Algebra: Introductory and Intermediate

Algebra: Introductory and Intermediate

Richard N. Aufmann
Palomar College, California

Vernon C. Barker
Palomar College, California

Joanne S. Lockwood
Plymouth State College, New Hampshire

HOUGHTON MIFFLIN COMPANY Boston   Toronto
Dallas   Geneva, Illinois   Palo Alto
Princeton, New Jersey

Cover Designer: Harold Burch, Harold Burch Design, New York City

Cover Image: Eiji Yanagi/PHOTONICA

Senior Sponsoring Editor: *Maureen O'Connor*
Associate Editor: *Dawn Nuttall*
Senior Project Editor: *Maria A. Morelli*
Editorial Assistant: *Christina Lillios*
Senior Production/Design Coordinator: *Carol Merrigan*
Senior Manufacturing Coordinator: *Marie Barnes*

Printed in the U.S.A.

ISBN:
  Text: 0-395-75579-4
  Exam Copy: 0-395-76484-X

23456789-B-99 98 97 96

Contents

· ·

Content and Format © 1996 HMCo.

6   Polynomials   305

7   Factoring   351

Content and Format © 1996 HMCo.

Preface

. . . .

Algebra: Introductory and Intermediate examines the fundamental ideas of algebra. Recognizing that the basic principles of geometry are a necessary part of mathematics, we have also included a separate chapter on geometry (Chapter 3) and have integrated geometry topics, where appropriate, throughout the text. Careful attention has been given to implementing the standards suggested by AMATYC and NCTM. At the end of each section are Applying the Concepts Exercises that include writing, synthesis, and challenge problems. Each chapter ends with Projects in Mathematics. The Projects in Mathematics feature is an extension or application of a concept that was covered in the chapter. These Projects can be used for cooperative learning activities or extra credit.

INSTRUCTIONAL FEATURES

Interactive Approach

Algebra: Introductory and Intermediate uses an interactive style that provides students with the opportunity to try a skill as it is presented. Each section is divided into objectives, and every objective contains one or more sets of matched-pair examples. The first example in each pair is worked out; the second example, labeled You Try It, is for the student to work. By solving this problem, the student practices concepts as they are presented in the text. There are *complete* worked-out solutions to these problems at the end of the book. By comparing their solutions to model solutions, students are able to obtain immediate feedback on and reinforcement of the concepts.

Emphasis on Problem-Solving Strategies

Algebra: Introductory and Intermediate features a carefully sequenced approach to application problems that emphasizes proven strategies to solve problems. Students are encouraged to develop their own strategies, draw diagrams, and to write their strategies as part of the solution to each application problem. In each case, model strategies are presented as guides for students to follow as they attempt the corresponding You Try It. Having students provide strategies is a natural way to incorporate writing into the math curriculum.

Emphasis on Applications

The traditional approach to teaching algebra covers only the straightforward manipulation of numbers and variables and thereby fails to teach students the practical value of algebra. By contrast, *Algebra: Introductory and Intermediate* contains an extensive collection of contemporary application problems. Wherever appropriate, the last objective of a section presents applications that require the student to use the skills covered in that section to solve practical problems. This carefully integrated applied approach generates awareness of the value of algebra as a real-life tool.

Completely Integrated Learning System
Organized by Objectives

Each chapter begins with a list of the learning objectives included within that chapter. Each objective is then restated in the chapter to remind the student of the current topic of discussion. The same objectives that organize the text are used as the structure for exercises, the testing programs, and the Computer Tutor. For every objective in the text there is a corresponding computer tutorial and a corresponding set of test questions associated with that objective.

THE INTERACTIVE APPROACH

Instructors have long realized the need for a text that requires students to use a skill as it is being taught. *Algebra: Introductory and Intermediate* uses an interactive technique that meets this need. Every objective, including the one shown below, contains at least one pair of examples. One of the examples is worked. The corresponding You Try It is not worked so that the student may "interact" with the text by solving a similar problem. So as to provide immediate feedback, a complete solution to this You Try It is provided in the Solutions

---

An explanatory passage begins each skill objective.

2.3 General Equations

Objective A   *To solve an equation of the form ax + b = c* ....

POINT OF INTEREST
Evariste Galois, despite being killed in a duel at the age of 21, made significant contributions to solving equations. In fact, there is a branch of mathematics called Galois Theory showing what kinds of equations can and cannot be solved.

In solving an equation of the form $ax + b = c$, the goal is to rewrite the equation in the form *variable = constant*. This requires the application of both the Addition and the Multiplication Properties of Equations.

➡ Solve: $\frac{3}{4}x - 2 = -11$

The goal is to write the equation in the form *variable = constant*.

$$\frac{3}{4}x - 2 = -11$$

$$\frac{3}{4}x - 2 + 2 = -11 + 2$$ • Add 2 to each side of the equation.

$$\frac{3}{4}x = -9$$ • Simplify.

$$\frac{4}{3} \cdot \frac{3}{4}x = \frac{4}{3}(-9)$$ • Multiply each side of the equation by $\frac{4}{3}$.

$$x = -12$$ • The equation is of the form *variable = constant*.

Check: • Check the solution.

$$\frac{3}{4}x - 2 = -11$$

$$\begin{array}{c|c} \frac{3}{4}(-12) - 2 & -11 \\ -9 - 2 & -11 \\ -11 = -11 \end{array}$$ • A true equation

The solution is -12.

Paired examples follow the explanatory passage.

The interactive key is the You Try It. It has not been worked so that the student may practice the skill, referring to the worked example at the left if necessary.

Reference to the Solution Section allows the student to check solutions immediately.

Example 1   Solve: $3x - 7 = -5$

Solution
$$3x - 7 = -5$$
$$3x - 7 + 7 = -5 + 7$$
$$3x = 2$$
$$\frac{3x}{3} = \frac{2}{3}$$
$$x = \frac{2}{3}$$

The solution is $\frac{2}{3}$.

You Try It 1   Solve: $5x + 7 = 10$

Your solution

Solution on p. A8

section. The benefit of this interactive style is that the student can verify that a new skill has been learned before attempting a homework assignment.

EMPHASIS ON APPLICATIONS

The solution of an application problem in *Algebra: Introductory and Intermediate* is presented in two parts: **Strategy** and **Solution.** The strategy is a written description of the steps that are necessary to solve the problem; the solution is the implementation of the strategy. Using this format provides students with a structure for problem solving. It also encourages students to write strategies for solving problems which, in turn, fosters organizing such strategies in a logical way. Having students write strategies is a natural way to incorporate writing into the math curriculum.

550   Chapter 11 / Functions and Relations

Objective B   *To solve application problems* ....

Linear functions can be used to model a variety of applications in science and business. For each application, data are collected and the independent and dependent variables are selected. Then a linear function is determined that models the data.

A strategy that the student may use in solving an application problem is stated.

This strategy is used in the solution of the worked example.

When students compare their solutions to model solutions in the appendix, they will see a complete solution of the problem along with a written strategy for solving the problem.

Example 2
Suppose a manufacturer has determined that at a price of $115, consumers will purchase 1 million portable CD players and that at a price of $90, consumers will purchase 1.25 million portable CD players. Describe this situation with a linear function. Use this function to predict how many portable CD players consumers will purchase if the price is $80.

Strategy
• Select the independent and dependent variables. Because you are trying to determine the number of CD players, that quantity is the *dependent* variable, y. The price of CD players is the *independent* variable, x. From the given data, two ordered pairs are (115, 1) and (90, 1.25). (The ordinates are in millions of units.) Use these ordered pairs to determine the linear function.
• Evaluate the function for $x = 80$ to predict how many CD players consumers will purchase if the price is $80.

Solution
Let $(x_1, y_1) = (115, 1)$ and $(x_2, y_2) = (90, 1.25)$.

$$m = \frac{y_2 - y_1}{x_2 - x_1} = \frac{1.25 - 1}{90 - 115} = -\frac{0.25}{25} = -0.01$$

$$y - y_1 = m(x - x_1)$$
$$y - 1 = -0.01(x - 115)$$
$$y = -0.01x + 2.15$$

The linear function is
$f(x) = -0.01x + 2.15$.

$f(80) = -0.01(80) + 2.15 = 1.35$

Consumers will purchase 1.35 million CD players at a price of $80.

You Try It 2
Gabriel Daniel Fahrenheit invented the mercury thermometer in 1717. In terms of readings on this thermometer, water freezes at 32°F and boils at 212°F. In 1742 Anders Celsius invented the Celsius temperature scale. On this scale, water freezes at 0°C and boils at 100°C. Determine a linear function that can be used to predict the Celsius temperature when the Fahrenheit temperature is known.

Your strategy

Your solution

Solution on p. A51

OBJECTIVE-SPECIFIC APPROACH

Many texts in mathematics are not organized in a manner that facilitates management of learning. Typically, students are left to wander through a maze of seemingly unrelated lessons, exercise sets, and tests. *Algebra: Introductory and Intermediate* solves this problem by organizing all lessons, exercise sets, computer tutorials and tests around a carefully constructed hierarchy of objectives. The advantage of this objective-by-objective organization is that it enables the student who is uncertain at any step in the learning process to refer easily to the original presentation and review that material.

The objective-specific approach also allows the instructor greater control over the management of student progress. The Computerized Testing Program and the Printed Testing Program are organized by the same objectives as the text. These references are provided with the answers to the test items. This allows the instructor to quickly determine those objectives on which a student may need additional instruction.

The Computer Tutor is also organized around the objectives of the text. As a result, supplemental instruction is available for any objectives that are troublesome for a student.

A numbered objective statement names the topic of each lesson.

 **General Equations**

Objective A *To solve an equation of the form ax + b = c* ....

The exercise sets correspond to the objectives in the text.

2.3 Exercises

Objective A

Solve and check.

1. $3x + 1 = 10$ **2.** $4y + 3 = 11$ **3.** $2a - 5 = 7$ **4.** $5m - 6 = 9$

The answers to the Chapter Review Exercises show the objective from which a problem was taken.

CHAPTER REVIEW EXERCISES *pages 137–138*

1. no [2.1A]   **2.** $x = 21$ [2.1B]   **3.** $a = 20$ [2.1C]   **4.** $x = -3$ [2.3A]
5. $x = -\frac{6}{7}$ [2.3A]   **6.** $y = \frac{1}{3}$ [2.3B]   **7.** $x = 4$ [2.3B]   **8.** $x = -2$ [2.3C]

The answers to the Cumulative Review Exercises also show the objective that relates to the exercise.

CUMULATIVE REVIEW EXERCISES *pages 139–140*

1. 6 [1.1C]   **2.** -48 [1.1D]   **3.** $-\frac{19}{48}$ [1.2C]   **4.** 54 [1.2E]   **5.** $\frac{49}{40}$ [1.3A]
6. 6 [1.4A]   **7.** $-17x$ [1.4B]   **8.** $-5a - 4b$ [1.4B]   **9.** $2x$ [1.4C]   **10.** $36y$ [1.4C]

ADDITIONAL LEARNING AIDS

Projects in Mathematics

The Projects in Mathematics feature occurs at the end of each chapter. These projects can be used as extra credit or cooperative learning activities. The projects cover various aspects of mathematics including the field of geometry, statistics, science and business.

Chapter Summaries

The Chapter Summaries are a useful guide to students as they review for a test. A Chapter Summary includes the Key Words and Essential Rules that were covered in the chapter.

Study Skills

To the Student, which follows the preface, suggests how to use this text and provides approaches to creating good study habits.

Computer Tutor

The Computer Tutor is an interactive computer tutorial that covers *every* objective in the text.

Glossary

There is a glossary at the back of the book that includes definitions of terms used in the text.

Graphing Calculator

There are various optional exercises where a graphing calculator is needed. To assist students with these exercises, a Graphing Calculator Appendix is included. This appendix is a short guide to the basic operation of graphing calculators from Texas Instruments, Sharp, and Casio.

EXERCISES

End-of-Section Exercises

Algebra: Introductory and Intermediate contains more than 6000 exercises. At the end of each section are exercise sets keyed to the corresponding learning objectives. The exercises are carefully developed to ensure that students can apply the concepts in the section to a variety of problem situations.

Applying the Concepts Exercises

The End-of-Section Exercises are followed by Applying the Concepts exercises. These contain a variety of exercise types which include:

- challenge problems
- graphing calculator exercises
- writing exercises
- problems that ask students to determine incorrect procedures
- problems that require a more in-depth analysis

Writing Exercises

Within the Applying the Concepts exercises are Writing Exercises denoted by **[W]**. These exercises ask students to write about a topic in the section or to research and report on a related topic. There are also writing exercises in some of the application problems. These exercises ask students to write a sentence that describes the meaning of their answer in the context of the problem.

Chapter Review Exercises

Review Exercises are found at the end of each chapter. These exercises are selected to help the student integrate all of the topics presented in the chapter. The answers to all review exercises are given in the Answer Section. Along with the answer, there is a reference to the objective that pertains to each exercise.

Cumulative Review Exercises

Cumulative Review Exercises, which appear at the end of each chapter (beginning with Chapter 2), help students maintain skills learned in previous chapters. The answers to all cumulative review exercises are given in the Answer Section. Along with the answers is a reference to the objective that pertains to each exercise.

SUPPLEMENTS FOR THE INSTRUCTOR

Instructor's Resource Manual with Solutions Manual

The Instructor's Resource Manual includes suggestions for course sequencing and outlines for the answers to the writing exercises. The Solutions Manual contains worked-out solutions for all end-of-section exercises, Applying the Concepts exercises, Chapter Review Exercises, and Cumulative Review Exercises.

Computerized Test Generator

The Computerized Test Generator is the first of three testing materials. The data base contains more than 3000 test items. These questions are unique to the Test Generator. The Test Generator is designed to provide an unlimited number of tests for each chapter, cumulative chapter tests, and a final exam. It is available for the IBM PC and compatible computers and the Macintosh. The IBM version provides new on-line testing and also contains algorithms, which produce an unlimited number of certain types of test questions.

Printed Test Bank with Chapter Tests

The Printed Test Bank, the second component of the testing material, is a printout of all items in the Computerized Test Generator. Instructors who do not have access to a computer can use the test bank to select items to include on a test being prepared by hand. Chapter Tests contain the printed testing program, which is the third source of testing material. Four printed tests, two free response and two multiple choice, are provided for each chapter.

SUPPLEMENTS FOR THE STUDENT

Student Solutions Manual

The Student Solutions Manual contains the complete solutions to all odd-numbered exercises in the text.

Computer Tutor

The Computer Tutor is an algorithmically based tutor that includes color and animation. The algorithmic feature essentially provides an infinite number of practice problems for students to attempt. The algorithms have been carefully crafted to present a variety of problem types from easy to difficult. A complete solution to each problem is available.

Each objective of the text is supported by a tutorial in the Computer Tutor. There is an interactive feature of the Computer Tutor that requires students to respond to questions about the topic in the current lesson. In this way, students can assess their understanding of concepts as they are presented. There is a Glossary which can be accessed at any time so that students can look up words whose definitions they may have forgotten.

When the student completes a lesson, a printed report is available. This optional report gives the student's name, the objectives studied, the number of problems attempted, the number of problems answered correctly, the percent correct, and the time spent working on the exercises in that objective.

The Computer Tutor can be used in several ways: (1) to cover material the student missed because of an absence; (2) to reinforce instruction on a concept that the student has not yet mastered; (3) to review material in preparation for exams. This tutorial is available for the Macintosh and IBM PC and compatible computers running Windows.

Videotapes

The videotape series contains lessons that accompany *Algebra: Introductory and Intermediate.* These lessons follow the format and style of the text and are closely tied to specific sections of the text. Each videotape begins with an application and then the necessary mathematics needed to solve that application is presented. The tape closes with the solution of the application problem.

ACKNOWLEDGMENTS

The authors would like to thank the following reviewers for their helpful suggestions: **John W. Coburn,** *St. Louis Community College;* **Ann Flamm,** *Reading Area Community College* (PA); **Larry Friesen,** *Butler County Community College* (KS); **Edith Hays,** *Texas Women's University;* **Patricia Lord,** *Bainbridge College* (GA); **Cindy Noftle; Thomas O'Keefe,** *Bucks County Community College* (PA); **John Searcy.**

To the Student

Many students feel that they will never understand math while others appear to do very well with little effort. Oftentimes what makes the difference is that successful students take an active role in the learning process.

Learning mathematics requires your *active* participation. Although doing homework is one way you can actively participate, it is not the only way. First, you must attend class regularly and become an active participant. Secondly, you must become actively involved with the textbook.

Algebra: Introductory and Intermediate was written and designed with you in mind as a participant. Here are some suggestions on how to use the features of this textbook.

There are 12 chapters in this text. Each chapter is divided into sections and each section is subdivided into learning objectives. Each learning objective is labeled with a letter from A–G.

First, read each objective statement carefully so you will understand the learning goal that is being presented. Next, read the objective material carefully, being sure to note each bold word. These words indicate important concepts that you should familiarize yourself with. Study each in-text example carefully, noting the techniques and strategies used to solve the example.

You will then come to the key learning feature of this text, the *boxed examples*. These examples have been designed to aid you in a very specific way. Notice that in each example box, the example on the left is completely worked out and the "You Try It" example on the right is not. *You* are expected to work the right-hand example (in the space provided) in order to immediately test your understanding of the material you have just studied.

You should study the worked-out example carefully by working through each step presented. This allows you to focus on each step and reinforces the technique for solving that type of problem. You can then use the worked-out example as a model for solving similar problems.

Solve the "You Try It" example using the problem-solving techniques that you have just studied. When you have completed your solution, check your work by turning to the page in the appendix where the complete solution can be found. The page number on which the solution appears is printed at the bottom of the example box in the right-hand corner. By checking your solution, you will know immediately whether or not you fully understand the skill just studied.

When you have completed studying an objective, do all of the exercises in the exercise set that correspond with that objective. The exercises will be labeled with the same letter as the objective. Algebra is a subject that needs to be learned in small sections and practiced continually in order to be mastered. Doing all of the exercises in each exercise set will help you master the problem solving techniques necessary for success.

Once you have completed the exercises to an objective, you should check your answers to the odd-numbered exercises with those found in the back of the book.

After completing a chapter, read the Chapter Summary. This summary highlights the important topics covered in the chapter. Following the Chapter Summary are Chapter Review Exercises, and a Cumulative Review (beginning with Chapter 2). Doing the review exercises is an important way of testing your understanding of the chapter. The answer to each review exercise is in an appendix at the back of the book. Each answer is followed by a reference that tells which objective that exercise was taken from. For example, (4.2B) means Section 4.2, Objective B. After checking your answers, restudy any objective that you missed. It may be very helpful to retry some of the exercises for that objective to reinforce your problem-solving techniques.

The Cumulative Review allows you to refresh the skills you have learned in previous chapters. This is very important in mathematics. By consistently reviewing previous materials, you will retain the previous skills as you build new ones.

Remember, to be successful, attend class regularly; read the textbook carefully; actively participate in class; work with your textbook using the "You Try It" examples for immediate feedback and reinforcement of each skill; do all the homework assignments; review constantly; and work carefully.

Index of Applications

· ·

Algebra: Introductory and Intermediate

Real Numbers and Variable Expressions

History of Variables

Prior to the 16th century, unknown quantities were represented by words. In Latin, the language in which most scholarly works were written, the word *res*, meaning "thing," was used. In Germany the word *zahl*, meaning "number," was used. In Italy the word *cosa*, also meaning "thing," was used.

Then in 1637, René Descartes, a French mathematician, began using the letters x, y, and z to represent variables. It is interesting to note, upon examining Descartes's work, that toward the end of the book the letters y and z were no longer used and x became the choice for a variable.

One explanation of why the letters y and z appeared less frequently has to do with the nature of printing presses during Descartes's time. A printer had a large tray that contained all the letters of the alphabet. There were many copies of each letter, especially those letters that are used frequently. For example, there were more e's than q's. Because the letters y and z do not occur frequently in French, a printer would have few of these letters on hand. Consequently, when Descartes started using these letters as variables, it quickly depleted the printer's supply and x's had to be used instead.

Today, x is used by most nations as the standard letter for a single unknown. In fact, x rays were so named because the scientists who discovered them did not know what they were and thus labeled them the "unknown rays" or x rays.

1.1 Integers

Objective A   *To use inequality symbols with integers* ....

It seems to be a human characteristic to put similar items in the same place. For instance, a biologist places similar animals in groups called *phyla*, and a geologist divides the history of the earth into *eras*.

Mathematicians likewise place objects with similar properties in *sets* and use braces to surround the objects in the set. The numbers that we use to count objects, such as the number of people at a baseball game or the number of horses on a ranch, have similar characteristics. These numbers are the *natural numbers*.

$$\textbf{Natural numbers} = \{1, 2, 3, 4, 5, 6, 7, 8, 9, 10, 11, \ldots\}$$

The natural numbers alone do not provide all the numbers that are useful in applications. For instance, a meteorologist needs numbers below zero and above zero.

$$\textbf{Integers} = \{\ldots, -5, -4, -3, -2, -1, 0, 1, 2, 3, 4, 5, \ldots\}$$

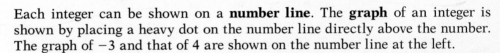

Negative integers   Zero   Positive integers

Each integer can be shown on a **number line**. The **graph** of an integer is shown by placing a heavy dot on the number line directly above the number. The graph of -3 and that of 4 are shown on the number line at the left.

The integers to the left of zero are **negative integers**. The integers to the right of zero are **positive integers**. Zero is neither a positive nor a negative integer.

Consider the sentences below.

> The quarterback threw the football and the receiver caught *it*.
> An accountant purchased a calculator and placed *it* in a briefcase.

In the first sentence, *it* means football; in the second sentence, *it* means calculator. In language, the word *it* can stand for many different objects. Similarly, in mathematics, a letter of the alphabet can be used to stand for a number. Such a letter is called a **variable**. Variables are used in the next definition.

Definition of Inequality Symbols

If a and b are two numbers and a is to the left of b on the number line, then a is **less than** b. This is written $a < b$.
If a and b are two numbers and a is to the right of b on the number line, then a is **greater than** b. This is written $a > b$.

There are also inequality symbols for **less than or equal to** ($\leq$) and **greater than or equal to** ($\geq$). For instance,

$$6 \leq 6 \text{ because } 6 = 6. \qquad 7 \leq 15 \text{ because } 7 < 15.$$

It is convenient to use a variable to represent, or stand for, any one of the elements of a set. For instance, the statement "*x* is an *element of* the set $\{0, 2, 4, 6\}$" means that *x* can be replaced by 0, 2, 4, or 6. The symbol for "is an element of" is $\in$; the symbol for "is not an element of" is $\notin$. For example,

$$2 \in \{0, 2, 4, 6\} \qquad 6 \in \{0, 2, 4, 6\} \qquad 7 \notin \{0, 2, 4, 6\}$$

Example 1

Let $x \in \{-6, -2, 0\}$. For which values of x is the inequality $x \leq -2$ a true statement?

Solution

Replace x by each element of the set and determine whether the inequality is true.

$$x \leq -2$$
$$-6 \leq -2 \quad \text{True.} \quad -6 < -2$$
$$-2 \leq -2 \quad \text{True.} \quad -2 = -2$$
$$0 \leq -2 \quad \text{False.}$$

The inequality is true for -6 and -2.

You Try It 1

Let $y \in \{-5, -1, 5\}$. For which values of y is the inequality $y > -1$ a true statement?

Your solution

Solution on p. A1

Objective B   To find the additive inverse and absolute value of a number ....

A number line with "5" and "5" marked above, from −5 to 5:
$$-5 \; -4 \; -3 \; -2 \; -1 \; 0 \; 1 \; 2 \; 3 \; 4 \; 5$$

CONSIDER THIS

From Fig. 2, the distance from 0 to 5 is 5; |5| = 5. The distance from 0 to −5 is 5; |−5| = 5.

POINT OF INTEREST

The definition of absolute value that we have given in the box is written in what is called "rhetorical style." That is, it is written without the use of variables. This is how *all* mathematics was written prior to the Renaissance. During that period, from the 14th to the 16th century, the idea of expressing a variable symbolically was developed.

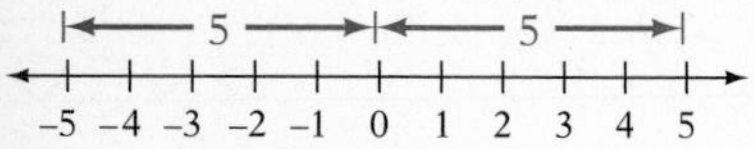

On the number line, the numbers 5 and −5 are the same distance from zero but on opposite sides of zero. The numbers 5 and −5 are called **opposites** or **additive inverses** of each other. (See the number line at the left.)

The opposite (or additive inverse) of 5 is −5. The opposite of −5 is 5. The symbol for opposite is −.

$-(5)$ means the opposite of *positive* 5. $-(5) = -5$
$-(-5)$ means the opposite of *negative* 5. $-(-5) = 5$

The **absolute value** of a number is its distance from zero on the number line. The symbol for absolute value is two vertical bars, | |.

> **Absolute Value**
>
> The absolute value of a positive number is the number itself. The absolute value of zero is zero. The absolute value of a negative number is the opposite of the negative number.

➡ Evaluate: $-|-12|$

$$-|-12| = -12$$

● The absolute value sign does not affect the negative sign in front of the absolute value sign.

Example 2

Let $a \in \{-12, 0, 4\}$. Find the additive inverse of a and the absolute value of a for each element of the set.

Solution

Replace a by each element of the set.

$-a$	$|a|$
$-(-12) = 12$	$|-12| = 12$
$-(0) = 0$	$|0| = 0$
$-(4) = -4$	$|4| = 4$

You Try It 2

Let $z \in \{-11, 0, 8\}$. Find the additive inverse of z and the absolute value of z for each element of the set.

Your solution

Solution on p. A1

Objective C   **To add or subtract integers** ....

A number can be represented anywhere along the number line by an arrow. A positive number is represented by an arrow pointing to the right, and a negative number is represented by an arrow pointing to the left. The size of the number is represented by the length of the arrow.

Addition of integers can be shown on the number line. To add integers, start at zero and draw an arrow representing the first number. At the tip of the first arrow, draw a second arrow representing the second number. The sum is below the tip of the second arrow.

CONSIDER THIS

There are several ways to model the addition of integers. The model at the right uses arrows. Another model uses money. For instance, if you are $8 in debt ($-8$) and you repay $5, then you are only $3 in debt ($-3$). This can also be related to credit card debt. If you owe $100 ($-100$) and charge $25 more ($-25$), then you owe $125 ($-125$).

$4 + 2 = 6$

$-4 + (-2) = -6$

$-4 + 2 = -2$

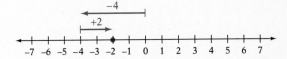

$4 + (-2) = 2$

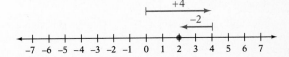

The pattern for the addition of integers shown on the number line can be summarized in the following rule.

> **Addition of Integers**
>
> - *Numbers with the same sign*
>   To add two numbers with the same sign, add the absolute values of the numbers. Then attach the sign of the addends.
>
> - *Numbers with different signs*
>   To add two numbers with different signs, find the absolute value of each number. Then subtract the smaller of these numbers from the larger one. Attach the sign of the number with the larger absolute value.

➡ Add: $(-9) + 8$

$|-9| = 9 \quad |8| = 8$ • The signs are different. Find the absolute value of each number.

$9 - 8 = 1$ • Subtract the smaller number from the larger.

$(-9) + 8 = -1$ • Attach the sign of the number with the larger absolute value. Because $|-9| > |8|$, use the sign of -9.

➡ Add: $(-23) + 47 + (-18) + 5$

To add more than two numbers, add the first two numbers. Then add the sum to the third number. Continue until all the numbers are added.

$$(-23) + 47 + (-18) + 5 = 24 + (-18) + 5$$
$$= 6 + 5$$
$$= 11$$

Look at the two expressions below and note that each expression equals the same number.

$$8 - 3 = 5 \qquad \text{8 minus 3 is 5.}$$
$$8 + (-3) = 5 \qquad \text{8 plus the opposite of 3 is 5.}$$

This example suggests that to subtract two numbers, we add the opposite of the second number to the first number.

first number	−	second number	=	first number	+	the opposite of the second number	
40	−	60	=	40	+	(-60)	$= -20$
−40	−	60	=	−40	+	(-60)	$= -100$
−40	−	(-60)	=	−40	+	60	$= 20$
40	−	(-60)	=	40	+	60	$= 100$

➡ Subtract: $-21 - (-40)$

Change this sign to plus.

$$-21 - (-40) = -21 + 40 = 19$$

Change −40 to the opposite of −40.

• Rewrite each subtraction as addition of the opposite. Then add.

➡ Subtract: $15 - 51$

Change this sign to plus.

$$15 - 51 = 15 + (-51) = -36$$

Change 51 to the opposite of 51.

• Rewrite each subtraction as addition of the opposite. Then add.

➡ Subtract: $-12 - (-21) - 15$

$$-12 - (-21) - 15 = -12 + 21 + (-15)$$
$$= 9 + (-15)$$
$$= -6$$

• Rewrite each subtraction as addition of the opposite. Then add.

Example 3   Add: $(-52) + (-39)$

Solution   The signs are the same.
Add the absolute values of
the numbers: $52 + 39 = 91$

Attach the sign of the addends:
$(-52) + (-39) = -91$

You Try It 3   Add: $100 + (-43)$

Your solution

Example 4   Add:
$37 + (-52) + (-21) + (-7)$

Solution $37 + (-52) + (-21) + (-7)$
$= -15 + (-21) + (-7)$
$= -36 + (-7)$
$= -43$

You Try It 4   Add:
$(-51) + 42 + 17 + (-102)$

Your solution

Example 5   Subtract: $-11 - 15$

Solution $-11 - 15 = -11 + (-15)$
$= -26$

You Try It 5   Subtract: $19 - (-32)$

Your solution

Example 6   Subtract:
$-14 - 18 - (-21) - 4$

Solution $-14 - 18 - (-21) - 4$
$= -14 + (-18) + 21 + (-4)$
$= -32 + 21 + (-4)$
$= -11 + (-4)$
$= -15$

You Try It 6   Subtract:
$-9 - (-12) - 17 - 4$

Your solution

Solutions on p. A1

Objective D   *To multiply or divide integers* ....

Multiplication is the repeated addition of the same number. The product 3×5 is shown on the number line below.

$$\underbrace{5 \qquad 5 \qquad 5}$$

```
 ├─┼─┼─┼─┼─┼─┼─┼─┼─┼─┼─┼─┼─┼─┼─►
 0 1 2 3 4 5 6 7 8 9 10 11 12 13 14 15
```

$\overbrace{\text{5 is added 3 times}}$
$3 \times 5 = 5 + 5 + 5 = 15$

To indicate multiplication, several different symbols are used.

$$3 \times 5 = 15 \qquad 3 \cdot 5 = 15 \qquad (3)(5) = 15 \qquad 5(3) = 15 \qquad (3)5 = 15$$

Note that when parentheses are used and there is no arithmetic operation symbol, the operation is multiplication. Each number in a product is called a **factor**. For instance, 3 and 5 are factors of the product $3 \cdot 5 = 15$.

Content and Format © 1996 HMCo.

POINT OF INTEREST

The cross × was first used as a symbol for multiplication in 1631 in a book titled *The Key to Mathematics.* Also in that year, another book, *Practice of the Analytical Art,* advocated the use of a dot to indicate multiplication.

Now consider the product of a positive and a negative number.

$$\overbrace{3(-5) = (-5) + (-5) + (-5)}^{-5 \text{ is added 3 times}} = -15 \qquad \bullet \text{ Multiplication is repeated addition.}$$

This suggests that the product of a positive number and a negative number is negative. Here are a few more examples.

$$4(-7) = -28 \qquad (-6)5 = -30 \qquad (-5) \cdot 7 = -35$$

To find the product of two negative numbers, look at the pattern at the right. As -5 multiplies a sequence of decreasing integers, the products increase by 5.

The pattern can be continued by requiring that the product of two negative numbers be positive.

These numbers decrease by 1.   These numbers increase by 5.

$$-5(3) = -15$$
$$-5(2) = -10$$
$$-5(1) = -5$$
$$-5(0) = 0$$
$$-5(-1) = 5$$
$$-5(-2) = 10$$
$$-5(-3) = 15$$

Multiplication of Integers

- *Numbers with the same sign*
  To multiply two numbers with the same sign, multiply the absolute values of the numbers. The product is positive.

- *Numbers with different signs*
  To multiply two numbers with different signs, multiply the absolute values of the numbers. The product is negative.

➡ Multiply: $-2(5)(-7)(-4)$

$$-2(5)(-7)(-4) = -10(-7)(-4)$$
$$= 70(-4) = -280$$

● To multiply more than two numbers, multiply the first two. Then multiply the product by the third number. Continue until all the numbers are multiplied.

For every division problem there is a related multiplication problem.

$$\frac{8}{2} = 4 \quad \text{because} \qquad 4 \cdot 2 = 8$$

Division   Related multiplication

This fact and the rules for multiplying integers can be used to illustrate the rules for dividing integers.

Note in the following examples that the quotient of two numbers with the same sign is positive.

$$\frac{12}{3} = 4 \text{ because } 4 \cdot 3 = 12 \qquad \frac{-12}{-3} = 4 \text{ because } 4 \cdot (-3) = -12$$

The next two examples illustrate that the quotient of two numbers with different signs is negative.

$$\frac{12}{-3} = -4 \text{ because } (-4)(-3) = 12 \qquad \frac{-12}{3} = -4 \text{ because } (-4) \cdot 3 = -12$$

> **Division of Integers**
>
> - *Numbers with the same sign*
>   To divide two numbers with the same sign, divide the absolute values of the numbers. The quotient is positive.
>
> - *Numbers with different signs*
>   To divide two numbers with different signs, divide the absolute values of the numbers. The quotient is negative.

➡ Simplify: $-\dfrac{-56}{7}$

$$-\frac{-56}{7} = -\left(\frac{-56}{7}\right) = -(-8) = 8$$

Note that $\dfrac{-12}{3} = -4$, $\dfrac{12}{-3} = -4$, and $-\dfrac{12}{3} = -4$. This suggests the following rule.

CONSIDER THIS

The symbol $\neq$ is read "is not equal to."

> If a and b are integers, and $b \neq 0$, then $\dfrac{-a}{b} = \dfrac{a}{-b} = -\dfrac{a}{b}$.

Properties of Zero and One in Division

- Zero divided by any number other than zero is zero.

  $\dfrac{0}{a} = 0$ because $0 \cdot a = 0$ For example, $\dfrac{0}{7} = 0$ because $0 \cdot 7 = 0$.

- Division by zero is not defined.

  To understand that division by zero is not permitted, suppose that $\dfrac{4}{0}$ were equal to n, where n is some number. Because each division problem has a related multiplication problem, $\dfrac{4}{0} = n$ means $n \cdot 0 = 4$. But $n \cdot 0 = 4$ is impossible because any number times 0 is 0. Therefore, division by 0 is not defined.

- Any number other than zero divided by itself is 1.

  $\dfrac{a}{a} = 1$, $a \neq 0$ For example, $\dfrac{-8}{-8} = 1$.

- Any number divided by one is the number.

  $\dfrac{a}{1} = a$ For example, $\dfrac{9}{1} = 9$.

Example 7	Multiply: $(-3)4(-5)$	**You Try It 7** Multiply: $8(-9)10$
Solution	$(-3)4(-5) = (-12)(-5) = 60$	**Your solution**

Solution on p. A1

Example 8 Multiply: $12(-4)(-3)(-5)$

Solution $12(-4)(-3)(-5)$
$= (-48)(-3)(-5)$
$= 144(-5) = -720$

You Try It 8 Multiply: $(-2)3(-8)7$

Your solution

Example 9 Divide: $(-120) \div (-8)$

Solution $(-120) \div (-8) = 15$

You Try It 9 Divide: $(-135) \div (-9)$

Your solution

Example 10 Divide: $\dfrac{95}{-5}$

Solution $\dfrac{95}{-5} = -19$

You Try It 10 Divide: $\dfrac{-72}{4}$

Your solution

Example 11 Divide: $-\dfrac{-81}{3}$

Solution $-\dfrac{-81}{3} = -(-27) = 27$

You Try It 11 Divide: $-\dfrac{36}{-12}$

Your solution

Solutions on p. A1

Objective E   *To solve application problems*

To solve an application problem, first read the problem carefully. The Strategy involves identifying the quantity to be found and planning the steps that are necessary to find that quantity. The Solution involves performing each operation stated in the Strategy and writing the answer.

Example 12
The daily high temperatures (in degrees Celsius) for six days in Anchorage, Alaska, were $-14°$, $3°$, $0°$, $-8°$, $2°$, $-1°$. Find the average daily high temperature.

Strategy
To find the average daily high temperature:

• Add the six temperature readings.
• Divide the sum by 6.

Solution
$-14 + 3 + 0 + (-8) + 2 + (-1) = -18$
$-18 \div 6 = -3$

The average daily high temperature was $-3°C$.

You Try It 12
The daily low temperatures (in degrees Celsius) during one week were recorded as follows: $-6°$, $-7°$, $0°$, $-5°$, $-8°$, $-1°$, $-1°$. Find the average daily low temperature.

Your strategy

Your solution

Solution on p. A1

1.1 Exercises

Objective A

Place the correct symbol, $<$ or $>$, between the two numbers.

1. 8 -6

2. -14 16

3. -12 1

4. 35   28

5. 42   19

6. -42 27

7. 0 -31

8. -17 0

9. 53 -46

10. -27 -39

Answer True or False.

11. $-13 > 0$

12. $-20 > 3$

13. $12 > -31$

14. $9 > 7$

15. $-5 > -2$

16. $-44 > -21$

17. $-4 > -120$

18. $0 > -8$

19. $-1 > 0$

20. $-10 > -88$

21. Let $x \in \{-23, -18, -8, 0\}$. For which values of x is the inequality $x < -8$ a true statement?

22. Let $w \in \{-33, -24, -10, 0\}$. For which values of w is the inequality $w < -10$ a true statement?

23. Let $a \in \{-33, -15, 21, 37\}$. For which values of a is the inequality $a > -10$ a true statement?

24. Let $v \in \{-27, -14, 14, 27\}$. For which values of v is the inequality $v > -15$ a true statement?

25. Let $n \in \{-23, -1, 0, 4, 29\}$. For which values of n is the inequality $-6 > n$ a true statement?

26. Let $m \in \{-33, -11, 0, 12, 45\}$. For which values of m is the inequality $-15 > m$ a true statement?

Objective B

Find the additive inverse.

27. 4

28. 8

29. -9

30. -12

31. -28

32. -36

Evaluate.

33. $-(-14)$

34. $-(-40)$

35. $-(77)$

36. $-(39)$

37. $-(0)$

38. $-(-13)$

39. $|-74|$

40. $|-96|$

41. $-|-82|$

42. $-|-53|$

43. $-|81|$

44. $-|38|$

Place the correct symbol, $<$ or $>$, between the values of the two numbers.

45. $|-83|$ $|58|$ **46.** $|22|$ $|-19|$ **47.** $|43|$ $|-52|$ **48.** $|-71|$ $|-92|$

49. $|-68|$ $|-42|$ **50.** $|12|$ $|-31|$ **51.** $|-45|$ $|-61|$ **52.** $|-28|$ $|43|$

53. Let $p \in \{-19, 0, 28\}$. Evaluate $-p$ for each element of the set.

54. Let $q \in \{-34, 0, 31\}$. Evaluate $-q$ for each element of the set.

55. Let $x \in \{-45, 0, 17\}$. Evaluate $-|x|$ for each element of the set.

56. Let $y \in \{-91, 0, 48\}$. Evaluate $-|y|$ for each element of the set.

Objective C

Add or subtract.

57. $-3 + (-8)$ **58.** $-6 + (-9)$ **59.** $-8 + 3$ **60.** $-9 + 2$

61. $-3 + (-80)$ **62.** $-12 + (-1)$ **63.** $-23 + (-23)$ **64.** $-12 + (-12)$

65. $16 + (-16)$ **66.** $-17 + 17$ **67.** $48 + (-53)$ **68.** $19 + (-41)$

69. $-17 + (-3) + 29$ **70.** $13 + 62 + (-38)$ **71.** $-3 + (-8) + 12$ **72.** $-27 + (-42) + (-18)$

73. $16 - 8$ **74.** $12 - 3$ **75.** $7 - 14$ **76.** $6 - 9$

77. $-7 - 2$ **78.** $-9 - 4$ **79.** $7 - (-2)$ **80.** $3 - (-4)$

81. $-6 - (-3)$ **82.** $-4 - (-2)$ **83.** $6 - (-12)$ **84.** $-12 - 16$

85. $13 + (-22) + 4 + (-5)$ **86.** $-14 + (-3) + 7 + (-21)$

87. $-22 + 20 + 2 + (-18)$ **88.** $-6 + (-8) + 14 + (-4)$

Add or subtract.

89. $-16 + (-17) + (-18) + 10$ **90.** $-25 + (-31) + 24 + 19$

91. $26 + (-15) + (-11) + (-12)$ **92.** $-32 + 40 + (-8) + (-19)$

93. $-17 + (-18) + 45 + (-10)$ **94.** $23 + (-15) + 9 + (-15)$

95. $46 + (-17) + (-13) + (-50)$ **96.** $-37 + (-17) + (-12) + (-15)$

97. $-14 + (-15) + (-11) + 40$ **98.** $28 + (-19) + (-8) + (-1)$

99. $-4 - 3 - 2$ **100.** $4 - 5 - 12$ **101.** $12 - (-7) - 8$

102. $-12 - (-3) - (-15)$ **103.** $-19 - (-19) - 18$ **104.** $-8 - (-8) - 14$

105. $-17 - (-8) - (-9)$ **106.** $7 - 8 - (-1)$ **107.** $-30 - (-65) - 29 - 4$

108. $42 - (-82) - 65 - 7$ **109.** $-16 - 47 - 63 - 12$ **110.** $42 - (-30) - 65 - (-11)$

111. $-47 - (-67) - 13 - 15$ **112.** $-18 - 49 - (-84) - 27$ **113.** $-19 - 17 - (-36) - 12$

114. $48 - 19 - 29 - 51$ **115.** $21 - (-14) - 43 - 12$ **116.** $17 - (-17) - 14 - 21$

Objective D

Multiply or divide.

117. $(14)3$ **118.** $(17)6$ **119.** $-7 \cdot 4$ **120.** $-8 \cdot 7$ **121.** $(-12)(-5)$ **122.** $(-13)(-9)$

123. $-11(23)$ **124.** $-8(21)$ **125.** $(-17)14$ **126.** $(-15)12$ **127.** $6(-19)$ **128.** $17(-13)$

129. $12 \div (-6)$ **130.** $18 \div (-3)$ **131.** $(-72) \div (-9)$ **132.** $(-64) \div (-8)$ **133.** $-42 \div 6$

134. $(-56) \div 8$ **135.** $(-144) \div 12$ **136.** $(-93) \div (-3)$ **137.** $48 \div (-8)$ **138.** $57 \div (-3)$

139. $\dfrac{-49}{7}$ **140.** $\dfrac{-45}{5}$ **141.** $\dfrac{-44}{-4}$ **142.** $\dfrac{-36}{-9}$ **143.** $\dfrac{98}{-7}$

144. $\dfrac{85}{-5}$ **145.** $-\dfrac{-120}{8}$ **146.** $-\dfrac{-72}{4}$ **147.** $-\dfrac{-80}{-5}$ **148.** $-\dfrac{-114}{-6}$

149. $0 \div (-9)$ **150.** $0 \div (-14)$ **151.** $\dfrac{-261}{9}$ **152.** $\dfrac{-128}{4}$ **153.** $9 \div 0$

154. $(-21) \div 0$ **155.** $\dfrac{132}{-12}$ **156.** $\dfrac{250}{-25}$ **157.** $\dfrac{0}{0}$ **158.** $\dfrac{-58}{0}$

159. $7(5)(-3)$ **160.** $(-3)(-2)8$ **161.** $9(-7)(-4)$ **162.** $(-2)(6)(-4)$

163. $16(-3)5$ **164.** $20(-4)3$ **165.** $-4(-3)8$ **166.** $-5(-9)6$

167. $-3(-8)(-9)$ **168.** $-7(-6)(-5)$ **169.** $(-9)7(5)$ **170.** $(-8)7(10)$

171. $7(-2)(5)(-6)$ **172.** $(-3)7(-2)8$ **173.** $-9(-4)(-8)(-10)$ **174.** $-11(-3)(-5)(-2)$

175. $7(9)(-11)4$ **176.** $-12(-4)7(-2)$ **177.** $(-14)9(-11)0$ **178.** $(-13)(15)(-19)0$

Objective E   *Application Problems*

The elevation, or height, of places on earth is measured in relation to sea level—that is, the average level of the ocean's surface. The following table shows height above sea level as a positive number and shows depth below sea level as a negative number.

Place	Elevation (in feet)
Mt. Everest	29,028
Mt. Aconcagua	23,035
Mt. McKinley	20,320
Mt. Kilimanjaro	19,340
Salinas Grandes	−131
Death Valley	−282
Qattara Depression	−436
Dead Sea	−1286

179. Use the table to find the difference in elevation between Mt. McKinley and Death Valley (the highest and lowest points in North America).

180. Use the table to find the difference in elevation between Mt. Kilimanjaro and the Qattara Depression (the highest and lowest points in Africa).

181. Use the table to find the difference in elevation between Mt. Everest and the Dead Sea (the highest and lowest points in Asia).

182. Use the table to find the difference in elevation between Mt. Aconcagua and Salinas Grandes (the highest and lowest points in South America).

A meteorologist may report a wind-chill temperature. This is the equivalent temperature, including the effects of wind and temperature, that a person would feel in calm air conditions. The table below gives the wind-chill temperature for various wind speeds and temperatures. For instance, when the temperature is 5°F and the wind is blowing at 15 mph, the wind-chill temperature is −25°F. Use this table for Exercises 183 to 185.

Wind Chill Factors

Wind		Calm Air Temperature (Fahrenheit)						
		15°F	*10°F*	*5°F*	*0°F*	*−5°F*	*−10°F*	*−15°F*
Speed	*5*	12	7	0	−5	−10	−15	−21
(Miles	*10*	−3	−9	−15	−22	−27	−34	−40
per	*15*	−11	−18	−25	−31	−38	−45	−51
Hour)	*20*	−17	−24	−31	−39	−46	−53	−60

183. What is the difference between a calm air temperature of −10°F and the wind-chill temperature when the wind is blowing 15 mph?

184. What is the difference between a calm air temperature of 10°F and the wind-chill temperature when the wind is blowing 10 mph?

185. For a wind speed of 15 mph, when the calm air temperature decreases by 5°F, does the wind-chill temperature decrease by the same number of degrees for each of the temperatures shown? If not, determine the largest decrease in temperature.

Solve.

186. On January 22, 1943, the temperature at Spearfish, South Dakota, rose from $-4°F$ to $45°F$ in two minutes. How many degrees did the temperature rise during those two minutes?

187. In a 24-hour period in January of 1916, the temperature in Browning, Montana, dropped from $44°F$ to $-56°F$. How many degrees did the temperature drop during that time?

188. The high temperatures for a six-day period in Barrow, Alaska, were $-23°F$, $-29°F$, $-21°F$, $-28°F$, $-28°F$, and $-27°F$. Calculate the average daily high temperature.

189. The low temperatures for a ten-day period in a midwestern city were $-4°F$, $-9°F$, $-5°F$, $-2°F$, $4°F$, $-1°F$, $-1°F$, $-2°F$, $-2°F$, and $2°F$. Calculate the average daily low temperature for this city.

190. To discourage random guessing on a multiple-choice exam, a professor assigns 5 points for a correct answer, -2 points for an incorrect answer, and 0 points for leaving the question blank. What is the score for a student who had 20 correct answers, had 13 incorrect answers, and left 7 questions blank?

191. To discourage random guessing on a multiple-choice exam, a professor assigns 7 points for a correct answer, -3 points for an incorrect answer, and -1 point for leaving the question blank. What is the score for a student who had 17 correct answers, had 8 incorrect answers, and left 2 questions blank?

APPLYING THE CONCEPTS

192. Let $z \in \{-15, -10, 0, 5, 10\}$. For which values of z is the inequality $-z > -|z|$ a true statement?

193. Let $z \in \{-5, 0, 6\}$, for which value of z is the expression $4 - z$ the smallest?

194. If $x \in \{-5, 0, 6\}$, for which value of x does the expression $2x$ have the smallest value?

195. If x represents a negative integer, then $-x$ represents a _____ integer.

196. Determine whether the statement is always true, sometimes true, or never true.
   a. If a and b are integers and $a < b$, then $|a| < |b|$.
   b. If x is an integer, then $|x| < -3$.
   c. If a and b are integers, then $|a + b| = |a| + |b|$.
   d. If a and b are integers, then $|a - b| = |a| - |b|$.

197.
[W] Explain why division by zero is not allowed.

198.
[W] If $-4x$ equals a positive integer, is x a positive or a negative integer? Explain your answer.

1.2 Rational and Irrational Numbers

Objective A *To write a rational number as a decimal* ....

A *rational number* is the quotient of two integers. A rational number written in this way is commonly called a fraction. Here are some examples of rational numbers.

$$\frac{3}{4}, \quad \frac{-4}{9}, \quad \frac{15}{-4}, \quad \frac{8}{1}, \quad -\frac{5}{6}$$

> **Rational Numbers**
>
> A **rational number** is a number that can be written in the form $\frac{a}{b}$, where a and b are integers and $b \neq 0$.

Because an integer can be written as the quotient of the integer and 1, every integer is a rational number. For instance,

$$\frac{6}{1} = 6 \qquad \frac{-8}{1} = -8$$

A number written in **decimal notation** is also a rational number.

three-tenths $0.3 = \frac{3}{10}$ forty-three thousandths $0.043 = \frac{43}{1000}$

A rational number written as a fraction can be written in decimal notation.

Write $\frac{5}{8}$ as a decimal.

The fraction bar can be read "÷".

$$\frac{5}{8} = 5 \div 8$$

$$\begin{array}{r} 0.625 \\ 8\overline{)5.000} \\ -4\,8 \\ \hline 20 \\ -16 \\ \hline 40 \\ -40 \\ \hline 0 \end{array}$$ ← This is called a **terminating decimal.**

0 ← The remainder is zero.

$$\frac{5}{8} = 0.625$$

Write $\frac{4}{11}$ as a decimal.

$$\begin{array}{r} 0.3636\ldots \\ 11\overline{)4.0000} \\ -3\,3 \\ \hline 70 \\ -66 \\ \hline 40 \\ -33 \\ \hline 70 \\ -66 \\ \hline 4 \end{array}$$ ← This is called a **repeating decimal.**

4 ← The remainder is never zero.

$$\frac{4}{11} = 0.\overline{36}$$ ← The bar over the digits 3 and 6 is used to show that these digits repeat.

Example 1   Write $\dfrac{8}{11}$ as a decimal. Place a bar over the repeating digits of the decimal.

You Try It 1   Write $\dfrac{4}{9}$ as a decimal. Place a bar over the repeating digits of the decimal.

Solution $\dfrac{8}{11} = 8 \div 11 = 0.7272\ldots = 0.\overline{72}$

Your solution

Solution on p. A2

Objective B   *To convert among percents, fractions, and decimals* ....

Percent means "parts of 100." Thus 27% means 27 parts of 100. In applied problems involving percent, it may be necessary to rewrite a percent as a fraction or decimal or to rewrite a fraction or decimal as a percent.

To write a percent as a fraction, remove the percent sign and multiply by $\dfrac{1}{100}$.

➡ Write 27% as a fraction.

$$27\% = 27\left(\dfrac{1}{100}\right) = \dfrac{27}{100}$$ • **Remove the percent sign and multiply by $\dfrac{1}{100}$.**

To write a percent as a decimal, remove the percent sign and multiply by 0.01.

$$33\% \quad = \quad 33(0.01) \quad = \quad 0.33$$

> Move the decimal point two places to the left. Then remove the percent sign.

A fraction or decimal can be written as a percent by multiplying by 100%.

$$\dfrac{5}{8} = \dfrac{5}{8}(100\%) = \dfrac{500}{8}\% = 62.5\%, \text{ or } 62\dfrac{1}{2}\%$$

$$0.82 \quad = \quad 0.82(100\%) \quad = \quad 82\%$$

> Move the decimal point two places to the right. Then write the percent sign.

Example 2
Write 130% as a fraction and as a decimal.

You Try It 2
Write 125% as a fraction and as a decimal.

Solution

$$130\% = 130\left(\dfrac{1}{100}\right) = \dfrac{130}{100} = 1\dfrac{3}{10}$$

$$130\% = 130(0.01) = 1.30$$

Your solution

Solution on p. A2

Example 3   Write $\dfrac{5}{6}$ as a percent.

Solution $\dfrac{5}{6} = \dfrac{5}{6}(100\%) = \dfrac{500}{6}\% = 83\dfrac{1}{3}\%$

You Try It 3   Write $\dfrac{1}{3}$ as a percent.

Your solution

Example 4   Write 0.092 as a percent.

Solution $0.092 = 0.092(100\%) = 9.2\%$

You Try It 4   Write 0.043 as a percent.

Your solution

Solutions on p. A2

Objective C   *To add or subtract rational numbers* ....

Fractions with the same denominator are added by adding the numerators and placing the sum over the common denominator.

> **Addition of Fractions**
>
> To add two fractions with the same denominator, add the numerators and place the sum over the common denominator.
>
> $$\dfrac{a}{c} + \dfrac{b}{c} = \dfrac{a + b}{c}$$

To add fractions with different denominators, first rewrite the fractions as equivalent fractions with a common denominator. Then add the fractions.

The common denominator is the **least common multiple** (LCM) of the denominators. This is the smallest number that is a multiple of each of the denominators.

➡ Add: $-\dfrac{5}{6} + \dfrac{3}{10}$

The LCM of 6 and 10 is 30. This is frequently called the **least common denominator.** Rewrite the fractions as equivalent fractions with the denominator 30. Then add the fractions.

$$-\dfrac{5}{6} + \dfrac{3}{10} = -\dfrac{25}{30} + \dfrac{9}{30} = \dfrac{-25 + 9}{30} = \dfrac{-16}{30} = -\dfrac{8}{15}$$

To subtract fractions, subtract the numerators and place the difference over the common denominator.

➡ Subtract: $-\dfrac{4}{9} - \left(-\dfrac{7}{12}\right)$

The LCM of 9 and 12 is 36. Rewrite the fractions as equivalent fractions with the denominator 36. Then subtract the fractions.

$$-\dfrac{4}{9} - \left(-\dfrac{7}{12}\right) = -\dfrac{16}{36} - \left(-\dfrac{21}{36}\right) = \dfrac{-16 - (-21)}{36} = \dfrac{-16 + 21}{36} = \dfrac{5}{36}$$

CONSIDER THIS

You can find the LCM by multiplying the denominators and then dividing by the *common factor* of the two denominators. In the case of 6 and 10, $6 \cdot 10 = 60$. Now divide by 2, the common factor of 6 and 10.

$60 \div 2 = 30.$

To add or subtract decimals, write the numbers so that the decimal points are in a vertical line. Then proceed as in the addition or subtraction of integers. Write the decimal point in the answer directly below the decimal points in the problem.

➡ Add: $-114.039 + 84.76$

$|-114.039| = 114.039$
$|84.76| = 84.76$

- The signs are different. Find the absolute value of each number.

$$\begin{array}{r} 114.039 \\ -84.76 \\ \hline 29.279 \end{array}$$

- Subtract the smaller of these numbers from the larger.

$-114.039 + 84.76 = -29.279$

- Attach the sign of the number with the larger absolute value. Because $|-114.039| > |84.76|$, use the sign of -114.039.

Example 5   Simplify: $-\dfrac{3}{4} + \dfrac{1}{6} - \dfrac{5}{8}$

Solution   The LCM of 4, 6, and 8 is 24.

$$-\frac{3}{4} + \frac{1}{6} - \frac{5}{8} = -\frac{18}{24} + \frac{4}{24} - \frac{15}{24}$$

$$= \frac{-18 + 4 - 15}{24}$$

$$= \frac{-29}{24} = -\frac{29}{24}$$

You Try It 5   Simplify: $-\dfrac{7}{8} - \dfrac{5}{6} + \dfrac{3}{4}$

Your solution

Example 6   Subtract: $42.987 - 98.61$

Solution $42.987 - 98.61$
$= 42.987 + (-98.61)$
$= -55.623$

You Try It 6   Subtract: $16.127 - 67.91$

Your solution

Solutions on p. A2

Objective D   ***To multiply or divide rational numbers*** ....

The product of two fractions is the product of the numerators divided by the product of the denominators.

➡ Multiply: $\dfrac{3}{8} \cdot \dfrac{12}{17}$

$$\frac{3}{8} \cdot \frac{12}{17} = \frac{3 \cdot 12}{8 \cdot 17}$$

- Multiply the numerators. Multiply the denominators.

$$= \frac{3 \cdot \overset{1}{\cancel{2}} \cdot \overset{1}{\cancel{2}} \cdot 3}{2 \cdot \underset{1}{\cancel{2}} \cdot \underset{1}{\cancel{2}} \cdot 17}$$

- Write the prime factorization of each factor. Divide by the common factors.

$$= \frac{9}{34}$$

- Multiply the factors in the numerator and in the denominator.

To divide fractions, invert the divisor. Then multiply the fractions.

➡ Divide: $\dfrac{3}{10} \div \left(-\dfrac{18}{25}\right)$

The signs are different. The quotient is negative.

$$\frac{3}{10} \div \left(-\frac{18}{25}\right) = -\left(\frac{3}{10} \div \frac{18}{25}\right) = -\left(\frac{3}{10} \cdot \frac{25}{18}\right) = -\left(\frac{3 \cdot 25}{10 \cdot 18}\right)$$

$$= -\left(\frac{\overset{1}{\cancel{3}} \cdot \overset{1}{\cancel{5}} \cdot 5}{2 \cdot \underset{1}{\cancel{5}} \cdot 2 \cdot \underset{1}{\cancel{3}} \cdot 3}\right) = -\frac{5}{12}$$

To multiply decimals, multiply as with integers. Write the decimal point in the product so that the number of decimal places in the product equals the sum of the decimal places in the factors.

➡ Multiply: $-6.89(0.00035)$

$$
\begin{array}{r}
6.89 \\
\times\, 0.00035 \\
\hline
3445 \\
2067 \\
\hline
0.0024115
\end{array}
$$

2 decimal places
5 decimal places   • **Multiply the absolute values.**

7 decimal places

$-6.89(0.00035) = -0.0024115$ • **The signs are different. The product is negative.**

To divide decimals, move the decimal point in the divisor to the right to make it a whole number. Move the decimal point in the dividend the same number of places to the right. Place the decimal point in the quotient directly over the decimal point in the dividend. Then divide as with whole numbers.

➡ Divide: $1.32 \div 0.27$. Round to the nearest tenth.

CONSIDER THIS
The symbol $\approx$ is used to indicate that the quotient is an approximate value that has been rounded off.

$$
\begin{array}{r}
4.88 \approx 4.9 \\
0.27\overline{)1.32.00} \\
-1\,08 \\
\hline
240 \\
-216 \\
\hline
240 \\
-216 \\
\hline
24
\end{array}
$$

• **Move the decimal point 2 places to the right in the divisor and then in the dividend. Place the decimal point in the quotient.**

Example 7   Divide: $-\dfrac{5}{8} \div \left(-\dfrac{5}{40}\right)$

Solution   The quotient is positive.

$$-\frac{5}{8} \div \left(-\frac{5}{40}\right) = \frac{5}{8} \div \frac{5}{40} = \frac{5}{8} \cdot \frac{40}{5} = \frac{5 \cdot 40}{8 \cdot 5}$$

$$= \frac{\overset{1}{\cancel{5}} \cdot \overset{1}{\cancel{2}} \cdot \overset{1}{\cancel{2}} \cdot \overset{1}{\cancel{2}} \cdot 5}{\underset{1}{\cancel{2}} \cdot \underset{1}{\cancel{2}} \cdot \underset{1}{\cancel{2}} \cdot \underset{1}{\cancel{5}}} = \frac{5}{1} = 5$$

You Try It 7   Divide: $-\dfrac{3}{8} \div \left(-\dfrac{5}{12}\right)$

Your solution

Solution on p. A2

Example 8 Multiply: $-4.29(8.2)$

Solution The product is negative.

$$
\begin{array}{r}
4.29 \\
\times 8.2 \\
\hline
858 \\
3432 \\
\hline
35.178
\end{array}
$$

$-4.29(8.2) = -35.178$

You Try It 8 Multiply: $-5.44(3.8)$

Your solution

Solution on p. A2

Objective E *To evaluate exponential expressions* ....

Repeated multiplication of the same factor can be written using an exponent.

POINT OF INTEREST
René Descartes (1596–1650)
was the first mathematician
to extensively use exponential
notation as it is used today.
However, for some unknown
reason, he always used xx
for x^2.

$$2 \cdot 2 \cdot 2 \cdot 2 \cdot 2 = 2^5 \leftarrow \textbf{exponent}$$
$$\underset{\textbf{base}}{\uparrow\!\!\!\!\!\_\_\_}$$

$$a \cdot a \cdot a \cdot a = a^4 \leftarrow \textbf{exponent}$$
$$\underset{\textbf{base}}{\uparrow\!\!\!\!\!\_\_\_}$$

The **exponent** indicates how many times the factor, called the **base,** occurs in the multiplication. The multiplication $2 \cdot 2 \cdot 2 \cdot 2 \cdot 2$ is in **factored form.** The exponential expression 2^5 is in **exponential form.**

2^1 is read "the first power of 2" or just 2. Usually the exponent 1 is not written.

2^2 is read "the second power of 2" or "2 squared."

2^3 is read "the third power of 2" or "2 cubed."

2^4 is read "the fourth power of 2."

a^4 is read "the fourth power of a."

There is a geometric interpretation of the first three natural-number powers.

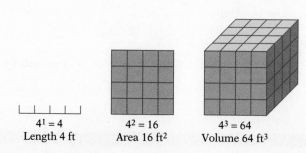

| $4^1 = 4$ | $4^2 = 16$ | $4^3 = 64$ |
| Length 4 ft | Area 16 ft² | Volume 64 ft³ |

To evaluate an exponential expression, write each factor as many times as indicated by the exponent. Then multiply.

➡ Evaluate $(-2)^4$.

$$(-2)^4 = (-2)(-2)(-2)(-2)$$ ● Write (-2) as a factor 4 times.
$$= 16$$ ● Multiply.

➡ Evaluate -2^4.

$$-2^4 = -(2 \cdot 2 \cdot 2 \cdot 2)$$ ● Write 2 as a factor 4 times.
$$= -16$$ ● Multiply.

From these last two examples, note the difference between $(-2)^4$ and -2^4.

$$(-2)^4 = 16$$
$$-2^4 = -(2^4) = -16$$

Example 9   Evaluate -5^3.

Solution $-5^3 = -(5 \cdot 5 \cdot 5) = -125$

You Try It 9   Evaluate -6^3.

Your solution

Example 10   Evaluate $(-4)^4$.

Solution $(-4)^4 = (-4)(-4)(-4)(-4)$
$$= 256$$

You Try It 10   Evaluate $(-3)^4$.

Your solution

Example 11   Evaluate $(-3)^2 \cdot 2^3$

Solution $(-3)^2 \cdot 2^3 = (-3)(-3) \cdot (2)(2)(2)$
$$= 9 \cdot 8 = 72$$

You Try It 11   Evaluate $(3^3)(-2)^3$.

Your solution

Example 12   Evaluate $\left(-\dfrac{2}{3}\right)^3$.

Solution $\left(-\dfrac{2}{3}\right)^3 = \left(-\dfrac{2}{3}\right)\left(-\dfrac{2}{3}\right)\left(-\dfrac{2}{3}\right)$
$$= -\dfrac{2 \cdot 2 \cdot 2}{3 \cdot 3 \cdot 3} = -\dfrac{8}{27}$$

You Try It 12   Evaluate $\left(-\dfrac{2}{5}\right)^2$.

Your solution

Example 13   Evaluate $-4(0.7)^2$.

Solution $-4(0.7)^2 = -4(0.7)(0.7)$
$$= -2.8(0.7) = -1.96$$

You Try It 13   Evaluate $-3(0.3)^3$.

Your solution

Solutions on p. A2

Objective F   *To simplify numerical radical expressions* ....

A **square root** of a positive number x is a number whose square is x.

A square root of 16 is 4 because $4^2 = 16$.
A square root of 16 is -4 because $(-4)^2 = 16$.

Every positive number has two square roots, one a positive and one a negative number. The symbol "$\sqrt{}$," called a **radical sign,** is used to indicate the positive or **principal square root** of a number. For example, $\sqrt{16} = 4$ and $\sqrt{25} = 5$. The number under the radical sign is called the **radicand.**

When the negative square root of a number is to be found, a negative sign is placed in front of the radical. For example, $-\sqrt{16} = -4$ and $-\sqrt{25} = -5$.

The square of an integer is a **perfect square.** 49, 81, and 144 are examples of perfect squares.

$$7^2 = 49$$
$$9^2 = 81$$
$$12^2 = 144$$

An integer that is a perfect square can be written as the product of prime factors, each of which has an even exponent when expressed in exponential form.

$$49 = 7 \cdot 7 = 7^2$$
$$81 = 3 \cdot 3 \cdot 3 \cdot 3 = 3^4$$
$$144 = 2 \cdot 2 \cdot 2 \cdot 2 \cdot 3 \cdot 3 = 2^4 3^2$$

➡ Simplify $\sqrt{625}$.

$\sqrt{625} = \sqrt{5^4}$ • Write the prime factorization of the radicand in exponential form.

$\qquad = 5^2$ • Remove the radical sign and multiply the exponent by $\frac{1}{2}$.

$\qquad = 25$ • Simplify.

If a number is not a perfect square, its square root can only be approximated. For example, 2 and 7 are not perfect squares. The square roots of these numbers are **irrational numbers.** Their decimal representations never terminate or repeat.

$$\sqrt{2} \approx 1.4142135\ldots \qquad \sqrt{7} \approx 2.6457513\ldots$$

Recall that rational numbers are fractions such as $-\frac{6}{7}$ or $\frac{10}{3}$, where the numerator and denominator are integers. Rational numbers are also represented by repeating decimals, such as $0.25767676\ldots$, and by terminating decimals, such as 1.73. An irrational number is neither a repeating nor a terminating decimal. For instance, $2.45445444544445\ldots$ is an irrational number.

Real Numbers

The rational numbers and the irrational numbers taken together are called the **real numbers.**

A radical expression is in simplest form when the radicand contains no factor greater than 1 that is a perfect square. The Product Property of Square Roots is used to simplify radical expressions.

The Product Property of Square Roots

If a and b are positive real numbers, then $\sqrt{ab} = \sqrt{a} \cdot \sqrt{b}$.

➡ Simplify: $\sqrt{360}$

$$\sqrt{360} = \sqrt{2^3 \cdot 3^2 \cdot 5}$$

  • Write the prime factorization of the radicand.

$$= \sqrt{(2^2 \cdot 3^2)(2 \cdot 5)}$$

  • Write the radicand as a product of a perfect square and factors that do not contain perfect-square factors.

$$= \sqrt{2^2 \cdot 3^2}\,\sqrt{2 \cdot 5}$$

  • Use the Product Property of Square Roots.

$$= (2 \cdot 3)\sqrt{10}$$

  • Take the square root of the perfect squares.

$$= 6\sqrt{10}$$

From the last example, note that $\sqrt{360} = 6\sqrt{10}$. The two expressions are different representations of the same number. Using a calculator, we find that $\sqrt{360} \approx 18.973666$ and $6\sqrt{10} \approx 6(3.1622777) = 18.973666$.

➡ Simplify: $\sqrt{-16}$

Because the square of any real number is positive, there is no real number whose square is -16. $\sqrt{-16}$ is not a real number.

Example 14   Simplify: $3\sqrt{90}$

Solution $3\sqrt{90} = 3\sqrt{2 \cdot 3^2 \cdot 5}$
$$= 3\sqrt{3^2(2 \cdot 5)}$$
$$= 3\sqrt{3^2}\,\sqrt{2 \cdot 5}$$
$$= 3 \cdot 3\sqrt{10} = 9\sqrt{10}$$

You Try It 14   Simplify: $-5\sqrt{32}$

Your solution

Example 15   Simplify: $\sqrt{252}$

Solution $\sqrt{252} = \sqrt{2^2 \cdot 3^2 \cdot 7}$
$$= \sqrt{2^2 \cdot 3^2}\,\sqrt{7}$$
$$= 2 \cdot 3\sqrt{7}$$
$$= 6\sqrt{7}$$

You Try It 15   Simplify: $\sqrt{216}$

Your solution

Solutions on pp. A2–A3

Objective G *To solve application problems* ....

An article in *Business Week* magazine reported that the U.S. budget deficit for 1993 was $236.4 billion. The number 236.4 billion means

$$236.4 \cdot \underbrace{1,000,000,000}_{\text{one billion}} = 236,400,000,000$$

Numbers such as 236.4 billion are used in many instances because they are easy to read and offer an approximation to the actual number. Such numbers are used in Example 16 and You Try It 16.

One of the applications of percent is to express a portion of a total as a percent. For instance, a recent survey of 450 mall shoppers found that 270 preferred the mall closer to their home even though it did not have the same store variety as a mall farther from home. The percent of shoppers who preferred the mall closer to home can be found by converting a fraction to a percent.

$$\frac{\text{number preferring the mall close to home}}{\text{total number surveyed}} = \frac{270}{450}$$
$$= 0.60 = 60\%$$

Example 16
The graph below shows the profit and loss of Hartmarx for the years 1990 through 1994.

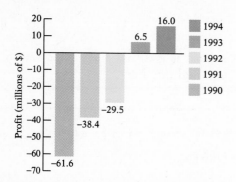

What was the difference between the profit in 1994 and the loss in 1990?

Strategy
To find the difference between the profit in 1994 and the loss in 1990, read the graph. Subtract the loss in 1990 (−$61.6 million) from the profit in 1994 ($16.0 million).

Solution
$16.0 − (−61.6) = 16.0 + 61.6 = 77.6$

The difference between the profit in 1994 and the loss in 1990 was $77.6 million.

You Try It 16
The circle graph below shows the amount spent in the United States for the most commonly prescribed drugs.

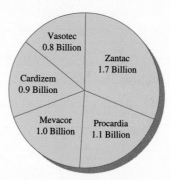

What percent of the total amount spent was spent for Procardia?

Your strategy

Your solution

Solution on p. A3

Content and Format © 1996 HMCo.

1.2 Exercises

Objective A

Write as a decimal. Place a bar over the repeating digits of a repeating decimal.

1. $\dfrac{1}{8}$ **2.** $\dfrac{7}{8}$ **3.** $\dfrac{2}{9}$ **4.** $\dfrac{8}{9}$ **5.** $\dfrac{1}{6}$

6. $\dfrac{5}{6}$ **7.** $\dfrac{9}{16}$ **8.** $\dfrac{15}{16}$ **9.** $\dfrac{7}{12}$ **10.** $\dfrac{11}{12}$

11. $\dfrac{6}{25}$ **12.** $\dfrac{14}{25}$ **13.** $\dfrac{9}{40}$ **14.** $\dfrac{21}{40}$ **15.** $\dfrac{5}{11}$

Objective B

Write as a fraction and as a decimal.

16. 75%   **17.** 40%   **18.** 64%   **19.** 88%   **20.** 125%

21. 160%   **22.** 19%   **23.** 87%   **24.** 5%   **25.** 450%

Write as a fraction.

26. $11\dfrac{1}{9}\%$ **27.** $4\dfrac{2}{7}\%$ **28.** $12\dfrac{1}{2}\%$ **29.** $37\dfrac{1}{2}\%$ **30.** $66\dfrac{2}{3}\%$

31. $\dfrac{1}{4}\%$ **32.** $\dfrac{1}{2}\%$ **33.** $6\dfrac{1}{4}\%$ **34.** $83\dfrac{1}{3}\%$ **35.** $5\dfrac{3}{4}\%$

Write as a decimal.

36. 7.3%   **37.** 9.1%   **38.** 15.8%   **39.** 16.7%   **40.** 0.3%

41. 0.9%   **42.** 9.9%   **43.** 9.15%   **44.** 121.2%   **45.** 18.23%

Write as a percent.

46. 0.15   **47.** 0.37   **48.** 0.05   **49.** 0.02   **50.** 0.175

51. 0.125   **52.** 1.15   **53.** 1.36   **54.** 0.008   **55.** 0.004

56. $\dfrac{27}{50}$ **57.** $\dfrac{83}{100}$ **58.** $\dfrac{1}{3}$ **59.** $\dfrac{3}{8}$ **60.** $\dfrac{5}{11}$

61. $\dfrac{4}{9}$ **62.** $\dfrac{7}{8}$ **63.** $\dfrac{9}{20}$ **64.** $1\dfrac{2}{3}$ **65.** $2\dfrac{1}{2}$

Objective C

Add or subtract.

66. $\dfrac{1}{2} + \dfrac{3}{8}$ **67.** $\dfrac{5}{8} - \dfrac{5}{6}$ **68.** $\dfrac{1}{9} - \dfrac{5}{27}$ **69.** $-\dfrac{5}{12} - \dfrac{3}{8}$

70. $-\dfrac{5}{6} - \dfrac{5}{9}$ **71.** $-\dfrac{6}{13} + \dfrac{17}{26}$ **72.** $-\dfrac{7}{12} + \dfrac{5}{8}$ **73.** $\dfrac{5}{8} - \left(-\dfrac{3}{4}\right)$

74. $\dfrac{3}{5} - \dfrac{11}{12}$ **75.** $\dfrac{11}{12} - \dfrac{5}{6}$ **76.** $-\dfrac{2}{3} - \left(-\dfrac{11}{18}\right)$ **77.** $-\dfrac{5}{8} - \left(-\dfrac{11}{12}\right)$

78. $\dfrac{1}{3} + \dfrac{5}{6} - \dfrac{2}{9}$ **79.** $\dfrac{1}{2} - \dfrac{2}{3} + \dfrac{1}{6}$ **80.** $-\dfrac{5}{16} + \dfrac{3}{4} - \dfrac{7}{8}$ **81.** $\dfrac{1}{2} - \dfrac{3}{8} - \left(-\dfrac{1}{4}\right)$

82. $\dfrac{3}{4} - \left(-\dfrac{7}{12}\right) - \dfrac{7}{8}$ **83.** $\dfrac{1}{3} - \dfrac{1}{4} - \dfrac{1}{5}$ **84.** $\dfrac{2}{3} - \dfrac{1}{2} + \dfrac{5}{6}$ **85.** $\dfrac{5}{16} + \dfrac{1}{8} - \dfrac{1}{2}$

Add or subtract.

86. $7.56 + 0.462$

87. $1.09 + 6.2$

88. $-32.1 - 6.7$

89. $5.13 - 8.179$

90. $-13.092 + 6.9$

91. $2.54 - 3.6$

92. $5.43 + 7.925$

93. $-16.92 - 6.925$

94. $-3.87 + 8.546$

95. $6.9027 - 17.692$

96. $2.09 - 6.72 - 5.4$

97. $-18.39 + 4.9 - 23.7$

98. $19 - (-3.72) - 82.75$

99. $-3.07 - (-2.97) - 17.4$

100. $16.4 - (-3.09) - 7.93$

101. $-3.09 - 4.6 - (-27.3)$

102. $2.66 - (-4.66) - 8.2$

Objective D

Multiply or divide.

103. $\dfrac{1}{2}\left(-\dfrac{3}{4}\right)$

104. $-\dfrac{2}{9}\left(-\dfrac{3}{14}\right)$

105. $\left(-\dfrac{3}{8}\right)\left(-\dfrac{4}{15}\right)$

106. $\left(-\dfrac{3}{4}\right)\left(-\dfrac{8}{27}\right)$

107. $-\dfrac{1}{2}\left(\dfrac{8}{9}\right)$

108. $\dfrac{5}{12}\left(-\dfrac{8}{15}\right)$

109. $\dfrac{5}{8}\left(-\dfrac{7}{12}\right)\dfrac{16}{25}$

110. $\left(\dfrac{5}{12}\right)\left(-\dfrac{8}{15}\right)\left(-\dfrac{1}{3}\right)$

111. $\dfrac{1}{2}\left(-\dfrac{3}{4}\right)\left(-\dfrac{5}{8}\right)$

112. $\dfrac{3}{8} \div \dfrac{1}{4}$

113. $\dfrac{5}{6} \div \left(-\dfrac{3}{4}\right)$

114. $-\dfrac{5}{12} \div \dfrac{15}{32}$

115. $-\dfrac{7}{8} \div \dfrac{4}{21}$

116. $\dfrac{7}{10} \div \dfrac{2}{5}$

117. $-\dfrac{15}{64} \div \left(-\dfrac{3}{40}\right)$

118. $\dfrac{1}{8} \div \left(-\dfrac{5}{12}\right)$

119. $-\dfrac{4}{9} \div \left(-\dfrac{2}{3}\right)$

120. $-\dfrac{6}{11} \div \dfrac{4}{9}$

Multiply or divide.

121. 1.2(3.47)

122. (−0.8)6.2

123. (−1.89)(−2.3)

124. (6.9)(−4.2)

125. 1.06(−3.8)

126. −2.7(−3.5)

127. 1.2(−0.5)(3.7)

128. −2.4(6.1)(0.9)

129. 2.3(−0.6)(0.8)

130. −1.2(−0.55)(1.9)

131. 0.44(−2.3)(−0.5)

132. −3.4(−22.1)(−0.5)

133. 1.8(0.33)(−0.4)

134. 4.5(−0.22)(−0.8)

135. −24.7 ÷ 0.09

Divide. Round to the nearest hundredth.

136. −1.27 ÷ (−1.7)

137. 9.07 ÷ (−3.5)

138. 0.0976 ÷ 0.042

139. −6.904 ÷ 1.35

140. −7.894 ÷ (−2.06)

141. −354.2086 ÷ 0.1719

Objective E

Evaluate.

142. 6^2

143. 7^4

144. -7^2

145. -4^3

146. $(-3)^2$

147. $(-2)^3$

148. $(-3)^4$

149. $(-5)^3$

150. $\left(\dfrac{1}{2}\right)^2$

151. $\left(-\dfrac{3}{4}\right)^3$

152. $(0.3)^2$

153. $(1.5)^3$

154. $\left(\dfrac{2}{3}\right)^2 \cdot 3^3$

155. $\left(-\dfrac{1}{2}\right)^3 \cdot 8$

156. $(0.3)^3 \cdot 2^3$

157. $(0.5)^2 \cdot 3^3$

158. $(-3) \cdot 2^2$

159. $(-5) \cdot 3^4$

160. $(-2) \cdot (-2)^3$

161. $(-2) \cdot (-2)^2$

162. $2^3 \cdot 3^3 \cdot (-4)$

163. $(-3)^3 \cdot 5^2 \cdot 10$

164. $(-7) \cdot 4^2 \cdot 3^2$

165. $(-2) \cdot 2^3 \cdot (-3)^2$

166. $\left(\dfrac{2}{3}\right)^2 \cdot \dfrac{1}{4} \cdot 3^3$

167. $\left(\dfrac{3}{4}\right)^2 \cdot (-4) \cdot 2^3$

168. $8^2 \cdot (-3)^5 \cdot 5$

Objective F

Simplify.

169. $\sqrt{16}$ **170.** $\sqrt{64}$ **171.** $\sqrt{49}$ **172.** $\sqrt{144}$ **173.** $\sqrt{32}$ **174.** $\sqrt{50}$

175. $\sqrt{8}$ **176.** $\sqrt{12}$ **177.** $6\sqrt{18}$ **178.** $-3\sqrt{48}$ **179.** $5\sqrt{40}$ **180.** $2\sqrt{28}$

181. $\sqrt{15}$ **182.** $\sqrt{21}$ **183.** $\sqrt{29}$ **184.** $\sqrt{13}$ **185.** $-9\sqrt{72}$ **186.** $11\sqrt{80}$

187. $\sqrt{45}$ **188.** $\sqrt{225}$ **189.** $\sqrt{0}$ **190.** $\sqrt{210}$ **191.** $6\sqrt{128}$ **192.** $9\sqrt{288}$

Find the decimal approximation rounded to the nearest thousandth.

193. $\sqrt{240}$ **194.** $\sqrt{300}$ **195.** $\sqrt{288}$ **196.** $\sqrt{600}$ **197.** $\sqrt{256}$ **198.** $\sqrt{324}$

199. $\sqrt{275}$ **200.** $\sqrt{450}$ **201.** $\sqrt{245}$ **202.** $\sqrt{525}$ **203.** $\sqrt{352}$ **204.** $\sqrt{363}$

Objective G   Application Problems

205. The table at the right shows the number of wins and losses for the Midwest League of the Western Conference of the National Basketball Association for 1994.

Team	Won	Lost
Houston	58	24
San Antonio	55	27
Utah	53	29
Denver	42	40
Minnesota	20	62
Dallas	13	69

  a. Which team came closest to winning $\frac{2}{3}$ of its games?

  b. Which teams lost more than $\frac{3}{5}$ of their games?

  c. What percent of its games did Dallas lose? Round to the nearest tenth of a percent.

206. The results of a survey conducted in part by Nintendo of 5000 children are shown in the circle graph at the right. In the survey, the children were asked which characteristic they would most like to have.

  a. What percent of the total number of students surveyed selected "smart" as the characteristic they would most like to have?

  b. What percent of the total number of students surveyed did not choose "wealthy" as the characteristic they would most like to have?

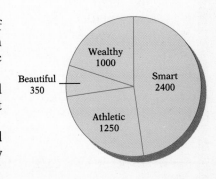

Wealthy 1000

Beautiful 350

Smart 2400

Athletic 1250

207. The net profits and losses for Rose Stores for the years 1990 through 1994 are shown in the graph at the right.
   a. What is the difference between the profit or loss in 1990 and that in 1992?
   b. What is the difference between the profit or loss in 1993 and that in 1991?

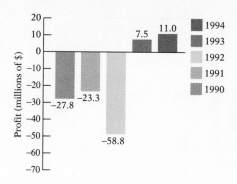

208. When a U.S. company does business with another country, it is necessary to convert U.S. currency into the currency of the other nation. These *exchange rates* are determined by various factors. The exchange rates for some currencies are shown in the table at the right.
   a. If you sold goods worth $1.2 million U.S. dollars to Australia, how many Australian dollars would you receive?
   b. How many U.S. dollars, to the nearest thousandth, are equivalent to 1 Irish punt?

Foreign Currency per U.S. Dollar

Australian dollar	1.5186
Belgian franc	34.80
Irish punt	0.6840

209. According to the Federal Highway Administration, the average car is driven approximately 10,300 mi per year and uses approximately 495 gal of gas. Assuming that the average cost of gasoline is $1.097 per gallon, which includes 44.3¢ for all taxes, and that there are 1.53 million cars on the road, determine how much total tax is paid for gasoline in one year. (You may not need all the data given in this problem.)

APPLYING THE CONCEPTS

210. Use a calculator to determine the decimal representations of $\frac{17}{99}, \frac{45}{99},$ and $\frac{73}{99}$. Make a conjecture as to the decimal representation of $\frac{83}{99}$. Does your conjecture work for $\frac{33}{99}$? What about $\frac{1}{99}$?

211. a. Find the two-digit perfect square that has exactly nine factors.
   b. Find two whole numbers such that their difference is 10, the smaller number is a perfect square, and the larger number is two less than a perfect square.

212. List the whole numbers between $\sqrt{8}$ and $\sqrt{90}$.

213. Describe in your own words how to simplify a radical expression.
[W]

214. Explain why $2\sqrt{2}$ is in simplest form and $\sqrt{8}$ is not in simplest form.
[W]

1.3 The Order of Operations Agreement

Objective A   *To use the Order of Operations Agreement to simplify expressions* ....

Let's evaluate $2 + 3 \cdot 5$.

There are two arithmetic operations, addition and multiplication, in this expression. The operations could be performed in different orders.

Multiply first.	$2 + \underline{3 \cdot 5}$	Add first.	$\underline{2 + 3} \cdot 5$
Then add.	$\underline{2 + 15}$	Then multiply.	$\underline{5 \cdot 5}$
	17		25

In order to prevent there being more than one answer for a numerical expression, an Order of Operations Agreement has been established.

The Order of Operations Agreement

> **Step 1** Perform operations inside grouping symbols. Grouping symbols include parentheses ( ), brackets [ ], braces { }, absolute value symbols | |, and the fraction bar.
> **Step 2** Simplify exponential expressions.
> **Step 3** Do multiplication and division as they occur from left to right.
> **Step 4** Do addition and subtraction as they occur from left to right.

➡ Evaluate $12 - 24(8 - 5) \div 2^2$.

$$12 - 24(8 - 5) \div 2^2 = 12 - 24(3) \div 2^2$$

• Perform operations inside grouping symbols.

$$= 12 - 24(3) \div 4$$

• Simplify exponential expressions.

$$= 12 - 72 \div 4$$

• Do multiplication and division as they occur from left to right.

$$= 12 - 18$$

$$= -6$$

• Do addition and subtraction as they occur from left to right.

One or more of the above steps may not be needed to evaluate an expression. In that case, proceed to the next step in the Order of Operations Agreement.

When an expression has grouping symbols inside grouping symbols, perform the operations inside the inner grouping symbols first.

➡ Evaluate $6 \div [4 - (6 - 8)] + 2^2$.

$6 \div [4 - (6 - 8)] + 2^2 = 6 \div [4 - (-2)] + 2^2$ • **Perform operations inside grouping symbols.**

$= 6 \div 6 + 2^2$

$= 6 \div 6 + 4$ • **Simplify exponential expressions.**

$= 1 + 4$ • **Do multiplication and division as they occur from left to right.**

$= 5$ • **Do addition and subtraction as they occur from left to right.**

Example 1
Evaluate $4 - 3[4 - 2(6 - 3)] \div 2$.

Solution
$4 - 3[4 - 2(6 - 3)] \div 2$
$= 4 - 3[4 - 2 \cdot 3] \div 2$
$= 4 - 3[4 - 6] \div 2$
$= 4 - 3[-2] \div 2$
$= 4 + 6 \div 2$
$= 4 + 3$
$= 7$

You Try It 1
Evaluate $18 - 5[8 - 2(2 - 5)] \div 10$.

Your solution

Example 2
Evaluate $27 \div (5 - 2)^2 + (-3)^2 \cdot 4$.

Solution
$27 \div (5 - 2)^2 + (-3)^2 \cdot 4$
$= 27 \div 3^2 + (-3)^2 \cdot 4$
$= 27 \div 9 + 9 \cdot 4$
$= 3 + 9 \cdot 4$
$= 3 + 36$
$= 39$

You Try It 2
Evaluate $36 \div (8 - 5)^2 - (-3)^2 \cdot 2$.

Your solution

Example 3
Evaluate $(1.75 - 1.3)^2 \div 0.025 + 6.1$.

Solution
$(1.75 - 1.3)^2 \div 0.025 + 6.1$
$= (0.45)^2 \div 0.025 + 6.1$
$= 0.2025 \div 0.025 + 6.1$
$= 8.1 + 6.1$
$= 14.2$

You Try It 3
Evaluate $(6.97 - 4.72)^2 \cdot 4.5 \div 0.05$.

Your solution

Solutions on p. A3

1.3 Exercises

· ·

___

Objective A

Evaluate by using the Order of Operations Agreement.

1. $4 - 8 \div 2$

2. $2^2 \cdot 3 - 3$

3. $2(3 - 4) - (-3)^2$

4. $16 - 32 \div 2^3$

5. $24 - 18 \div 3 + 2$

6. $8 - (-3)^2 - (-2)$

7. $8 - 2(3)^2$

8. $16 - 16 \cdot 2 \div 4$

9. $12 + 16 \div 4 \cdot 2$

10. $16 - 2 \cdot 4^2$

11. $27 - 18 \div (-3^2)$

12. $4 + 12 \div 3 \cdot 2$

13. $16 + 15 \div (-5) - 2$

14. $14 - 2^2 - (4 - 7)$

15. $14 - 2^2 - |4 - 7|$

16. $10 - |5 - 8| + 2^3$

17. $3 - 2[8 - (3 - 2)]$

18. $-2^2 + 4[16 \div (3 - 5)]$

19. $6 + \dfrac{16 - 4}{2^2 + 2} - 2$

20. $24 \div \dfrac{3^2}{8 - 5} - (-5)$

21. $18 \div |9 - 2^3| + (-3)$

22. $96 \div 2[12 + (6 - 2)] - 3^2$

23. $4[16 - (7 - 1)] \div 10$

24. $18 \div 2 - 4^2 - (-3)^2$

25. $20 \div (10 - 2^3) + (-5)$

26. $16 - 3(8 - 3)^2 \div 5$

27. $4(-8) \div [2(7 - 3)^2]$

28. $\dfrac{(-10) + (-2)}{6^2 - 30} \div |2 - 4|$

29. $16 - 4 \cdot \dfrac{3^3 - 7}{2^3 + 2} - (-2)^2$

30. $(0.2)^2 \cdot (-0.5) + 1.72$

31. $0.3(1.7 - 4.8) + (1.2)^2$

32. $(1.8)^2 - 2.52 \div 1.8$

33. $(1.65 - 1.05)^2 \div 0.4 + 0.8$

Evaluate by using the Order of Operations Agreement.

34. $\dfrac{3}{8} \div \left(\dfrac{5}{6} + \dfrac{2}{3} \right)$

35. $\left(\dfrac{5}{12} - \dfrac{9}{16} \right) \dfrac{3}{7}$

36. $\left(\dfrac{3}{4} \right)^2 - \left(\dfrac{1}{2} \right)^3 \div \dfrac{3}{5}$

APPLYING THE CONCEPTS

37. Find two fractions between $\dfrac{2}{3}$ and $\dfrac{3}{4}$. (There is more than one answer to this question.)

38. A magic square is one in which the numbers in every row, column, and diagonal sum to the same number. Complete the magic square at the right.

$\dfrac{2}{3}$		
	$\dfrac{1}{6}$	$\dfrac{5}{6}$
		$-\dfrac{1}{3}$

39. For each part below, find a rational number, r, that satisfies the condition.
  a. $r^2 < r$ **b.** $r^2 = r$ **c.** $r^2 > r$

40. In a survey of consumers, approximately 43% said they would be willing to pay between $1000 and $2000 more for a new car if the car had an EPA rating of 80 mpg. If your car now gets 28 mpg and you drive approximately 10,000 mi per year, in how many months would your savings on gasoline pay for the increased cost of such a car? Assume the average cost for gasoline is $1.06 per gallon.

41. Find three natural numbers a, b, and c such that $\dfrac{1}{a} + \dfrac{1}{b} + \dfrac{1}{c}$ is a natural number.

42. The following was offered as the simplification of $6 + 2(4 - 9)$.
[W]

$$6 + 2(4 - 9) = 6 + 2(-5)$$
$$= 8(-5)$$
$$= -40$$

Is this a correct simplification? Explain your answer.

43. The following was offered as the simplification of $2 \cdot 3^3$.
[W]

$$2 \cdot 3^3 = 6^3 = 216$$

Is this is a correct simplification? Explain your answer.

44. If a and b are rational numbers and $a < b$, is it always possible to find
[W] a rational number c such that $a < c < b$? If not, explain why. If so, show how to find one.

1.4 Variable Expressions

Objective A   *To evaluate a variable expression* ....

POINT OF INTEREST

Historical manuscripts indicate that mathematics is at least 4000 years old. Yet it was only 400 years ago that mathematicians started using variables to stand for numbers. The idea that a letter can stand for some number was a critical turning point in mathematics.

Often we discuss a quantity without knowing its exact value—for example, the price of gold next month, the cost of a new automobile next year, or the tuition cost for next semester. Recall that a letter of the alphabet can be used to stand for a quantity that is unknown or that can change, or *vary*. Such a letter is called a variable. An expression that contains one or more variables is called a **variable expression.**

A variable expression is shown at the right. The expression can be re-written by writing subtraction as the addition of the opposite.

$$3x^2 - 5y + 2xy - x - 7$$

$$3x^2 + (-5y) + 2xy + (-x) + (-7)$$

Note that the expression has 5 addends. The **terms** of a variable expression are the addends of the expression. The expression has 5 terms.

$$\underbrace{\overbrace{3x^2 \;-\; 5y \;+\; 2xy \;-\; x}^{\text{5 terms}}}_{\text{variable terms}} \quad \underbrace{-\; 7}_{\substack{\text{constant} \\ \text{term}}}$$

The terms $3x^2$, $-5y$, $2xy$, and $-x$ are **variable terms.**

The term -7 is a **constant term,** or simply a **constant.**

Each variable term is composed of a **numerical coefficient** and a **variable part** (the variable or variables and their exponents).

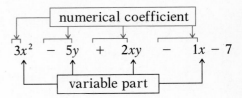

When the numerical coefficient is 1 or -1, the 1 is usually not written ($x = 1x$ and $-x = -1x$).

Replacing each variable by its value and then simplifying the resulting numerical expression is called **evaluating the variable expression.**

➡ Evaluate $ab - b^2$ when $a = 2$ and $b = -3$.

$ab - b^2$

$2(-3) - (-3)^2$ • Replace each variable in the expression by its value.

$= 2(-3) - 9$ • Use the Order of Operations Agreement to simplify the resulting numerical expression.

$= -6 - 9$

$= -15$

Example 1 Evaluate $\dfrac{a^2 - b^2}{a - b}$ when $a = 3$ and $b = -4$.

Solution $\dfrac{a^2 - b^2}{a - b}$

$$\dfrac{3^2 - (-4)^2}{3 - (-4)} = \dfrac{9 - 16}{3 - (-4)}$$

$$= \dfrac{-7}{7} = -1$$

You Try It 1 Evaluate $\dfrac{a^2 + b^2}{a + b}$ when $a = 5$ and $b = -3$.

Your solution

Example 2 Evaluate $x^2 - 3(x - y) - z^2$ when $x = 2$, $y = -1$, and $z = 3$.

Solution $x^2 - 3(x - y) - z^2$

$2^2 - 3[2 - (-1)] - 3^2$

$$= 2^2 - 3(3) - 3^2$$

$$= 4 - 3(3) - 9$$

$$= 4 - 9 - 9$$

$$= -5 - 9$$

$$= -14$$

You Try It 2 Evaluate $x^3 - 2(x + y) + z^2$ when $x = 2$, $y = -4$, and $z = -3$.

Your solution

Solutions on p. A4

Objective B *To simplify a variable expression using the Properties of Addition* ....

Like terms of a variable expression are terms with the same variable part. (Because $x^2 = x \cdot x$, x^2 and x are not like terms.)

Constant terms are like terms. 4 and 9 are like terms.

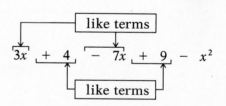

To simplify a variable expression, use the Distributive Property to combine like terms by adding the numerical coefficients. The variable part remains unchanged.

> **Distributive Property**
>
> If *a*, *b*, and *c* are real numbers, then $a(b + c) = ab + ac$.

The Distributive Property can also be written as $ba + ca = (b + c)a$. This form is used to simplify a variable expression.

➡ Simplify: $2x + 3x$

Use the Distributive Property to add the numerical coefficients of the like variable terms. This is called **combining like terms.**

$$2x + 3x = (2 + 3)x$$
$$= 5x$$

● Use the Distributive Property.

➡ Simplify: $5y - 11y$

$$5y - 11y = \boxed{(5 - 11)y}$$
$$= -6y$$

● Use the Distributive Property. This step is usually done mentally.

➡ Simplify: $5 + 7p$

The terms 5 and $7p$ are not like terms. The expression $5 + 7p$ is in simplest form.

In simplifying variable expressions, the following Properties of Addition are used.

CONSIDER THIS

To simplify an expression means to combine like terms. A constant term (5) and a variable term ($7p$) are not like terms and therefore cannot be combined. Combining like terms can be related to many everyday experiences. For instance, 5 bricks plus 7 bricks is 12 bricks. But 5 bricks plus 7 nails is 5 bricks plus 7 nails.

The Associative Property of Addition

If a, b, and c are real numbers, then $(a + b) + c = a + (b + c)$.

When three or more like terms are added, the terms can be grouped (with parentheses, for example) in any order. The sum is the same. For example,

$$(3x + 5x) + 9x = 3x + (5x + 9x)$$
$$8x + 9x = 3x + 14x$$
$$17x = 17x$$

The Commutative Property of Addition

If a and b are real numbers, then $a + b = b + a$.

When two like terms are added, the terms can be added in either order. The sum is the same. For example,

$$2x + (-4x) = -4x + 2x$$
$$-2x = -2x$$

The Addition Property of Zero

If a is a real number, then $a + 0 = 0 + a = a$.

The sum of a term and zero is the term. For example,

$$5x + 0 = 0 + 5x = 5x$$

> **The Inverse Property of Addition**
>
> If a is a real number, then $a + (-a) = (-a) + a = 0$.

The sum of a term and its opposite is zero. The opposite of a number is called its **additive inverse.**

$$7x + (-7x) = -7x + 7x = 0$$

➡ Simplify: $8x + 4y - 8x + y$

Use the Commutative and Associative Properties of Addition to rearrange and group like terms. Then combine like terms.

$$8x + 4y - 8x + y = \boxed{(8x - 8x) + (4y + y)} \quad \bullet \text{ This step is usually done mentally.}$$
$$= 0 + 5y$$
$$= 5y$$

➡ Simplify: $4x^2 + 5x - 6x^2 - 2x + 1$

Use the Commutative and Associative Properties of Addition to rearrange and group like terms. Then combine like terms.

$$4x^2 + 5x - 6x^2 - 2x + 1 = (4x^2 - 6x^2) + (5x - 2x) + 1$$
$$= -2x^2 + 3x + 1$$

Example 3   Simplify: $3x + 4y - 10x + 7y$ **You Try It 3**   Simplify: $3a - 2b - 5a + 6b$

Solution $3x + 4y - 10x + 7y = -7x + 11y$ **Your solution**

Example 4   Simplify: $x^2 - 7 + 4x^2 - 16$ **You Try It 4**   Simplify: $-3y^2 + 7 + 8y^2 - 14$

Solution $x^2 - 7 + 4x^2 - 16 = 5x^2 - 23$ **Your solution**

Solutions on p. A4

Objective C   *To simplify a variable expression using the Properties of Multiplication*

In simplifying variable expressions, the following Properties of Multiplication are used.

> **The Associative Property of Multiplication**
>
> If a, b, and c are real numbers, then $(a \cdot b) \cdot c = a \cdot (b \cdot c)$.

When three or more factors are multiplied, the factors can be grouped in any order. The product is the same. For example,

$$2(3x) = (2 \cdot 3)x = 6x$$

> **The Commutative Property of Mulitplication**
>
> If a and b are real numbers, then $a \cdot b = b \cdot a$.

Two factors can be multiplied in either order. The product is the same. For example,

$$(2x) \cdot 3 = 3 \cdot (2x) = 6x$$

> **The Multiplication Property of One**
>
> If a is a real number, then $a \cdot 1 = 1 \cdot a = a$.

The product of a term and one is the term. For example,

$$(8x)(1) = (1)(8x) = 8x$$

> **The Inverse Property of Multiplication**
>
> If a is a real number, and a is not equal to zero, then
>
> $a \cdot \dfrac{1}{a} = \dfrac{1}{a} \cdot a = 1.$

$\dfrac{1}{a}$ is called the **reciprocal** of a. $\dfrac{1}{a}$ is also called the **multiplicative inverse** of a.

The product of a number and its reciprocal is one. For example,

$$7 \cdot \frac{1}{7} = \frac{1}{7} \cdot 7 = 1$$

The multiplication properties just discussed are used to simplify variable expressions.

➡ Simplify: $2(-x)$

$$\begin{aligned} 2(-x) &= 2(-1 \cdot x) \\ &= [2(-1)]x \\ &= -2x \end{aligned}$$

• Use the Associative Property of Multiplication to group factors.

➡ Simplify: $\dfrac{3}{2}\left(\dfrac{2x}{3}\right)$

Use the Associative Property of Multiplication to group factors.

$$\begin{aligned} \frac{3}{2}\left(\frac{2x}{3}\right) &= \frac{3}{2}\left(\frac{2}{3}x\right) \\ &= \left(\frac{3}{2} \cdot \frac{2}{3}\right)x \\ &= 1 \cdot x \\ &= x \end{aligned}$$

• Note that $\dfrac{2x}{3} = \dfrac{2}{3}x$.

• The steps in the dashed box are usually done mentally.

➡ Simplify: $(16x)2$

Use the Commutative and Associative Properties of Multiplication to rearrange and group factors.

$$(16x)2 = \boxed{2(16x)}$$
$$= \boxed{(2 \cdot 16)x}$$
$$= 32x$$

● The steps in the dashed box are usually done mentally.

Example 5 Simplify: $-2(3x^2)$

Solution $-2(3x^2) = -6x^2$

You Try It 5 Simplify: $-5(4y^2)$

Your solution

Example 6 Simplify: $-5(-10x)$

Solution $-5(-10x) = 50x$

You Try It 6 Simplify: $-7(-2a)$

Your solution

Example 7 Simplify: $(6x)(-4)$

Solution $(6x)(-4) = -24x$

You Try It 7 Simplify: $(-5x)(-2)$

Your solution

Solutions on p. A4

Objective D *To simplify a variable expression using the Distributive Property* ....

Recall that the Distributive Property states that if a, b, and c are real numbers, then

$$a(b + c) = ab + ac$$

The Distributive Property is used to remove parentheses from a variable expression.

➡ Simplify: $3(2x + 7)$

$$3(2x + 7) = \boxed{3(2x) + 3(7)}$$
$$= 6x + 21$$

● Use the Distributive Property.
  Do this step mentally.

➡ Simplify: $-5(4x + 6)$

$$-5(4x + 6) = \boxed{-5(4x) + (-5) \cdot 6}$$
$$= -20x - 30$$

● Use the Distributive Property.
  Do this step mentally.

➡ Simplify: $-(2x - 4)$

$$-(2x - 4) = \boxed{\begin{array}{l} -1(2x - 4) \\ -1(2x) - (-1)(4) \end{array}}$$

● Use the Distributive Property.
  Do these steps mentally.

$$= -2x + 4$$

Note: When a negative sign immediately precedes the parentheses, the sign of each term inside the parentheses is changed.

➡ Simplify: $-\dfrac{1}{2}(8x - 12y)$

$$-\frac{1}{2}(8x - 12y) = \boxed{-\frac{1}{2}(8x) - \left(-\frac{1}{2}\right)(12y)}$$

● Use the Distributive Property.
  Do this step mentally.

$$= -4x + 6y$$

➡ Simplify: $4(x - y) - 2(-3x + 6y)$

$$4(x - y) - 2(-3x + 6y) = 4x - 4y + 6x - 12y$$

● Use the Distributive Property twice.

$$= 10x - 16y$$

● Combine like terms.

An extension of the Distributive Property is used when an expression inside parentheses contains more than two terms.

➡ Simplify: $3(4x - 2y - z)$

$$3(4x - 2y - z) = \boxed{3(4x) - 3(2y) - 3(z)}$$

● Use the Distributive Property
  Do this step mentally.

$$= 12x - 6y - 3z$$

Example 8   Simplify: $-3(-5a + 7b)$

Solution $-3(-5a + 7b) = 15a - 21b$

You Try It 8   Simplify: $-8(-2a + 7b)$

Your solution

Example 9   Simplify: $(2x - 6)2$

Solution $(2x - 6)2 = 4x - 12$

You Try It 9   Simplify: $(3a - 1)5$

Your solution

Solutions on p. A4

Example 10   Simplify: $3(x^2 - x - 5)$

Solution $3(x^2 - x - 5) = 3x^2 - 3x - 15$

You Try It 10   Simplify: $2(x^2 - x + 7)$

Your solution

Example 11   Simplify: $2x - 3(2x - 7y)$

Solution $2x - 3(2x - 7y) = 2x - 6x + 21y$
$$= -4x + 21y$$

You Try It 11   Simplify: $3y - 2(y - 7x)$

Your solution

Example 12   Simplify:
$7(x - 2y) - (-x - 2y)$

Solution $7(x - 2y) - (-x - 2y)$
$$= 7x - 14y + x + 2y$$
$$= 8x - 12y$$

You Try It 12   Simplify:
$-2(x - 2y) - (-x + 3y)$

Your solution

Example 13   Simplify:
$2x - 3[2x - 3(x + 7)]$

Solution $2x - 3[2x - 3(x + 7)]$
$$= 2x - 3[2x - 3x - 21]$$
$$= 2x - 3[-x - 21]$$
$$= 2x + 3x + 63$$
$$= 5x + 63$$

You Try It 13   Simplify:
$3y - 2[x - 4(2 - 3y)]$

Your solution

Solutions on p. A4

Objective E   *To translate a verbal expression into a variable expression* ....

One of the major skills required in applied mathematics is to translate a verbal expression into a variable expression. This requires recognizing the verbal phrases that translate into mathematical operations. A partial list of the verbal phrases used to indicate the different mathematical operations follows.

Addition	added to	6 added to y	$y + 6$
	more than	8 more than x	$x + 8$
	the sum of	the sum of x and z	$x + z$
	increased by	t increased by 9	$t + 9$
	the total of	the total of 5 and y	$5 + y$

Content and Format © 1996 HMCo.

POINT OF INTEREST

The way in which expressions are symbolized has changed over time. Here is how some of the expressions shown at the right may have appeared in the early 16th century.

R p. 9 for $x + 9$. The symbol R was used for a variable to the first power. The symbol p. was used for plus.

R m. 3 for $x - 3$. The symbol R is still used for the variable. The symbol m. was used for minus.

The square of a variable was designated by Q and the cube was designated by C. The expression $x^3 + x^2$ was written **C p. Q.**

Subtraction	minus	x minus 2	$x - 2$
	less than	7 less than t	$t - 7$
	decreased by	m decreased by 3	$m - 3$
	the difference between	the difference between y and 4	$y - 4$
Multiplication	times	10 times t	$10t$
	of	one half of x	$\dfrac{1}{2}x$
	the product of	the product of y and z	yz
	multiplied by	y multiplied by 11	$11y$
Division	divided by	x divided by 12	$\dfrac{x}{12}$
	the quotient of	the quotient of y and z	$\dfrac{y}{z}$
	the ratio of	the ratio of t to 9	$\dfrac{t}{9}$
Power	the square of	the square of x	x^2
	the cube of	the cube of a	a^3

⟹ Translate "14 less than the cube of x" into a variable expression.

14 *less than* the *cube* of x • Identify the words that indicate the mathematical operations.

$x^3 - 14$ • Use the identified operations to write the variable expression.

In most applications that involve translating phrases into variable expressions, the variable to be used is not given. To translate these phrases, a variable must be assigned to an unknown quantity before the variable expression can be written.

⟹ Translate "the sum of two consecutive integers" into a variable expression. Then simplify.

the first integer: n • Assign a variable to one of the unknown quantities.

the next consecutive integer: $n + 1$ • Use the assigned variable to write an expression for any other unknown quantity.

$n + (n + 1)$ • Use the assigned variable to write the variable expression.

$(n + n) + 1$ • Simplify the variable expression.

$2n + 1$

Many of the applications of mathematics require that you identify an unknown quantity, assign a variable to that quantity, and then attempt to express another unknown quantity in terms of the variable.

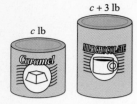

$c + 3$ lb

c lb

➡ A confectioner makes a mixture of candy that contains 3 lb more of milk chocolate than of caramel. Express the amount of milk chocolate in the mixture in terms of the amount of caramel in the mixture.

Amount of caramel
in the mixture: c

- Assign a variable to the amount of caramel in the mixture.

Amount of milk chocolate
in the mixture: $c + 3$

- Express the amount of milk chocolate in the mixture in terms of c.

Example 14
Translate "four times the sum of half of a number and fourteen" into a variable expression. Then simplify.

Solution
the unknown number: n

half of the number: $\frac{1}{2}n$

the sum of half of the number and

fourteen: $\frac{1}{2}n + 14$

$4\left(\frac{1}{2}n + 14\right)$

$2n + 56$

You Try It 14
Translate "five times the difference between a number and sixty" into a variable expression. Then simplify.

Your solution

$5(x - 60)$

$5x - 300$

Example 15
The length of a swimming pool is 4 ft less than two times the width. Express the length of the pool in terms of the width.

Solution
the width of the pool: w
the length is 4 ft less than two times the width: $2w - 4$

You Try It 15
The speed of a new printer is twice the speed of an older model. Express the speed of the new model in terms of the speed of the older model.

Your solution

s old model
$2s$ new model

$2s$

Example 16
A banker divided $5000 between two accounts, one paying 10% annual interest and the second paying 8% annual interest. Express the amount invested in the 10% account in terms of the amount invested in the 8% account.

Solution
the amount invested at 8%: x
the amount invested at 10%: $5000 - x$

You Try It 16
A guitar string 6 ft long was cut into two pieces. Express the length of the shorter piece in terms of the length of the longer piece.

Your solution

shorter $6 - x$
longer x

$6 - x$

Solutions on p. A4

1.4 Exercises

Objective A

Evaluate the variable expression when $a = 2$, $b = 3$, and $c = -4$.

1. $6b \div (-a)$

2. $bc \div (2a)$

3. $b^2 - 4ac$

4. $a^2 - b^2$

5. $b^2 - c^2$

6. $(a + b)^2$

7. $a^2 + b^2$

8. $2a - (c + a)^2$

9. $(b - a)^2 + 4c$

10. $b^2 - \dfrac{ac}{8}$

11. $\dfrac{5ab}{6} - 3cb$

12. $(b - 2a)^2 + bc$

$3 - 2(2)^2 + 3(-4)$
$3 - 4^2 + 3(-4)$
$(-1^2) + (-12)$
$1 - 12 = -11$

Evaluate the variable expression when $a = -2$, $b = 4$, $c = -1$, and $d = 3$.

13. $\dfrac{b + c}{d}$

14. $\dfrac{d - b}{c}$

15. $\dfrac{2d + b}{-a}$

16. $\dfrac{b + 2d}{b}$

17. $\dfrac{b - d}{c - a}$

18. $\dfrac{2c - d}{-ad}$

19. $(b + d)^2 - 4a$

20. $(d - a)^2 - 3c$

21. $(d - a)^2 \div 5$

22. $3(b - a) - bc$

23. $\dfrac{b - 2a}{bc^2 - d}$

24. $\dfrac{b^2 - a}{ad + 3c}$

25. $\dfrac{1}{3}d^2 - \dfrac{3}{8}b^2$

26. $\dfrac{5}{8}a^4 - c^2$

27. $\dfrac{-4bc}{2a - b}$

28. $-\dfrac{3}{4}b + \dfrac{1}{2}(ac + bd)$

29. $-\dfrac{2}{3}d - \dfrac{1}{5}(bd - ac)$

30. $(b - a)^2 - (d - c)^2$

31. $(b + c)^2 + (a + d)^2$

32. $4ac + (2a)^2$

33. $3dc - (4c)^2$

Objective B

Simplify.

34. $6x + 8x$

35. $12x + 13x$

36. $9a - 4a$

37. $12a - 3a$

38. $4y + (-10y)$

39. $8y + (-6y)$

40. $-3b - 7$

41. $-12y - 3$

42. $-12a + 17a$

43. $-3a + 12a$

44. $5ab - 7ab$

45. $9ab - 3ab$

46. $-12xy + 17xy$

47. $-15xy + 3xy$

48. $-3ab + 3ab$

49. $-7ab + 7ab$

50. $-\dfrac{1}{2}x - \dfrac{1}{3}x$

51. $-\dfrac{2}{5}y + \dfrac{3}{10}y$

52. $\dfrac{3}{8}x^2 - \dfrac{5}{12}x^2$

53. $\dfrac{2}{3}y^2 - \dfrac{4}{9}y^2$

54. $3x + 5x + 3x$

55. $8x + 5x + 7x$

56. $5a - 3a + 5a$

57. $10a - 17a + 3a$

58. $-5x^2 - 12x^2 + 3x^2$

59. $-y^2 - 8y^2 + 7y^2$

60. $7x + (-8x) + 3y$

61. $8y + (-10x) + 8x$

62. $7x - 3y + 10x$

63. $8y + 8x - 8y$

64. $3a + (-7b) - 5a + b$

65. $-5b + 7a - 7b + 12a$

66. $3x + (-8y) - 10x + 4x$

67. $3y + (-12x) - 7y + 2y$

68. $x^2 - 7x + (-5x^2) + 5x$

69. $3x^2 + 5x - 10x^2 - 10x$

Objective C

Simplify.

70. $4(3x)$ **71.** $12(5x)$ **72.** $-3(7a)$ **73.** $-2(5a)$ **74.** $-2(-3y)$

75. $-5(-6y)$ **76.** $(4x)2$ **77.** $(6x)12$ **78.** $(3a)(-2)$ **79.** $(7a)(-4)$

80. $(-3b)(-4)$ **81.** $(-12b)(-9)$ **82.** $-5(3x^2)$ **83.** $-8(7x^2)$ **84.** $\frac{1}{3}(3x^2)$

85. $\frac{1}{6}(6x^2)$ **86.** $\frac{1}{5}(5a)$ **87.** $\frac{1}{8}(8x)$ **88.** $-\frac{1}{2}(-2x)$ **89.** $-\frac{1}{4}(-4a)$

90. $-\frac{1}{7}(-7n)$ **91.** $-\frac{1}{9}(-9b)$ **92.** $(3x)\left(\frac{1}{3}\right)$ **93.** $(12x)\left(\frac{1}{12}\right)$ **94.** $(-6y)\left(-\frac{1}{6}\right)$

95. $(-10n)\left(-\frac{1}{10}\right)$ **96.** $\frac{1}{3}(9x)$ **97.** $\frac{1}{7}(14x)$ **98.** $-\frac{1}{5}(10x)$ **99.** $-\frac{1}{8}(16x)$

100. $-\frac{2}{3}(12a^2)$ **101.** $-\frac{5}{8}(24a^2)$ **102.** $-\frac{1}{2}(-16y)$ **103.** $-\frac{3}{4}(-8y)$ **104.** $(16y)\left(\frac{1}{4}\right)$

105. $(33y)\left(\frac{1}{11}\right)$ **106.** $(-6x)\left(\frac{1}{3}\right)$ **107.** $(-10x)\left(\frac{1}{5}\right)$ **108.** $(-8a)\left(-\frac{3}{4}\right)$ **109.** $(21y)\left(-\frac{3}{7}\right)$

Objective D

Simplify.

110. $-(x + 2)$ **111.** $-(x + 7)$ **112.** $2(4x - 3)$ **113.** $5(2x - 7)$

114. $-2(a + 7)$ **115.** $-5(a + 16)$ **116.** $-3(2y - 8)$ **117.** $-5(3y - 7)$

Simplify.

118. $(5 - 3b)7$ **119.** $(10 - 7b)2$ **120.** $\frac{1}{3}(6 - 15y)$ **121.** $\frac{1}{2}(-8x + 4y)$

122. $3(5x^2 + 2x)$ **123.** $6(3x^2 + 2x)$ **124.** $-2(-y + 9)$ **125.** $-5(-2x + 7)$

126. $(-3x - 6)5$ **127.** $(-2x + 7)7$ **128.** $2(-3x^2 - 14)$ **129.** $5(-6x^2 - 3)$

130. $-3(2y^2 - 7)$ **131.** $-8(3y^2 - 12)$ **132.** $3(x^2 - y^2)$ **133.** $5(x^2 + y^2)$

134. $-\frac{2}{3}(6x - 18y)$ **135.** $-\frac{1}{2}(x - 4y)$ **136.** $-(6a^2 - 7b^2)$

137. $3(x^2 + 2x - 6)$ **138.** $4(x^2 - 3x + 5)$ **139.** $-2(y^2 - 2y + 4)$

140. $\frac{1}{2}(2x - 6y + 8)$ **141.** $-\frac{1}{3}(6x - 9y + 1)$ **142.** $4(-3a^2 - 5a + 7)$

143. $-5(-2x^2 - 3x + 7)$ **144.** $-3(-4x^2 + 3x - 4)$ **145.** $3(2x^2 + xy - 3y^2)$

146. $5(2x^2 - 4xy - y^2)$ **147.** $-(3a^2 + 5a - 4)$ **148.** $-(8b^2 - 6b + 9)$

149. $4x - 2(3x + 8)$ **150.** $6a - (5a + 7)$ **151.** $9 - 3(4y + 6)$

152. $10 - (11x - 3)$ **153.** $5n - (7 - 2n)$ **154.** $8 - (12 + 4y)$

Simplify.

155. $3(x + 2) - 5(x - 7)$ **156.** $2(x - 4) - 4(x + 2)$ **157.** $12(y - 2) + 3(7 - 3y)$

158. $6(2y - 7) - (3 - 2y)$ **159.** $3(a - b) - (a + b)$ **160.** $2(a + 2b) - (a - 3b)$

161. $4[x - 2(x - 3)]$ **162.** $2[x + 2(x + 7)]$ **163.** $-2[3x + 2(4 - x)]$

164. $-5[2x + 3(5 - x)]$ **165.** $-3[2x - (x + 7)]$ **166.** $-2[3x - (5x - 2)]$

167. $2x - 3[x - (4 - x)]$ **168.** $-7x + 3[x - (3 - 2x)]$ **169.** $-5x - 2[2x - 4(x + 7)] - 6$

Objective E

Translate into a variable expression. Then simplify.

170. twelve minus a number

171. a number divided by eighteen

172. two-thirds of a number

173. twenty more than a number

174. the quotient of twice a number and nine

175. ten times the difference between a number and fifty

176. eight less than the product of eleven and a number

177. the sum of five-eighths of a number and six

178. nine less than the total of a number and two

179. the difference between a number and three more than the number

Translate into a variable expression. Then simplify.

180. the quotient of seven and the total of five and a number

181. four times the sum of a number and nineteen

182. five increased by one-half of the sum of a number and three

183. the quotient of fifteen and the sum of a number and twelve

184. a number added to the difference between twice the number and four

185. the product of two-thirds and the sum of a number and seven

186. the product of five less than a number and seven

187. the difference between forty and the quotient of a number and twenty

188. the quotient of five more than twice a number and the number

189. the sum of the square of a number and twice the number

190. a number decreased by the difference between three times the number and eight

191. the sum of eight more than a number and one-third of the number

192. a number added to the product of three and the number

193. a number increased by the total of the number and nine

194. five more than the sum of a number and six

195. a number decreased by the difference between eight and the number

196. a number minus the sum of the number and ten

197. the difference between one-third of a number and five-eighths of the number

198. the sum of one-sixth of a number and four-ninths of the number

199. two more than the total of a number and five

200. the sum of a number divided by three and the number

201. twice the sum of six times a number and seven

Write a variable expression.

202. A propeller-driven plane flies at a rate that is half that of a jet plane. Express the rate of the propeller plane in terms of the rate of the jet plane.

203. The length of a football field is 30 yd more than the width. Express the length of the field in terms of the width.

204. The diameter of a basketball is approximately 4 times the diameter of a baseball. Express the diameter of the basketball in terms of the diameter of the baseball.

205. An 18-carat gold chain has 6 g less than twice the amount of gold in a 12-carat gold chain. Express the amount of gold in the 18-carat gold chain in terms of the amount of gold in the 12-carat gold chain.

206. The highest percent income tax bracket is 3 percent more than twice the lowest percent income tax bracket. Express the highest percent income tax bracket in terms of the lowest percent income tax bracket.

207. A batter can swing a bat at a rate that is two-thirds the rate at which a pitcher can throw the ball. Express the rate of the bat in terms of the rate of the ball.

208. Twenty gallons of crude oil were poured into two containers of different size. Express the amount of crude oil poured into the smaller container in terms of the amount poured into the larger container.

209. A rope was cut into two pieces in such a way that the length of the longer piece was 2 ft less than 3 times the length of the shorter piece. Express the length of the longer piece in terms of the length of the shorter piece.

210. A new model of a laser printer can print pages at a rate that is seven pages more than one-half the speed of the older model. Express the speed of the newer model in terms of that of the older model.

211. Two cars are traveling in opposite directions and at different rates. Two hours later the cars are 200 mi apart. Express the distance traveled by the slower car in terms of the distance traveled by the faster car.

APPLYING THE CONCEPTS

212. Does every number have an additive inverse? If not, which real numbers do not have an additive inverse?

213. Does every number have a multiplicative inverse? If not, which real numbers do not have a multiplicative inverse?

214. The chemical formula for glucose (sugar) is $C_6H_{12}O_6$. This formula means that there are twelve hydrogen atoms, six carbon atoms, and six oxygen atoms in each molecule of glucose. If x represents the number of atoms of oxygen in a pound of sugar, express the number of hydrogen atoms in the pound of sugar.

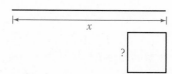

215. Determine whether the statement is true or false. If the statement is false, give an example that illustrates that it is false.
  a. Division is a commutative operation.
  b. Division is an associative operation.
  c. Subtraction is an associative operation.
  d. Subtraction is a commutative operation.
  e. Addition is a commutative operation.

216. A wire whose length is given as x in. is bent into a square. Express the length of a side of the square in terms of x.

217. For each of the following, determine the first natural number x, greater than 2, for which the second expression is larger than the first.
  a. $x^3, 3^x$ **b.** $x^4, 4^x$ **c.** $x^5, 5^x$ **d.** $x^6, 6^x$

On the basis of your answers, make a conjecture that appears to be true about the expressions x^n and n^x, where $n = 3, 4, 5, 6, 7, \ldots$ and x is a natural number greater than 2.

218. A block-and-tackle system is designed so that pulling five feet on one end of a rope will move a weight on the other end a distance of three feet. If x represents the distance the rope is pulled, express the distance the weight moves in terms of x.

219. Give examples of two operations that occur in everyday experience
[W]   that are not commutative (for example, putting on socks and then shoes).

220. Choose any number a. Evaluate the expressions $6a^2 + 2a - 10$ and
[W] $2a(3a - 4) + 10(a - 1)$. Now choose a different number and evaluate the expressions again. Repeat this two more times with different numbers. What conclusions might you draw from your evaluations?

1.5   Sets

Objective A   *To write a set using the roster method* ....

Recall that a *set* is a collection of objects, which are called the *elements* of the set.

The **roster method** of writing a set encloses a list of the elements in braces.

The set of the last three letters of the alphabet is written $\{x, y, z\}$.

The set of the positive integers less than 5 is written $\{1, 2, 3, 4\}$.

➡ Use the roster method to write the set of integers between 0 and 10.

$$A = \{1, 2, 3, 4, 5, 6, 7, 8, 9\}$$

A set can be designated by a capital letter. Note that 0 and 10 are not elements of set A.

➡ Use the roster method to write the set of natural numbers.

$A = \{1, 2, 3, 4, \ldots\}$ • **The three dots mean that the pattern of numbers continues without end.**

The **empty set,** or **null set,** is the set that contains no elements. The symbol $\varnothing$ or { } is used to represent the empty set.

The set of people who have run a two-minute mile is the empty set.

The **union** of two sets, written $A \cup B$, is the set that contains the elements of A and the elements of B.

➡ Find $A \cup B$, given $A = \{1, 2, 3, 4\}$ and $B = \{3, 4, 5, 6\}$.

$A \cup B = \{1, 2, 3, 4, 5, 6\}$ • **The union of *A* and *B* contains all the elements of *A* and all the elements of *B*. Any elements that are in both *A* and *B* are listed only once.**

The **intersection** of two sets, written $A \cap B$, is the set that contains the elements that are common to both A and B.

➡ Find $A \cap B$, given $A = \{1, 2, 3, 4\}$ and $B = \{3, 4, 5, 6\}$.

$A \cap B = \{3, 4\}$ • **The intersection of *A* and *B* contains the elements common to *A* and *B*.**

Example 1
Use the roster method to write the set of the odd positive integers less than 12.

Solution
$A = \{1, 3, 5, 7, 9, 11\}$

You Try It 1
Use the roster method to write the set of the odd negative integers greater than -10.

Your solution

Solution on p. A5

Content and Format © 1996 HMCo.

Example 2
Use the roster method to write the set of the even positive integers.

Solution
$A = \{2, 4, 6, \ldots\}$

You Try It 2
Use the roster method to write the set of the odd positive integers.

Your solution

Example 3
Find $D \cup E$, given $D = \{6, 8, 10, 12\}$ and $E = \{-8, -6, 10, 12\}$.

Solution
$D \cup E = \{-8, -6, 6, 8, 10, 12\}$

You Try It 3
Find $A \cup B$, given $A = \{-2, -1, 0, 1, 2\}$ and $B = \{0, 1, 2, 3, 4\}$.

Your solution

Example 4
Find $A \cap B$, given $A = \{5, 6, 9, 11\}$ and $B = \{5, 9, 13, 15\}$.

Solution
$A \cap B = \{5, 9\}$

You Try It 4
Find $C \cap D$, given $C = \{10, 12, 14, 16\}$ and $D = \{10, 16, 20, 26\}$.

Your solution

Example 5
Find $A \cap B$, given $A = \{1, 2, 3, 4\}$ and $B = \{8, 9, 10, 11\}$.

Solution
$A \cap B = \varnothing$

You Try It 5
Find $A \cap B$, given $A = \{-5, -4, -3, -2\}$ and $B = \{2, 3, 4, 5\}$.

Your solution

Solutions on p. A5

Objective B   *To write a set using set-builder notation* ....

POINT OF INTEREST
The symbol $\in$ was first used in the book *Arithmeticae Principia,* published in 1889. It was the first letter of the Greek word $\varepsilon\sigma\tau\iota$, which means "is." The symbols for union and intersection were also introduced at that time.

Another method of representing sets is called **set-builder notation.** Using set-builder notation, the set of all positive integers less than 10 is as follows:

$\{x \mid x < 10, x \in \text{positive integers}\}$, which is read "the set of all x such that x is less than 10 and x is an element of the positive integers."

➡ Use set-builder notation to write the set of real numbers greater than 4.

$\{x \mid x > 4, x \in \text{ real numbers}\}$

- "$x \in$ real numbers" is read "x is an element of the real numbers."

Example 6

Use set-builder notation to write the set of negative integers greater than -100.

Solution

$\{x \mid x > -100, x \in \text{negative integers}\}$

You Try It 6

Use set-builder notation to write the set of positive even integers less than 59.

Your solution

Example 7

Use set-builder notation to write the set of real numbers less than 60.

Solution

$\{x \mid x < 60, x \in \text{real numbers}\}$

You Try It 7

Use set-builder notation to write the set of real numbers greater than -3.

Your solution

Solutions on p. A5

Objective C   ***To graph an inequality on the number line*** ....

An expression that contains the symbol $>$, $<$, $\geq$, or $\leq$ is called an **inequality.** An inequality expresses the relative order of two mathematical expressions. The expressions can be either numerical or variable.

$$\left.\begin{array}{l} 4 > 2 \\ 3x \leq 7 \\ x^2 - 2x > y + 4 \end{array}\right\} \text{Inequalities}$$

An **inequality** can be graphed on the number line.

➡ Graph: $x > 1$

The graph is the real numbers greater than 1. The parenthesis at 1 indicates that 1 is not included in the graph.

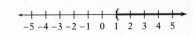

➡ Graph: $x \geq 1$

The bracket at 1 indicates that 1 is included in the graph.

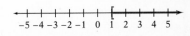

➡ Graph: $-1 > x$

$-1 > x$ is equivalent to $x < -1$. The numbers less than -1 are to the left of -1 on the number line.

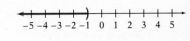

The union of two sets is the set that contains all the elements of each set.

➡ Graph: $\{x \mid x > 4\} \cup \{x \mid x < 1\}$

The graph is the numbers greater than 4 and the numbers less than 1.

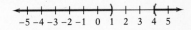

The intersection of two sets is the set that contains the elements common to both sets.

➡ Graph: $\{x \mid x > -1\} \cap \{x \mid x < 2\}$

The graph is the numbers between -1 and 2.

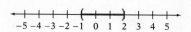

Example 8
Graph: $x < 5$

Solution
The graph is the numbers less than 5.

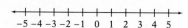

You Try It 8
Graph: $-2 < x$

Your solution

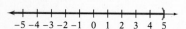

Example 9
Graph: $\{x \mid x > -2\} \cap \{x \mid x < 1\}$

Solution
The graph is the numbers between -2 and 1.

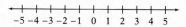

You Try It 9
Graph: $\{x \mid x \leq 4\} \cap \{x \mid x \geq -4\}$

Your solution

Example 10
Graph: $\{x \mid x \leq 5\} \cup \{x \mid x \geq -3\}$

Solution
The graph is the real numbers.

You Try It 10
Graph: $\{x \mid x < 5\} \cup \{x \mid x \geq -2\}$

Your solution

Example 11
Graph: $\{x \mid x > 3\} \cup \{x \mid x < 1\}$

Solution
The graph is the numbers greater than 3 and the numbers less than 1.

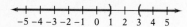

You Try It 11
Graph: $\{x \mid x > -1\} \cup \{x \mid x < -3\}$

Your solution

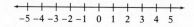

Solutions on p. A5

1.5 Exercises

Objective A

Use the roster method to write the set.

1. the integers between 15 and 22

2. the integers between -10 and -4

3. the odd integers between 8 and 18

4. the even integers between -11 and -1

5. the letters of the alphabet between a and d

6. the letters of the alphabet between p and v

Find $A \cup B$.

7. $A = \{3, 4, 5\}$ $B = \{4, 5, 6\}$

8. $A = \{-3, -2, -1\}$ $B = \{-2, -1, 0\}$

9. $A = \{-10, -9, -8\}$ $B = \{8, 9, 10\}$

10. $A = \{a, b, c\}$ $B = \{x, y, z\}$

11. $A = \{a, b, d, e\}$ $B = \{c, d, e, f\}$

12. $A = \{m, n, p, q\}$ $B = \{m, n, o\}$

13. $A = \{1, 3, 7, 9\}$ $B = \{7, 9, 11, 13\}$

14. $A = \{-3, -2, -1\}$ $B = \{-1, 1, 2\}$

Find $A \cap B$.

15. $A = \{3, 4, 5\}$ $B = \{4, 5, 6\}$

16. $A = \{-4, -3, -2\}$ $B = \{-6, -5, -4\}$

17. $A = \{-4, -3, -2\}$ $B = \{2, 3, 4\}$

18. $A = \{1, 2, 3, 4\}$ $B = \{1, 2, 3, 4\}$

19. $A = \{a, b, c, d, e\}$ $B = \{c, d, e, f, g\}$

20. $A = \{m, n, o, p\}$ $B = \{k, l, m, n\}$

Objective B

Use set-builder notation to write the set.

21. the negative integers greater than -5

22. the positive integers less than 5

23. the integers greater than 30

24. the integers less than -70

25. the even integers greater than 5

26. the odd integers less than -2

27. the real numbers greater than 8

28. the real numbers less than 57

Objective C

Graph.

29. $x > 2$

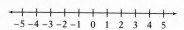

30. $x \geq -1$

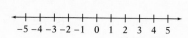

31. $0 \geq x$

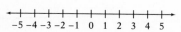

32. $4 > x$

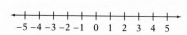

33. $\{x \,|\, x > -2\} \cup \{x \,|\, x < -4\}$

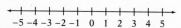

34. $\{x \,|\, x > 4\} \cup \{x \,|\, x < -2\}$

35. $\{x \,|\, x > -2\} \cap \{x \,|\, x < 4\}$

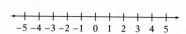

36. $\{x \,|\, x > -3\} \cap \{x \,|\, x < 3\}$

37. $\{x \,|\, x \geq -2\} \cup \{x \,|\, x < 4\}$

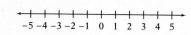

38. $\{x \,|\, x > 0\} \cup \{x \,|\, x \leq 4\}$

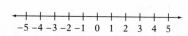

APPLYING THE CONCEPTS

39. Determine whether the statement is always true, sometimes true, or never true.
   a. Given that $a > 0$ and $b < 0$, then $ab > 0$.
   b. Given that $a < 0$, then $a^2 > 0$.
   c. Given that $a > 0$ and $b < 0$, then $a^2 > b$.

40. By trying various sets, make a conjecture as to whether the union of two sets is
   a. a commutative operation
   b. an associative operation

41. By trying various sets, make a conjecture as to whether the intersection of two sets is
   a. a commutative operation
   b. an associative operation

42. Explain how to find the union of two sets.
[W]

43. Explain how to find the intersection of two sets.
[W]

Projects in Mathematics

Prime and Composite Numbers

A **prime number** is a natural number greater than 1 whose only natural-number factors are itself and 1. The number 11 is a prime number because the only natural-number factors of 11 are 11 and 1.

Eratosthenes, a Greek philosopher and astronomer who lived from 270 to 190 B.C., devised a method of identifying prime numbers. It is called the **Sieve of Eratosthenes.** The procedure is illustrated below.

1	②	③	4	⑤	6	⑦	8	9	10
⑪	12	⑬	14	15	16	⑰	18	⑲	20
21	22	㉓	24	25	26	27	28	㉙	30
㉛	32	33	34	35	36	㊲	38	39	40
㊶	42	㊸	44	45	46	㊷	48	49	50
51	52	㊼	54	55	56	57	58	㊾	60
㉛	62	63	64	65	66	㊿	68	69	70
71	72	73	74	75	76	77	78	79	80
81	82	83	84	85	86	87	88	89	90
91	92	93	94	95	96	97	98	99	100

List all the natural numbers from 1 to 100. Cross out the number 1, because it is not a prime number. The number 2 is prime; circle it. Cross out all the other multiples of 2 (4, 6, 8, ...), because they are not prime. The number 3 is prime; circle it. Cross out all the other multiples of 3 (6, 9, 12, ...) that are not already crossed out. The number 4, the next consecutive number in the list, has already been crossed out. The number 5 is prime; circle it. Cross out all the other multiples of 5 that are not already crossed out. Continue in this manner until all the prime numbers less than 100 are circled.

A **composite number** is a natural number greater than 1 that has a natural-number factor other than itself and 1. The number 21 is a composite number because it has factors of 3 and 7. All the numbers crossed out in the table above, except the number 1, are composite numbers.

Solve.

1. Use the Sieve of Eratosthenes to find the prime numbers between 100 and 200.

2. How many prime numbers are even numbers?

3. **a.** List two prime numbers that are consecutive natural numbers.
   b. Can there by any other pairs of prime numbers that are consecutive natural numbers?

4. Find the "twin primes" between 1 and 200. Twin primes are two prime numbers whose difference is 2. For instance, 3 and 5 are twin primes; 5 and 7 are also twin primes.

5. Some primes are the sum of a square and 1. For example, $5 = 2^2 + 1$. Find another prime p such that $p = n^2 + 1$, where n is a natural number.

Using the $\boxed{+/-}$ Key on a Scientific Calculator

Using your calculator to simplify numerical expressions sometimes requires use of the $\boxed{+/-}$ key or, on some calculators, the negative key, which is frequently shown as $\boxed{(-)}$. These keys change the sign of the number currently in the display. To enter -4:

- For those calculators with $\boxed{+/-}$, press 4 and then $\boxed{+/-}$.
- For those calculators with $\boxed{(-)}$, press $\boxed{(-)}$ and then 4.

Here are the keystrokes for evaluating the expression $3(-4) - (-5)$.

Calculators with $\boxed{+/-}$ key: $3\boxed{\times}4\boxed{+/-}\boxed{-}5\boxed{+/-}\boxed{=}$

Calculators with $\boxed{(-)}$ key: $3\boxed{\times}\boxed{(-)}4\boxed{-}\boxed{(-)}5\boxed{=}$

This example illustrates that calculators make a distinction between negative and minus. To perform the operation $3 - (-3)$, you cannot enter $3\boxed{-}\boxed{-}3$. This would result in 0, which is not the correct answer. You must enter

$$3\boxed{-}3\boxed{+/-}\boxed{=} \quad \text{or} \quad 3\boxed{-}\boxed{(-)}3\boxed{=}$$

Use your calculator to evaluate the expressions.

1. $8 - (-3) + (-9)$ **2.** $5(-7) - (2 - 6)$

3. $-12 \div (-6)(-2)$ **4.** $\dfrac{(-4) + (-10)}{-7} + (-1)$

Chapter Summary

· ·

Key Words   A *set* is a collection of objects. The objects in the set are called the *elements* of the set. The *roster method* of writing sets encloses a list of the elements in braces.

The set of *natural numbers* is {1, 2, 3, 4, 5, 6, 7, ...}. The set of *integers* is {..., -4, -3, -2, -1, 0, 1, 2, 3, 4, ...}.

The *empty set* or *null set*, written ∅ or { }, is the set that contains no elements.

A number *a is less than* another number *b*, written $a < b$, if *a* is to the left of *b* on the number line. A number *a is greater than* another number *b*, written $a > b$, if *a* is to the right of *b* on the number line. The symbol ≤ means *is less than or equal to*. The symbol ≥ means *is greater than or equal to*. An expression that contains the symbol >, <, ≥, or ≤ is an *inequality*.

Two numbers that are the same distance from zero on the number line but on opposite sides of zero are *opposite numbers*, or *opposites*. The *additive inverse* of a number is the opposite of the number.

The *multiplicative inverse* of a number is the reciprocal of the number.

The *absolute value* of a number is its distance from zero on the number line.

A *rational number* is a number that can be written in the form $\frac{a}{b}$, where a and b are integers and $b \neq 0$. An *irrational number* is a number that has a decimal representation that never terminates or repeats. The rational numbers and the irrational numbers taken together are called the *real numbers.*

Percent means "parts of 100."

An expression of the form a^n is in *exponential form,* where a is the base and n is the exponent.

A *square root* of a positive number x is a n⟍   ⟍are is x. The *principal square root* of a number is the positiv   ⟍mbol $\sqrt{}$ is called a *radical sign* and is used to indicate   t of a number. The *radicand* is the number under th⟍

The square of an integer is a *perfe⟍   ⟍are, its square root can only be app⟍

A *variable* is a letter th⟍   ⟍known or that varies. A *variabl⟍   one or more variables.

The *terms⟍   ⟍expression. A *variable term* is⟍   able part. *Like terms* of a vari⟍   ⟍le part.

⟍t that contains all the elements
that are in both set A and set B are
⟍ sets, written $A \cap B$, is the set that
⟍ to both A and B.

Ess⟍   ⟍ame sign, add the absolute values of the num-
⟍e addends.

⟍ith different signs,** find the absolute value of each
⟍ smaller of these from the larger. Attach the sign of the
⟍ger absolute value.

⟍wo numbers,** add the opposite of the second number to the first

⟍ultiply two numbers with the same sign,** multiply the absolute values
⟍the factors. The product is positive.

To multiply two numbers with different signs, multiply the absolute values of the factors. The product is negative.

To divide two numbers with the same sign, divide the absolute values of the numbers. The quotient is positive.

To divide two numbers with different signs, divide the absolute values of the numbers. The quotient is negative.

To convert a percent to a decimal, remove the percent sign and multiply by 0.01.

To convert a percent to a fraction, remove the percent sign and multiply by $\frac{1}{100}$.

To convert a decimal or a fraction to a percent, multiply by 100%.

Product Property of Square Roots	$\sqrt{ab} = \sqrt{a} \cdot \sqrt{b}$
Addition Property of Zero	$a + 0 = a$ or $0 + a = a$
Commutative Property of Addition	$a + b = b + a$
Associative Property of Addition	$(a + b) + c = a + (b + c)$
Inverse Property of Addition	$a + (-a) = 0$ or $-a + a = 0$
Multiplication Property of Zero	$a \cdot 0 = 0$ or $0 \cdot a = 0$
Multiplication Property of One	$a \cdot 1 = a$ or $1 \cdot a = a$
Commutative Property of Multiplication	$a \cdot b = b \cdot a$
Associative Property of Multiplication	$(a \cdot b) \cdot c = a \cdot (b \cdot c)$
Inverse Property of Multiplication	$a \cdot \dfrac{1}{a} = \dfrac{1}{a} \cdot a = 1, a \neq 0$
Division Properties of Zero and One	If $a \neq 0, 0 \div a = 0$. If $a \neq 0, a \div a = 1$. $a \div 1 = a$ $a \div 0$ is undefined.
Distributive Property	$a(b + c) = ab + ac$

The Order of Operations Agreement

Step 1   Perform operations inside grouping symbols. The grouping symbols include parentheses, brackets, braces, absolute value symbols, and the fraction bar.

Step 2   Simplify exponential expressions.

Step 3   Do multiplication and division as they occur from left to right.

Step 4   Do addition and subtraction as they occur from left to right.

Chapter Review Exercises

1. Let $x \in \{-4, 0, 11\}$. For what values of x is the inequality $x < 1$ a true statement?

2. Find the additive inverse of -4.

3. Evaluate $-|-5|$.

4. Add: $-3 + (-12) + 6 + (-4)$

5. Subtract: $16 - (-3) - 18$

6. Multiply: $(-6)(7)$

7. Divide: $-100 \div 5$

8. Write $\frac{7}{25}$ as a decimal.

9. Write 6.2% as a decimal.

10. Write $\frac{5}{8}$ as a percent.

11. Simplify: $\frac{1}{3} - \frac{1}{6} + \frac{5}{12}$

12. Subtract: $5.17 - 6.238$

13. Divide: $-\frac{18}{35} \div \frac{17}{28}$

14. Multiply: $4.32(-1.07)$

15. Evaluate $\left(-\frac{2}{3}\right)^4$.

16. Simplify: $2\sqrt{36}$

17. Simplify: $-3\sqrt{120}$

18. Evaluate $-3^2 + 4[18 + (12 - 20)]$.

19. Evaluate $(b - a)^2 + c$ when $a = -2$, $b = 3$, and $c = 4$.

20. Simplify: $6a - 4b + 2a$

21. Simplify: $-3(-12y)$

22. Simplify: $5(2x - 7)$

23. Simplify: $-4(2x - 9) + 5(3x + 2)$

24. Simplify: $5[2 - 3(6x - 1)]$

25. Use the roster method to write the set of odd positive integers less than 8.

26. Find $A \cap B$, given $A = \{1, 5, 9, 13\}$ and $B = \{1, 3, 5, 7, 9\}$.

27. Graph: $x > 3$

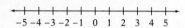

28. Graph: $\{x \mid x < 2\} \cup \{x \mid x > 5\}$

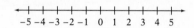

29. To discourage random guessing on a multiple-choice exam, a professor assigns 6 points for a correct answer, -4 points for an incorrect answer, and -2 points for leaving a question blank. What is the score for a student who had 21 correct answers, had 5 incorrect answers, and left 4 questions blank?

30. The exchange rates for some currencies are shown at the right. If you are in Sweden and purchase a car for 64,000 krona, what is the value of the purchase in U.S. dollars?

Foreign Currency per U.S. Dollar	
Hong Kong dollar	7.7350
Swedish krona	8.0414
Japanese yen	103.85

31. Translate "the difference between twice a number and one-half of the number" into a variable expression. Then simplify.

32. A baseball card collection contains five times as many National League players' cards as American League players' cards. Express the number of National League players' cards in terms of the number of American League players' cards.

33. A club treasurer has some five-dollar bills and some ten-dollar bills. The treasurer has a total of 35 bills. Express the number of five-dollar bills in terms of the number of ten-dollar bills.

First-Degree Equations and Inequalities

Objectives

2.1 Introduction to Equations
A To determine whether a given number is a solution of an equation
B To solve an equation of the form $x + a = b$
C To solve an equation of the form $ax = b$

2.2 Applications: The Basic Percent Equation
A To use the basic percent equation
B To solve application problems

2.3 General Equations
A To solve an equation of the form $ax + b = c$
B To solve an equation of the form $ax + b = cx + d$
C To solve an equation containing parentheses
D To translate a sentence into an equation and solve

2.4 Mixture, Investment, and Motion Problems
A To solve value mixture problems
B To solve percent mixture problems
C To solve investment problems
D To solve uniform motion problems

2.5 Inequalities in One Variable
A To solve an inequality in one variable
B To solve a compound inequality
C To solve application problems

2.6 Absolute Value Equations and Inequalities
A To solve an absolute value equation
B To solve an absolute value inequality
C To solve application problems

Mersenne Primes

A prime number that can be written in the form $2^n - 1$, where n is also prime, is called a Mersenne prime. The table below shows some Mersenne primes.

$$3 = 2^2 - 1$$
$$7 = 2^3 - 1$$
$$31 = 2^5 - 1$$
$$127 = 2^7 - 1$$

Not every prime number is a Mersenne prime. For example, 5 is a prime number but not a Mersenne prime. Also, not all numbers in the form $2^n - 1$, where n is prime, yield a prime number. For example, $2^{11} - 1 = 2047$ is not a prime number.

The search for Mersenne primes has been quite extensive, especially since the advent of the computer. One reason for the extensive research into large prime numbers (not only Mersenne primes) has to do with cryptology.

Cryptology is the study of making or breaking secret codes. One method for making a code that is difficult to break is called public key cryptology. For this method to work, it is necessary to use very large prime numbers. To keep anyone from breaking the code, each prime should have at least 200 digits.

Today the largest known Mersenne prime is $2^{216091} - 1$. This number has 65,050 digits in its representation.

Another Mersenne prime got special recognition in a postage-meter stamp! It is the number $2^{11213} - 1$. This number has 3276 digits in its representation.

2.1 Introduction to Equations

Objective A   *To determine whether a given number is a solution of an equation*

An **equation** expresses the equality of two mathematical expressions. The expressions can be either numerical or variable expressions.

$$\left.\begin{array}{l} 9 + 3 = 12 \\ 3x - 2 = 10 \\ y^2 + 4 = 2y - 1 \\ z = 2 \end{array}\right\} \text{Equations}$$

The equation at the right is true if the variable is replaced by 5.

$$x + 8 = 13$$
$$5 + 8 = 13 \quad \text{A true equation}$$

The equation is false if the variable is replaced by 7.

$$7 + 8 = 13 \quad \text{A false equation}$$

A **solution** of an equation is a number that, when substituted for the variable, results in a true equation. 5 is a solution of the equation $x + 8 = 13$. 7 is not a solution of the equation $x + 8 = 13$.

➡ Is -2 a solution of $2x + 5 = x^2 - 3$?

To determine whether -2 is a solution of $2x + 5 = x^2 - 3$, replace x by -2. Then simplify and compare the results. If the results are equal, -2 is a solution of the equation. If the results are not equal, -2 is not a solution of the equation.

$$\begin{array}{c|c} \multicolumn{2}{c}{2x + 5 = x^2 - 3} \\ \hline 2(-2) + 5 & (-2)^2 - 3 \\ -4 + 5 & 4 - 3 \\ \multicolumn{2}{c}{1 = 1} \end{array}$$

The results are equal. Therefore, -2 is a solution of $2x + 5 = x^2 - 3$.

Example 1   Is $\dfrac{2}{3}$ a solution of $12x - 2 = 6x + 2$?

Solution
$$\begin{array}{c|c} \multicolumn{2}{c}{12x - 2 = 6x + 2} \\ \hline 12\left(\dfrac{2}{3}\right) - 2 & 6\left(\dfrac{2}{3}\right) + 2 \\ 8 - 2 & 4 + 2 \\ \multicolumn{2}{c}{6 = 6} \end{array}$$

Yes, $\dfrac{2}{3}$ is a solution.

You Try It 1   Is $\dfrac{1}{4}$ a solution of $5 - 4x = 8x + 2$?

Your solution
$5 - 4(\tfrac{1}{4}) = 8(\tfrac{1}{4}) + 2$
$5 - 1 = 2 + 2$
$4 = 4$

Example 2   Is -4 a solution of $4 + 5x = x^2 - 2x$?

Solution
$$\begin{array}{c|c} \multicolumn{2}{c}{4 + 5x = x^2 - 2x} \\ \hline 4 + 5(-4) & (-4)^2 - 2(-4) \\ 4 + (-20) & 16 - (-8) \\ \multicolumn{2}{c}{-16 \neq 24} \end{array}$$

No, -4 is not a solution.

You Try It 2   Is 5 a solution of $10x - x^2 = 3x - 10$?

Your solution
$10(5) - 5^2 = 3(5) - 10$
$50 - 25 = 15 - 10$
$25 \neq 5$

Solutions on p. A6

Objective B To solve an equation of the form $x + a = b$

POINT OF INTEREST

Finding solutions of equations has been a principal aim of mathematics for thousands of years. However, the equals sign did not occur in any text until 1557.

To **solve** an equation means to find a solution of the equation. The simplest equation to solve is an equation of the form *variable = constant,* because the constant is the solution.

The solution of the equation $x = 5$ is 5 because $5 = 5$ is a true equation.

The solution of the equation at the right is 7 because $7 + 2 = 9$ is a true equation.

$$x + 2 = 9$$
$$7 + 2 = 9$$

Note that when 4 is added to each side of the equation $x + 2 = 9$, the solution is still 7.

$$x + 2 = 9$$
$$x + 2 + 4 = 9 + 4$$
$$x + 6 = 13$$
$$7 + 6 = 13$$

When -5 is added to each side of the equation $x + 2 = 9$, the solution is still 7.

$$x + 2 = 9$$
$$x + 2 + (-5) = 9 + (-5)$$
$$x - 3 = 4$$
$$7 - 3 = 4$$

Equations that have the same solution are **equivalent equations.** The equations $x + 2 = 9$, $x + 6 = 13$, and $x - 3 = 4$ are equivalent equations; each equation has 7 as its solution. These examples suggest that adding the same number to each side of an equation produces equivalent equations. This is called the *Addition Property of Equations.*

Addition Property of Equations

The same number can be added to each side of an equation without changing its solution. In symbols, the equation $a = b$ has the same solution as the equation $a + c = b + c.$

CONSIDER THIS

Think of an equation as a balance scale. If the same weights are not added to each side of an equation, the "pans" no longer balance.

In solving an equation, the goal is to rewrite the given equation in the form *variable = constant.* The Addition Property of Equations is used to remove a term from one side of the equation by adding the opposite of that term to each side of the equation.

➡ Solve: $x - 4 = 2$

$$x - 4 = 2$$ • The goal is to rewrite the equation as **variable = constant.**
$$x - 4 + 4 = 2 + 4$$ • Add 4 to each side of the equation.
$$x + 0 = 6$$ • Simplify.
$$x = 6$$ • The equation is in the form **variable = constant.**

Check: $x - 4 = 2$
$$6 - 4 \mid 2$$
$$2 = 2 \quad \text{A true equation}$$

The solution is 6.

Because subtraction is defined in terms of addition, the Addition Property of Equations also makes it possible to subtract the same number from each side of an equation without changing the solution of the equation.

⇒ Solve: $y + \dfrac{3}{4} = \dfrac{1}{2}$

$$y + \frac{3}{4} = \frac{1}{2}$$

- The goal is to rewrite the equation in the form *variable = constant*.

$$y + \frac{3}{4} - \frac{3}{4} = \frac{1}{2} - \frac{3}{4}$$

- Subtract $\dfrac{3}{4}$ from each side of the equation.

$$y + 0 = \frac{2}{4} - \frac{3}{4}$$

- Simplify.

$$y = -\frac{1}{4}$$

- The equation is in the form *variable = constant*.

The solution is $-\dfrac{1}{4}$. You should check this solution.

Example 3 Solve: $x + \dfrac{3}{4} = \dfrac{1}{3}$

Solution

$$x + \frac{3}{4} = \frac{1}{3}$$

$$x + \frac{3}{4} - \frac{3}{4} = \frac{1}{3} - \frac{3}{4}$$

$$x = -\frac{5}{12} \quad \bullet \quad \frac{1}{3} - \frac{3}{4} = \frac{4}{12} - \frac{9}{12}$$

The solution is $-\dfrac{5}{12}$.

You Try It 3 Solve: $\dfrac{5}{6} = y - \dfrac{3}{8}$

Your solution

Solution on p. A6

Objective C   *To solve an equation of the form ax = b*

POINT OF INTEREST

Equations of the form $ax = b$ have appeared in algebra texts for a long time. The problem below is an adaptation from Fibonacci's text *Liber Abaci,* which dates from 1202.

A merchant purchased 7 eggs for 1 denarius and sold them at a rate of 5 eggs for 1 denarius. The merchant's profit was 18 denarii. How much did the merchant invest?

The resulting equation is

$$\frac{7}{5}x - x = 18$$

The solution is 45 denarii.

The solution of the equation at the right is 3 because $2 \cdot 3 = 6$ is a true equation.

$$2x = 6$$
$$2 \cdot 3 = 6$$

Note that when each side of $2x = 6$ is multiplied by 5, the solution is still 3.

$$2x = 6$$
$$5(2x) = 5 \cdot 6$$
$$10x = 30$$
$$10 \cdot 3 = 30$$

When each side of $2x = 6$ is multiplied by -4, the solution is still 3.

$$2x = 6$$
$$(-4)(2x) = (-4) \cdot 6$$
$$-8x = -24$$
$$-8 \cdot 3 = -24$$

The equations $2x = 6$, $10x = 30$, and $-8x = -24$ are equivalent equations; each equation has 3 as its solution. These examples suggest that multiplying each side of an equation by the same number produces equivalent equations.

> **Multiplication Property of Equations**
>
> Each side of an equation can be multiplied by the same *nonzero* number without changing the solution of the equation. In symbols, if $c \neq 0$, then the equation $a = b$ has the same solutions as the equation $ac = bc$.

Recall that the goal of solving an equation is to rewrite the equation in the form *variable = constant*. The Multiplication Property of Equations is used to remove a coefficient by multiplying each side of the equation by the reciprocal of the coefficient.

➡ Solve: $\frac{3}{4}z = 9$

$$\frac{3}{4}z = 9$$ • The goal is to rewrite the equation in the form *variable = constant*.

$$\frac{4}{3} \cdot \frac{3}{4}z = \frac{4}{3} \cdot 9$$ • Multiply each side of the equation by $\frac{4}{3}$.

$$1 \cdot z = 12$$ • Simplify.

$$z = 12$$ • The equation is in the form *variable = constant*.

The solution is 12. You should check this solution.

Because division is defined in terms of multiplication, each side of an equation can be divided by the same nonzero number without changing the solution of the equation.

➡ Solve: $6x = 14$

$$6x = 14$$ • The goal is to rewrite the equation in the form *variable = constant*.

$$\frac{6x}{6} = \frac{14}{6}$$ • Divide each side of the equation by 6.

$$1 \cdot x = \frac{7}{3}$$ • Simplify.

$$x = \frac{7}{3}$$ • The equation is in the form *variable = constant*.

The solution is $\frac{7}{3}$.

CONSIDER THIS

Remember to check the solution.

Check: $\dfrac{6x = 14}{6\left(\dfrac{7}{3}\right) \bigg| 14}$

$$14 = 14$$

When using the Multiplication Property of Equations, multiply each side of the equation by the reciprocal of the coefficient when the coefficient is a fraction. Divide each side of the equation by the coefficient when the coefficient is an integer or a decimal.

Example 4   Solve: $\frac{3x}{4} = -9$

Solution

$$\frac{3x}{4} = -9$$

$$\frac{4}{3} \cdot \frac{3}{4}x = \frac{4}{3}(-9) \quad \bullet \left[\frac{3x}{4} = \frac{3}{4}x\right]$$

$$x = -12$$

The solution is -12.

You Try It 4   Solve: $-\frac{2x}{5} = 6$

Your solution

Example 5   Solve: $5x - 9x = 12$

Solution

$$5x - 9x = 12$$
$$-4x = 12 \quad \bullet \text{ Combine like terms.}$$
$$\frac{-4x}{-4} = \frac{12}{-4}$$
$$x = -3$$

The solution is -3.

You Try It 5   Solve: $4x - 8x = 16$

Your solution

Solutions on p. A6

Objective C

Solve and check.

73. $5x = -15$

74. $4y = -28$

75. $3b = 0$

76. $2a = 0$

77. $-3x = 6$

78. $-5m = 20$

79. $-3x = -27$

80. $-\dfrac{1}{6}n = -30$

81. $20 = \dfrac{1}{4}c$

82. $18 = 2t$

83. $-32 = 8w$

84. $-56 = 7x$

85. $0 = -5x$

86. $0 = -8a$

87. $-32 = -4y$

88. $-54 = 6c$

89. $49 = -7t$

90. $\dfrac{x}{3} = 2$

91. $\dfrac{x}{4} = 3$

92. $-\dfrac{y}{2} = 5$

93. $-\dfrac{b}{3} = 6$

94. $\dfrac{3}{4}y = 9$

95. $\dfrac{2}{5}x = 6$

96. $-\dfrac{2}{3}d = 8$

97. $-\dfrac{3}{5}m = 12$

98. $\dfrac{2n}{3} = 0$

99. $\dfrac{5x}{6} = 0$

100. $\dfrac{-3z}{8} = 9$

101. $\dfrac{-4x}{5} = -12$

102. $-6 = -\dfrac{2}{3}y$

103. $-15 = -\dfrac{1}{5}x$

104. $\dfrac{2}{5}a = 3$

105. $\dfrac{3x}{4} = 2$

106. $\dfrac{3}{4}c = \dfrac{3}{5}$

107. $\dfrac{2}{9} = \dfrac{2}{3}y$

108. $-\dfrac{6}{7} = -\dfrac{3}{4}b$

109. $\dfrac{1}{5}x = -\dfrac{1}{10}$

110. $-\dfrac{2}{3}y = -\dfrac{8}{9}$

111. $-1 = \dfrac{2n}{3}$

112. $-\dfrac{3}{4} = \dfrac{a}{8}$

Solve and check.

113. $-\dfrac{2}{5}m = -\dfrac{6}{7}$ **114.** $5x + 2x = 14$ **115.** $3n + 2n = 20$ **116.** $d - 4d = 9$

117. $10y - 3y = 21$ **118.** $2x - 5x = 9$ **119.** $\dfrac{x}{1.46} = 3.25$ **120.** $\dfrac{z}{2.95} = -7.88$

121. $3.47a = 7.1482$ **122.** $2.31m = 2.4255$ **123.** $-3.7x = 7.881$ **124.** $\dfrac{n}{2.65} = 9.08$

APPLYING THE CONCEPTS

125. Solve the equation $ax = b$ for x. Is the solution you have written valid for all real numbers a and b?

126. Make up an equation of the form $x + a = b$ that has 2 as a solution.

127. Make up an equation of the form $ax = b$ that has -1 as a solution.

128. Solve.

    a. $\dfrac{3}{\dfrac{1}{x}} = 5$ **b.** $\dfrac{2}{\dfrac{1}{y}} = -2$

129. Solve.

    a. $\dfrac{3x + 2x}{3} = 2$ **b.** $\dfrac{4a - 7a}{3} = -2$

130. One-half of a certain number equals two-thirds of the same number. What is the number?

131. [W] To solve the equation $x + 7 = 10$, you can either add -7 to each side of the equation or subtract 7 from each side. Explain why this is possible.

132. [W] Write out the steps for solving the equation $\dfrac{1}{2}x = -3$. Identify each Property of Real Numbers or Property of Equations as you use it.

133. [W] In your own words, state the Addition Property of Equations and the Multiplication Property of Equations.

2.2 Applications: The Basic Percent Equation

Objective A   *To use the basic percent equation* ....

A real estate broker receives a payment that is 6% of a $175,000 sale. To find the amount the broker receives requires answering the question "6% of $175,000 is what?" This sentence can be written using mathematical symbols and then solved for the unknown number. The word *of* is written as · (times), *is* is written as = (equals), and *what* is written as a variable (the unknown number).

6%	of	$175,000	is	what?
↓	↓	↓	↓	↓
Percent	·	base	=	amount
6%		$175,000		A

$$0.06 \cdot \$175{,}000 = A$$
$$\$10{,}500 = A$$

The broker receives a payment of $10,500.

The solution was found by solving the basic percent equation for amount.

The Basic Percent Equation

Percent · base = amount
$$P \cdot B = A$$

➡ Find 2.5% of 800.

Percent · base = amount
$$0.025 \cdot 800 = A$$
$$20 = A$$

• Use the basic percent equation.
  Percent = 2.5% = 0.025, base = 800,
  amount = A

2.5% of 800 is 20.

A recent promotional game at a grocery store listed the probability of winning a prize as "1 chance in 2." A percent can be used to describe the chance of winning. This requires answering the question "What percent of 2 is 1?"

The chance of winning can be found by solving the basic percent equation for percent.

What	percent	of	2	is	1?
	↓	↓	↓	↓	↓
	percent	·	base	=	amount
	P		2		1

$$P \cdot 2 = 1$$
$$P = \frac{1}{2}$$

Write the fraction as a percent. $P = \dfrac{1}{2}(100\%) = 50\%$

There is a 50% chance of winning a prize.

➡ 32 is what percent of 20?

Percent · base = amount
$$P \cdot 20 = 32$$
$$P = \frac{32}{20}$$
$$P = 1.6$$
$$P = 160\%$$

• Use the basic percent equation. Percent = P, base = 20, amount = 32

• Write 1.6 as a percent.

32 is 160% of 20.

Each year an investor receives a payment that equals 8% of the value of an investment. This year that payment amounted to $640. To find the value of the investment this year, we must answer the question "8% of what value is $640?"

The value of the investment can be found by solving the basic percent equation for the base.

8%	of	what	is	$640?
↓	↓	↓	↓	↓
Percent	·	base	=	amount
8%		B		640

$$0.08 \cdot B = 640$$
$$B = \frac{640}{0.08}$$
$$B = 8000$$

This year the investment is worth $8000.

➡ 62% of what is 800? Round to the nearest tenth.

Percent · base = amount
$$0.62 \cdot B = 800$$
$$B = \frac{800}{0.62}$$
$$B \approx 1290.3$$

• Use the basic percent equation. Percent = 62% = 0.62, base = B, amount = 800

62% of 1290.3 is approximately 800.

Note from the previous problems that when any two parts of the basic percent equation are given, the third part can be found. **In percent problems, the base usually follows the word *of*.** Some percent problems may use the word *find*. In this case, we can substitute *what is* for *find*.

Example 1 Find 9.4% of 240.

Strategy To find the amount, solve the basic percent equation. Percent = 9.4% = 0.094, base = 240, amount = A

Solution Percent · base = amount
$$0.094 \cdot 240 = A$$
$$22.56 = A$$

22.56 is 9.4% of 240.

You Try It 1 Find $33\frac{1}{3}\%$ of 45.

Your strategy

Your solution

Solution on p. A6

Example 2 What percent of 30 is 12?

Strategy To find the percent, solve the basic percent equation. Percent = P, base = 30, amount = 12

Solution Percent · base = amount
$$P \cdot 30 = 12$$
$$P = \frac{12}{30} = 0.4$$
$$P = 40\%$$

12 is 40% of 30.

You Try It 2 25 is what percent of 40?

Your strategy

Your solution

Example 3 60 is 2.5% of what?

Strategy To find the base, solve the basic percent equation. Percent = 2.5% = 0.025, base = B, amount = 60

Solution Percent · base = amount
$$0.025 \cdot B = 60$$
$$B = \frac{60}{0.025}$$
$$B = 2400$$

60 is 2.5% of 2400.

You Try It 3 $16\frac{2}{3}\%$ of what is 15?

Your strategy

Your solution

Solutions on pp. A6–A7

Objective B   *To solve application problems* ·································

➡ A computer programmer receives a weekly wage of $650, of which $110.50 is deducted for income tax. Find the percent of the computer programmer's salary deducted for income tax.

Percent · base = amount   • **Use the basic percent equation.**
$$P \cdot 650 = 110.50$$ **Percent = *P*, base = 650,**
$$P = \frac{110.50}{650}$$ **amount = 110.50**
$$P = 0.17$$
$$P = 17\%$$

17% of the computer programmer's salary is deducted for income tax.

⇒ A department store has 250 employees and must hire an additional 18% for the holiday season. What is the total number of employees needed for the holiday season?

Percent · base = amount

$$0.18 \cdot 250 = A$$
$$45 = A$$

$$250 + 45 = 295$$

- Use the basic percent equation to find the number of additional employees needed. Percent = 18% = 0.18, base = 250, amount = A

- Add the number of employees hired to the present number of employees.

The department store will employ 295 employees for the holiday season.

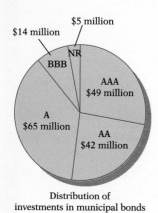

Distribution of investments in municipal bonds

The circle graph at the left represents the amounts of money a mutual fund company has invested in different municipal bonds. The complete circle represents the total amount invested in all bonds, $175 million. Each sector of the circle represents the amount of money invested in a different quality of bonds. To find the percent of the total money invested in AAA-rated bonds, solve the basic percent equation for percent. Percent = P, base = 175 million, amount = 49 million

Percent · base = amount

$$P \cdot 175 = 49$$

$$P = \frac{49}{175}$$

$$P = 0.28$$

$$P = 28\%$$

28% of the total amount invested is invested in AAA-rated bonds.

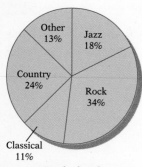

Total sales of compact disks for 1 month

In some circle graphs, each sector represents a percent of the total. The circle graph at the left represents the total sales in dollars of compact disks for a record store for one month. The complete circle represents 100% of the $55,000 in compact disks sold. Each sector expresses the percent of sales for a particular type of music. To find the dollar sales for jazz compact disks for the one-month period, solve the basic percent equation for amount. Percent = 18% = 0.18, base = 55,000, amount = A

Percent · base = amount

$$0.18 \cdot 55,000 = A$$

$$9900 = A$$

The dollar sales for jazz compact disks were $9900.

Example 4
The circle graph below shows the result of a survey of 300 people who were asked to name their favorite sport. Find the percent of people surveyed who listed tennis as their favorite spectator sport.

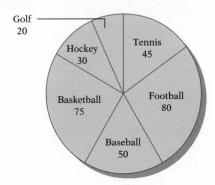

Strategy
To find the percent, solve the basic percent equation for percent.
Percent = P, base = 300,
amount = 45

Solution
Percent · base = amount
$$P \cdot 300 = 45$$
$$P = \frac{45}{300}$$
$$P = 0.15 = 15\%$$

15% of the people surveyed selected tennis as their favorite spectator sport.

Example 5
Twelve percent of a company's $60,000 budget is used for advertising. Find the amount of the company's budget spent for advertising.

Strategy
To find the amount, use the basic percent equation.
Percent = 12% = 0.12, base = 60,000,
amount = A

Solution
Percent · base = amount
$$0.12 \cdot 60,000 = A$$
$$7200 = A$$

$7200 is spent for advertising.

You Try It 4
The circle graph below represents the percent of a family's annual income of $26,000 that is budgeted for various expenses. Find the annual amount budgeted for food.

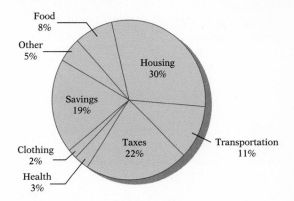

Your strategy

Your solution

You Try It 5
An instructor receives a monthly salary of $2165, and $324.75 is deducted for income tax. Find the percent of the instructor's salary deducted for income tax.

Your strategy

Your solution

Solutions on p. A7

Example 6

A taxpayer pays 35% of his net income for state and federal taxes. The taxpayer has a net income of $37,500. Find the amount of state and federal taxes paid by the taxpayer.

Strategy

To find the amount, solve the basic percent equation.
Percent = 35% = 0.35, base = 37,500, amount = A

Solution

Percent · base = amount
$$0.35 \cdot 37,500 = A$$
$$13,125 = A$$

The amount of taxes paid is $13,125.

You Try It 6

Seventy percent of the people polled in a public opinion survey approved of the way the mayor was performing the duties of government. If 210 persons approved, how many were polled?

Your strategy

Your solution

Example 7

A department store has a blue blazer on sale for $114, which is 60% of the original price. What is the difference between the original price and the sale price?

Strategy

To find the difference between the original price and the sale price:

• Find the original price. Solve the basic percent equation.
  Percent = 60% = 0.60, amount = 114, base = B
• Subtract the sale price from the original price.

Solution

Percent · base = amount
$$0.60 \cdot B = 114$$
$$B = \frac{114}{0.60}$$
$$B = 190$$

$$190 - 114 = 76$$

The difference in price is $76.

You Try It 7

An electrician's wage this year is $20.01 per hour, which is 115% of last year's wage. What was the increase in the hourly wage over last year?

Your strategy

Your solution

Solutions on p. A7

2.2 Exercises

· ·

___

Objective A

Solve.

1. 8% of 100 is what?

2. 16% of 50 is what?

3. 0.05% of 150 is what?

4. 0.075% of 625 is what?

5. 15 is what percent of 90?

6. 24 is what percent of 60?

7. What percent of 16 is 6?

8. What percent of 24 is 18?

9. 10 is 10% of what?

10. 37 is 37% of what?

11. 2.5% of what is 30?

12. 10.4% of what is 52?

13. Find 10.7% of 485.

14. Find 12.8% of 625.

15. 80% of 16.25 is what?

16. 26% of 19.5 is what?

17. 54 is what percent of 2000?

18. 8 is what percent of 2500?

19. 16.4 is what percent of 4.1?

20. 5.3 is what percent of 50?

21. 18 is 240% of what?

22. 24 is 320% of what?

23. 25.6 is 12.8% of what?

24. 45.014 is 63.4% of what?

25. 1 is what percent of 40?

26. 0.3 is what percent of 20?

27. What percent of 48 is 18?

28. What percent of 11 is 88?

29. 0.7% of what is 0.56?

30. 0.25% of what is 1?

31. 30% of what is 2.7?

32. 78% of what is 3.9?

Objective B

Solve.

33. A computer programmer's salary increased $19.52 per week. By what percent did the programmer's salary increase if the salary was $244 before the raise?

34. A charity organization spent $2940 for administrative expenses. This amount is 12% of the money it collected. What is the total amount the charity organization collected?

35. A mechanic estimates that the brakes of an RV still have 6000 mi of wear. This amount is 12% of the estimated safe-life use of the brakes. What is the estimated life of the brakes?

36. A company spends $4500 of its $90,000 budget for advertising. What percent of the budget is spent for advertising?

37. A sales clerk receives a salary of $2240 per month, and 18% of this amount is deducted for income tax. Find the amount deducted for income tax.

38. An antiques shop owner expects to receive $16\frac{2}{3}$% of the shop's sales as profit. What is the expected profit in a month when the total sales are $24,000?

39. Last month a thrift store brought in an income of $2812.50. The rent for the store is $900 per month. What percent of last month's income was spent for rent?

40. A used mobile home was purchased for $18,000. This amount was 64% of the new mobile home cost. What is the cost of a new mobile home?

The circle graph represents the number of playground injuries suffered by children on various pieces of playground equipment.

41. What percent of the total injuries happen on slides? Round to the nearest tenth of a percent.

42. What percent of the total injuries occur on monkey bars? Round to the nearest tenth of a percent.

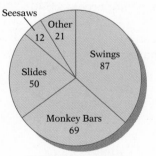

Number of playground
injuries in thousands

Solve.

43. A ski resort had a snowfall of 198 in. during the year. This amount was 120% of the previous year's snowfall. What was the previous year's snowfall?

44. A fire department received 24 false alarms out of a total of 200 alarms received. What percent of the alarms received were false alarms?

45. A store advertised a scientific calculator for $55.80. This amount was 112% of the cost at a competing store. What was the price at the competitor's store?

46. A calculator can be purchased for $28.50. This amount is 40% of the cost of the calculator 8 years ago. What was the cost of the calculator 8 years ago?

47. A city has a population of 42,000. This amount is 75% of what the population was 5 years ago. What was the city's population 5 years ago?

48. A car is sold for $8900. The buyer of the car pays a down payment of $1780. What percent of the selling price is the down payment?

49. A farmer is given an income tax credit of 15% of the cost of some farm machinery. What tax credit would the farmer receive on farm equipment that cost $85,000?

An accounting major recorded the number of units required in each discipline to graduate with a degree in accounting. The results are shown in the circle graph.

50. What percent of the units required to graduate are taken in accounting?

51. What percent of the units required to graduate are taken in mathematics?

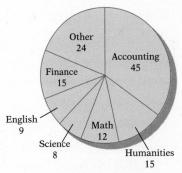

Number of units required to graduate with an accounting degree

52. A customer bought a car for $8500 and paid a sales tax of 5.5% of the cost. Find the total cost of the car, including the sales tax.

53. During a quality control test, a manufacturer of computer boards found that 56 boards were defective. This amount was 0.7% of the computer boards tested. How many computer boards tested were not defective?

Solve.

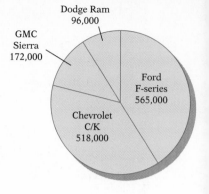

54. The circle graph at the right shows the sales of full-size pickup trucks for different manufacturers in a recent year. What percent of the total number of pickups sold by these companies were Ford F-series pickups?

55. A baseball player with a salary of $1,650,000 is offered a new contract that contains an 8% salary cut. What is the amount of money offered in the new contract?

56. To receive a license to sell insurance, an insurance account executive must answer 70% of the 250 questions on a test correctly. An account executive answered 177 questions correctly. Did the account executive pass the test?

57. To receive a B− grade in a history class, a student must give 75 correct responses on a test of 90 questions. What percent of the total number of questions must a student answer correctly to receive a B− grade?

APPLYING THE CONCEPTS

58. A $10,000, 8% bond has a quoted price of $10,500. Look up in a business math book the meaning of current yield. Then find the current yield on the $10,000 bond.

59. A welder earning $12 per hour is given a 10% raise. To find the new wage, we can multiply $12 by 0.10 and add the product to $12. Can the new wage be found by multiplying $12 by 1.10? Try both methods and compare your answers.

60. A wide-screen TV with a price of $3000 was on sale for 30% off. When the TV didn't sell, an additional 10% of the sale price was taken off the sale price. Calculate the sale price after the two discounts. Compare the two successive discounts of 30% and 10% with a 40% discount off the original price. What single discount would be equivalent to the successive discounts?

61. If a quantity increases by 100%, how many times its original value is the new value?

62. Find 10% of a number and subtract it from the original number. Now
[W] take 10% of the new number and subtract it from the new number. Is this the same as taking 20% of the original number? Explain.

63. Increase a number by 10%. Now decrease the new number by 10%. Is
[W] the result the original number? Explain.

64. Explain how the term *percentage* is used in different situations.
[W]

65. Explain the meaning of *per millage,* and explain its relation to
[W] percent.

66. Visit a savings and loan institution or credit union to research, and to
[W] write a report on, the meaning of points as related to a loan.

2.3 General Equations

Objective A *To solve an equation of the form ax + b = c* ....

POINT OF INTEREST
Evariste Galois, despite being killed in a duel at the age of 21, made significant contributions to solving equations. In fact, there is a branch of mathematics called Galois Theory showing what kinds of equations can and cannot be solved.

In solving an equation of the form $ax + b = c$, the goal is to rewrite the equation in the form *variable = constant*. This requires the application of both the Addition and the Multiplication Properties of Equations.

➡ Solve: $\dfrac{3}{4}x - 2 = -11$

The goal is to write the equation in the form *variable = constant*.

$$\frac{3}{4}x - 2 = -11$$

$$\frac{3}{4}x - 2 + 2 = -11 + 2$$ • Add 2 to each side of the equation.

$$\frac{3}{4}x = -9$$ • Simplify.

$$\frac{4}{3} \cdot \frac{3}{4}x = \frac{4}{3}(-9)$$ • Multiply each side of the equation by $\dfrac{4}{3}$.

$$x = -12$$ • The equation is of the form *variable = constant*.

Check: • Check the solution.

$$\frac{3}{4}x - 2 = -11$$

$$\begin{array}{c|c} \frac{3}{4}(-12) - 2 & -11 \\ -9 - 2 & -11 \\ -11 = -11 \end{array}$$ • A true equation

The solution is -12.

Example 1 Solve: $3x - 7 = -5$

Solution
$$3x - 7 = -5$$
$$3x - 7 + 7 = -5 + 7$$
$$3x = 2$$
$$\frac{3x}{3} = \frac{2}{3}$$
$$x = \frac{2}{3}$$

The solution is $\dfrac{2}{3}$.

You Try It 1 Solve: $5x + 7 = 10$

Your solution

Solution on p. A8

Example 2

Solve: $5 = 9 - 2x$

Solution

$$5 = 9 - 2x$$
$$5 - 9 = 9 - 9 - 2x$$
$$-4 = -2x$$
$$\frac{-4}{-2} = \frac{-2x}{-2}$$
$$2 = x$$

The solution is 2.

You Try It 2

Solve: $2 = 11 + 3x$

Your solution

$2 - 11 = 3x$

$9 = 3x$

$9 \div 3 = x$

$\dfrac{9}{3} = x$

$-3 = x$

Example 3

Solve: $2x + 4 - 5x = 10$

Solution

$$2x + 4 - 5x = 10 \qquad \bullet \text{ Combine like terms.}$$
$$-3x + 4 = 10$$
$$-3x + 4 - 4 = 10 - 4$$
$$-3x = 6$$
$$\frac{-3x}{-3} = \frac{6}{-3}$$
$$x = -2$$

The solution is -2.

You Try It 3

Solve: $x - 5 + 4x = 25$

Your solution

$5x - 5 = 25$

$5x = 25 + 5$

$5x - 30$

$x = \dfrac{30}{5} = 6$

Solutions on p. A8

Objective B   To solve an equation of the form $ax + b = cx + d$ ....

In solving an equation of the form $ax + b = cx + d$, the goal is to rewrite the equation in the form *variable = constant*. Begin by rewriting the equation so that there is only one variable term in the equation. Then rewrite the equation so that there is only one constant term.

➡ Solve: $2x + 3 = 5x - 9$

$$2x + 3 = 5x - 9$$

$$2x - 5x + 3 = 5x - 5x - 9 \qquad \bullet \textbf{ Subtract } 5x \textbf{ from each side of the equation.}$$

$$-3x + 3 = -9 \qquad \bullet \textbf{ Simplify.}$$

$$-3x + 3 - 3 = -9 - 3 \qquad \bullet \textbf{ Subtract 3 from each side of the equation.}$$

$$-3x = -12 \qquad \bullet \textbf{ Simplify.}$$

$$\frac{-3x}{-3} = \frac{-12}{-3} \qquad \bullet \textbf{ Divide each side of the equation by } -3.$$

$$x = 4 \qquad \bullet \textbf{ The equation is in the form } \textit{variable = constant.}$$

The solution is 4. You should verify this by checking this solution.

Example 4

Solve: $4x - 5 = 8x - 7$

Solution

$$4x - 5 = 8x - 7$$

$$4x - 8x - 5 = 8x - 8x - 7$$

$$-4x - 5 = -7$$

$$-4x - 5 + 5 = -7 + 5$$

$$-4x = -2$$

$$\frac{-4x}{-4} = \frac{-2}{-4}$$

$$x = \frac{1}{2}$$

The solution is $\frac{1}{2}$.

You Try It 4

Solve: $5x + 4 = 6 + 10x$

Your solution

Example 5

Solve: $3x + 4 - 5x = 2 - 4x$

Solution

$$3x + 4 - 5x = 2 - 4x$$

$$-2x + 4 = 2 - 4x$$

$$-2x + 4x + 4 = 2 - 4x + 4x$$

$$2x + 4 = 2$$

$$2x + 4 - 4 = 2 - 4$$

$$2x = -2$$

$$\frac{2x}{2} = \frac{-2}{2}$$

$$x = -1$$

The solution is -1.

You Try It 5

Solve: $5x - 10 - 3x = 6 - 4x$

Your solution

Solutions on p. A8

Objective C   *To solve an equation containing parentheses* ....

When an equation contains parentheses, one of the steps in solving the equation requires the use of the Distributive Property. The Distributive Property is used to remove parentheses from a variable expression.

⇒ Solve: $4 + 5(2x - 3) = 3(4x - 1)$

$4 + 5(2x - 3) = 3(4x - 1)$

$4 + 10x - 15 = 12x - 3$ • **Use the Distributive Property. Then simplify.**

$10x - 11 = 12x - 3$

$10x - 12x - 11 = 12x - 12x - 3$ • **Subtract 12x from each side of the equation.**

$-2x - 11 = -3$ • **Simplify.**

$-2x - 11 + 11 = -3 + 11$ • **Add 11 to each side of the equation.**

$-2x = 8$ • **Simplify.**

$\dfrac{-2x}{-2} = \dfrac{8}{-2}$ • **Divide each side of the equation by −2.**

$x = -4$ • **The equation is in the form** *variable = constant.*

The solution is −4. You should verify this by checking this solution.

Example 6
Solve: $3x - 4(2 - x) = 3(x - 2) - 4$

Solution
$3x - 4(2 - x) = 3(x - 2) - 4$
$3x - 8 + 4x = 3x - 6 - 4$
$7x - 8 = 3x - 10$
$7x - 3x - 8 = 3x - 3x - 10$
$4x - 8 = -10$
$4x - 8 + 8 = -10 + 8$
$4x = -2$
$\dfrac{4x}{4} = \dfrac{-2}{4}$
$x = -\dfrac{1}{2}$

The solution is $-\dfrac{1}{2}$.

You Try It 6
Solve: $5x - 4(3 - 2x) = 2(3x - 2) + 6$

Your solution

$5x - 12 - 8x = 6x - 4 + 6$
$13x - 12 = 6x + 2$
$13x =$

Solution on p. A8

Example 7

Solve: $3[2 - 4(2x - 1)] = 4x - 10$

Solution

$$3[2 - 4(2x - 1)] = 4x - 10$$
$$3[2 - 8x + 4] = 4x - 10$$
$$3[6 - 8x] = 4x - 10$$
$$18 - 24x = 4x - 10$$
$$18 - 24x - 4x = 4x - 4x - 10$$
$$18 - 28x = -10$$
$$18 - 18 - 28x = -10 - 18$$
$$-28x = -28$$
$$\frac{-28x}{-28} = \frac{-28}{-28}$$
$$x = 1$$

The solution is 1.

You Try It 7

Solve: $-2[3x - 5(2x - 3)] = 3x - 8$

Your solution

$$-2[3x - 10x + 15] = 3x - 8$$
$$-2[-7x + 15] = 3x - 8$$
$$14x + 30 = 3x - 8$$
$$14x - 3x = -8 + 30$$
$$11x = 22$$
$$x = 22 \div 11 = 2$$

Solution on p. A9

Objective D   *To translate a sentence into an equation and solve*

An equation states that two mathematical expressions are equal. Therefore, to translate a sentence into an equation requires recognition of the words or phrases that mean "equals." Some of these words and phrases are listed below.

equals   *is*   *totals*   *is the same as*

POINT OF INTEREST

Number puzzle problems similar to the one on this page have appeared in textbooks for hundreds of years. Here is one from a 1st-century Chinese textbook: "When a number is divided by 3, the remainder is 2; when it is divided by 5, the remainder is 3; when it is divided by 7, the remainder is 2. Find the number." There are actually an infinite number of solutions to this problem. See if you can find one of them.

➡ Translate "five less than four times a number is four more than the number" into an equation and solve.

the unknown number: n

five less than four times a number	is	four more than the number

- Assign a variable to the unknown number.
- Find two verbal expressions for the same value.

$$4n - 5 = n + 4$$
$$4n - n - 5 = n - n + 4$$
$$3n - 5 = 4$$
$$3n - 5 + 5 = 4 + 5$$
$$3n = 9$$
$$\frac{3n}{3} = \frac{9}{3}$$
$$n = 3$$

- Write an equation.
- Solve the equation.
- Check:

5 less than 4 times 3	4 more than 3
$4 \cdot 3 - 5$	$3 + 4$
$12 - 5$	7
$7 =$	7

The number is 3.

Example 8

Translate "eight less than three times a number equals five times the number" into an equation and solve.

Solution

the unknown number: x

eight less than three times a number	equals	five times the number

$$3x - 8 = 5x$$
$$3x - 3x - 8 = 5x - 3x$$
$$-8 = 2x$$
$$\frac{-8}{2} = \frac{2x}{2}$$
$$-4 = x$$

-4 checks as the solution.

The number is -4.

Example 9

The sum of two numbers is 9. Eight times the smaller number is five less than three times the larger number. Find the numbers.

Solution

the smaller number: p
the larger number: $9 - p$

eight times the smaller number	is	five less than three times the larger number

$$8p = 3(9 - p) - 5$$
$$8p = 27 - 3p - 5$$
$$8p = 22 - 3p$$
$$8p + 3p = 22 - 3p + 3p$$
$$11p = 22$$
$$\frac{11p}{11} = \frac{22}{11}$$
$$p = 2$$
$$9 - p = 9 - 2 = 7$$

These numbers check as solutions.

The smaller number is 2.
The larger number is 7.

You Try It 8

Translate "seven less than a number is equal to five more than three times the number" into an equation and solve.

Your solution

$$n - 7 = 3n + 5$$
$$-5 - 7 = 3n - n$$
$$-12 = 2n$$
$$12 \div 2 = n$$
$$-6 = n$$

You Try It 9

The sum of two numbers is 14. One more than three times the smaller equals the sum of the larger number and three. Find the two numbers.

Your solution

$$n = smaller$$
$$14 - n = larger$$

$$3n + 1 = 14 - n + 3$$
$$3n + 1 = 17 - n$$
$$3n + n = 17 - 1$$
$$4n = 16$$
$$n = 4$$

Solutions on p. A9

2.3 Exercises

Objective A

Solve and check.

1. $3x + 1 = 10$ **2.** $4y + 3 = 11$ **3.** $2a - 5 = 7$ **4.** $5m - 6 = 9$

5. $5 = 4x + 9$ **6.** $2 = 5b + 12$ **7.** $2x - 5 = -11$ **8.** $3n - 7 = -19$

9. $4 - 3w = -2$ **10.** $5 - 6x = -13$ **11.** $8 - 3t = 2$ **12.** $12 - 5x = 7$

13. $4a - 20 = 0$ **14.** $3y - 9 = 0$ **15.** $6 + 2b = 0$ **16.** $10 + 5m = 0$

17. $-2x + 5 = -7$ **18.** $-5d + 3 = -12$ **19.** $-12x + 30 = -6$ **20.** $-13 = -11y + 9$

21. $2 = 7 - 5a$ **22.** $3 = 11 - 4n$ **23.** $-35 = -6b + 1$ **24.** $-8x + 3 = -29$

25. $-3m - 21 = 0$ **26.** $-5x - 30 = 0$ **27.** $-4y + 15 = 15$ **28.** $-3x + 19 = 19$

29. $9 - 4x = 6$ **30.** $3t - 2 = 0$ **31.** $9x - 4 = 0$ **32.** $7 - 8z = 0$

33. $1 - 3x = 0$ **34.** $9d + 10 = 7$ **35.** $12w + 11 = 5$ **36.** $6y - 5 = -7$

37. $8b - 3 = -9$ **38.** $5 - 6m = 2$ **39.** $7 - 9a = 4$ **40.** $9 = -12c + 5$

Solve and check.

41. $10 = -18x + 7$ **42.** $2y + \dfrac{1}{3} = \dfrac{7}{3}$ **43.** $4a + \dfrac{3}{4} = \dfrac{19}{4}$ **44.** $2n - \dfrac{3}{4} = \dfrac{13}{4}$

45. $3x - \dfrac{5}{6} = \dfrac{13}{6}$ **46.** $5y + \dfrac{3}{7} = \dfrac{3}{7}$ **47.** $9x + \dfrac{4}{5} = \dfrac{4}{5}$ **48.** $8 = 7d - 1$

49. $8 = 10x - 5$ **50.** $4 = 7 - 2w$ **51.** $7 = 9 - 5a$ **52.** $8t + 13 = 3$

53. $12x + 19 = 3$ **54.** $-6y + 5 = 13$ **55.** $-4x + 3 = 9$ **56.** $\dfrac{1}{2}a - 3 = 1$

57. $\dfrac{1}{3}m - 1 = 5$ **58.** $\dfrac{2}{5}y + 4 = 6$ **59.** $\dfrac{3}{4}n + 7 = 13$ **60.** $-\dfrac{2}{3}x + 1 = 7$

61. $-\dfrac{3}{8}b + 4 = 10$ **62.** $\dfrac{x}{4} - 6 = 1$ **63.** $\dfrac{y}{5} - 2 = 3$ **64.** $\dfrac{2x}{3} - 1 = 5$

65. $\dfrac{3c}{7} - 1 = 8$ **66.** $4 - \dfrac{3}{4}z = -2$ **67.** $3 - \dfrac{4}{5}w = -9$ **68.** $5 + \dfrac{2}{3}y = 3$

69. $17 + \dfrac{5}{8}x = 7$ **70.** $17 = 7 - \dfrac{5}{6}t$ **71.** $9 = 3 - \dfrac{2x}{7}$ **72.** $3 = \dfrac{3a}{4} + 1$

73. $7 = \dfrac{2x}{5} + 4$ **74.** $5 - \dfrac{4c}{7} = 8$ **75.** $7 - \dfrac{5}{9}y = 9$ **76.** $6a + 3 + 2a = 11$

77. $5y + 9 + 2y = 23$ **78.** $7x - 4 - 2x = 6$ **79.** $11z - 3 - 7z = 9$ **80.** $2x - 6x + 1 = 9$

81. $b - 8b + 1 = -6$ **82.** $3 = 7x + 9 - 4x$ **83.** $-1 = 5m + 7 - m$ **84.** $8 = 4n - 6 + 3n$

Objective B

Solve and check.

85. $8x + 5 = 4x + 13$

86. $6y + 2 = y + 17$

87. $5x - 4 = 2x + 5$

88. $13b - 1 = 4b - 19$

89. $15x - 2 = 4x - 13$

90. $7a - 5 = 2a - 20$

91. $3x + 1 = 11 - 2x$

92. $n - 2 = 6 - 3n$

93. $2x - 3 = -11 - 2x$

94. $4y - 2 = -16 - 3y$

95. $2b + 3 = 5b + 12$

96. $m + 4 = 3m + 8$

97. $4y - 8 = y - 8$

98. $5a + 7 = 2a + 7$

99. $6 - 5x = 8 - 3x$

100. $10 - 4n = 16 - n$

101. $5 + 7x = 11 + 9x$

102. $3 - 2y = 15 + 4y$

103. $2x - 4 = 6x$

104. $2b - 10 = 7b$

105. $8m = 3m + 20$

106. $9y = 5y + 16$

107. $8b + 5 = 5b + 7$

108. $6y - 1 = 2y + 2$

109. $7x - 8 = x - 3$

110. $2y - 7 = -1 - 2y$

111. $2m - 1 = -6m + 5$

Objective C

Solve and check.

112. $5x + 2(x + 1) = 23$

113. $6y + 2(2y + 3) = 16$

114. $9n - 3(2n - 1) = 15$

115. $12x - 2(4x - 6) = 28$

116. $7a - (3a - 4) = 12$

117. $9m - 4(2m - 3) = 11$

Solve and check.

118. $5(3 - 2y) + 4y = 3$ **119.** $4(1 - 3x) + 7x = 9$ **120.** $5y - 3 = 7 + 4(y - 2)$

121. $5 + 2(3b + 1) = 3b + 5$ **122.** $6 - 4(3a - 2) = 2(a + 5)$ **123.** $7 - 3(2a - 5) = 3a + 10$

124. $2a - 5 = 4(3a + 1) - 2$ **125.** $5 - (9 - 6x) = 2x - 2$ **126.** $7 - (5 - 8x) = 4x + 3$

127. $3[2 - 4(y - 1)] = 3(2y + 8)$ **128.** $5[2 - (2x - 4)] = 2(5 - 3x)$

129. $3a + 2[2 + 3(a - 1)] = 2(3a + 4)$ **130.** $5 + 3[1 + 2(2x - 3)] = 6(x + 5)$

131. $-2[4 - (3b + 2)] = 5 - 2(3b + 6)$ **132.** $-4[x - 2(2x - 3)] + 1 = 2x - 3$

Objective D

Translate into an equation and solve.

133. The sum of a number and twelve is twenty. Find the number.

134. The difference between nine and a number is seven. Find the number.

135. Three-fifths of a number is negative thirty. Find the number.

136. The quotient of a number and six is twelve. Find the number.

137. Four more than three times a number is thirteen. Find the number.

138. The sum of twice a number and five is fifteen. Find the number.

139. The difference between nine times a number and six is twelve. Find the number.

Translate into an equation and solve.

140. Six less than four times a number is twenty-two. Find the number.

141. The sum of a number and twice the number is nine. Find the number.

142. Eleven more than negative four times a number is three. Find the number.

143. Seventeen less than the product of five and a number is two. Find the number.

144. Eight less than the product of eleven and a number is negative nineteen. Find the number.

145. Seven more than the product of six and a number is eight less than the product of three and the number. Find the number.

146. Fifteen less than the product of four and a number is the difference between six times the number and eleven. Find the number.

147. Thirty equals nine less than the product of seven and a number. Find the number.

148. Twenty-three equals the difference between eight and the product of five and a number. Find the number.

149. The sum of two numbers is twenty-one. Twice the smaller number is three more than the larger number. Find the two numbers.

150. The sum of two numbers is thirty. Three times the smaller number is twice the larger number. Find the two numbers.

151. The sum of two numbers is twenty-three. The larger number is five more than twice the smaller number. Find the two numbers.

152. The sum of two numbers is twenty-five. The larger number is seven less than four times the smaller number. Find the two numbers.

APPLYING THE CONCEPTS

153. If $2x - 3 = 7$, evaluate $3x + 4$.

154. If $3x + 5 = -4$, evaluate $2x - 5$.

155. If $5x = 3x - 8$, evaluate $4x + 2$.

156. If $7x + 3 = 5x - 7$, evaluate $3x - 2$.

157. Solve: $x \div 28 = 1481$ remainder 25

158. If $3 + 2(4a - 3) = 5$ and $4 - 3(2 - 3b) = 11$, which is larger, a or b?

159. Does the sentence "Solve $2x - 3(4x + 1)$" make sense? Why or why not?

Solve. If the equation has no solution, write "No solution."

160. $3(2x - 1) - (6x - 4) = -9$

161. $7(3x + 6) - 4(3 + 5x) = 13 + x$

162. $\dfrac{1}{5}(25 - 10a) + 4 = \dfrac{1}{3}(12a - 15) + 14$

163. $5[m + 2(3 - m)] = 3[2(4 - m) - 5]$

164. Solve the equation $x + a = b$ for x. Is the solution you have written valid for all real numbers a and b?

165.
[W] Explain in your own words the steps you would take to solve the equation $\dfrac{2}{3}x - 4 = 10$. State the Property of Real Numbers or the Property of Equations that is used at each step.

166.
[W] The equation $x = x + 1$ has no solution, whereas the solution of the equation $2x + 3 = 3$ is zero. Is there a difference between no solution and a solution of zero? Explain your answer.

167.
[W] Explain the difference between the word *equation* and the word *expression*.

168.
[W] The following problem does not contain enough information for us to find only one solution. Supply some additional information so that the problem has exactly one solution. Then write and solve an equation.
"The sum of two numbers is 15. Find the numbers."

Mixture, Investment, and Motion Problems

Objective A   *To solve value mixture problems*

A value mixture problem involves combining two ingredients that have different prices into a single blend. For example, a coffee merchant may blend two types of coffee into a single blend, or a candy manufacturer may combine two types of candy to sell as a "variety pack."

The solution of a value mixture problem is based on the equation $V = AC$, where V is the value of an ingredient, A is the amount of the ingredient, and C is the cost per unit of the ingredient.

➡ A coffee merchant wants to make 6 lb of a blend of coffee costing $5 per pound. The blend is made using a $6-per-pound grade and a $3-per-pound grade of coffee. How many pounds of each of these grades should be used?

Strategy for Solving a Value Mixture Problem

1. For each ingredient in the mixture, write a numerical or variable expression for the amount of the ingredient used, the unit cost of the ingredient, and the value of the amount used. For the blend, write a numerical or variable expression for the amount, the unit cost of the blend, and the value of the amount. The results can be recorded in a table.

Amount of $6 coffee: x • The sum of the amounts is 6 lb.
Amount of $3 coffee: $6 - x$

CONSIDER THIS
Use the information given in the problem to fill in the "Amount" and "Unit cost" columns of the table. Fill in the "Value" column by multiplying the two expressions you wrote in each row. Use the expressions in the last column to write the equation.

	Amount, A	·	Unit Cost, C	=	Value, V
$6 grade	x	·	6	=	$6x$
$3 grade	$6 - x$	·	3	=	$3(6 - x)$
$5 blend	6	·	5	=	$5(6)$

---

2. Determine how the values of each ingredient are related. Use the fact that the sum of the values of all the ingredients is equal to the value of the blend.

$6x + 3(6 - x) = 5(6)$ • The sum of the values of the $6 grade
$6x + 18 - 3x = 30$ and the $3 grade is equal to the value
$3x + 18 = 30$ of the $5 blend.
$3x = 12$
$x = 4$

$6 - x = 6 - 4 = 2$ • Find the amount of $3 coffee.

The merchant must use 4 lb of the $6 coffee and 2 lb of the $3 coffee.

Example 1

How many ounces of a silver alloy that costs $4 an ounce must be mixed with 10 oz of an alloy that costs $6 an ounce to make a mixture that costs $4.32 an ounce?

Strategy

• Ounces of $4 alloy: x

	Amount	Cost	Value
$4 alloy	x	4	$4x$
$6 alloy	10	6	6(10)
$4.32 mixture	$10 + x$	4.32	$4.32(10 + x)$

• The sum of the values before mixing equals the value after mixing.

Solution

$$4x + 6(10) = 4.32(10 + x)$$

$$4x + 60 = 43.2 + 4.32x$$

$$-0.32x + 60 = 43.2$$

$$-0.32x = -16.8$$

$$x = 52.5$$

52.5 oz of the $4 silver alloy must be used.

You Try It 1

A gardener has 20 lb of a lawn fertilizer that costs $.80 per pound. How many pounds of a fertilizer that costs $.55 per pound should be mixed with this 20 lb of lawn fertilizer to produce a mixture that costs $.75 per pound?

Your strategy

Your solution

Solution on p. A10

Objective B *To solve percent mixture problems* ....

The amount of a substance in a solution can be given as a percent of the total solution. For example, a 5% salt water solution means that 5% of the total solution is salt. The remaining 95% is water.

The solution of a percent mixture problem is based on the equation $Q = Ar$, where Q is the quantity of a substance in the solution, r is the percent of concentration, and A is the amount of solution.

➡ A 500-milliliter bottle contains a 4% solution of hydrogen peroxide. Find the amount of hydrogen peroxide in the solution.

$Q = Ar$ • Given: $A = 500$; $r = 4\% = 0.04$
$Q = 500(0.04)$
$Q = 20$

The bottle contains 20 ml of hydrogen peroxide.

➡ How many gallons of a 20% salt solution must be mixed with 6 gal of a 30% salt solution to make a 22% salt solution?

> **Strategy for Solving a Percent Mixture Problem**
>
> **1.** For each solution, write a numerical or variable expression for the amount of solution, the percent of concentration, and the quantity of the substance in the solution. The results can be recorded in a table.

CONSIDER THIS

Use the information given in the problem to fill in the "Amount" and "Percent" columns of the table. Fill in the "Quantity" column by multiplying the two expressions you wrote in each row. Use the expressions in the last column to write the equation.

The unknown quantity of 20% solution: x

	Amount of Solution, A	·	Percent of Concentration, r	=	Quantity of Substance, Q
20% solution	x	·	0.20	=	$0.20x$
30% solution	6	·	0.30	=	$0.30(6)$
22% solution	$x + 6$	·	0.22	=	$0.22(x + 6)$

> **2.** Determine how the quantities of the substances in each solution are related. Use the fact that the sum of the quantities of the substances being mixed is equal to the quantity of the substance after mixing.

$$0.20x + 0.30(6) = 0.22(x + 6)$$
$$0.20x + 1.80 = 0.22x + 1.32$$
$$-0.02x + 1.80 = 1.32$$
$$-0.02x = -0.48$$
$$x = 24$$

• The sum of the quantities of the substances in the 20% solution and the 30% solution is equal to the quantity of the substance in the 22% solution.

24 gal of the 20% solution are required.

Example 2

A chemist wishes to make 2 L of an 8% acid solution by mixing a 10% acid solution and a 5% acid solution. How many liters of each solution should the chemist use?

Strategy

- Liters of 10% solution: x
  Liters of 5% solution: $2 - x$

	Amount	*Percent*	*Quantity*
10%	x	0.10	$0.10x$
5%	$2 - x$	0.05	$0.05(2 - x)$
8%	2	0.08	$0.08(2)$

- The sum of the quantities before mixing is equal to the quantity after mixing.

Solution

$$0.10x + 0.05(2 - x) = 0.08(2)$$

$$0.10x + 0.10 - 0.05x = 0.16$$

$$0.05x + 0.10 = 0.16$$

$$0.05x = 0.06$$

$$x = 1.2$$

$$2 - x = 2 - 1.2 = 0.8$$

The chemist needs 1.2 L of the 10% solution and 0.8 L of the 5% solution.

You Try It 2

A pharmacist dilutes 5 L of a 12% solution with a 6% solution. How many liters of the 6% solution are added to make an 8% solution?

Your strategy

Your solution

Solution on p. A2

Objective C   *To solve investment problems* ....

POINT OF INTEREST
You may be familiar with the simple interest formula $I = Prt$. If so, you know that t represents time. In the problems in this section, time is always 1 (one year), so the formula $I = Prt$ simplifies to
$$I = Pr(1)$$
$$I = Pr$$

The annual simple interest that an investment earns is given by the equation $I = Pr$, where I is the simple interest, P is the principal, or the amount invested, and r is the simple interest rate.

➡ The annual interest rate on a $2500 investment is 8%. Find the annual simple interest earned on the investment.

$$I = Pr$$ • **Given:** $P = \$2500$; $r = 8\% = 0.08$
$$I = 2500(0.08)$$
$$I = 200$$

The annual simple interest is $200.

➡ An investor has a total of $10,000 deposited in two simple interest accounts. On one account, the annual simple interest rate is 6%. On the second account, the annual simple interest rate is 10%. How much is invested in the 6% account if the total annual interest earned is $900?

> **Strategy for Solving a Problem Involving Money Deposited in Two Simple Interest Accounts**
>
> **1.** For each amount invested, write a numerical or variable expression for the principal, the interest rate, and the interest earned. The results can be recorded in a table.

CONSIDER THIS
Use the information given in the problem to fill in the "Principal" and "Interest Rate" columns of the table. Fill in the "Interest Earned" column by multiplying the two expressions you wrote in each row.

Amount invested at 6%: x
Amount invested at 10%: $\$10{,}000 - x$

• The sum of the amounts invested is $10,000.

	Principal, P	·	*Interest Rate, r*	=	*Interest Earned, I*
Amount at 6%	x	·	0.06	=	$0.06x$
Amount at 10%	$10{,}000 - x$	·	0.10	=	$0.10(10{,}000 - x)$

> **2.** Determine how the amounts of interest earned on each amount are related. For example, the total interest earned by both accounts may be known, or it may be known that the interest earned on one account is equal to the interest earned on the other account.

$$0.06x + 0.10(10{,}000 - x) = 900$$ • The sum of the interest earned
$$0.06x + 1000 - 0.10x = 900$$ on the two accounts is $900.
$$-0.04x + 1000 = 900$$
$$-0.04x = -100$$
$$x = 2500$$

The amount invested at 6% is $2500.

Example 3

An investment counselor invested 75% of a client's money in a 9% annual simple interest money market fund. The remainder was invested in 7% annual simple interest government securities. Find the amount invested in each if the total annual interest earned is $3825.

Strategy

- Amount invested: x

  Amount invested at 7%: $0.25x$

  Amount invested at 9%: $0.75x$

	Principal	Rate	Interest
Amount at 7%	$0.25x$	0.07	$0.0175x$
Amount at 9%	$0.75x$	0.09	$0.0675x$

- The sum of the interest earned by the two investments equals the total annual interest earned ($3825).

You Try It 3

An investment of $5000 is made at an annual simple interest rate of 8%. How much additional money must be invested at 11% so that the total interest earned will be 9% of the total investment?

Your strategy

Solution

$$0.0175x + 0.0675x = 3825$$
$$0.085x = 3825$$
$$x = 45,000$$

$$0.25x = 0.25(45,000) = 11,250$$

$$0.75x = 0.75(45,000) = 33,750$$

The amount invested at 7% is $11,250.
The amount invested at 9% is $33,750.

Your solution

Solution on p. A10

Objective D   ***To solve uniform motion problems*** ....

A train that travels constantly in a straight line at 50 mph is in *uniform motion*. **Uniform motion** means that the speed or direction of an object does not change.

The solution of a uniform motion problem is based on the equation $d = rt$, where d is the distance traveled, r is the rate of travel, and t is the time traveled.

A train traveled at a speed of 55 mph for 3 h. The distance traveled by the train can be found by the equation $d = rt$.

$$d = rt$$
$$d = (55)(3)$$
$$d = 165$$

The distance traveled is 165 mi.

➡ A car leaves a town traveling at 40 mph. Two hours later, a second car leaves the same town, on the same road, traveling at 60 mph. In how many hours will the second car pass the first car?

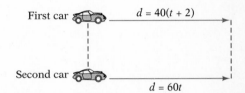

Strategy for Solving a Uniform Motion Problem

1. For each object, write a numerical or variable expression for the distance, rate, and time. The results can be recorded in a table.

Unknown time for the second car: t
Time for the first car: $t + 2$

● **The first car traveled 2 h longer than the second car.**

	Rate, r	·	*Time, t*	=	*Distance, d*
First car	40	·	$t + 2$	=	$40(t + 2)$
Second car	60	·	t	=	$60t$

---

2. Determine how the distances traveled by each object are related. For example, the total distance traveled by both objects may be known, or it may be known that the two objects traveled the same distance.

$$40(t + 2) = 60t$$ ● **The two cars travel the same distance.**
$$40t + 80 = 60t$$
$$80 = 20t$$
$$4 = t$$

The second car will pass the first car in 4 h.

Example 4

Two cars, one traveling 10 mph faster than the other car, start at the same time from the same point and travel in opposite directions. In 3 h they are 300 mi apart. Find the rate of each car.

Strategy

- Rate of 1st car: r
  Rate of 2nd car: $r + 10$

	Rate	Time	Distance
1st car	r	3	$3r$
2nd car	$r + 10$	3	$3(r + 10)$

- The total distance traveled by the two cars is 300 mi.

Solution

$$3r + 3(r + 10) = 300$$
$$3r + 3r + 30 = 300$$
$$6r + 30 = 300$$
$$6r = 270$$
$$r = 45$$

$$r + 10 = 45 + 10 = 55$$

The first car is traveling 45 mph.
The second car is traveling 55 mph.

Example 5

How far can the members of a bicycling club ride out into the country at a speed of 12 mph and return over the same road at 8 mph if they travel a total of 10 h?

Strategy

- Time spent riding out: t
  Time spent riding back: $10 - t$

	Rate	Time	Distance
Out	12	t	$12t$
Back	8	$10 - t$	$8(10 - t)$

- The distance out equals the distance back.

Solution

$$12t = 8(10 - t)$$
$$12t = 80 - 8t$$
$$20t = 80$$
$$t = 4 \quad \text{(The time is 4 h.)}$$

The distance out $= 12t = 12(4) = 48$ mi.
The club can ride 48 mi into the country.

You Try It 4

Two trains, one traveling at twice the speed of the other, start at the same time on parallel tracks from stations that are 288 mi apart and travel toward each other. In 3 h, the trains pass each other. Find the rate of each train.

Your strategy

Your solution

You Try It 5

A pilot flew out to a parcel of land and back in 5 h. The rate out was 150 mph, and the rate returning was 100 mph. How far away was the parcel of land?

Your strategy

Your solution

Solutions on p. A11

2.4 Exercises

· ·

___

Objective A *Application Problems*

Solve.

1. A high-protein diet supplement that costs $6.75 per pound is mixed with a vitamin supplement that costs $3.25 per pound. How many pounds of each should be used to make 5 lb of a mixture that costs $4.65 per pound?

2. A 20-ounce alloy of platinum that costs $220 per ounce is mixed with an alloy that costs $400 per ounce. How many ounces of the $400 alloy should be mixed with the $220 alloy to make an alloy that costs $300 per ounce?

3. Find the cost per pound of a coffee mixture made from 8 lb of coffee that costs $9.20 per pound and 12 lb of coffee that costs $5.50 per pound.

4. How many pounds of tea that costs $4.20 per pound must be mixed with 12 lb of tea that costs $2.25 per pound to make a mixture that costs $3.40 per pound?

5. A goldsmith combined an alloy that costs $4.30 per ounce with an alloy that costs $1.80 per ounce. How many ounces of each were used to make a mixture of 200 oz costing $2.50 per ounce?

6. How many liters of a solvent that costs $80 per liter must be mixed with 6 L of a solvent that costs $25 per liter to make a solvent that costs $36 per liter?

7. Find the cost per pound of a trail mix made from 40 lb of raisins that cost $4.40 per pound and 100 lb of granola that costs $2.30 per pound.

8. Find the cost per ounce of a mixture of 200 oz of cologne that costs $5.50 per ounce and 500 oz of cologne that costs $2.00 per ounce.

9. How many kilograms of hard candy that costs $7.50 per kilogram must be mixed with 24 kg of jelly beans that cost $3.25 per kilogram to make a mixture that costs $4.50 per kilogram?

10. A grocery store offers a cheese and fruit sampler that combines cheddar cheese that costs $8 per kilogram with kiwis that cost $3 per kilogram. How many kilograms of each were used to make a 5-kilogram mixture that costs $4.50 per kilogram?

Solve.

11. A ground meat mixture is formed by combining meat that costs $2.20 per pound with meat that costs $4.20 per pound. How many pounds of each were used to make a 50-pound mixture that costs $3.00 per pound?

12. A lumber company combined oak wood chips that cost $3.10 per pound with pine wood chips that cost $2.50 per pound. How many pounds of each were used to make an 80-pound mixture that costs $2.65 per pound?

13. A caterer makes an ice cream punch by combining fruit juice that costs $2.25 per gallon with ice cream that costs $3.25 per gallon. How many gallons of each should be used to make 100 gal of punch that costs $2.50 per gallon?

14. The manager of a specialty food store combined almonds that cost $4.50 per pound with walnuts that cost $2.50 per pound. How many pounds of each were used to make a 100-pound mixture that costs $3.24 per pound?

15. Find the cost per gallon of a carbonated fruit drink made from 12 gal of fruit juice that costs $4.00 per gallon and 30 gal of carbonated water that costs $2.25 per gallon.

16. Find the cost per pound of a sugar-coated breakfast cereal made from 40 lb of sugar that costs $1.00 per pound and 120 lb of corn flakes that cost $.60 per pound.

Objective B   *Application Problems*

Solve.

17. A chemist wants to make 50 ml of a 16% acid solution. How many milliliters each of a 13% acid solution and an 18% acid solution should be mixed to produce the desired solution?

18. A blend of coffee was made by combining some coffee that was 40% java beans with 80 lb of coffee that was 30% java beans to make a mixture that is 32% java. How many pounds of the 40% java coffee were used?

19. Thirty ounces of pure silver are added to 50 oz of a silver alloy that is 20% silver. What is the percent concentration of silver in the resulting mixture?

Solve.

20. Two hundred liters of a punch that contains 35% fruit juice are mixed with 300 L of another punch. The resulting fruit punch is 20% fruit juice. Find the percent of fruit juice in the 300 L of punch.

21. The manager of a garden shop mixes grass seed that is 60% rye grass with 70 lb of grass seed that is 80% rye grass to make a mixture that is 74% rye grass. How much of the 60% mixture is used?

22. Ten grams of sugar are added to a 40-gram serving of a breakfast cereal that is 30% sugar. What is the percent concentration of sugar in the resulting mixture?

23. A dermatologist mixes 50 g of a cream that is 0.5% hydrocortisone with 150 g of another hydrocortisone cream. The resulting mixture is 0.68% hydrocortisone. Find the percent of hydrocortisone in the 150 g of cream.

24. A carpet manufacturer blends two fibers, one 20% wool and the second 50% wool. How many pounds of each fiber should be woven together to produce 500 lb of a fabric that is 35% wool?

25. A hair dye is made by blending some 7% hydrogen peroxide solution with some 4% hydrogen peroxide solution. How many milliliters of each should be mixed to make a 300-milliliter solution that is 5% hydrogen peroxide?

26. How many grams of pure salt must be added to 40 g of a 20% salt solution to make a solution that is 36% salt?

27. How many ounces of pure water must be added to 50 oz of a 15% saline solution to make a saline solution that is 10% salt?

28. A paint is blended by using a paint that contains 21% green dye and a paint that contains 15% green dye. How many gallons of each must be mixed to produce 60 gal of paint that is 19% green dye?

29. A goldsmith mixes 8 oz of a 30% gold alloy with 12 oz of a 25% gold alloy. What is the percent concentration of the resulting alloy?

30. A physicist mixes 40 L of oxygen with 50 L of air that contains 64% oxygen. What is the percent concentration of oxygen in the resulting air?

31. A 50-ounce box of cereal is 40% bran flakes. How many ounces of pure bran flakes must be added to this box to produce a mixture that is 50% bran flakes?

Solve.

32. A pastry chef has 150 ml of a chocolate topping that is 50% chocolate. How many milliliters of pure chocolate must be added to this topping to make a topping that is 75% chocolate?

Objective C   Application Problems

Solve.

33. An investment of $3000 is made at an annual simple interest rate of 5%. How much additional money must be invested at an annual simple interest rate of 9% so that the total annual interest earned is 7.5% of the total investment?

34. A total of $6000 is invested into two simple interest accounts. The annual simple interest rate on one account is 9%; on the second account, the annual simple interest rate is 6%. How much should be invested in each account so that both accounts earn the same amount of annual interest?

35. An engineer invested a portion of $15,000 in a 7% annual simple interest account and the remainder in a 6.5% annual simple interest government bond. The amount of interest earned for one year was $1020. How much was invested in each account?

36. An investment club invested part of $20,000 in preferred stock that pays 8% annual simple interest and the remainder in a municipal bond that pays 7% annual simple interest. The amount of interest earned each year is $1520. How much was invested in each account?

37. A grocery checker deposited an amount of money into a high-yield mutual fund that returns a 9% annual simple interest rate. A second deposit, $2500 more than the first, was placed in a certificate of deposit that returns a 5% annual simple interest rate. The total interest earned on both investments for one year was $475. How much money was deposited in the mutual fund?

38. A deposit was made into a 7% annual simple interest account. Another deposit, $1500 less than the first deposit, was placed in a 9% annual simple interest certificate of deposit. The total interest earned on both accounts for one year was $505. How much money was deposited in the certificate of deposit?

39. A corporation gave a university $300,000 to support product safety research. The university deposited some of the money in a 10% simple interest account and the remainder in an 8.5% simple interest account. How much should be deposited in each account so that the annual interest earned is $28,500?

Solve.

40. A financial consultant advises a client to invest part of $30,000 in municipal bonds that earn 6.5% annual simple interest and the remainder of the money in 8.5% corporate bonds. How much should be invested in each account so that the total annual interest earned each year is $2190?

41. To provide for retirement income, an auto mechanic purchases a $5000 bond that earns 7.5% annual simple interest. How much money must be invested in additional bonds that have an interest rate of 8% so that the total annual interest earned from the two investments is $615?

42. The portfolio manager for an investment group invested $40,000 in a certificate of deposit that earns 7.25% annual simple interest. How much money must be invested in additional certificates that have an interest rate of 8.5% so that the total annual interest earned from the two investments is $5025?

43. A charity deposited a total of $54,000 into two simple interest accounts. The annual simple interest rate on one account is 8%. The annual simple interest rate on the second account is 12%. How much was invested in each account if the total interest earned is 9% of the total investment?

44. A college sports foundation deposited a total of $24,000 into two simple interest accounts. The annual simple interest rate on one account is 7%. The annual simple interest rate on the second account is 11%. How much is invested in each account if the total annual interest earned is 10% of the total investment?

45. An investment banker invested 55% of the bank's available cash in an account that earns 8.25% annual simple interest. The remainder of the cash was placed in an account that earns 10% annual simple interest. The interest earned in one year was $58,743.75. Find the total amount invested.

46. A financial planner recommended that 40% of a client's cash account be placed in preferred stock that earns 9% annual simple interest. The remainder of the client's cash was placed in treasury bonds that earn 7% annual interest. The total annual interest earned from the two investments was $2496. What was the total amount invested?

47. The manager of a mutual fund placed 30% of the fund's available cash in a 6% simple interest account, 25% in 8% corporate bonds, and the remainder in a money market fund that earns 7.5% annual simple interest. The total annual interest from the investments was $35,875. What was the total amount invested?

Solve.

48. The manager of a trust decided to invest 30% of a client's cash in government bonds that earn 6.5% annual simple interest. Another 30% was placed in utility stocks that earn 7% annual simple interest. The remainder of the cash was placed in an account earning 8% annual simple interest. The total annual interest earned from the investments was $5437.50. What was the total amount invested?

Objective D   *Application Problems*

Solve.

49. A 555-mile, 5-hour plane trip was flown at two speeds. For the first part of the trip, the average speed was 105 mph. For the remainder of the trip, the average speed was 115 mph. For how long did the plane fly at each speed?

50. An executive drove from home at an average speed of 30 mph to an airport where a helicopter was waiting. The executive boarded the helicopter and flew to the corporate offices at an average speed of 60 mph. The entire distance was 150 mi. The entire trip took 3 h. Find the distance from the airport to the corporate offices.

51. After a saiboat had been on the water for 3 h, a change in wind direction reduced the average speed of the boat by 5 mph. The entire distance sailed was 57 mi. The total time spent sailing was 6 h. How far did the sailboat travel in the first 3 h?

52. A car and a bus set out at 2 P.M. from the same point headed in the same direction. The average speed of the car is 30 mph slower than twice the speed of the bus. In 2 h the car is 20 mi ahead of the bus. Find the rate of the car.

53. A passenger train leaves a train depot 2 h after a freight train leaves the same depot. The freight train is traveling 20 mph slower than the passenger train. Find the rate of each train if the passenger train overtakes the freight train in 3 h.

54. Two cyclists start at the same time from opposite ends of a course that is 45 mi long. One cyclist is riding at 14 mph, and the second cyclist is riding at 16 mph. How long after they begin will they meet?

55. A cyclist and a jogger set out at 11 A.M. from the same point headed in the same direction. The average speed of the cyclist is twice the average speed of the jogger. In 1 h the cyclist is 8 mi ahead of the jogger. Find the rate of the cyclist.

Solve.

56. Two cyclists start from the same point and ride in opposite directions. One cyclist rides twice as fast as the other. In 3 h they are 72 mi apart. Find the rate of each cyclist.

57. Two small planes start from the same point and fly in opposite directions. The first plane is flying 25 mph slower than the second plane. In 2 h the planes are 430 mi apart. Find the rate of each plane.

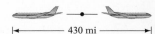

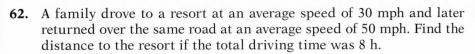

58. A motorboat leaves a harbor and travels at an average speed of 8 mph toward a small island. Two hours later a cabin cruiser leaves the same harbor and travels at an average speed of 16 mph toward the same island. In how many hours after the cabin cruiser leaves will it be alongside the motorboat?

59. Two joggers start at the same time from opposite ends of a 10-mile course. One jogger is running at 4 mph, and the other is running at 6 mph. How long after they begin will they meet?

60. On a 195-mile trip, a car traveled at an average speed of 45 mph and then reduced its speed to an average of 30 mph for the remainder of the trip. The trip took a total of 5 h. How long did the car travel at each speed?

61. A long-distance runner started on a course running at an average speed of 6 mph. One hour later, a second runner began the same course at an average speed of 8 mph. How long after the second runner started will the second runner overtake the first runner?

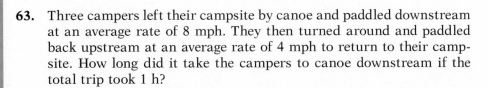

62. A family drove to a resort at an average speed of 30 mph and later returned over the same road at an average speed of 50 mph. Find the distance to the resort if the total driving time was 8 h.

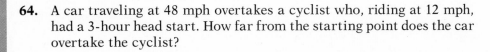

63. Three campers left their campsite by canoe and paddled downstream at an average rate of 8 mph. They then turned around and paddled back upstream at an average rate of 4 mph to return to their campsite. How long did it take the campers to canoe downstream if the total trip took 1 h?

64. A car traveling at 48 mph overtakes a cyclist who, riding at 12 mph, had a 3-hour head start. How far from the starting point does the car overtake the cyclist?

APPLYING THE CONCEPTS

Solve.

65. Find the cost per ounce of a mixture of 30 oz of an alloy that costs $4.50 per ounce, 40 oz of an alloy that costs $3.50 per ounce, and 30 oz of an alloy that costs $3.00 per ounce.

Solve.

66. A grocer combined walnuts that cost $1.60 per pound and cashews that cost $2.50 per pound with 20 lb of peanuts that cost $1.00 per pound. Find the amount of walnuts and the amount of cashews used to make a 50-pound mixture that costs $1.72 per pound.

67. How many ounces of water must be evaporated from 50 oz of a 12% salt solution to produce a 15% salt solution?

68. A chemist mixed pure acid with water to make 10 L of a 30% acid solution. How much pure acid and how much water did the chemist use?

69. A radiator contains 15 gal of a 20% antifreeze solution. How many gallons must be drained from the radiator and replaced by pure anti-freeze so that the radiator will contain 15 gal of a 40% antifreeze solution?

70. A sales representative invests in a stock paying 9% dividends. A re-search consultant invests $5000 more than the sales representative in bonds paying 8% annual simple interest. The research consultant's income from the investment is equal to the sales representative's. Find the amount of the research consultant's investment.

71. A financial manager invested 20% of a client's money in bonds paying 9% annual simple interest, 35% in an 8% simple interest account, and the remainder in 9.5% corporate bonds. Find the amount invested in each if the total annual interest earned is $5325.

CORPORATION BONDS				
Volume, $46,830,000				
Bonds	Cur Yld	Vol	Close	Net Chg.
AMR 9s16	8.9	20	$101\frac{1}{2}$	$- \frac{1}{4}$
ANR $10\frac{5}{8}$95	10.4	10	102	$+ \frac{1}{2}$
Advst 9s08	cv	15	89	...
AetnLf $8\frac{1}{8}$07	8.2	75	$98\frac{3}{4}$	$- \frac{1}{8}$
AlaP $7\frac{7}{8}$s02	7.9	8	$99\frac{1}{2}$	$+ \frac{1}{2}$
AlaP $7\frac{3}{4}$s02	7.9	3	$97\frac{3}{4}$	$+ \frac{3}{8}$
AlaP $8\frac{7}{8}$s03	8.7	8	102	...
AlaP $8\frac{1}{4}$s03	8.0	22	$102\frac{3}{4}$	...
AlaP $9\frac{3}{4}$s04	9.3	1	105	$+ \frac{1}{8}$
AlaP $8\frac{7}{8}$06	8.5	31	104	$+ 1\frac{5}{8}$
AlaP $8\frac{1}{4}$07	8.6	31	102	$+ \frac{1}{4}$
AlaP $9\frac{1}{2}$08	9.1	4	$104\frac{3}{4}$	$+ \frac{5}{8}$
AlskAr $6\frac{7}{8}$14	cv	34	$79\frac{1}{2}$	$- \frac{1}{2}$
AlskAr zr06	...	22	$33\frac{1}{4}$	...
AlldC zr98	...	20	$59\frac{3}{8}$	$- \frac{5}{8}$
AlldC zr92	...	5	$97\frac{17}{32}$	$- \frac{1}{32}$
AlldC zr96	...	4	$76\frac{3}{4}$	...
AlldC zr2000	...	4	$50\frac{1}{2}$	$+ \frac{5}{8}$
AlldC zr9	...	10	$91\frac{7}{8}$	$+ \frac{5}{8}$
AlldC zr95	...	50	$78\frac{1}{4}$	$- \frac{1}{8}$
AlldC zr01	...	10	$45\frac{3}{8}$	$- \frac{1}{8}$
AlldC zr 03	...	15	$38\frac{1}{4}$	$- \frac{1}{2}$
Allwst $7\frac{1}{4}$14	cv	13	87	$+ \frac{1}{4}$
AmStor 01	cv	12	$98\frac{3}{4}$	$+ \frac{1}{4}$
ATT $5\frac{5}{8}$95	5.7	170	$97\frac{7}{8}$	$+ \frac{1}{4}$
ATT $5\frac{1}{2}$97	5.9	50	$93\frac{3}{4}$	$- 1\frac{1}{4}$

72. A plant manager invested $3000 more in stocks than in bonds. The stocks paid 8% annual simple interest, and the bonds paid 9.5% an-nual simple interest. Both investments yielded the same income. Find the total annual interest received on both investments.

73. At 10 A.M., two campers left their campsite by canoe and paddled downstream at an average speed of 12 mph. They then turned around and paddled back upstream at an average rate of 4 mph. The total trip took 1 h. At what time did the campers turn around downstream?

74. At 7 A.M., two joggers start from opposite ends of an 8-mile course. One jogger is running at a rate of 4 mph, and the other is running at a rate of 6 mph. At what time will the joggers meet?

75. A bicyclist rides for 2 h at a speed of 10 mph and then returns at a speed of 20 mph. Find the cyclist's average speed for the trip.

76. A car travels a 1-mile track at an average speed of 30 mph. At what average speed must the car travel the next mile so that the average speed for the 2 mi is 60 mph?

2.5

Inequalities in One Variable

Objective A   *To solve an inequality in one variable*

The **solution set of an inequality** is a set of numbers, each element of which, when substituted for the variable, results in a true inequality.

The inequality at the right is true if the variable is replaced by (for instance) 3, -1.98, or $\frac{2}{3}$.

$$x - 1 < 4$$
$$3 - 1 < 4$$
$$-1.98 - 1 < 4$$
$$\frac{2}{3} - 1 < 4$$

There are many values of the variable x that will make the inequality $x - 1 < 4$ true. The solution set of the inequality is any number less than 5. The solution set can be written in set-builder notation as $\{x \mid x < 5\}$.

The graph of the solution set of $x - 1 < 4$ is shown at the right.

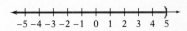

In solving an inequality, we use the Addition and Multiplication Properties of Inequalities to rewrite the inequality in the form *variable < constant* or in the form *variable > constant*.

> **The Addition Property of Inequalities**
>
> If $a > b$, then $a + c > b + c$.
> If $a < b$, then $a + c < b + c$.

The Addition Property of Inequalities states that the same number can be added to each side of an inequality without changing the solution set of the inequality. This property is also true for an inequality that contains the symbol $\leq$ or $\geq$.

The Addition Property of Inequalities is used to remove a term from one side of an inequality by adding the additive inverse of that term to each side of the inequality. Because subtraction is defined in terms of addition, the same number can be subtracted from each side of an inequality without changing the solution set of the inequality.

➡ Solve: $x + 2 \geq 4$

$$x + 2 \geq 4$$
$$x + 2 - 2 \geq 4 - 2 \qquad \bullet \text{ Subtract 2 from each side of the inequality.}$$
$$x \geq 2 \qquad \bullet \text{ Simplify.}$$

The solution set is $\{x \mid x \geq 2\}$.

➡ Solve: $3x - 4 < 2x - 1$

$$3x - 4 < 2x - 1$$

$$3x - 4 - 2x < 2x - 1 - 2x$$ ● Subtract $2x$ from each side of the inequality.

$$x - 4 < -1$$

$$x - 4 + 4 < -1 + 4$$ ● Add 4 to each side of the inequality.

$$x < 3$$

The solution set is $\{x \mid x < 3\}$.

The Multiplication Property of Inequalities is used to remove a coefficient from one side of an inequality by multiplying each side of the inequality by the reciprocal of the coefficient.

CONSIDER THIS

$c > 0$ means c is a positive number.

$c < 0$ means c is a negative number.

> **The Multiplication Property of Inequalities**
>
> Rule 1   If $a > b$ and $c > 0$, then $ac > bc$.
>   If $a < b$ and $c > 0$, then $ac < bc$.
>
> Rule 2   If $a > b$ and $c < 0$, then $ac < bc$.
>   If $a < b$ and $c < 0$, then $ac > bc$.

Here are some examples of this property.

| **Rule 1** | **Rule 2** |

$3 > 2$	$2 < 5$	$3 > 2$	$2 < 5$
$3(4) > 2(4)$	$2(4) < 5(4)$	$3(-4) < 2(-4)$	$2(-4) > 5(-4)$
$12 > 8$	$8 < 20$	$-12 < -8$	$-8 > -20$

Rule 1 states that **when each side of an inequality is multiplied by a positive number, the inequality symbol remains the same.** However, Rule 2 states that **when each side of an inequality is multiplied by a negative number, the inequality symbol must be reversed.** Because division is defined in terms of multiplication, **when each side of an inequality is divided by a positive number, the inequality symbol remains the same. But when each side of an inequality is divided by a negative number, the inequality symbol must be reversed.**

The Multiplication Property of Inequalities is also true for the symbols $\leq$ and $\geq$.

CONSIDER THIS

Each side of the inequality is *divided by* a negative number; the inequality symbol must be reversed.

➡ Solve: $-3x > 9$

$$-3x > 9$$

$$\frac{-3x}{-3} < \frac{9}{-3}$$ ● Divide each side of the inequality by the coefficient -3. Because -3 is a negative number, the inequality symbol must be reversed.

$$x < -3$$

The solution set is $\{x \mid x < -3\}$.

CONSIDER THIS

Any time an inequality is multiplied or divided by a negative number, the inequality symbol must be reversed. Compare the next two examples.

$2x < -4$ Divide each side by *positive* 2.

$\dfrac{2x}{2} < \dfrac{-4}{2}$ Inequality *is not* reversed.

$x < -2$

$-2x < 4$ Divide each side by *negative* 2.

$\dfrac{-2x}{-2} > \dfrac{4}{-2}$ Inequality *is* reversed.

$x > -2$

➡ Solve: $3x + 2 < -4$

$3x + 2 < -4$

$3x < -6$ • Subtract 2 from each side of the inequality.

$\dfrac{3x}{3} < \dfrac{-6}{3}$ • Divide each side of the inequality by the coefficient 3.

$x < -2$

The solution set is $\{x \mid x < -2\}$.

➡ Solve: $2x - 9 > 4x + 5$

$2x - 9 > 4x + 5$

$-2x - 9 > 5$ • Subtract 4x from each side of the inequality.

$-2x > 14$ • Add 9 to each side of the inequality.

$\dfrac{-2x}{-2} < \dfrac{14}{-2}$ • Divide each side of the inequality by the coefficient −2. Reverse the inequality symbol.

$x < -7$

The solution set is $\{x \mid x < -7\}$.

CONSIDER THIS

Solving these inequalities is similar to solving the equations solved in Section 3 *except* that when you multiply or divide the inequality by a negative number, you must reverse the inequality symbol.

➡ Solve: $5(x - 2) \geq 9x - 3(2x - 4)$

$5(x - 2) \geq 9x - 3(2x - 4)$

$5x - 10 \geq 9x - 6x + 12$ • Use the Distributive Property to remove parentheses.

$5x - 10 \geq 3x + 12$ • Combine like terms.

$2x - 10 \geq 12$ • Subtract 3x from each side of the inequality.

$2x \geq 22$ • Add 10 to each side of the inequality.

$\dfrac{2x}{2} \geq \dfrac{22}{2}$ • Divide each side of the inequality by the coefficient 2.

$x \geq 11$

The solution set is $\{x \mid x \geq 11\}$.

Example 1
Solve: $x + 3 > 4x + 6$

Solution

$x + 3 > 4x + 6$

$-3x + 3 > 6$ • Subtract 4x from each side.

$-3x > 3$ • Subtract 3 from each side.

$\dfrac{-3x}{-3} < \dfrac{3}{-3}$ • Divide each side by −3.

$x < -1$

The solution set is $\{x \mid x < -1\}$.

You Try It 1
Solve: $2x - 1 < 6x + 7$

Your solution

Solution on p. A11

Example 2   Solve:
$$3x - 5 \leq 3 - 2(3x + 1)$$

You Try It 2   Solve:
$$5x - 2 \leq 4 - 3(x - 2)$$

Solution
$$3x - 5 \leq 3 - 2(3x + 1)$$
$$3x - 5 \leq 3 - 6x - 2$$
$$3x - 5 \leq 1 - 6x$$
$$9x - 5 \leq 1$$
$$9x \leq 6$$
$$\frac{9x}{9} \leq \frac{6}{9}$$
$$x \leq \frac{2}{3}$$
$$\left\{ x \,\middle|\, x \leq \frac{2}{3} \right\}$$

Your solution

Solution on p. A11

Objective B   *To solve a compound inequality*

A **compound inequality** is formed by joining two inequalities with a connective word such as *and* or *or*. The inequalities at the right are compound inequalities.

$$2x < 4 \text{ and } 3x - 2 > -8$$

$$2x + 3 > 5 \text{ or } x + 2 < 5$$

The solution set of a compound inequality with the connective word *and* is the set of all elements that appear in the solution sets of both inequalities. Therefore, it is the intersection of the solution sets of the two inequalities.

➡ Solve: $2x < 6$ and $3x + 2 > -4$

$$\begin{array}{lll} 2x < 6 & \text{and} & 3x + 2 > -4 \\ x < 3 & & 3x > -6 \qquad \bullet \text{ Solve each inequality.} \\ \{x \,|\, x < 3\} & & x > -2 \\ & & \{x \,|\, x > -2\} \end{array}$$

The solution of a compound inequality with *and* is the intersection of the solution sets of the two inequalities.

$$\{x \,|\, x < 3\} \cap \{x \,|\, x > -2\} = \{x \,|\, -2 < x < 3\}$$

➡ Solve: $-3 < 2x + 1 < 5$

This inequality is equivalent to the compound inequality shown at the right.

$$-3 < 2x + 1 \text{ and } 2x + 1 < 5$$

$$\begin{array}{lll} -3 < 2x + 1 & \text{and} & 2x + 1 < 5 \\ -4 < 2x & & 2x < 4 \qquad \bullet \text{ Solve each inequality.} \\ -2 < x & & x < 2 \\ \{x \,|\, x > -2\} & & \{x \,|\, x < 2\} \end{array}$$

$$\{x \,|\, x > -2\} \cap \{x \,|\, x < 2\} = \{x \,|\, -2 < x < 2\}$$

There is an alternative method for solving the inequality in the last example.

➡ Solve: $-3 < 2x + 1 < 5$

$$-3 < 2x + 1 < 5$$

$$-3 - 1 < 2x + 1 - 1 < 5 - 1$$ ● Subtract 1 from each of the three parts of the inequality.

$$-4 < 2x < 4$$

$$\frac{-4}{2} < \frac{2x}{2} < \frac{4}{2}$$ ● Divide each of the three parts of the inequality by the coefficient 2.

$$-2 < x < 2$$

The solution set is $\{x \mid -2 < x < 2\}$.

The solution set of a compound inequality with the connective word *or* is the union of the solution sets of the two inequalities.

➡ Solve: $2x + 3 > 7$ or $4x - 1 < 3$

$$2x + 3 > 7 \quad \text{or} \quad 4x - 1 < 3$$

$$2x > 4 \qquad\qquad 4x < 4$$ ● Solve each inequality.

$$x > 2 \qquad\qquad\quad x < 1$$

$$\{x \mid x > 2\} \qquad \{x \mid x < 1\}$$

$$\{x \mid x > 2\} \cup \{x \mid x < 1\} = \{x \mid x > 2 \text{ or } x < 1\}$$ ● Find the union of the solution sets.

Example 3
Solve: $1 < 3x - 5 < 4$

Solution

$$1 < 3x - 5 < 4$$

$$1 + 5 < 3x - 5 + 5 < 4 + 5$$

$$6 < 3x < 9$$

$$\frac{6}{3} < \frac{3x}{3} < \frac{9}{3}$$

$$2 < x < 3$$

$$\{x \mid 2 < x < 3\}$$

You Try It 3
Solve: $-2 \le 5x + 3 \le 13$

Your solution

Example 4
Solve: $11 - 2x > -3$ and $7 - 3x < 4$

Solution

$$11 - 2x > -3 \quad \text{and} \quad 7 - 3x < 4$$

$$-2x > -14 \qquad\qquad -3x < -3$$

$$x < 7 \qquad\qquad\quad x > 1$$

$$\{x \mid x < 7\} \qquad\qquad \{x \mid x > 1\}$$

$$\{x \mid x < 7\} \cap \{x \mid x > 1\} = \{x \mid 1 < x < 7\}$$

You Try It 4
Solve: $2 - 3x > 11$ or $5 + 2x > 7$

Your solution

Solutions on p. A11

Objective C   *To solve application problems* ....

Example 5

A rectangle is 10 ft wide and $(2x + 4)$ ft long. Express as an integer the maximum length of the rectangle when the area is less than 200 ft^2. (The area of a rectangle is equal to its length times its width.)

You Try It 5

Company A rents cars for $8 a day and 10¢ for every mile driven. Company B rents cars for $10 a day and 8¢ per mile driven. You want to rent a car for one week. What is the maximum number of miles you can drive a Company A car if it is to cost you less than a Company B car?

Strategy

$(2x + 4)$ ft

10 ft

To find the maximum length:

- Replace the variables in the area formula by the given values and solve for x.

- Replace the variable in the expression $2x + 4$ with the value found for x.

Your strategy

Solution

Length times width	is less than	200 ft^2

$$(2x + 4)10 < 200$$
$$20x + 40 < 200$$
$$20x + 40 - 40 < 200 - 40$$
$$20x < 160$$
$$\frac{20x}{20} < \frac{160}{20}$$
$$x < 8$$

The length is $(2x + 4)$ ft. Because $x < 8$, $2x + 4 < 2(8) + 4 = 20$. Therefore, the length is less than 20 ft.

The maximum length is 19 ft.

Your solution

Solution on p. A12

2.5 Exercises

Objective A

Solve.

1. $x - 3 < 2$

2. $x + 4 \geq 2$

3. $4x \leq 8$

4. $6x > 12$

5. $-2x > 8$

6. $-3x \leq -9$

7. $3x - 1 > 2x + 2$

8. $5x + 2 \geq 4x - 1$

9. $2x - 1 > 7$

10. $3x + 2 < 8$

11. $5x - 2 \leq 8$

12. $4x + 3 \leq -1$

13. $6x + 3 > 4x - 1$

14. $7x + 4 < 2x - 6$

15. $8x + 1 \geq 2x + 13$

16. $5x - 4 < 2x + 5$

17. $4 - 3x < 10$

18. $2 - 5x > 7$

19. $7 - 2x \geq 1$

20. $3 - 5x \leq 18$

21. $-3 - 4x > -11$

22. $-2 - x < 7$

23. $4x - 2 < x - 11$

24. $6x + 5 \leq x - 10$

25. $x + 7 \geq 4x - 8$

26. $3x + 1 \leq 7x - 15$

27. $3x + 2 \leq 7x + 4$

28. $3x - 5 \geq -2x + 5$

29. $\dfrac{3}{5}x - 2 < \dfrac{3}{10} - x$

30. $\dfrac{5}{6}x - \dfrac{1}{6} < x - 4$

Solve.

31. $\dfrac{2}{3}x - \dfrac{3}{2} < \dfrac{7}{6} - \dfrac{1}{3}x$ **32.** $\dfrac{7}{12}x - \dfrac{3}{2} < \dfrac{2}{3}x + \dfrac{5}{6}$ **33.** $\dfrac{1}{2}x - \dfrac{3}{4} < \dfrac{7}{4}x - 2$

34. $6 - 2(x - 4) \le 2x + 10$ **35.** $4(2x - 1) > 3x - 2(3x - 5)$

36. $2(1 - 3x) - 4 > 10 + 3(1 - x)$ **37.** $2 - 5(x + 1) \ge 3(x - 1) - 8$

38. $7 + 2(4 - x) < 9 - 3(6 + x)$ **39.** $3(4x + 3) \le 7 - 4(x - 2)$

40. $2 - 2(7 - 2x) < 3(3 - x)$ **41.** $3 + 2(x + 5) \ge x + 5(x + 1) + 1$

42. $10 - 13(2 - x) < 5(3x - 2)$ **43.** $3 - 4(x + 2) \le 6 + 4(2x + 1)$

44. $3x - 2(3x - 5) \le 2 - 5(x - 4)$ **45.** $12 - 2(3x - 2) \ge 5x - 2(5 - x)$

Objective B

Solve.

46. $3x < 6$ and $x + 2 > 1$ **47.** $x - 3 \le 1$ and $2x \ge -4$

48. $x + 2 \ge 5$ or $3x \le 3$ **49.** $2x < 6$ or $x - 4 > 1$

50. $-2x > -8$ and $-3x < 6$ **51.** $\dfrac{1}{2}x > -2$ and $5x < 10$

Solve.

52. $\frac{1}{3}x < -1$ or $2x > 0$

53. $\frac{2}{3}x > 4$ or $2x < -8$

54. $x + 4 \geq 5$ and $2x \geq 6$

55. $3x < -9$ and $x - 2 < 2$

56. $-5x > 10$ and $x + 1 > 6$

57. $7x < 14$ and $1 - x < 4$

58. $2x - 3 > 1$ and $3x - 1 < 2$

59. $4x + 1 < 5$ and $4x + 7 > -1$

60. $3x + 7 < 10$ or $2x - 1 > 5$

61. $6x - 2 < -14$ or $5x + 1 > 11$

62. $-5 < 3x + 4 < 16$

63. $5 < 4x - 3 < 21$

64. $0 < 2x - 6 < 4$

65. $-2 < 3x + 7 < 1$

66. $4x - 1 > 11$ or $4x - 1 \leq -11$

67. $3x - 5 > 10$ or $3x - 5 < -10$

68. $2x - 3 \geq 5$ and $3x - 1 > 11$

69. $6x - 2 < 5$ or $7x - 5 < 16$

70. $9x - 2 < 7$ and $3x - 5 > 10$

71. $8x + 2 \leq -14$ and $4x - 2 > 10$

72. $3x - 11 < 4$ or $4x + 9 \geq 1$

73. $5x + 12 \geq 2$ or $7x - 1 \leq 13$

Solve.

74. $-6 \leq 5x + 14 \leq 24$

75. $3 \leq 7x - 14 \leq 31$

76. $3 - 2x > 7$ and $5x + 2 > -18$

77. $1 - 3x < 16$ and $1 - 3x > -16$

78. $5 - 4x > 21$ or $7x - 2 > 19$

79. $6x + 5 < -1$ or $1 - 2x < 7$

80. $3 - 7x \leq 31$ and $5 - 4x > 1$

81. $9 - x \geq 7$ and $9 - 2x < 3$

Objective C   *Application Problems*

Solve.

82. Five times the difference between a number and two is greater than the quotient of two times the number and three. Find the smallest integer that will satisfy the inequality.

83. Two times the difference between a number and eight is less than or equal to five times the sum of the number and four. Find the smallest number that will satisfy the inequality.

84. The length of a rectangle is 2 ft more than four times the width. Express as an integer the maximum width of the rectangle when the perimeter is less than 34 ft.

85. The length of a rectangle is 5 cm less than twice the width. Express as an integer the maximum width of the rectangle when the perimeter is less than 60 cm.

86. A cellular phone company offers its customers a rate of $99.00 for up to 200 min per month of cellular phone time or a rate of $35.00 per month plus $.40 for each minute of cellular phone time. For how many minutes per month can a customer who chooses the second option use a cellular phone before the charges exceed those the customer would have paid under the first option?

Solve.

87. A cellular phone company offers its customers a rate of $36.20 per month plus $.40 for each minute of cellular phone time or a rate of $20 per month plus $.76 for each minute of cellular phone time. For how many minutes can a customer who chooses the second option use a cellular phone before the charges exceed those the customer would have paid under the first option?

88. You can rent a car from Company A for $15 a day and 10¢ a mile or from Company B for $10 a day and 24¢ a mile. You want to rent a car for one week. How many miles can you drive a Company B car during the week if it is to cost you less than a Company A car?

89. Woolf Rental rents cars for $11 a day and 12¢ for every mile driven. Moreno Agency rents cars for $8 a day and 18¢ for every mile driven. You want to rent a car for one week. How many miles can you drive a Moreno Agency car during the week if it is to cost you less than a Woolf Rental car?

90. The temperature range for a week is between 14°F and 77°F. Find the temperature range in Celsius degrees. $\left[F = \dfrac{9}{5}C + 32 \right]$

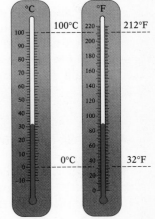

91. The temperature range for a week in a mountain town was between 0°C and 30°C. Find the temperature range in Fahrenheit degrees. $\left[C = \dfrac{5}{9}(F - 32) \right]$

92. You are a sales account executive earning $1200 per month plus a 6% commission on the amount of sales. Your goal is to earn a minimum of $6000 per month. What amount of sales will enable you to earn $6000 or more per month?

93. George Stoia earns $1000 per month plus a 5% commission on the amount of sales. George's goal is to earn a minimum of $3200 per month. What amount of sales will enable George to earn $3200 or more per month?

94. You have a choice of two types of checking accounts. One account has a charge of $6 per month plus 2¢ per check. The second account has a charge of $2 per month plus 7¢ per check. If you choose the second account, how many checks can you write if it is to cost less than the first type of account?

Solve.

95. A bank offers two types of checking accounts. One account has a charge of $4 per month plus 4¢ per check. The second account has a charge of $1 per month plus 10¢ per check. How many checks can a customer who has the second type of account write if it is to cost the customer less than the first type of account?

96. An average score of 90 or above in a history class receives an A grade. You have grades of 95, 89, and 81 on three exams. Find the range of scores on the fourth exam that will give you an A grade for the course.

97. An average of 70 to 79 in a mathematics class receives a C grade. A student has grades of 56, 91, 83, and 62 on four tests. Find the range of scores on the fifth test that will give the student a C for the course.

98. Grade A hamburger cannot contain more than 20% fat. How much fat can a butcher mix with 300 lb of lean meat to meet the 20% requirement?

99. A shuttle service taking skiers to a ski area charges $8 per person each way. Four skiers are debating whether to take the shuttle bus or rent a car for $45 plus $.25 per mile. The skiers will share the cost of the car, and they want the least expensive method of transportation. How far away is the ski area if they choose the shuttle service?

APPLYING THE CONCEPTS

100. Determine whether the statement is always true, sometimes true, or never true, given that a, b, and c are real numbers.
   a. If $a > b$, then $-a > -b$.
   b. If $a < b$, then $ac < bc$.
   c. If $a > b$, then $a + c > b + c$.
   d. If $a \neq 0$, $b \neq 0$, and $a > b$, then $\frac{1}{a} > \frac{1}{b}$.

101. Use the roster method to list the set of positive integers that are solutions of the inequality $7 - 2b \leq 15 - 5b$.

102. Determine the solution set of $2 - 3(x + 4) < 5 - 3x$.

103. Determine the solution set of $3x + 2(x - 1) > 5(x + 1)$.

104. In your own words, state the Multiplication Property of Inequalities.
[W]

Absolute Value Equations and Inequalities

Objective A *To solve an absolute value equation*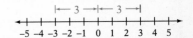

The **absolute value** of a number is its distance from zero on the number line. Distance is always a positive number or zero. Therefore, the absolute value of a number is always a positive number or zero.

The distance from 0 to 3 or from 0 to -3 is 3 units.

$$|3| = 3 \qquad\qquad |-3| = 3$$

An equation that contains an absolute value symbol is called an **absolute value equation.** The solution of an absolute value equation is based on the following property:

If $a \geq 0$ and $|x| = a$, then $x = a$ or $x = -a$.

For instance, given $|x| = 3$, then $x = 3$ or $x = -3$, because $|3| = 3$ and $|-3| = 3$.

⮕ Solve: $|x + 2| = 8$

$$|x + 2| = 8$$

$x + 2 = 8 \qquad x + 2 = -8$ • Remove the absolute value sign and rewrite as two equations.

$x = 6 \qquad\quad x = -10$ • Solve each equation.

Check:
$$\begin{array}{c|c} |x + 2| = 8 \\ \hline |6 + 2| & 8 \\ |8| & 8 \\ 8 = 8 \end{array} \qquad \begin{array}{c|c} |x + 2| = 8 \\ \hline |-10 + 2| & 8 \\ |-8| & 8 \\ 8 = 8 \end{array}$$

The solutions are 6 and -10.

⮕ Solve: $|5 - 3x| - 8 = -4$

$$|5 - 3x| - 8 = -4$$
$$|5 - 3x| = 4$$ • Solve for the absolute value.

$5 - 3x = 4 \qquad 5 - 3x = -4$ • Remove the absolute value sign and rewrite as two equations.

$-3x = -1 \qquad\quad -3x = -9$ • Solve each equation.

$x = \dfrac{1}{3} \qquad\qquad x = 3$

Check:
$$\begin{array}{c|c} |5 - 3x| - 8 = -4 \\ \hline \left|5 - 3\left(\frac{1}{3}\right)\right| - 8 & -4 \\ |5 - 1| - 8 & -4 \\ 4 - 8 & -4 \\ -4 = -4 \end{array} \qquad \begin{array}{c|c} |5 - 3x| - 8 = -4 \\ \hline |5 - 3(3)| - 8 & -4 \\ |5 - 9| - 8 & -4 \\ 4 - 8 & -4 \\ -4 = -4 \end{array}$$

The solutions are $\dfrac{1}{3}$ and 3.

CONSIDER THIS

Because the absolute value of $x + 2$ is 8, the number $x + 2$ is 8 units from 0 on the number line. Therefore, $x + 2$ is equal to 8 or -8.

Example 1

Solve: $|2 - x| = 12$

Solution

$$|2 - x| = 12$$

$$2 - x = 12 \qquad 2 - x = -12$$
$$\quad -x = 10 \qquad \quad -x = -14$$
$$\quad\; x = -10 \qquad \quad\; x = 14$$

The solutions are -10 and 14.

You Try It 1

Solve: $|2x - 3| = 5$

Your solution

Example 2

Solve: $3 - |2x - 4| = -5$

Solution

$$3 - |2x - 4| = -5$$
$$\quad -|2x - 4| = -8$$
$$\qquad |2x - 4| = 8$$

$$2x - 4 = 8 \qquad 2x - 4 = -8$$
$$\quad 2x = 12 \qquad \quad 2x = -4$$
$$\quad\; x = 6 \qquad \qquad x = -2$$

The solutions are 6 and -2.

You Try It 2

Solve: $5 - |3x + 5| = 3$

Your solution

Solutions on p. A12

Objective B *To solve an absolute value inequality* ....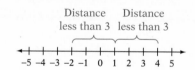

Recall that absolute value represents the distance between two points. For example, the solutions of the absolute value equation $|x - 1| = 3$ are the numbers whose distance from 1 is 3. Therefore, the solutions are -2 and 4.

The solutions of the absolute value inequality $|x - 1| < 3$ are the numbers whose distance from 1 is less than 3. Therefore, the solutions are the numbers greater than -2 and less than 4. The solution set is $\{x \,|\, -2 < x < 4\}$.

Distance less than 3   Distance less than 3

$-5 \; -4 \; -3 \; -2 \; -1 \;\; 0 \;\; 1 \;\; 2 \;\; 3 \;\; 4 \;\; 5$

To solve an absolute value inequality of the form $|ax + b| < c$, solve the equivalent compound inequality $-c < ax + b < c$.

CONSIDER THIS

Because the absolute value of $3x - 1$ is less than 5, the number $3x - 1$ is less than 5 units from 0 on the number line.

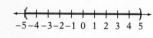

$-5\,-4\,-3\,-2\,-1\,\ 0\ \ 1\ \ 2\ \ 3\ \ 4\ \ 5$

Therefore, $3x - 1$ is between -5 and 5.

➡ Solve: $|3x - 1| < 5$

$$|3x - 1| < 5$$

$$-5 < 3x - 1 < 5$$

$$-5 + 1 < 3x - 1 + 1 < 5 + 1$$

$$-4 < 3x < 6$$

$$\frac{-4}{3} < \frac{3x}{3} < \frac{6}{3}$$

$$-\frac{4}{3} < x < 2$$

$$\left\{ x \,\middle|\, -\frac{4}{3} < x < 2 \right\}$$

• **Solve the equivalent compound inequality.**

The solutions of the absolute value inequality $|x + 1| > 2$ are the numbers whose distance from -1 is greater than 2. Therefore, the solutions are the numbers that are less than -3 or greater than 1. The solution set of $|x + 1| > 2$ is $\{x \mid x < -3 \text{ or } x > 1\}$.

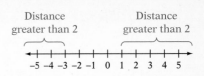

CONSIDER THIS

Because the absolute value of $3 - 2x$ is greater than 1, the number $3 - 2x$ is more than 1 unit from 0 on the number line.

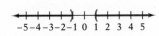

Therefore, $3 - 2x$ is less than -1 or greater than 1.

To solve an absolute value inequality of the form $|ax + b| > c$, solve the equivalent compound inequality $ax + b < -c$ or $ax + b > c$.

⇒ Solve: $|3 - 2x| > 1$

$$3 - 2x < -1 \quad \text{or} \quad 3 - 2x > 1$$
$$-2x < -4 \qquad\qquad -2x > -2$$
$$x > 2 \qquad\qquad\quad x < 1$$
$$\{x \mid x > 2\} \qquad\quad \{x \mid x < 1\}$$

• Solve each inequality.

$$\{x \mid x > 2\} \cup \{x \mid x < 1\}$$
$$= \{x \mid x > 2 \text{ or } x < 1\}$$

• Find the union of the solution sets of the two inequalities.

Example 3   Solve: $|4x - 3| < 5$

Solution   Solve the equivalent compound inequality.

$$-5 < 4x - 3 < 5$$
$$-5 + 3 < 4x - 3 + 3 < 5 + 3$$
$$-2 < 4x < 8$$
$$\frac{-2}{4} < \frac{4x}{4} < \frac{8}{4}$$
$$-\frac{1}{2} < x < 2$$
$$\left\{ x \,\middle|\, -\frac{1}{2} < x < 2 \right\}$$

You Try It 3   Solve: $|3x + 2| < 8$

Your solution

Example 4   Solve: $|2x - 1| > 7$

Solution   Solve the equivalent compound inequality.

$$2x - 1 < -7 \quad \text{or} \quad 2x - 1 > 7$$
$$2x < -6 \qquad\qquad 2x > 8$$
$$x < -3 \qquad\qquad\, x > 4$$
$$\{x \mid x < -3\} \qquad\quad \{x \mid x > 4\}$$

$$\{x \mid x < -3\} \cup \{x \mid x > 4\}$$
$$= \{x \mid x < -3 \text{ or } x > 4\}$$

You Try It 4   Solve: $|5x + 3| > 8$

Your solution

Solutions on p. A13

Objective C   *To solve application problems* ⋯⋯⋯⋯⋯⋯⋯⋯⋯⋯⋯⋯⋯⋯⋯⋯

The **tolerance** of a component, or part, is the acceptable amount by which the component may vary from a given measurement. For example, the diameter of a piston may vary from the given measurement of 9 cm by 0.001 cm. This is written 9 cm ± 0.001 cm and is read "9 centimeters plus or minus 0.001 centimeters." The maximum diameter, or **upper limit,** of the piston is 9 cm + 0.001 cm = 9.001 cm. The minimum diameter, or **lower limit,** is 9 cm − 0.001 cm = 8.999 cm.

The lower and upper limits of the diameter of the piston could also be found by solving the absolute value inequality $|d - 9| \leq 0.001$, where d is the diameter of the piston.

$$|d - 9| \leq 0.001$$
$$-0.001 \leq d - 9 \leq 0.001$$
$$-0.001 + 9 \leq d - 9 + 9 \leq 0.001 + 9$$
$$8.999 \leq d \leq 9.001$$

The lower and upper limits of the diameter of the piston are 8.999 cm and 9.001 cm.

Example 5

A doctor has prescribed 2 cc of medication for a patient. The tolerance is 0.03 cc. (In the medical field, cubic centimeter is usually abbreviated cc.) Find the lower and upper limits of the amount of medication to be given.

Strategy

Let p represent the prescribed amount of medication, T the tolerance, and m the given amount of medication. Solve the absolute value inequality $|m - p| \leq T$ for m.

Solution

$$|m - p| \leq T$$
$$|m - 2| \leq 0.03$$

$$-0.03 \leq m - 2 \leq 0.03$$
$$-0.03 + 2 \leq m - 2 + 2 \leq 0.03 + 2$$
$$1.97 \leq m \leq 2.03$$

The lower and upper limits of the amount of medication to be given are 1.97 cc and 2.03 cc.

You Try It 5

A machinist must make a bushing that has a diameter of 2.55 in. The tolerance of the bushing is 0.003 in. Find the lower and upper limits of the diameter of the bushing.

Your strategy

Your solution

Solution on p. A13

2.6 Exercises

_____

Objective A

Solve.

1. $|x| = 7$ **2.** $|a| = 2$ **3.** $|b| = 4$ **4.** $|c| = 12$

5. $|-y| = 6$ **6.** $|-t| = 3$ **7.** $|-a| = 7$ **8.** $|-x| = 3$

9. $|x| = -4$ **10.** $|y| = -3$ **11.** $|-t| = -3$ **12.** $|-y| = -2$

13. $|x + 2| = 3$ **14.** $|x + 5| = 2$ **15.** $|y - 5| = 3$ **16.** $|y - 8| = 4$

17. $|a - 2| = 0$ **18.** $|a + 7| = 0$ **19.** $|x - 2| = -4$ **20.** $|x + 8| = -2$

21. $|3 - 4x| = 9$ **22.** $|2 - 5x| = 3$ **23.** $|2x - 3| = 0$ **24.** $|5x + 5| = 0$

25. $|3x - 2| = -4$ **26.** $|2x + 5| = -2$ **27.** $|x - 2| - 2 = 3$

28. $|x - 9| - 3 = 2$ **29.** $|3a + 2| - 4 = 4$ **30.** $|2a + 9| + 4 = 5$

31. $|2 - y| + 3 = 4$ **32.** $|8 - y| - 3 = 1$ **33.** $|2x - 3| + 3 = 3$

34. $|4x - 7| - 5 = -5$ **35.** $|2x - 3| + 4 = -4$ **36.** $|3x - 2| + 1 = -1$

Solve.

37. $|6x - 5| - 2 = 4$

38. $|4b + 3| - 2 = 7$

39. $|3t + 2| + 3 = 4$

40. $|5x - 2| + 5 = 7$

41. $3 - |x - 4| = 5$

42. $2 - |x - 5| = 4$

43. $8 - |2x - 3| = 5$

44. $8 - |3x + 2| = 3$

45. $|2 - 3x| - 7 = 2$

46. $|1 - 5a| + 2 = 3$

47. $|8 - 3x| - 3 = 2$

48. $|6 - 5b| - 4 = 3$

49. $|2x - 8| + 12 = 2$

50. $|3x - 4| + 8 = 3$

51. $2 + |3x - 4| = 5$

52. $5 + |2x + 1| = 8$

53. $5 - |2x + 1| = 5$

54. $3 - |5x + 3| = 3$

55. $6 - |2x + 4| = 3$

56. $8 - |3x - 2| = 5$

57. $8 - |1 - 3x| = -1$

58. $3 - |3 - 5x| = -2$

59. $5 + |2 - x| = 3$

60. $6 + |3 - 2x| = 2$

Objective B

Solve.

61. $|x| > 3$

62. $|x| < 5$

63. $|x + 1| > 2$

64. $|x - 2| > 1$

65. $|x - 5| \leq 1$

66. $|x - 4| \leq 3$

Solve.

67. $|2 - x| \geq 3$

68. $|3 - x| \geq 2$

69. $|2x + 1| < 5$

70. $|3x - 2| < 4$

71. $|5x + 2| > 12$

72. $|7x - 1| > 13$

73. $|4x - 3| \leq 2$

74. $|5x + 1| \leq 4$

75. $|2x + 7| > 5$

76. $|3x - 1| > 4$

77. $|4 - 3x| \geq 5$

78. $|7 - 2x| > 9$

79. $|5 - 4x| \leq 13$

80. $|3 - 7x| < 17$

81. $|6 - 3x| \leq 0$

82. $|10 - 5x| \geq 0$

83. $|2 - 9x| > 20$

84. $|5x - 1| < 16$

85. $|2x - 3| + 2 < 8$

86. $|3x - 5| + 1 < 7$

87. $|2 - 5x| - 4 > -2$

88. $|4 - 2x| - 9 > -3$

89. $8 - |2x - 5| < 3$

90. $12 - |3x - 4| > 7$

Objective C   *Application Problems*

Solve.

91. A doctor has prescribed 4 cc of medication for a patient. The tolerance is 0.05 cc. Find the lower and upper limits of the amount of medication to be given.

92. A doctor has prescribed 3 cc of medication for a patient. The tolerance is 0.04 cc. Find the lower and upper limits of the amount of medication to be given.

Solve.

93. A machinist must make a bushing that has a tolerance of 0.002 in. The diameter of the bushing is 2.65 in. Find the lower and upper limits of the diameter of the bushing.

94. The diameter of a bushing is 2.45 in. The bushing has a tolerance of 0.001 in. Find the lower and upper limits of the diameter of the bushing.

95. A piston rod for an automobile is $9\frac{5}{8}$ in. long with a tolerance of $\frac{1}{32}$ in. Find the lower and upper limits of the length of the piston rod.

96. A piston rod for an automobile is $9\frac{3}{8}$ in. long with a tolerance of $\frac{1}{64}$ in. Find the lower and upper limits of the length of the piston rod.

The tolerance of the resistors used in electronics is given as a percent.

97. Find the lower and upper limits of a 29,000-ohm resistor with a 2% tolerance.

98. Find the lower and upper limits of a 15,000-ohm resistor with a 10% tolerance.

99. Find the lower and upper limits of a 25,000-ohm resistor with a 5% tolerance.

100. Find the lower and upper limits of a 56-ohm resistor with a 5% tolerance.

APPLYING THE CONCEPTS

101. Determine whether the following statements are always true, sometimes true, or never true.
  a. An absolute value equation has two solutions.
  b. A solution of an absolute value equation is a negative number.

102. Write an absolute value inequality to represent all real numbers within 5 units of 2.

103. Replace the question mark with $\le$, $\ge$, or $=$.
  a. $|x + y| \; ? \; |x| + |y|$
  b. $|x - y| \; ? \; |x| - |y|$
  c. $||x| - |y|| \; ? \; |x| - |y|$
  d. $|xy| \; ? \; |x||y|$

Projects in Mathematics

. . . .

Counterexamples   Some of the exercises in this text ask you to determine whether a statement is true or false. For instance, the statement "Every real number has a reciprocal" is false because 0 is a real number and 0 does not have a reciprocal.

Finding an example, such as the fact that 0 has no reciprocal, to show that a statement is not always true is called *finding a counterexample*. A counterexample is an example that shows that a statement is not always true.

Consider the statement "The product of two numbers is greater than either factor." A counterexample to this statement is the product of the numbers $\frac{2}{3}$ and $\frac{3}{4}$. The product of these numbers is $\frac{1}{2}$, and $\frac{1}{2}$ is *smaller* than $\frac{2}{3}$ or $\frac{3}{4}$. There are many other counterexamples to the given statement.

Here are some counterexamples to the statement "The square of a number is always larger than the number."

$$\left(\frac{1}{2}\right)^2 = \frac{1}{4} \text{ but } \frac{1}{4} < \frac{1}{2} \qquad 1^2 = 1 \text{ but } 1 = 1$$

For each of the following five statements, find at least one counterexample to show that the statement, or conjecture, is false.

1. The product of two integers is always a positive number.

2. The sum of two prime numbers is never a prime number.

3. For all real numbers, $|x + y| = |x| + |y|$.

4. If x and y are nonzero real numbers and $x > y$, then $x^2 > y^2$.

5. The quotient of any two nonzero real numbers is less than either one of the numbers.

When a problem is posed, it may not be known whether the problem statement is true or false. For instance, Christian Goldbach (1690–1764) stated that every even integer greater than 2 can be written as the sum of two prime numbers. No one has been able to find a counterexample to this statement, but neither has anyone been able to prove that it is always true.

In the next set of problems, answer true if the statement is always true. Give a counterexample if the statement is false.

6. The reciprocal of a positive number is always smaller than the number.

7. If $x < 0$, then $|x| = -x$.

8. For any two real numbers x and y, $x + y > x - y$.

9. For any positive integer n, $n^2 + n + 17$ is a prime number.

10. The list of numbers 1, 11, 111, 1111, 11111, ... contains infinitely many composit numbers. *Hint:* A number is divisible by 3 if the sum of the digits of the number is divisible by 3.

Chapter Summary

· ·

Key Words An *equation* expresses the equality of two mathematical expressions. A *solution* of an equation is a number that, when substituted for the variable, results in a true equation. To *solve* an equation means to find a solution of the equation.

An *inequality* is an expression that contains the symbol >, <, ≥, or ≤. The *solution set of an inequality* is a set of numbers, each element of which, when substituted for the variable, results in a true inequality.

A *compound inequality* is formed by joining two inequalities with a connective word such as *and* or *or*.

The *absolute value* of a number is its distance from zero on the number line. An equation that contains an absolute value symbol is an *absolute value equation.*

Essential Rules **Addition Property of Equations**   If $a = b$, then $a + c = b + c$.

Multiplication Property of Equations   If $a = b$ and $c \neq 0$, then $ac = bc$.

Addition Property of Inequalities   If $a > b$, then $a + c > b + c$.
                                                                         If $a < b$, then $a + c < b + c$.

Multiplication Property of Inequalities   **Rule 1**
                                                                              If $a > b$ and $c > 0$, then $ac > bc$.
                                                                              If $a < b$ and $c > 0$, then $ac < bc$.
                                                                              Rule 2
                                                                              If $a > b$ and $c < 0$, then $ac < bc$.
                                                                              If $a < b$ and $c < 0$, then $ac > bc$.

Basic Percent Equation   Percent · base = amount
                                                                     $$P \cdot B = A$$

Value Mixture Equation   Amount · unit cost = value
                                                                     $$A \cdot C = V$$

Percent Mixture Equation $\dfrac{\text{Amount of}}{\text{solution}} \cdot \dfrac{\text{Percent of}}{\text{concentration}} = \dfrac{\text{Quantity of}}{\text{substance}}$
                                                                     $$A \cdot r = Q$$

Annual Simple Interest Equation   Principal $\cdot \dfrac{\text{interest}}{\text{rate}} = \dfrac{\text{interest}}{\text{earned}}$
                                                                     $$P \cdot r = I$$

Uniform Motion Equation   Rate · time = distance
                                                                     $$r \cdot t = d$$

To solve an absolute value inequality of the form $|ax + b| < c$, solve the equivalent compound inequality $-c < ax + b < c$.

To solve an absolute value inequality of the form $|ax + b| > c$, solve the equivalent compound inequality $ax + b < -c$ or $ax + b > c$.

Chapter Review Exercises

1. Is 3 a solution of $5x - 2 = 4x + 5$?

2. Solve: $x + 3 = 24$

3. Solve: $\dfrac{3}{5}a = 12$

4. Solve: $-4x - 2 = 10$

5. Solve: $14x + 7x + 8 = -10$

6. Solve: $12y - 1 = 3y + 2$

7. Solve: $-6x + 16 = -2x$

8. Solve: $6x + 3(2x - 1) = -27$

9. Solve: $x + 5(3x - 20) = 10(x - 4)$

10. Solve: $3x - 7 > -2$

11. Solve: $4 - 3(x + 2) < 2(2x + 3) - 1$

12. Solve: $3x < 4$ and $x + 2 > -1$

13. Solve: $3x - 2 > x - 4$ or $7x - 5 < 3x + 3$

14. Solve: $|3 - 5x| = 12$

15. Solve: $|x - 4| - 8 = -3$

16. Solve: $|2x - 5| < 3$

17. Solve: $|4x - 5| \geq 3$

18. 30 is what percent of 12?

19. $\dfrac{1}{2}\%$ of what is 8?

20. Translate "four less than the product of five and a number is sixteen" into an equation and solve.

21. The sum of two numbers is twenty-one. Three times the smaller number is two less than twice the larger number. Translate into an equation. Then find the two numbers.

22. An airline knowingly overbooks certain flights by selling 18% more tickets than there are available seats. How many tickets would this airline sell for an airplane that has 150 seats?

23. An auto manufacturer offers a rebate of $1000 on each car sold by a dealership. A customer bought a car from the dealership for $16,500. What percent of the cost is the $1000 rebate? Round to the nearest tenth of a percent.

24. A health food store combined cranberry juice that costs $1.79 per quart with apple juice that costs $1.19 per quart. How many quarts of each were used to make 10 qt of a cranapple juice mixture that costs $1.61 per quart?

25. Find the cost per pound of a meatloaf mixture made from 3 lb of ground beef that costs $1.99 per pound and 1 lb of ground turkey that costs $1.39 per pound.

26. A dairy mixed 5 gal of cream that is 30% butterfat with 8 gal of milk that is 4% butterfat. What is the percent concentration of butterfat in the resulting mixture?

27. An alloy containing 30% tin is mixed with an alloy containing 70% tin. How many pounds of each were used to make 500 lb of an alloy containing 40% tin?

28. An investment banker invested 45% of the bank's available cash in an account earning 8.5% annual simple interest. The remainder of the cash was placed in an account earning 10% annual simple interest. The interest earned in one year was $41,962.50. What was the total amount invested?

29. A club treasurer deposited $2400 into two simple interest accounts. On one account the annual simple interest rate was 6.75%. The annual simple interest rate on the other account was 9.45%. How much was deposited in each account if both accounts earned the same amount of interest?

30. A jet plane traveling at 600 mph overtakes a propeller-driven plane that had a 2-hour headstart. The propeller-driven plane is traveling at 200 mph. How far from the starting point did the jet overtake the propeller-driven plane?

31. A bus traveled on a level road for 2 h at an average speed of 20 mph faster than it traveled on a winding road. The time spent on the winding road was 3 h. Find the average speed on the winding road if the total trip was 200 mi.

32. An average score of 80 to 90 in a psychology class receives a B grade. A student has grades of 92, 66, 72, and 88 on four tests. Find the range of scores on the fifth test that will earn the student a B for the course.

33. A doctor has prescribed 2 cc of medication for a patient. The tolerance is 0.25 cc. Find the lower and upper limits of the amount of medication to be given.

Content and Format © 1996 HMCo.

Cumulative Review Exercises

. . . .

1. Subtract: $-6 - (-20) - 8$

2. Multiply: $(-2)(-6)(-4)$

3. Subtract: $-\dfrac{5}{6} - \left(-\dfrac{7}{16}\right)$

4. Simplify: $-4^2 \cdot \left(-\dfrac{3}{2}\right)^3$

5. Simplify: $\dfrac{5}{8} - \left(\dfrac{1}{2}\right)^2 \div \left(\dfrac{1}{3} - \dfrac{3}{4}\right)$

6. Evaluate $3(a - c) - 2ab$ when $a = 2$, $b = 3$, and $c = -4$.

7. Simplify: $3x - 8x + (-12x)$

8. Simplify: $2a - (-b) - 7a - 5b$

9. Simplify: $(16x)\left(\dfrac{1}{8}\right)$

10. Simplify: $-4(-9y)$

11. Simplify: $-2(-x^2 - 3x + 2)$

12. Simplify: $-2(x - 3) + 2(4 - x)$

13. Simplify: $-3[2x - 4(x - 3)] + 2$

14. Find $A \cap B$ given $A = \{-4, -2, 0, 2\}$ and $B = \{-4, 0, 4, 8\}$.

15. Graph: $\{x \mid x \le 3\} \cap \{x \mid x > -1\}$.

16. Is -3 a solution of $x^2 + 6x + 9 = x + 3$?

17. Solve: $\dfrac{3}{5}x = -15$

18. Solve: $7x - 8 = -29$

19. Solve: $13 - 9x = -14$

20. Solve: $5x - 8 = 12x + 13$

21. Solve: $11 - 4x = 2x + 8$

22. Solve: $8x - 3(4x - 5) = -2x - 11$

23. Solve: $3 - 2(2x - 1) \geq 3(2x - 2) + 1$

24. Solve: $3x + 2 \leq 5$ and $x + 5 \geq 1$

25. Solve: $|3 - 2x| = 5$

26. Solve: $|3x - 1| > 5$

27. Write 55% as a fraction.

28. Write 1.03 as a percent.

29. 25% of what number is 30?

30. Translate "the sum of six times a number and thirteen is five less than the product of three and the number" into an equation and solve.

31. How many pounds of an oat flour that costs $.80 per pound must be mixed with 40 lb of a wheat flour that costs $.50 per pound to make a blend that costs $.60 per pound?

32. How many grams of pure gold must be added to 100 g of a 20% gold alloy to make an alloy that is 36% gold?

33. A sprinter ran to the end of a track at an average rate of 8 m/s and then jogged back to the starting point at an average rate of 3 m/s. The sprinter took 55 s to run to the end of the track and jog back. Find the length of the track.

Chapter

Geometry

Objectives

3.1 Introduction to Geometry

A To solve problems involving lines and angles

B To solve problems involving angles formed by intersecting lines

C To solve problems involving the angles of a triangle

3.2 Plane Geometric Figures

A To solve problems involving the perimeter of geometric figures

B To solve problems involving the area of geometric figures

3.3 Solids

A To solve problems involving the volume of a solid

B To solve problems involving the surface area of a solid

Mobius Strips and Klein Bottles

Some geometric shapes have very unusual characteristics. Among these figures are Mobius strips and Klein bottles.

A mobius strip is formed by taking a long strip of paper and twisting it one-half turn. The resulting figure is called a one-sided surface. It is one-sided in the sense that if you tried to paint the strip in one continuous motion beginning at one spot, the entire surface would be painted the same color. This is not possible with a strip that has not been twisted.

Another remarkable fact of being one-sided can be demonstrated by making a Mobius strip and sealing the junction with tape. Now cut the entire strip along the center. Try this; you will be amazed at the result.

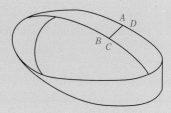

A second interesting surface is called a Klein bottle, which is a one-sided surface with no edges and no "inside" or "outside." A Klein bottle is formed by pulling the small open end of a tapering tube through the side of the tube and joining the ends of the small open end to the ends of the larger open end.

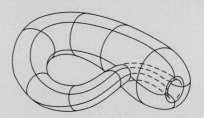

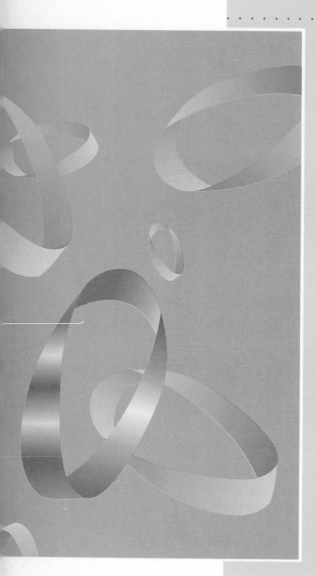

Introduction to Geometry

Objective A   *To solve problems involving lines and angles* ....

The word *geometry* comes from the Greek words for "earth" and "measure." The original purpose of geometry was to measure land. Today geometry is used in many fields, such as physics, medicine, and geology, and is applied in such areas as mechanical drawing and astronomy. Geometric forms are also used in art and design.

Three basic concepts of geometry are the point, line, and plane. A **point** is symbolized by drawing a dot. A **line** is determined by two distinct points and extends indefinitely in both directions, as the arrows on the line shown at the right indicate. This line contains points A and B and is represented by $\overleftrightarrow{AB}$. A line can also be represented by a single letter, such as ℓ.

A **ray** starts at a point and extends indefinitely in *one* direction. The point at which a ray starts is called the **endpoint** of the ray. The ray shown at the right is denoted by $\overrightarrow{AB}$. Point A is the endpoint of the ray.

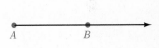

A **line segment** is part of a line and has two endpoints. The line segment shown at the right is denoted by $\overline{AB}$.

The distance between the endpoints of $\overline{AC}$ is denoted by AC. If B is a point on $\overline{AC}$, then AC (the distance from A to C) is the sum of AB (the distance from A to B) and BC (the distance from B to C).

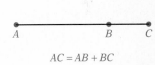

$$AC = AB + BC$$

➡ Given $AB = 22$ cm and $AC = 31$ cm, find BC.

$AC = AB + BC$	• Write an equation for the distances between points on the line segment.
$31 = 22 + BC$	• Substitute the given distances for AB and AC into the equation.
$9 = BC$	• Solve for BC.
$BC = 9$ cm	

In this section we will be discussing figures that lie in a plane. A **plane** is a flat surface and can be pictured as a table top or blackboard that extends in all directions. Figures that lie in a plane are called **plane figures.**

Lines in a plane can be intersecting or parallel. **Intersecting lines** cross at a point in the plane. **Parallel lines** never meet. The distance between them is always the same.

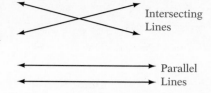

The symbol ∥ means "is parallel to." In the figure at the right, $j \parallel k$ and $\overline{AB} \parallel \overline{CD}$. Note that j contains $\overline{AB}$ and k contains $\overline{CD}$. Parallel lines contain parallel line segments.

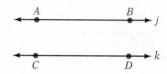

An **angle** is formed by two rays with the same endpoint. The **vertex** of the angle is the point at which the two rays meet. The rays are called the **sides** of the angle.

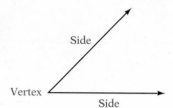

If A and C are points on rays r_1 and r_2, and B is the vertex, then the angle is called $\angle B$ or $\angle ABC$, where $\angle$ is the symbol for angle. Note that either the angle is named by the vertex, or the vertex is the second point listed when the angle is named by giving three points. $\angle ABC$ could also be called $\angle CBA$.

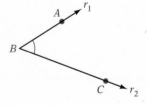

An angle can also be named by a variable written between the rays close to the vertex. In the figure at the right, $\angle x = \angle QRS$ and $\angle y = \angle SRT$. Note that in this figure, more than two rays meet at R. In this case, the vertex cannot be used to name an angle.

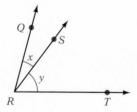

POINT OF INTEREST

The first woman mathematician for which documented evidence exists is Hypatia (370–415). She lived in Alexandria, Egypt, and lectured at the Museum, the forerunner of our modern university. She made important contributions in mathematics, astronomy, and philosophy.

An angle is measured in **degrees.** The symbol for degrees is a small raised circle, °. Probably because early Babylonians believed that Earth revolves around the sun in approximately 360 days, the angle formed by a circle has a measure of 360° (360 degrees).

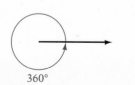

A **protractor** is used to measure an angle. Place the center of the protractor at the vertex of the angle with the edge of the protractor along a side of the angle. The angle shown in the figure below measures 58°.

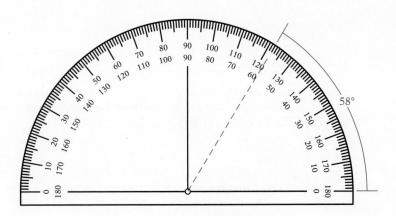

CONSIDER THIS

The corner of a page of this book is a good example of a 90° angle.

A 90° is called a **right angle.** The symbol ⌐ represents a right angle.

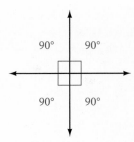

Perpendicular lines are intersecting lines that form right angles.

The symbol ⊥ means "is perpendicular to." In the figure at the right, $p \perp q$ and $\overline{AB} \perp \overline{CD}$. Note that line p contains $\overline{AB}$ and line q contains $\overline{CD}$. Perpendicular lines contain perpendicular line segments.

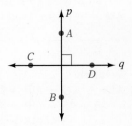

Complementary angles are two angles whose measures have the sum 90°.

$$\angle A + \angle B = 70° + 20° = 90°$$

$\angle A$ and $\angle B$ are complementary angles.

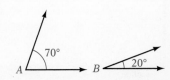

A 180° angle is called a **straight angle.**

∠AOB is a straight angle.

Supplementary angles are two angles whose measures have the sum 180°.

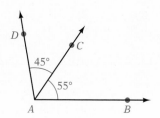

$$\angle A + \angle B = 130° + 50° = 180°$$

∠A and ∠B are supplementary angles.

An **acute angle** is an angle whose measure is between 0° and 90°. ∠B above is an acute angle. An **obtuse angle** is an angle whose measure is between 90° and 180°. ∠A above is an obtuse angle.

Two angles that share a common side are **adjacent angles.** In the figure at the right, ∠DAC and ∠CAB are adjacent angles. ∠DAC = 45° and ∠CAB = 55°.

$$\angle DAB = \angle DAC + \angle CAB$$
$$= 45° + 55° = 100°$$

➡ In the figure at the right, ∠EDG = 80°. ∠FDG is three times the measure of ∠EDF. Find the measure of ∠EDF.

CONSIDER THIS
Answers to application problems must have units, such as degrees, feet, dollars, or hours.

Let x = the measure of ∠EDF. Then $3x$ = the measure of ∠FDG. Write an equation and solve for x, the measure of ∠EDF.

$$\angle EDF + \angle FDG = \angle EDG$$
$$x + 3x = 80$$
$$4x = 80$$
$$x = 20$$

$$\angle EDF = 20°$$

Example 1
Given MN = 15 mm, NO = 18 mm, and MP = 48 mm, find OP.

Solution
$$MN + NO + OP = MP$$
$$15 + 18 + OP = 48$$
$$33 + OP = 48$$
$$OP = 15$$

$$OP = 15 \text{ mm}$$

You Try It 1
Given QR = 24 cm, ST = 17 cm, and QT = 62 cm, find RS.

Q   R   S   T

Your solution

Solution on p. A14

Example 2

Given $XY = 9$ m and YZ is twice XY, find XZ.

Solution

$XZ = XY + YZ$
$XZ = XY + 2(XY)$
$XZ = 9 + 2(9)$
$XZ = 9 + 18$
$XZ = 27$

$XZ = 27$ m

You Try It 2

Given $BC = 16$ ft and $AB = \frac{1}{4}(BC)$, find AC.

Your solution

Example 3

Find the complement of a 38° angle.

Strategy

Complementary angles are two angles whose sum is 90°. To find the complement, let x represent the complement of a 38° angle. Write an equation and solve for x.

Solution

$x + 38° = 90°$
$x = 52°$

The complement of a 38° angle is a 52° angle.

You Try It 3

Find the supplement of a 129° angle.

Your strategy

Your solution

Example 4

Find the measure of $\angle x$.

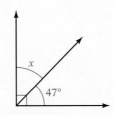

Strategy

To find the measure of $\angle x$, write an equation using the fact that the sum of the measure of $\angle x$ and 47° is 90°. Solve for $\angle x$.

Solution

$\angle x + 47° = 90°$
$\angle x = 43°$

The measure of $\angle x$ is 43°.

You Try It 4

Find the measure of $\angle a$.

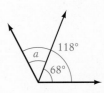

Your strategy

Your solution

Solutions on p. A14

Objective B **To solve problems involving angles formed by intersecting lines** ....

Four angles are formed by the intersection of two lines. If the two lines are perpendicular, each of the four angles is a right angle. If the two lines are not perpendicular, then two of the angles formed are acute angles and two of the angles are obtuse angles. The two acute angles are always opposite each other, and the two obtuse angles are always opposite each other.

In the figure at the right, $\angle w$ and $\angle y$ are acute angles. $\angle x$ and $\angle z$ are obtuse angles.

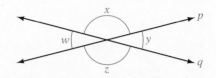

Two angles that are on opposite sides of the intersection of two lines are called **vertical angles.** Vertical angles have the same measure. $\angle w$ and $\angle y$ are vertical angles. $\angle x$ and $\angle z$ are vertical angles.

Vertical angles have the same measure.

$$\angle w = \angle y$$
$$\angle x = \angle z$$

Two angles that share a common side are called **adjacent angles.** For the figure shown above, $\angle x$ and $\angle y$ are adjacent angles, as are $\angle y$ and $\angle z$, $\angle z$ and $\angle w$, and $\angle w$ and $\angle x$. Adjacent angles of intersecting lines are supplementary angles.

Adjacent angles of intersecting lines are supplementary angles.

$$\angle x + \angle y = 180°$$
$$\angle y + \angle z = 180°$$
$$\angle z + \angle w = 180°$$
$$\angle w + \angle x = 180°$$

➡ Given that $\angle c = 65°$, find the measures of angles a, b, and d.

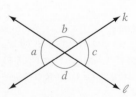

$\angle a = 65°$

* $\angle a = \angle c$ because $\angle a$ and $\angle c$ are vertical angles.

$\angle b + \angle c = 180°$
$\angle b + 65° = 180°$
$\angle b = 115°$

* $\angle b$ is supplementary to $\angle c$ because $\angle b$ and $\angle c$ are adjacent angles of intersecting lines.

$\angle d = 115°$

* $\angle d = \angle b$ because $\angle d$ and $\angle b$ are vertical angles.

A line that intersects two other lines at different points is called a **transversal.**

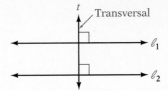

If the lines cut by a transversal t are parallel lines and the transversal is perpendicular to the parallel lines, all eight angles formed are right angles.

If the lines cut by a transversal t are parallel lines and the transversal is not perpendicular to the parallel lines, all four acute angles have the same measure and all four obtuse angles have the same measure. For the figure at the right,

$$\angle b = \angle d = \angle x = \angle z$$
$$\angle a = \angle c = \angle w = \angle y$$

Alternate interior angles are two angles that are on opposite sides of the transversal and lie between the lines. In the figure above, $\angle c$ and $\angle w$ are alternate interior angles; $\angle d$ and $\angle x$ are alternate interior angles. Alternate interior angles have the same measure.

Alternate interior angles have the same measure.

$$\angle c = \angle w$$
$$\angle d = \angle x$$

Alternate exterior angles are two angles that are on opposite sides of the transversal and lie outside the parallel lines. In the figure above, $\angle a$ and $\angle y$ are alternate exterior angles; $\angle b$ and $\angle z$ are alternate exterior angles. Alternate exterior angles have the same measure.

Alternate exterior angles have the same measure.

$$\angle a = \angle y$$
$$\angle b = \angle z$$

Corresponding angles are two angles that are on the same side of the transversal and are both acute angles or are both obtuse angles. For the figure above, the following pairs of angles are corresponding angles: $\angle a$ and $\angle w$, $\angle d$ and $\angle z$, $\angle b$ and $\angle x$, and $\angle c$ and $\angle y$. Corresponding angles have the same measure.

Corresponding angles have the same measure.

$$\angle a = \angle w$$
$$\angle d = \angle z$$
$$\angle b = \angle x$$
$$\angle c = \angle y$$

➡ Given that $\ell_1 \parallel \ell_2$ and $\angle c = 58°$,
find the measures of $\angle f$, $\angle h$, and $\angle g$.

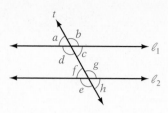

$\angle f = \angle c = 58°$ • $\angle c$ and $\angle f$ are alternate interior angles.

$\angle h = \angle c = 58°$ • $\angle c$ and $\angle h$ are corresponding angles.

$\angle g + \angle h = 180°$ • $\angle g$ is supplementary to $\angle h$.
$\angle g + 58° = 180°$
$\angle g = 122°$

Example 5
Find x.

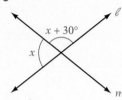

Strategy
The angles labeled are adjacent angles of intersecting lines and are, therefore, supplementary angles. To find x, write an equation and solve for x.

Solution
$x + (x + 30°) = 180°$
$2x + 30° = 180°$
$2x = 150°$
$x = 75°$

You Try It 5
Find x.

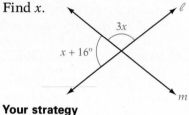

Your strategy

Your solution

Example 6
Given $\ell_1 \parallel \ell_2$, find x.

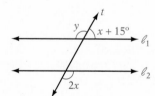

Strategy
$2x = y$ because alternate exterior angles have the same measure.
$(x + 15°) + y = 180°$ because adjacent angles of intersecting lines are supplementary angles. Substitute $2x$ for y and solve for x.

Solution
$(x + 15°) + 2x = 180°$
$3x + 15° = 180°$
$3x = 165°$
$x = 55°$

You Try It 6
Given $\ell_1 \parallel \ell_2$, find x.

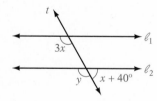

Your strategy

Your solution

Solutions on p. A14

Objective C   ***To solve problems involving the angles of a triangle*** ....

If the lines cut by a transversal are not parallel lines, the three lines will intersect at three points. In the figure at the right, the transversal *t* intersects lines *p* and *q*. The three lines intersect at points *A*, *B*, and *C*. These three points define three line segments: $\overline{AB}$, $\overline{BC}$, and $\overline{AC}$. The plane figure formed by these three line segments is called a **triangle.**

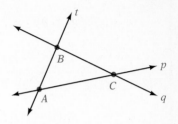

Each of the three points of intersection is the vertex of four angles. The angles within the region enclosed by the triangle are called **interior angles.** In the figure at the right, angles *a*, *b*, and *c* are interior angles. The sum of the measures of the interior angles of a triangle is 180°.

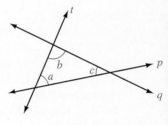

$\angle a + \angle b + \angle c = 180°$

> **The Sum of the Measures of the Interior Angles of a Triangle**
>
> The sum of the measures of the interior angles of a triangle is 180°.

An angle adjacent to an interior angle is an **exterior angle.** In the figure at the right, angles *m* and *n* are exterior angles for angle *a*. The sum of the measures of an interior and an exterior angle is 180°.

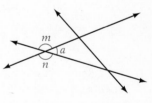

$\angle a + \angle m = 180°$
$\angle a + \angle n = 180°$

➡ Given that $\angle c = 40°$ and $\angle d = 100°$, find the measure of $\angle e$.

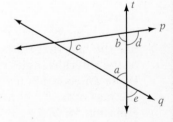

$\angle d$ and $\angle b$ are supplementary angles.

$$\angle d + \angle b = 180°$$
$$100° + \angle b = 180°$$
$$\angle b = 80°$$

The sum of the interior angles is 180°.

$$\angle c + \angle b + \angle a = 180°$$
$$40° + 80° + \angle a = 180°$$
$$120° + \angle a = 180°$$
$$\angle a = 60°$$

$\angle a$ and $\angle e$ are vertical angles.

$$\angle e = \angle a = 60°$$

Example 7

Given that $\angle y = 55°$, find the measures of angles a, b, and d.

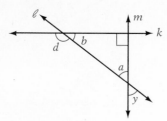

Strategy

- To find the measure of angle a, use the fact that $\angle a$ and $\angle y$ are vertical angles.
- To find the measure of angle b, use the fact that the sum of the measures of the interior angles of a triangle is 180°.
- To find the measure of angle d, use the fact that the sum of an interior and an exterior angle is 180°.

Solution

$\angle a = \angle y = 55°$

$\angle a + \angle b + 90° = 180°$
$55° + \angle b + 90° = 180°$
$\quad\quad \angle b + 145° = 180°$
$\quad\quad\quad\quad\quad \angle b = 35°$

$\angle d + \angle b = 180°$
$\angle d + 35° = 180°$
$\quad\quad \angle d = 145°$

Example 8

Two angles of a triangle measure 53° and 78°. Find the measure of the third angle.

Strategy

To find the measure of the third angle, use the fact that the sum of the measures of the interior angles of a triangle is 180°. Write an equation using x to represent the measure of the third angle. Solve the equation for x.

Solution

$x + 53° + 78° = 180°$
$\quad\quad x + 131° = 180°$
$\quad\quad\quad\quad\quad x = 49°$

The measure of the third angle is 49°.

You Try It 7

Given that $\angle a = 45°$ and $\angle x = 100°$, find the measures of angles b, c, and y.

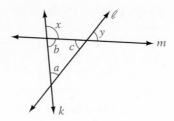

Your strategy

Your solution

You Try It 8

One angle in a triangle is a right angle, and one angle measures 34°. Find the measure of the third angle.

Your strategy

Your solution

Solutions on p. A14

3.1 Exercises

- -

Objective A

Use a protractor to measure the angle. State whether the angle is acute, obtuse, or right.

1.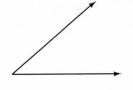

2.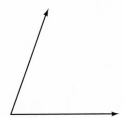

3.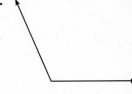

4.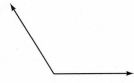

5.

6.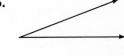

Solve.

7. Find the complement of a 62° angle.

8. Find the complement of a 31° angle.

9. Find the supplement of a 162° angle.

10. Find the supplement of a 72° angle.

11. Given $AB = 12$ cm, $CD = 9$ cm, and $AD = 35$ cm, find the length of BC.

12. Given $AB = 21$ mm, $BC = 14$ mm, and $AD = 54$ mm, find the length of CD.

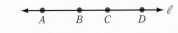

13. Given $QR = 7$ ft and RS is three times the length of QR, find the length of QS.

14. Given $QR = 15$ in. and RS is twice the length of QR, find the length of QS.

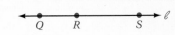

15. Given $EF = 20$ m and FG is $\frac{1}{2}$ the length of EF, find the length of EG.

Solve.

16. Given *EF* = 18 cm and *FG* is $\frac{1}{3}$ the length of *EF*, find the length of *EG*.

17. Given ∠*LOM* = 53° and ∠*LON* = 139°, find the measure of ∠*MON*.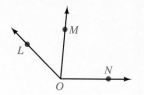

18. Given ∠*MON* = 38° and ∠*LON* = 85°, find the measure of ∠*LOM*.

Find the measure of ∠*x*.

19.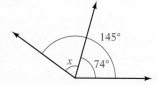

20.

Given that ∠*LON* is a right angle, find the measure of ∠*x*.

21.

22.

23.

24.

Find the measure of ∠*a*.

25.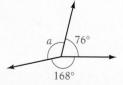

26.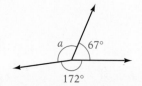

27.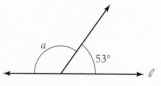

28.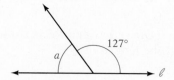

Find *x*.

29.

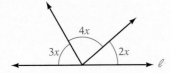

30.

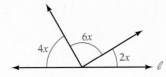

31.

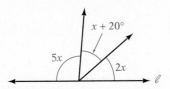

32.

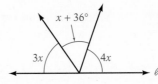

33.

34.

Solve.

35. Given ∠*a* = 51°, find the measure of ∠*b*.

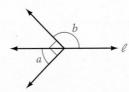

36. Given ∠*a* = 38°, find the measure of ∠*b*.

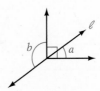

Objective B

Find the measure of ∠*x*.

37.

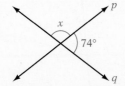

38.

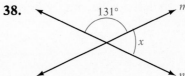

Find *x*.

39.
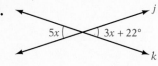
$5x$ $3x + 22°$

40.
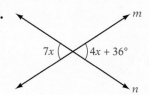
$7x$ $4x + 36°$

Given that $\ell_1 \parallel \ell_2$, find the measures of angles *a* and *b*.

41.
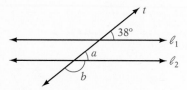
$38°$ a b

42.
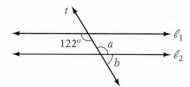
$122°$ a b

43.
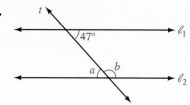
$47°$ a b

44.
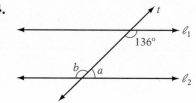
$136°$ b a

Given that $\ell_1 \parallel \ell_2$, find *x*.

45.
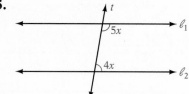
$5x$ $4x$

46.
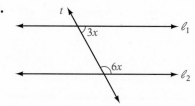
$3x$ $6x$

47.
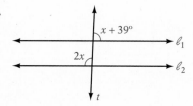
$x + 39°$ $2x$

48.
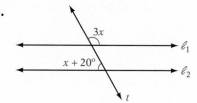
$3x$ $x + 20°$

Objective C

Solve.

49. Given that $\angle a = 95°$ and $\angle b = 70°$, find the measures of angles *x* and *y*.

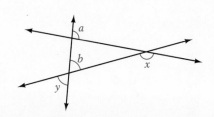

Solve.

50. Given that $\angle a = 35°$ and $\angle b = 55°$, find the measures of angles x and y.

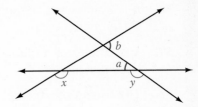

51. Given that $\angle y = 45°$, find the measures of angles a and b.

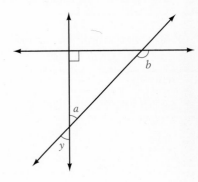

52. Given that $\angle y = 130°$, find the measures of angles a and b.

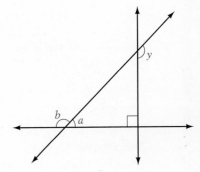

53. Given that $\overline{AO} \perp \overline{OB}$, express in terms of x the number of degrees in $\angle BOC$.

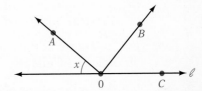

54. Given that $\overline{AO} \perp \overline{OB}$, express in terms of x the number of degrees in $\angle AOC$.

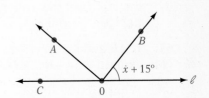

55. One angle in a triangle is a right angle, and one angle is equal to 30°. What is the measure of the third angle?

Solve.

56. A triangle has a 45° angle and a right angle. Find the measure of the third angle.

57. Two angles of a triangle measure 42° and 103°. Find the measure of the third angle.

58. Two angles of a triangle measure 62° and 45°. Find the measure of the third angle.

59. A triangle has a 13° angle and a 65° angle. What is the measure of the third angle?

60. A triangle has a 105° angle and a 32° angle. What is the measure of the third angle?

APPLYING THE CONCEPTS

61. **a.** What is the smallest possible whole number of degrees in an angle of a triangle?
   b. What is the largest possible whole number of degrees in an angle of a triangle?

62. Cut out a triangle and then tear off two of the angles, as shown at the right. Position the pieces you tore off so that angle a is adjacent to angle b and angle c is adjacent to angle b (on the other side). Describe what you observe. What does this demonstrate?

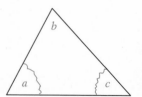

63. Construct a triangle with the given angle measures.
   a. 45°, 45°, and 90°   **b.** 30°, 60°, and 90°   **c.** 40°, 40°, and 100°

64. Determine whether the statement is always true, sometimes true, or never true.
   a. Two lines that are parallel to a third line are parallel to each other.
   b. A triangle contains two acute angles.
   c. Vertical angles are complementary angles.

65. For the figure at the right, find the sum of the measures of angles x, y, and z.

66. For the figure at the right, explain why $\angle a + \angle b = \angle x$.
[W] Write a rule that describes the relationship between an exterior angle of a triangle and the opposite interior angles. Use the rule to write an equation involving angles a, c, and z.

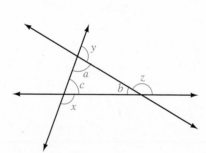

67. If $\overline{AB}$ and $\overline{CD}$ intersect at point O, and $\angle AOC = \angle BOC$,
[W] explain why $\overline{AB} \perp \overline{CD}$.

68. What are the meanings of the words *acute* and *obtuse*
[W] when they are used to describe a person?

69. Do some research on the principle of reflection. Explain
[W] how this principle applies to the operation of a periscope and to the game of billiards.

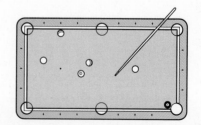

Plane Geometric Figures

Objective A   *To solve problems involving the perimeter of geometric figures* ....

A **polygon** is a closed figure determined by three or more line segments that lie in a plane. The line segments that form the polygon are called its **sides.** The figures below are examples of polygons.

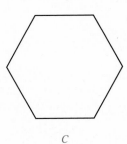
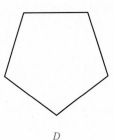
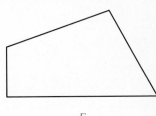

    A   B   C   D   E

POINT OF INTEREST

Although a polygon is defined in terms of its *sides* (see the definition above), the word actually comes from the Latin word *polygonum,* which means "having many *angles.*" This is certainly the case for a polygon.

A **regular polygon** is one in which each side has the same length and each angle has the same measure. The polygons in Figures *A*, *C*, and *D* above are regular polygons.

The name of a polygon is based on the number of its sides. The table below lists the names of polygons that have from 3 to 10 sides.

Number of Sides	Name of the Polygon
3	Triangle
4	Quadrilateral
5	Pentagon
6	Hexagon
7	Heptagon
8	Octagon
9	Nonagon
10	Decagon

Triangles and quadrilaterals are two of the most common types of polygons. Triangles are distinguished by the number of equal sides and also by the measures of their angles.

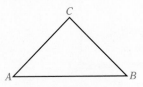
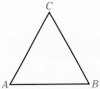
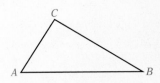

An **isosceles triangle** has two sides of equal length. The angles opposite the equal sides are of equal measure.
$AC = BC$
$\angle A = \angle B$

The three sides of an **equilateral triangle** are of equal length. The three angles are of equal measure.
$AB = BC = AC$
$\angle A = \angle B = \angle C$

A **scalene triangle** has no two sides of equal length. No two angles are of equal measure.

An **acute triangle** has three acute angles.

An **obtuse triangle** has one obtuse angle.

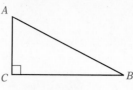

A **right triangle** has a right angle.

CONSIDER THIS

The diagram below shows the relationships among all quadrilaterals. The description of each quadrilateral is within an example of that quadrilateral.

Quadrilaterals are also distinguished by their sides and angles, as shown below. Note that a rectangle, a square, and a rhombus are different forms of a parallelogram.

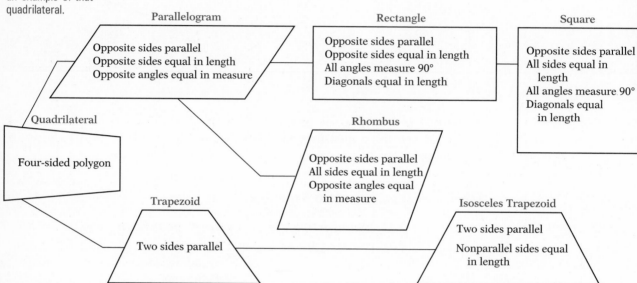

Parallelogram
Opposite sides parallel
Opposite sides equal in length
Opposite angles equal in measure

Rectangle
Opposite sides parallel
Opposite sides equal in length
All angles measure 90°
Diagonals equal in length

Square
Opposite sides parallel
All sides equal in length
All angles measure 90°
Diagonals equal in length

Quadrilateral
Four-sided polygon

Rhombus
Opposite sides parallel
All sides equal in length
Opposite angles equal in measure

Trapezoid
Two sides parallel

Isosceles Trapezoid
Two sides parallel
Nonparallel sides equal in length

The **perimeter** of a plane geometric figure is a measure of the distance around the figure. Perimeter is used in buying fencing for a lawn or determining how much baseboard is needed for a room.

The perimeter of a triangle is the sum of the lengths of the three sides.

> **Perimeter of a Triangle**
>
> Let a, b, and c be the lengths of the sides of a triangle. The perimeter, P, of the triangle is given by $P = a + b + c$.

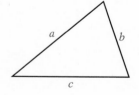

$P = a + b + c$

➡ Find the perimeter of the triangle shown at the right.

$P = 5 + 7 + 10 = 22$

The perimeter is 22 ft.

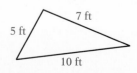

The perimeter of a quadrilateral is the sum of the lengths of its four sides.

A rectangle is a quadrilateral with opposite sides of equal length. Usually the length, *L*, of a rectangle refers to the length of one of the longer sides of the rectangle, and the width, *W*, refers to the length of one of the shorter sides. The perimeter can then be represented $P = L + W + L + W$.

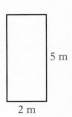

$P = L + W + L + W$

The formula for the perimeter of a rectangle is derived by combining like terms.

$$P = 2L + 2W$$

Perimeter of a Rectangle

Let *L* represent the length and *W* the width of a rectangle. The perimeter, *P*, of the rectangle is given by $P = 2L + 2W$.

➡ Find the perimeter of the rectangle shown at the right.

$P = 2L + 2W$

$P = 2(5) + 2(2)$ • **The length is 5 m. Substitute 5 for *L*. The width is 2 m. Substitute 2 for *W*.**

$P = 10 + 4$ • **Solve for *P*.**

$P = 14$

The perimeter is 14 m.

5 m

2 m

A square is a rectangle in which each side has the same length. Let *s* represent the length of each side of a square. Then the perimeter of a square can be represented $P = s + s + s + s$.

The formula for the perimeter of a square is derived by combining like terms.

$P = s + s + s + s$

$P = 4s$

Perimeter of a Square

Let *s* represent the length of a side of a square. The perimeter, *P*, of the square is given by $P = 4s$.

➡ Find the perimeter of the square shown at the right.

$P = 4s = 4(8) = 32$

The perimeter is 32 in.

8 in.

A **circle** is a plane figure in which all points are the same distance from point O, which is called the **center** of the circle.

The **diameter** of a circle is a line segment across the circle through point O. AB is a diameter of the circle at the right. The variable d is used to designate the diameter of a circle.

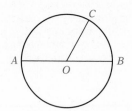

The **radius** of a circle is a line segment from the center of the circle to a point on the circle. OC is a radius of the circle at the right. The variable r is used to designate a radius of a circle.

The length of the diameter is twice the length of the radius.

$$d = 2r \text{ or } r = \frac{1}{2}d$$

The distance around a circle is called the **circumference**. The circumference, C, of a circle is equal to the product of π (pi) and the diameter.

$$C = \pi d$$

Because $d = 2r$, the formula for the circumference can be written in terms of r.

$$C = 2\pi r$$

The Circumference of a Circle

The circumference, C, of a circle with diameter d and radius r is given by $C = \pi d$ or $C = 2\pi r$.

POINT OF INTEREST
Archimedes (c. 287–212 B.C.) was the mathematician who gave us the approximate value of π as $\frac{22}{7} = 3\frac{1}{7}$. He actually showed that π was between $3\frac{10}{71}$ and $3\frac{1}{7}$. The approximation $3\frac{10}{71}$ is closer to the exact value of π, but it is more difficult to use.

The formula for circumference uses the number π, which is an irrational number. The value of π can be approximated by a fraction or by a decimal.

$$\pi \approx \frac{22}{7} \text{ or } \pi \approx 3.14$$

The π key on a scientific calculator gives a closer approximation of π than 3.14. Use a scientific calculator to find approximate values in calculations involving π.

➡ Find the circumference of a circle with a diameter of 6 in.

$C = \pi d$ • The diameter of the circle is given.
$C = \pi(6)$ Use the circumference formula that involves the diameter. $d = 6$.

$C = 6\pi$ • The exact circumference of the circle is 6π in.

$C \approx 18.85$ • An approximate measure is found by using the π key on a calculator.

The circumference is approximately 18.85 in.

Example 1

A carpenter is designing a square patio with a perimeter of 44 ft. What is the length of each side?

Strategy

To find the length of each side, use the formula for the perimeter of a square. Substitute 44 for P and solve for s.

Solution

$P = 4s$
$44 = 4s$
$11 = s$

The length of each side of the patio is 11 ft.

You Try It 1

The infield for a softball field is a square with each side of length 60 ft. Find the perimeter of the infield.

Your strategy

Your solution

Example 2

The dimensions of a triangular sail are 18 ft, 11 ft, and 15 ft. What is the perimeter of the sail?

Strategy

To find the perimeter, use the formula for the perimeter of a triangle. Substitute 18 for a, 11 for b, and 15 for c. Solve for P.

Solution

$P = a + b + c$
$P = 18 + 11 + 15$
$P = 44$

The perimeter of the sail is 44 ft.

You Try It 2

What is the perimeter of a standard piece of typing paper that measures $8\frac{1}{2}$ in. by 11 in.?

Your strategy

Your solution

Example 3

Find the circumference of a circle with a radius of 15 cm. Round to the nearest hundredth.

Strategy

To find the circumference, use the circumference formula that involves the radius. An approximation is asked for; use the π key on a calculator. $r = 15$.

Solution

$C = 2\pi r = 2\pi(15) = 30\pi \approx 94.25$

The circumference is 94.25 cm.

You Try It 3

Find the circumference of a circle with a diameter of 9 in. Give the exact measure.

Your strategy

Your solution

Solutions on p. A15

Objective B   *To solve problems involving the area of geometric figures* ....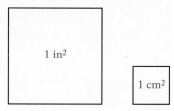

Area is the amount of surface in a region. Area can be used to describe the size of a rug, a parking lot, a farm, or a national park. Area is measured in square units.

A square that measures 1 in. on each side has an area of 1 square inch, written 1 in^2.

A square that measures 1 cm on each side has an area of 1 square centimeter, written 1 cm^2.

Larger areas can be measured in square feet (ft^2), square meters (m^2), square miles (mi^2), acres (43,560 ft^2), or any other square unit.

The area of a geometric figure is the number of squares that are necessary to cover the figure. In the figures below, two rectangles have been drawn and covered with squares. In the figure on the left, 12 squares, each of area 1 cm^2, were used to cover the rectangle. The area of the rectangle is 12 cm^2. In the figure on the right, 6 squares, each of area 1 in^2, were used to cover the rectangle. The area of the rectangle is 6 in^2.

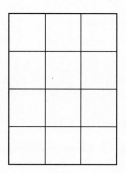

The area of the rectangle is 12 cm^2.

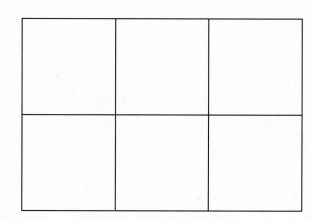

The area of the rectangle is 6 in^2.

Note from the above figures that the area of a rectangle can be found by multiplying the length of the rectangle by its width.

> **Area of a Rectangle**
>
> Let *L* represent the length and *W* the width of a rectangle. The area, *A*, of the rectangle is given by $A = LW$.

➡ Find the area of the rectangle shown at the right.

$A = LW = 11(7) = 77$

The area is 77 m^2.

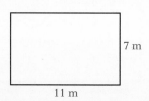

A square is a rectangle in which all sides are the same length. Therefore, both the length and the width of a square can be represented by s, and $A = LW = s \cdot s = s^2$.

Area of a Square

Let s represent the length of a side of a square. The area, A, of the square is given by $A = s^2$.

$A = s \cdot s = s^2$

➡ Find the area of the square shown at the right.

$$A = s^2 = 9^2 = 81$$

The area is 81 mi^2.

9 mi

Figure $ABCD$ is a parallelogram. BC is the **base**, b, of the parallelogram. AE, perpendicular to the base, is the **height**, h, of the parallelogram.

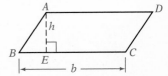

Any side of a parallelogram can be designated as the base. The corresponding height is found by drawing a line segment perpendicular to the base from the opposite side.

A rectangle can be formed from a parallelogram by cutting a right triangle from one end of the parallelogram and attaching it to the other end. The area of the resulting rectangle will equal the area of the original parallelogram.

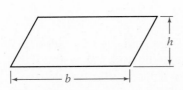

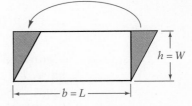

Area of a Parallelogram

Let b represent the length of the base and h the height of a parallelogram. The area, A, of the parallelogram is given by $A = bh$.

➡ Find the area of the parallelogram shown at the right.

$$A = bh = 12 \cdot 6 = 72$$

The area is 72 m^2.

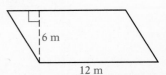

6 m

12 m

Figure *ABC* is a triangle. *AB* is the **base**, *b*, of the triangle. *CD*, perpendicular to the base, is the **height**, *h*, of the triangle.

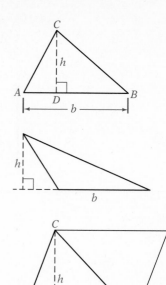

Any side of a triangle can be designated as the base. The corresponding height is found by drawing a line segment perpendicular to the base from the vertex opposite the base.

Consider the triangle with base *b* and height *h* shown at the right. By extending a line from *C* parallel to the base *AB* and equal in length to the base, and extending a line from *B* parallel to *AC* and equal in length to *AC*, a parallelogram is formed. The area of the parallelogram is *bh* and is twice the area of the triangle. Therefore, the area of the triangle is one-half the area of the parallelogram, or $\frac{1}{2}bh$.

Area of a Triangle

Let *b* represent the length of the base and *h* the height of a triangle.

The area, *A*, of the triangle is given by $A = \frac{1}{2}bh$.

➡ Find the area of a triangle with a base of 18 cm and a height of 6 cm.

$$A = \frac{1}{2}bh = \frac{1}{2} \cdot 18 \cdot 6 = 54$$

The area is 54 cm².

Figure *ABCD* is a trapezoid. *AB* is one **base**, b_1, of the trapezoid, and *CD* is the other base, b_2. *AE*, perpendicular to the two bases, is the **height**, *h*.

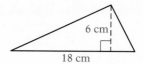

In the trapezoid at the right, the line segment *BD* divides the trapezoid into two triangles, *ABD* and *BCD*. In triangle *ABD*, b_1 is the base and *h* is the height. In triangle *BCD*, b_2 is the base and *h* is the height. The area of the trapezoid is the sum of the areas of the two triangles.

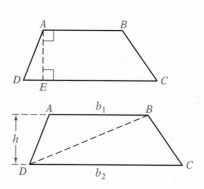

Area of trapezoid *ABCD* = area of triangle *ABD* + area of triangle *BCD*

$$= \frac{1}{2}b_1 h + \frac{1}{2}b_2 h = \frac{1}{2}h(b_1 + b_2)$$

> **Area of a Trapezoid**
>
> Let b_1 and b_2 represent the lengths of the bases and h the height of a trapezoid. The area, A, of the trapezoid is given by
>
> $A = \dfrac{1}{2}h(b_1 + b_2)$.

➡ Find the area of a trapezoid that has bases measuring 15 in. and 5 in. and a height of 8 in.

$$A = \frac{1}{2}h(b_1 + b_2)$$

$$= \frac{1}{2} \cdot 8(15 + 5) = 4(20) = 80$$

The area is 80 in^2.

The area of a circle is equal to the product of π and the square of the radius.

$A = \pi r^2$

> **The Area of a Circle**
>
> The area, A, of a circle with radius r is given by $A = \pi r^2$.

➡ Find the area of a circle that has a radius of 6 cm.

$A = \pi r^2$ • **Use the formula for the area of a**
$A = \pi(6)^2$ **circle. $r = 6$**
$A = \pi(36)$

$A = 36\pi$ • **The exact area of the circle is 36π cm^2.**

$A \approx 113.10$ • **An approximate measure is found by using the π key on a calculator.**

The approximate area of the circle is 113.10 cm^2.

For your reference, all of the formulas for the perimeters and areas of the geometric figures presented in this section are listed in the Chapter Summary on page 187.

Example 4

The Parks and Recreation Department of a city plans to plant grass seed in a playground that has the shape of a trapezoid, as shown below. Each bag of grass seed will seed 1500 ft². How many bags of grass seed should the department purchase?

You Try It 4

An interior designer decides to wallpaper two walls of a room. Each roll of wallpaper will cover 30 ft². Each wall measures 8 ft by 12 ft. How many rolls of wallpaper should be purchased?

Strategy

To find the number of bags to be purchased:

• Use the formula for the area of a trapezoid to find the area of the playground.
• Divide the area of the playground by the area one bag will seed (1500).

Your strategy

Solution

$$A = \frac{1}{2}h(b_1 + b_2)$$

$$A = \frac{1}{2} \cdot 64(80 + 115)$$

$A = 6240$ • The area of the playground is 6240 ft².

$6240 \div 1500 = 4.16$

Because a portion of a fifth bag is needed, 5 bags of grass seed should be purchased.

Your solution

Example 5

Find the area of a circle with a diameter of 5 ft. Give the exact measure.

You Try It 5

Find the area of a circle with a radius of 11 cm. Round to the nearest hundredth.

Strategy

To find the area:

• Find the radius of the circle.
• Use the formula for the area of a circle. Leave the answer in terms of π.

Your strategy

Solution

$$r = \frac{1}{2}d = \frac{1}{2}(5) = 2.5$$

$$A = \pi r^2 = \pi(2.5)^2 = \pi(6.25) = 6.25\pi$$

The area of the circle is 6.25π ft².

Your solution

Solutions on p. A15

3.2 Exercises

. . . .

Objective A

Name each polygon.

1.    **2.**    **3.** 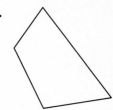   **4.**

Classify the triangle as isosceles, equilateral, or scalene.

5. 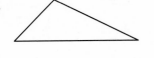   **6.**   **7.** 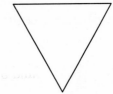   **8.**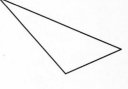

Classify the triangle as acute, obtuse, or right.

9.    **10.** 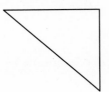   **11.**    **12.**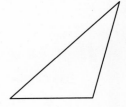

Find the perimeter of the figure.

13.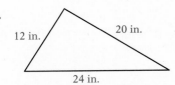

12 in.   20 in.

24 in.

14.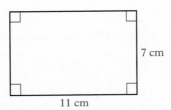

7 cm

11 cm

15.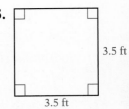

3.5 ft

3.5 ft

16.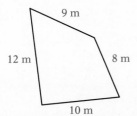

9 m

12 m   8 m

10 m

17.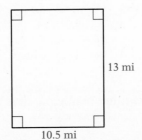

13 mi

10.5 mi

18.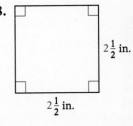

$2\frac{1}{2}$ in.

$2\frac{1}{2}$ in.

Find the circumference of the figure.
Give both the exact value and an approximation to the nearest hundredth.

19.

4 cm

20.

12 m

21.

5.5 mi

22.

18 in.

23.

17 ft

24.

6.6 km

Solve.

25. The lengths of the three sides of a triangle are 3.8 cm, 5.2 cm, and 8.4 cm. Find the perimeter of the triangle.

26. The lengths of the three sides of a triangle are 7.5 m, 6.1 m, and 4.9 m. Find the perimeter of the triangle.

27. The length of each of two sides of an isosceles triangle is $2\frac{1}{2}$ cm. The third side measures 3 cm. Find the perimeter of the triangle.

28. The length of each side of an equilateral triangle is $4\frac{1}{2}$ in. Find the perimeter of the triangle.

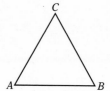

29. A rectangle has a length of 8.5 m and a width of 3.5 m. Find the perimeter of the rectangle.

30. Find the perimeter of a rectangle that has a length of $5\frac{1}{2}$ ft and a width of 4 ft.

31. The length of each side of a square is 12.2 cm. Find the perimeter of the square.

32. Find the perimeter of a square that measures 0.5 m on each side.

33. Find the perimeter of a regular pentagon that measures 3.5 in. on each side.

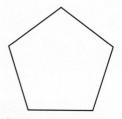

Solve.

34. What is the perimeter of a regular hexagon that measures 8.5 cm on each side?

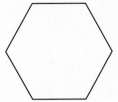

35. The radius of a circle is 4.2 cm. Find the length of a diameter of the circle.

36. The diameter of a circle is 0.56 m. Find the length of a radius of the circle.

37. Find the circumference of a circle that has a diameter of 1.5 in. Give the exact value.

38. The diameter of a circle is 4.2 ft. Find the circumference of the circle. Round to the nearest hundredth.

39. The radius of a circle is 36 cm. Find the circumference of the circle. Round to the nearest hundredth.

40. Find the circumference of a circle that has a radius of 2.5 m. Give the exact value.

41. How many feet of fencing should be purchased for a rectangular garden that is 18 ft long and 12 ft wide?

42. How many meters of binding are required to bind the edge of a rectangular quilt that measures 3.5 m by 8.5 m?

43. Wall-to-wall carpeting is installed in a room that is 12 ft long and 10 ft wide. The edges of the carpet are nailed to the floor. Along how many feet must the carpet be nailed down?

44. The length of a rectangular park is 55 yd. The width is 47 yd. How many yards of fencing are needed to surround the park?

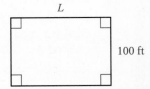

45. The perimeter of a rectangular playground is 440 ft. If the width is 100 ft, what is the length of the playground?

46. A rectangular vegetable garden has a perimeter of 64 ft. The length of the garden is 20 ft. What is the width of the garden?

47. Each of two sides of a triangular banner measures 18 in. If the perimeter of the banner is 46 in., what is the length of the third side of the banner?

48. The perimeter of an equilateral triangle is 13.2 cm. What is the length of each side of the triangle?

Solve.

49. The perimeter of a square picture frame is 48 in. Find the length of each side of the frame.

50. A square rug has a perimeter of 32 ft. Find the length of each edge of the rug.

51. The circumference of a circle is 8 cm. Find the length of a diameter of the circle. Round to the nearest hundredth.

52. The circumference of a circle is 15 in. Find the length of a radius of the circle. Round to the nearest hundredth.

53. Find the length of molding needed to put around a circular table that is 4.2 ft in diameter. Round to the nearest hundredth.

54. How much binding is needed to bind the edge of a circular rug that is 3 m in diameter? Round to the nearest hundredth.

55. A bicycle tire has a diameter of 24 in. How many feet does the bicycle travel when the wheel makes eight revolutions? Round to the nearest hundredth.

24 in.

56. A tricycle tire has a diameter of 12 in. How many feet does the tricycle travel when the wheel makes twelve revolutions? Round to the nearest hundredth.

57. The distance from the surface of Earth to its center is 6356 km. What is the circumference of Earth? Round to the nearest hundredth.

58. Bias binding is to be sewed around the edge of a rectangular tablecloth measuring 72 in. by 45 in. If the bias binding comes in packages containing 15 ft of binding, how many packages of bias binding are needed for the tablecloth?

Objective B

Find the area of the figure.

59.

5 ft

12 ft

60.

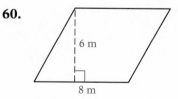

6 m

8 m

61.

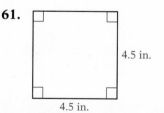

4.5 in.

4.5 in.

Find the area of the figure.

62.

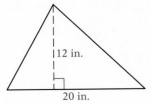

63.

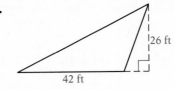

64.

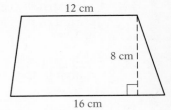

Find the area of the figure.
Give both the exact value and an appxoximation to the nearest hundredth.

65.

66.

67.

68.

69.

70.

Solve.

71. The length of a side of a square is 12.5 cm. Find the area of the square.

72. Each side of a square measure $3\frac{1}{2}$ in. Find the area of the square.

73. The length of a rectangle is 38 in., and the width is 15 in. Find the area of the rectangle.

74. Find the area of a rectangle that has a length of 6.5 m and a width of 3.8 m.

75. The length of the base of a parallelogram is 16 in., and the height is 12 in. Find the area of the parallelogram.

76. The height of a parallelogram is 3.4 m, and the length of the base is 5.2 m. Find the area of the parallelogram.

Solve.

77. The length of the base of a triangle is 6 ft. The height is 4.5 ft. Find the area of the triangle.

78. The height of a triangle is 4.2 cm. The length of the base is 5 cm. Find the area of the triangle.

79. The length of one base of a trapezoid is 35 cm, and the length of the other base is 20 cm. If the height is 12 cm, what is the area of the trapezoid?

80. The height of a trapezoid is 5 in. The bases measure 16 in. and 18 in. Find the area of the trapezoid.

81. The radius of a circle is 5 in. Find the area of the circle. Give the exact value.

82. Find the area of a circle with a radius of 14 m. Round to the nearest hundredth.

83. Find the area of a circle that has a diameter of 3.4 ft. Round to the nearest hundredth.

84. The diameter of a circle is 6.5 m. Find the area of the circle. Give the exact value.

85. The Hale telescope at Mount Palomar, California, has a diameter of 200 in. Find its area. Give the exact value.

86. An irrigation system waters a circular field that has a 50-foot radius. Find the area watered by the irrigation system. Give the exact value.

87. Find the area of a rectangular flower garden that measures 14 ft by 9 ft.

88. What is the area of a square patio that measures 8.5 m on each side?

89. Artificial turf is being used to cover a playing field. If the field is rectangular with a length of 100 yd and a width of 75 yd, how much artificial turf must be purchased to cover the field?

90. A fabric wall hanging is to fill a space that measures 5 m by 3.5 m. Allowing for 0.1 m of the fabric to be folded back along each edge, how much fabric must be purchased for the wall hanging?

91. The area of a rectangle is 300 in². If the length of the rectangle is 30 in., what is the width?

30 in.

W

Solve.

92. The width of a rectangle is 12 ft. If the area is 312 ft^2, what is the length of the rectangle?

93. The height of a triangle is 5 m. The area of the triangle is 50 m^2. Find the length of the base of the triangle.

94. The area of a parallelogram is 42 m^2. If the height of the parallelogram is 7 m, what is the length of the base?

95. You plan to stain the wooden deck attached to your house. The deck measures 10 ft by 8 ft. If a quart of stain will cover 50 ft^2, how many quarts of stain should you buy?

96. You want to tile your kitchen floor. The floor measures 12 ft by 9 ft. How many tiles, each a square with side $1\frac{1}{2}$ ft, should you purchase for the job?

97. You are wallpapering two walls of a child's room, one measuring 9 ft by 8 ft and the other measuring 11 ft by 8 ft. The wallpaper costs $18.50 per roll, and each roll of the wallpaper will cover 40 ft^2. What will it cost to wallpaper the two walls?

98. An urban renewal project involves reseeding a park that is in the shape of a square 60 ft on each side. Each bag of grass seed costs $5.75 and will seed 1200 ft^2. How much money should be budgeted for buying grass seed for the park?

99. A circle has a radius of 8 in. Find the increase in area when the radius is increased by 2 in. Round to the nearest hundredth.

100. A circle has a radius of 6 cm. Find the increase in area when the radius is doubled. Round to the nearest hundredth.

101. You want to install wall-to-wall carpeting in your living room, which measures 15 ft by 24 ft. If the cost of the carpet you would like to purchase is $15.95 per square yard, what will be the cost of the carpeting for your living room? (*Hint:* 9 ft^2 = 1 yd^2)

102. You want to paint the walls of your bedroom. Two walls measure 15 ft by 9 ft, and the other two walls measure 12 ft by 9 ft. The paint you wish to purchase costs $12.98 per gallon, and each gallon will cover 400 ft^2 of wall. Find the total amount you will spend on paint.

103. A walkway 2 m wide surrounds a rectangular plot of grass. The plot is 30 m long and 20 m wide. What is the area of the walkway?

Solve.

104. Pleated draperies for a window must be twice as wide as the width of the window. Draperies are being made for four windows, each 2 ft wide and 4 ft high. Because the drapes will fall slightly below the window sill, and because extra fabric will be needed for hemming the drapes, 1 ft must be added to the height of the window. How much material must be purchased to make the drapes?

APPLYING THE CONCEPTS

105. Find the ratio of the areas of two squares if the ratio of the lengths of their sides is 2:3.

106. If both the length and the width of a rectangle are doubled, how many times larger is the area of the resulting rectangle?

107. A **hexagram** is a six-pointed star, formed by extending each of the sides of a regular hexagon into an equilateral triangle. A hexagram is shown at the right. Use a pencil, paper, protractor, and ruler to create a hexagram.

108. If the formula $C = \pi d$ is solved for π, the resulting equation is $\pi = \dfrac{C}{d}$. Therefore, π is the ratio of the circumference of a circle to the length of its diameter. Use several circular objects, such as coins, plates, tin cans, and wheels, to show that the ratio of the circumference of each object to its diameter is approximately equal to 3.14.

109. Derive a formula for the area of a circle in terms of the diameter of the circle.

110. Determine whether the statement is always true, sometimes true, or never true.
a. If two triangles have the same perimeter, then they have the same area.
b. If two rectangles have the same area, then they have the same perimeter.
c. If two squares have the same area, then the sides of the squares have the same length.
d. An equilateral triangle is also an isosceles triangle.
e. All the radii (plural of radius) of a circle are equal.
f. All the diameters of a circle are equal.

111.
[W] Suppose a circle is cut into 16 equal pieces, which are then arranged as shown at the right. The figure formed resembles a parallelogram. What variable expression could describe the base of the parallelogram? What variable could describe its height? Explain how the formula for the area of a circle is derived from this approach.

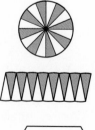

112.
[W] Prepare a report on the history of quilts in America. Find examples of quilt patterns that incorporate regular polygons. Use pieces of cardboard to create the shapes needed for one block of one of the quilt patterns you learned about.

113.
[W] The **apothem** of a regular polygon is the distance from the center of the polygon to a side. Explain how to derive a formula for the area of a regular polygon using the apothem. (*Hint:* Use the formula for the area of a triangle.)

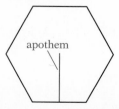

apothem

3.3 Solids

Objective A   *To solve problems involving the volume of a solid*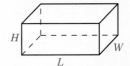

Geometric solids are figures in space. Five common geometric solids are the rectangular solid, the sphere, the cylinder, the cone, and the pyramid.

A **rectangular solid** is one in which all six sides, called **faces**, are rectangles. The variable L is used to represent the length of a rectangular solid, W its width, and H its height.

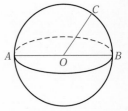

A **sphere** is a solid in which all points are the same distance from point O, which is called the **center** of the sphere. The **diameter**, d, of a sphere is a line across the sphere going through point O. The **radius**, r, is a line from the center to a point on the sphere. AB is a diameter and OC is a radius of the sphere shown at the right.

$$d = 2r \text{ or } r = \frac{1}{2}d$$

The most common cylinder, called a **right circular cylinder,** is one in which the bases are circles and are perpendicular to the height of the cylinder. The variable r is used to represent the radius of a base of a cylinder, and h represents the height. In this text, only right circular cylinders are discussed.

A **right circular cone** is obtained when one base of a right circular cylinder is shrunk to a point, called a **vertex**, V. The variable r is used to represent the radius of the base of the cone, and h represents the height. The variable ℓ is used to represent the **slant height,** which is the distance from a point on the circumference of the base to the vertex. In this text, only right circular cones are discussed.

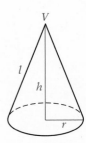

The base of a **regular pyramid** is a regular polygon, and the sides are isosceles triangles. The height, h, is the distance from the vertex, V, to the base and is perpendicular to the base. The variable ℓ is used to represent the **slant height,** which is the height of one of the isosceles triangles on the face of the pyramid. The regular square pyramid at the right has a square base. This is the only type of pyramid discussed in this text.

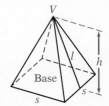

A **cube** is a special type of rectangular solid. Each of the six faces of a cube is a square. The variable *s* is used to represent the length of one side of a cube.

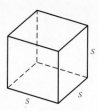

Volume is a measure of the amount of space inside a figure in space. Volume can be used to describe the amount of heating gas used for cooking, the amount of concrete delivered for the foundation of a house, or the amount of water in storage for a city's water supply.

A cube that is 1 ft on each side has a volume of 1 cubic foot, which is written 1 ft^3. A cube that measures 1 cm on each side has a volume of 1 cubic centimeter, which is written 1 cm^3.

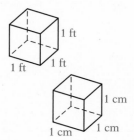

The volume of a solid is the number of cubes that are necessary to exactly fill the solid. The volume of the rectangular solid at the right is 24 cm^3 because it will hold exactly 24 cubes, each 1 cm on a side. Note that the volume can be found by multiplying the length times the width times the height.

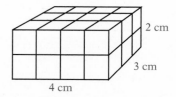

$4 \cdot 3 \cdot 2 = 24$

The formulas for the volumes of the geometric solids described above are given below.

Volumes of Geometric Solids

The volume, *V*, of a **rectangular solid** with length *L*, width *W*, and height *H* is given by $V = LWH$.

The volume, *V*, of a **cube** with side *s* is given by $V = s^3$.

The volume, *V*, of a **sphere** with radius *r* is given by $V = \frac{4}{3}\pi r^3$.

The volume, *V*, of a **right circular cylinder** is given by $V = \pi r^2 h$, where *r* is the radius of the base and *h* is the height.

The volume, *V*, of a **right circular cone** is given by $V = \frac{1}{3}\pi r^2 h$, where *r* is the radius of the circular base and *h* is the height.

The volume, *V*, of a **regular square pyramid** is given by $V = \frac{1}{3}s^2 h$, where *s* is the length of a side of the base and *h* is the height.

➡ Find the volume of a sphere with a diameter of 6 in.

$r = \dfrac{1}{2}d = \dfrac{1}{2}(6) = 3$ ● First find the radius of the sphere.

$V = \dfrac{4}{3}\pi r^3$ ● Use the formula for the volume of a sphere.

$V = \dfrac{4}{3}\pi(3)^3$

$V = \dfrac{4}{3}\pi(27)$

$V = 36\pi$ ● The exact volume of the sphere is 36π in³.

$V \approx 113.10$ ● An approximate measure can be found by using the π key on a calculator.

The approximate volume is 113.10 in³.

Example 1

The length of a rectangular solid is 5 m, the width is 3.2 m, and the height is 4 m. Find the volume of the solid.

Strategy

To find the volume, use the formula for the volume of a rectangular solid. $L = 5$, $W = 3.2$, $H = 4$.

Solution

$V = LWH = 5(3.2)(4) = 64$

The volume of the rectangular solid is 64 m³.

Example 2

The radius of the base of a cone is 8 cm. The height is 12 cm. Find the volume of the cone. Round to the nearest hundredth.

Strategy

To find the volume, use the formula for the volume of a cone. An approximation is asked for; use the π key on a calculator. $r = 8$, $h = 12$.

Solution

$V = \dfrac{1}{3}\pi r^2 h$

$V = \dfrac{1}{3}\pi(8)^2(12) = \dfrac{1}{3}\pi(64)(12) = 256\pi$

≈ 804.25

The volume is approximately 804.25 cm³.

You Try It 1

Find the volume of a cube that measures 2.5 m on a side.

Your strategy

Your solution

You Try It 2

The diameter of the base of a cylinder is 8 ft. The height of the cylinder is 22 ft. Find the exact volume of the cylinder.

Your strategy

Your solution

Solutions on p. A16

Objective B **To solve problems involving the surface area of a solid** ....

The **surface area** of a solid is the total area on the surface of the solid.

When a rectangular solid is cut open and flattened out, each face is a rectangle. The surface area, *SA*, of the rectangular solid is the sum of the areas of the six rectangles:

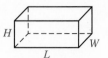

$$SA = LW + LH + WH + LW + WH + LH$$

which simplifies to

$$SA = 2LW + 2LH + 2WH$$

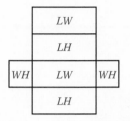

The surface area of a cube is the sum of the areas of the six faces of the cube. The area of each face is s^2. Therefore, the surface area, *SA*, of a cube is given by the formula $SA = 6s^2$.

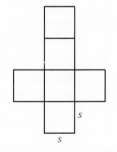

When a cylinder is cut open and flattened out, the top and bottom of the cylinder are circles. The side of the cylinder flattens out to a rectangle. The length of the rectangle is the circumference of the base, which is $2\pi r$; the width is h, the height of the cylinder. Therefore, the area of the rectangle is $2\pi rh$. The surface area, *SA*, of the cylinder is

$$SA = \pi r^2 + 2\pi rh + \pi r^2$$

which simplifies to

$$SA = 2\pi r^2 + 2\pi rh$$

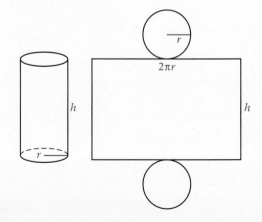

The surface area of a pyramid is the area of the base plus the area of the four isosceles triangles. A side of the square base is s; therefore, the area of the base is s^2. The slant height, ℓ, is the height of each triangle, and s is the base of each triangle. The surface area, SA, of a pyramid is

$$SA = s^2 + 4\left(\frac{1}{2}s\ell\right)$$

which simplifies to

$$SA = s^2 + 2s\ell$$

Formulas for the surface areas of geometric solids are given below.

Surface Areas of Geometric Solids

The surface area, SA, of a **rectangular solid** with length L, width W, and height H is given by $SA = 2LW + 2LH + 2WH$.

The surface area, SA, of a **cube** with side s is given by $SA = 6s^2$.

The surface area, SA, of a **sphere** with radius r is given by $SA = 4\pi r^2$.

The surface area, SA, of a **right circular cylinder** is given by $SA = 2\pi r^2 + 2\pi rh$, where r is the radius of the base and h is the height.

The surface area, SA, of a **right circular cone** is given by $SA = \pi r^2 + \pi r\ell$, where r is the radius of the circular base and ℓ is the slant height.

The surface area, SA, of a **regular pyramid** is given by $SA = s^2 + 2s\ell$, where s is the length of a side of the base and ℓ is the slant height.

➡ Find the surface area of a sphere with a diameter of 18 cm.

$r = \dfrac{1}{2}d = \dfrac{1}{2}(18) = 9$ • First find the radius of the sphere.

$SA = 4\pi r^2$

$SA = 4\pi(9)^2$ • Use the formula for the surface area of a sphere.

$SA = 4\pi(81)$

$SA = 324\pi$ • The exact surface area of the sphere is 324π cm².

$SA \approx 1017.88$ • An approximate measure can be found by using the π key on a calculator.

The approximate surface area is 1017.88 cm².

Example 3

The diameter of the base of a cone is 5 m, and the slant height is 4 m. Find the surface area of the cone. Give the exact measure.

Strategy

To find the surface area of the cone:

• Find the radius of the base of the cone.
• Use the formula for the surface area of a cone. Leave the answer in terms of π.

Solution

$$r = \frac{1}{2}d = \frac{1}{2}(5) = 2.5$$

$$SA = \pi r^2 + \pi r \ell$$
$$SA = \pi (2.5)^2 + \pi (2.5)(4)$$
$$SA = \pi (6.25) + \pi (2.5)(4)$$
$$SA = 6.25\pi + 10\pi$$
$$SA = 16.25\pi$$

The surface area of the cone is 16.25π m^2.

Example 4

Find the area of a label used to cover a soup can that has a radius of 4 cm and a height of 12 cm. Round to the nearest hundredth.

Strategy

To find the area of the label, use the fact that the surface area of the sides of a cylinder is given by $2\pi rh$. An approximation is asked for; use the π key on a calculator. $r = 4$, $h = 12$.

Solution

Area of the label $= 2\pi rh$
Area of the label $= 2\pi (4)(12) = 96\pi$
≈ 301.59

The area is approximately 301.59 cm^2.

You Try It 3

The diameter of the base of a cylinder is 6 ft, and the height is 8 ft. Find the surface area of the cylinder. Round to the nearest hundredth.

Your strategy

Your solution

You Try It 4

Which has a larger surface area, a cube with a side measuring 10 cm or a sphere with a diameter measuring 8 cm?

Your strategy

Your solution

3.3 Exercises

· ·

Objective A

Find the volume of the figure. For calculations involving π, give both the exact value and an approximation to the nearest hundredth.

1.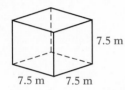
6 in.
14 in.   10 in.

2.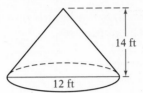
14 ft
12 ft

3.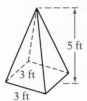
5 ft
3 ft
3 ft

4.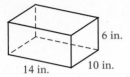
7.5 m
7.5 m   7.5 m

5.
3 cm

6.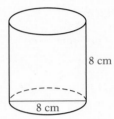
8 cm
8 cm

Solve.

7. A rectangular solid has a length of 6.8 m, a width of 2.5 m, and a height of 2 m. Find the volume of the solid.

8. Find the volume of a rectangular solid that has a length of 4.5 ft, a width of 3 ft, and a height of 1.5 ft.

9. Find the volume of a cube whose side measures 2.5 in.

10. The length of a side of a cube is 7 cm. Find the volume of the cube.

11. The diameter of a sphere is 6 ft. Find the volume of the sphere. Give the exact measure.

12. Find the volume of a sphere that has a radius of 1.2 m. Round to the nearest tenth.

13. The diameter of the base of a cylinder is 24 cm. The height of the cylinder is 18 cm. Find the volume of the cylinder. Round to the nearest hundredth.

14. The height of a cylinder is 7.2 m. The radius of the base is 4 m. Find the volume of the cylinder. Give the exact measure.

15. The radius of the base of a cone is 5 in. The height of the cone is 9 in. Find the volume of the cone. Give the exact measure.

Solve.

16. The height of a cone is 15 cm. The diameter of the cone is 10 cm. Find the volume of the cone. Round to the nearest hundredth.

17. The length of a side of the base of a pyramid is 6 in., and the height is 10 in. Find the volume of the pyramid.

18. The height of a pyramid is 8 m, and the length of a side of the base is 9 m. What is the volume of the pyramid?

19. The volume of a freezer with a length of 7 ft and a height of 3 ft is 52.5 ft³. Find the width of the freezer.

20. The length of an aquarium is 18 in., and the width is 12 in. If the volume of the aquarium is 1836 in³, what is the height of the aquarium?

21. The volume of a cylinder with a height of 10 in. is 502.4 in³. Find the radius of the base of the cylinder. Round to the nearest hundredth.

22. The diameter of the base of a cylinder is 14 cm. If the volume of the cylinder is 2310 cm³, find the height of the cylinder. Round to the nearest hundredth.

23. A rectangular solid has a square base and a height of 5 in. If the volume of the solid is 125 in³, find the length and the width.

24. The volume of a rectangular solid is 864 m³. The rectangular solid has a square base and a height of 6 m. Find the dimensions of the solid.

25. An oil storage tank, which is in the shape of a cylinder, is 4 m high and has a diameter of 6 m. The oil tank is two-thirds full. Find the number of cubic meters of oil in the tank. Round to the nearest hundredth.

26. A silo, which is in the shape of a cylinder, is 16 ft in diameter and has a height of 30 ft. The silo is three-fourths full. Find the volume of the portion of the silo that is not being used for storage. Round to the nearest hundredth.

Objective B

Find the surface area of the figure.

27.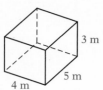

3 m
5 m
4 m

28.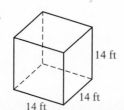

14 ft
14 ft
14 ft

29.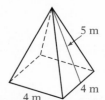

5 m
4 m
4 m

Find the surface area of the figure. Give both the exact value and an approximation to the nearest hundredth.

30.

2 cm

31.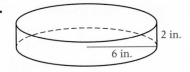

2 in.

6 in.

32.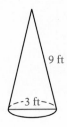

9 ft

3 ft

Solve.

33. The height of a rectangular solid is 5 ft. The length is 8 ft, and the width is 4 ft. Find the surface area of the solid.

34. The width of a rectangular solid is 32 cm. The length is 60 cm, and the height is 14 cm. What is the surface area of the solid?

35. The side of a cube measures 3.4 m. Find the surface area of the cube.

36. Find the surface area of a cube that has a side measuring 1.5 in.

37. Find the surface area of a sphere with a diameter of 15 cm. Give the exact value.

38. The radius of a sphere is 2 in. Find the surface area of the sphere. Round to the nearest hundredth.

39. The radius of the base of a cylinder is 4 in. The height of the cylinder is 12 in. Find the surface area of the cylinder. Round to the nearest hundredth.

40. The diameter of the base of a cylinder is 1.8 m. The height of the cylinder is 0.7 m. Find the surface area of the cylinder. Give the exact value.

41. The slant height of a cone is 2.5 ft. The radius of the base is 1.5 ft. Find the surface area of the cone. Give the exact value.

42. The diameter of the base of a cone is 21 in. The slant height is 16 in. What is the surface area of the cone? Round to the nearest hundredth.

43. The length of a side of the base of a pyramid is 9 in., and the slant height is 12 in. Find the surface area of the pyramid.

44. The slant height of a pyramid is 18 m, and the length of a side of the base is 16 m. What is the surface area of the pyramid?

Solve.

45. The surface area of a rectangular solid is 108 cm². The height of the solid is 4 cm, and the length is 6 cm. Find the width of the rectangular solid.

46. The length of a rectangular solid is 12 ft. The width is 3 ft. If the surface area is 162 ft², find the height of the rectangular solid.

47. A can of paint will cover 300 ft². How many cans of paint should be purchased in order to paint a cylinder that has a height of 30 ft and a radius of 12 ft?

48. A hot air balloon is in the shape of a sphere. Approximately how much fabric was used to construct the balloon if its diameter is 32 ft? Round to the nearest whole number.

49. How much glass is needed to make a fish tank that is 12 in. long, 8 in. wide, and 9 in. high? The fish tank is open at the top.

50. Find the area of a label used to cover a can of juice that has a diameter of 16.5 cm and a height of 17 cm. Round to the nearest hundredth.

51. The length of a side of the base of a pyramid is 5 cm, and the slant height is 8 cm. How much larger is the surface area of this pyramid than the surface area of a cone with a diameter of 5 cm and a slant height of 8 cm? Round to the nearest hundredth.

APPLYING THE CONCEPTS

52. Half of a sphere is called a **hemisphere**. Derive formulas for the volume and surface area of a hemisphere.

53. Determine whether the statement is always true, sometimes true, or never true.
  a. The slant height of a regular pyramid is longer than the height.
  b. The slant height of a cone is shorter than the height.
  c. The four triangular faces of a regular pyramid are equilateral triangles.

54. **a.** What is the effect on the surface area of a rectangular solid when the width and height are doubled?
  b. What is the effect on the volume of a rectangular solid when both the length and the width are doubled?
  c. What is the effect on the volume of a cube when the length of each side of the cube is doubled?
  d. What is the effect on the surface area of a cylinder when the radius and height are doubled?

55.
[W] Explain how you could cut through a cube so that the face of the resulting solid is
  a. a square
  b. an equilateral triangle
  c. a trapezoid
  d. a hexagon

56.
[W] Prepare a report on the rods and cones at the back of the retina. Describe the differences between the ways in which rods and cones perceive color.

Projects in Mathematics

Lines of Symmetry

Look at the letter A printed at the left. If the letter were folded along line ℓ, the two sides of the letter would match exactly. This letter has **symmetry** with respect to line ℓ. Line ℓ is called the **axis of symmetry.**

Now consider the letter H printed below at the left. Both line ℓ_1 and line ℓ_2 are axes of symmetry for this letter; the letter could be folded along either line and the two sides would match exactly. Does the letter A have more than one axis of symmetry? Find axes of symmetry for other capital letters of the alphabet. Which lowercase letters have one axis of symmetry? Do any of the lowercase letters have more than one axis of symmetry?

Find the number of axes of symmetry for each of the plane geometric figures presented in this chapter.

There are other types of symmetry. Look up the meaning of point symmetry and rotational symmetry. Which plane geometric figures provide examples of these types of symmetry?

Find examples of symmetry in nature, art, and architecture.

Chapter Summary

Key Words

A *line* is determined by two distinct points and extends indefinitely in both directions. A *line segment* is part of a line that has two endpoints. *Parallel lines* never meet; the distance between them is always the same. *Perpendicular lines* are intersecting lines that form right angles.

A *ray* starts at a point and extends indefinitely in one direction. The point at which a ray starts is the *endpoint* of the ray. An *angle* is formed by two rays with the same endpoint. The *vertex* of an angle is the point at which the two rays meet. An angle is measured in *degrees*. A 90° angle is a *right angle*. A 180° angle is a *straight angle*. An *acute angle* is an angle whose measure is between 0° and 90°. An *obtuse angle* is an angle whose measure is between 90° and 180°. *Complementary angles* are two angles whose measures have the sum 90°. *Supplementary angles* are two angles whose measures have the sum 180°.

Two angles that are on opposite sides of the intersection of two lines are *vertical angles;* vertical angles have the same measure. Two angles that share a common side are *adjacent angles;* adjacent angles of intersecting lines are supplementary angles.

A line that intersects two other lines at two different points is a *transversal.* If the lines cut by a transversal are parallel lines, equal angles are formed: *alternate interior angles, alternate exterior angles,* and *corresponding angles.*

A *polygon* is a closed figure determined by three or more line segments. The line segments that form the polygon are its *sides.* A *regular polygon* is one in which each side has the same length and each angle has the same measure. Polygons are classified by the number of sides.

A *triangle* is a plane figure formed by three line segments. An *isosceles triangle* has two sides of equal length. The three sides of an *equilateral triangle* are of equal length. A *scalene triangle* has no two sides of equal length. An *acute triangle* has three acute angles. An *obtuse triangle* has one obtuse angle. A *right triangle* has a right angle.

A *quadrilateral* is a four-sided polygon. A parallelogram, a rectangle, a square, a rhombus, and a trapezoid are all quadrilaterals.

A *circle* is a plane figure in which all points are the same distance from the center of the circle. A *diameter* of a circle is a line segment across the circle through the center. A *radius* of a circle is a line segment from the center of the circle to a point on the circle.

The *perimeter* of a plane geometric figure is a measure of the distance around the figure. The distance around a circle is called the *circumference*. *Area* is the amount of surface in a region. *Volume* is a measure of the amount of space inside a figure in space. The *surface area* of a solid is the total area on the surface of the solid.

Essential Rules

Triangles
Sum of the measures of the interior angles = 180°
Sum of an interior and corresponding exterior angle = 180°

Perimeter

Triangle:	$P = a + b + c$
Rectangle:	$P = 2L + 2W$
Square:	$P = 4s$
Circle:	$C = \pi d$ or $C = 2\pi r$

Area

Triangle:	$A = \dfrac{1}{2}bh$
Rectangle:	$A = LW$
Square:	$A = s^2$
Circle:	$A = \pi r^2$
Parallelogram:	$A = bh$
Trapezoid:	$A = \dfrac{1}{2}h(b_1 + b_2)$

Volume

Rectangular solid:	$V = LWH$
Cube:	$V = s^3$
Sphere:	$V = \dfrac{4}{3}\pi r^3$
Right circular cylinder:	$V = \pi r^2 h$
Right circular cone:	$V = \dfrac{1}{3}\pi r^2 h$
Regular pyramid:	$V = \dfrac{1}{3}s^2 h$

Surface Area

Rectangular solid:	$SA = 2LW + 2LH + 2WH$
Cube:	$SA = 6s^2$
Sphere:	$SA = 4\pi r^2$
Right circular cylinder:	$SA = 2\pi r^2 + 2\pi rh$
Right circular cone:	$SA = \pi r^2 + \pi r\ell$
Regular pyramid:	$SA = s^2 + 2s\ell$

Chapter Review Exercises

1. Given that $\angle a = 74°$ and $\angle b = 52°$, find the measures of angles x and y.

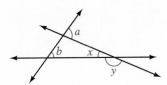

2. Find the measure of $\angle x$.

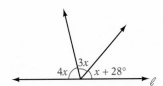

3. Given that $BC = 11$ cm and AB is three times the length of BC, find the length of AC.

4. Find x.

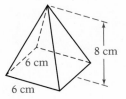

5. Find the volume of the figure.

8 cm
6 cm
6 cm

6. Given that $\ell_1 \| \ell_2$, find the measures of angles a and b.

138° a ℓ_1
b ℓ_2
t

7. Find the surface area of the figure.

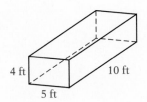

4 ft   10 ft
5 ft

8. Find the supplement of a 32° angle.

9. Determine the area of a rectangle with a length of 12 cm and a width of 6.5 cm.

10. Determine the area of a triangle whose base is 9 m and whose height is 14 m.

11. Find the volume of a rectangular solid with a length of 6.5 ft, a width of 2 ft, and a height of 3 ft.

12. Two angles of a triangle measure 37° and 48°. Find the measure of the third angle.

13. The height of a triangle is 7 cm. The area of the triangle is 28 cm². Find the length of the base of the triangle.

14. Find the volume of a sphere that has a diameter of 12 mm. Find the exact value.

15. Determine the exact volume of a right circular cone whose radius is 7 cm and whose height is 16 cm.

16. The perimeter of a square picture frame is 86 cm. Find the length of each side of the frame.

17. A can of paint will cover 200 ft². How many cans of paint should be purchased in order to paint a cylinder that has a height of 15 ft and a radius of 6 ft?

18. The length of a rectangular park is 56 yd. The width is 48 yd. How many yards of fencing are needed to surround the park?

19. What is the area of a square patio that measures 9.5 m on each side?

20. A walkway 2 m wide surrounds a rectangular plot of grass. The plot is 40 m long and 25 m wide. What is the area of the walkway?

Cumulative Review Exercises

1. Let $x \in \{-3, 0, 1\}$. For what values of x is the inequality $x \leq 1$ a true statement?

2. Write 8.9% as a decimal.

3. Write $\dfrac{7}{20}$ as a percent.

4. Divide: $-\dfrac{4}{9} \div \dfrac{2}{3}$

5. Multiply: $5.7(-4.3)$

6. Simplify: $-\sqrt{125}$

7. Evaluate $5 - 3[10 + (5 - 6)^2]$.

8. Evaluate $a(b - c)^3$ when $a = -1$, $b = -2$, and $c = -4$.

9. Simplify: $5m + 3n - 8m$

10. Simplify: $-7(-3y)$

11. Simplify: $4(3x + 2) - (5x - 1)$

12. Use the roster method to write the set of negative integers greater than or equal to -2.

13. Find $C \cup D$, given $C = \{0, 10, 20, 30\}$ and $D = \{-10, 0, 10\}$.

14. Graph: $x > -2$

15. Solve: $4x + 2 = 6x - 8$

16. Solve: $3(2x + 5) = 18$

17. Solve: $4y - 3 \geq 6y + 5$

18. Solve: $8 - 4(3x + 5) \leq 6(x - 8)$

19.   Solve: $2x - 3 > 5$ or $x + 4 < 1$

20.   Solve: $-3 \le 2x - 7 \le 5$

21.   Solve: $|3x - 1| = 2$

22.   Solve: $|x - 8| \le 2$

23.   Find the measure of $\angle x$.

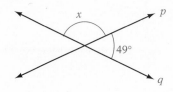

24.   Translate "the difference between four times a number and ten is two" into an equation and solve.

25.   Two angles of a triangle measure 37° and 21°. Find the measure of the third angle of the triangle.

26.   An engineering consultant invested $14,000 in a 5.5% annual simple interest account. How much additional money is deposited in an account that pays 9.5% annual simple interest if the total interest earned on both accounts is $1245?

27.   Two sides of an isosceles triangle measure 7.5 m. The perimeter of the triangle is 19.5 m. Find the measure of the third side of the triangle.

28.   Of the 375 students in a high school graduating class, 300 students went on to college. What percent of the students in the class went on to college?

29.   Find the exact area of a circle that has a diameter of 9 cm.

30.   The volume of a box is 144 ft³. The length of the box is 12 ft, and the width is 4 ft. Find the height of the box.

Chapter

4

Linear Equations and Inequalities in Two Variables

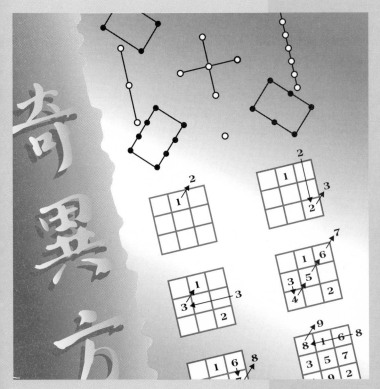

Objectives

Magic Squares

A magic square is a square array of distinct integers so arranged that the numbers along every row, column, and main diagonal have the same sum. An example of a magic square is shown at the right.

8	3	4
1	5	9
6	7	2

The oldest known example of a magic square comes from China. Estimates are that this magic square is over 4000 years old. It is shown at the left.

There is a simple way to produce a magic square with an odd number of cells. Start by writing a 1 in the top middle cell. The rule then is to proceed diagonally upward to the right with the successive integers.

When the rule takes you outside the square, write the number by shifting either across the square from right to left or down the square from top to bottom, as the case may be. For example, in Fig. B the second number (2) is outside the square above a column. Because the 2 is above a column, it should be shifted down to the bottom cell in that column. In Fig. C, the 3 is outside the square to the right of a column and should therefore be shifted all the way to the left.

If the rule takes you to a square that is already filled (as shown in Fig. D), then write the number in the cell directly below the last number written. Continue until the entire square is filled.

It is possible to begin a magic square with any integer and proceed by using the above rule and consecutive integers.

For an odd magic square beginning with 1, the sum of a row, column or diagonal is $\frac{n(n^2 + 1)}{2}$, where n is the number of rows.

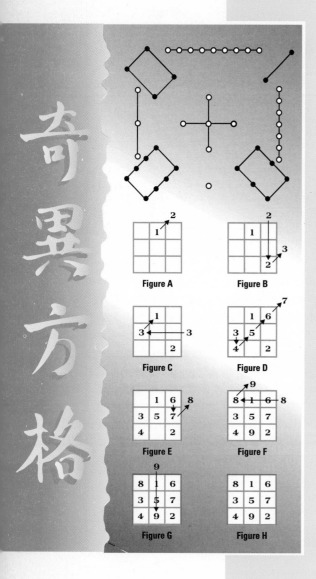

Figure A

Figure B

Figure C

Figure D

Figure E

Figure F

Figure G

Figure H

The Rectangular Coordinate System

Objective A *To graph points in a rectangular coordinate system* ....

Before the 15th century, geometry and algebra were considered separate branches of mathematics. That all changed when René Descartes, a French mathematician who lived from 1596 to 1650, founded **analytic geometry.** In this geometry, a *coordinate system* is used to study relationships between variables.

A **rectangular coordinate system** is formed by two number lines, one horizontal and one vertical, that intersect at the zero point of each line. The point of intersection is called the **origin.** The two lines are called **coordinate axes,** or simply **axes.** Generally, the horizontal axis is labeled the *x*-axis and the vertical axis is labeled the *y*-axis.

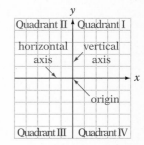

The axes determine a **plane,** which can be thought of as a large, flat sheet of paper. The two axes divide the plane into four regions called **quadrants,** which are numbered counterclockwise from I to IV starting from the upper right.

Each point in the plane can be identified by a pair of numbers called an **ordered pair.** The first number of the ordered pair measures a horizontal distance and is called the **abscissa,** or *x*-**coordinate.** The second number of the pair measures a vertical distance and is called the **ordinate,** or *y*-**coordinate.** The ordered pair (x, y) associated with a point is also called the **coordinates** of the point.

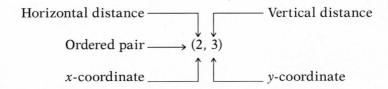

To **graph** or **plot** a point in the plane, place a dot at the location given by the ordered pair. The **graph of an ordered pair** is the dot drawn at the coordinates of the point in the plane. The points whose coordinates are $(3, 4)$ and $(-2.5, -3)$ are graphed in the figures below.

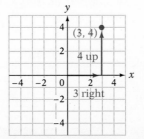

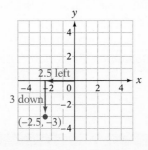

CONSIDER THIS

This is very important. An **ordered pair** is a *pair* of coordinates, and the *order* in which the coordinates appear is important.

The points whose coordinates are (3, −1) and (−1, 3) are graphed at the right. Note that the graphed points are in different locations. *The order of the coordinates of an ordered pair is important.*

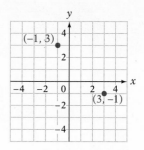

Each point in the plane is associated with an ordered pair, and each ordered pair is associated with a point in the plane. Although only the labels for integers are given on a coordinate grid, the graph of any ordered pair can be approximated. For example, the points whose coordinates are (−2.3, 4.1) and (π, 1) are shown on the graph at the right.

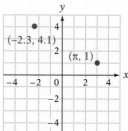

Example 1 Graph the ordered pairs (−2, −3), (3, −2), (0, −2), and (3, 0).

Solution

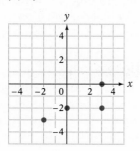

You Try It 1 Graph the ordered pairs (−4, 1), (3, −3), (0, 4), and (−3, 0).

Your solution

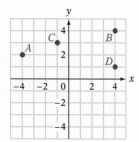

Example 2 Give the coordinates of each of the points.

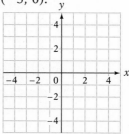

Solution The coordinates of *A* are (−4, 2).
The coordinates of *B* are (4, 4).
The coordinates of *C* are (−1, 3).
The coordinates of *D* are (4, 1).

You Try It 2 Give the coordinates of each of the points.

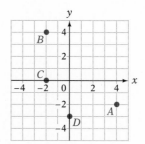

Your solution

Solutions on p. A17

Objective B   *To determine ordered-pair solutions of an equation in two variables* ....

A coordinate system is used to study the relationship between two variables. Frequently this relationship is given by an equation. Examples of equations in two variables include

$$y = 2x - 3 \qquad 3x + 2y = 6 \qquad x^2 - y = 0$$

A **solution of an equation in two variables** is an ordered pair (x, y) whose coordinates make the equation a true statement.

CONSIDER THIS

An ordered pair is of the form (x, y). For the ordered pair $(-3, 7)$, -3 is the x value and 7 is the y value. Substitute -3 for x and 7 for y.

➡ Is the ordered pair $(-3, 7)$ a solution of the equation $y = -2x + 1$?

$$\begin{array}{c|l} y = & -2x + 1 \\ \hline 7 & -2(-3) + 1 \\ 7 & 6 + 1 \\ 7 = 7 \end{array}$$

- Replace x by -3 and y by 7.
- Simplify.
- Compare the results. If the resulting equation is a true statement, the ordered pair is a solution of the equation. If it is not a true statement, the ordered pair is not a solution of the equation.

Yes, the ordered pair $(-3, 7)$ is a solution of the equation.

Besides $(-3, 7)$, there are many other ordered-pair solutions of $y = -2x + 1$. For example, $(0, 1)$, $\left(-\dfrac{3}{2}, 4\right)$, and $(4, -7)$ are also solutions. In general, an equation in two variables has an infinite number of solutions. By choosing any value of x and substituting that value into the equation, we can calculate a corresponding value of y.

➡ Find the ordered-pair solution of $y = \dfrac{2}{3}x - 3$ that corresponds to $x = 6$.

$$y = \frac{2}{3}x - 3$$

$$= \frac{2}{3}(6) - 3 \qquad \text{• Replace } x \text{ by 6.}$$

$$= 4 - 3 \qquad\qquad \text{• Solve for } y.$$

$$= 1$$

The ordered-pair solution is $(6, 1)$.

The solution of an equation in two variables can be graphed in an xy-coordinate system.

➡ Graph the ordered-pair solutions of $y = -2x + 1$ when $x = -2, -1, 0, 1,$ and 2.

Use the values of x to determine ordered-pair solutions of the equation. It is convenient to record these in a table.

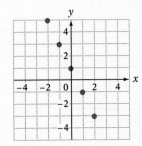

x	$y = -2x + 1$	y	(x, y)
-2	$-2(-2) + 1$	5	$(-2, 5)$
-1	$-2(-1) + 1$	3	$(-1, 3)$
0	$-2(0) + 1$	1	$(0, 1)$
1	$-2(1) + 1$	-1	$(1, -1)$
2	$-2(2) + 1$	-3	$(2, -3)$

Example 3

Is $(3, -2)$ a solution of $3x - 4y = 15$?

Solution

$$3x - 4y = 15$$

$3(3) - 4(-2)$	15
$9 + 8$	
	$17 \neq 15$

• Replace x by 3 and y by -2.

No, $(3, -2)$ is not a solution of $3x - 4y = 15$.

Example 4

Find the ordered-pair solution of $y = \dfrac{x}{x - 2}$ corresponding to $x = 4$.

Solution

Replace x by 4 and solve for y.

$$y = \frac{x}{x - 2} = \frac{4}{4 - 2} = \frac{4}{2} = 2$$

The ordered-pair solution is $(4, 2)$.

Example 5

Graph the ordered-pair solutions of $y = \dfrac{2}{3}x - 2$ when $x = -3, 0, 3, 6$.

Solution

Replace x in $y = \dfrac{2}{3}x - 2$ by $-3, 0, 3$, and 6. For each value of x, determine the value of y.

x	$y = \dfrac{2}{3}x - 2$	y	(x, y)
-3	$\dfrac{2}{3}(-3) - 2$	-4	$(-3, -4)$
0	$\dfrac{2}{3}(0) - 2$	-2	$(0, -2)$
3	$\dfrac{2}{3}(3) - 2$	0	$(3, 0)$
6	$\dfrac{2}{3}(6) - 2$	2	$(6, 2)$

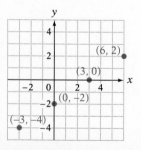

You Try It 3

Is $(-2, 4)$ a solution of $x - 3y = -14$?

Your solution

You Try It 4

Find the ordered-pair solution of $y = \dfrac{3x}{x + 1}$ corresponding to $x = -2$.

Your solution

You Try It 5

Graph the ordered-pair solutions of $y = -\dfrac{1}{2}x + 2$ when $x = -4, -2, 0, 2$.

Your solution

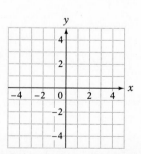

Solutions on p. A17

Objective C   **To determine whether a set of ordered pairs is a function** ....

Discovering a relationship between two variables is an important task in the application of mathematics. Here are some examples.

- Botanists study the relationship between the number of bushels of wheat yielded per acre and the amount of watering per acre.
- Environmental scientists study the relationship between the incidence of skin cancer and the amount of ozone in the atmosphere.
- Business analysts study the relationship between the price of a product and the number of products that are sold at that price.

Each of these relationships can be described by a set of ordered pairs.

Definition of a Relation

A **relation** is any set of ordered pairs.

The following table shows the number of hours that each of 9 students spent studying for a midterm exam and the grade that each of these 9 students received.

Hours	3	3.5	2.75	2	4	4.5	3	2.5	5
Grade	78	75	70	65	85	85	80	75	90

This information can be written as the relation

{(3, 78), (3.5, 75), (2.75, 70), (2, 65), (4, 85), (4.5, 85), (3, 80), (2.5, 75), (5, 90)}

where the first coordinate of the ordered pair is the hours spent studying and the second coordinate is the score on the midterm.

The **domain** of a relation is the set of first coordinates of the ordered pairs; the **range** is the set of second coordinates. For the relation above,

Domain = {2, 2.5, 2.75, 3, 3.5, 4, 4.5, 5}   Range = {65, 70, 75, 78, 80, 85, 90}

The **graph of a relation** is the graph of the ordered pairs that belong to the relation. The graph of the relation given above is shown at the right. The horizontal axis represents the hours spent studying (the domain); the vertical axis represents the test score (the range). The axes could be labeled *H* for hours studied and *S* for test score.

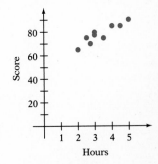

A *function* is a special type of relation for which no two ordered pairs have the same first coordinate.

Definition of a Function

A **function** is a relation in which no two ordered pairs that have the same first coordinate have different second coordinates.

The table at the right is the grading scale for a 50-point test. This table defines a relationship between the *score* on the test and a *letter grade*. Some of the ordered pairs of this function are (38, C), (47, A), (44, B), and (42, B).

Score	Grade
46–50	A
40–45	B
33–39	C
28–32	D
0–27	F

The grading-scale table defines a function, because no two ordered pairs can have the *same* first coordinate and *different* second coordinates. For instance, it is not possible to have the ordered pairs (32, C) and (32, B)—same first coordinate (test score) but different second coordinate (test grade). The domain of this function is {0, 1, 2, . . . , 49, 50}. The range is {A, B, C, D, F}.

The example of hours spent studying and test score given earlier is *not* a function, because (3, 78) and (3, 80) are ordered pairs of the relation that have the *same* first coordinate but *different* second coordinates.

Consider again the grading-scale example. Note that (44, B) and (42, B) are ordered pairs of the function. Ordered pairs of a function may have the same *second* coordinates but not the same first coordinates.

Although relations and functions are given by tables, they are frequently given by an equation in two variables.

The equation $y = 2x$ expresses the relationship between a number, x, and twice the number, y. For instance, if $x = 3$, then $y = 6$, which is twice 3. To indicate exactly which ordered pairs are determined by the equation, the domain (values of x) is specified. If $x \in \{-2, -1, 0, 1, 2\}$, then the ordered pairs determined by the equation are {(−2, −4), (−1, −2), (0, 0), (1, 2), (2, 4)}. This relation is a function because no two ordered pairs have the same first coordinate.

The graph of the function is shown at the right. The horizontal axis (domain) is labeled x; the vertical axis (range) is labeled y.

The domain {−2, −1, 0, 1, 2} was chosen arbitrarily. Other domains could have been selected. The type of application usually influences the choice of the domain.

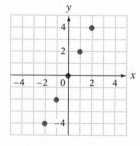

For the equation $y = 2x$, we say that "y is a function of x" because the set of ordered pairs is a function.

Not all equations, however, define a function. For instance, the equation $|y| = x + 2$ does not define y as a function of x. The ordered pairs (2, 4) and (2, −4) both satisfy the equation. Thus there are two ordered pairs with the same first coordinate but different second coordinates.

Example 6

To test a heart medicine, Dr. Nabib measures the heart rate of 5 patients before and after they take the medicine. The results are given in the table below.

| Before medicine | 85 | 80 | 85 | 75 | 90 |
| After medicine | 75 | 70 | 70 | 80 | 80 |

Write a relation where the first coordinate is the heart rate before taking the medicine and the second coordinate is the heart rate after taking the medicine. Is the relation a function?

Solution

$\{(85, 75), (80, 70), (85, 70), (75, 80), (90, 80)\}$

The two ordered pairs (85, 75) and (85, 70) have the same first coordinate but different second coordinates.

No, the relation is not a function.

You Try It 6

Professor Kaplan records the midterm and final exam scores for 5 graduate students. The results are given in the table below.

| Midterm | 82 | 78 | 81 | 87 | 81 |
| Final Exam | 91 | 86 | 96 | 79 | 87 |

Write a relation where the first coordinate is the score on the midterm and the second coordinate is the score on the final exam. Is the relation a function?

Your solution

Example 7

Does $y = x^2 + 3$, where $x \in \{-2, -1, 1, 3\}$, define y as a function of x?

Solution

Determine the ordered pairs defined by the equation. Replace x in $y = x^2 + 3$ by the given values and solve for y.

$\{(-2, 7), (-1, 4), (1, 4), (3, 12)\}$

No two ordered pairs have the same first coordinate. Therefore, the relation is a function, and the equation $y = x^2 + 3$ defines y as a function of x.

Note that $(-1, 4)$ and $(1, 4)$ are ordered pairs that belong to this function. Ordered pairs of a function may have the same *second* coordinates but not the same *first* coordinates.

You Try It 7

Does $y = \frac{1}{2}x + 1$, where $x \in \{-4, 0, 2\}$, define y as a function of x?

Your solution

Solutions on p. A17

Objective D *To evaluate a function written in functional notation* ....

When an equation defines y as a function of x, **functional notation** is frequently used to emphasize that the relation is a function. In this case, it is common to use the notation $f(x)$, where

$$f(x) \text{ is read "} f \text{ of } x \text{" or "the value of } f \text{ at } x \text{"}$$

For instance, the equation $y = x^2 + 3$ from Example 7 defined y as a function of x. The equation can also be written in functional notation as

$$f(x) = x^2 + 3$$

where y has been replaced by $f(x)$.

The symbol $f(x)$ is called the **value of the function** at x because it is the result of evaluating a variable expression. For instance, $f(4)$ means to replace x by 4 and then simplify the resulting numerical expression.

$$f(x) = x^2 + 3$$
$$f(4) = 4^2 + 3 \qquad \bullet \text{ Replace } x \text{ by 4.}$$
$$= 16 + 3 = 19$$

This process is called **evaluating the function.**

➡ Given $f(x) = x^2 + x - 3$, find $f(-2)$.

$$f(x) = x^2 + x - 3$$
$$f(-2) = (-2)^2 + (-2) - 3 \qquad \bullet \text{ Replace } x \text{ by } -2.$$
$$= 4 - 2 - 3 = -1$$
$$f(-2) = -1$$

In this example, $f(-2)$ is the second coordinate of an ordered pair of the function; the first coordinate is -2. Therefore, an ordered pair of this function is $(-2, f(-2))$, which simplifies to $(-2, -1)$.

For the functions given by $y = f(x) = x^2 + 3$ and $y = f(x) = x^2 + x - 3$, y is called the **dependent variable** because its value depends on the value of x. The **independent variable** is x.

Functions can be written using other letters or even combinations of letters. For instance, some calculators use $ABS(x)$ for the absolute-value function. Thus the equation $y = |x|$ would be written $ABS(x) = |x|$, where $ABS(x)$ replaces y.

Example 8

Given $G(t) = \dfrac{3t}{t + 4}$, find $G(1)$.

Solution

$$G(t) = \dfrac{3t}{t + 4}$$

$$G(1) = \dfrac{3(1)}{1 + 4} \qquad \bullet \text{ Replace } t \text{ by 1. Then simplify.}$$

$$G(1) = \dfrac{3}{5}$$

You Try It 8

Given $H(x) = \dfrac{x}{x - 4}$, find $H(8)$.

Your solution

Solution on p. A17

4.1 Exercises

Objective A

1. Graph (−2, 1), (3, −5), (−2, 4), and (0, 3).

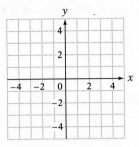

2. Graph (5, −1), (−3, −3), (−1, 0), and (1, −1).

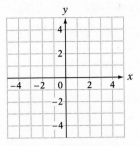

3. Graph (0, 0), (0, −5), (−3, 0), and (0, 2).

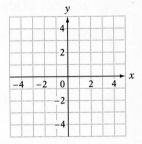

4. Graph (−4, 5), (−3, 1), (3, −4), and (5, 0).

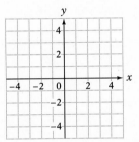

5. Graph (−1, 4), (−2, −3), (0, 2), and (4, 0).

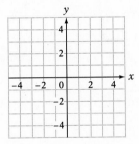

6. Graph (5, 2), (−4, −1), (0, 0), and (0, 3).

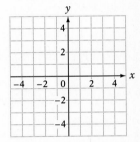

7. Find the coordinates of each of the points.

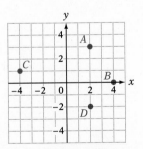

8. Find the coordinates of each of the points.

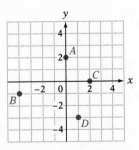

9. Find the coordinates of each of the points.

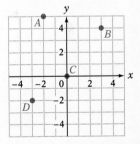

10. Find the coordinates of each of the points.

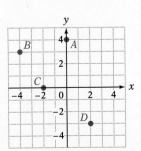

11. Find the coordinates of each of the points.

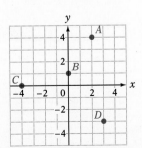

12. Find the coordinates of each of the points.

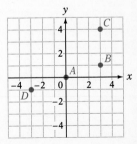

Objective B

13. Is $(3, 4)$ a solution of $y = -x + 7$?

14. Is $(2, -3)$ a solution of $y = x + 5$?

15. Is $(-1, 2)$ a solution of $y = \frac{1}{2}x - 1$?

16. Is $(1, -3)$ a solution of $y = -2x - 1$?

17. Is $(4, 1)$ a solution of $2x - 5y = 4$?

18. Is $(-5, 3)$ a solution of $3x - 2y = 9$?

19. Is $(0, 4)$ a solution of $3x - 4y = -4$?

20. Is $(-2, 0)$ a solution of $x + 2y = -1$?

21. Find the ordered-pair solution of $y = 3x - 2$ corresponding to $x = 3$.

22. Find the ordered-pair solution of $y = 4x + 1$ corresponding to $x = -1$.

23. Find the ordered-pair solution of $y = \frac{2}{3}x - 1$ corresponding to $x = 6$.

24. Find the ordered-pair solution of $y = \frac{3}{4}x - 2$ corresponding to $x = 4$.

25. Find the ordered-pair solution of $y = -3x + 1$ corresponding to $x = 0$.

26. Find the ordered-pair solution of $y = \frac{2}{5}x - 5$ corresponding to $x = 0$.

27. Find the ordered-pair solution of $y = \frac{2}{5}x + 2$ corresponding to $x = -5$.

28. Find the ordered-pair solution of $y = -\frac{1}{6}x - 2$ corresponding to $x = 12$.

Graph the ordered-pair solutions for the given values of x.

29. $y = 2x; x = -2, -1, 0, 2$ **30.** $y = -2x; x = -2, -1, 0, 2$ **31.** $y = x + 2; x = -4, -2, 0, 3$

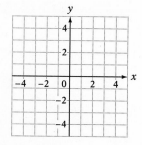
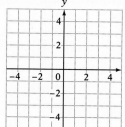
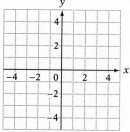

32. $y = \frac{1}{2}x - 1;$ $x = -2, 0, 2, 4$

33. $y = \frac{2}{3}x + 1; x = -3, 0, 3$

34. $y = -\frac{1}{3}x - 2; x = -3, 0, 3$

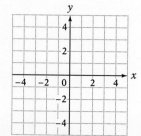
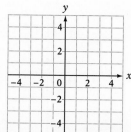
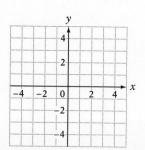

Objective C

35. The runs scored by a baseball team and whether the team won, W, or lost, L, its game are recorded in the following table. Write a relation where the first component is the runs scored and the second component is a win or a loss. Is the relation a function?

Runs scored	4	6	4	2	1	6
Win or loss	L	W	W	W	L	W

36. The number of aces served by a tennis player during a set and the number of games that player won in the set are given in the following table. Write a relation where the first component is the aces served and the second component is the number of games won. Is the relation a function?

Aces served	6	8	4	10	6
Games won	6	5	3	6	4

37. The number of concerts given in one year by five musical groups and the number of hit records that group had that year are given in the following table. Write a relation where the first component is the number of hit records and the second component is the number of concerts. Is the relation a function?

Hit records	11	9	7	12	12
Number of concerts	200	150	200	175	250

38. The monthly salary (in thousands) of an employee and the number of years of college that employee completed are given in the following table. Write a relation where the first component is the years of college and the second component is the monthly salary. Is the relation a function?

Years of college	4	4	5	4	6	8
Salary	3	3.5	4	2.5	5	5.5

39. Does $y = -2x - 3$, where $x \in \{-2, -1, 0, 3\}$, define y as a function of x?

40. Does $y = 2x + 3$, where $x \in \{-2, -1, 1, 4\}$, define y as a function of x?

41. Does $|y| = x - 1$, where $x \in \{1, 2, 3, 4\}$, define y as a function of x?

42. Does $|y| = x + 2$, where $x \in \{-2, -1, 0, 3\}$, define y as a function of x?

43. Does $y = x^2$, where $x \in \{-2, -1, 0, 1, 2\}$, define y as a function of x?

44. Does $y = x^2 - 1$, where $x \in \{-2, -1, 0, 1, 2\}$, define y as a function of x?

Objective D

45. Given $f(x) = 3x - 4$, find $f(4)$.

46. Given $f(x) = 5x + 1$, find $f(2)$.

47. Given $f(x) = x^2$, find $f(3)$.

48. Given $f(x) = x^2 - 1$, find $f(1)$.

49. Given $G(x) = x^2 + x$, find $G(-2)$.

50. Given $H(x) = x^2 - x$, find $H(-2)$.

51. Given $s(t) = \dfrac{3}{t - 1}$, find $s(-2)$.

52. Given $P(x) = \dfrac{4}{2x + 1}$, find $P(-2)$.

53. Given $h(x) = 3x^2 - 2x + 1$, find $h(3)$.

54. Given $Q(r) = 4r^2 - r - 3$, find $Q(2)$.

55. Given $f(x) = \dfrac{x}{x + 5}$, find $f(-3)$.

56. Given $v(t) = \dfrac{2t}{2t + 1}$, find $v(3)$.

57. Given $g(x) = x^3 - x^2 + 2x - 7$, find $g(0)$.

58. Given $F(z) = \dfrac{z}{z^2 + 1}$, find $F(0)$.

APPLYING THE CONCEPTS

59.
[W] Suppose you are helping a student who is having trouble graphing ordered pairs. The work of the student is at the right. What can you say to this student to correct the error that is being made?

60.
[W] Write a few sentences that describe the similarities and differences between relations and functions.

61.
[W] The graph of $y^2 = x$, where $x \in \{0, 1, 4, 9\}$, is shown at the right. Is this the graph of a function? Explain your answer.

62.
[W] Is it possible to evaluate $f(x) = \dfrac{5}{x - 1}$ when $x = 1$? If so, what is $f(1)$? If not, explain why not.

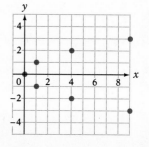

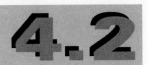

Linear Equations in Two Variables

Objective A   *To graph an equation of the form y = mx + b*

POINT OF INTEREST
The Project in Mathematics at the end of this chapter contains information on using calculators to graph an equation.

The **graph of an equation in two variables** is a graph of the ordered-pair solutions of the equation.

Consider $y = 2x + 1$. Choosing $x = -2$, -1, 0, 1, and 2 and determining the corresponding values of y produces some of the ordered pairs of the equation. These are recorded in the table at the right. See the graph of the ordered pairs in Fig. 1.

x	$y = 2x + 1$	y	(x, y)
-2	$2(-2) + 1$	-3	$(-2, -3)$
-1	$2(-1) + 1$	-1	$(-1, -1)$
0	$2(0) + 1$	1	$(0, 1)$
1	$2(1) + 1$	3	$(1, 3)$
2	$2(2) + 1$	5	$(2, 5)$

Choosing values of x that are not integers produces more ordered pairs to graph, such as $\left(-\frac{5}{2}, -4\right)$ and $\left(\frac{3}{2}, 4\right)$, as shown in Fig. 2. Choosing still other values of x would result in more and more ordered pairs being graphed. The result would be so many dots that the graph would appear as the straight line shown in Fig. 3, which is the graph of $y = 2x + 1$.

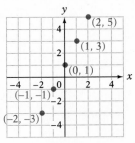

Fig. 1

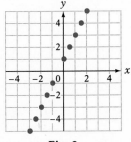

Fig. 2

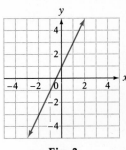

Fig. 3

Equations in two variables have characteristic graphs. The equation $y = 2x + 1$ is an example of a *linear equation*, or *linear function*, because its graph is a straight line.

> **Linear Equation in Two Variables**
>
> Any equation of the form $y = mx + b$, where m is the coefficient of x and b is a constant, is a **linear equation in two variables**. The graph of a linear equation in two variables is a straight line.

Examples of linear equations are shown at the right. These equations represent linear functions because there is only one possible y for each x.

$$y = 2x + 1 \qquad (m = 2, b = 1)$$
$$y = x - 4 \qquad (m = 1, b = -4)$$
$$y = -\frac{3}{4}x \qquad \left(m = -\frac{3}{4}, b = 0\right)$$

The equation $y = x^2 + 4x + 3$ is not a linear equation in two variables because there is a term with a variable squared. The equation $y = \frac{3}{x - 4}$ is not a linear equation because a variable occurs in the denominator of a fraction.

To graph a linear equation, find ordered-pair solutions of the equation. Do this by choosing any value of x and finding the corresponding value of y. Repeat this procedure, choosing different values for x, until you have found the number of solutions desired.

Because the graph of a linear equation in two variables is a straight line, and a straight line is determined by two points, it is necessary to find only two solutions. However, it is recommended that at least three points be used to ensure accuracy.

CONSIDER THIS

If the three points you graph do not lie on a straight line, then either you have made an arithmetic error in calculating a point or you have plotted a point incorrectly.

➡ Graph $y = 2x + 1$.

Choose any values of x, and then find corresponding values of y. The numbers 0, 2, and -1 were chosen arbitrarily for x. It is convenient to record these solutions in a table.

Graph the ordered-pair solutions $(0, 1)$, $(2, 5)$, and $(-1, -1)$. Draw a line through the ordered-pair solutions.

x	$y =$	$2x + 1$	y
0		$2(0) + 1$	1
2		$2(2) + 1$	5
-1		$2(-1) + 1$	-1

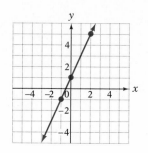

Remember that a graph is a drawing of the ordered-pair solutions of the equation. Therefore, every point on the graph is a solution of the equation, and every solution of the equation is a point on the graph.

The graph at the right is the graph of $y = x + 2$. Note that $(-4, -2)$ and $(1, 3)$ are points on the graph and that these points are solutions of $y = x + 2$. The point whose coordinates are $(4, 1)$ is not a point on the graph and is not a solution of the equation.

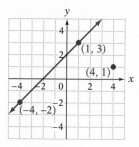

When m is a fraction in the equation $y = mx + b$, choose values of x that will simplify the evaluation.

➡ Graph $y = \frac{1}{3}x - 1$.

m is a fraction. $\left(m = \frac{1}{3}\right)$

Choose values of x that are multiples of the denominator. The numbers 0, 3, and -3 are used here.

x	y
0	-1
3	0
-3	-2

Example 1   Graph $y = 3x - 2$.

Solution

x	y
0	-2
-1	-5
2	4

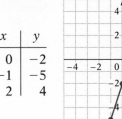

You Try It 1   Graph $y = 3x + 1$.

Your solution

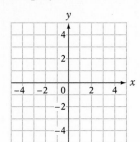

Example 2   Graph $y = 2x$.

Solution

x	y
0	0
2	4
-2	-4

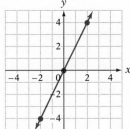

You Try It 2   Graph $y = -2x$.

Your solution

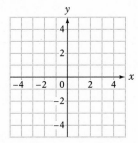

Example 3   Graph $y = \frac{1}{2}x - 1$.

Solution

x	y
0	-1
2	0
-2	-2

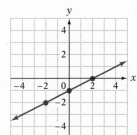

You Try It 3   Graph $y = \frac{1}{3}x - 3$.

Your solution

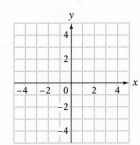

Solutions on pp. A17–A18

Objective B   To graph an equation of the form $Ax + By = C$ ....

An equation in the form $Ax + By = C$ is also a linear equation. Examples of these equations are shown below.

$$2x + 3y = 6 \qquad (A = 2, B = 3, \quad C = 6)$$
$$x - 2y = -4 \qquad (A = 1, B = -2, C = -4)$$
$$2x + y = 0 \qquad (A = 2, B = 1, \quad C = 0)$$
$$4x - 5y = 2 \qquad (A = 4, B = -5, C = 2)$$

One method of graphing an equation of the form $Ax + By = C$ involves first solving the equation for y and then following the same procedure used for graphing an equation of the form $y = mx + b$. To solve the equation for y means to rewrite the equation so that y is alone on one side of the equation and the term containing x and the constant are on the other side of the equation. The Addition and Multiplication Properties of Equations are used to rewrite an equation of the form $Ax + By = C$ in the form $y = mx + b$.

➡ Solve the equation $3x + 2y = 4$ for y.

$$3x + 2y = 4$$

- The equation is in the form $Ax + By = C$.

$$3x - 3x + 2y = -3x + 4$$

- Use the Addition Property of Equations to subtract the term $3x$ from each side of the equation.

$$2y = -3x + 4$$

- Simplify. Note that on the right side of the equation, the term containing x is first, followed by the constant.

$$\frac{1}{2} \cdot 2y = \frac{1}{2}(-3x + 4)$$

- Use the Multiplication Property of Equations to multiply each side of the equation by the reciprocal of the coefficient of y. (The coefficient of y is 2; the reciprocal of 2 is $\frac{1}{2}$.)

$$y = \frac{1}{2}(-3x) + \frac{1}{2}(4)$$

- Simplify. Use the Distributive Property on the right side of the equation.

$$y = -\frac{3}{2}x + 2$$

- The equation is now in the form $y = mx + b$, with $m = -\frac{3}{2}$ and $b = 2$.

In solving the equation $3x + 2y = 4$ for y, where we multiplied both sides of the equation by $\frac{1}{2}$, we could have divided both sides of the equation by 2, as shown at the right. In simplifying the right side after dividing both sides by 2, be sure to divide *each term* by 2.

$$2y = -3x + 4$$
$$\frac{2y}{2} = \frac{-3x + 4}{2}$$
$$y = \frac{-3x}{2} + \frac{4}{2}$$
$$y = -\frac{3}{2}x + 2$$

➡ Graph $3x + 4y = 12$.

$$3x + 4y = 12$$

- Solve the equation for y.

$$4y = -3x + 12$$

- Subtract $3x$ from each side of the equation.

$$y = -\frac{3}{4}x + 3$$

- Divide each side of the equation by 4.

x	y
0	3
4	0
-4	6

- Find three ordered-pair solutions of the equation.

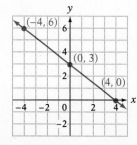

- Graph the ordered pairs and then draw a line through the points.

The graph of the equation $2x + 3y = 6$ is shown at the right. The graph crosses the x-axis at the point $(3, 0)$. This point is called the **x-intercept.** The graph also crosses the y-axis at the point $(0, 2)$. This point is called the **y-intercept.**

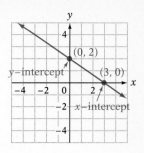

We can find the x-intercept and the y-intercept of the graph of the equation $2x + 3y = 6$ algebraically.

CONSIDER THIS

Any point on the x-axis has y-coordinate 0. Any point on the y-axis has x-coordinate 0.

To find the x-intercept, let $y = 0$.
(Any point on the x-axis has y-coordinate 0.)

$$2x + 3y = 6$$
$$2x + 3(0) = 6$$
$$2x = 6$$
$$x = 3$$

The x-intercept is $(3, 0)$.

To find the y-intercept, let $x = 0$.
(Any point on the y-axis has x-coordinate 0.)

$$2x + 3y = 6$$
$$2(0) + 3y = 6$$
$$3y = 6$$
$$y = 2$$

The y-intercept is $(0, 2)$.

Another method of graphing an equation of the form $Ax + By = C$ is to find the x- and y-intercepts, plot both intercepts, and then draw a line through the two points. This method of graphing the equation $3x + 4y = 12$ is shown below. Note that this is the same equation graphed at the bottom of page 210.

➡ Graph $3x + 4y = 12$ by using the x- and y-intercepts.

x-intercept: $3x + 4y = 12$
$$3x + 4(0) = 12$$
$$3x = 12$$
$$x = 4$$

- To find the x-intercept, let $y = 0$.

- The x-intercept is $(4, 0)$.

y-intercept: $3x + 4y = 12$
$$3(0) + 4y = 12$$
$$4y = 12$$
$$y = 3$$

- To find the y-intercept, let $x = 0$.

- The y-intercept is $(0, 3)$.

- Graph the ordered pairs $(4, 0)$, and $(0, 3)$. Draw a straight line through the points.

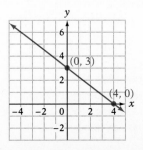

The graph of an equation with one of the variables missing is either a horizontal line or a vertical line.

The equation $y = 2$ could be written $0x + y = 2$. Because $0x = 0$ for any value of x, the value of y is always 2 no matter what value of x is chosen. For instance, replace x by -4, -1, 0, or 3. In each case, $y = 2$.

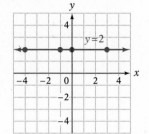

$$0x + y = 2$$

$0(-4) + y = 2$	$(-4, 2)$ is a solution.
$0(-1) + y = 2$	$(-1, 2)$ is a solution.
$0(0) + y = 2$	$(0, 2)$ is a solution.
$0(3) + y = 2$	$(3, 2)$ is a solution.

The solutions are plotted in the graph above, and a line is drawn through the plotted points. Note that the line is horizontal.

Graph of a Horizontal Line

The graph of $y = b$ is a horizontal line passing through $(0, b)$.

The equation $x = -2$ could be written $x + 0y = -2$. Because $0y = 0$ for any value of y, the value of x is always -2 no matter what value of y is chosen. For instance, replace y by -2, 0, 2, or 3. In each case, $x = -2$.

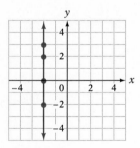

$$x + 0y = -2$$

$x + 0(-2) = -2$	$(-2, -2)$ is a solution.
$x + 0(0) = -2$	$(-2, 0)$ is a solution.
$x + 0(2) = -2$	$(-2, 2)$ is a solution.
$x + 0(3) = -2$	$(-2, 3)$ is a solution.

The solutions are plotted in the graph at the right, and a line is drawn through the plotted points. Note that the line is vertical.

Graph of a Vertical Line

The graph of $x = a$ is a vertical line passing through $(a, 0)$.

⇒ Graph $x = -3$ and $y = 2$ in the same coordinate grid.

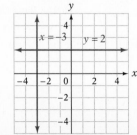

- The graph of $x = -3$ is a vertical line passing through $(-3, 0)$.
- The graph of $y = 2$ is a horizontal line passing through $(0, 2)$.

Example 4

Solve the equation $2x - 5y = 10$ for y. Then graph the equation.

Solution

$2x - 5y = 10$
$ -5y = -2x + 10$
$ y = \dfrac{2}{5}x - 2$

x	y
0	-2
5	0
-5	-4

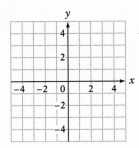

You Try It 4

Solve the equation $5x - 2y = 10$ for y. Then graph the equation.

Your solution

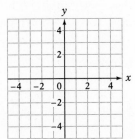

Example 5

Find the x- and y-intercepts of $x - 2y = 4$. Graph the line.

Solution

x-intercept:
$x - 2y = 4$
$x - 2(0) = 4$
$ x = 4$
$(4, 0)$

y-intercept:
$x - 2y = 4$
$0 - 2y = 4$
$ -2y = 4$
$ y = -2$
$(0, -2)$

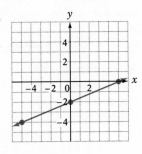

You Try It 5

Find the x- and y-intercepts of $x - 4y = -4$. Graph the line.

Your solution

Example 6

Graph $y = -2$.

Solution

The graph of an equation of the form $y = b$ is a horizontal line passing through the point $(0, b)$.

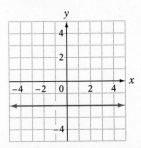

You Try It 6

Graph $x = -4$.

Your solution

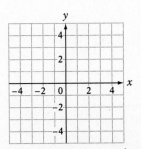

Solutions on p. A18

Objective C *To solve application problems*

There are a variety of applications of linear functions.

➡ Solve: An installer of marble kitchen countertops charges $250 plus $180 per foot of countertop. The equation that describes the total cost, *C*, to have *x* feet of countertop installed is $C = 180x + 250$.

a. Graph this equation for $0 \le x \le 25$. (*Note:* In many applications, the domain of the variable is given so that the equation makes sense. For instance, it would not be sensible to have values of *x* that are less than 0. This would mean negative countertop! The choice of 25 is somewhat arbitrary, but most kitchens have less than 25 ft of counter space.)

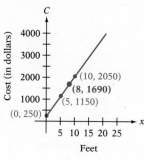

● Choosing *x* = 0, 5, and 10, you find that the corresponding ordered pairs are (0, 250), (5, 1150), and (10, 2050). Plot these points and draw a line through them.

b. The point whose coordinates are (8, 1690) is on the graph. Write a sentence that describes this ordered pair.

The point whose coordinates are (8, 1690) means that 8 ft of countertop costs $1690 to install.

Example 7

The value, *y*, of an investment of $2500 at an annual simple interest rate of 6% is given by the equation $y = 150x + 2500$, where *x* is the number of years the investment is held. Graph this equation for $0 \le x \le 10$. The point whose coordinates are (5, 3250) is on the graph. Write a sentence that describes this ordered pair.

Solution

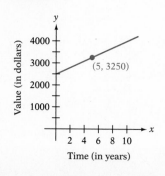

The ordered pair (5, 3250) means that in 5 years, the value of the investment will be $3250.

You Try It 7

A car is traveling at a uniform speed of 40 mph. The distance, *d*, the car travels in *t* hours is given by $d = 40t$. Graph this equation for $0 \le t \le 5$. The point whose coordinates are (3, 120) is on the graph. Write a sentence that describes this ordered pair.

Your solution

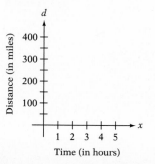

Solution on p. A18

4.2 Exercises

Objective A

Graph.

1. $y = 2x - 3$

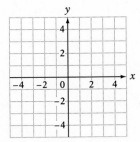

2. $y = -2x + 2$

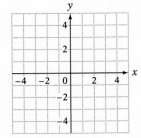

3. $y = \dfrac{1}{3}x$

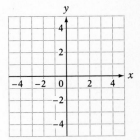

4. $y = -3x$

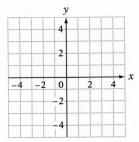

5. $y = \dfrac{2}{3}x - 1$

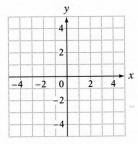

6. $y = \dfrac{3}{4}x + 2$

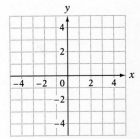

7. $y = -\dfrac{1}{4}x + 2$

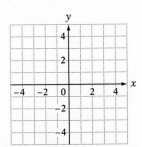

8. $y = -\dfrac{1}{3}x + 1$

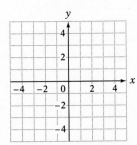

9. $y = -\dfrac{2}{5}x + 1$

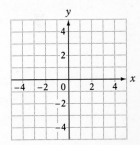

10. $y = -\dfrac{1}{2}x + 3$

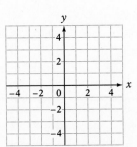

11. $y = 2x - 4$

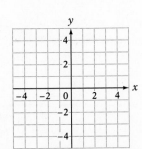

12. $y = 3x - 4$

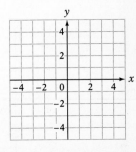

Graph.

13. $y = x - 3$

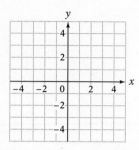

14. $y = x + 2$

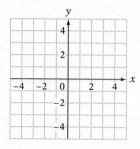

15. $y = -x + 2$

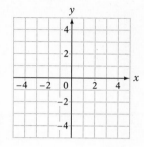

16. $y = -x - 1$

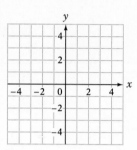

17. $y = -\dfrac{2}{3}x + 1$

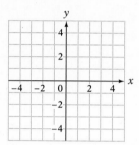

18. $y = 5x - 4$

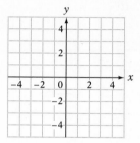

Objective B

Find the *x*- and *y*-intercepts.

19. $x - y = 3$

20. $3x + 4y = 12$

21. $x - 5y = 10$

22. $3x + 2y = 12$

23. $2x - 3y = 0$

24. $3x + 4y = 0$

Graph.

25. $3x + y = 3$

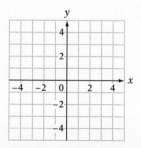

26. $2x + y = 4$

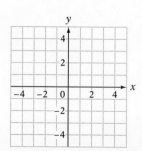

27. $2x + 3y = 6$

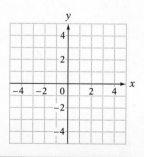

Graph.

28. $3x + 2y = 4$

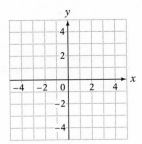

29. $x - 2y = 4$

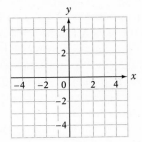

30. $x - 3y = 6$

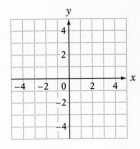

31. $2x - 3y = 6$

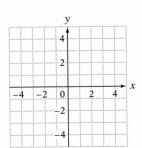

32. $3x - 2y = 8$

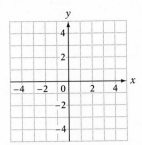

33. $2x + 5y = 10$

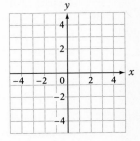

34. $3x + 4y = 12$

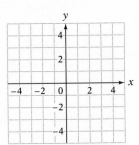

35. $x = 3$

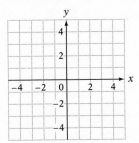

36. $y = -4$

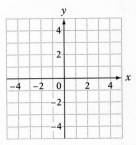

37. $x + 4y = 4$

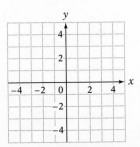

38. $4x - 3y = 12$

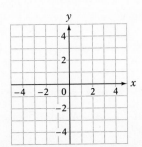

39. $y = 4$

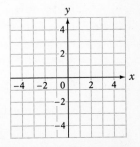

Objective C   *Application Problems*

Solve.

40. Depreciation is the declining value of an asset. For instance, a company that purchases a truck for $20,000 has an asset worth $20,000. In five years, however, the value of the truck will have declined and it may be worth only $4000. An equation that represents this decline is $V = 20,000 - 3200x$, where V is the value of the truck after x years. Graph this equation for $0 \leq x \leq 5$. The point (3, 10,400) is on the graph (Fig. 1). Write a sentence that describes the meaning of this ordered pair.

41. A company uses the equation $V = 30,000 - 5000x$ to estimate the depreciated value (see Exercise 40) of a computer. Graph this equation for $0 \leq x \leq 4$. The point (1, 25,000) is on the graph (Fig. 2). Write a sentence that describes the meaning of this ordered pair.

42. An architect charges a fee of $500 plus $2.65 per square foot to design a house. The equation that represents the architect's fee is given by $F = 2.65x + 500$, where F is the fee and x is the number of square feet in the house. Graph this equation for $0 \leq x \leq 5000$. The point (3500, 9775) is on the graph (Fig. 3). Write a sentence that describes the meaning of this ordered pair.

43. A rental car company charges a "drop-off" fee of $50 to return a car to a location different from that from which it was rented. In addition, a fee of $.18 per mile the car is driven is charged. An equation that represents the total cost to rent a car with a drop-off fee is $C = 0.18x + 50$, where C is the total cost and x is the number of miles the car is driven. Graph this equation for $0 \leq x \leq 1000$. The point (500, 140) is on the graph (Fig. 4). Write a sentence that describes the meaning of this ordered pair.

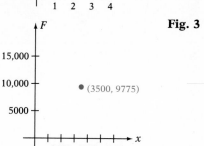

Fig. 1

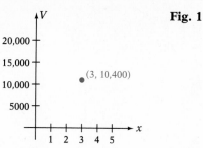

Fig. 2

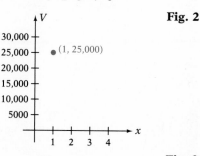

Fig. 3

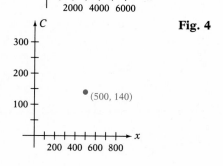

Fig. 4

APPLYING THE CONCEPTS

44. For the equation $y = 3x + 2$, when the value of x changes from 1 to 2, does the value of y increase or decrease? What is the change in y? Suppose that the value of x changes from 13 to 14. What is the change in y?

45. For the equation $y = -2x + 1$, when the value of x changes from 1 to 2, does the value of y increase or decrease? What is the change in y? Suppose the value of x changes from 13 to 14. What is the change in y?

46.
[W] Graph $y = 2x - 2$, $y = 2x$, and $y = 2x + 3$ on the same coordinate system in Fig. 5. What observation can you make about the graphs?

47.
[W] Graph $y = x + 3$, $y = 2x + 3$, and $y = -\dfrac{1}{2}x + 3$ on the same coordinate system in Fig. 6. What observation can you make about the graphs?

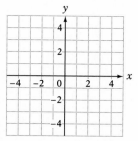

Fig. 5

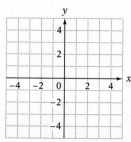

Fig. 6

4.3

Slopes of Straight Lines

Objective A   *To find the slope of a straight line* ....

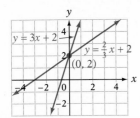

The graphs of $y = 3x + 2$ and $y = \frac{2}{3}x + 2$ are shown at the left. Each graph crosses the y-axis at the point $(0, 2)$, but the graphs have different slants. The **slope** of a line is a measure of the slant of a line. The symbol for slope is m.

The slope of a line containing two points is the ratio of the change in the y values between the two points to the change in the x values. The line containing the points whose coordinates are $(-1, -3)$ and $(5, 2)$ is shown below.

The change in the y values is the difference between the y-coordinates of the two points.

$$\text{Change in } y = 2 - (-3) = 5$$

The change in the x values is the difference between the x-coordinates of the two points.

$$\text{Change in } x = 5 - (-1) = 6$$

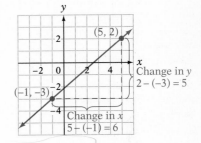

The slope of the line between the two points is the ratio of the change in y to the change in x.

$$\text{Slope} = m = \frac{\text{change in } y}{\text{change in } x} = \frac{5}{6} \qquad m = \frac{2 - (-3)}{5 - (-1)} = \frac{5}{6}$$

In general, if $P_1(x_1, y_1)$ and $P_2(x_2, y_2)$ are two points on a line, then

$$\text{Change in } y = y_2 - y_1 \qquad\qquad \text{Change in } x = x_2 - x_1$$

Using these ideas, we can state a formula for slope.

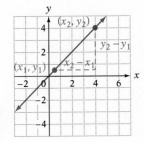

> **Slope Formula**
>
> The slope of the line containing the two points $P_1(x_1, y_1)$ and $P_2(x_2, y_2)$ is given by
>
> $$m = \frac{y_2 - y_1}{x_2 - x_1}, \; x_1 \neq x_2$$

Frequently, the Greek letter Δ is used to designate the change in a variable. Using this notation, we can write the equations for the change in y and change in x as follows:

$$\text{Change in } y = \Delta y = y_2 - y_1 \qquad\qquad \text{Change in } x = \Delta x = x_2 - x_1$$

With this notation, the slope formula is written $m = \frac{\Delta y}{\Delta x}$.

➡ Find the slope of the line containing the points whose coordinates are $(-2, 0)$ and $(4, 5)$.

Let $P_1 = (-2, 0)$ and $P_2 = (4, 5)$. (It does not matter which point is named P_1 or P_2; the slope will be the same.)

$$m = \frac{y_2 - y_1}{x_2 - x_1} = \frac{5 - 0}{4 - (-2)} = \frac{5}{6}$$

A line that slants upward to the right always has a **positive slope.**

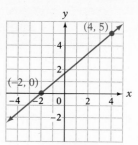

Positive slope

CONSIDER THIS

Positive slope means that the value of y increases as the value of x increases.

➡ Find the slope of the line containing the points whose coordinates are $(-3, 4)$ and $(4, 2)$.

Let $P_1 = (-3, 4)$ and $P_2 = (4, 2)$.

$$m = \frac{y_2 - y_1}{x_2 - x_1} = \frac{2 - 4}{4 - (-3)} = \frac{-2}{7} = -\frac{2}{7}$$

A line that slants downward to the right always has a **negative slope.**

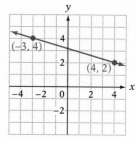

Negative slope

CONSIDER THIS

Negative slope means that the value of y decreases as the value of x increases. Compare this to positive slope.

➡ Find the slope of the line containing the points whose coordinates are $(-2, 2)$ and $(4, 2)$.

Let $P_1 = (-2, 2)$ and $P_2 = (4, 2)$.

$$m = \frac{y_2 - y_1}{x_2 - x_1} = \frac{2 - 2}{4 - (-2)} = \frac{0}{6} = 0$$

A horizontal line has **zero slope.**

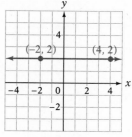

Zero slope

➡ Find the slope of the line containing the points whose coordinates are $(1, -2)$ and $(1, 3)$.

Let $P_1 = (1, -2)$ and $P_2 = (1, 3)$.

$$m = \frac{y_2 - y_1}{x_2 - x_1} = \frac{3 - (-2)}{1 - 1} = \frac{5}{0} \quad \text{Not a real number}$$

The slope of a vertical line is **undefined.**

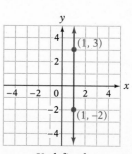

Undefined

There are many applications of the concept of slope. Here are two possibilities.

In 1988, when Florence Griffith-Joyner set the world record for the 100-meter dash, her average rate of speed was approximately 9.5 m/s. The graph at the right shows the distance she ran during her record-setting run. From the graph, note that after 4 s she had traveled 38 m and that after 6 s she had traveled 57 m. The slope of the line between these two points is

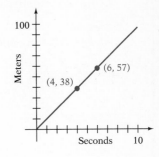

$$m = \frac{57 - 38}{6 - 4} = \frac{19}{2} = 9.5$$

Note that the slope of the line is the same as the rate she was running, 9.5 m/s. The average speed of an object is related to slope.

Here is another example, this one related to economics.

During the 1980s the value of a 1957 Chevrolet increased dramatically to around $26,000 by 1990. Since then, however, the value has been decreasing at a rate of approximately $3500 per year. The graph at the right shows the value of the car for the years 1990 through 1994. From the graph, note that the value of the car in 1991 was $22,500 and that its value in 1993 was $15,500. The slope of the line between these two points is

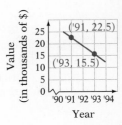

$$m = \frac{15,500 - 22,500}{1993 - 1991} = \frac{-7000}{2} = -3500$$

Note that if we interpret negative slope as decreasing, then the slope of the line is the same as the rate at which the value of the car is decreasing, $3500 per year.

In general, any quantity that is expressed by using the word *per* is represented mathematically as slope. In the first example, slope was 9.5 meters *per* second; in the second example, slope was −$3500 *per* year.

Example 1

Find the slope of the line containing the points $(2, -5)$ and $(-4, 2)$.

Solution

Let $P_1 = (2, -5)$ and $P_2 = (-4, 2)$.

$$m = \frac{y_2 - y_1}{x_2 - x_1} = \frac{2 - (-5)}{-4 - 2} = \frac{7}{-6}$$

The slope is $-\frac{7}{6}$.

You Try It 1

Find the slope of the line containing the points $(4, -3)$ and $(2, 7)$.

Your solution

Example 2

Find the slope of the line containing the points $(-3, 4)$ and $(5, 4)$.

Solution

Let $P_1 = (-3, 4)$ and $P_2 = (5, 4)$.

$$m = \frac{y_2 - y_1}{x_2 - x_1} = \frac{4 - 4}{5 - (-3)} = \frac{0}{8} = 0$$

The slope of the line is zero.

You Try It 2

Find the slope of the line containing the points $(6, -1)$ and $(6, 7)$.

Your solution

Example 3

The graph below shows the relationship between the cost of an item and the sales tax. Find the slope of the line between the two points shown on the graph. Write a sentence that states the meaning of the slope.

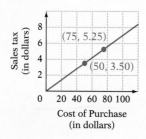

Solution

$$m = \frac{5.25 - 3.50}{75 - 50} = \frac{1.75}{25} = 0.07$$

A slope of 0.07 means that the sales tax is $.07 per dollar.

You Try It 3

The graph below shows the decrease in the value of a printing press for a period of six years. Find the slope of the line between the two points shown on the graph. Write a sentence that states the meaning of the slope.

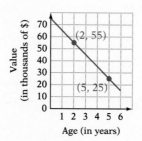

Your solution

Solutions on p. A18–A19

Objective B   **To graph a line using the slope and the y-intercept** ································

Recall that we can find the y-intercept of a linear equation by letting $x = 0$.

➡ Find the y-intercept of $y = 3x + 4$.

$$y = 3x + 4 = 3(0) + 4 = 4 \qquad \bullet \text{ Let } x = 0.$$

The y-intercept is $(0, 4)$.

For any equation of the form $y = mx + b$, the y-intercept is $(0, b)$.

The graph of the equation $y = \dfrac{2}{3}x + 1$ is shown at the right. The points $(-3, -1)$ and $(3, 3)$ are on the graph. The slope of the line between the two points is

$$m = \frac{3 - (-1)}{3 - (-3)} = \frac{4}{6} = \frac{2}{3}$$

Observe that the slope of the line is the coefficient of x in the equation $y = \dfrac{2}{3}x + 1$.

Slope–Intercept Form of a Straight Line

An equation of the form $y = mx + b$ is called the **slope–intercept** form of a straight line. The slope of the line is m, the coefficient of x. The y-intercept is $(0, b)$, where b is the constant term of the equation.

The following equations are written in slope–intercept form.

$y = 2x - 3$	Slope = 2, y-intercept = $(0, -3)$
$y = -x + 2$	Slope = -1 $(-x = -1x)$, y-intercept = $(0, 2)$
$y = \dfrac{1}{2}x$	Slope = $\dfrac{1}{2}$, y-intercept = $(0, 0)$

When the equation of a straight line is in the form $y = mx + b$, the graph can be drawn by using the slope and y-intercept. First locate the y-intercept. Use the slope to find a second point on the line. Then draw a line through the two points.

➡ Graph $y = \dfrac{5}{3}x - 4$ by using the slope and y-intercept.

The slope is the coefficient of x:

$$m = \frac{5}{3} = \frac{\text{change in } y}{\text{change in } x}$$

The y-intercept is $(0, -4)$.

Beginning at the y-intercept $(0, -4)$, move right 3 units (change in x) and then up 5 units (change in y).

The point whose coordinates are $(3, 1)$ is a second point on the graph. Draw a line through the points $(0, -4)$ and $(3, 1)$.

Content and Format © 1996 HMCo.

➡ Graph $y = 2x - 3$.

y-intercept $= (0, b) = (0, -3)$

$m = 2 = \dfrac{2}{1} = \dfrac{\text{change in } y}{\text{change in } x}$

Beginning at the y-intercept, move right 1 unit (change in x) and then up 2 units (change in y).

$(1, -1)$ is a second point on the graph.

Draw a line through the two points $(0, -3)$ and $(1, -1)$.

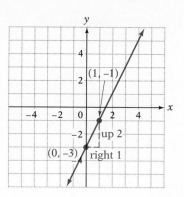

Example 4   Graph $y = -\dfrac{2}{3}x + 1$ by using the slope and y-intercept.

Solution y-intercept $= (0, b) = (0, 1)$

$m = -\dfrac{2}{3} = \dfrac{-2}{3}$

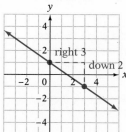

You Try It 4   Graph $y = -\dfrac{1}{4}x - 1$ by using the slope and y-intercept.

Your solution

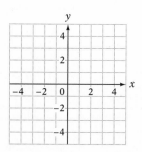

Example 5   Graph $2x - 3y = 6$ by using the slope and y-intercept.

Solution   Solve the equation for y.

$2x - 3y = 6$

$-3y = -2x + 6$

$y = \dfrac{2}{3}x - 2$

y-intercept $= (0, -2)$; $m = \dfrac{2}{3}$

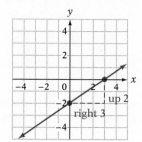

You Try It 5   Graph $x - 2y = 4$ by using the slope and y-intercept.

Your solution

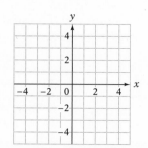

Solutions on p. A19

4.3 Exercises

· ·

Objective A

Find the slope of the line containing the points.

1. $P_1(1, 3)$, $P_2(3, 1)$

2. $P_1(2, 3)$, $P_2(5, 1)$

3. $P_1(-1, 4)$, $P_2(2, 5)$

4. $P_1(3, -2)$, $P_2(1, 4)$

5. $P_1(-1, 3)$, $P_2(-4, 5)$

6. $P_1(-1, -2)$, $P_2(-3, 2)$

7. $P_1(0, 3)$, $P_2(4, 0)$

8. $P_1(-2, 0)$, $P_2(0, 3)$

9. $P_1(2, 4)$, $P_2(2, -2)$

10. $P_1(4, 1)$, $P_2(4, -3)$

11. $P_1(2, 5)$, $P_2(-3, -2)$

12. $P_1(4, 1)$, $P_2(-1, -2)$

13. $P_1(2, 3)$, $P_2(-1, 3)$

14. $P_1(3, 4)$, $P_2(0, 4)$

15. $P_1(0, 4)$, $P_2(-2, 5)$

16. $P_1(-2, 3)$, $P_2(-2, 5)$

17. $P_1(-3, -1)$, $P_2(-3, 4)$

18. $P_1(-2, -5)$, $P_2(-4, -1)$

19. The graph below shows the relationship between the distance traveled by a motorist and the time of travel. Find the slope of the line between the two points shown on the graph. Write a sentence that states the meaning of the slope.

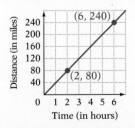

20. The graph below shows the relationship between the cost of a building and the depreciation allowed for income tax purposes. Find the slope of the line between the two points shown on the graph. Write a sentence that states the meaning of the slope.

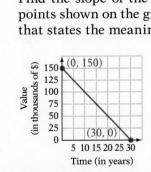

21. The graph below shows the relationship between the amount of tax and the amount of taxable income between $22,101 and $54,500. Find the slope of the line between the two points shown on the graph. Write a sentence that states the meaning of the slope.

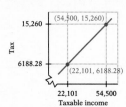

22. The graph below shows the relationship between the payment on a mortgage and the amount of the mortgage. Find the slope of the line between the two points shown on the graph. Write a sentence that states the meaning of the slope.

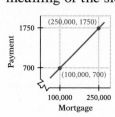

23. The graph below shows the relationship between the distance and the time for the 5000-meter run for the world record by Said Aouita in 1987. Find the slope of the line between the two points shown on the graph. Round to the nearest tenth. Write a sentence that states the meaning of the slope.

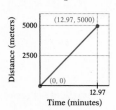

24. The graph below shows the relationship between the distance and the time for the 10,000-meter run for the world record by Arturo Barrios in 1989. Find the slope of the line between the two points shown on the graph. Write a sentence that states the meaning of the slope.

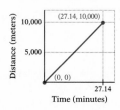

Objective B

Graph by using the slope and y-intercept.

25. $y = \dfrac{1}{2}x + 2$

26. $y = \dfrac{2}{3}x - 3$

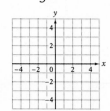

27. $y = -\dfrac{3}{2}x$

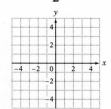

28. $y = \dfrac{3}{4}x$

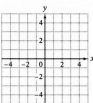

29. $y = \dfrac{3}{4}x + 1$

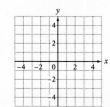

30. $y = -\dfrac{1}{2}x + 2$

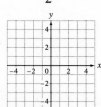

Graph by using the slope and *y*-intercept.

31. $y = 3x + 1$

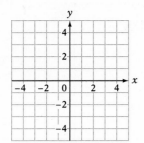

32. $y = -2x - 1$

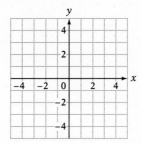

33. $y = \dfrac{2}{5}x - 2$

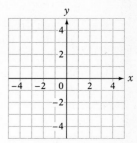

34. $y = \dfrac{1}{2}x$

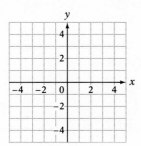

35. $y = -x + 1$

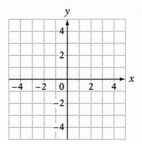

36. $y = -x - 3$

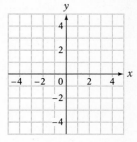

37. $y = \dfrac{2}{3}x$

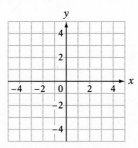

38. $y = -4x + 2$

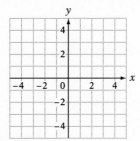

39. $2x + y = 3$

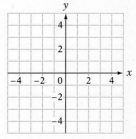

40. $3x - y = -1$

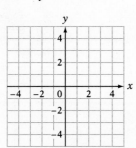

41. $x - 2y = 4$

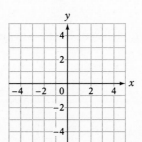

42. $x + 3y = 6$

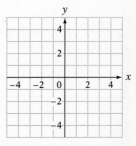

APPLYING THE CONCEPTS

43. Do all straight lines have a *y*-intercept? If not, give an example of one that does not.

44. If two lines have the same slope and same *y*-intercept, must the graphs of the lines be the same? If not, give an example.

Complete the following sentences.

45. If a line has a slope of 2, then the value of *y* increases/decreases by _____ as the value of *x* increases by 1.

46. If a line has a slope of −3, then the value of *y* increases/decreases by _____ as the value of *x* increases by 1.

47. If a line has a slope of −2, then the value of *y* increases/decreases by _____ as the value of *x* decreases by 1.

48. If a line has a slope of 3, then the value of *y* increases/decreases by _____ as the value of *x* decreases by 1.

49. If a line has a slope of $\frac{1}{2}$, then the value of *y* increases/decreases by _____ as the value of *x* increases by 1.

50. If a line has a slope of $-\frac{2}{3}$, then the value of *y* increases/decreases by _____ as the value of *x* increases by 1.

Determine the value of *k* such that the points whose coordinates are given below lie on the same line.

51. (3, 2), (4, 6), (5, *k*)   **52.** (−2, 3), (1, 0), (*k*, 2)

53. (*k*, 1), (0, −1), (2, −2)   **54.** (4, −1), (3, −4), (*k*, *k*)

55. What does the highway sign at the right have to do with slope?
[W]

56. Explain how to graph the equation of a straight line by using its slope and a point on the line.
[W]

57. Explain why $y = mx + b$ is called the slope–intercept form of the equation of a straight line.
[W]

Grade
6%

4.4 Equations of Straight Lines

Objective A   *To find the equation of a line given a point and the slope*

When the slope of a line and a point on the line are known, the equation of the line can be determined. If the particular point is the *y*-intercept, use the slope–intercept form, $y = mx + b$, to find the equation.

➡ Find the equation of the line that contains the point (0, 3) and has slope $\frac{1}{2}$.

The known point is the *y*-intercept, (0, 3).

$$y = mx + b$$ • Use the slope–intercept form.

$$y = \frac{1}{2}x + 3$$ • Replace *m* with $\frac{1}{2}$, the given slope.

    Replace *b* with 3, the *y*-coordinate of the *y*-intercept.

The equation of the line is $y = \frac{1}{2}x + 3$.

One method of finding the equation of a line when the slope and *any* point on the line are known involves using the *point–slope* formula. This formula is derived from the formula for the slope of a line.

Let (x_1, y_1) be the given point on the line, and let (x, y) be any other point on the line.

$$\frac{y - y_1}{x - x_1} = m$$ • Use the formula for the slope of a line.

$$\frac{y - y_1}{x - x_1}(x - x_1) = m(x - x_1)$$ • Multiply each side by $(x - x_1)$.

$$y - y_1 = m(x - x_1)$$ • Simplify.

Point–Slope Formula

If (x_1, y_1) is a point on a line with slope *m*, then $y - y_1 = m(x - x_1)$.

➡ Find the equation of the line that contains the point (4, −1) and has slope $-\frac{3}{4}$.

$$y - y_1 = m(x - x_1)$$ • Use the point–slope formula.

$$y - (-1) = \left(-\frac{3}{4}\right)(x - 4)$$ • $m = -\frac{3}{4}$, $(x_1, y_1) = (4, -1)$

$$y + 1 = -\frac{3}{4}x + 3$$

$$y = -\frac{3}{4}x + 2$$ • Write the equation in slope–intercept form.

The equation of the line is $y = -\frac{3}{4}x + 2$.

⇒ Find the equation of the line that passes through the point (4, 3) and whose slope is undefined.

Because the slope is undefined, the point–slope formula cannot be used to find the equation. Instead, recall that when the slope is undefined, the line is vertical and that the equation of a vertical line is $x = a$, where a is the x-coordinate of the x-intercept. Because the line is vertical and passes through (4, 3), the x-intercept is (4, 0).

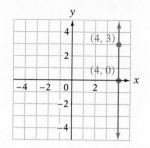

The equation of the line is $x = 4$.

Example 1
Find the equation of the line whose slope is $-\frac{2}{3}$ and whose y-intercept is (0, −1).

Solution
Because the slope and y-intercept are known, use the slope–intercept formula, $y = mx + b$.

$$y = -\frac{2}{3}x - 1 \quad \bullet \; m = -\frac{2}{3}; \; b = -1$$

You Try It 1
Find the equation of the line whose slope is $\frac{5}{3}$ and whose y-intercept is (0, 2).

Your solution

Example 2
Find the equation of the line that contains the point (3, 0) and has slope −4.

Solution
$m = -4 \quad (x_1, y_1) = (3, 0)$

$$y - y_1 = m(x - x_1)$$
$$y - 0 = -4(x - 3)$$
$$y = -4x + 12$$

The equation of the line is $y = -4x + 12$.

You Try It 2
Find the equation of the line that contains the point (−3, −2) and has slope $-\frac{1}{3}$.

Your solution

Example 3
Find the equation of the line that contains the point (−2, 4) and has slope 2.

Solution
$m = 2 \quad (x_1, y_1) = (-2, 4)$

$$y - y_1 = m(x - x_1)$$
$$y - 4 = 2[x - (-2)]$$
$$y - 4 = 2(x + 2)$$
$$y - 4 = 2x + 4$$
$$y = 2x + 8$$

The equation of the line is $y = 2x + 8$.

You Try It 3
Find the equation of the line that contains the point (4, −3) and has slope −3.

Your solution

Solutions on pp. A19–A20

Objective B   ***To find the equation of a line given two points*** ....

The point–slope formula is used to find the equation of a line when a point on the line and the slope of the line are known. But this formula can also be used to find the equation of a line given two points on the line. In this case:

1. Use the slope formula to determine the slope of the line between the points.
2. Use the point–slope formula, the slope you just calculated, and one of the given points to find the equation of the line.

⇒ Find the equation of the line that passes through the points $(-3, -1)$ and $(3, 3)$.

CONSIDER THIS

It does not matter which ordered pair you use in the point–slope formula. The point $(-1, -3)$ is used at the right. The point $(3, 3)$ is used below. In each case, the result is the same.

$$y - 3 = \frac{2}{3}(x - 3)$$

$$y - 3 = \frac{2}{3}x - 2$$

$$y = \frac{2}{3}x + 1$$

$$m = \frac{y_2 - y_1}{x_2 - x_1} = \frac{3 - (-1)}{3 - (-3)} = \frac{4}{6} = \frac{2}{3}$$

$$y - y_1 = m(x - x_1)$$

$$y - (-1) = \frac{2}{3}[x - (-3)]$$

$$y + 1 = \frac{2}{3}(x + 3)$$

$$y + 1 = \frac{2}{3}x + 2$$

$$y = \frac{2}{3}x + 1$$

- Use the formula for slope to find the slope of the line between the points.
- Use the point–slope formula.

The equation of the line is $y = \frac{2}{3}x + 1$.

Example 4
Find the equation of the line that passes through the points $(-4, 0)$ and $(2, -3)$.

Solution

$$m = \frac{y_2 - y_1}{x_2 - x_1} = \frac{-3 - 0}{2 - (-4)} = \frac{-3}{6} = -\frac{1}{2}$$

$$y - y_1 = m(x - x_1) \qquad \bullet \text{ Point–slope formula}$$

$$y - 0 = -\frac{1}{2}[x - (-4)] \qquad \bullet \ m = -\frac{1}{2}; (x_1, y_1) = (-4, 0)$$

$$y = -\frac{1}{2}(x + 4)$$

$$y = -\frac{1}{2}x - 2$$

The equation of the line is $y = -\frac{1}{2}x - 2$.

You Try It 4
Find the equation of the line that passes through the points $(-6, -1)$ and $(3, 1)$.

Your solution

Solution on p. A20

Objective C   To solve application problems ....

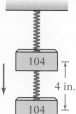

A **linear model** describes a relationship between quantities. In many cases, a linear model is used to approximate collected data. The data are graphed as points in a coordinate system, and then a line is drawn that approximates the data. The graph of the points is called a **scatter diagram;** the line is called a **line of best fit.**

Consider an experiment to determine the amount of weight that is necessary to stretch a spring a certain distance. Data from such an experiment are shown in the table below. Distance is in inches; weight is in pounds.

Distance	2.5	4	2	3.5	1	4.5
Weight	63	104	47	85	27	115

The accompanying graph shows the scatter diagram, which is the plotted points, and the line of best fit, which is the line that approximately goes through the plotted points. The equation of the line of best fit is $y = 25.7x - 1.3$, where x is the number of inches the spring is stretched and y is the weight in pounds.

The table below shows the values that the model would predict, to the nearest tenth. Good linear models should predict values that are close to the actual values. A more thorough analysis of lines of best fit is undertaken in statistics courses.

Distance, x	2.5	4	2	3.5	1	4.5
Weight predicted using $y = 25.7x - 1.3$	63.0	101.5	50.1	88.7	24.4	114.4

Example 5

The data in the table below show the size of a house in square feet and the cost to build the house. The line of best fit is $y = 70.3x + 41{,}100$, where x is the number of square feet and y is the cost of the house.

Square ft	1250	1400	1348	2675	2900
Cost	128,000	140,000	136,100	233,450	241,500

Graph the data and the line of best fit in the coordinate system below. Write a sentence that describes the meaning of the slope of the line of best fit.

Solution

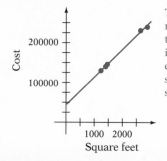

The slope of the line means that the cost to build the house increases $70.30 for each additional square foot in the size of the house.

You Try It 5

The data in the table below show a reading test grade and the final exam grade in a history class. The line of best fit is $y = 8.3x - 7.8$, where x is the reading test score and y is the history test score.

Reading	8.5	9.4	10.0	11.4	12.0
History	64	68	76	87	92

Graph the data and the line of best fit in the coordinate system below. Write a sentence that describes the meaning of the slope of the line of best fit.

Your solution

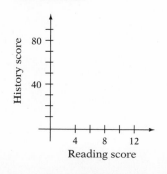

Solution on p. A20

Content and Format © 1996 HMCo.

4.4 Exercises

. . . .

Objective A

Find the equation of the line that contains the given point and has the given slope.

1. Point $(0, 5)$, $m = 2$

2. Point $(0, 3)$, $m = 1$

3. Point $(2, 3)$, $m = \frac{1}{2}$

4. Point $(5, 1)$, $m = \frac{2}{3}$

5. Point $(-1, 4)$, $m = \frac{5}{4}$

6. Point $(-2, 1)$, $m = \frac{3}{2}$

7. Point $(3, 0)$, $m = -\frac{5}{3}$

8. Point $(-2, 0)$, $m = \frac{3}{2}$

9. Point $(2, 3)$, $m = -3$

10. Point $(1, 5)$, $m = -\frac{4}{5}$

11. Point $(-1, 7)$, $m = -3$

12. Point $(-2, 4)$, $m = -4$

13. Point $(-1, -3)$, $m = \frac{2}{3}$

14. Point $(-2, -4)$, $m = \frac{1}{4}$

15. Point $(0, 0)$, $m = \frac{1}{2}$

16. Point $(0, 0)$, $m = \frac{3}{4}$

17. Point $(2, -3)$, $m = 3$

18. Point $(4, -5)$, $m = 2$

19. Point $(3, 5)$, $m = -\frac{2}{3}$

20. Point $(5, 1)$, $m = -\frac{4}{5}$

21. Point $(0, -3)$, $m = -1$

22. Point $(2, 0)$, $m = \frac{5}{6}$

23. Point $(1, -4)$, $m = \frac{7}{5}$

24. Point $(3, 5)$, $m = -\frac{3}{7}$

25. Point $(4, -1)$, $m = -\frac{2}{5}$

26. Point $(-3, 5)$, $m = -\frac{1}{4}$

27. Point $(3, -4)$, slope is undefined

28. Point $(-2, 5)$, slope is undefined

29. Point $(-2, -5)$, $m = -\frac{5}{4}$

30. Point $(-3, -2)$, $m = -\frac{2}{3}$

Find the equation of the line that contains the given point and has the given slope.

31. Point $(-2, -3)$, $m = 0$ **32.** Point $(-3, -2)$, $m = 0$ **33.** Point $(4, -5)$, $m = -2$

34. Point $(-3, 5)$, $m = 3$ **35.** Point $(-5, -1)$, slope is undefined   **36.** Point $(0, 4)$, slope is undefined

---

Objective B

Find the equation of the line that contains the given points.

37. $P_1(0, 2)$, $P_2(3, 5)$ **38.** $P_1(0, 4)$, $P_2(1, 5)$ **39.** $P_1(0, -3)$, $P_2(-4, 5)$

40. $P_1(0, -2)$, $P_2(-3, 4)$ **41.** $P_1(-1, 1)$, $P_2(-4, -1)$ **42.** $P_1(4, 1)$, $P_2(6, 3)$

43. $P_1(-1, 3)$, $P_2(2, 4)$ **44.** $P_1(-1, 1)$, $P_2(4, 4)$ **45.** $P_1(-1, -2)$, $P_2(3, 4)$

46. $P_1(-3, -1)$, $P_2(2, 4)$ **47.** $P_1(0, 3)$, $P_2(2, 0)$ **48.** $P_1(0, 4)$, $P_2(2, 0)$

49. $P_1(-3, -1)$, $P_2(2, -1)$ **50.** $P_1(-3, -5)$, $P_2(4, -5)$ **51.** $P_1(-2, -3)$, $P_2(-1, -2)$

52. $P_1(4, 1)$, $P_2(3, -2)$ **53.** $P_1(-2, 3)$, $P_2(2, -1)$ **54.** $P_1(3, 1)$, $P_2(-3, -2)$

55. $P_1(2, 3)$, $P_2(5, 5)$ **56.** $P_1(7, 2)$, $P_2(4, 4)$ **57.** $P_1(2, 0)$, $P_2(0, -1)$

58. $P_1(0, 4)$, $P_2(-2, 0)$ **59.** $P_1(3, -4)$, $P_2(-2, -4)$ **60.** $P_1(-3, 3)$, $P_2(-2, 3)$

Find the equation of the line that contains the given points.

61. $P_1(0, 0), P_2(4, 3)$

62. $P_1(2, -5), P_2(0, 0)$

63. $P_1(2, -1), P_2(-1, 3)$

64. $P_1(3, -5), P_2(-2, 1)$

65. $P_1(-2, 5), P_2(-2, -5)$

66. $P_1(3, 2), P_2(3, -4)$

67. $P_1(2, 1), P_2(-2, -3)$

68. $P_1(-3, -2), P_2(1, -4)$

69. $P_1(-4, -3), P_2(2, 5)$

70. $P_1(4, 5), P_2(-4, 3)$

71. $P_1(0, 3), P_2(3, 0)$

72. $P_1(1, -3), P_2(-2, 4)$

Objective C   *Application Problems*

Solve.

73. The data in the table below show the tread depth of a tire and the number of miles that have been driven on that tire. The line of best fit is $y = -0.2x + 10$, where x is the number of miles driven, in thousands, and y is the depth of the tread, in millimeters.

Miles driven, x	25	35	40	20	45
Tread depth, y	4.8	3.5	2.1	5.5	1.0

Graph the data and the line of best fit in the coordinate system at the right. Write a sentence that describes the meaning of the slope of the line of best fit.

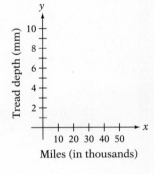

74. The data in the table below are estimates of the number of employed persons per retired person during the decade from 2020 to 2030. Information like this is important to the Social Security Administration as it plans for paying future retirement benefits. The line of best fit is $y = -0.13x + 3.35$, where x is the year (with 2020 as 0) and y is the number of employed persons per retired person.

Year, x	2 (2022)	4 (2024)	6 (2026)	8 (2028)	10 (2030)
Employed per retired, y	3.1	2.9	2.5	2.4	2.1

Graph the data and the line of best fit in the coordinate system at the right. Write a sentence that describes the meaning of the slope of the line of best fit.

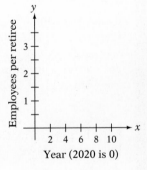

Solve.

75. The data in the table below show the recorded and estimated sales of microprocessors by Texas Instruments for 1991 through 1995. The line of best fit is $y = 1.6x + 2.4$, where x is the year (with 1991 as 1) and y is the revenue in billions.

Year, x	1 (1991)	2 (1992)	3 (1993)	4 (1994)	5 (1995)
Revenue, y	3.9	5.5	7.7	8.7	10.3

Graph the data and the line of best fit in the coordinate system at the right. Write a sentence that describes the meaning of the slope of the line of best fit.

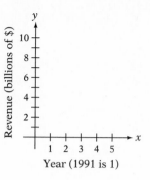

76. The data in the table below show the numbers of ATMs (automated teller machines) in service from 1985 through 1995. The line of best fit is $y = 3.6x + 61.6$, where x is the year (with 1985 as 0) and y is the number of ATMs, in thousands.

Year, x	0 (1985)	3 (1988)	5 (1990)	7 (1992)	10 (1995)
ATMs, y	61	73	81	86	98

Graph the data and the line of best fit in the coordinate system at the right. Write a sentence that describes the meaning of the slope of the line of best fit.

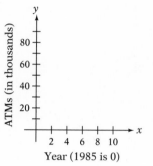

APPLYING THE CONCEPTS

77. If $(-2, 4)$ are the coordinates of a point on the line whose equation is $y = mx + 1$, what is the slope of the line?

78. If $(3, 1)$ are the coordinates of a point on the line whose equation is $y = mx - 3$, what is the slope of the line?

The formula $y - y_1 = \frac{y_2 - y_1}{x_2 - x_1}(x - x_1)$, where $x_1 \neq x_2$, is called the **two-point formula** for a straight line. This formula can be used to find the equation of a line given two points. Use this formula for Exercises 79 and 80.

79. Find the equation of the line passing through $(-2, 3)$ and $(4, -1)$.

80. Find the equation of the line passing through $(3, -1)$ and $(4, -3)$.

81. Explain why the condition $x_1 \neq x_2$ is placed on the two-point formula
[W] given above.

82. Explain how the two-point formula given above can be derived from
[W] the point–slope formula.

4.5 Parallel and Perpendicular Lines

Objective A   *To find the equations of parallel and perpendicular lines* ....

Two lines that have the same slope do not intersect and are called **parallel lines**.

The slope of each of the lines at the right is $\frac{2}{3}$.

The lines are parallel.

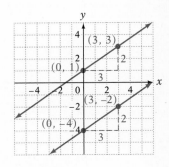

Slopes of Parallel Lines

Two nonvertical lines with slopes of m_1 and m_2 are parallel if and only if $m_1 = m_2$. Any two vertical lines are parallel.

➡ Is the line that contains the points $(-2, 1)$ and $(-5, -1)$ parallel to the line that contains the points $(1, 0)$ and $(4, 2)$?

$$m_1 = \frac{-1 - 1}{-5 - (-2)} = \frac{-2}{-3} = \frac{2}{3}$$
- Find the slope of the line through $(-2, 1)$ and $(-5, -1)$.

$$m_2 = \frac{2 - 0}{4 - 1} = \frac{2}{3}$$
- Find the slope of the line through $(1, 0)$ and $(4, 2)$.

Because $m_1 = m_2$, the lines are parallel.

➡ Find the equation of the line that contains the point $(2, 3)$ and is parallel to the line $y = \frac{1}{2}x - 4$.

The slope of the given line is $\frac{1}{2}$. Because parallel lines have the same slope, the slope of the unknown line is also $\frac{1}{2}$.

$$y - y_1 = m(x - x_1)$$
- Use the point–slope formula.

$$y - 3 = \frac{1}{2}(x - 2)$$
- $m = \frac{1}{2}$, $(x_1, y_1) = (2, 3)$

$$y - 3 = \frac{1}{2}x - 1$$
- Write in the form $y = mx + b$.

$$y = \frac{1}{2}x + 2$$

The equation of the line is $y = \frac{1}{2}x + 2$.

➡ Find the equation of the line that contains the point $(-1, 4)$ and is parallel to the line $2x - 3y = 5$.

Because the lines are parallel, the slope of the unknown line is the same as the slope of the given line. Solve $2x - 3y = 5$ for y and determine its slope.

$$2x - 3y = 5$$
$$-3y = -2x + 5$$
$$y = \frac{2}{3}x - \frac{5}{3}$$

The slope of the given line is $\frac{2}{3}$. Because the lines are parallel, this is the slope of the unknown line. Use the point–slope formula to determine the equation.

$$y - y_1 = m(x - x_1)$$ • Use the point–slope formula.

$$y - 4 = \frac{2}{3}[x - (-1)]$$ • $m = \frac{2}{3}$, $(x_1, y_1) = (-1, 4)$

$$y - 4 = \frac{2}{3}x + \frac{2}{3}$$ • Write in the form $y = mx + b$.

$$y = \frac{2}{3}x + \frac{14}{3}$$

The equation of the line is $y = \frac{2}{3}x + \frac{14}{3}$.

Two lines that intersect at right angles are **perpendicular lines.**

Any horizontal line is perpendicular to any vertical line. For example, $x = 3$ is perpendicular to $y = -2$.

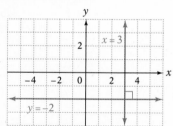

Slopes of Perpendicular Lines

If m_1 and m_2 are the slopes of two lines, neither of which is vertical, then the lines are perpendicular if and only if $m_1 \cdot m_2 = -1$.

A vertical line is perpendicular to a horizontal line.

Solving $m_1 \cdot m_2 = -1$ for m_1 gives $m_1 = -\dfrac{1}{m_2}$. This last equation states that the slopes of perpendicular lines are *negative reciprocals* of each other.

➡ Is the line that contains the points $(4, 2)$ and $(-2, 5)$ perpendicular to the line that contains the points $(-4, 3)$ and $(-3, 5)$?

$$m_1 = \frac{5 - 2}{-2 - 4} = \frac{3}{-6} = -\frac{1}{2}$$ • Find the slope of the line through $(4, 2)$ and $(-2, 5)$.

$$m_2 = \frac{5 - 3}{-3 - (-4)} = \frac{2}{1} = 2$$ • Find the slope of the line through $(-4, 3)$ and $(-3, 5)$.

$$m_1 \cdot m_2 = -\frac{1}{2}(2) = -1$$ • Find the product of the two slopes.

Because $m_1 \cdot m_2 = -1$, the lines are perpendicular.

➡ Are the graphs of the equations $3x + 4y = 8$ and $8x + 6y = 5$ perpendicular?

To determine whether the lines are perpendicular, solve each equation for y and find the slope of each line. Then use the equation $m_1 \cdot m_2 = -1$.

$$3x + 4y = 8 \qquad\qquad\qquad 8x + 6y = 5$$
$$4y = -3x + 8 \qquad\qquad\qquad 6y = -8x + 5$$
$$y = -\frac{3}{4}x + 2 \quad m_1 = -\frac{3}{4} \qquad y = -\frac{4}{3}x + \frac{5}{6} \quad m_2 = -\frac{4}{3}$$

$$m_1 \cdot m_2 = \left(-\frac{3}{4}\right)\left(-\frac{4}{3}\right) = 1$$

Because $m_1 \cdot m_2 = 1 \neq -1$, the lines are not perpendicular.

➡ Find the equation of the line that contains the point $(-2, 1)$ and is perpendicular to the line $y = -\frac{2}{3}x + 2$.

The slope of the given line is $-\frac{2}{3}$. The slope of the line perpendicular to the given line is the negative reciprocal of $-\frac{2}{3}$, which is $\frac{3}{2}$.

$$y - y_1 = m(x - x_1)$$ • Use the point–slope formula.

$$y - 1 = \frac{3}{2}[x - (-2)]$$ • $m = \frac{3}{2}$, $(x_1, y_1) = (-2, 1)$

$$y - 1 = \frac{3}{2}x + 3$$ • Write in the form $y = mx + b$.

$$y = \frac{3}{2}x + 4$$

The equation of the perpendicular line is $y = \frac{3}{2}x + 4$.

➡ Find the equation of the line that contains the point $(3, -4)$ and is perpendicular to the line $2x - y = -3$.

$$2x - y = -3$$ • Find the slope of the given line.
$$-y = -2x - 3$$
$$y = 2x + 3$$ • The slope is 2. The slope of the line perpendicular to this line is $-\frac{1}{2}$.

$$y - y_1 = m(x - x_1)$$ • Use the point–slope formula.

$$y - (-4) = -\frac{1}{2}(x - 3)$$ • $m = -\frac{1}{2}$, $(x_1, y_1) = (3, -4)$

$$y + 4 = -\frac{1}{2}x + \frac{3}{2}$$ • Write in the form $y = mx + b$.

$$y = -\frac{1}{2}x - \frac{5}{2}$$

The equation of the perpendicular line is $y = -\frac{1}{2}x - \frac{5}{2}$.

Example 1

Is the line that contains the points $(-4, 2)$ and $(1, 6)$ parallel to the line that contains the points $(2, -4)$ and $(7, 0)$?

Solution

$$m_1 = \frac{6 - 2}{1 - (-4)} = \frac{4}{5}$$

$$m_2 = \frac{0 - (-4)}{7 - 2} = \frac{4}{5}$$

$$m_1 = m_2 = \frac{4}{5}$$

The lines are parallel.

You Try It 1

Is the line that contains the points $(-2, -3)$ and $(7, 1)$ perpendicular to the line that contains the points $(4, 1)$ and $(6, -5)$?

Your solution

Example 2

Are the lines $4x - y = -2$ and $x + 4y = -12$ perpendicular?

Solution

$4x - y = -2$
$\quad -y = -4x - 2$
$\qquad y = 4x + 2 \qquad$ • $m_1 = 4$

$x + 4y = -12$
$\quad 4y = -x - 12$
$\qquad y = -\frac{1}{4}x - 3 \qquad$ • $m_2 = -\frac{1}{4}$

$$m_1 \cdot m_2 = 4\left(-\frac{1}{4}\right) = -1$$

The lines are perpendicular.

You Try It 2

Are the lines $5x + 2y = 2$ and $5x + 2y = -6$ parallel?

Your solution

Example 3

Find the equation of the line that contains the point $(3, -1)$ and is parallel to the line $y = \frac{3}{2}x - 2$.

Solution

$$y - y_1 = m(x - x_1)$$

$$y - (-1) = \frac{3}{2}(x - 3) \qquad \bullet \ m = \frac{3}{2}$$

$$y + 1 = \frac{3}{2}x - \frac{9}{2}$$

$$y = \frac{3}{2}x - \frac{11}{2}$$

The equation of the line is $y = \frac{3}{2}x - \frac{11}{2}$.

You Try It 3

Find the equation of the line that contains the point $(-2, 2)$ and is perpendicular to the line $y = \frac{1}{4}x - 3$.

Your solution

Solutions on p. A20

4.5 Exercises

. . . .

Objective A

1. Is the line $x = -2$ perpendicular to the line $y = 3$?

2. Is the line $y = \frac{1}{2}$ perpendicular to the line $y = -4$?

3. Is the line $x = -3$ parallel to the line $y = \frac{1}{3}$?

4. Is the line $x = 4$ parallel to the line $x = -4$?

5. Is the line $y = \frac{2}{3}x - 4$ parallel to the line $y = -\frac{3}{2}x - 4$?

6. Is the line $y = -2x + \frac{2}{3}$ parallel to the line $y = -2x + 3$?

7. Is the line $y = \frac{4}{3}x - 2$ perpendicular to the line $y = -\frac{3}{4}x + 2$?

8. Is the line $y = \frac{1}{2}x + \frac{3}{2}$ perpendicular to the line $y = -\frac{1}{2}x + \frac{3}{2}$?

9. Are the lines $2x + 3y = 2$ and $2x + 3y = -4$ parallel?

10. Are the lines $2x - 4y = 3$ and $2x + 4y = -3$ parallel?

11. Are the lines $x - 4y = 2$ and $4x + y = 8$ perpendicular?

12. Are the lines $4x - 3y = 2$ and $4x + 3y = -7$ perpendicular?

13. Is the line that contains the points $(3, 2)$ and $(1, 6)$ parallel to the line that contains the points $(-1, 3)$ and $(-1, -1)$?

14. Is the line that contains the points $(4, -3)$ and $(2, 5)$ parallel to the line that contains the points $(-2, -3)$ and $(-4, 1)$?

15. Is the line that contains the points $(-3, 2)$ and $(4, -1)$ perpendicular to the line that contains the points $(1, 3)$ and $(-2, -4)$?

16. Is the line that contains the points $(-1, 2)$ and $(3, 4)$ perpendicular to the line that contains the points $(-1, 3)$ and $(-4, 1)$?

17. Is the line that contains the points $(-5, 0)$ and $(0, 2)$ parallel to the line that contains the points $(5, 1)$ and $(0, -1)$?

18. Is the line that contains the points $(3, 5)$ and $(-3, 3)$ perpendicular to the line that contains the points $(2, -5)$ and $(-4, 4)$?

19. Find the equation of the line that contains the point $(4, -3)$ and is parallel to the line $y = -\frac{3}{4}x + 2$.

20. Find the equation of the line that contains the point $(-5, 4)$ and is parallel to the line $y = -\frac{3}{5}x - 7$.

21. Find the equation of the line that contains the point $(-2, -4)$ and is parallel to the line $2x - 3y = 2$.

22. Find the equation of the line that contains the point $(3, 2)$ and is parallel to the line $3x + y = -3$.

23. Find the equation of the line that contains the point $(4, 1)$ and is perpendicular to the line $y = -3x + 4$.

24. Find the equation of the line that contains the point $(2, -5)$ and is perpendicular to the line $y = \frac{5}{2}x - 4$.

25. Find the equation of the line that contains the point $(4, -3)$ and is perpendicular to the line $y = 4x$.

26. Find the equation of the line that contains the point $(-2, 5)$ and is perpendicular to the line $y = \frac{2}{3}x - 5$.

27. Find the equation of the line that contains the point $(-1, -3)$ and is perpendicular to the line $3x - 5y = 2$.

28. Find the equation of the line that contains the point $(-1, 3)$ and is perpendicular to the line $2x + 4y = -1$.

APPLYING THE CONCEPTS

Find the value of k such that the line containing P_1 and P_2 is parallel to the line containing P_3 and P_4.

29. $P_1(3, 4)$, $P_2(-2, -1)$, $P_3(4, 1)$, $P_4(0, k)$

30. $P_1(6, 2)$, $P_2(-3, -1)$, $P_3(1, 5)$, $P_4(k, 4)$

Find the value of k such that the line containing P_1 and P_2 is perpendicular to the line containing P_3 and P_4.

31. $P_1(2, 5)$, $P_2(6, 4)$, $P_3(-2, 1)$, $P_4(-3, k)$

32. $P_1(-1, 4)$, $P_2(2, 5)$, $P_3(6, 1)$, $P_4(k, 4)$

33. The graphs of $y = -\frac{1}{2}x + 2$ and $y = \frac{2}{3}x - 5$ intersect at the point whose coordinates are $(6, -1)$. Find the equation of a line whose graph intersects the graphs of the given lines to form a right triangle. (*Hint:* There is more than one answer to this question.)

34. Explain how to determine whether the graphs of two lines are [W] parallel.

35. Explain how to determine whether the graphs of two lines are [W] perpendicular.

4.6   Linear Inequalities in Two Variables

Objective A   *To graph an inequality in two variables* ....

The graph of the linear equation $y = x - 2$ separates a plane into three sets:

the set of points on the line
the set of points above the line
the set of points below the line

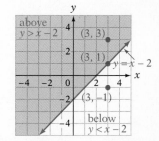

The point $(3, 1)$ is a solution of $y = x - 2$.

$$y = x - 2$$
$$\begin{array}{c|c} 1 & 3 - 2 \\ \end{array}$$
$$1 = 1$$

The point $(3, 3)$ is a solution of $y > x - 2$.

$$y > x - 2$$
$$\begin{array}{c|c} 3 & 3 - 2 \\ \end{array}$$
$$3 > 1$$

Any point above the line is a solution of $y > x - 2$.

The point $(3, -1)$ is a solution of $y < x - 2$.

$$y < x - 2$$
$$\begin{array}{c|c} -1 & 3 - 2 \\ \end{array}$$
$$-1 < 1$$

Any point below the line is a solution of $y < x - 2$.

The solution set of $y = x - 2$ is all points on the line. The solution set of $y > x - 2$ is all points above the line. The solution set of $y < x - 2$ is all points below the line. The solution set of an inequality in two variables is a **half-plane.**

The following illustrates the procedure for graphing a linear inequality.

➡ Graph the solution set of $2x + 3y \leq 6$.

Solve the inequality for y.

$$2x + 3y \leq 6$$
$$2x - 2x + 3y \leq -2x + 6$$
$$3y \leq -2x + 6$$
$$y \leq -\frac{2}{3}x + 2$$

Change the inequality to an equality and graph the line. If the inequality is $\geq$ or $\leq$, the line is in the solution set and is shown by a **solid line.** If the inequality is $>$ or $<$, the line is not a part of the solution set and is shown by a **dashed line.**

$$y = -\frac{2}{3}x + 2$$

If the inequality is in the form $y > mx + b$ or $y \geq mx + b$, shade the **upper half-plane.** If the inequality is in the form $y < mx + b$ or $y \leq mx + b$, shade the **lower half-plane.**

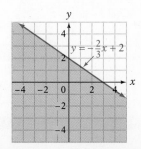

The inequality $2x + 3y \leq 6$ can also be graphed as shown below.

$2x + 3y = 6$ • **Change the inequality to an equality.**

$2x + 3(0) = 6$ • **Find the x- and y-intercepts of the equation.**
$\qquad 2x = 6$
$\qquad\ \ x = 3$ • **The x-intercept is (3, 0).**

$2(0) + 3y = 6$
$\qquad\quad 3y = 6$
$\qquad\quad\ y = 2$ • **The y-intercept is (0, 2).**

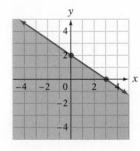

Graph the ordered pairs $(3, 0)$ and $(0, 2)$. Draw a solid line through the points because the inequality is $\leq$. The point $(0, 0)$ can be used to determine which region to shade. If $(0, 0)$ is a solution of the inequality, then shade the region that includes the point $(0, 0)$. If $(0, 0)$ is not a solution of the inequality, then shade the region that does not include the point $(0, 0)$. For this example, $(0, 0)$ is a solution of the inequality. The region that contains the point $(0, 0)$ is shaded.

CONSIDER THIS

Any ordered pair is of the form (x, y). For the point $(0, 0)$, substitute 0 for x and 0 for y in the inequality.

$2x + 3y \leq 6$
$2(0) + 3(0) \leq 6$
$\qquad\qquad 0 \leq 6$ True

If the line passes through point $(0, 0)$, another point must be used to determine which region to shade. For example, use the point $(1, 0)$.

It is important to note that every point in the shaded region is a solution of the inequality and that every solution of the inequality is a point in the shaded region. No point outside the shaded region is a solution of the inequality.

Example 1
Graph the solution set of $3x + y > -2$.

Solution
$$3x + y > -2$$
$$3x - 3x + y > -3x - 2$$
$$y > -3x - 2$$

Graph $y = -3x - 2$ as a dashed line.
Shade the upper half-plane.

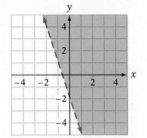

You Try It 1
Graph the solution set of $x - 3y < 2$.

Your solution

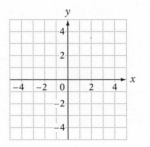

Solution on p. A21

4.6 Exercises

Objective A

Graph the solution set.

1. $y > -x + 4$

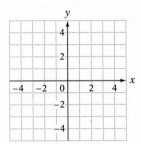

2. $y < x + 3$

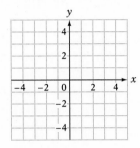

3. $y > 2x + 3$

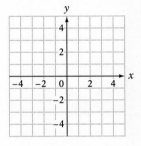

4. $y \geq -2x + 4$

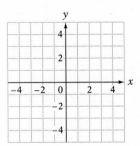

5. $y > \dfrac{3}{2}x - 4$

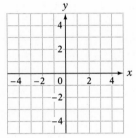

6. $y > -\dfrac{5}{4}x + 1$

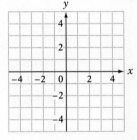

7. $y \leq -\dfrac{3}{4}x - 1$

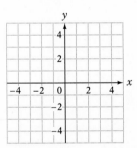

8. $y \leq -\dfrac{5}{2}x - 4$

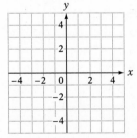

9. $y \leq -2$

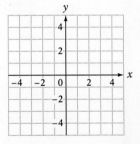

10. $y > 3$

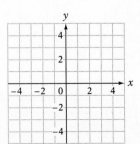

11. $y < \dfrac{4}{5}x + 3$

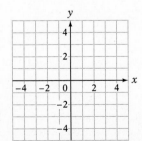

12. $y \leq -\dfrac{6}{5}x - 2$

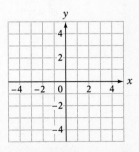

Graph the solution set.

13. $3x - y < 9$

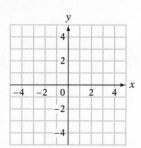

14. $3x + y \geq 6$

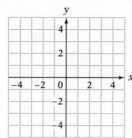

15. $2x + 2y \leq -4$

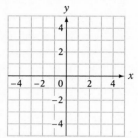

16. $-4x + 3y < -12$

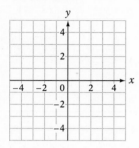

17. $-2x + 3y \leq 6$

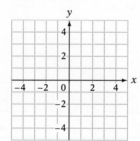

18. $3x - 4y > 12$

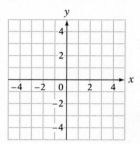

APPLYING THE CONCEPTS

Graph the solution set.

19. $\dfrac{x}{4} + \dfrac{y}{2} > 1$

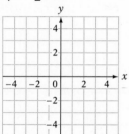

20. $2x - 3(y + 1) > y - (4 - x)$

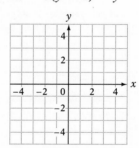

Write the inequality given its graph.

21.

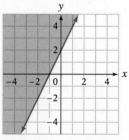

22.

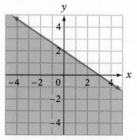

23. Are there any points whose coordinates satisfy both $y \leq x + 3$ and
[W] $y \geq -\dfrac{1}{2}x + 1$? If so, given the coordinates of three such points. If not, explain why not.

24. Are there any points whose coordinates satisfy both $y \leq x - 1$ and
[W] $y \geq x + 2$? If so, give the coordinates of three such points. If not, explain why not.

Projects in Mathematics

Graphing Linear Equations with a Graphing Utility

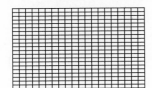

A computer or graphing calculator screen is divided into *pixels*. There are approximately 6000 to 790,000 pixels available on the screen (depending on the computer or calculator). The greater the number of pixels, the smoother a graph will appear. A portion of a screen is shown at the left. Each little rectangle represents one pixel.

The graphing utilities that are used by computers or calculators to graph an equation do basically what we have shown in the text: They choose values of x and, for each, calculate the corresponding value of y. The pixel corresponding to the ordered pair is then turned on. The graph is jagged because pixels are much larger than the dots we draw on paper.

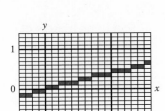

The graph of $y = 0.45x$ is shown at the left as the calculator drew it. The x- and y-axes have been chosen so that each pixel represents $\frac{1}{10}$ of a unit. Consider the region of the graph where $x = 1$, 1.1, and 1.2.

The corresponding values of y are 0.45, 0.495, and 0.54. Because the y-axis is in tenths, the numbers 0.45, 0.495, and 0.54 are rounded to the nearest tenth before plotting. Rounding 0.45, 0.495 and 0.54 to the nearest tenth results in 0.5 for each number. Thus the ordered pairs $(1, 0.45)$, $(1.1, 0.495)$, and $(1.2, 0.54)$ are graphed as $(1, 0.5)$, $(1.1, 0.5)$, and $(1.2, 0.5)$. These points appear as three illuminated horizontal pixels. The graph of the line appears horizontal. However, if you use the TRACE feature of the calculator (see Appendix A), the actual y-coordinate for each value of x is displayed.

Here are the keystrokes to graph $y = \frac{2}{3}x + 1$. First the domain of the function (Xmin to Xmax) and the range (Ymin to Ymax) are entered. This is called the **viewing window** for the graph. Then the equation is entered. By changing the keystrokes that are shaded in color, you can graph different equations.

TI-82

ZOOM 6 Y= CLEAR 2 X,T,θ ÷ 3 + 1 GRAPH

SHARP EL-9300

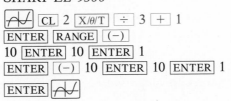 CL 2 X/θ/T ÷ 3 + 1 ENTER RANGE (−) 10 ENTER 10 ENTER 1 ENTER (−) 10 ENTER 10 ENTER 1 ENTER

CASIO fx-7700GB

SHIFT F5 EXE Range SHIFT (−) 10 EXE 10 EXE EXE SHIFT (−) 10 EXE 10 EXE Range Range Graph 2 X,θ,T ÷ 3 + 1 EXE

Graph using a graphing utility.

1. $y = 2x + 1$ For $2x$, you may enter $2 \times x$ or just $2x$. The times sign $\times$ is not necessary on many graphing calculators.

2. $y = -\frac{1}{2}x - 2$ Use the (−) key to enter a negative sign.

3. $3x + 2y = 6$ Solve for y. Then enter the equation.

4. $4x + 3y = 75$ You must adjust the viewing window. *Suggestion:* Xmin = −25, Xmax = 25, Xscl = 5, Ymin = −35, Ymax = 35, Yscl = 5.

Chapter Summary

Key Words   A *rectangular coordinate system* is formed by two number lines, one horizontal and one vertical, that intersect at the zero point of each line. A rectangular coordinate system divides the plane into four regions called *quadrants*. The number lines that make up a rectangular coordinate system are called the *coordinate axes*, or simply the *axes*. The *origin* is the point of intersection of the two coordinate axes.

Every point in the plane can be identified by an ordered pair (x, y). The first number in an ordered pair is called the *abscissa* or *x*-coordinate. The second number is called the *ordinate* or *y*-coordinate. The *coordinates* of a point are the numbers in the ordered pair associated with the point.

A *relation* is any set of ordered pairs. The *domain* of a relation is the set of first coordinates of the ordered pairs. The *range* is the set of second coordinates of the ordered pairs. A *function* is a relation in which no two ordered pairs have the same first coordinate and different second coordinates. A function designated by $f(x)$ is written in *functional notation*. The *value* of the function at x is $f(x)$.

An equation of the form $y = mx + b$, where m is the coefficient of x and b is a constant, is a *linear equation in two variables*. m is the slope of the line, and $(0, b)$ is the *y*-intercept. An equation of the form $Ax + By = C$ is also a linear equation in two variables. A *solution* of a linear equation in two variables is an ordered pair (x, y) that makes the equation a true statement. The *graph of a linear equation* is a straight line.

The point at which a graph crosses the *x*-axis is called the *x-intercept*. The point at which a graph crosses the *y*-axis is called the *y-intercept*.

The *slope* of a line is a measure of the slant of the line. The symbol for slope is m. A line that slants upward to the right has a *positive slope*. A line that slants downward to the right has a *negative slope*. A horizontal line has *zero slope*. The slope of a vertical line is *undefined*.

Two lines that have the same slope do not intersect and are *parallel lines*. Two lines that intersect at right angles are *perpendicular lines*.

An inequality of the form $y > mx + b$ or $Ax + By > C$ is a *linear inequality in two variables*. The symbol $>$ here could be replaced by $\geq$, $<$, or $\leq$. The solution set of an inequality in two variables is a *half-plane*.

Essential Rules   **To find the *x*-intercept,** let $y = 0$.
To find the *y*-intercept, let $x = 0$.

Slope of a Linear Equation	Slope $= m = \dfrac{y_2 - y_1}{x_2 - x_1}$, $x_1 \neq x_2$
Slope–Intercept Form of a Straight Line	$y = mx + b$
Point–Slope Formula	$y - y_1 = m(x - x_1)$
Slopes of Parallel Lines	$m_1 = m_2$
Slopes of Perpendicular Lines	$m_1 \cdot m_2 = -1$

Chapter Review Exercises

1. Graph $(0, -1)$, $(2, 0)$, $(3, 2)$, and $(-1, 4)$.

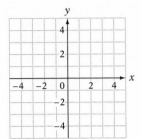

2. Graph the ordered-pair solutions of $y = -\frac{1}{2}x - 2$ when $x = -4, -2, 0, 2$.

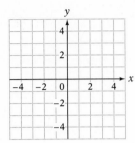

3. Find the ordered-pair solution of $y = 2x + 6$ that corresponds to $x = -3$.

4. Does $y = -x + 3$, where $x \in \{-2, 0, 3, 5\}$, define y as a function of x?

5. Given $f(x) = x^2 - 2$, find $f(-1)$.

6. Find the x- and y-intercepts of $3x - 2y = 24$.

7. Graph $y = \frac{1}{4}x + 3$.

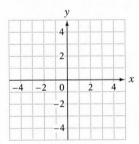

8. Graph $3x - 2y = -6$.

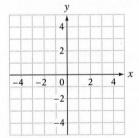

9. Graph the line that has slope 2 and y-intercept $(0, -4)$.

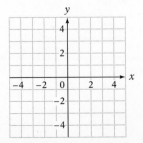

10. Graph the solution set of $y < \frac{3}{4}x - 2$.

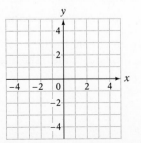

11. Find the slope of the line containing the points (9, 8) and (−2, 1).

12. Find the equation of the line that has slope −3 and y-intercept (0, 5).

13. Find the equation of the line that passes through the point (−3, 4) and has slope $\frac{2}{3}$.

14. Find the equation of the line that contains the points (−2, 5) and (4, 1).

15. Find the equation of the line that contains the point (1, 2) and is parallel to the line $y = -\frac{3}{2}x - 6$.

16. Find the equation of the line that contains the point (−2, −3) and is perpendicular to the line $y = -\frac{1}{2}x - 3$.

17. A computer bulletin board charges an access fee of $10 per month plus $.20 per minute to use the service. An equation that represents the monthly cost to use this bulletin board is $C = 0.20x + 10$, where C is the monthly cost and x is the number of minutes. Graph this equation for $0 \leq x \leq 200$. The point (100, 30) is on the graph. Write a sentence that describes the meaning of this ordered pair.

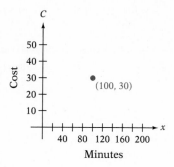

18. The height and weight of 8 seventh-grade students are shown in the following table. Write a relation where the first coordinate is the height, in inches, of the student and the second coordinate is the weight, in pounds, of the student. Is the relation a function?

Height	55	57	53	57	60	61	58	54
Weight	95	101	94	98	100	105	97	95

19. The graph at the right shows the increase in the cost of tuition for a college for the years 1990 through 1995 (with 1990 as 0). Find the slope of the line. Write a sentence that states the meaning of the slope.

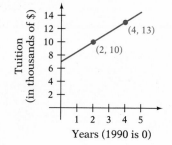

20. The data in the table below show the age of a machine, in years, and its down time, in hours per month. (Down time is the number of hours a machine is not available because of a malfunction.) The line of best fit is $y = 10.4x - 1.1$, where x is the age of the machine in years and y is the down time each month.

Age, x	1.0	2.7	4.1	1.2	2.5	1.9
Down time, y	10	30	40	9	25	19

Graph the data and the line in the coordinate system at the right. Write a sentence that describes the meaning of the slope of the line of best fit.

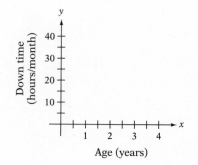

Cumulative Review Exercises

. . . .

1. Let $x \in \{-5, -3, -1\}$. For what values of x is the inequality $x \leq -3$ a true statement?

2. Write $\frac{17}{20}$ as a decimal.

3. Simplify: $3\sqrt{45}$

4. Simplify $12 - 18 \div 3(-2)^2$.

5. Evaluate $\frac{a-b}{a^2-c}$ when $a = -2$, $b = 3$, and $c = -4$.

6. Simplify: $3d - 9 - 7d$

7. Simplify: $4(-8z)$

8. Simplify: $2(x + y) - 5(3x - y)$

9. Graph: $\{x \mid x < -2\} \cup \{x \mid x > 0\}$

$$\begin{array}{c}\xleftarrow{\hspace{0.3cm}} \hspace{-0.2cm}+\hspace{-0.1cm}+\hspace{-0.1cm}+\hspace{-0.1cm}+\hspace{-0.1cm}+\hspace{-0.1cm}+\hspace{-0.1cm}+\hspace{-0.1cm}+\hspace{-0.1cm}+\hspace{-0.1cm}+\hspace{-0.1cm}+\hspace{-0.2cm}\xrightarrow{\hspace{0.3cm}}\\ \hspace{-0.2cm}-5\hspace{0.1cm}-4\hspace{0.1cm}-3\hspace{0.1cm}-2\hspace{0.1cm}-1\hspace{0.15cm}0\hspace{0.2cm}1\hspace{0.2cm}2\hspace{0.2cm}3\hspace{0.2cm}4\hspace{0.2cm}5\end{array}$$

10. Solve: $2x - \frac{2}{3} = \frac{7}{3}$

11. Solve: $3x - 2(10x - 6) = x - 6$

12. Solve: $4x - 3 < 9x + 2$

13. Solve: $3x - 1 < 4$ and $x - 2 > 2$

14. Solve: $|3x - 5| < 5$

15. Given $f(t) = t^2 + t$, find $f(2)$.

16. Find the slope of the line containing the points $(2, -3)$ and $(4, 1)$.

17. Graph $y = 3x + 1$.

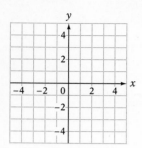

18. Graph $x = -3$.

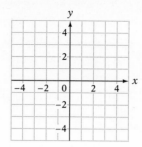

19. Graph the line that has slope $-\frac{2}{3}$ and y-intercept $(0, 4)$.

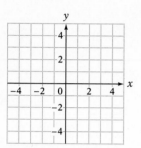

20. Graph the solution set of $3x - 2y \geq 6$.

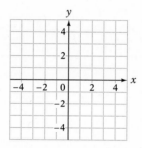

21. Find the equation of the line that passes through the points $(6, -4)$ and $(-3, -1)$.

22. Find the equation of the line that contains the point $(2, 4)$ and is parallel to the line $y = -\frac{3}{2}x + 2$.

23. Two planes are 1800 mi apart and traveling toward each other. The first plane is traveling at twice the speed of the second plane. The planes meet in 3 h. Find the speed of each plane.

24. A grocer combines coffee that costs $8 per pound with coffee that costs $3 per pound. How many pounds of each should be used to make 80 lb of a blend that costs $5 per pound?

25. The graph at the right shows the relationship between the cost of a rental house and the depreciation allowed for income tax purposes. Find the slope of the line between the two points on the graph. Write a sentence that states the meaning of the slope.

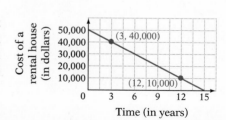

Systems of Linear Equations and Inequalities

Input–Output Analysis

The economies of the industrial nations are very complex; they comprise hundreds of different industries, and each industry supplies other industries with goods and services needed in the production process. For example, the steel industry requires coal to produce steel, and the coal industry requires steel (in the form of machinery) to mine and transport coal.

Wassily Leontief, a Russian-born economist, developed a method of describing mathematically the interactions of an economic system. His technique was to examine various sectors of an economy (steel industry, oil, farms, autos, and so on) and determine how each sector interacted with the others. More than five hundred sectors of the economy were studied.

The interactions of every sector with the others were expressed as a series of equations. This series of equations is called a *system of equations*. Using a computer, economists searched for a solution to the system of equations that would determine the output levels various sectors would have to meet to satisfy the requests from other sectors. The method is called input–output analysis.

Input–output analysis has many applications. For example, it is used today to predict the production needs of large corporations and to determine the effect of price changes on the economy. In recognition of the importance of his ideas, Leontief was awarded the Nobel Prize in Economics in 1973.

This chapter begins the study of systems of equations.

Solving Systems of Linear Equations by Graphing

Objective A   *To solve a system of linear equations by graphing* ....

A **system of equations** is two or more equations considered together. The system at the right is a system of two linear equations in two variables. The graphs of the equations are straight lines.

$$3x + 4y = 7$$
$$2x - 3y = 6$$

A **solution of a system of equations in two variables** is an ordered pair that is a solution of each equation of the system.

➡ Is $(3, -2)$ a solution of the system $2x - 3y = 12$
 $5x + 2y = 11$?

$2x - 3y = 12$	
$2(3) - 3(-2)$	12
$6 - (-6)$	
	$12 = 12$ True

$5x + 2y = 11$	
$5(3) + 2(-2)$	11
$15 + (-4)$	
	$11 = 11$ True

• Replace *x* by 3 and *y* by −2.

Yes, because $(3, -2)$ is a solution of each equation, it is a solution of the system of equations.

A solution of a system of linear equations can be found by graphing the lines of the system on the same coordinate axes. The point of intersection of the lines is the ordered pair that lies on both lines. It is the solution of the system of equations.

➡ Solve by graphing: $x + 2y = 4$
 $2x + y = -1$

Graph each line.

Find the point of intersection.

The ordered pair $(-2, 3)$ lies on each line.

The solution is $(-2, 3)$.

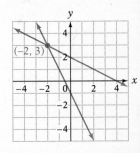

When the graphs of a system of equations intersect at only one point, the system of equations is called an **independent system of equations.**

➡ Solve by graphing: $2x + 3y = 6$
 $4x + 6y = -12$

Graph each line.

The lines are parallel and therefore do not intersect. The system of equations has no solution.

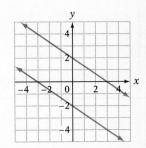

When a system of equations has no solution, it is called an **inconsistent system of equations.**

➡ Solve by graphing: $\quad x - 2y = 4$
$\qquad\qquad\qquad\qquad 2x - 4y = 8$

Graph each line.

The two equations represent the same line. This is a **dependent system of equations.** When a system of two equations is dependent, solve (if necessary) one of the equations of the system for y. The solutions of the dependent system are $(x, mx + b)$, where $y = mx + b$. For this system,

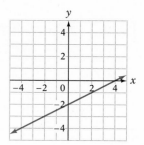

$$x - 2y = 4$$
$$-2y = -x + 4$$
$$y = \frac{1}{2}x - 2$$

The solutions are the ordered pairs $\left(x, \dfrac{1}{2}x - 2\right)$.

CONSIDER THIS

Keep in mind the ways in which independent, dependent, and inconsistent systems of equations differ. You should be able to express your understanding of these terms by using graphs.

Example 1 Solve by graphing:
$2x - y = 3$
$3x + y = 2$

Solution

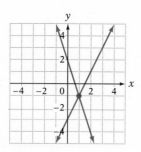

The solution is $(1, -1)$.

You Try It 1 Solve by graphing:
$x + y = 1$
$2x + y = 0$

Your solution

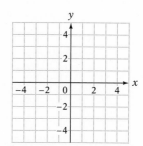

Example 2 Solve by graphing:
$2x + 3y = 6$
$y = -\dfrac{2}{3}x + 1$

Solution

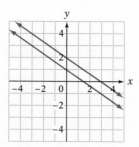

The lines are parallel and therefore do not intersect. The system of equations has no solution. It is inconsistent.

You Try It 2 Solve by graphing:
$3x - 4y = 12$
$y = \dfrac{3}{4}x - 3$

Your solution

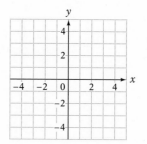

Solutions on p. A22

5.1 Exercises

. . . .

Objective A

1. Is $(2, 3)$ a solution of the system
$$3x + 4y = 18$$
$$2x - y = 1?$$

2. Is $(2, -1)$ a solution of the system
$$x - 2y = 4$$
$$2x + y = 3?$$

3. Is $(1, -2)$ a solution of the system
$$3x - y = 5$$
$$2x + 5y = -8?$$

4. Is $(-1, -1)$ a solution of the system
$$x - 4y = 3$$
$$3x + y = 2?$$

5. Is $(4, 3)$ a solution of the system
$$5x - 2y = 14$$
$$x + y = 8?$$

6. Is $(2, 5)$ a solution of the system
$$3x + 2y = 16$$
$$2x - 3y = 4?$$

7. Is $(-1, 3)$ a solution of the system
$$4x - y = -5$$
$$2x + 5y = 13?$$

8. Is $(4, -1)$ a solution of the system
$$x - 4y = 9$$
$$2x - 3y = 11?$$

9. Is $(0, 0)$ a solution of the system
$$4x + 3y = 0$$
$$2x - y = 1?$$

10. Is $(2, 0)$ a solution of the system
$$3x - y = 6$$
$$x + 3y = 2?$$

11. Is $(2, -3)$ a solution of the system
$$y = 2x - 7$$
$$3x - y = 9?$$

12. Is $(-1, -2)$ a solution of the system
$$3x - 4y = 5$$
$$y = x - 1?$$

13. Is $(5, 2)$ a solution of the system
$$y = 2x - 8$$
$$y = 3x - 13?$$

14. Is $(-4, 3)$ a solution of the system
$$y = 2x + 11$$
$$y = 5x - 19?$$

15. Is $(-2, -3)$ a solution of the system
$$3x - 4y = 6$$
$$2x - 7y = 17?$$

16. Is $(0, 0)$ a solution of the system
$$y = 2x$$
$$3x + 5y = 0?$$

17. Is $(0, -3)$ a solution of the system
$$4x - 3y = 9$$
$$2x + 5y = 15?$$

18. Is $(4, 0)$ a solution of the system
$$2x + 3y = 8$$
$$x - 5y = 4?$$

Solve by graphing.

19. $x + y = 2$
$x - y = 4$

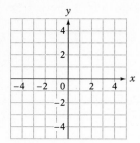

20. $x + y = 1$
$3x - y = -5$

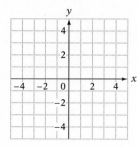

21. $x - y = -2$
$x + 2y = 10$

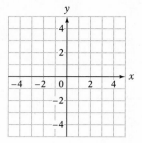

22. $2x - y = 5$
$3x + y = 5$

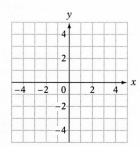

23. $3x - 2y = 6$
$y = 3$

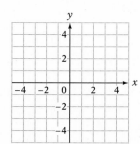

24. $x = 4$
$3x - 2y = 4$

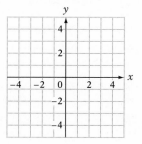

25. $x = 4$
$y = -1$

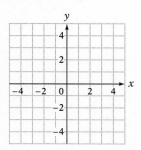

26. $x + 2 = 0$
$y - 1 = 0$

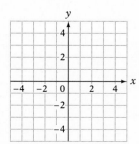

27. $2x + y = 3$
$x - 2 = 0$

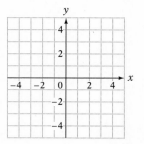

Solve by graphing.

28. $x - 3y = 6$
$y + 3 = 0$

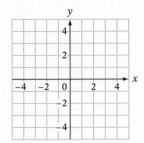

29. $x - y = 4$
$x + y = 2$

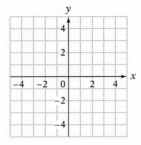

30. $2x + y = 2$
$-x + y = 5$

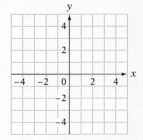

31. $y = x - 5$
$2x + y = 4$

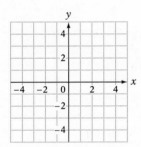

32. $2x - 5y = 4$
$y = x + 1$

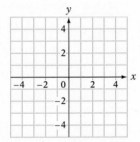

33. $y = \dfrac{1}{2}x - 2$
$x - 2y = 8$

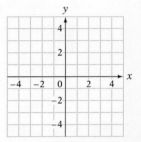

34. $2x + 3y = 6$
$y = -\dfrac{2}{3}x + 1$

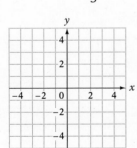

35. $2x - 5y = 10$
$y = \dfrac{2}{5}x - 2$

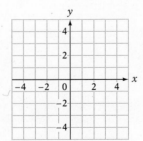

36. $3x - 2y = 6$
$y = \dfrac{3}{2}x - 3$

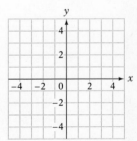

Solve by graphing.

37. $3x - 4y = 12$
$5x + 4y = -12$

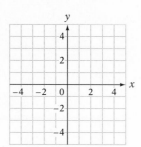

38. $2x - 3y = 6$
$2x - 5y = 10$

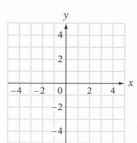

39. $2x - 3y = 2$
$5x + 4y = 5$

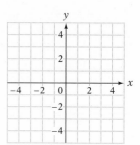

APPLYING THE CONCEPTS

40. Determine whether the statement is always true, sometimes true, or never true.
  a. A solution of a system of two equations with two variables is a point in the plane.
  b. Two parallel lines have the same slope.
  c. Two different lines with the same y-intercept are parallel.
  d. Two different lines with the same slope are parallel.

Use a graphing calculator to solve each of the following systems of equations. Round answers to the nearest hundredth. See the Projects in Mathematics feature on page 299 for assistance.

41. $y = -\dfrac{1}{2}x + 2$
$y = 2x - 1$

42. $y = 1.2x + 2$
$y = -1.3x - 3$

43. $y = \sqrt{2}x - 1$
$y = -\sqrt{3}x + 1$

44. $y = \pi x - \dfrac{2}{3}$
$y = -x + \dfrac{\pi}{2}$

45. Write three different systems of equations: (a) one that has $(-3, 5)$ as its only solution, (b) one for which there is no solution, and (c) one that is a dependent system of equations.

46. Explain how you can determine from the graph of a system of two
[W] equations in two variables whether it is an independent system of equations. Explain how you can determine whether it is an inconsistent system of equations.

47. The graph below shows the life expectancy at birth for males and
[W] females. Write an essay describing your interpretation of the data presented. Be sure to include in your discussion an interpretation of the point at which the two lines intersect.

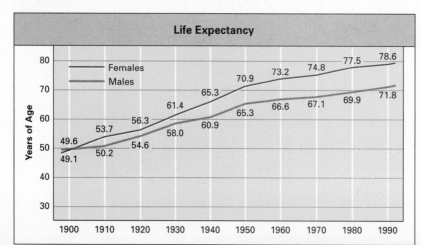

5.2 Solving Systems of Linear Equations by the Substitution Method

Objective A   *To solve a system of linear equations by the substitution method*

A graphical solution of a system of equations is based on approximating the coordinates of a point of intersection. An algebraic method called the **substitution method** can be used to find an exact solution of a system of equations.

➡ Solve by the substitution method: (1) $2x + 5y = -11$
(2) $y = 3x - 9$

CONSIDER THIS
When you evaluate a variable expression, you replace a variable with a constant. Here you are replacing a variable with a variable expression.

Equation (2) states that $y = 3x - 9$. Substitute $3x - 9$ for y in Equation (1).

$2x + 5y = -11$ • This is Equation (1).
$2x + 5(3x - 9) = -11$ • From Equation (2), substitute $3x - 9$ for y.
$2x + 15x - 45 = -11$ • Solve for x.
$17x - 45 = -11$
$17x = 34$
$x = 2$

Now substitute the value of x into Equation (2) and solve for y.

$y = 3x - 9$ • This is Equation (2).
$y = 3(2) - 9$ • Substitute 2 for x.
$y = 6 - 9 = -3$

The solution is the ordered pair $(2, -3)$.

The graph of the equations in this system of equations is shown at the right. Note that the lines intersect at the point whose coordinates are $(2, -3)$, which is the algebraic solution we determined by the substitution method.

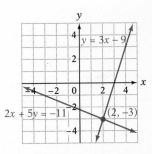

To solve a system of equations by the substitution method, sometimes we must solve one of the equations in the system of equations for one of its variables. For instance, the first step in solving the system of equations

(1) $x + 2y = -3$
(2) $2x - 3y = 5$

is to solve an equation of the system for one of its variables. Either equation can be used.

Solving Equation (1) for x: Solving Equation (2) for x:

$x + 2y = -3$ $2x - 3y = 5$
$x = -2y - 3$ $2x = 3y + 5$
$x = \dfrac{3y + 5}{2}$

Because solving Equation (1) for x does not result in fractions, it is the easier of the two equations to use.

Here is the solution of the system of equations given on the previous page.

➡ Solve by the substitution method: (1) $x + 2y = -3$
                                    (2) $2x - 3y = 5$

To use the substitution method, we must solve an equation for one of its variables. Equation (1) is used here because solving it for x does not result in fractions.

$$x + 2y = -3$$
(3) $$x = -2y - 3$$ • Solve for x. This is Equation (3).

Now substitute $-2y - 3$ for x in Equation (2) and solve for y.

$$2x - 3y = 5$$ • This is Equation (2).
$$2(-2y - 3) - 3y = 5$$ • From Equation (3), substitute $-2y - 3$ for x.
$$-4y - 6 - 3y = 5$$ • Solve for y.
$$-7y - 6 = 5$$
$$-7y = 11$$
$$y = -\frac{11}{7}$$

Substitute the value of y into Equation (3) and solve for x.

$$x = -2y - 3$$ • This is Equation (3).
$$= -2\left(-\frac{11}{7}\right) - 3$$ • Substitute $-\frac{11}{7}$ for y.
$$= \frac{22}{7} - 3 = \frac{22}{7} - \frac{21}{7} = \frac{1}{7}$$

The solution is $\left(\frac{1}{7}, -\frac{11}{7}\right)$.

The graph of the system of equations given above is shown at the right. It would be difficult to determine the exact solution of this system of equations from the graphs of the equations.

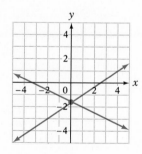

➡ Solve by the substitution method: (1) $y = 3x - 1$
                                    (2) $y = -2x - 6$

$$y = -2x - 6$$
$$3x - 1 = -2x - 6$$ • Substitute $3x - 1$ for y in Equation (2).
$$5x = -5$$ • Solve for x.
$$x = -1$$

Substitute this value of x into Equation (1) or Equation (2) and solve for y. Equation (1) is used here.

$$y = 3x - 1$$
$$y = 3(-1) - 1 = -4$$

The solution is $(-1, -4)$.

The substitution method can be used on inconsistent and dependent systems of equations.

➡ Solve by the substitution method: (1) $2x + 3y = 3$

  (2) $y = -\dfrac{2}{3}x + 3$

$2x + 3y = 3$ • This is Equation (1).

$2x + 3\left(-\dfrac{2}{3}x + 3\right) = 3$ • From Equation (2), replace y with $-\dfrac{2}{3}x + 3$.

$2x - 2x + 9 = 3$ • Solve for x.

$9 = 3$ • This is not a true equation.

Because $9 = 3$ is not a true equation, the system of equations has no solution.

Solving Equation (1) above for y, we have $y = -\dfrac{2}{3}x + 1$. Comparing this with Equation (2) reveals that the slopes are equal and the y-intercepts are different. The graphs of the equations that make up this system of equations are parallel and thus never intersect. Because the graphs do not intersect, there are no solutions of the system of equations. The system of equations is inconsistent.

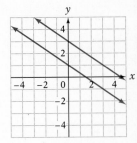

➡ Solve by the substitution method: (1) $x = 2y + 3$
  (2) $4x - 8y = 12$

$4x - 8y = 12$ • This is Equation (2).

$4(2y + 3) - 8y = 12$ • From Equation (1), replace x by $2y + 3$.

$8y + 12 - 8y = 12$ • Solve for y.

$12 = 12$ • This is a true equation.

The true equation $12 = 12$ indicates that any ordered pair (x, y) that satisfies one equation of the system satisfies the other equation. Therefore, the system of equations has an infinite number of solutions.

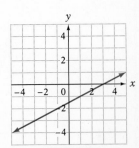

If we write Equation (1) and Equation (2) in slope–intercept form, we have

$$x = 2y + 3 \qquad\qquad 4x - 8y = 12$$

$$y = \dfrac{1}{2}x - \dfrac{3}{2} \qquad\qquad y = \dfrac{1}{2}x - \dfrac{3}{2}$$

The slope–intercept forms of the equations are the same, and therefore the graphs are the same. If we graph these two equations, we essentially graph one over the other. Accordingly, the graphs intersect at an infinite number of points. The solutions are the ordered pairs $\left(x, \dfrac{1}{2}x - \dfrac{3}{2}\right)$.

Example 1 Solve by substitution:
(1) $3x + 4y = -2$
(2) $-x + 2y = 4$

Solution Solve Equation (2) for x.
$$-x + 2y = 4$$
$$-x = -2y + 4$$
$$x = 2y - 4$$

Substitute in Equation (1).
(1) $\qquad 3x + 4y = -2$
$$3(2y - 4) + 4y = -2$$
$$6y - 12 + 4y = -2$$
$$10y - 12 = -2$$
$$10y = 10$$
$$y = 1$$

Substitute in Equation (2).
(2) $\qquad -x + 2y = 4$
$$-x + 2(1) = 4$$
$$-x + 2 = 4$$
$$-x = 2$$
$$x = -2$$

The solution is $(-2, 1)$.

You Try It 1 Solve by substitution:
$$7x - y = 4$$
$$3x + 2y = 9$$

Your solution

Example 2 Solve by substitution:
$$4x + 2y = 5$$
$$y = -2x + 1$$

Solution
$$4x + 2y = 5$$
$$4x + 2(-2x + 1) = 5$$
$$4x - 4x + 2 = 5$$
$$2 = 5$$

This is not a true equation. The system of equations is inconsistent and therefore does not have a solution.

You Try It 2 Solve by substitution:
$$3x - y = 4$$
$$y = 3x + 2$$

Your solution

Example 3 Solve by substitution:
$$y = 3x - 2$$
$$6x - 2y = 4$$

Solution
$$6x - 2y = 4$$
$$6x - 2(3x - 2) = 4$$
$$6x - 6x + 4 = 4$$
$$4 = 4$$

This is a true equation. The system of equations is dependent. The solutions are the ordered pairs $(x, 3x - 2)$.

You Try It 3 Solve by substitution:
$$y = -2x + 1$$
$$6x + 3y = 3$$

Your solution

Solutions on pp. A22–A23

5.2 Exercises

· ·

Objective A

Solve by substitution.

1. $2x + 3y = 7$
$\quad\quad x = 2$

2. $\quad\quad y = 3$
$\quad 3x - 2y = 6$

3. $\quad\quad y = x - 3$
$\quad x + y = 5$

4. $\quad\quad y = x + 2$
$\quad x + y = 6$

5. $\quad\quad x = y - 2$
$\quad x + 3y = 2$

6. $\quad\quad x = y + 1$
$\quad x + 2y = 7$

7. $\quad\quad y = 4 - 3x$
$\quad 3x + y = 5$

8. $\quad\quad y = 2 - 3x$
$\quad 6x + 2y = 7$

9. $\quad\quad x = 3y + 3$
$\quad 2x - 6y = 12$

10. $\quad\quad x = 2 - y$
$\quad 3x + 3y = 6$

11. $3x + 5y = -6$
$\quad\quad x = 5y + 3$

12. $\quad\quad y = 2x + 3$
$\quad 4x - 3y = 1$

13. $3x + y = 4$
$\quad 4x - 3y = 1$

14. $x - 4y = 9$
$\quad 2x - 3y = 11$

15. $3x - y = 6$
$\quad x + 3y = 2$

16. $4x - y = -5$
$\quad 2x + 5y = 13$

17. $3x - y = 5$
$\quad 2x + 5y = -8$

18. $3x + 4y = 18$
$\quad 2x - y = 1$

19. $4x + 3y = 0$
$\quad 2x - y = 0$

20. $5x + 2y = 0$
$\quad x - 3y = 0$

21. $2x - y = 2$
$\quad 6x - 3y = 6$

Solve by substitution.

22. $3x + y = 4$
$9x + 3y = 12$

23. $x = 3y + 2$
$y = 2x + 6$

24. $x = 4 - 2y$
$y = 2x - 13$

25. $y = 2x + 11$
$y = 5x - 19$

26. $y = 2x - 8$
$y = 3x - 13$

27. $y = -4x + 2$
$y = -3x - 1$

28. $x = 3y + 7$
$x = 2y - 1$

29. $x = 4y - 2$
$x = 6y + 8$

30. $x = 3 - 2y$
$x = 5y - 10$

APPLYING THE CONCEPTS

For what value of k does the system of equations have no solution?

31. $2x - 3y = 7$
$kx - 3y = 4$

32. $8x - 4y = 1$
$2x - ky = 3$

33. $\qquad x = 4y + 4$
$kx - 8y = 4$

34. The following was offered as a solution to the system of equations

(1) $\qquad y = \dfrac{1}{2}x + 2$

(2) $\quad 2x + 5y = 10$

$2x + 5y = 10 \qquad$ • This is Equation (2).

$2x + 5\left(\dfrac{1}{2}x + 2\right) = 10 \qquad$ • Substitute $\dfrac{1}{2}x + 2$ for y.

$2x + \dfrac{5}{2}x + 10 = 10 \qquad$ • Solve for x.

$\dfrac{9}{2}x = 0$

$x = 0$

At this point the student stated that because $x = 0$, the system of equations has no solution. If this assertion is correct, is the system of equations independent, dependent, or inconsistent? If the assertion is not correct, what is the correct solution?

35. Describe in your own words the process of solving a system of equa-
[W] tions by the substitution method.

36. When you solve a system of equations by the substitution method,
[W] how do you determine whether the system of equations is dependent?

37. When you solve a system of equations by the substitution method,
[W] how do you determine whether the system of equations is inconsistent?

Solving Systems of Linear Equations by the Addition Method

Objective A   *To solve a system of two linear equations in two variables by the addition method* ....

The **addition method** is an alternative method for solving a system of equations. This method is based on the Addition Property of Equations. Use the addition method when it is not convenient to solve one equation for one variable in terms of the other variable.

CONSIDER THIS

Equation (1) states $5x - 3y$ equals 14, and Equation (2) states $2x + 3y$ equals -7. Thus adding Equations (1) and (2) is like adding

$$14 = 14$$
$$\underline{-7 = -7}$$
$$7 = 7 \quad \text{A true equation}$$

The addition method of solving a system of equations is based on adding the same number to each side of the equation.

Note, for the system of equations at the right, the effect of adding Equation (2) to Equation (1). Because $-3y$ and $3y$ are additive inverses, adding the equations results in an equation with only one variable.

(1) $\quad 5x - 3y = 14$
(2) $\quad 2x + 3y = -7$
$\qquad 7x + 0y = 7$
$\qquad 7x = 7$

The solution of the resulting equation is the first component of the ordered-pair solution of the system.

$\qquad 7x = 7$
$\qquad x = 1$

The second component is found by substituting the value of x into Equation (1) or (2) and then solving for y. Equation (1) is used here.

(1) $\quad 5x - 3y = 14$
$\qquad 5(1) - 3y = 14$
$\qquad 5 - 3y = 14$
$\qquad -3y = 9$
$\qquad y = -3$

The solution is $(1, -3)$.

Sometimes each equation of the system of equations must be multiplied by a constant so that the coefficients of one of the variables are opposites.

→ Solve by the addition method: (1) $3x + 4y = 2$
$\qquad\qquad\qquad\qquad\qquad\quad$ (2) $2x + 5y = -1$

To eliminate x, multiply Equation (1) by 2 and Equation (2) by -3. Note at the right how the constants are chosen.

$$2(3x + 4y) = 2 \cdot 2$$
$$-3(2x + 5y) = -3(-1)$$

● The negative is used so that the coefficients will be opposites.

$$\begin{aligned} 6x + 8y &= 4 \\ \underline{-6x - 15y} &= 3 \\ -7y &= 7 \\ y &= -1 \end{aligned}$$

● 2 times Equation (1)
● -3 times Equation (2)
● Add the equations.
● Solve for y.

Substitute the value of y into Equation (1) or Equation (2) and solve for x. Equation (1) will be used here.

(1) $\qquad 3x + 4y = 2$
$\qquad 3x + 4(-1) = 2$
$\qquad 3x - 4 = 2$
$\qquad 3x = 6$
$\qquad x = 2$

● Substitute -1 for y.
● Solve for x.

The solution is $(2, -1)$.

POINT OF INTEREST

There are records of Babylonian mathematicians solving systems of equations 3600 years ago. Here is a system of equations from that time (in our modern notation):

$$\frac{2}{3}x = \frac{1}{2}y - 500$$
$$x - y = 1800$$

We say *modern notation* for many reasons. Foremost is the fact that using variables did not become widespread until the 17th century. There are many other reasons, however. The equals sign had not been invented, 2 and 3 did not look like they do today, and zero had not even been considered as a possible number.

➡ Solve by the addition method: (1) $\dfrac{2}{3}x + \dfrac{1}{2}y = 4$

(2) $\dfrac{1}{4}x - \dfrac{3}{8}y = -\dfrac{3}{4}$

Clear fractions. Multiply each equation by the LCM of the denominators.

$$6\left(\frac{2}{3}x + \frac{1}{2}y\right) = 6(4)$$

$$8\left(\frac{1}{4}x - \frac{3}{8}y\right) = 8\left(-\frac{3}{4}\right)$$

$$4x + 3y = 24$$
$$2x - 3y = -6$$
$$6x = 18$$ • **Eliminate y. Add the equations.**
$$x = 3$$ • **Solve for x.**

Substitute the value of x into Equation (1) and solve for y.

$$\frac{2}{3}x + \frac{1}{2}y = 4$$ • **This is Equation (1).**

$$\frac{2}{3}(3) + \frac{1}{2}y = 4$$ • **$x = 3$**

$$2 + \frac{1}{2}y = 4$$

$$\frac{1}{2}y = 2$$

$$y = 4$$

The solution is (3, 4).

➡ Solve by the addition method: (1) $2x - y = 3$
(2) $4x - 2y = 6$

Eliminate y. Multiply Equation (1) by -2.

(1) $-2(2x - y) = -2(3)$ • **Multiply both sides of Equation (1) by -2.**
(3) $-4x + 2y = -6$ • **This is Equation (3).**

Add Equation (3) to Equation (2).

(2) $4x - 2y = 6$
(3) $\underline{-4x + 2y = -6}$
$0 = 0$ • **This is a true equation.**

The equation $0 = 0$ indicates that the system of equations is dependent. This means that the graphs of the two lines are the same. Therefore, the solutions of the system of equations are the ordered-pair solutions of the equation of the line. Solve Equation (1) for y.

$$2x - y = 3$$
$$-y = -2x + 3$$
$$y = 2x - 3$$

The ordered-pair solutions are $(x, 2x - 3)$.

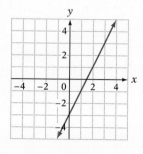

Example 1 Solve by the addition method:
(1) $3x - 2y = 2x + 5$
(2) $2x + 3y = -4$

Solution Write Equation (1) in the form $Ax + By = C$.

$$3x - 2y = 2x + 5$$
$$x - 2y = 5$$

Solve the system:

$$x - 2y = 5$$
$$2x + 3y = -4$$

Eliminate x.

$$-2(x - 2y) = -2(5)$$
$$2x + 3y = -4$$

$$-2x + 4y = -10$$
$$2x + 3y = -4$$

Add the equations.

$$7y = -14$$
$$y = -2$$

Replace y in Equation (2).

$$2x + 3y = -4$$
$$2x + 3(-2) = -4$$
$$2x - 6 = -4$$
$$2x = 2$$
$$x = 1$$

The solution is $(1, -2)$.

You Try It 1 Solve by the addition method:
$2x + 5y = 6$
$3x - 2y = 6x + 2$

Your solution

Example 2 Solve by the addition method:
(1) $4x - 8y = 36$
(2) $3x - 6y = 27$

Solution Eliminate x.

$$3(4x - 8y) = 3(36)$$
$$-4(3x - 6y) = -4(27)$$

$$12x - 24y = 108$$
$$-12x + 24y = -108$$

Add the equations.

$$0 = 0$$

The system of equations is dependent. The solutions are the ordered pairs $\left(x, \frac{1}{2}x - \frac{9}{2}\right)$.

You Try It 2 Solve by the addition method:
$2x + y = 5$
$4x + 2y = 6$

Your solution

Solutions on p. A23

Objective B   *To solve a system of three linear equations in three*
variables by the addition method ·······································

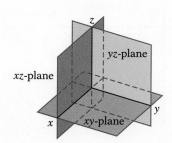

An equation of the form $Ax + By + Cz = D$, where A, B, and C are coefficients and D is a constant, is a **linear equation in three variables.** Examples of these equations are shown at the right. The graph of a linear equation in three variables is a plane.

$$2x + 4y - 3z = 7$$
$$x - 6y + z = -3$$

Graphing an equation in three variables requires a third coordinate axis perpendicular to the xy-plane. The third axis is commonly called the z-axis. The result is a three-dimensional coordinate system called the xyz-coordinate system. To help visualize a three-dimensional coordinate system, think of a corner of a room: the floor is the xy-plane, one wall is the yz-plane, and the other wall is the xz-plane. A three-dimensional coordinate system is shown at the right.

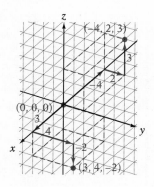

Graphing an ordered triple requires three moves, the first along the x-axis, the second parallel to the y-axis, and the third parallel to the z-axis. The graphs of the points $(-4, 2, 3)$ and $(3, 4, -2)$ are shown at the right.

The graph of a linear equation in three variables is a plane. That is, if all the solutions of a linear equation in three variables were plotted in an xyz-coordinate system, the graph would look like a large piece of paper extending infinitely. The graph of $x + y + z = 3$ is shown at the right.

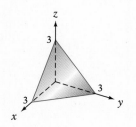

There are different ways in which three planes can be oriented in an *xyz*-coordinate system. The systems of equations represented by the planes below are inconsistent.

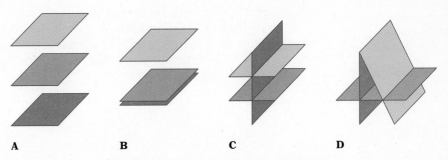

A   B   C   D

Graphs of Inconsistent Systems of Equations

For a system of three equations in three variables to have a solution, the graphs of the planes must intersect at a single point, they must intersect along a common line, or all equations must have a graph that is the same plane. These situations are shown in the figures below.

The three planes shown in Figure E intersect at a point. A system of equations represented by planes that intersect at a point is independent.

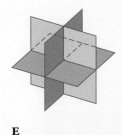

E

An Independent System
of Equations

The planes shown in Figures F and G intersect along a common line. The system of equations represented by the planes in Figure H has a graph that is the same plane. The systems of equations represented by these three graphs are dependent.

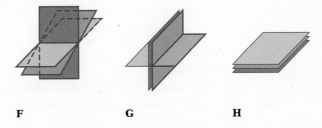

F   G   H

Dependent Systems of Equations

Just as a solution of an equation in two variables is an ordered pair (x, y), a **solution of an equation in three variables** is an ordered triple (x, y, z). For example, $(2, 1, -3)$ is a solution of the equation $2x - y - 2z = 9$. The ordered triple $(1, 3, 2)$ is not a solution.

A **system of linear equations in three variables** is shown at the right. A **solution of a system of equations in three variables** is an ordered triple that is a solution of each equation of the system.

$$x - 2y + z = 6$$
$$3x + y - 2z = 2$$
$$2x - 3y + 5z = 1$$

A system of linear equations in three variables can be solved by using the addition method. First, eliminate one variable from any two of the given equations. Then eliminate the same variable from any other two equations. The result will be a system of two equations in two variables. Solve this system by the addition method.

➡ Solve: (1) $\quad x + 4y - z = 10$
(2) $\quad 3x + 2y + z = 4$
(3) $\quad 2x - 3y + 2z = -7$

Eliminate z from Equations (1) and (2) by adding the two equations.

$$x + 4y - z = 10$$
$$3x + 2y + z = 4$$

(4) $\quad 4x + 6y = 14 \qquad$ • **Add the equations. This is Equation (4).**

Eliminate z from Equations (1) and (3). Multiply Equation (1) by 2 and add to Equation (3).

$$2x + 8y - 2z = 20 \qquad$$ • **2 times Equation (1)**
$$2x - 3y + 2z = -7 \qquad$$ • **This is Equation (3).**

(5) $\quad 4x + 5y = 13 \qquad$ • **Add the equations. This is Equation (5).**

Using Equations (4) and (5), solve the system of two equations in two variables.

(4) $\quad 4x + 6y = 14$
(5) $\quad 4x + 5y = 13$

Eliminate x. Multiply Equation (5) by -1 and add to Equation (4).

$$4x + 6y = 14 \qquad$$ • **This is Equation (4).**
$$-4x - 5y = -13 \qquad$$ • **−1 times Equation (5)**
$$y = 1 \qquad$$ • **Add the equations.**

Substitute the value of y into Equation (4) or Equation (5) and solve for x. Equation (4) is used here.

$$4x + 6y = 14 \qquad$$ • **This is Equation (4).**
$$4x + 6(1) = 14 \qquad$$ • $y = 1$
$$4x + 6 = 14 \qquad$$ • **Solve for x.**
$$4x = 8$$
$$x = 2$$

Substitute the value of y and the value of x into one of the equations in the original system. Equation (2) is used here.

$$3x + 2y + z = 4$$
$$3(2) + 2(1) + z = 4 \qquad$$ • $x = 2, y = 1$
$$6 + 2 + z = 4$$
$$8 + z = 4$$
$$z = -4$$

The solution is $(2, 1, -4)$.

➡ Solve: (1) $\qquad 2x - 3y - z = 1$
    (2) $\qquad x + 4y + 3z = 2$
    (3) $\qquad 4x - 6y - 2z = 5$

Eliminate x from Equations (1) and (2).

$\begin{aligned} 2x - 3y - z &= 1 \\ -2x - 8y - 6z &= -4 \\ \hline -11y - 7z &= -3 \end{aligned}$
    ● This is Equation (1).
    ● -2 times Equation (2).
    ● Add the equations.

Eliminate x from Equations (1) and (3).

$\begin{aligned} -4x + 6y + 2z &= -2 \\ 4x - 6y - 2z &= 5 \\ \hline 0 &= 3 \end{aligned}$
    ● -2 times Equation (1).
    ● This is Equation (3).
    ● Add the equations.

The equation $0 = 3$ is not a true equation. The system of equations is inconsistent and therefore has no solution.

➡ Solve: (1) $\qquad 3x - z = -1$
    (2) $\qquad 2y - 3z = 10$
    (3) $\qquad x + 3y - z = 7$

Eliminate x from Equations (1) and (3). Multiply Equation (3) by -3 and add to Equation (1).

$\begin{aligned} 3x - z &= -1 \\ -3x - 9y + 3z &= -21 \\ \hline -9y + 2z &= -22 \end{aligned}$
    ● This is Equation (1).
    ● -3 times Equation (3).
    ● Add the equations.

(4) marks the line $-9y + 2z = -22$.

Use Equations (2) and (4) to form a system of equations in two variables.

(2) $\qquad 2y - 3z = 10$
(4) $\qquad -9y + 2z = -22$

Eliminate z. Multiply Equation (2) by 2 and Equation (4) by 3.

$\begin{aligned} 4y - 6z &= 20 \\ -27y + 6z &= -66 \\ \hline -23y &= -46 \\ y &= 2 \end{aligned}$
    ● 2 times Equation (2).
    ● 3 times Equation (4).
    ● Add the equations.
    ● Solve for y.

Substitute the value of y into Equation (2) or Equation (4) and solve for z. Equation (2) is used here.

(2) $\begin{aligned} 2y - 3z &= 10 \\ 2(2) - 3z &= 10 \\ 4 - 3z &= 10 \\ -3z &= 6 \\ z &= -2 \end{aligned}$
    ● This is Equation (2).
    ● $y = 2$
    ● Solve for z.

Substitute the value of z into Equation (1) and solve for x.

(1) $\begin{aligned} 3x - z &= -1 \\ 3x - (-2) &= -1 \\ 3x + 2 &= -1 \\ 3x &= -3 \\ x &= -1 \end{aligned}$
    ● This is Equation (1).
    ● $z = -2$
    ● Solve for x.

The solution is $(-1, 2, -2)$.

Example 3 Solve: (1) $3x - y + 2z = 1$
(2) $2x + 3y + 3z = 4$
(3) $x + y - 4z = -9$

You Try It 3 Solve: $x - y + z = 6$
$2x + 3y - z = 1$
$x + 2y + 2z = 5$

Solution Eliminate y. Add
Equations (1) and (3).

$$3x - y + 2z = 1$$
$$x + y - 4z = -9$$
$$4x \quad - 2z = -8$$

Multiply each side of the
equation by $\frac{1}{2}$.

(4) $2x - z = -4$

Multiply Equation (1) by 3
and add to Equation (2).

$$9x - 3y + 6z = 3$$
$$2x + 3y + 3z = 4$$
(5) $\quad 11x \quad + 9z = 7$

Solve the system of two
equations.

(4) $\quad 2x - z = -4$
(5) $\quad 11x + 9z = 7$

Multiply Equation (4) by 9
and add to Equation (5).

$$18x - 9z = -36$$
$$11x + 9z = 7$$
$$29x \quad = -29$$
$$x = -1$$

Replace x by -1 in Equation
(4).

$$2x - z = -4$$
$$2(-1) - z = -4$$
$$-2 - z = -4$$
$$-z = -2$$
$$z = 2$$

Replace x by -1 and z by 2 in
Equation (3).

$$x + y - 4z = -9$$
$$-1 + y - 4(2) = -9$$
$$-9 + y = -9$$
$$y = 0$$

The solution is $(-1, 0, 2)$.

Your solution

Solution on p. A24

5.3 Exercises

. . . .

Objective A

Solve by the addition method.

1. $x - y = 5$
$x + y = 7$

2. $x + y = 1$
$2x - y = 5$

3. $3x + y = 4$
$x + y = 2$

4. $x - 3y = 4$
$x + 5y = -4$

5. $3x + y = 7$
$x + 2y = 4$

6. $x - 2y = 7$
$3x - 2y = 9$

7. $2x + 3y = -1$
$x + 5y = 3$

8. $x + 5y = 7$
$2x + 7y = 8$

9. $3x - y = 4$
$6x - 2y = 8$

10. $x - 2y = -3$
$-2x + 4y = 6$

11. $2x + 5y = 9$
$4x - 7y = -16$

12. $8x - 3y = 21$
$4x + 5y = -9$

13. $4x - 6y = 5$
$2x - 3y = 7$

14. $3x + 6y = 7$
$2x + 4y = 5$

15. $3x - 5y = 7$
$x - 2y = 3$

16. $3x + 4y = 25$
$2x + y = 10$

17. $x + 3y = 7$
$-2x + 3y = 22$

18. $2x - 3y = 14$
$5x - 6y = 32$

19. $3x + 2y = 16$
$2x - 3y = -11$

20. $2x - 5y = 13$
$5x + 3y = 17$

21. $4x + 4y = 5$
$2x - 8y = -5$

Solve by the addition method.

22. $3x + 7y = 16$
$4x - 3y = 9$

23. $5x + 4y = 0$
$3x + 7y = 0$

24. $3x - 4y = 0$
$4x - 7y = 0$

25. $5x + 2y = 1$
$2x + 3y = 7$

26. $3x + 5y = 16$
$5x - 7y = -4$

27. $3x - 6y = 6$
$9x - 3y = 8$

28. $\dfrac{2}{3}x - \dfrac{1}{2}y = 3$
$\dfrac{1}{3}x - \dfrac{1}{4}y = \dfrac{3}{2}$

29. $\dfrac{3}{4}x + \dfrac{1}{3}y = -\dfrac{1}{2}$
$\dfrac{1}{2}x - \dfrac{5}{6}y = -\dfrac{7}{2}$

30. $\dfrac{2}{5}x - \dfrac{1}{3}y = 1$
$\dfrac{3}{5}x + \dfrac{2}{3}y = 5$

31. $\dfrac{5x}{6} + \dfrac{y}{3} = \dfrac{4}{3}$
$\dfrac{2x}{3} - \dfrac{y}{2} = \dfrac{11}{6}$

32. $\dfrac{3x}{4} + \dfrac{2y}{5} = -\dfrac{3}{20}$
$\dfrac{3x}{2} - \dfrac{y}{4} = \dfrac{3}{4}$

33. $\dfrac{2x}{5} - \dfrac{y}{2} = \dfrac{13}{2}$
$\dfrac{3x}{4} - \dfrac{y}{5} = \dfrac{17}{2}$

34. $\dfrac{x}{2} + \dfrac{y}{3} = \dfrac{5}{12}$
$\dfrac{x}{2} - \dfrac{y}{3} = \dfrac{1}{12}$

35. $\dfrac{3x}{2} - \dfrac{y}{4} = -\dfrac{11}{12}$
$\dfrac{x}{3} - y = -\dfrac{5}{6}$

36. $\dfrac{3x}{4} - \dfrac{2y}{3} = 0$
$\dfrac{5x}{4} - \dfrac{y}{3} = \dfrac{7}{12}$

37. $4x - 5y = 3y + 4$
$2x + 3y = 2x + 1$

38. $5x - 2y = 8x - 1$
$2x + 7y = 4y + 9$

39. $2x + 5y = 5x + 1$
$3x - 2y = 3y + 3$

40. $4x - 8y = 5$
$8x + 2y = 1$

41. $5x + 2y = 2x + 1$
$2x - 3y = 3x + 2$

42. $3x + 3y = y + 1$
$x + 3y = 9 - x$

Objective B

Solve by the addition method.

43. $x + 2y - z = 1$
$2x - y + z = 6$
$x + 3y - z = 2$

44. $x + 3y + z = 6$
$3x + y - z = -2$
$2x + 2y - z = 1$

45. $2x - y + 2z = 7$
$x + y + z = 2$
$3x - y + z = 6$

Solve by the addition method.

46. $x - 2y + z = 6$
$x + 3y + z = 16$
$3x - y - z = 12$

47. $3x + y = 5$
$3y - z = 2$
$x + z = 5$

48. $2y + z = 7$
$2x - z = 3$
$x - y = 3$

49. $x - y + z = 1$
$2x + 3y - z = 3$
$-x + 2y - 4z = 4$

50. $2x + y - 3z = 7$
$x - 2y + 3z = 1$
$3x + 4y - 3z = 13$

51. $2x + 3z = 5$
$3y + 2z = 3$
$3x + 4y = -10$

52. $3x + 4z = 5$
$2y + 3z = 2$
$2x - 5y = 8$

53. $2x + 4y - 2z = 3$
$x + 3y + 4z = 1$
$x + 2y - z = 4$

54. $x - 3y + 2z = 1$
$x - 2y + 3z = 5$
$2x - 6y + 4z = 3$

55. $2x + y - z = 5$
$x + 3y + z = 14$
$3x - y + 2z = 1$

56. $3x - y - 2z = 11$
$2x + y - 2z = 11$
$x + 3y - z = 8$

57. $3x + y - 2z = 2$
$x + 2y + 3z = 13$
$2x - 2y + 5z = 6$

58. $4x + 5y + z = 6$
$2x - y + 2z = 11$
$x + 2y + 2z = 6$

59. $2x - y + z = 6$
$3x + 2y + z = 4$
$x - 2y + 3z = 12$

60. $3x + 2y - 3z = 8$
$2x + 3y + 2z = 10$
$x + y - z = 2$

Solve by the addition method.

61. $3x - 2y + 3z = -4$
$2x + y - 3z = 2$
$3x + 4y + 5z = 8$

62. $3x - 3y + 4z = 6$
$4x - 5y + 2z = 10$
$x - 2y + 3z = 4$

63. $3x - y + 2z = 2$
$4x + 2y - 7z = 0$
$2x + 3y - 5z = 7$

64. $2x + 2y + 3z = 13$
$-3x + 4y - z = 5$
$5x - 3y + z = 2$

65. $2x - 3y + 7z = 0$
$x + 4y - 4z = -2$
$3x + 2y + 5z = 1$

66. $5x + 3y - z = 5$
$3x - 2y + 4z = 13$
$4x + 3y + 5z = 22$

APPLYING THE CONCEPTS

67. The point of intersection of the graphs of the equations $Ax + 2y = 2$ and $2x + By = 10$ is $(2, -2)$. Find A and B.

68. The point of intersection of the graphs of the equations $Ax - 4y = 9$ and $4x + By = -1$ is $(-1, -3)$. Find A and B.

69. For what value of k is the system of equations dependent?

a. $2x + 3y = 7$
$4x + 6y = k$

b. $y = \dfrac{2}{3}x - 3$
$y = kx - 3$

c. $x = ky - 1$
$y = 2x + 2$

70. For what values of k is the system of equations independent?

a. $x + y = 7$
$kx + y = 3$

b. $x + 2y = 4$
$kx + 3y = 2$

c. $2x + ky = 1$
$x + 2y = 2$

71. Given that the graphs of the equations $2x - y = 6$, $3x - 4y = 4$, and $Ax - 2y = 0$ all intersect at the same point, find A.

72. Given that the graphs of the equations $3x - 2y = -2$, $2x - y = 0$, and $Ax + y = 8$ all intersect at the same point, find A.

73. Describe in your own words the process of solving a system of two
[W] linear equations in two variables by the addition method.

74. Explain, graphically, the following situations when they are related
[W] to a system of three linear equations in three variables.
a. The system of equations has no solution.
b. The system of equations has exactly one solution.
c. The system of equations has infinitely many solutions.

Solving Systems of Equations by Using Determinants

Objective A   *To evaluate a determinant* ....

A **matrix** is a rectangular array of numbers. Each number in the matrix is called an **element** of the matrix. The matrix at the right, with three rows and four columns, is called a 3×4 (read "3 by 4") matrix.

$$A = \begin{pmatrix} 1 & -3 & 2 & 4 \\ 0 & 4 & -3 & 2 \\ 6 & -5 & 4 & -1 \end{pmatrix}$$

A matrix of m rows and n columns is said to be of **order** $m \times n$. The matrix above has order 3×4. The notation a_{ij} refers to the element of a matrix in the ith row and the jth column. For matrix A, $a_{23} = -3$, $a_{31} = 6$, and $a_{13} = 2$.

A **square matrix** is one that has the same number of rows as columns. A 2×2 matrix and a 3×3 matrix are shown at the right.

$$\begin{pmatrix} -1 & 3 \\ 5 & 2 \end{pmatrix} \qquad \begin{pmatrix} 4 & 0 & 1 \\ 5 & -3 & 7 \\ 2 & 1 & 4 \end{pmatrix}$$

Associated with every square matrix is a number called its **determinant.**

> **Determinant of a 2 × 2 Matrix**
>
> The determinant of a 2×2 matrix $\begin{pmatrix} a_{11} & a_{12} \\ a_{21} & a_{22} \end{pmatrix}$ is written $\begin{vmatrix} a_{11} & a_{12} \\ a_{21} & a_{22} \end{vmatrix}$.
> The value of this determinant is given by the formula
>
> $$\begin{vmatrix} a_{11} & a_{12} \\ a_{21} & a_{22} \end{vmatrix} = a_{11}a_{22} - a_{21}a_{12}$$

Note that vertical bars are used to represent the determinant and that parentheses are used to represent the matrix.

➡ Find the value of the determinant $\begin{vmatrix} 3 & 4 \\ -1 & 2 \end{vmatrix}$.

$$\begin{vmatrix} 3 & 4 \\ -1 & 2 \end{vmatrix} = 3 \cdot 2 - (-1)(4) = 6 - (-4) = 10$$

The value of the determinant is 10.

For a square matrix whose order is 3×3 or greater, the value of the determinant of that matrix is found by using 2×2 determinants, called *minors*. The **minor of an element** in a 3×3 determinant is the 2×2 determinant that is obtained by eliminating the row and column that contain that element.

➡ Find the minor of -3 for the determinant $\begin{vmatrix} 2 & -3 & 4 \\ 0 & 4 & 8 \\ -1 & 3 & 6 \end{vmatrix}$.

The minor of -3 is the 2×2 determinant created by eliminating the row and column that contain -3.

$$\begin{vmatrix} 2 & -3 & 4 \\ 0 & 4 & 8 \\ -1 & 3 & 6 \end{vmatrix}$$

The minor of -3 is $\begin{vmatrix} 0 & 8 \\ -1 & 6 \end{vmatrix}$.

CONSIDER THIS

The only difference between the cofactor and the minor is one of sign. The definition at the right can be stated symbolically as follows: If C_{ij} is the cofactor and M_{ij} is the minor of the matrix element a_{ij}, then $C_{ij} = (-1)^{i+j}M_{ij}$. If $i + j$ is an even number, then $(-1)^{i+j} = 1$ and $C_{ij} = M_{ij}$. If $i + j$ is an odd number, then $(-1)^{i+j} = -1$ and $C_{ij} = -M_{ij}$.

> **Definition of a Cofactor**
>
> The **cofactor** of an element of a determinant is $(-1)^{i+j}$ times the minor of that element, where i is the row number of the element and j is the column number of the element.

➡ For the determinant $\begin{vmatrix} 3 & -2 & 1 \\ 2 & -5 & -4 \\ 0 & 3 & 1 \end{vmatrix}$, find the cofactor of -2 and of -5.

Because -2 is in the first row and the second column, $i = 1$ and $j = 2$. Therefore, $i + j = 1 + 2 = 3$, and $(-1)^{i+j} = (-1)^3 = -1$.

The cofactor of -2 is $(-1)\begin{vmatrix} 2 & -4 \\ 0 & 1 \end{vmatrix}$.

Because -5 is in the second row and the second column, $i = 2$ and $j = 2$. Therefore, $i + j = 2 + 2 = 4$, and $(-1)^{i+j} = (-1)^4 = 1$.

The cofactor of -5 is $1 \cdot \begin{vmatrix} 3 & 1 \\ 0 & 1 \end{vmatrix}$.

Note from this example that the cofactor of an element is -1 times the minor of that element or 1 times the minor of that element, depending on whether the sum $i + j$ is an odd or even integer.

The value of a 3×3 or larger determinant can be found by **expanding by cofactors** of any row or any column. The result of expanding by cofactors of the first row of a 3×3 determinant is shown below.

$$\begin{vmatrix} a_{11} & a_{12} & a_{13} \\ a_{21} & a_{22} & a_{23} \\ a_{31} & a_{32} & a_{33} \end{vmatrix} = a_{11}(-1)^{1+1}\begin{vmatrix} a_{22} & a_{23} \\ a_{32} & a_{33} \end{vmatrix} + a_{12}(-1)^{1+2}\begin{vmatrix} a_{21} & a_{23} \\ a_{31} & a_{33} \end{vmatrix} + a_{13}(-1)^{1+3}\begin{vmatrix} a_{21} & a_{22} \\ a_{31} & a_{32} \end{vmatrix}$$

$$= a_{11}\begin{vmatrix} a_{22} & a_{23} \\ a_{32} & a_{33} \end{vmatrix} - a_{12}\begin{vmatrix} a_{21} & a_{23} \\ a_{31} & a_{33} \end{vmatrix} + a_{13}\begin{vmatrix} a_{21} & a_{22} \\ a_{31} & a_{32} \end{vmatrix}$$

➡ Find the value of the determinant $\begin{vmatrix} 2 & -3 & 2 \\ 1 & 3 & -1 \\ 0 & -2 & 2 \end{vmatrix}$.

Expand by cofactors of the first row.

$$\begin{vmatrix} 2 & -3 & 2 \\ 1 & 3 & -1 \\ 0 & -2 & 2 \end{vmatrix} = 2\begin{vmatrix} 3 & -1 \\ -2 & 2 \end{vmatrix} - (-3)\begin{vmatrix} 1 & -1 \\ 0 & 2 \end{vmatrix} + 2\begin{vmatrix} 1 & 3 \\ 0 & -2 \end{vmatrix}$$

$$= 2(6 - 2) - (-3)(2 - 0) + 2(-2 - 0)$$
$$= 2(4) - (-3)(2) + 2(-2) = 8 - (-6) + (-4)$$
$$= 10$$

To illustrate a statement made earlier, the value of this determinant will now be found by expanding by cofactors of the second column.

$$\begin{vmatrix} 2 & -3 & 2 \\ 1 & 3 & -1 \\ 0 & -2 & 2 \end{vmatrix} = -3(-1)^{1+2}\begin{vmatrix} 1 & -1 \\ 0 & 2 \end{vmatrix} + 3(-1)^{2+2}\begin{vmatrix} 2 & 2 \\ 0 & 2 \end{vmatrix} + (-2)(-1)^{3+2}\begin{vmatrix} 2 & 2 \\ 1 & -1 \end{vmatrix}$$

$$= -3(-1)\begin{vmatrix} 1 & -1 \\ 0 & 2 \end{vmatrix} + 3 \cdot 1\begin{vmatrix} 2 & 2 \\ 0 & 2 \end{vmatrix} + (-2)(-1)\begin{vmatrix} 2 & 2 \\ 1 & -1 \end{vmatrix}$$

$$= 3(2 - 0) + 3(4 - 0) + 2(-2 - 2)$$
$$= 3(2) + 3(4) + 2(-4) = 6 + 12 + (-8)$$
$$= 10$$

Note that the value of the determinant is the same whether the first row or the second column is used to expand by cofactors. *Any row or column* can be used to evaluate a determinant by expanding by cofactors.

Example 1

Find the value of $\begin{vmatrix} 3 & -2 \\ 6 & -4 \end{vmatrix}$.

Solution

$\begin{vmatrix} 3 & -2 \\ 6 & -4 \end{vmatrix} = 3(-4) - 6(-2) = -12 + 12 = 0$

The value of the determinant is 0.

You Try It 1

Find the value of $\begin{vmatrix} -1 & -4 \\ 3 & -5 \end{vmatrix}$.

Your solution

Example 2

Find the value of $\begin{vmatrix} -2 & 3 & 1 \\ 4 & -2 & 0 \\ 1 & -2 & 3 \end{vmatrix}$.

Solution

Expand by cofactors of the first row.

$\begin{vmatrix} -2 & 3 & 1 \\ 4 & -2 & 0 \\ 1 & -2 & 3 \end{vmatrix}$

$= -2 \begin{vmatrix} -2 & 0 \\ -2 & 3 \end{vmatrix} - 3 \begin{vmatrix} 4 & 0 \\ 1 & 3 \end{vmatrix} + 1 \begin{vmatrix} 4 & -2 \\ 1 & -2 \end{vmatrix}$

$= -2(-6 - 0) - 3(12 - 0) + 1(-8 + 2)$
$= -2(-6) - 3(12) + 1(-6)$
$= 12 - 36 - 6$
$= -30$

The value of the determinant is -30.

You Try It 2

Find the value of $\begin{vmatrix} 1 & 4 & -2 \\ 3 & 1 & 1 \\ 0 & -2 & 2 \end{vmatrix}$.

Your solution

Example 3

Find the value of $\begin{vmatrix} 0 & -2 & 1 \\ 1 & 4 & 1 \\ 2 & -3 & 4 \end{vmatrix}$.

Solution

$\begin{vmatrix} 0 & -2 & 1 \\ 1 & 4 & 1 \\ 2 & -3 & 4 \end{vmatrix}$

$= 0 \begin{vmatrix} 4 & 1 \\ -3 & 4 \end{vmatrix} - (-2) \begin{vmatrix} 1 & 1 \\ 2 & 4 \end{vmatrix} + 1 \begin{vmatrix} 1 & 4 \\ 2 & -3 \end{vmatrix}$

$= 0 - (-2)(4 - 2) + 1(-3 - 8)$
$= 2(2) + 1(-11)$
$= 4 - 11$
$= -7$

The value of the determinant is -7.

You Try It 3

Find the value of $\begin{vmatrix} 3 & -2 & 0 \\ 1 & 4 & 2 \\ -2 & 1 & 3 \end{vmatrix}$.

Your solution

Solutions on p. A25

Objective B *To solve a system of equations by using Cramer's Rule* ....

The connection between determinants and systems of equations can be understood by solving a general system of linear equations.

Solve: (1) $a_1x + b_1y = c_1$
(2) $a_2x + b_2y = c_2$

Eliminate y. Multiply Equation (1) by b_2 and Equation (2) by $-b_1$.

$a_1b_2x + b_1b_2y = c_1b_2$ • b_2 times Equation (1)
$-a_2b_1x - b_1b_2y = -c_2b_1$ • $-b_1$ times Equation (2)

$a_1b_2x - a_2b_1x = c_1b_2 - c_2b_1$ • Add the equations.

$(a_1b_2 - a_2b_1)x = c_1b_2 - c_2b_1$ • Solve for x, assuming $a_1b_2 - a_2b_1 \neq 0$.

$$x = \frac{c_1b_2 - c_2b_1}{a_1b_2 - a_2b_1}$$

The denominator $a_1b_2 - a_2b_1$ is the determinant of the coefficients of x and y. This is called the **coefficient determinant.**

$$a_1b_2 - a_2b_1 = \begin{vmatrix} a_1 & b_1 \\ a_2 & b_2 \end{vmatrix}$$

coefficients of x ⟶
coefficients of y ⟶

The numerator $c_1b_2 - c_2b_1$ is the determinant obtained by replacing the first column in the coefficient determinant by the constants c_1 and c_2. This is called a **numerator determinant.**

$$c_1b_2 - c_2b_1 = \begin{vmatrix} c_1 & b_1 \\ c_2 & b_2 \end{vmatrix}$$

constants of
the equations ⟶

Following a similar procedure and eliminating x, it is possible also to express the y-component of the solution in determinant form. These results are summarized in Cramer's Rule.

Cramer's Rule for a System of Two Equations in Two Variables
The solution of the system of equations $\begin{array}{c}a_1x + b_1y = c_1\\a_2x + b_2y = c_2\end{array}$ is given by
$x = \dfrac{D_x}{D}$ and $y = \dfrac{D_y}{D}$, where $D = \begin{vmatrix} a_1 & b_1 \\ a_2 & b_2 \end{vmatrix}$, $D_x = \begin{vmatrix} c_1 & b_1 \\ c_2 & b_2 \end{vmatrix}$,
$D_y = \begin{vmatrix} a_1 & c_1 \\ a_2 & c_2 \end{vmatrix}$, and $D \neq 0$.

➡ Solve by using Cramer's Rule: $3x - 2y = 1$
$2x + 5y = 3$

$$D = \begin{vmatrix} 3 & -2 \\ 2 & 5 \end{vmatrix} = 19$$ • **Find the value of the coefficient determinant.**

$$D_x = \begin{vmatrix} 1 & -2 \\ 3 & 5 \end{vmatrix} = 11, \quad D_y = \begin{vmatrix} 3 & 1 \\ 2 & 3 \end{vmatrix} = 7$$ • **Find the value of each of the numerator determinants.**

$$x = \frac{D_x}{D} = \frac{11}{19}, \quad y = \frac{D_y}{D} = \frac{7}{19}$$ • **Use Cramer's Rule to write the solution.**

The solution is $\left(\dfrac{11}{19}, \dfrac{7}{19}\right)$.

A procedure similar to that followed for a system of two equations in two variables can be used to extend Cramer's Rule to a system of three equations in three variables.

Cramer's Rule for a System of Three Equations in Three Variables

The solution of the system of equations
$$\begin{aligned} a_1 x + b_1 y + c_1 z &= d_1 \\ a_2 x + b_2 y + c_2 z &= d_2 \\ a_3 x + b_3 y + c_3 z &= d_3 \end{aligned}$$

is given by $x = \dfrac{D_x}{D}$, $y = \dfrac{D_y}{D}$, and $z = \dfrac{D_z}{D}$, where

$$D = \begin{vmatrix} a_1 & b_1 & c_1 \\ a_2 & b_2 & c_2 \\ a_3 & b_3 & c_3 \end{vmatrix}, \quad D_x = \begin{vmatrix} d_1 & b_1 & c_1 \\ d_2 & b_2 & c_2 \\ d_3 & b_3 & c_3 \end{vmatrix}, \quad D_y = \begin{vmatrix} a_1 & d_1 & c_1 \\ a_2 & d_2 & c_2 \\ a_3 & d_3 & c_3 \end{vmatrix},$$

$$D_z = \begin{vmatrix} a_1 & b_1 & d_1 \\ a_2 & b_2 & d_2 \\ a_3 & b_3 & d_3 \end{vmatrix}, \text{ and } D \neq 0.$$

➡️ Solve by using Cramer's Rule:
$$\begin{aligned} 2x - y + z &= 1 \\ x + 3y - 2z &= -2 \\ 3x + y + 3z &= 4 \end{aligned}$$

Find the value of the coefficient determinant.

$$D = \begin{vmatrix} 2 & -1 & 1 \\ 1 & 3 & -2 \\ 3 & 1 & 3 \end{vmatrix} = 2\begin{vmatrix} 3 & -2 \\ 1 & 3 \end{vmatrix} - (-1)\begin{vmatrix} 1 & -2 \\ 3 & 3 \end{vmatrix} + 1\begin{vmatrix} 1 & 3 \\ 3 & 1 \end{vmatrix}$$

$$= 2(11) + 1(9) + 1(-8)$$
$$= 23$$

Find the value of each of the numerator determinants.

$$D_x = \begin{vmatrix} 1 & -1 & 1 \\ -2 & 3 & -2 \\ 4 & 1 & 3 \end{vmatrix} = 1\begin{vmatrix} 3 & -2 \\ 1 & 3 \end{vmatrix} - (-1)\begin{vmatrix} -2 & -2 \\ 4 & 3 \end{vmatrix} + 1\begin{vmatrix} -2 & 3 \\ 4 & 1 \end{vmatrix}$$

$$= 1(11) + 1(2) + 1(-14)$$
$$= -1$$

$$D_y = \begin{vmatrix} 2 & 1 & 1 \\ 1 & -2 & -2 \\ 3 & 4 & 3 \end{vmatrix} = 2\begin{vmatrix} -2 & -2 \\ 4 & 3 \end{vmatrix} - 1\begin{vmatrix} 1 & -2 \\ 3 & 3 \end{vmatrix} + 1\begin{vmatrix} 1 & -2 \\ 3 & 4 \end{vmatrix}$$

$$= 2(2) - 1(9) + 1(10)$$
$$= 5$$

$$D_z = \begin{vmatrix} 2 & -1 & 1 \\ 1 & 3 & -2 \\ 3 & 1 & 4 \end{vmatrix} = 2\begin{vmatrix} 3 & -2 \\ 1 & 4 \end{vmatrix} - (-1)\begin{vmatrix} 1 & -2 \\ 3 & 4 \end{vmatrix} + 1\begin{vmatrix} 1 & 3 \\ 3 & 1 \end{vmatrix}$$

$$= 2(14) + 1(10) + 1(-8)$$
$$= 30$$

Use Cramer's Rule to write the solution.

$$x = \frac{D_x}{D} = \frac{-1}{23}, \qquad y = \frac{D_y}{D} = \frac{5}{23}, \qquad z = \frac{D_z}{D} = \frac{30}{23}$$

The solution is $\left(-\dfrac{1}{23}, \dfrac{5}{23}, \dfrac{30}{23}\right)$.

Example 4

Solve by using Cramer's Rule:

$6x - 9y = 5$
$4x - 6y = 4$

Solution

$$D = \begin{vmatrix} 6 & -9 \\ 4 & -6 \end{vmatrix} = -36 + 36 = 0$$

Because $D = 0$, $\dfrac{D_x}{D}$ is undefined.

Therefore, the system is dependent or inconsistent.

It is not possible to solve this system by Cramer's Rule.

You Try It 4

Solve by using Cramer's Rule:

$3x - y = 4$
$6x - 2y = 5$

Your solution

Example 5

Solve by using Cramer's Rule:

$3x - y + z = 5$
$x + 2y - 2z = -3$
$2x + 3y + z = 4$

Solution

$$D = \begin{vmatrix} 3 & -1 & 1 \\ 1 & 2 & -2 \\ 2 & 3 & 1 \end{vmatrix} = 28$$

$$D_x = \begin{vmatrix} 5 & -1 & 1 \\ -3 & 2 & -2 \\ 4 & 3 & 1 \end{vmatrix} = 28$$

$$D_y = \begin{vmatrix} 3 & 5 & 1 \\ 1 & -3 & -2 \\ 2 & 4 & 1 \end{vmatrix} = 0$$

$$D_z = \begin{vmatrix} 3 & -1 & 5 \\ 1 & 2 & -3 \\ 2 & 3 & 4 \end{vmatrix} = 56$$

$x = \dfrac{D_x}{D} = \dfrac{28}{28} = 1,\quad y = \dfrac{D_y}{D} = \dfrac{0}{28} = 0,$

$z = \dfrac{D_z}{D} = \dfrac{56}{28} = 2$

The solution is $(1, 0, 2)$.

You Try It 5

Solve by using Cramer's Rule:

$2x - y + z = -1$
$3x + 2y - z = 3$
$x + 3y + z = -2$

Your solution

Solutions on p. A25

5.4 Exercises

Objective A

Evaluate the determinant.

1. $\begin{vmatrix} 2 & -1 \\ 3 & 4 \end{vmatrix}$

2. $\begin{vmatrix} 5 & 1 \\ -1 & 2 \end{vmatrix}$

3. $\begin{vmatrix} 6 & -2 \\ -3 & 4 \end{vmatrix}$

4. $\begin{vmatrix} -3 & 5 \\ 1 & 7 \end{vmatrix}$

5. $\begin{vmatrix} 3 & 6 \\ 2 & 4 \end{vmatrix}$

6. $\begin{vmatrix} 5 & -10 \\ 1 & -2 \end{vmatrix}$

7. $\begin{vmatrix} 1 & -1 & 2 \\ 3 & 2 & 1 \\ 1 & 0 & 4 \end{vmatrix}$

8. $\begin{vmatrix} 4 & 1 & 3 \\ 2 & -2 & 1 \\ 3 & 1 & 2 \end{vmatrix}$

9. $\begin{vmatrix} 3 & -1 & 2 \\ 0 & 1 & 2 \\ 3 & 2 & -2 \end{vmatrix}$

10. $\begin{vmatrix} 4 & 5 & -2 \\ 3 & -1 & 5 \\ 2 & 1 & 4 \end{vmatrix}$

11. $\begin{vmatrix} 4 & 2 & 6 \\ -2 & 1 & 1 \\ 2 & 1 & 3 \end{vmatrix}$

12. $\begin{vmatrix} 3 & 6 & -3 \\ 4 & -1 & 6 \\ -1 & -2 & 3 \end{vmatrix}$

Objective B

Solve by using Cramer's Rule.

13. $2x - 5y = 26$
$5x + 3y = 3$

14. $3x + 7y = 15$
$2x + 5y = 11$

15. $x - 4y = 8$
$3x + 7y = 5$

16. $5x + 2y = -5$
$3x + 4y = 11$

17. $2x + 3y = 4$
$6x - 12y = -5$

18. $5x + 4y = 3$
$15x - 8y = -21$

19. $2x + 5y = 6$
$6x - 2y = 1$

20. $7x + 3y = 4$
$5x - 4y = 9$

21. $-2x + 3y = 7$
$4x - 6y = 9$

22. $9x + 6y = 7$
$3x + 2y = 4$

23. $2x - 5y = -2$
$3x - 7y = -3$

24. $8x + 7y = -3$
$2x + 2y = 5$

Solve by using Cramer's Rule.

25. $2x - y + 3z = 9$
    $x + 4y + 4z = 5$
    $3x + 2y + 2z = 5$

26. $3x - 2y + z = 2$
    $2x + 3y + 2z = -6$
    $3x - y + z = 0$

27. $3x - y + z = 11$
    $x + 4y - 2z = -12$
    $2x + 2y - z = -3$

28. $x + 2y + 3z = 8$
    $2x - 3y + z = 5$
    $3x - 4y + 2z = 9$

29. $4x - 2y + 6z = 1$
    $3x + 4y + 2z = 1$
    $2x - y + 3z = 2$

30. $x - 3y + 2z = 1$
    $2x + y - 2z = 3$
    $3x - 9y + 6z = -3$

31. $5x - 4y + 2z = 4$
    $3x - 5y + 3z = -4$
    $3x + y - 5z = 12$

32. $2x + 4y + z = 7$
    $x + 3y - z = 1$
    $3x + 2y - 2z = 5$

33. $3x - 2y + 2z = 5$
    $6x + 3y - 4z = -1$
    $3x - y + 2z = 4$

APPLYING THE CONCEPTS

34. Determine whether the following statements are always true, sometimes true, or never true.
   a. The determinant of a matrix is a positive number.
   b. A determinant can be evaluated by expanding about any row or column of the matrix.
   c. Cramer's Rule can be used to solve a system of linear equations in three variables.

Complete.

35. If all the elements in one row or one column of a 2×2 matrix are zeros, the value of the determinant of the matrix is _____ .

36. If all the elements in one row or one column of a 3×3 matrix are zeros, the value of the determinant of the matrix is _____ .

37. a. The value of the determinant $\begin{vmatrix} x & x & a \\ y & y & b \\ z & z & c \end{vmatrix}$ is _____ .

   b. If two columns of a 3×3 matrix contain identical elements, the value of the determinant is _____ .

5.5

Application Problems in Two Variables

....

Objective A   *To solve rate-of-wind or rate-of-current problems* ....

Motion problems that involve an object moving with or against a wind or current normally require two variables to solve.

➡ A motorboat traveling with the current can go 24 mi in 2 h. Against the current, it takes 3 h to go the same distance. Find the rate of the motorboat in calm water and the rate of the current.

> **Strategy for Solving Rate-of-Wind or Rate-of-Current Problems**
>
> 1. Choose one variable to represent the rate of the object in calm conditions and a second variable to represent the rate of the wind or current. Using these variables, express the rate of the object with and against the wind or current. Use the equation $rt = d$ to write expressions for the distance traveled by the object. The results can be recorded in a table.

CONSIDER THIS

The boat travels faster when it travels with the current than when it travels against the current. Note that the boat's rate is *increased* by the rate of the current when it is traveling with the current, and the boat's rate is *decreased* by the rate of the current when it is traveling against the current.

Rate of the boat in calm water: x
Rate of the current: y

	Rate	·	*Time*	=	*Distance*
With the current	$x + y$	·	2	=	$2(x + y)$
Against the current	$x - y$	·	3	=	$3(x - y)$

> 2. Determine how the expressions for distance are related.

The distance traveled with the current is 24 mi: $2(x + y) = 24$
The distance traveled against the current is 24 mi: $3(x - y) = 24$

$2(x + y) = 24$ ● Solve this system of equations.
$3(x - y) = 24$

$x + y = 12$ ● Divide each side of the first equation by 2.
$x - y = 8$ ● Divide each side of the second equation by 3.

$2x = 20$ ● Add the equations.
$x = 10$ ● Solve the equation for x.

$x + y = 12$
$10 + y = 12$ ● Replace x by 10 in the equation $x + y = 12$
$y = 2$ and solve for y.

The rate of the boat in calm water is 10 mph.
The rate of the current is 2 mph.

Example 1

Flying with the wind, a plane flew 1000 mi in 5 h. Flying against the wind, the plane could fly only 500 mi in the same amount of time. Find the rate of the plane in calm air and the rate of the wind.

Strategy

- Rate of the plane in still air: p
  Rate of the wind: w

	Rate	Time	Distance
With wind	$p + w$	5	$5(p + w)$
Against wind	$p - w$	5	$5(p - w)$

- The distance traveled with the wind is 1000 mi.
  The distance traveled against the wind is 500 mi.

Solution

$5(p + w) = 1000$
$5(p - w) = 500$

$p + w = 200$ • Divide each side of
$p - w = 100$ each equation by 5.

$2p = 300$
$p = 150$

$p + w = 200$
$150 + w = 200$
$w = 50$

The rate of the plane in calm air is 150 mph. The rate of the wind is 50 mph.

You Try It 1

A rowing team rowing with the current traveled 18 mi in 2 h. Against the current, the team rowed 10 mi in 2 h. Find the rate of the rowing team in calm water and the rate of the current.

Your strategy

Your solution

Solution on p. A26

Objective B   *To solve application problems using two variables* ·······················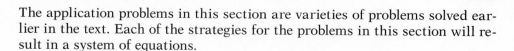

The application problems in this section are varieties of problems solved earlier in the text. Each of the strategies for the problems in this section will result in a system of equations.

➡ A store owner purchased twenty 60-watt light bulbs and 30 fluorescent bulbs for a total cost of $40. A second purchase, at the same prices, included thirty 60-watt light bulbs and 10 fluorescent bulbs for a total cost of $25. Find the cost of a 60-watt bulb and that of a fluorescent bulb.

Strategy for Solving an Application Problem in Two Variables

1. Choose one variable to represent one of the unknown quantities and a second variable to represent the other unknown quantity. Write numerical or variable expressions for all the remaining quantities. These results can be recorded in tables.

Cost of a 60-watt bulb: b
Cost of a fluorescent bulb: f

First purchase

	Amount	·	Unit Cost	=	Value
60-watt	20	·	b	=	$20b$
Fluorescent	30	·	f	=	$30f$

Second purchase

	Amount	·	Unit Cost	=	Value
60-watt	30	·	b	=	$30b$
Fluorescent	10	·	f	=	$10f$

2. Determine a system of equations. The strategies presented in Chapter 2 can be used to determine the relationships between the expressions in the tables.

The total of the first purchase was $40: $\quad 20b + 30f = 40$
The total of the second purchase was $25: $30b + 10f = 25$

Solve the system of equations: (1) $\quad 20b + 30f = 40$
(2) $\quad 30b + 10f = 25$

$$
\begin{aligned}
60b + 90f &= 120 \quad \bullet \ \textbf{3 times Equation (1)} \\
-60b - 20f &= -50 \quad \bullet \ \textbf{-2 times Equation (2)} \\
70f &= 70 \\
f &= 1
\end{aligned}
$$

Replace f by 1 in Equation (1). Solve for b.

$$
\begin{aligned}
20b + 30f &= 40 \\
20b + 30(1) &= 40 \\
20b + 30 &= 40 \\
20b &= 10 \\
b &= 0.5
\end{aligned}
$$

The cost of a 60-watt bulb was $.50.
The cost of a fluorescent bulb was $1.00.

Example 2

A total of 260 tickets were sold for a softball game. Adult tickets sold for $6 each, and children's tickets sold for $2 each. If the total receipts were $1220, how many adult tickets and how many children's tickets were sold?

Strategy

• Number of adult tickets sold: A
  Number of children's tickets sold: C

Ticket	Price	Number Sold	Total Receipts
Adult	6	A	$6A$
Children	2	C	$2C$

• The total number of tickets sold was 260.
  The total receipts were $1220.

You Try It 2

An investment adviser invested $15,000 in two accounts. One investment earned 8% annual simple interest. The other investment earned 7% annual simple interest. The total interest in one year was $1170. How much was invested in each account?

Your strategy

Solution

$$A + C = 260$$
$$6A + 2C = 1220$$

$$-2A - 2C = -520$$
$$6A + 2C = 1220$$
$$4A = 700$$
$$A = 175$$

$$A + C = 260$$
$$175 + C = 260$$
$$C = 85$$

There were 175 adult tickets sold.
There were 85 children's tickets sold.

Your solution

Solution on p. A27

5.5 Exercises

- -

Objective A   *Application Problems*

Solve.

1. A motorboat traveling with the current went 36 mi in 2 h. Against the current, it took 3 h to travel the same distance. Find the rate of the boat in calm water and the rate of the current.

2. A cabin cruiser traveling with the current went 45 mi in 3 h. Against the current, it took 5 h to travel the same distance. Find the rate of the cabin cruiser in calm water and the rate of the current.

3. A jet plane flying with the wind went 2200 mi in 4 h. Against the wind, the plane could fly only 1820 mi in the same amount of time. Find the rate of the plane in calm air and the rate of the wind.

4. Flying with the wind, a small plane flew 300 mi in 2 h. Against the wind, the plane could fly only 270 mi in the same amount of time. Find the rate of the plane in calm air and the rate of the wind.

5. A rowing team rowing with the current traveled 20 km in 2 h. Rowing against the current, the team rowed 12 km in the same amount of time. Find the rate of the rowing team in calm water and the rate of the current.

6. A motorboat traveling with the current went 72 km in 3 h. Against the current, the boat could go only 48 km in the same amount of time. Find the rate of the boat in calm water and the rate of the current.

7. A turbo-prop plane flying with the wind flew 800 mi in 4 h. Flying against the wind, the plane required 5 h to travel the same distance. Find the rate of the wind and the rate of the plane in calm air.

8. Flying with the wind, a pilot flew 600 mi between two cities in 4 h. The return trip against the wind took 5 h. Find the rate of the plane in calm air and the rate of the wind.

Solve.

9. A plane flying with a tailwind flew 600 mi in 5 h. Against the wind, the plane required 6 h to fly the same distance. Find the rate of the plane in calm air and the rate of the wind.

10. Flying with the wind, a plane flew 720 mi in 3 h. Against the wind, the plane required 4 h to fly the same distance. Find the rate of the plane in calm air and the rate of the wind.

11. A motorboat traveling with the current went 48 mi in 3 h. Against the current, it took 4.8 h to travel the same distance. Find the rate of the boat in calm water and the rate of the current.

12. A plane traveling with the wind flew 3625 mi in 6.25 h. Against the wind, the plane required 7.25 h to fly the same distance. Find the rate of the plane in calm air and the rate of the wind.

13. A cabin cruiser traveling with the current went 45 mi in 2.5 h. Against the current, the boat could go only 30 mi in the same amount of time. Find the rate of the cabin cruiser and the rate of the current.

14. Flying with the wind, a plane flew 450 mi in 3 h. Against the wind, the plane could fly only 270 mi in the same amount of time. Find the rate of the plane in calm air and the rate of the wind.

Objective B   *Application Problems*

Solve.

15. A merchant mixed 10 lb of a cinnamon tea with 5 lb of spice tea. The 15-pound mixture cost $40. A second mixture included 12 lb of the cinnamon tea and 8 lb of the spice tea. The 20-pound mixture cost $54. Find the cost per pound of the cinnamon tea and the spice tea.

16. A carpenter purchased 60 ft of redwood and 80 ft of pine for a total cost of $27. A second purchase, at the same prices, included 100 ft of redwood and 60 ft of pine for a total of $34. Find the cost per foot of redwood and the cost per foot of pine.

Solve.

17. A contractor buys 16 yd of nylon carpet and 20 yd of wool carpet for $920. A second purchase, at the same prices, includes 18 yd of nylon carpet and 25 yd of wool carpet for $1100. Find the cost per yard of the wool carpet.

18. During one month, a homeowner used 500 units of electricity and 100 units of gas for a total cost of $88. The next month, 400 units of electricity and 150 units of gas were used, for a total cost of $76. Find the cost per unit of gas.

19. A company manufactures both 10-speed and standard model bicycles. The cost of materials for a 10-speed bicycle is $35, and the cost of materials for a standard bicycle is $25. The cost of labor to manufacture a 10-speed bicycle is $40, and the cost of labor to manufacture a standard bicycle is $20. During a week when the company has budgeted $1250 for materials and $1300 for labor, how many 10-speed bicycles does the company plan to manufacture?

20. A company manufactures both color and black-and-white television sets. The cost of materials for a black-and-white TV is $20, and the cost of materials for a color TV is $80. The cost of labor to manufacture a black-and-white TV is $30, and the cost of labor to manufacture a color TV is $50. During a week when the company has budgeted $4200 for materials and $2800 for labor, how many color TVs does the company plan to manufacture?

21. A chemist has two alloys, one of which is 10% gold and 15% lead and the other of which is 30% gold and 40% lead. How many grams of each of the two alloys should be used to make an alloy that contains 60 g of gold and 88 g of lead?

22. A pharmacist has two vitamin-supplement powders. The first powder is 20% vitamin B_1 and 10% vitamin B_2. The second is 15% vitamin B_1 and 20% vitamin B_2. How many milligrams of each of the two powders should the pharmacist use to make a mixture that contains 130 mg of vitamin B_1 and 80 mg of vitamin B_2?

23. Two angles are complementary. The larger angle is 9° more than eight times the measure of the smaller angle. Find the measure of the two angles. (Complementary angles are two angles whose sum is 90°.)

Solve.

24. Two angles are supplementary. The larger angle is 40° more than three times the measure of the smaller angle. Find the measure of the two angles. (Supplementary angles are two angles whose sum is 180°.)

25. The total receipts from 245 tickets sold to a play were $1045. Adult tickets sold for $5 each, and student tickets sold for $3 each. Find the number of adult tickets and the number of student tickets sold.

26. The income from a theater production was $4150. The price of a student ticket was $6, and the price of a nonstudent ticket was $10. Find the number of student tickets and the number of nonstudent tickets sold.

27. You have a total of $5000 invested in two simple interest accounts. On one account, the annual simple interest rate is 6%. On the second account, the annual simple interest rate is 8%. The total annual interest earned in one year is $370. How much is invested in each account?

28. Two investments earn annual interest of $970. One investment is in a 7.5% annual simple interest account. The other investment is in a 9.5% annual simple interest account. The total in the two investments is $12,000. How much is invested in each account?

APPLYING THE CONCEPTS

29. The sum of the digits of a two-digit number equals $\frac{1}{7}$ of the number.

If the digits of the number are reversed, the new number is equal to 36 less than the original number. Find the original number.

30. The sum of the digits of a two-digit number equals $\frac{1}{5}$ of the number.

If the digits of the number are reversed, the new number is equal to 9 more than the original number. Find the original number.

31. A coin bank contains only nickels and dimes. The total value of the coins in the bank is $2.50. If the nickels were dimes and the dimes were nickels, the total value of the coins would be $3.50. Find the number of nickels in the bank.

32. The total value of the quarters and dimes in a coin bank is $5.75. If the quarters were dimes and the dimes were quarters, the total value of the coins would be $6.50. Find the number of quarters in the bank.

Solving Systems of Linear Inequalities

Objective A   *To graph the solution set of a system of linear inequalities* ....

POINT OF INTEREST

Large systems of linear inequalities containing over 100 inequalities have been used to solve application problems in such diverse areas as providing health care and hardening a nuclear missile silo.

Two or more inequalities considered together are called a **system of inequalities.** The **solution set of a system of inequalities** is the intersection of the solution sets of the individual inequalities. To graph the solution set of a system of inequalities, first graph the solution set of each inequality. The solution set of the system of inequalities is the region of the plane represented by the intersection of the two shaded areas.

➡ Graph the solution set: $2x - y < 3$
$3x + 2y \geq 8$

Solve each equation for y.

$$2x - y < 3$$
$$-y < -2x + 3$$
$$y > 2x - 3$$

$$3x + 2y \geq 8$$
$$2y \geq -3x + 8$$
$$y \geq -\frac{3}{2}x + 4$$

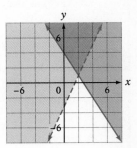

Graph $y = 2x - 3$ as a dashed line. Because the inequality is $>$, shade above the line.

Graph $y = -\frac{3}{2}x + 4$ as a solid line. Because the inequality is $\geq$, shade above the line.

The solution set of the system is the region of the plane represented by the intersection of the solution sets of the individual inequalities.

➡ Graph the solution set: $-x + 2y \geq 4$
$x - 2y \geq 6$

Solve each equation for y.

$$-x + 2y \geq 4$$
$$2y \geq x + 4$$
$$y \geq \frac{1}{2}x + 2$$

$$x - 2y \geq 6$$
$$-2y \geq -x + 6$$
$$y \leq \frac{1}{2}x - 3$$

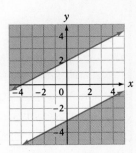

Shade above the solid line $y = \frac{1}{2}x + 2$.

Shade below the solid line $y = \frac{1}{2}x - 3$.

Because the solution sets of the two inequalities do not intersect, the solution of the system is the empty set.

Example 1
Graph the solution set: $y \geq x - 1$
$\qquad\qquad\qquad\qquad y < -2x$

Solution
Shade above the solid line $y = x - 1$.
Shade below the dashed line $y = -2x$.

The solution of the system is the
intersection of the solution sets of the
individual inequalities.

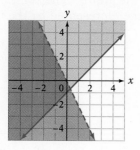

You Try It 1
Graph the solution set: $y \geq 2x - 3$
$\qquad\qquad\qquad\qquad y > -3x$

Your solution

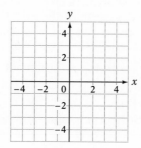

Example 2
Graph the solution set: $2x + 3y > 9$
$\qquad\qquad\qquad\qquad y < -\frac{2}{3}x + 1$

Solution
$2x + 3y > 9$
$\qquad 3y > -2x + 9$
$\qquad\quad y > -\frac{2}{3}x + 3$

Graph above the dashed line $y = -\frac{2}{3}x + 3$.

Graph below the dashed line $y = -\frac{2}{3}x + 1$.

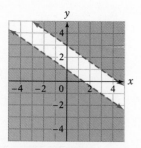

The intersection of the system is the
empty set, because the solution sets of
the two inequalities do not intersect.

You Try It 2
Graph the solution set: $3x + 4y > 12$
$\qquad\qquad\qquad\qquad y < \frac{3}{4}x - 1$

Your solution

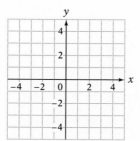

Solutions on p. A28

Content and Format © 1996 HMCo.

5.6 Exercises

· ·

Objective A

Graph the solution set.

1. $x - y \geq 3$
$\quad x + y \leq 5$

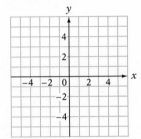

2. $2x - y < 4$
$\quad x + y < 5$

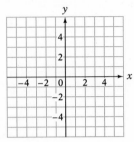

3. $3x - y < 3$
$\quad 2x + y \geq 2$

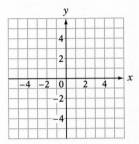

4. $x + 2y \leq 6$
$\quad x - y \leq 3$

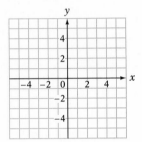

5. $2x + y \geq -2$
$\quad 6x + 3y \leq 6$

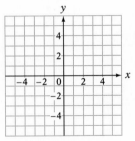

6. $x + y \geq 5$
$\quad 3x + 3y \leq 6$

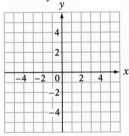

7. $3x - 2y < 6$
$\quad\quad y \leq 3$

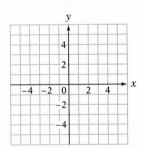

8. $\quad\quad x \leq 2$
$\quad 3x + 2y > 4$

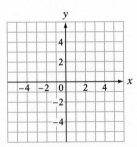

9. $\quad y > 2x - 6$
$\quad x + y < 0$

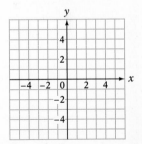

10. $x < 3$
$\quad y < -2$

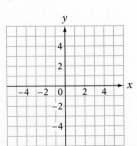

11. $x + 1 \geq 0$
$\quad y - 3 \leq 0$

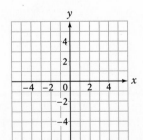

12. $5x - 2y \geq 10$
$\quad 3x + 2y \geq 6$

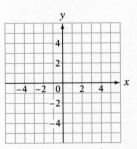

Graph the solution set.

13. $2x + y \geq 4$
$3x - 2y < 6$

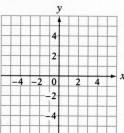

14. $3x - 4y < 12$
$x + 2y < 6$

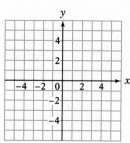

15. $x - 2y \leq 6$
$2x + 3y \leq 6$

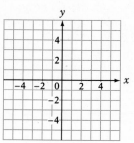

16. $x - 3y > 6$
$2x + y > 5$

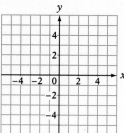

17. $x - 2y \leq 4$
$3x + 2y \leq 8$

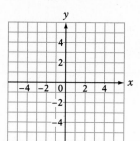

18. $3x - 2y < 0$
$5x + 3y > 9$

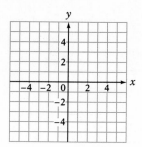

APPLYING THE CONCEPTS

Graph the solution set.

19. $2x + 3y \leq 15$
$3x - y \leq 6$
$y \geq 0$

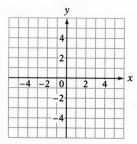

20. $x + y \leq 6$
$x - y \leq 2$
$x \geq 0$

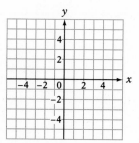

21. $x - y \leq 5$
$2x - y \geq 6$
$y \geq 0$

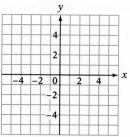

22. $x - 3y \leq 6$
$5x - 2y \geq 4$
$y \geq 0$

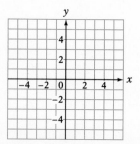

23. $2x - y \leq 4$
$3x + y < 1$
$y \leq 0$

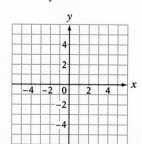

24. $x - y \leq 4$
$2x + 3y > 6$
$x \geq 0$

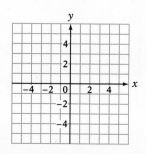

Projects in Mathematics

. . . .

Using a Graphing Calculator to Solve a System of Equations

A graphing calculator can be used to solve a system of equations. For this procedure to work on most calculators, it is necessary that the point of intersection be on the screen. This means that you may have to experiment with Xmin, Xmax, Ymin, and Ymax values until the graphs intersect on the screen.

To solve a system of equations graphically, solve each equation for y. Then graph the equations of the system. Their point of intersection is the solution.

The keystrokes to solve the system of equations

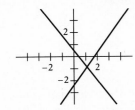

$$4x - 3y = 7$$
$$5x + 4y = 2$$

are given below. We are using a viewing window $[-4.7, 4.7]$ by $[-3.1, 3.1]$. The approximate solution is $(1.096744, -0.870967)$.

TI-82	SHARP EL-9300	CASIO fx-7700GB
[Y=] [CLEAR] 4 [X,T,θ] [÷] 3 [−] 7 [÷] 3 [ENTER] [CLEAR] [(-)] 5 [X,T,θ] [÷] 4 [+] 1 [÷] 2 [ZOOM] 4 [Graph]	[↰↲] [CL] 4 [X/θ/T] [÷] 3 [−] 7 [÷] 3 [ENTER] [CLEAR] [(-)] 5 [X/θ/T] [÷] 4 [+] 1 [÷] 2 [ENTER] [RANGE] [MENU] [ENTER] [↰↲]	[SHIFT] [F5] [EXE] [Range] [F1] [Range] [Range] [Graph] 4 [X,θ,T] [÷] 3 [−] 7 [÷] 3 [EXE] [Graph] [SHIFT] [(-)] 5 [X,θ,T] [÷] 4 [+] 1 [÷] 2 [EXE]

Once the calculator has drawn the graphs, use the TRACE feature and move the cursor to the approximate point of intersection. This will give you an approximate solution of the system of equations. The method by which a more accurate solution can be determined depends on the type of calculator you use. Consult the user's manual under "systems of equations."

Some of the exercises in the first section of this chapter asked you to solve a system of equations by graphing. Try those exercises again, this time using your graphing calculator.

Current models of calculators do not allow you to solve graphically a system of equations in three variables. However, these calculators do have matrix and determinant operations that can be used to solve these systems.

Chapter Summary

. . . .

Key Words   Equations considered together are called a *system of equations*.

A *solution of a system of equations* in two variables is an ordered pair that is a solution of each equation of the system.

When the graphs of a system of equations intersect at only one point, the system is called an *independent system of equations.* When the graphs of a system of equations coincide, the system is called a *dependent system of equations.* When a system of equations has no solution, it is called an *inconsistent system of equations.*

An equation of the form $Ax + By + Cz = D$ is a *linear equation in three variables.* A *solution of a system of equations in three variables* is an ordered triple that is a solution of each equation of the system.

A *matrix* is a rectangular array of numbers. Each number in the matrix is called an *element* of the matrix. A matrix of m rows and n columns is said to be of *order* $m \times n$. A *square matrix* has the same number of rows as columns.

A *determinant* is a number associated with a square matrix.

The *minor of an element* in a 3×3 determinant is the 2×2 determinant obtained by eliminating the row and column that contain that element. The *cofactor* of an element of a matrix is $(-1)^{i+j}$ times the minor of that element, where i is the row number of the element and j is its column number.

The evaluation of the determinant of a 3×3 or larger matrix is accomplished by *expanding by cofactors.*

Inequalities considered together are called a *system of inequalities.* The *solution set of a system of inequalities* is the intersection of the solution sets of the individual inequalities.

Essential Rules

A system of equations can be solved by a graphing method, by the substitution method, or by the addition method.

Cramer's Rule

The solution of the system of equations $\begin{aligned} a_1x + b_1y &= c_1 \\ a_2x + b_2y &= c_2 \end{aligned}$ is

given by $x = \dfrac{D_x}{D}$ and $y = \dfrac{D_y}{D}$, where $D = \begin{vmatrix} a_1 & b_1 \\ a_2 & b_2 \end{vmatrix}$,

$D_x = \begin{vmatrix} c_1 & b_1 \\ c_2 & b_2 \end{vmatrix}$, $D_y = \begin{vmatrix} a_1 & c_1 \\ a_2 & c_2 \end{vmatrix}$, and $D \neq 0$.

The solution of the system of equations

$\begin{aligned} a_1x + b_1y + c_1z &= d_1 \\ a_2x + b_2y + c_2z &= d_2 \\ a_3x + b_3y + c_3z &= d_3 \end{aligned}$ is given by $x = \dfrac{D_x}{D}$, $y = \dfrac{D_y}{D}$, and

$z = \dfrac{D_z}{D}$, where $D = \begin{vmatrix} a_1 & b_1 & c_1 \\ a_2 & b_2 & c_2 \\ a_3 & b_3 & c_3 \end{vmatrix}$, $D_x = \begin{vmatrix} d_1 & b_1 & c_1 \\ d_2 & b_2 & c_2 \\ d_3 & b_3 & c_3 \end{vmatrix}$,

$D_y = \begin{vmatrix} a_1 & d_1 & c_1 \\ a_2 & d_2 & c_2 \\ a_3 & d_3 & c_3 \end{vmatrix}$, $D_z = \begin{vmatrix} a_1 & b_1 & d_1 \\ a_2 & b_2 & d_2 \\ a_3 & b_3 & d_3 \end{vmatrix}$, and $D \neq 0$.

Chapter Review Exercises

. . . .

1. Solve by graphing: $2x - 3y = -6$
$2x - y = 2$

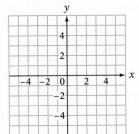

2. Solve by graphing: $x - 2y = -5$
$3x + 4y = -15$

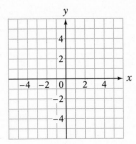

3. Graph the solution set: $2x - y < 3$
$4x + 3y < 11$

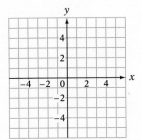

4. Graph the solution set: $x + y > 2$
$2x - y < -1$

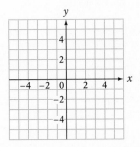

5. Solve by substitution: $3x + 2y = 4$
$x = 2y - 1$

6. Solve by substitution: $5x + 2y = -23$
$2x + y = -10$

7. Solve by substitution: $y = 3x - 7$
$y = -2x + 3$

8. Solve by the addition method:
$3x + 4y = -2$
$2x + 5y = 1$

9. Solve by the addition method:
$4x - 6y = 5$
$6x - 9y = 4$

10. Solve by the addition method:
$3x - y = 2x + y - 1$
$5x + 2y = y + 6$

11. Solve by the addition method:
$2x + 4y - z = 3$
$x + 2y + z = 5$
$4x + 8y - 2z = 7$

12. Solve by the addition method:
$x - y - z = 5$
$2x + z = 2$
$3y - 2z = 1$

13. Evaluate the determinant: $\begin{vmatrix} 3 & -1 \\ -2 & 4 \end{vmatrix}$

14. Evaluate the determinant: $\begin{vmatrix} 1 & -2 & 3 \\ 3 & 1 & 1 \\ 2 & -1 & -2 \end{vmatrix}$

15. Solve by using Cramer's Rule:
$x - y = 3$
$2x + y = -4$

16. Solve by using Cramer's Rule:
$5x + 2y = 9$
$3x + 5y = -7$

17. Solve by using Cramer's Rule:
$x - y + z = 2$
$2x - y - z = 1$
$x + 2y - 3z = -4$

18. Solve by using Cramer's Rule:
$3x + 2y + 2z = 2$
$x - 2y - z = 1$
$2x - 3y - 3z = -3$

19. A plane flying with the wind went 350 mi in 2 h. The return trip, flying against the wind, took 2.8 h. Find the rate of the plane in calm air and the rate of the wind.

20. A clothing manufacturer purchased 60 yd of cotton and 90 yd of wool for a total cost of $900. Another purchase, at the same prices, included 80 yd of cotton and 20 yd of wool for a total cost of $500. Find the cost per yard of the cotton and of the wool.

Cumulative Review Exercises

1. Simplify: $-2\sqrt{90}$

2. Solve: $3(x - 5) = 2x + 7$

3. Simplify: $3[x - 2(5 - 2x) - 4x] + 6$

4. Evaluate $a + bc \div 2$ when $a = 4$, $b = 8$, and $c = -2$.

5. Solve: $2x - 3 < 9$ or $5x - 1 < 4$

6. Solve: $|x - 2| - 4 < 2$

7. Solve: $|2x - 3| > 5$

8. Given $F(x) = x^2 - 3$, find $F(2)$.

9. Graph the solution set of $\{x \,|\, x \le 2\} \cap \{x \,|\, x > -3\}$.

$\xleftarrow{\;+\;+\;+\;+\;+\;+\;+\;+\;+\;+\;+\;}\rightarrow$
$\quad -5\ -4\ -3\ -2\ -1\ \ 0\ \ 1\ \ 2\ \ 3\ \ 4\ \ 5$

10. Find the equation of the line that contains the point $(-2, 3)$ and has slope $-\dfrac{2}{3}$.

11. Find the equation of the line that contains the points $(2, -1)$ and $(3, 4)$.

12. Find the equation of the line that contains the point $(-1, 2)$ and is perpendicular to the line $2x - 3y = 7$.

13. Graph $2x - 5y = 10$ by using the slope and y-intercept.

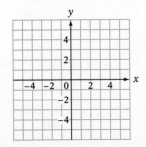

14. Graph the solution set of the inequality $3x - 4y \ge 8$.

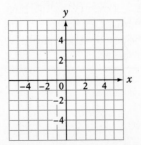

15. Solve by graphing:
$5x - 2y = 10$
$3x + 2y = 6$

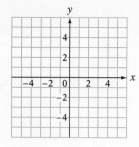

16. Solve by substitution:
$3x - 2y = 7$
$y = 2x - 1$

17. Solve by the addition method:
$3x + 2z = 1$
$2y - z = 1$
$x + 2y = 1$

18. Evaluate the determinant.
$$\begin{vmatrix} 2 & -5 & 1 \\ 3 & 1 & 2 \\ 6 & -1 & 4 \end{vmatrix}$$

19. Solve by using Cramer's Rule:
$4x - 3y = 17$
$3x - 2y = 12$

20. Graph the solution set.
$3x - 2y \geq 4$
$x + y < 3$

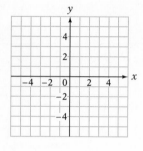

21. How many milliliters of pure water must be added to 100 ml of a 4% salt solution to make a 2.5% salt solution?

22. Flying with the wind, a small plane required 2 h to fly 150 mi. Against the wind, it took 3 h to fly the same distance. Find the rate of the wind.

23. A restaurant manager buys 100 lb of hamburger and 50 lb of steak for a total cost of $270. A second purchase, at the same prices, includes 150 lb of hamburger and 100 lb of steak. The total cost is $480. Find the price of 1 lb of steak.

24. Find the lower and upper limits of a 12,000-ohm resistor with a 15% tolerance.

25. The graph shows the relationship between the monthly income and the sales of an account executive. Find the slope of the line between the two points shown on the graph. Write a sentence that states the meaning of the slope.

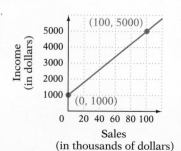

Chapter

Polynomials

Objectives

Origins of the Word *Algebra*

The word *algebra* has its origins in an Arabic book written around A.D. 825, *Hisab al-jabr w' almuqa-balah* by al-Khowarizmi. The word *al-jabr,* which literally translated means "reunion," was written as the word *algebra* in Latin translations of al-Khowarizmi's work and became synonymous with equations and the solutions of equations. It is interesting to note that an early meaning of the Spanish word *algebrista* was "bonesetter" or "reuniter of broken bones."

This same al-Khowarizmi also made a second contribution to our language of mathematics. One of the translations of his work into Latin shortened his name to *Algoritmi.* A further modification of this word gives us our present word *algorithm.* An algorithm is a procedure or set of instructions used to solve different types of problems. Computer scientists use algorithms when writing computer programs.

A further historical note: Omar Khayyám, a Persian who probably read al-Khowarizmi's work, is especially noted as a poet and the author of the *Rubáiyat.* However, he was also an excellent mathematician and astronomer and made many contributions to mathematics.

6.1 Exponential Expressions

Objective A   *To multiply monomials*

A **monomial** is a number, a variable, or a product of a number and variables.

POINT OF INTEREST

Around A.D. 250, the monomial $3x^2$ shown at the right would have been written $\Delta^{Y}3$ or at least approximately like that. In A.D. 250, the symbol for 3 was not the one we use today.

The examples at the right are monomials. The **degree of a monomial** is the sum of the exponents of the variables.

x	degree 1 ($x = x^1$)
$3x^2$	degree 2
$4x^2y$	degree 3
$6x^3y^4z^2$	degree 9

In this chapter, the variable n is considered a positive integer when used as an exponent.

x^n degree n

The degree of a nonzero constant term is zero.

6   degree 0

The expression $5\sqrt{x}$ is not a monomial because $\sqrt{x}$ cannot be written as a product of variables. The expression $\dfrac{x}{y}$ is not a monomial because it is a quotient of variables.

The expression x^4 is an exponential expression. The exponent, 4, indicates the number of times the base, x, occurs as a factor.

The product of exponential expressions with the *same* base can be simplified by writing each expression in factored form and writing the result with an exponent.

$$x^3 \cdot x^4 = \overbrace{(x \cdot x \cdot x)}^{3 \text{ factors}} \cdot \overbrace{(x \cdot x \cdot x \cdot x)}^{4 \text{ factors}}$$
$$\underbrace{}_{7 \text{ factors}}$$
$$= x^7$$

Note that adding the exponents results in the same product.

$$x^3 \cdot x^4 = x^{3+4} = x^7$$

Rule for Multiplying Exponential Expressions

If m and n are positive integers, then $x^m \cdot x^n = x^{m+n}$.

CONSIDER THIS

The Rule for Multiplying Exponential Expressions requires the bases to be the same. The expression x^6y^5 cannot be simplified.

⇒ Multiply: $(-4x^5y^3)(3xy^2)$

$$(-4x^5y^3)(3xy^2) = (-4 \cdot 3)(x^5 \cdot x)(y^3 \cdot y^2)$$

- Use the Commutative and Associative Properties of Multiplication to rearrange and group factors.

$$= -12(x^{5+1})(y^{3+2})$$

- Multiply variables with the same base by adding their exponents. Do these steps mentally.

$$= -12x^6y^5$$

- Simplify.

POINT OF INTEREST
One of the first symbolic representations of powers was given by Diophantus (c. A.D. 250) in his book *Arithmetica*. He used Δ^γ for x^2 and κ^γ for x^3. The symbol Δ^γ was the first two letters of the Greek word *dunamis*, meaning "power"; κ^γ was from the Greek word *kubos*, meaning "cube." He also combined these symbols to denote higher powers. For instance, $\Delta\kappa^\gamma$ was the symbol for x^5.

The power of a monomial can be simplified by writing the power in factored form and then using the Rule for Multiplying Exponential Expressions.

$$(x^4)^3 = x^4 \cdot x^4 \cdot x^4 \qquad (a^2b^3)^2 = (a^2b^3)(a^2b^3)$$
$$= x^{4+4+4} \qquad\qquad\qquad = a^{2+2}b^{3+3}$$
$$= x^{12} \qquad\qquad\qquad\quad = a^4b^6$$

- Write in factored form.
- Use the Rule for Multiplying Exponential Expressions.

Note that multiplying each exponent inside the parentheses by the exponent outside the parentheses results in the same product.

$$(x^4)^3 = x^{4\cdot3} = x^{12} \qquad (a^2b^3)^2 = a^{2\cdot2}b^{3\cdot2} = a^4b^6$$

- Multiply each exponent inside the parentheses by the exponent outside the parentheses.

Rule for Simplifying Powers of Exponential Expressions

If m and n are positive integers, then $(x^m)^n = x^{mn}$.

Rule for Simplifying Powers of Products

If m, n, and p are positive integers, then $(x^m y^n)^p = x^{mp} y^{np}$.

➡ Simplify: $(x^4)^5$

Multiply the exponents.

$$(x^4)^5 = x^{4\cdot5}$$
$$= x^{20}$$

- Use the Rule for Simplifying Powers of Exponential Expressions.

➡ Simplify: $(2a^3b^4)^3$

Multiply each exponent inside the parentheses by the exponent outside the parentheses.

$$(2a^3b^4)^3 = 2^{1\cdot3}a^{3\cdot3}b^{4\cdot3}$$
$$= 2^3a^9b^{12}$$
$$= 8a^9b^{12}$$

- Use the Rule for Simplifying Powers of Products. Note that $2 = 2^1$.

Example 1
Simplify: $(2xy^2)(-3xy^4)^3$

Solution
$$(2xy^2)(-3xy^4)^3 = (2xy^2)[(-3)^3x^3y^{12}]$$
$$= (2xy^2)(-27x^3y^{12})$$
$$= -54x^4y^{14}$$

You Try It 1
Simplify: $(-3a^2b^4)(-2ab^3)^4$

Your solution

Example 2
Simplify: $(x^{n+2})^5$

Solution
$$(x^{n+2})^5 = x^{5n+10}$$

You Try It 2
Simplify: $(y^{n-3})^2$

Your solution

Solutions on p. A29

Example 3

Simplify: $[(2xy^2)^2]^3$

Solution

$[(2xy^2)^2]^3 = [2^2x^2y^4]^3$
$= 2^6x^6y^{12} = 64x^6y^{12}$

You Try It 3

Simplify: $[(ab^3)^3]^4$

Your solution

Solution on p. A29

Objective B   *To divide monomials and simplify expressions with negative exponents* ....

The quotient of two exponential expressions with the same base can be simplified by writing each expression in factored form, dividing by the common factors, and then writing the result with an exponent.

$$\frac{x^5}{x^2} = \frac{\overset{1}{\cancel{x}} \cdot \overset{1}{\cancel{x}} \cdot x \cdot x \cdot x}{\underset{1}{\cancel{x}} \cdot \underset{1}{\cancel{x}}} = x^3$$

Note that subtracting the exponents gives the same result.

$$\frac{x^5}{x^2} = x^{5-2} = x^3$$

To divide two monomials with the same base, subtract the exponents of the like bases.

➡ Simplify: $\frac{z^8}{z^2}$

$$\frac{z^8}{z^2} = \boxed{z^{8-2}}$$

• The bases are the same. Subtract the exponents. This step is often done mentally.

$$= z^6$$

➡ Simplify: $\frac{a^5b^9}{a^4b}$

$$\frac{a^5b^9}{a^4b} = \boxed{a^{5-4}b^{9-1}}$$

• Subtract the exponents of the like bases. This step is often done mentally.

$$= ab^8$$

Consider the expression $\frac{x^4}{x^4}, x \neq 0$. This expression can be simplified, as shown below, by (1) subtracting exponents and (2) dividing by common factors.

$$\frac{x^4}{x^4} = x^{4-4} = x^0 \qquad\qquad \frac{x^4}{x^4} = \frac{\overset{1}{\cancel{x}} \cdot \overset{1}{\cancel{x}} \cdot \overset{1}{\cancel{x}} \cdot \overset{1}{\cancel{x}}}{\underset{1}{\cancel{x}} \cdot \underset{1}{\cancel{x}} \cdot \underset{1}{\cancel{x}} \cdot \underset{1}{\cancel{x}}} = 1$$

The equations $\frac{x^4}{x^4} = x^0$ and $\frac{x^4}{x^4} = 1$ suggest the following definition of x^0.

Definition of Zero as an Exponent

If $x \neq 0$, then $x^0 = 1$. The expression 0^0 is not defined.

➡ Simplify: $(16z^5)^0, z \neq 0$

$$(16z^5)^0 = 1$$

• Any nonzero expression to the zero power is 1.

➡ Simplify: $-(7x^4y^3)^0$

$$-(7x^4y^3)^0 = -(1) = -1$$

• The negative outside the parentheses is not affected by the exponent.

Consider the expression $\dfrac{x^4}{x^6}$, $x \neq 0$. This expression can be simplified, as shown below, by (1) subtracting exponents and (2) dividing by common factors.

$$\frac{x^4}{x^6} = x^{4-6} = x^{-2} \qquad \frac{x^4}{x^6} = \frac{\overset{1}{\cancel{x}} \cdot \overset{1}{\cancel{x}} \cdot \overset{1}{\cancel{x}} \cdot \overset{1}{\cancel{x}}}{\underset{1}{\cancel{x}} \cdot \underset{1}{\cancel{x}} \cdot \underset{1}{\cancel{x}} \cdot \underset{1}{\cancel{x}} \cdot x \cdot x} = \frac{1}{x^2}$$

The equations $\dfrac{x^4}{x^6} = x^{-2}$ and $\dfrac{x^4}{x^6} = \dfrac{1}{x^2}$ suggest that $x^{-2} = \dfrac{1}{x^2}$.

Definition of a Negative Exponent

If $x \neq 0$ and n is a positive integer, then

$$x^{-n} = \frac{1}{x^n} \quad \text{and} \quad \frac{1}{x^{-n}} = x^n$$

➡ Evaluate: 2^{-4}

$$2^{-4} = \frac{1}{2^4}$$

• Use the Definition of a Negative Exponent.

$$= \frac{1}{16}$$

• Evaluate the expression.

A power of the quotient of two exponential expressions can be simplified by multiplying each exponent in the quotient by the exponent outside the parentheses.

Rule for Simplifying Powers of Quotients

If m, n, and p are integers and $y \neq 0$, then $\left(\dfrac{x^m}{y^n}\right)^p = \dfrac{x^{mp}}{y^{np}}$.

➡ Simplify: $\left(\dfrac{a^2}{b^3}\right)^{-2}$

$$\left(\frac{a^2}{b^3}\right)^{-2} = \frac{a^{2(-2)}}{b^{3(-2)}}$$

• Use the Rule for Simplifying Powers of Quotients.

$$= \frac{a^{-4}}{b^{-6}} = \frac{b^6}{a^4}$$

• Use the Definition of a Negative Exponent.

Simplest Form of an Exponential Expression

An exponential expression is in simplest form when it is written with only positive exponents.

CONSIDER THIS

The exponent on n is -5 (*negative* five). The n^{-5} is written in the denominator as n^5. The exponent on 3 is 1 (*positive* one). The 3 remains in the numerator.

➡ Simplify: $3n^{-5}$

$$3n^{-5} = 3 \cdot \frac{1}{n^5} = \frac{3}{n^5}$$

- **Use the Definition of Negative Exponents to rewrite the expression with a positive exponent.**

➡ Simplify: $\dfrac{2}{5a^{-4}}$

$$\frac{2}{5a^{-4}} = \frac{2}{5} \cdot \frac{1}{a^{-4}} = \frac{2}{5} \cdot a^4 = \frac{2a^4}{5}$$

- **Use the Definition of Negative Exponents to rewrite the expression with a positive exponent.**

Now that zero and negative exponents have been defined, a rule for dividing exponential expressions can be stated.

Rule for Dividing Exponential Expressions

If m and n are integers and $x \neq 0$, then $\dfrac{x^m}{x^n} = x^{m-n}$.

➡ Simplify: $\dfrac{x^4}{x^9}$

$$\frac{x^4}{x^9} = \boxed{x^{4-9}}$$

- **Use the Rule for Dividing Exponential Expressions. This step is often done mentally.**

$$= x^{-5}$$

- **Subtract the exponents.**

$$= \frac{1}{x^5}$$

- **Use the Definition of Negative Exponents to rewrite the expression with a positive exponent.**

The rules for simplifying exponential expressions and powers of exponential expressions are true for all integer exponents. These rules are restated here for your convenience.

Rules of Exponents

If m, n, and p are integers, then

$$x^m \cdot x^n = x^{m+n} \qquad (x^m)^n = x^{mn} \qquad (x^m y^n)^p = x^{mp} y^{np}$$

$$\frac{x^m}{x^n} = x^{m-n},\ x \neq 0 \qquad \left(\frac{x^m}{y^n}\right)^p = \frac{x^{mp}}{y^{np}},\ y \neq 0 \qquad x^{-n} = \frac{1}{x^n},\ x \neq 0$$

$$x^0 = 1,\ x \neq 0$$

➡ Simplify: $(3ab^{-4})(-2a^{-3}b^7)$

$$(3ab^{-4})(-2a^{-3}b^7) = \boxed{[3 \cdot (-2)](a^{1+(-3)}b^{-4+7})}$$

- **When multiplying expressions, add the exponents on like bases. Do this step mentally.**

$$= -6a^{-2}b^3$$

$$= -\frac{6b^3}{a^2}$$

➡ Simplify: $\dfrac{4a^{-2}b^5}{6a^5b^2}$

$$\dfrac{4a^{-2}b^5}{6a^5b^2} = \dfrac{\cancel{2} \cdot 2a^{-2}b^5}{\cancel{2} \cdot 3a^5b^2} = \dfrac{2a^{-2}b^5}{3a^5b^2}$$

- **Divide the coefficients by their common factor.**

$$= \dfrac{2a^{-2-5}b^{5-2}}{3}$$

- **Use the Rule for Dividing Exponential Expressions.**

$$= \dfrac{2a^{-7}b^3}{3} = \dfrac{2b^3}{3a^7}$$

- **Use the Definition of Negative Exponents to rewrite the expression with a positive exponent.**

➡ Simplify: $\left[\dfrac{6m^2n^3}{8m^7n^2}\right]^{-3}$

$$\left[\dfrac{6m^2n^3}{8m^7n^2}\right]^{-3} = \left[\dfrac{3m^{2-7}n^{3-2}}{4}\right]^{-3}$$

- **Simplify inside the brackets.**

$$= \left[\dfrac{3m^{-5}n}{4}\right]^{-3}$$

- **Subtract the exponents.**

$$= \dfrac{3^{-3}m^{15}n^{-3}}{4^{-3}}$$

- **Use the Rule for Simplifying Powers of Quotients.**

$$= \dfrac{4^3m^{15}}{3^3n^3} = \dfrac{64m^{15}}{27n^3}$$

- **Use the Definition of Negative Exponents to rewrite the expression with positive exponents. Then simplify.**

Example 4

Simplify: $\dfrac{-28x^6z^{-3}}{42x^{-1}z^4}$

Solution

$$\dfrac{-28x^6z^{-3}}{42x^{-1}z^4} = -\dfrac{14 \cdot 2x^{6-(-1)}z^{-3-4}}{14 \cdot 3}$$

$$= -\dfrac{2x^7z^{-7}}{3} = -\dfrac{2x^7}{3z^7}$$

You Try It 4

Simplify: $\dfrac{20r^{-2}t^{-5}}{-16r^{-3}s^{-2}}$

Your solution

Example 5

Simplify: $\dfrac{(3a^{-1}b^4)^{-3}}{(6^{-1}a^{-3}b^{-4})^3}$

Solution

$$\dfrac{(3a^{-1}b^4)^{-3}}{(6^{-1}a^{-3}b^{-4})^3} = \dfrac{3^{-3}a^3b^{-12}}{6^{-3}a^{-9}b^{-12}} = 3^{-3} \cdot 6^3a^{12}b^0$$

$$= \dfrac{6^3a^{12}}{3^3} = \dfrac{216a^{12}}{27} = 8a^{12}$$

You Try It 5

Simplify: $\dfrac{(9u^{-6}v^4)^{-1}}{(6u^{-3}v^{-2})^{-2}}$

Your solution

Example 6

Simplify: $\dfrac{x^{4n-2}}{x^{2n-5}}$

Solution

$$\dfrac{x^{4n-2}}{x^{2n-5}} = x^{4n-2-(2n-5)}$$

$$= x^{4n-2-2n+5} = x^{2n+3}$$

You Try It 6

Simplify: $\dfrac{a^{2n+1}}{a^{n+3}}$

Your solution

Solutions on p. A29

Objective C   *To write a number using scientific notation* ....

Integer exponents are used to represent the very large and very small numbers encountered in the fields of science and engineering. For example, the mass of the electron is 0.00000000000000000000000000009 g. Numbers such as this are difficult to read and write, so a more convenient system for writing such numbers has been developed. It is called **scientific notation.**

To express a number in scientific notation, write the number as the product of a number between 1 and 10 and a power of 10. The form for scientific notation is $\boldsymbol{a} \times \boldsymbol{10^n}$, where $1 \le a < 10$.

For numbers greater than 10, move the decimal point to the right of the first digit. The exponent n is positive and equal to the number of places the decimal point has been moved.

$$965{,}000 = 9.65 \times 10^5$$
$$3{,}600{,}000 = 3.6 \times 10^6$$
$$92{,}000{,}000{,}000 = 9.2 \times 10^{10}$$

For numbers less than 1, move the decimal point to the right of the first nonzero digit. The exponent n is negative. The absolute value of the exponent is equal to the number of places the decimal point has been moved.

$$0.0002 = 2 \times 10^{-4}$$
$$0.0000000974 = 9.74 \times 10^{-8}$$
$$0.000000000086 = 8.6 \times 10^{-11}$$

Converting a number written in scientific notation to decimal notation requires moving the decimal point.

When the exponent is positive, move the decimal point to the right the same number of places as the exponent.

$$1.32 \times 10^4 = 13{,}200$$
$$1.4 \times 10^8 = 140{,}000{,}000$$

When the exponent is negative, move the decimal point to the left the same number of places as the absolute value of the exponent.

$$1.32 \times 10^{-2} = 0.0132$$
$$1.4 \times 10^{-4} = 0.00014$$

Numerical calculations involving numbers that have more digits than the hand-held calculator is able to handle can be performed using scientific notation.

➡ Simplify: $\dfrac{220{,}000 \times 0.000000092}{0.0000011}$

$$\frac{220{,}000 \times 0.000000092}{0.0000011} = \frac{2.2 \times 10^5 \times 9.2 \times 10^{-8}}{1.1 \times 10^{-6}}$$

- Write the numbers in scientific notation.

$$= \frac{(2.2)(9.2) \times 10^{5+(-8)-(-6)}}{1.1}$$

- Simplify.

$$= 18.4 \times 10^3 = 18{,}400$$

Example 7   Write 0.000041 in scientific notation.

Solution $0.000041 = 4.1 \times 10^{-5}$

You Try It 7   Write 942,000,000 in scientific notation.

Your solution

Solution on p. A29

Example 8   Write 3.3×10^7 in decimal notation.

Solution $3.3 \times 10^7 = 33,000,000$

You Try It 8   Write 2.7×10^{-5} in decimal notation.

Your solution

Example 9   Simplify:
$$\frac{2,400,000,000 \times 0.0000063}{0.00009 \times 480}$$

Solution
$$\frac{2,400,000,000 \times 0.0000063}{0.00009 \times 480}$$

$$= \frac{2.4 \times 10^9 \times 6.3 \times 10^{-6}}{9 \times 10^{-5} \times 4.8 \times 10^2}$$

$$= \frac{(2.4)(6.3) \times 10^{9+(-6)-(-5)-2}}{(9)(4.8)}$$

$$= 0.35 \times 10^6 = 350,000$$

You Try It 9   Simplify:
$$\frac{5,600,000 \times 0.000000081}{900 \times 0.000000028}$$

Your solution

Solutions on p. A29

Objective D   To solve application problems ....

Example 10

How many miles does light travel in one day? The speed of light is 186,000 mi/s. Write the answer in scientific notation.

Strategy

To find the distance traveled:

• Write the speed of light in scientific notation.
• Write the number of seconds in one day in scientific notation.
• Use the equation $d = rt$, where r is the speed of light and t is the number of seconds in one day.

Solution

$r = 186,000 = 1.86 \times 10^5$

$t = 24 \cdot 60 \cdot 60 = 86,400 = 8.64 \times 10^4$

$d = rt$
$d = (1.86 \times 10^5)(8.64 \times 10^4)$
$\quad = 1.86 \times 8.64 \times 10^9$
$\quad = 16.0704 \times 10^9$
$\quad = 1.60704 \times 10^{10}$

Light travels 1.60704×10^{10} mi in one day.

You Try It 10

A computer can do an arithmetic operation in 1×10^{-7} s. In scientific notation, how many arithmetic operations can the computer perform in 1 min?

Your strategy

Your solution

Solution on p. A29

6.1 Exercises

Objective A

Simplify.

1. $(ab^3)(a^3b)$

2. $(-2ab^4)(-3a^2b^4)$

3. $(9xy^2)(-2x^2y^2)$

4. $(x^2y)^2$

5. $(x^2y^4)^4$

6. $(-2ab^2)^3$

7. $(-3x^2y^3)^4$

8. $(2^2a^2b^3)^3$

9. $(3^3a^5b^3)^2$

10. $(xy)(x^2y)^4$

11. $(x^2y^2)(xy^3)^3$

12. $[(2x)^4]^2$

13. $[(3x)^3]^2$

14. $[(x^2y)^4]^5$

15. $[(ab)^3]^6$

16. $[(2ab)^3]^2$

17. $[(2xy)^3]^4$

18. $[(3x^2y^3)^2]^2$

19. $[(2a^4b^3)^3]^2$

20. $y^n \cdot y^{2n}$

21. $x^n \cdot x^{n+1}$

22. $y^{2n} \cdot y^{4n+1}$

23. $y^{3n} \cdot y^{3n-2}$

24. $(a^n)^{2n}$

25. $(a^{n-3})^{2n}$

26. $(y^{2n-1})^3$

27. $(x^{3n+2})^5$

28. $(b^{2n-1})^n$

29. $(2xy)(-3x^2yz)(x^2y^3z^3)$

30. $(x^2z^4)(2xyz^4)(-3x^3y^2)$

31. $(3b^5)(2ab^2)(-2ab^2c^2)$

32. $(-c^3)(-2a^2bc)(3a^2b)$

33. $(-2x^2y^3z)(3x^2yz^4)$

34. $(2a^2b)^3(-3ab^4)^2$

35. $(-3ab^3)^3(-2^2a^2b)^2$

36. $(4ab)^2(-2ab^2c^3)^3$

37. $(-2ab^2)(-3a^4b^5)^3$

Objective B

Simplify.

38. 2^{-3}

39. $\dfrac{1}{3^{-5}}$

40. $\dfrac{1}{x^{-4}}$

41. $\dfrac{1}{y^{-3}}$

Simplify.

42. $\dfrac{2x^{-2}}{y^4}$ **43.** $\dfrac{a^3}{4b^{-2}}$ **44.** $x^{-3}y$ **45.** xy^{-4}

46. $-5x^0$ **47.** $\dfrac{1}{2x^0}$ **48.** $\dfrac{(2x)^0}{-2^3}$ **49.** $\dfrac{-3^{-2}}{(2y)^0}$

50. $\dfrac{y^{-7}}{y^{-8}}$ **51.** $\dfrac{y^{-2}}{y^6}$ **52.** $(x^2y^{-4})^2$ **53.** $(x^3y^5)^{-2}$

54. $\dfrac{x^{-2}y^{-11}}{xy^{-2}}$ **55.** $\dfrac{x^4y^3}{x^{-1}y^{-2}}$ **56.** $\dfrac{a^{-1}b^{-3}}{a^4b^{-5}}$ **57.** $\dfrac{a^6b^{-4}}{a^{-2}b^5}$

58. $(2a^{-1})^{-2}(2a^{-1})^4$ **59.** $(3a)^{-3}(9a^{-1})^{-2}$ **60.** $(x^{-2}y)^2(xy)^{-2}$ **61.** $(x^{-1}y^2)^{-3}(x^2y^{-4})^{-3}$

62. $\dfrac{50b^{10}}{70b^5}$ **63.** $\dfrac{x^3y^6}{x^6y^2}$ **64.** $\dfrac{x^{17}y^5}{-x^7y^{10}}$ **65.** $\dfrac{-6x^2y}{12x^4y}$

66. $\dfrac{2x^2y^4}{(3xy^2)^3}$ **67.** $\dfrac{-3ab^2}{(9a^2b^4)^3}$ **68.** $\left(\dfrac{-12a^2b^3}{9a^5b^9}\right)^3$ **69.** $\left(\dfrac{12x^3y^2z}{18xy^3z^4}\right)^4$

70. $\dfrac{(4x^2y)^2}{(2xy^3)^3}$ **71.** $\dfrac{(3a^2b)^3}{(-6ab^3)^2}$ **72.** $\dfrac{(-4x^2y^3)^2}{(2xy^2)^3}$ **73.** $\dfrac{(-3a^2b^3)^2}{(-2ab^4)^3}$

74. $\dfrac{(-4xy^3)^3}{(-2x^7y)^4}$ **75.** $\dfrac{(-8x^2y^2)^4}{(16x^3y^7)^2}$ **76.** $\dfrac{a^{5n}}{a^{3n}}$ **77.** $\dfrac{b^{6n}}{b^{10n}}$

78. $\dfrac{-x^{5n}}{x^{2n}}$ **79.** $\dfrac{y^{2n}}{-y^{8n}}$ **80.** $\dfrac{x^{2n-1}}{x^{n-3}}$ **81.** $\dfrac{y^{3n+2}}{y^{2n+4}}$

82. $\dfrac{a^{3n}b^n}{a^nb^{2n}}$ **83.** $\dfrac{x^ny^{3n}}{x^ny^{5n}}$ **84.** $\dfrac{a^{3n-2}b^{n+1}}{a^{2n+1}b^{2n+2}}$ **85.** $\dfrac{x^{2n-1}y^{n-3}}{x^{n+4}y^{n+3}}$

86. $\left(\dfrac{4^{-2}xy^{-3}}{x^{-3}y}\right)^3\left(\dfrac{8^{-1}x^{-2}y}{x^4y^{-1}}\right)^{-2}$ **87.** $\left(\dfrac{9ab^{-2}}{8a^{-2}b}\right)^{-2}\left(\dfrac{3a^{-2}b}{2a^2b^{-2}}\right)^3$ **88.** $\left(\dfrac{2ab^{-1}}{ab}\right)^{-1}\left(\dfrac{3a^{-2}b}{a^2b^2}\right)^{-2}$

89. $1 + (1 + (1 + 2^{-1})^{-1})^{-1}$ **90.** $2 - (2 - (2 - 2^{-1})^{-1})^{-1}$

Objective C

Write in scientific notation.

91. 0.00000467

92. 0.00000005

93. 0.00000000017

94. 4,300,000

95. 200,000,000,000

96. 9,800,000,000

Write in decimal notation.

97. 1.23×10^{-7}

98. 6.2×10^{-12}

99. 8.2×10^{15}

100. 6.34×10^{5}

101. 3.9×10^{-2}

102. 4.35×10^{9}

Simplify.

103. $(3 \times 10^{-12})(5 \times 10^{16})$

104. $(8.9 \times 10^{-5})(3.2 \times 10^{-6})$

105. $(0.0000065)(3,200,000,000,000)$

106. $(480,000)(0.0000000096)$

107. $\dfrac{9 \times 10^{-3}}{6 \times 10^{5}}$

108. $\dfrac{2.7 \times 10^{4}}{3 \times 10^{-6}}$

109. $\dfrac{0.0089}{500,000,000}$

110. $\dfrac{4800}{0.00000024}$

111. $\dfrac{0.00056}{0.000000000004}$

112. $\dfrac{0.000000346}{0.0000005}$

113. $\dfrac{(3.3 \times 10^{-11})(2.7 \times 10^{15})}{8.1 \times 10^{-3}}$

114. $\dfrac{(6.9 \times 10^{27})(8.21 \times 10^{-13})}{4.1 \times 10^{15}}$

115. $\dfrac{(0.00000004)(84,000)}{(0.0003)(1,400,000)}$

116. $\dfrac{(720)(0.0000000039)}{(26,000,000,000)(0.018)}$

Objective D   Application Problems

Solve. Write the answer in scientific notation.

117. How many kilometers does light travel in one day? The speed of light is 300,000 km/s.

118. How many meters does light travel in 8 h? The speed of light is 300,000,000 m/s.

Solve. Write the answer in scientific notation.

119. A computer can do an arithmetic operation in 9×10^{-7} s. How many arithmetic operations can the computer perform in 1 h?

120. The national debt is 4.3×10^{12} dollars. How much would each American citizen have to pay in order to pay off the national debt? Use two hundred fifty million as the number of citizens.

121. A high-speed centrifuge makes 9×10^7 revolutions each minute. Find the time in seconds for the centrifuge to make one revolution.

122. How long does it take light to travel to the earth from the sun? The sun is 9.3×10^7 mi from the earth, and light travels 1.86×10^5 mi/s.

123. The mass of the earth is 5.9×10^{27} g. The mass of the sun is 2×10^{33} g. How many times heavier is the sun than the earth?

124. The distance to the sun is 9.3×10^7 mi. A satellite leaves the earth traveling at a constant speed of 1×10^5 mph. How long does it take for the satellite to reach the sun?

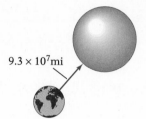

9.3×10^7 mi

125. The weight of 31 million orchid seeds is one ounce. Find the weight of one orchid seed.

126. One light-year, an astronomical unit of distance, is the distance that light will travel in one year. Light travels 1.86×10^5 mi/s. Find the measure of one light-year in miles. Use a 365-day year.

127. The light from a nearby star, Alpha Centauri, takes 4.3 years to reach the earth. Light travels at 1.86×10^5 mi/s. How far is Alpha Centauri from the earth? Use a 365-day year.

APPLYING THE CONCEPTS

128. Which is larger, $(2^3)^3$ or $2^{(3^3)}$?

129.
[W] Correct the error in each of the following expressions. Explain which rule or property was used incorrectly.
   a. $2x + 3x = 5x^2$
   b. $a - (b - c) = a - b - c$
   c. $x^0 = 0$
   d. $(x^4)^5 = x^9$
   e. $x^2 \cdot x^3 = x^6$
   f. $b^{m+n} = b^m + b^n$

6.2 Introduction to Polynomials

Objective A *To evaluate polynomial functions*

A **polynomial** is a variable expression in which the terms are monomials.

A polynomial of one term is a **monomial.** $5x$

A polynomial of two terms is a **binomial.** $5x^2y + 6x$

A polynomial of three terms is a **trinomial.** $3x^2 + 9xy - 5y$

Polynomials with more than three terms do not have special names.

The **degree of a polynomial** is the greatest of the degrees of any of its terms.

$3x + 2$	degree 1
$3x^2 + 2x - 4$	degree 2
$4x^3y^2 + 6x^4$	degree 5
$3x^{2n} - 5x^n - 2$	degree $2n$

The terms of a polynomial in one variable are usually arranged so that the exponents of the variable decrease from left to right. This is called **descending order.**

$2x^2 - x + 8$

$3y^3 - 3y^2 + y - 12$

For a polynomial in more than one variable, descending order may refer to any one of the variables.

The polynomial at the right is shown first in descending order of the x variable and then in descending order of the y variable.

$2x^2 + 3xy + 5y^2$

$5y^2 + 3xy + 2x^2$

Polynomial functions have many applications in mathematics. The **linear function** given by $f(x) = mx + b$ is an example of a polynomial function. It is a polynomial function of degree one. A second-degree polynomial function, called a **quadratic function,** is given by the equation $f(x) = ax^2 + bx + c$, $a \neq 0$. A third-degree polynomial function is called a **cubic function.** In general, a **polynomial function** is an expression whose terms are monomials.

To evaluate a polynomial function, replace the variable by its value and simplify.

➡ Given $P(x) = x^3 - 3x^2 + 4$, evaluate $P(-3)$.

$P(x) = x^3 - 3x^2 + 4$

$P(-3) = (-3)^3 - 3(-3)^2 + 4$ • Substitute −3 for *x* and simplify.

$\quad\quad = -27 - 3(9) + 4$

$\quad\quad = -27 - 27 + 4$

$\quad\quad = -50$

The **leading coefficient** of a polynomial function is the coefficient of the variable with the largest exponent. The constant term is the term without a variable.

➡ Find the leading coefficient, the constant term, and the degree of the polynomial function $P(x) = 7x^4 - 3x^2 + 2x - 4$.

The leading coefficient is 7, the constant term is -4, and the degree is 4.

The three equations below do not represent polynomial functions.

$f(x) = 3x^2 + 2x^{-1}$ A polynomial function does not have a variable raised to a negative power.

$g(x) = 2\sqrt{x} - 3$ A polynomial function does not have a variable expression within a radical.

$h(x) = \dfrac{x}{x - 1}$ A polynomial function does not have a variable in the denominator of a fraction.

The graph of a linear function is a straight line and can be found by plotting just two points. The graph of a polynomial function of degree greater than one is a curve. Consequently, many points may have to be found before an accurate graph can be drawn.

Evaluating the quadratic function given by the equation $f(x) = x^2 - x - 6$ when $x = -3, -2, -1, 0, 1, 2, 3,$ and 4 gives the points shown in Fig. 1 below. For instance, $f(-3) = 6$, so $(-3, 6)$ is graphed; $f(2) = -4$, so $(2, -4)$ is graphed; and $f(4) = 6$, so $(4, 6)$ is graphed. Evaluating the function when x is not an integer, such as $x = -\dfrac{3}{2}$ and $x = \dfrac{5}{2}$, produces more points to graph, as shown in Fig. 2. Connecting the points with a smooth curve results in Fig. 3, which is the graph of f.

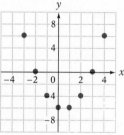

Fig. 1

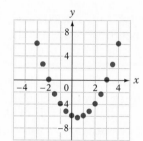

Fig. 2

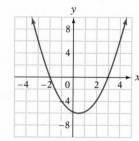

Fig. 3

Here is an example of graphing a cubic function, $P(x) = x^3 - 2x^2 - 5x + 6$. Evaluating the function when $x = -2, -1, 0, 1, 2, 3,$ and 4 gives the graph in Fig. 4 below. Evaluating at some noninteger values gives the graph in Fig. 5. Finally, connecting the dots with a smooth curve gives the graph in Fig. 6.

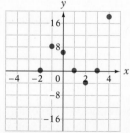

Fig. 4

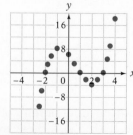

Fig. 5

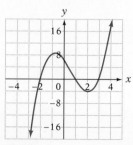

Fig. 6

Example 1

Given $P(x) = x^3 + 3x^2 - 2x + 8$, evaluate $P(-2)$.

Solution

$$P(x) = x^3 + 3x^2 - 2x + 8$$
$$P(-2) = (-2)^3 + 3(-2)^2 - 2(-2) + 8$$
$$= (-8) + 3(4) - 2(-2) + 8$$
$$= -8 + 12 + 4 + 8$$
$$= 16$$

You Try It 1

Given $R(x) = -2x^4 - 5x^3 + 2x - 8$, evaluate $R(2)$.

Your solution

Example 2

Find the leading coefficient, the constant term, and the degree of the polynomial $P(x) = 5x^6 - 4x^5 - 3x^2 + 7$.

Solution

The leading coefficient is 5, the constant term is 7, and the degree is 6.

You Try It 2

Find the leading coefficient, the constant term, and the degree of the polynomial $R(x) = -3x^4 + 3x^3 + 3x^2 - 2x - 12$.

Your solution

Example 3

Which of the following is a polynomial function?

a. $P(x) = 3x^{\frac{1}{2}} + 2x^2 - 3$
b. $T(x) = 3\sqrt{x} - 2x^2 - 3x + 2$
c. $R(x) = 14x^3 - \pi x^2 + 3x + 2$

Solution

a. This is not a polynomial function. A polynomial function does not have a variable raised to a fractional power.
b. This is not a polynomial function. A polynomial function does not have a variable expression within a radical.
c. This is a polynomial function.

You Try It 3

Which of the following is a polynomial function?

a. $R(x) = 5x^{14} - 5$
b. $V(x) = -x^{-1} + 2x - 7$
c. $P(x) = 2x^4 - 3\sqrt{x} - 3$

Your solution

Example 4

Graph $f(x) = x^2 - 2$.

Solution

x	$y = f(x)$
-3	7
-2	2
-1	-1
0	-2
1	-1
2	2
3	7

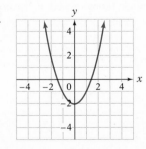

You Try It 4

Graph $f(x) = 2x^2 + 1$.

Your solution

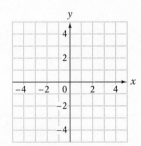

Solutions on p. A30

Example 5

Graph $f(x) = x^3 - 1$.

Solution

x	$y = f(x)$
-2	-9
-1	-2
0	-1
1	0
2	7

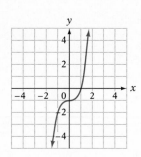

You Try It 5

Graph $f(x) = -x^3 + 1$.

Your solution

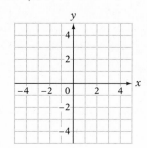

Solution on p. A30

Objective B   *To add or subtract polynomials* ....

Polynomials can be added by combining like terms. Either a vertical or a horizontal format can be used.

➡ Add $(3x^3 - 7x + 2) + (7x^2 + 2x - 7)$. Use a horizontal format.

$(3x^3 - 7x + 2) + (7x^2 + 2x - 7)$
$= 3x^3 + 7x^2 + (-7x + 2x) + (2 - 7)$

• Use the Commutative and Associative Properties of Addition to rearrange and group like terms.

$= 3x^3 + 7x^2 - 5x - 5$

• Combine like terms.

➡ Add $(4x^2 + 5x - 3) + (7x^3 - 7x + 1) + (2x - 3x^2 + 4x^3 + 1)$. Use a vertical format.

$$\begin{array}{l} \quad\;\; 4x^2 + 5x - 3 \\ 7x^3 \qquad\;\; - 7x + 1 \\ \underline{4x^3 - 3x^2 + 2x + 1} \\ 11x^3 + \;\; x^2 \qquad\;\, - 1 \end{array}$$

• Arrange the terms of each polynomial in descending order with like terms in the same column.

• Add the terms in each column.

CONSIDER THIS

The opposite of a polynomial is the polynomial with the sign of every term changed.

The additive inverse of the polynomial $x^2 + 5x - 4$ is $-(x^2 + 5x - 4)$.

To simplify the additive inverse of a polynomial, change the sign of every term inside the parentheses.

$-(x^2 + 5x - 4) = -x^2 - 5x + 4$

To subtract two polynomials, add the additive inverse of the second polynomial to the first.

⇒ Subtract $(3x^2 - 7xy + y^2) - (-4x^2 + 7xy - 3y^2)$. Use a horizontal format.

$(3x^2 - 7xy + y^2) - (-4x^2 + 7xy - 3y^2)$
$= (3x^2 - 7xy + y^2) + (4x^2 - 7xy + 3y^2)$ • **Rewrite the subtraction as the addition of the additive inverse.**
$= 7x^2 - 14xy + 4y^2$ • **Combine like terms.**

⇒ Subtract $(6x^3 - 3x + 7) - (3x^2 - 5x + 12)$. Use a vertical format.

$(6x^3 - 3x + 7) - (3x^2 - 5x + 12)$
$= (6x^3 - 3x + 7) + (-3x^2 + 5x - 12)$ • **Rewrite subtraction as the addition of the additive inverse.**

$$
\begin{array}{l}
6x^3 \qquad\quad -3x + 7 \\
\underline{\quad -3x^2 + 5x - 12} \\
6x^3 - 3x^2 + 2x - 5
\end{array}
$$

• **Arrange the terms of each polynomial in descending order with like terms in the same column.**

• **Combine the terms in each column.**

Functional notation can be used when adding or subtracting polynomials.

⇒ Given $P(x) = 3x^2 - 2x + 4$ and $R(x) = -5x^3 + 4x + 7$, find $P(x) + R(x)$.

$P(x) + R(x) = (3x^2 - 2x + 4) + (-5x^3 + 4x + 7)$
$= -5x^3 + 3x^2 + 2x + 11$

⇒ Given $P(x) = -5x^2 + 8x - 4$ and $R(x) = -3x^2 - 5x + 9$, find $P(x) - R(x)$.

$P(x) - R(x) = (-5x^2 + 8x - 4) - (-3x^2 - 5x + 9)$
$= (-5x^2 + 8x - 4) + (3x^2 + 5x - 9)$
$= -2x^2 + 13x - 13$

⇒ Given $P(x) = 3x^2 - 5x + 6$ and $R(x) = 2x^2 - 5x - 7$, find $S(x)$, the sum of the two polynomials.

$S(x) = P(x) + R(x) = (3x^2 - 5x + 6) + (2x^2 - 5x - 7)$
$= 5x^2 - 10x - 1$

Note that evaluating $P(x) = 3x^2 - 5x + 6$ and $R(x) = 2x^2 - 5x - 7$ at, for example, $x = 3$, and then adding the values, is the same as evaluating $S(x) = 5x^2 - 10x - 1$ at 3.

$P(3) = 3(3)^2 - 5(3) + 6 = 27 - 15 + 6 = 18$

$R(3) = 2(3)^2 - 5(3) - 7 = 18 - 15 - 7 = -4$

$P(3) + R(3) = 18 + (-4) = 14$

$S(3) = 5(3)^2 - 10(3) - 1 = 45 - 30 - 1 = 14$

Example 6
Add
$(4x^2 - 3xy + 7y^2) + (-3x^2 + 7xy + y^2)$.
Use a vertical format.

Solution

$$
\begin{array}{r}
4x^2 - 3xy + 7y^2 \\
-3x^2 + 7xy + y^2 \\
\hline
x^2 + 4xy + 8y^2
\end{array}
$$

You Try It 6
Add
$(-3x^2 - 4x + 9) + (-5x^2 - 7x + 1)$.
Use a vertical format.

Your solution

Example 7
Subtract
$(3x^2 - 2x + 4) - (7x^2 + 3x - 12)$.
Use a vertical format.

Solution
Add the additive inverse of $7x^2 + 3x - 12$
to $3x^2 - 2x + 4$.

$$
\begin{array}{r}
3x^2 - 2x + 4 \\
-7x^2 - 3x + 12 \\
\hline
-4x^2 - 5x + 16
\end{array}
$$

You Try It 7
Subtract
$(-5x^2 + 2x - 3) - (6x^2 + 3x - 7)$.
Use a vertical format.

Your solution

Example 8
Given $P(x) = -3x^2 + 2x - 6$ and
$R(x) = 4x^3 - 3x + 4$, find
$S(x) = P(x) + R(x)$. Evaluate $S(-2)$.

Solution
$$
\begin{aligned}
S(x) &= P(x) + R(x) \\
&= (-3x^2 + 2x - 6) + (4x^3 - 3x + 4) \\
&= 4x^3 - 3x^2 - x - 2
\end{aligned}
$$

$$
\begin{aligned}
S(-2) &= 4(-2)^3 - 3(-2)^2 - (-2) - 2 \\
&= 4(-8) - 3(4) - (-2) - 2 \\
&= -32 - 12 + 2 - 2 \\
&= -44
\end{aligned}
$$

You Try It 8
Given $P(x) = 4x^3 - 3x^2 + 2$ and
$R(x) = -2x^2 + 2x - 3$, find
$S(x) = P(x) + R(x)$. Evaluate $S(-1)$.

Your solution

Example 9
Given $P(x) = 2x^{2n} - 3x^n + 7$ and
$R(x) = 3x^{2n} + 3x^n + 5$, find
$D(x) = P(x) - R(x)$.

Solution
$$
\begin{aligned}
D(x) &= P(x) - R(x) \\
&= (2x^{2n} - 3x^n + 7) - (3x^{2n} + 3x^n + 5) \\
&= (2x^{2n} - 3x^n + 7) + (-3x^{2n} - 3x^n - 5) \\
&= -x^{2n} - 6x^n + 2
\end{aligned}
$$

You Try It 9
Given $P(x) = 5x^{2n} - 3x^n - 7$ and
$R(x) = -2x^{2n} - 5x^n + 8$, find
$D(x) = P(x) - R(x)$.

Your solution

Solutions on p. A30

6.2 Exercises

Objective A

1. Given $P(x) = 3x^2 - 2x - 8$, evaluate $P(3)$.

2. Given $P(x) = -3x^2 - 5x + 8$, evaluate $P(-5)$.

3. Given $R(x) = 2x^3 - 3x^2 + 4x - 2$, evaluate $R(2)$.

4. Given $R(x) = -x^3 + 2x^2 - 3x + 4$, evaluate $R(-1)$.

5. Given $f(x) = x^4 - 2x^2 - 10$, evaluate $f(-1)$.

6. Given $f(x) = x^5 - 2x^3 + 4x$, evaluate $f(2)$.

Which of the following define a polynomial function? For those that are polynomial functions, identify (a) the leading coefficient, (b) the constant term, and (c) the degree.

7. $P(x) = -x^2 + 3x + 8$

8. $P(x) = 3x^4 - 3x - 7$

9. $R(x) = \dfrac{x}{x + 1}$

10. $R(x) = \dfrac{3x^2 - 2x + 1}{x}$

11. $f(x) = \sqrt{x} - x^2 + 2$

12. $f(x) = x^2 - \sqrt{x + 2} - 8$

13. $g(x) = 3x^5 - 2x^2 + \pi$

14. $g(x) = -4x^5 - 3x^2 + x - \sqrt{7}$

15. $P(x) = 3x^2 - 5x^3 + 2$

16. $P(x) = x^2 - 5x^4 - x^6$

17. $R(x) = 14$

18. $R(x) = \dfrac{1}{x} + 2$

Graph.

19. $P(x) = x^2 - 1$

20. $P(x) = 2x^2 + 3$

21. $R(x) = x^3 + 2$

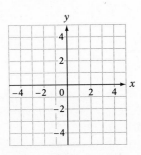

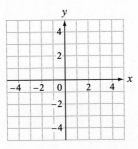

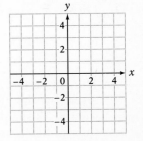

22. $R(x) = x^3 - 2x$

23. $f(x) = x^2 - 4x + 3$

24. $f(x) = x^2 - x - 2$

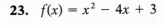

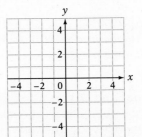

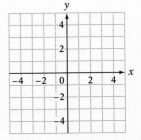

Objective B

Add or subtract. Use a vertical format.

25. $(5x^2 + 2x - 7) + (x^2 - 8x + 12)$

26. $(3x^2 - 2x + 7) + (-3x^2 + 2x - 12)$

27. $(x^2 - 3x + 8) - (2x^2 - 3x + 7)$

28. $(2x^2 + 3x - 7) - (5x^2 - 8x - 1)$

Add or subtract. Use a horizontal format for Exercises 29–32.

29. $(3y^2 - 7y) + (2y^2 - 8y + 2)$

30. $(-2y^2 - 4y - 12) + (5y^2 - 5y)$

31. $(2a^2 - 3a - 7) - (-5a^2 - 2a - 9)$

32. $(3a^2 - 9a) - (-5a^2 + 7a - 6)$

33. Given $P(x) = x^2 - 3xy + y^2$ and $R(x) = 2x^2 - 3y^2$, find $P(x) + R(x)$.

34. Given $P(x) = x^{2n} + 7x^n - 3$ and $R(x) = -x^{2n} + 2x^n + 8$, find $P(x) + R(x)$.

35. Given $P(x) = 3x^2 + 2y^2$ and $R(x) = -5x^2 + 2xy - 3y^2$, find $P(x) - R(x)$.

36. Given $P(x) = 2x^{2n} - x^n - 1$ and $R(x) = 5x^{2n} + 7x^n + 1$, find $P(x) - R(x)$.

37. Given $P(x) = 3x^4 - 3x^3 - x^2$ and $R(x) = 3x^3 - 7x^2 + 2x$, find $S(x) = P(x) + R(x)$. Evaluate $S(2)$.

38. Given $P(x) = 3x^4 - 2x + 1$ and $R(x) = 3x^5 - 5x - 8$, find $S(x) = P(x) + R(x)$. Evaluate $S(-1)$.

APPLYING THE CONCEPTS

39. For what value of k is the given equation an identity?
  a. $(2x^3 + 3x^2 + kx + 5) - (x^3 + 2x^2 + 3x + 7) = x^3 + x^2 + 5x - 2$
  b. $(6x^3 + kx^2 - 2x - 1) - (4x^3 - 3x^2 + 1) = 2x^3 - x^2 - 2x - 2$

40. The deflection D (in inches) of a beam that is uniformly loaded is given by the polynomial function $D(x) = 0.005x^4 - 0.1x^3 + 0.5x^2$, where x is the distance from one end of the beam. See the figure at the right. The maximum deflection occurs when x is the midpoint of the beam. Determine the maximum deflection for the beam in the diagram.

41.
[W] If $P(x)$ is a third-degree polynomial and $Q(x)$ is a fourth-degree polynomial, what can be said about the degree of $P(x) + Q(x)$? Give some examples of polynomials that support your answer.

42.
[W] If $P(x)$ is a fifth-degree polynomial and $Q(x)$ is a fourth-degree polynomial, what can be said about the degree of $P(x) - Q(x)$? Give some examples of polynomials that support your answer.

Multiplication of Polynomials

Objective A *To multiply a polynomial by a monomial* ··········

To multiply a polynomial by a monomial, use the Distributive Property and the Rule for Multiplying Exponential Expressions.

➡ Multiply: $-3a(4a^2 - 5a + 6)$

$$-3a(4a^2 - 5a + 6) = \boxed{-3a(4a^2) - (-3a)(5a) + (-3a)(6)}$$
$$= -12a^3 + 15a^2 - 18a$$

● Use the Distributive Property. This step is frequently done mentally.

Example 1
Multiply: $(5x + 4)(-2x)$

Solution
$(5x + 4)(-2x) = -10x^2 - 8x$

You Try It 1
Multiply: $(-2y + 3)(-4y)$

Your solution

Example 2
Multiply: $2a^2b(4a^2 - 2ab + b^2)$

Solution
$2a^2b(4a^2 - 2ab + b^2)$
$= 8a^4b - 4a^3b^2 + 2a^2b^3$

You Try It 2
Multiply: $-a^2(3a^2 + 2a - 7)$

Your solution

Solutions on p. A31

Objective B *To multiply two polynomials* ··········

Multiplication of two polynomials requires the repeated application of the Distributive Property.

$$(y - 2)(y^2 + 3y + 1) = (y - 2)(y^2) + (y - 2)(3y) + (y - 2)(1)$$
$$= y^3 - 2y^2 + 3y^2 - 6y + y - 2$$
$$= y^3 + y^2 - 5y - 2$$

A convenient method of multiplying two polynomials is to use a vertical format similar to that used for multiplication of whole numbers.

$$
\begin{array}{r}
y^2 + 3y + 1 \\
y - 2 \\
\hline
-2y^2 - 6y - 2 \\
y^3 + 3y^2 + y \\
\hline
y^3 + y^2 - 5y - 2
\end{array}
$$

● Multiply $-2(y^2 + 3y + 1)$.
● Multiply $y(y^2 + 3y + 1)$.
● Add the terms in each column.

➡ Multiply: $(2a^3 + a - 3)(a + 5)$

$$
\begin{array}{r}
2a^3 + a - 3 \\
a + 5 \\
\hline
10a^3 + 5a - 15 \\
2a^4 + a^2 - 3a \\
\hline
2a^4 + 10a^3 + a^2 + 2a - 15
\end{array}
$$

- Note that spaces are provided in each product so that like terms are in the same column.
- Add the terms in each column.

Example 3
Multiply: $(2b^3 - b + 1)(2b + 3)$

Solution

$$
\begin{array}{r}
2b^3 - b + 1 \\
2b + 3 \\
\hline
6b^3 - 3b + 3 \\
4b^4 + - 2b^2 + 2b \\
\hline
4b^4 + 6b^3 - 2b^2 - b + 3
\end{array}
$$

You Try It 3
Multiply: $(2y^3 + 2y^2 - 3)(3y - 1)$

Your solution

Solution on p. A31

Objective C   *To multiply two binomials* ....

It is frequently necessary to find the product of two binomials. The product can be found using a method called **FOIL**, which is based on the Distributive Property. The letters of FOIL stand for **F**irst, **O**uter, **I**nner, and **L**ast.

CONSIDER THIS
FOIL is not really a different way of multiplying. It is based on the Distributive Property.

$(2x + 3)(x + 5)$
$= 2x(x + 5) + 3(x + 5)$
   F   O   I   L
$= 2x^2 + 10x + 3x + 15$
$= 2x^2 + 13x + 15$

FOIL is an efficient way to remember how to do binomial multiplication.

➡ Multiply: $(2x + 3)(x + 5)$

Multiply the **F**irst terms.	$(2x + 3)(x + 5)$	$2x \cdot x = 2x^2$
Multiply the **O**uter terms.	$(2x + 3)(x + 5)$	$2x \cdot 5 = 10x$
Multiply the **I**nner terms.	$(2x + 3)(x + 5)$	$3 \cdot x = 3x$
Multiply the **L**ast terms.	$(2x + 3)(x + 5)$	$3 \cdot 5 = 15$

$$
\begin{array}{cccc}
\textbf{F} & \textbf{O} & \textbf{I} & \textbf{L}
\end{array}
$$

Add the products. $(2x + 3)(x + 5)$ $= 2x^2 + 10x + 3x + 15$

Combine like terms. $= 2x^2 + 13x + 15$

➡ Multiply: $(4x - 3)(3x - 2)$

$$(4x - 3)(3x - 2) = \boxed{4x(3x) + 4x(-2) + (-3)(3x) + (-3)(-2)}$$
$$= 12x^2 - 8x - 9x + 6$$
$$= 12x^2 - 17x + 6$$

- Do this step mentally.

➡ Multiply: $(3x - 2y)(x + 4y)$

$$(3x - 2y)(x + 4y) = \boxed{3x(x) + 3x(4y) + (-2y)(x) + (-2y)(4y)}$$
$$= 3x^2 + 12xy - 2xy - 8y^2$$
$$= 3x^2 + 10xy - 8y^2$$

- Do this step mentally.

Example 4

Multiply: $(2a - 1)(3a - 2)$

Solution

$(2a - 1)(3a - 2) = 6a^2 - 4a - 3a + 2$
$= 6a^2 - 7a + 2$

You Try It 4

Multiply: $(4y - 5)(2y - 3)$

Your solution

Example 5

Multiply: $(3x - 2)(4x + 3)$

Solution

$(3x - 2)(4x + 3) = 12x^2 + 9x - 8x - 6$
$= 12x^2 + x - 6$

You Try It 5

Multiply: $(3b + 2)(3b - 5)$

Your solution

Solutions on p. A31

Objective D   *To multiply binomials that have special products* ....

Using FOIL, it is possible to find a pattern for the product of the sum and difference of two terms and for the square of a binomial.

The Sum and Difference of Two Terms

$$(a + b)(a - b) = a^2 - ab + ab - b^2$$
$$= a^2 - b^2$$

Square of first term ───────
Square of second term ───────

The Square of a Binomial

$$(a + b)^2 = (a + b)(a + b) = a^2 + ab + ab + b^2$$
$$= a^2 + 2ab + b^2$$

Square of first term ───────
Twice the product of the two terms ───────
Square of the last term ───────

➡ Multiply: $(2x + 3)(2x - 3)$

$(2x + 3)(2x - 3)$ is the sum and difference of two terms.

$(2x + 3)(2x - 3) = \boxed{(2x)^2 - 3^2}$ • **Do this step mentally.**
$= 4x^2 - 9$

➡ Multiply: $(3x - 2)^2$

$(3x - 2)^2$ is the square of a binomial.

$(3x - 2)^2 = \boxed{(3x)^2 + 2(3x)(-2) + (-2)^2}$ • **Do this step mentally.**
$= 9x^2 - 12x + 4$

Example 6
Multiply: $(4z - 2w)(4z + 2w)$

Solution
$(4z - 2w)(4z + 2w) = 16z^2 - 4w^2$

You Try It 6
Multiply: $(2a + 5c)(2a - 5c)$

Your solution

Example 7
Simplify: $(2r - 3s)^2$

Solution
$(2r - 3s)^2 = 4r^2 - 12rs + 9s^2$

You Try It 7
Simplify: $(3x + 2y)^2$

Your solution

Solutions on p. A31

Objective E *To solve application problems*

Example 8
The length of a rectangle is $(2x + 3)$ ft.
The width is $(x - 5)$ ft. Find the area of
the rectangle in terms of the variable x.

$x - 5$ ▭
$2x + 3$

Strategy
To find the area, replace the variables L
and W in the equation $A = L \cdot W$ by the
given values and solve for A.

Solution
$A = L \cdot W$
$A = (2x + 3)(x - 5)$
$\quad = 2x^2 - 10x + 3x - 15$
$\quad = 2x^2 - 7x - 15$

The area is $(2x^2 - 7x - 15)$ ft^2.

You Try It 8
The base of a triangle is $(2x + 6)$ ft. The
height is $(x - 4)$ ft. Find the area of the
triangle in terms of the variable x.

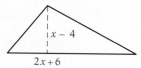
$x - 4$
$2x + 6$

Your strategy

Your solution

Solution on p. A31

6.3 Exercises

Objective A

Multiply.

1. $x(x - 2)$

2. $y(3 - y)$

3. $-x(x + 7)$

4. $-y(7 - y)$

5. $3a^2(a - 2)$

6. $4b^2(b + 8)$

7. $-5x^2(x^2 - x)$

8. $-6y^2(y + 2y^2)$

9. $-x^3(3x^2 - 7)$

10. $-y^4(2y^2 - y^6)$

11. $2x(6x^2 - 3x)$

12. $3y(4y - y^2)$

13. $(2x - 4)3x$

14. $(3y - 2)y$

15. $(3x + 4)x$

16. $(2x + 1)2x$

17. $-xy(x^2 - y^2)$

18. $-x^2y(2xy - y^2)$

19. $x(2x^3 - 3x + 2)$

20. $y(-3y^2 - 2y + 6)$

21. $-a(-2a^2 - 3a - 2)$

22. $-b(5b^2 + 7b - 35)$

23. $x^2(3x^4 - 3x^2 - 2)$

24. $y^3(-4y^3 - 6y + 7)$

25. $2y^2(-3y^2 - 6y + 7)$

26. $4x^2(3x^2 - 2x + 6)$

27. $(a^2 + 3a - 4)(-2a)$

28. $(b^3 - 2b + 2)(-5b)$

29. $-3y^2(-2y^2 + y - 2)$

30. $-5x^2(3x^2 - 3x - 7)$

31. $xy(x^2 - 3xy + y^2)$

32. $ab(2a^2 - 4ab - 6b^2)$

Objective B

Multiply.

33. $(x^2 + 3x + 2)(x + 1)$

34. $(x^2 - 2x + 7)(x - 2)$

35. $(a^2 - 3a + 4)(a - 3)$

Multiply.

36. $(x^2 - 3x + 5)(2x - 3)$ **37.** $(-2b^2 - 3b + 4)(b - 5)$ **38.** $(-a^2 + 3a - 2)(2a - 1)$

39. $(-2x^2 + 7x - 2)(3x - 5)$ **40.** $(-a^2 - 2a + 3)(2a - 1)$ **41.** $(x^2 + 5)(x - 3)$

42. $(y^2 - 2y)(2y + 5)$ **43.** $(x^3 - 3x + 2)(x - 4)$ **44.** $(y^3 + 4y^2 - 8)(2y - 1)$

45. $(5y^2 + 8y - 2)(3y - 8)$ **46.** $(3y^2 + 3y - 5)(4y - 3)$ **47.** $(5a^3 - 5a + 2)(a - 4)$

48. $(3b^3 - 5b^2 + 7)(6b - 1)$ **49.** $(y^3 + 2y^2 - 3y + 1)(y + 2)$ **50.** $(2a^3 - 3a^2 + 2a - 1)(2a - 3)$

Objective C

Multiply.

51. $(x + 1)(x + 3)$ **52.** $(y + 2)(y + 5)$ **53.** $(a - 3)(a + 4)$ **54.** $(b - 6)(b + 3)$

55. $(y + 3)(y - 8)$ **56.** $(x + 10)(x - 5)$ **57.** $(y - 7)(y - 3)$ **58.** $(a - 8)(a - 9)$

59. $(2x + 1)(x + 7)$ **60.** $(y + 2)(5y + 1)$ **61.** $(3x - 1)(x + 4)$ **62.** $(7x - 2)(x + 4)$

63. $(4x - 3)(x - 7)$ **64.** $(2x - 3)(4x - 7)$ **65.** $(3y - 8)(y + 2)$ **66.** $(5y - 9)(y + 5)$

67. $(3x + 7)(3x + 11)$ **68.** $(5a + 6)(6a + 5)$ **69.** $(7a - 16)(3a - 5)$ **70.** $(5a - 12)(3a - 7)$

71. $(3a - 2b)(2a - 7b)$ **72.** $(5a - b)(7a - b)$ **73.** $(a - 9b)(2a + 7b)$

Multiply.

74. $(2a + 5b)(7a - 2b)$ **75.** $(10a - 3b)(10a - 7b)$ **76.** $(12a - 5b)(3a - 4b)$

77. $(5x + 12y)(3x + 4y)$ **78.** $(11x + 2y)(3x + 7y)$ **79.** $(2x - 15y)(7x + 4y)$

80. $(5x + 2y)(2x - 5y)$ **81.** $(8x - 3y)(7x - 5y)$ **82.** $(2x - 9y)(8x - 3y)$

Objective D

Multiply.

83. $(y - 5)(y + 5)$ **84.** $(y + 6)(y - 6)$ **85.** $(2x + 3)(2x - 3)$ **86.** $(4x - 7)(4x + 7)$

87. $(x + 1)^2$ **88.** $(y - 3)^2$ **89.** $(3a - 5)^2$ **90.** $(6x - 5)^2$

91. $(3x - 7)(3x + 7)$ **92.** $(9x - 2)(9x + 2)$ **93.** $(2a + b)^2$ **94.** $(x + 3y)^2$

95. $(x - 2y)^2$ **96.** $(2x - 3y)^2$ **97.** $(4 - 3y)(4 + 3y)$

98. $(4x - 9y)(4x + 9y)$ **99.** $(5x + 2y)^2$ **100.** $(2a - 9b)^2$

Objective E   Application Problems

Solve.

101. The length of a rectangle is $5x$ ft. The width is $(2x - 7)$ ft. Find the area of the rectangle in terms of the variable x.

102. The width of a rectangle is $(x - 6)$ m. The length is $(2x + 3)$ m. Find the area of the rectangle in terms of the variable x.

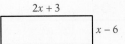

Solve.

103. The length of a side of a square is $(2x + 1)$ km. Find the area of the square in terms of the variable x.

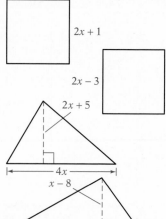

104. The length of a side of a square is $(2x - 3)$ yd. Find the area of the square in terms of the variable x.

105. The base of a triangle is $4x$ m, and the height is $(2x + 5)$ m. Find the area of the triangle in terms of the variable x.

106. The base of a triangle is $(2x + 6)$ in., and the height is $(x - 8)$ in. Find the area of the triangle in terms of the variable x.

107. The width of a rectangle is $(3x + 1)$ in. The length of the rectangle is twice the width. Find the area of the rectangle in terms of the variable x.

108. The width of a rectangle is $(4x - 3)$ cm. The length of the rectangle is twice the width. Find the area of the rectangle in terms of the variable x.

109. A softball diamond has dimensions 45 ft by 45 ft. A base path border x ft wide lies on both the first-base side and the third-base side of the diamond. Express the total area of the softball diamond and the base path in terms of the variable x.

110. An athletic field has dimensions 30 yd by 100 yd. An end zone that is w yd wide borders each end of the field. Express the total area of the field and the end zones in terms of the variable w.

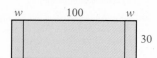

APPLYING THE CONCEPTS

Simplify.

111. $(a + b)^2 - (a - b)^2$ **112.** $(x^2 + x - 3)^2$ **113.** $(a + 3)^3$

114. **a.** What polynomial has quotient $3x - 4$ when divided by $4x + 5$?
　　b. What polynomial has quotient $x^2 + 2x - 1$ when divided by $x + 3$?

115. **a.** Add $x^2 + 2x - 3$ to the product of $2x - 5$ and $3x + 1$.
　　b. Subtract $4x^2 - x - 5$ from the product of $x^2 + x + 3$ and $x - 4$.

Division of Polynomials

Objective A   *To divide polynomials* ....

Some rational expressions cannot be simplified by factoring and dividing by the common factors. In these cases, long division of polynomials is used.

To divide two polynomials, use a method similar to that used for division of whole numbers. To check division of polynomials, use

Dividend = (quotient × divisor) + remainder

➡ Divide: $(x^2 + 5x - 7) \div (x + 3)$

Step 1

$$
\begin{array}{r}
x \\
x + 3 \overline{)x^2 + 5x - 7} \\
\underline{x^2 + 3x} \quad\downarrow \\
2x - 7
\end{array}
$$

Think: $x\overline{)x^2} = \dfrac{x^2}{x} = x$

Multiply: $x(x + 3) = x^2 + 3x$

Subtract: $(x^2 + 5x) - (x^2 + 3x) = 2x$

Bring down the -7.

Step 2

$$
\begin{array}{r}
x + 2 \\
x + 3 \overline{)x^2 + 5x - 7} \\
\underline{x^2 + 3x} \\
2x - 7 \\
\underline{2x + 6} \\
-13
\end{array}
$$

Think: $x\overline{)2x} = \dfrac{2x}{x} = 2$

Multiply: $2(x + 3) = 2x + 6$

Subtract: $(2x - 7) - (2x + 6) = -13$

The remainder is -13.

Check: (Quotient)(divisor) + remainder

$$= (x + 2)(x + 3) + (-13) = x^2 + 3x + 2x + 6 - 13 = x^2 + 5x - 7$$

$$(x^2 + 5x - 7) \div (x + 3) = x + 2 - \frac{13}{x + 3}$$

➡ Divide: $\dfrac{6 - 6x^2 + 4x^3}{2x + 3}$

Arrange the terms in descending order. Note that there is no term of x in $4x^3 - 6x^2 + 6$. Insert a zero for the missing term so that like terms will be in the same columns.

$$
\begin{array}{r}
2x^2 - 6x + 9 \\
2x + 3 \overline{)4x^3 - 6x^2 + 0x + 6} \\
\underline{4x^3 + 6x^2} \\
-12x^2 + 0x \\
\underline{-12x^2 - 18x} \\
18x + 6 \\
\underline{18x + 27} \\
-21
\end{array}
$$

$$\frac{4x^3 - 6x^2 + 6}{2x + 3} = 2x^2 - 6x + 9 - \frac{21}{2x + 3}$$

Example 1

Divide: $\dfrac{12x^2 - 11x + 10}{4x - 5}$

Solution

$$
\begin{array}{r}
3x + 1 \\
4x - 5 \overline{\smash{\big)}\ 12x^2 - 11x + 10} \\
\underline{12x^2 - 15x} \\
4x + 10 \\
\underline{4x - 5} \\
15
\end{array}
$$

$\dfrac{12x^2 - 11x + 10}{4x - 5} = 3x + 1 + \dfrac{15}{4x - 5}$

You Try It 1

Divide: $\dfrac{15x^2 + 17x - 20}{3x + 4}$

Your solution

Example 2

Divide: $\dfrac{x^3 + 1}{x + 1}$

Solution

$$
\begin{array}{r}
x^2 - x + 1 \\
x + 1 \overline{\smash{\big)}\ x^3 + 0x^2 + 0x + 1} \\
\underline{x^3 + x^2} \\
- x^2 + 0x \\
\underline{- x^2 - x} \\
x + 1 \\
\underline{x + 1} \\
0
\end{array}
$$

● **Insert zeros for the missing terms.**

$\dfrac{x^3 + 1}{x + 1} = x^2 - x + 1$

You Try It 2

Divide: $\dfrac{3x^3 + 8x^2 - 6x + 2}{3x - 1}$

Your solution

Example 3

Divide:
$(2x^4 - 7x^3 + 3x^2 + 4x - 5) \div (x^2 - 2x - 2)$

Solution

$$
\begin{array}{r}
2x^2 - 3x + 1 \\
x^2 - 2x - 2 \overline{\smash{\big)}\ 2x^4 - 7x^3 + 3x^2 + 4x - 5} \\
\underline{2x^4 - 4x^3 - 4x^2} \\
- 3x^3 + 7x^2 + 4x \\
\underline{- 3x^3 + 6x^2 + 6x} \\
x^2 - 2x - 5 \\
\underline{x^2 - 2x - 2} \\
- 3
\end{array}
$$

$(2x^4 - 7x^3 + 3x^2 + 4x - 5) \div (x^2 - 2x - 2)$

$= 2x^2 - 3x + 1 - \dfrac{3}{x^2 - 2x - 2}$

You Try It 3

Divide:
$(3x^4 - 11x^3 + 16x^2 - 16x + 8) \div (x^2 - 3x + 2)$

Your solution

Solutions on p. A32

Objective B   *To divide polynomials using synthetic division* ....

Synthetic division is a shorter method of dividing a polynomial by a binomial of the form $x - a$.

➡ Divide $(3x^2 - 4x + 6) \div (x - 2)$ by using long division.

$$
\begin{array}{r}
3x + 2 \\
x - 2 \overline{\smash{)}3x^2 - 4x + 6} \\
\underline{3x^2 - 6x} \\
2x + 6 \\
\underline{2x - 4} \\
10
\end{array}
$$

$$(3x^2 - 4x + 6) \div (x - 2) = 3x + 2 + \frac{10}{x - 2}$$

The variables can be omitted because the position of a term indicates the power of the term.

$$
\begin{array}{r}
3 \quad\ 2 \\
-2\overline{\smash{)}3 \ \ -4 \ \ \ \ 6} \\
\underline{3 \ \ -6} \\
2 \quad\ 6 \\
\underline{2 \ \ -4} \\
10
\end{array}
$$

Each number shown in color above is exactly the same as the number above it. Removing the colored numbers condenses the vertical spacing.

$$
\begin{array}{r}
3 \quad\ 2 \\
-2\overline{\smash{)}3 \ \ -4 \ \ \ \ 6} \\
\underline{-6 \ \ -4} \\
2 \quad 10
\end{array}
$$

The number in color on the top row is the same as the one in the bottom row. Writing the 3 from the top row in the bottom row allows the spacing to be condensed even further.

$$
\begin{array}{c|ccc}
-2 & 3 & -4 & 6 \\
 & & -6 & -4 \\
\hline
 & 3 & 2 & 10
\end{array}
$$

$$\underbrace{\qquad 3 \qquad 2 \qquad}_{\substack{\text{Terms of} \\ \text{the quotient}}} \quad \underbrace{10}_{\text{Remainder}}$$

Because the degree of the dividend $(3x^2 - 4x + 6)$ is 2 and the degree of the divisor $(3x + 2)$ is 1, the degree of the quotient is $2 - 1 = 1$. This means that, using the terms of the quotient given above, the quotient is $3x + 2$. The remainder is 10.

In general, the degree of the quotient of two polynomials is the difference between the degree of the dividend and the degree of the divisor.

By replacing the constant term in the divisor by its additive inverse, we may add rather than subtract terms. This is illustrated in the following example.

➡ Divide: $(3x^3 + 6x^2 - x - 2) \div (x + 3)$

The additive inverse of the binomial constant

Coefficients of the polynomial

$$
\begin{array}{c|cccc}
-3 & 3 & 6 & -1 & -2 \\
   & \downarrow & & & \\
\hline
   & 3 & & &
\end{array}
$$

• Bring down the 3.

$$
\begin{array}{c|cccc}
-3 & 3 & 6 & -1 & -2 \\
   &   & -9 & & \\
\hline
   & 3 & -3 & &
\end{array}
$$

• Multiply $-3(3)$ and add the product to 6.

$$
\begin{array}{c|cccc}
-3 & 3 & 6 & -1 & -2 \\
   &   & -9 & 9 & \\
\hline
   & 3 & -3 & 8 &
\end{array}
$$

• Multiply $-3(-3)$ and add the product to -1.

$$
\begin{array}{c|cccc}
-3 & 3 & 6 & -1 & -2 \\
   &   & -9 & 9 & -24 \\
\hline
   & 3 & -3 & 8 & -26
\end{array}
$$

• Multiply $-3(8)$ and add the product to -2.

Terms of the quotient   Remainder

The degree of the dividend is 3 and the degree of the divisor is 1. Therefore, the degree of the quotient is $3 - 1 = 2$.

$$(3x^3 + 6x^2 - x - 2) \div (x + 3) = 3x^2 - 3x + 8 - \frac{26}{x + 3}$$

➡ Divide: $(2x^3 - x + 2) \div (x - 2)$

The additive inverse of the binomial constant

Coefficients of the polynomial

$$
\begin{array}{c|cccc}
2 & 2 & 0 & -1 & 2 \\
  & \downarrow & & & \\
\hline
  & 2 & & &
\end{array}
$$

• Insert a 0 for the missing term and bring down the 2.

$$
\begin{array}{c|cccc}
2 & 2 & 0 & -1 & 2 \\
  &   & 4 & & \\
\hline
  & 2 & 4 & &
\end{array}
$$

• Multiply $2(2)$ and add the product to 0.

$$
\begin{array}{c|cccc}
2 & 2 & 0 & -1 & 2 \\
  &   & 4 & 8 & \\
\hline
  & 2 & 4 & 7 &
\end{array}
$$

• Multiply $2(4)$ and add the product to -1.

$$
\begin{array}{c|cccc}
2 & 2 & 0 & -1 & 2 \\
  &   & 4 & 8 & 14 \\
\hline
  & 2 & 4 & 7 & 16
\end{array}
$$

• Multiply $2(7)$ and add the product to 2.

Terms of the quotient   Remainder

$$(2x^3 - x + 2) \div (x - 2) = 2x^2 + 4x + 7 + \frac{16}{x - 2}$$

Example 4

Divide by using synthetic division:
$(7 - 3x + 5x^2) \div (x - 1)$

Solution

Arrange the coefficients in decreasing powers of x.

$$
\begin{array}{r|rrr}
1 & 5 & -3 & 7 \\
  &   & 5 & 2 \\
\hline
  & 5 & 2 & 9
\end{array}
$$

$(5x^2 - 3x + 7) \div (x - 1) = 5x + 2 + \dfrac{9}{x - 1}$

You Try It 4

Divide by using synthetic division:
$(8x + 6x^2 - 5) \div (x + 2)$

Your solution

Example 5

Divide by using synthetic division:
$(2x^3 + 4x^2 - 3x + 12) \div (x + 4)$

Solution

$$
\begin{array}{r|rrrr}
-4 & 2 & 4 & -3 & 12 \\
   &   & -8 & 16 & -52 \\
\hline
   & 2 & -4 & 13 & -40
\end{array}
$$

$(2x^3 + 4x^2 - 3x + 12) \div (x + 4)$

$\quad = 2x^2 - 4x + 13 - \dfrac{40}{x + 4}$

You Try It 5

Divide by using synthetic division:
$(5x^3 - 12x^2 - 8x + 16) \div (x - 2)$

Your solution

Example 6

Divide by using synthetic division:
$(3x^4 - 8x^2 + 2x + 1) \div (x + 2)$

Solution

Insert a zero for the missing term.

$$
\begin{array}{r|rrrrr}
-2 & 3 & 0 & -8 & 2 & 1 \\
   &   & -6 & 12 & -8 & 12 \\
\hline
   & 3 & -6 & 4 & -6 & 13
\end{array}
$$

$(3x^4 - 8x^2 + 2x + 1) \div (x + 2)$

$\quad = 3x^3 - 6x^2 + 4x - 6 + \dfrac{13}{x + 2}$

You Try It 6

Divide by using synthetic division:
$(2x^4 - 3x^3 - 8x^2 - 2) \div (x - 3)$

Your solution

Solutions on p. A32

Objective C   *To evaluate a polynomial using synthetic division* ....

A polynomial can be evaluated by using synthetic division. Consider the polynomial $P(x) = 2x^4 - 3x^3 + 4x^2 - 5x + 1$. One way to evaluate the polynomial when $x = 2$ is to replace x by 2 and then simplify the numerical expression.

$$P(x) = 2x^4 - 3x^3 + 4x^2 - 5x + 1$$

$$\begin{aligned} P(2) &= 2(2)^4 - 3(2)^3 + 4(2)^2 - 5(2) + 1 \\ &= 2(16) - 3(8) + 4(4) - 5(2) + 1 \\ &= 32 - 24 + 16 - 10 + 1 \\ &= 15 \end{aligned}$$

Now use synthetic division to divide $(2x^4 - 3x^3 + 4x^2 - 5x + 1) \div (x - 2)$.

$$
\begin{array}{r|rrrrr}
2 & 2 & -3 & 4 & -5 & 1 \\
  &   & 4 & 2 & 12 & 14 \\
\hline
  & 2 & 1 & 6 & 7 & 15
\end{array}
$$

$\underbrace{\qquad\qquad\qquad}_{\text{Terms of the quotient}}$ $\underbrace{\quad}_{\text{Remainder}}$

Note that the remainder is 15, which is the same value as $P(2)$. This is not a coincidence. The following theorem states that this situation is always true.

> **Remainder Theorem**
>
> If the polynomial $P(x)$ is divided by $x - a$, the remainder is $P(a)$.

➡ Use the Remainder Theorem to evaluate $P(x) = x^4 - 3x^2 + 4x - 5$ when $x = -2$.

The value at which the polynomial is evaluated

$$
\begin{array}{r|rrrrr}
-2 & 1 & 0 & -3 & 4 & -5 \\
   &   & -2 & 4 & -2 & -4 \\
\hline
   & 1 & -2 & 1 & 2 & -9
\end{array}
$$

● A 0 is inserted for the x^3 term.

⟵ The remainder

$$P(-2) = -9$$

Example 7
Use the Remainder Theorem to evaluate $P(x) = -x^4 + 3x^3 + 2x^2 - x - 5$ when $x = -2$.

Solution

$$
\begin{array}{r|rrrrr}
-2 & -1 & 3 & 2 & -1 & -5 \\
   &   & 2 & -10 & 16 & -30 \\
\hline
   & -1 & 5 & -8 & 15 & -35
\end{array}
$$

$$P(-2) = -35$$

You Try It 7
Use the Remainder Theorem to evaluate $P(x) = 2x^3 - 5x^2 + 7$ when $x = -3$.

Your solution

Solution on p. A32

6.4 Exercises

Objective A

Divide by using long division.

1. $(x^2 + 3x - 40) \div (x - 5)$

2. $(x^2 - 14x + 24) \div (x - 2)$

3. $(x^3 - 3x^2 + 2) \div (x - 3)$

4. $(x^3 + 4x^2 - 8) \div (x + 4)$

5. $(6x^2 + 13x + 8) \div (2x + 1)$

6. $(12x^2 + 13x - 14) \div (3x - 2)$

7. $(10x^2 + 9x - 5) \div (2x - 1)$

8. $(18x^2 - 3x + 2) \div (3x + 2)$

9. $(8x^3 - 9) \div (2x - 3)$

10. $(64x^3 + 4) \div (4x + 2)$

11. $(6x^4 - 13x^2 - 4) \div (2x^2 - 5)$

12. $(12x^4 - 11x^2 + 10) \div (3x^2 + 1)$

13. $\dfrac{-10 - 33x + 3x^3 - 8x^2}{3x + 1}$

14. $\dfrac{10 - 49x + 38x^2 - 8x^3}{1 - 4x}$

15. $\dfrac{x^3 - 5x^2 + 7x - 4}{x - 3}$

16. $\dfrac{2x^3 - 3x^2 + 6x + 4}{2x + 1}$

Divide by using long division.

17. $\dfrac{16x^2 - 13x^3 + 2x^4 + 20 - 9x}{x - 5}$

18. $\dfrac{x - x^2 + 5x^3 + 3x^4 - 2}{x + 2}$

19. $\dfrac{2x^3 + 4x^2 - x + 2}{x^2 + 2x - 1}$

20. $\dfrac{3x^3 - 2x^2 + 5x - 4}{x^2 - x + 3}$

21. $\dfrac{x^4 + 2x^3 - 3x^2 - 6x + 2}{x^2 - 2x - 1}$

22. $\dfrac{x^4 - 3x^3 + 4x^2 - x + 1}{x^2 + x - 3}$

23. $\dfrac{x^4 + 3x^2 - 4x + 5}{x^2 + 2x + 3}$

24. $\dfrac{x^4 + 2x^3 - x + 2}{x^2 - x - 1}$

Objective B

Divide by using synthetic division.

25. $(2x^2 - 6x - 8) \div (x + 1)$

26. $(3x^2 + 19x + 20) \div (x + 5)$

27. $(3x^2 - 14x + 16) \div (x - 2)$

28. $(4x^2 - 23x + 28) \div (x - 4)$

29. $(3x^2 - 4) \div (x - 1)$

30. $(4x^2 - 8) \div (x - 2)$

31. $(2x^3 - x^2 + 6x + 9) \div (x + 1)$

32. $(3x^3 + 10x^2 + 6x - 4) \div (x + 2)$

Divide by using synthetic division.

33. $(18 + x - 4x^3) \div (2 - x)$

34. $(12 - 3x^2 + x^3) \div (x + 3)$

35. $(2x^3 + 5x^2 - 5x + 20) \div (x + 4)$

36. $(5x^3 + 3x^2 - 17x + 6) \div (x + 2)$

37. $\dfrac{5 + 5x - 8x^2 + 4x^3 - 3x^4}{2 - x}$

38. $\dfrac{3 - 13x - 5x^2 + 9x^3 - 2x^4}{3 - x}$

39. $\dfrac{3x^4 + 3x^3 - x^2 + 3x + 2}{x + 1}$

40. $\dfrac{4x^4 + 12x^3 - x^2 - x + 2}{x + 3}$

41. $\dfrac{2x^4 - x^2 + 2}{x - 3}$

42. $\dfrac{x^4 - 3x^3 - 30}{x + 2}$

___

Objective C

Use the Remainder Theorem to evaluate the polynomial.

43. $P(x) = 2x^2 - 3x - 1;\ P(3)$

44. $Q(x) = 3x^2 - 5x - 1;\ Q(2)$

45. $R(x) = x^3 - 2x^2 + 3x - 1;\ R(4)$

46. $F(x) = x^3 + 4x^2 - 3x + 2;\ F(3)$

47. $P(z) = 2z^3 - 4z^2 + 3z - 1;\ P(-2)$

48. $R(t) = 3t^3 + t^2 - 4t + 2;\ R(-3)$

Use the Remainder Theorem to evaluate the polynomial.

49. $Z(p) = 2p^3 - p^2 + 3; Z(-3)$

50. $P(y) = 3y^3 + 2y^2 - 5; P(-2)$

51. $Q(x) = x^4 + 3x^3 - 2x^2 + 4x - 9; Q(2)$

52. $Y(z) = z^4 - 2z^3 - 3z^2 - z + 7; Y(3)$

53. $F(x) = 2x^4 - x^3 + 2x - 5; F(-3)$

54. $Q(x) = x^4 - 2x^3 + 4x - 2; Q(-2)$

55. $P(x) = x^3 - 3; P(5)$

56. $S(t) = 4t^3 + 5; S(-4)$

57. $R(t) = 4t^4 - 3t^2 + 5; R(-3)$

58. $P(z) = 2z^4 + z^2 - 3; P(-4)$

59. $Q(x) = x^5 - 4x^3 - 2x^2 + 5x - 2; Q(2)$

60. $T(x) = 2x^5 + 4x^4 - x^2 + 4; T(3)$

61. $R(x) = 2x^5 - x^3 + 4x - 1; R(-2)$

62. $P(x) = x^5 - x^3 + 4x + 1; P(-3)$

APPLYING THE CONCEPTS

63. Divide by using long division.

   a. $\dfrac{a^3 + b^3}{a + b}$
   b. $\dfrac{x^5 + y^5}{x + y}$
   c. $\dfrac{x^6 - y^6}{x + y}$

64. For what value of k will the remainder be zero?

   a. $(x^3 - x^2 - 3x + k) \div (x + 3)$
   b. $(2x^3 - x + k) \div (x - 1)$

65. Divide.

   a. $(2x^3 + 7x^2 + 2x - 8) \div (4x + 8)$
   b. $(4x^3 + 13x^2 - 22x + 24) \div (6x - 12)$

   c. $(2x^4 - 3x^3 + 4x^2 + x - 10) \div (x^2 - x + 1)$
   d. $(x^4 + 4x^3 + 2x^2 - x + 5) \div (x^2 - 2x - 3)$

66.
[W] Show how synthetic division can be modified so that the divisor can be of the form $ax + b$.

Projects in Mathematics

Pascal's Triangle   Simplifying the power of a binomial is called *expanding the binomial*. The expansion of the first three powers of a binomial is shown below.

$$(a + b)^1 = a + b$$

$$(a + b)^2 = (a + b)(a + b) = a^2 + 2ab + b^2$$

$$(a + b)^3 = (a + b)^2(a + b) = (a^2 + 2ab + b^2)(a + b) = a^3 + 3a^2b + 3ab^2 + b^3$$

Find $(a + b)^4$. [*Hint:* $(a + b)^4 = (a + b)^3(a + b)$]

Find $(a + b)^5$. [*Hint:* $(a + b)^5 = (a + b)^4(a + b)$]

If we continue in this way, we find that the result for $(a + b)^6$ is

$$(a + b)^6 = a^6 + 6a^5b + 15a^4b^2 + 20a^3b^3 + 15a^2b^4 + 6ab^5 + b^6$$

Now expand $(a + b)^8$. Before you begin, see if you can find a pattern that will help you write the expansion of $(a + b)^8$ without having to multiply it out. Here are some hints.

1. Write out the variable terms of each binomial expansion from $(a + b)^1$ through $(a + b)^6$. Observe how the exponents of the variables change.

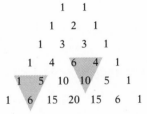

2. Write out the coefficients of each term without the variable part. It will be helpful to make a triangular arrangement as shown at the left. Note that each row begins and ends with a 1. Also note, in the two shaded regions, that any number in a row is the sum of the two closest numbers above it. For instance, $1 + 5 = 6$ and $6 + 4 = 10$.

The triangle of numbers shown at the left is called **Pascal's Triangle.** To find the expansion of $(a + b)^8$, you will need to find the eighth row of Pascal's Triangle. First find row seven. Then find row eight.

Use the patterns you have observed to write the expansion $(a + b)^8$.

Pascal's Triangle has been the subject of extensive analysis, and many patterns have been found. See if you can find some of them.

Chapter Summary

Key Words   A *monomial* is a number, a variable, or a product of a number and variables. The *degree of a monomial* is the sum of the exponents of the variables.

An *exponential expression is in simplest form* when it is written with only positive exponents.

A number written in *scientific notation* is a number written in the form $a \times 10^n$, where $1 \le a < 10$.

A *polynomial* is a variable expression in which the terms are monomials. A polynomial of two terms is a *binomial*. A polynomial of three terms is a *trinomial*. The *degree of a polynomial* is the greatest of the degrees of any of its terms. A polynomial in one variable is usually written in *descending order*, which means that the terms are arranged so that the exponents of the variable decrease from left to right.

A *polynomial function* is an expression whose terms are monomials. A *linear function* is given by the equation $f(x) = mx + b$. A second-degree polynomial function, called a *quadratic function*, is given by the equation $f(x) = ax^2 + bx + c$, $a \neq 0$. A third-degree polynomial function is called a *cubic function*. The *leading coefficient* of a polynomial function is the coefficient of the variable with the largest exponent.

The product of two binomials can be found using the *FOIL* method, which is based on the Distributive Property. The letters of FOIL stand for First, Outer, Inner, and Last.

Synthetic division is a shorter method of dividing a polynomial by a binomial of the form $x - a$. This method uses only the coefficients of the variable terms.

Essential Rules

Rule for Multiplying Exponential Expressions $x^m \cdot x^n = x^{m+n}$

Rule for Simplifying Powers of Exponential Expressions $(x^m)^n = x^{mn}$

Rule for Simplifying Powers of Products $(x^m y^n)^p = x^{mp} y^{np}$

Definition of Zero as an Exponent   If $x \neq 0$, then $x^0 = 1$.

Definition of a Negative Exponent $x^{-n} = \dfrac{1}{x^n}$ and $\dfrac{1}{x^{-n}} = x^n$, $x \neq 0$

Rule for Dividing Exponential Expressions $\dfrac{x^m}{x^n} = x^{m-n}$, $x \neq 0$

Rule for Simplifying Powers of Quotients $\left(\dfrac{x^m}{y^n}\right)^p = \dfrac{x^{mp}}{y^{np}}$, $y \neq 0$

The Sum and Difference of Two Terms $(a + b)(a - b) = a^2 - b^2$

The Square of a Binomial $(a + b)^2 = a^2 + 2ab + b^2$

Remainder Theorem   If the polynomial $P(x)$ is divided by $x - a$, then the remainder is $P(a)$.

Dividend = (quotient × divisor) + remainder

Chapter Review Exercises

. . . .

1. Multiply: $(-2a^2b^4)(3ab^2)$

2. Simplify: $(-3x^2y^3)^2$

3. Simplify: $\dfrac{(2a^4b^{-3}c^2)^3}{(2a^3b^2c^{-1})^4}$

4. Write 2.54×10^{-3} in decimal notation.

5. Given $P(x) = 2x^3 - x + 7$, find $P(-2)$.

6. Subtract: $(5x^2 - 8xy + 2y^2) - (x^2 - 3y^2)$

7. Simplify: $(2x^{-1}y^2z^5)(-3x^3yz^{-3})^2$

8. Multiply: $(2a^{12}b^3)(-9b^2c^6)(3ac)$

9. Graph: $f(x) = x^2 + 1$

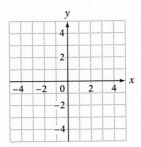

10. Identify (a) the leading coefficient, (b) the constant term, and (c) the degree of the polynomial $P(x) = 3x^5 - 3x^2 + 7x + 8$.

11. Use the Remainder Theorem to evaluate $P(x) = x^3 - 2x^2 + 3x - 5$ when $x = 2$.

12. Multiply: $-2x(4x^2 + 7x - 9)$

13. Multiply: $2ab^3(4a^2 - 2ab + 3b^2)$

14. Multiply: $(3y^2 + 4y - 7)(2y + 3)$

15. Multiply: $(5a - 7)(2a + 9)$

16. Write 0.000000127 in scientific notation.

17. Simplify: $(5y - 7)^2$

18. Divide: $\dfrac{15x^2 + 2x - 2}{3x - 2}$

19. Divide: $\dfrac{4x^3 + 27x^2 + 10x + 2}{x + 6}$

20. Divide: $\dfrac{x^4 - 4}{x - 4}$

21. Write 8.1×10^9 in decimal notation.

22. Simplify: $\dfrac{-18a^6b}{27a^3b^4}$

23. Multiply: $(5a + 2b)(5a - 2b)$

24. Use the Remainder Theorem to evaluate $P(x) = -2x^3 + 2x^2 - 4$ when $x = -3$.

25. Add: $(12y^2 + 17y - 4) + (9y^2 - 13y + 3)$

26. Multiply: $(6b^3 - 2b^2 - 5)(2b^2 - 1)$

27. Multiply: $(a + 7)(a - 7)$

28. Write 765,000,000,000 in scientific notation.

29. The mass of the moon is 3.7×10^{-8} times the mass of the sun. The mass of the sun is 2.19×10^{27} tons. Find the mass of the moon. Write the answer in scientific notation.

30. The most distant object visible from the earth without the aid of a telescope is the Great Galaxy of Andromeda. It takes light from this galaxy 2.2×10^6 years to travel to the earth. Light travels about 6.7×10^8 mph. How far from the earth is the Great Galaxy of Andromeda? Use a 365-day year.

31. Write the number of seconds in one week in scientific notation.

32. The length of a side of a square checkerboard is $(3x - 2)$ in. Express the area of the checkerboard in terms of the variable x.

33. The length of a rectangle is $(5x + 3)$ cm. The width is $(2x - 7)$ cm. Find the area of the rectangle in terms of the variable x.

Cumulative Review Exercises

1. Let $x \in \{-8, -3, 3\}$. For what values of x is the inequality $x \geq -3$ a true statement?

2. Find the additive inverse of 83.

3. Simplify: $8 - 2[-3 - (-1)]^2 \div 4$

4. Evaluate $\frac{2a - b}{b - c}$ when $a = 4$, $b = -2$, and $c = 6$.

5. Simplify: $-5\sqrt{300}$

6. Identify the property that justifies the statement $2x + (-2x) = 0$.

7. Simplify: $2x - 4[x - 2(3 - 23x) + 4]$

8. Solve: $\frac{2}{3} - y = \frac{5}{6}$

9. Solve: $8x - 3 - x = -6 + 3x - 8$

10. Solve: $3 - |2 - 3x| = -2$

11. Given $P(x) = 3x^2 - 2x + 2$, find $P(-2)$.

12. Is the relation $\{(-1, 0), (0, 0), (1, 0)\}$ a function?

13. Find the slope of the line containing the points $(-2, 3)$ and $(4, 2)$.

14. Find the equation of the line that contains the point $(-1, 2)$ and has slope $-\frac{3}{2}$.

15. Find the equation of the line that contains the point $(-2, 4)$ and is perpendicular to the line $3x + 2y = 4$.

16. Solve by using Cramer's Rule:
$$2x - 3y = 2$$
$$x + y = -3$$

17. Solve by the addition method:
$$x - y + z = 0$$
$$2x + y - 3z = -7$$
$$-x + 2y + 2z = 5$$

18. Simplify: $-2x - (-xy) + 7x - 4xy$

19. Multiply: $(2x + 3)(2x^2 - 3x + 1)$

20. Write the number 0.00000501 in scientific notation.

21. Graph: $3x - 4y = 12$

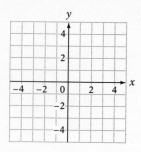

22. Graph: $-3x + 2y < 6$

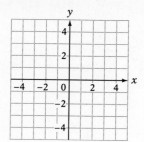

23. Solve by graphing:
$$x - 2y = 3$$
$$-2x + y = -3$$

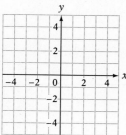

24. Graph the solution set:
$$2x + y < 3$$
$$-6x + 3y \geq 4$$

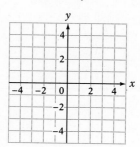

25. Simplify: $(4a^{-2}b^3)(2a^3b^{-1})^{-2}$

26. Simplify: $\dfrac{(5x^3y^{-3}z)^{-2}}{y^4z^{-2}}$

27. The sum of two integers is twenty-four. The difference between four times the smaller integer and nine is three less than twice the larger integer. Find the integers.

28. How many ounces of pure gold that costs $360 per ounce must be mixed with 80 oz of an alloy that costs $120 per ounce to make a mixture that costs $200 per ounce?

29. Two bicycles are 25 mi apart and traveling toward each other. One cyclist is traveling at 1.5 times the rate of the other cyclist. They meet in 2 h. Find the rate of each cyclist.

30. If $3000 is invested at an annual simple interest rate of 7.5%, how much additional money must be invested at an annual simple interest rate of 10% so that the total interest earned in one year is 9% of the total investment?

31. The graph shows the relationship between the distance traveled and the time of travel. Find the slope of the line between the two points on the graph. Write a sentence that states the meaning of the slope.

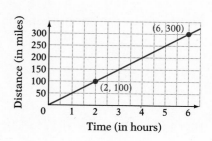

32. The width of a rectangle is 40% of the length. The perimeter of the rectangle is 42 m. Find the length and width of the rectangle.

33. The length of a side of a square is $(2x + 3)$ m. Find the area of the square in terms of the variable x.

Chapter 7

Factoring

Objectives

Algebra from Geometry

The early Babylonians made substantial progress in both algebra and geometry. Often the progress they made in algebra was based on geometric concepts.

Here are some geometric proofs of algebraic identities the Babylonians understood.

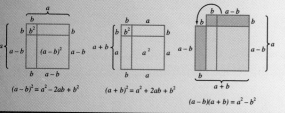

$$(a-b)^2 = a^2 - 2ab + b^2$$

$$(a+b)^2 = a^2 + 2ab + b^2$$

$$(a-b)(a+b) = a^2 - b^2$$

7.1

Common Factors

Objective A   *To factor a monomial from a polynomial* ....

The **greatest common factor (GCF)** of two or more monomials is the product of the GCF of the coefficients and the common variable factors.

$6x^3y = 2 \cdot 3 \cdot x \cdot x \cdot x \cdot y$
$8x^2y^2 = 2 \cdot 2 \cdot 2 \cdot x \cdot x \cdot y \cdot y$
$GCF = 2 \cdot x \cdot x \cdot y = 2x^2y$

Note that the exponent of each variable in the GCF is the same as the *smallest* exponent of that variable in either of the monomials.

The GCF of $6x^3y$ and $8x^2y^2$ is $2x^2y$.

➡ Find the GCF of $12a^4b$ and $18a^2b^2c$.

The common variable factors are a^2 and b; c is not a common variable factor.

$12a^4b = 2 \cdot 2 \cdot 3 \cdot a^4 \cdot b$
$18a^2b^2c = 2 \cdot 3 \cdot 3 \cdot a^2 \cdot b^2 \cdot c$
$GCF = 2 \cdot 3 \cdot a^2 \cdot b = 6a^2b$

CONSIDER THIS
$2x^2 + 10x$ is a sum.
$2x(x + 5)$ is a product.
When we factor a polynomial, we write it as a product.

To **factor** a polynomial means to write the polynomial as a product of other polynomials.

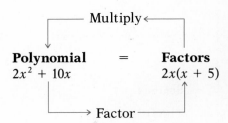

In the example above, $2x$ is the GCF of the terms $2x^2$ and $10x$. It is a **common factor** of the terms.

➡ Factor: $5x^3 - 35x^2 + 10x$

Find the GCF of the terms of the polynomial.

$5x^3 = 5 \cdot x^3$ $35x^2 = 5 \cdot 7 \cdot x^2$ $10x = 2 \cdot 5 \cdot x$

The GCF is $5x$.

CONSIDER THIS
The factors in color are determined by dividing each term of the trinomial by the GCF, $5x$. For instance,
$\left(\dfrac{5x^3}{5x}\right) = x^2,$
$\left(\dfrac{-35x^2}{5x}\right) = -7x,$ and
$\left(\dfrac{10x}{5x}\right) = 2.$

Rewrite the polynomial, expressing each term as a product with the GCF as one of the factors.

$5x^3 - 35x^2 + 10x = 5x(x^2) + 5x(-7x) + 5x(2)$
$= 5x(x^2 - 7x + 2)$

● Use the Distributive Property to write the polynomial as a product of factors.

Example 1

Factor: $8x^2 + 2xy$

Solution

The GCF is $2x$.

$8x^2 + 2xy = 2x(4x) + 2x(y) = 2x(4x + y)$

You Try It 1

Factor: $14a^2 - 21a^4b$

Your solution

Example 2

Factor: $n^3 - 5n^2 + 2n$

Solution

The GCF is n.

$n^3 - 5n^2 + 2n = n(n^2) + n(-5n) + n(2)$
$\qquad\qquad\qquad = n(n^2 - 5n + 2)$

You Try It 2

Factor: $27b^2 + 18b + 9$

Your solution

Example 3

Factor: $16x^2y + 8x^4y^2 - 12x^4y^5$

Solution

The GCF is $4x^2y$.

$16x^2y + 8x^4y^2 - 12x^4y^5$
$\quad = 4x^2y(4) + 4x^2y(2x^2y) + 4x^2y(-3x^2y^4)$
$\quad = 4x^2y(4 + 2x^2y - 3x^2y^4)$

You Try It 3

Factor: $6x^4y^2 - 9x^3y^2 + 12x^2y^4$

Your solution

Example 4

Factor: $x^{5n} - 2x^{3n}$

Solution

The GCF is x^{3n}.

$x^{5n} - 2x^{3n} = x^{3n}(x^{2n}) - x^{3n}(2)$
$\qquad\qquad\quad = x^{3n}(x^{2n} - 2)$

You Try It 4

Factor: $a^{n+4} + a^2$

Your solution

Solutions on p. A33

Objective B   ***To factor by grouping*** ....

In the examples at the right, the binomials in parentheses are called **binomial factors.**

$$2a(a + b)^2$$
$$3xy(x - y)$$

The Distributive Property is used to factor a common binomial factor from an expression.

The common binomial factor of the expression $6x(x - 3) + y^2(x - 3)$ is $(x - 3)$. To factor that expression, use the Distributive Property to write the expression as a product of factors.

$$6x(x - 3) + y^2(x - 3) = (x - 3)(6x + y^2)$$

Consider the following simplification of $-(a - b)$.

$$-(a - b) = -1(a - b) = -a + b = b - a$$

Thus, $$b - a = -(a - b)$$

This equation is sometimes used to factor a common binomial from an expression.

➡ Factor: $2x(x - y) + 5(y - x)$

$$\begin{aligned} 2x(x - y) + 5(y - x) &= 2x(x - y) - 5(x - y) \\ &= (x - y)(2x - 5) \end{aligned}$$

- $5(y - x) = 5[(-1)(x - y)]$
  $= -5(x - y)$

Some polynomials can be factored by grouping terms in such a way that a common binomial factor is found.

➡ Factor: $ax + bx - ay - by$

$$\begin{aligned} ax + bx - ay - by &= (ax + bx) - (ay + by) \\ &= x(a + b) - y(a + b) \\ &= (a + b)(x - y) \end{aligned}$$

- Group the first two terms and the last two terms. Note that $-ay - by = -(ay + by)$.
- Factor the GCF from each group.

➡ Factor: $6x^2 - 9x - 4xy + 6y$

$$\begin{aligned} 6x^2 - 9x - 4xy + 6y &= (6x^2 - 9x) - (4xy - 6y) \\ &= 3x(2x - 3) - 2y(2x - 3) \\ &= (2x - 3)(3x - 2y) \end{aligned}$$

- Group the first two terms and the last two terms. Note that $-4xy + 6y = -(4xy - 6y)$.
- Factor the GCF from each group.

Example 5
Factor: $4x(3x - 2) - 7(3x - 2)$

You Try It 5
Factor: $2y(5x - 2) - 3(2 - 5x)$

Solution

$4x(3x - 2) - 7(3x - 2)$
$\qquad = (3x - 2)(4x - 7)$

Your solution

Example 6
Factor: $9x^2 - 15x - 6xy + 10y$

You Try It 6
Factor: $a^2 - 3a + 2ab - 6b$

Solution

$9x^2 - 15x - 6xy + 10y$
$\qquad = (9x^2 - 15x) - (6xy - 10y)$
$\qquad = 3x(3x - 5) - 2y(3x - 5)$
$\qquad = (3x - 5)(3x - 2y)$

Your solution

Example 7
Factor: $3x^2y - 4x - 15xy + 20$

You Try It 7
Factor: $2mn^2 - n + 8mn - 4$

Solution

$3x^2y - 4x - 15xy + 20$
$\qquad = (3x^2y - 4x) - (15xy - 20)$
$\qquad = x(3xy - 4) - 5(3xy - 4)$
$\qquad = (3xy - 4)(x - 5)$

Your solution

Solutions on p. A33

7.1 Exercises

Objective A

Factor.

1. $5a + 5$ **2.** $7b - 7$ **3.** $16 - 8a^2$ **4.** $12 + 12y^2$ **5.** $8x + 12$

6. $16a - 24$ **7.** $30a - 6$ **8.** $20b + 5$ **9.** $7x^2 - 3x$ **10.** $12y^2 - 5y$

11. $3a^2 + 5a^5$ **12.** $9x - 5x^2$ **13.** $14y^2 + 11y$ **14.** $6b^3 - 5b^2$ **15.** $2x^4 - 4x$

16. $3y^4 - 9y$ **17.** $10x^4 - 12x^2$ **18.** $12a^5 - 32a^2$ **19.** $8a^8 - 4a^5$ **20.** $16y^4 - 8y^7$

21. $x^2y^2 - xy$ **22.** $a^2b^2 + ab$ **23.** $3x^2y^4 - 6xy$ **24.** $12a^2b^5 - 9ab$ **25.** $x^2y - xy^3$

26. $3x^3 + 6x^2 + 9x$ **27.** $5y^3 - 20y^2 + 10y$ **28.** $2x^4 - 4x^3 + 6x^2$ **29.** $3y^4 - 9y^3 - 6y^2$

30. $2x^3 + 6x^2 - 14x$ **31.** $3y^3 - 9y^2 + 24y$ **32.** $2y^5 - 3y^4 + 7y^3$ **33.** $6a^5 - 3a^3 - 2a^2$

34. $x^3y - 3x^2y^2 + 7xy^3$ **35.** $2a^2b - 5a^2b^2 + 7ab^2$ **36.** $5y^3 + 10y^2 - 25y$

37. $4b^5 + 6b^3 - 12b$ **38.** $3a^2b^2 - 9ab^2 + 15b^2$ **39.** $8x^2y^2 - 4x^2y + x^2$

40. $x^{2n} - x^n$ **41.** $2a^{5n} + a^{2n}$ **42.** $x^{3n} - x^{2n}$

43. $y^{4n} + y^{2n}$ **44.** $a^{2n+2} + a^2$ **45.** $b^{n+5} - b^5$

Objective B

Factor.

46. $x(b + 4) + 3(b + 4)$

47. $y(a + z) + 7(a + z)$

48. $a(y - x) - b(y - x)$

49. $3r(a - b) + s(a - b)$

50. $x(x - 2) + y(2 - x)$

51. $t(m - 7) + 7(7 - m)$

52. $2x(7 + b) - y(b + 7)$

53. $2y(4a - b) - (b - 4a)$

54. $8c(2m - 3n) + (3n - 2m)$

55. $x^2 + 2x + 2xy + 4y$

56. $x^2 - 3x + 4ax - 12a$

57. $p^2 - 2p - 3rp + 6r$

58. $t^2 + 4t - st - 4s$

59. $ab + 6b - 4a - 24$

60. $xy - 5y - 2x + 10$

61. $2z^2 - z + 2yz - y$

62. $2y^2 - 10y + 7xy - 35x$

63. $8v^2 - 12vy + 14v - 21y$

64. $21x^2 + 6xy - 49x - 14y$

65. $2x^2 - 5x - 6xy + 15y$

66. $4a^2 + 5ab - 10b - 8a$

67. $3y^2 - 6y - ay + 2a$

68. $2ra + a^2 - 2r - a$

69. $3xy - y^2 - y + 3x$

70. $2ab - 3b^2 - 3b + 2a$

71. $3st + t^2 - 2t - 6s$

72. $4x^2 + 3xy - 12y - 16x$

APPLYING THE CONCEPTS

A whole number is a perfect number if it is the sum of all of its factors less than itself. For example, 6 is a perfect number because all the factors of 6 that are less than 6 are 1, 2, and 3, and $1 + 2 + 3 = 6$.

73. Find the one perfect number between 20 and 30.

74. Find the one perfect number between 490 and 500.

75. In the equation $P = 2L + 2W$, what is the effect on P when the quantity $L + W$ doubles?

76. Write the area of the shaded portion of each diagram in factored form.

a.
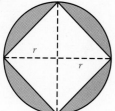

b.
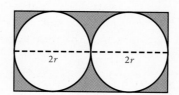

c.

7.2

Factoring Polynomials of the Form $x^2 + bx + c$

Objective A   *To factor a trinomial of the form $x^2 + bx + c$* ....

Trinomials of the form $x^2 + bx + c$, where b and c are integers, are shown at the right.

$x^2 + 8x + 12;\ b = 8,\ c = 12$
$x^2 - 7x + 12;\ b = -7,\ c = 12$
$x^2 - 2x - 15;\ b = -2,\ c = -15$

To factor a trinomial of this form means to express the trinomial as the product of two binomials.

Trinomials expressed as the product of binomials are shown at the right.

$x^2 + 8x + 12 = (x + 6)(x + 2)$
$x^2 - 7x + 12 = (x - 3)(x - 4)$
$x^2 - 2x - 15 = (x + 3)(x - 5)$

The method by which factors of a trinomial are found is based on FOIL. Consider the following binomial products, noting the relationship between the constant terms of the binomials and the terms of the trinomials.

Signs in the binomials are the same.

$(x + 6)(x + 2) = x^2 + 2x + 6x + (6)(2)\quad = x^2 + 8x + 12$
sum of 6 and 2 ⎯⎯⎯⎯⎯⎯⎯⎯⎯⎯
product of 6 and 2 ⎯⎯⎯⎯⎯⎯⎯⎯⎯⎯

$(x - 3)(x - 4) = x^2 - 4x - 3x + (-3)(-4)\ = x^2 - 7x + 12$
sum of -3 and -4 ⎯⎯⎯⎯⎯⎯⎯⎯
product of -3 and -4 ⎯⎯⎯⎯⎯⎯⎯⎯

Signs in the binomials are opposite.

$(x + 3)(x - 5) = x^2 - 5x + 3x + (3)(-5)\quad = x^2 - 2x - 15$
sum of 3 and -5 ⎯⎯⎯⎯⎯⎯⎯⎯⎯
product of 3 and -5 ⎯⎯⎯⎯⎯⎯⎯⎯⎯

$(x - 4)(x + 6) = x^2 + 6x - 4x + (-4)(6)\quad = x^2 + 2x - 24$
sum of -4 and 6 ⎯⎯⎯⎯⎯⎯⎯⎯⎯
product of -4 and 6 ⎯⎯⎯⎯⎯⎯⎯⎯⎯

IMPORTANT RELATIONSHIPS

1. When the constant term of the trinomial is positive, the constant terms of the binomials have the same sign. They are both positive when the coefficient of the x term in the trinomial is positive. They are both negative when the coefficient of the x term in the trinomial is negative.

2. When the constant term of the trinomial is negative, the constant terms of the binomials have opposite signs.

3. In the trinomial, the coefficient of x is the sum of the constant terms of the binomials.

4. In the trinomial, the constant term is the product of the constant terms of the binomials.

➡ Factor: $x^2 - 7x + 10$

Because the constant term is positive and the coefficient of x is negative, the binomial constants will be negative. Find two negative factors of 10 whose sum is -7. The results can be recorded in a table.

Negative Factors of 10	Sum	
$-1, -10$	-11	
$-2, -5$	-7	• These are the correct factors.

$x^2 - 7x + 10 = (x - 2)(x - 5)$ • Write the trinomial as a product of its factors.

Check: $(x - 2)(x - 5) = x^2 - 5x - 2x + 10$ • Check the proposed
$= x^2 - 7x + 10$ factorization by multiplying
the two binomials.

CONSIDER THIS

Always check your proposed factorization to ensure accuracy.

➡ Factor: $x^2 - 9x - 36$

The constant term is negative. The binomial constants will have opposite signs. Find two factors of -36 whose sum is -9.

Factors of -36	Sum	
$+1, -36$	-35	
$-1, +36$	35	
$+2, -18$	-16	
$-2, +18$	16	
$+3, -12$	-9	• Once the correct factors are found, it is not necessary to try the remaining factors.

$x^2 - 9x - 36 = (x + 3)(x - 12)$ • Write the trinomial as a product of its factors.

➡ Factor: $x^2 + 7x + 8$

Because the constant term is positive and the coefficient of x is positive, the binomial constants will be positive. Find two positive factors of 8 whose sum is 7.

Positive Factors of 8	Sum
1, 8	9
2, 4	6

There are no positive integer factors of 8 whose sum is 7. The trinomial $x^2 + 7x + 8$ is said to be **nonfactorable over the integers.** Just as 17 is a prime number, $x^2 + 7x + 8$ is a **prime polynomial.** Binomials of the form $x - a$ and $x + a$ are also prime polynomials.

Example 1
Factor: $x^2 - 8x + 15$

Solution

Factors	Sum	
		• Find two negative factors of 15 whose sum is -8.
$-1, -15$	-16	
$-3, -5$	-8	

$x^2 - 8x + 15 = (x - 3)(x - 5)$

You Try It 1
Factor: $x^2 + 9x + 20$

Your solution

Solution on p. A33

Example 2

Factor: $x^2 + 6x - 27$

Solution

Factors	Sum
+1, −27	−26
−1, +27	26
+3, −9	−6
−3, +9	6

● Find two factors of −27 whose sum is 6.

$x^2 + 6x - 27 = (x - 3)(x + 9)$

You Try It 2

Factor: $x^2 + 7x - 18$

Your solution

Solution on p. A33

Objective B   *To factor completely* ....

A polynomial is factored completely when it is written as a product of factors that are nonfactorable over the integers.

CONSIDER THIS

The first step in *any* factoring problem is to determine whether the terms of the polynomial have a **common factor.** If they do, factor it out first.

➡ Factor: $4y^3 - 4y^2 - 24y$

$$4y^3 - 4y^2 - 24y = \overline{4y(y^2) - 4y(y) - 4y(6)}$$

$$= 4y(y^2 - y - 6)$$

$$= 4y(y + 2)(y - 3)$$

● The GCF is 4*y*. Do this step mentally.
● Use the Distributive Property to factor out the GCF.
● Factor $y^2 - y - 6$. The two factors of −6 whose sum is −1 are 2 and −3.

It is always possible to check the proposed factorization by multiplying the polynomials. Here is the check for the last example.

Check: $4y(y + 2)(y - 3) = 4y(y^2 - 3y + 2y - 6)$
$$= 4y(y^2 - y - 6)$$
$$= 4y^3 - 4y^2 - 24y$$

● This is the original polynomial.

➡ Factor: $5x^2 + 60xy + 100y^2$

$$5x^2 + 60xy + 100y^2 = \overline{5(x^2) + 5(12xy) + 5(20y^2)}$$

$$= 5(x^2 + 12xy + 20y^2)$$

$$= 5(x + 2y)(x + 10y)$$

● The GCF is 5. Do this step mentally.
● Use the Distributive Property to factor out the GCF.
● Factor $x^2 + 12xy + 20y^2$. The two factors of 20 whose sum is 12 are 2 and 10.

CONSIDER THIS

$2y$ and $10y$ are placed in the binomials. This is necessary so that the middle term contains xy and the last term contains y^2.

Note that $2y$ and $10y$ were placed in the binomials. The following check shows that this was necessary.

Check: $5(x + 2y)(x + 10y) = 5(x^2 + 10xy + 2xy + 20y^2)$
$$= 5(x^2 + 12xy + 20y^2)$$
$$= 5x^2 + 60xy + 100y^2$$

● The original polynomial

➡ Factor: $15 - 2x - x^2$

CONSIDER THIS

When the coefficient of the highest power in a polynomial is negative, consider factoring out a negative GCF. Example 3 below is another example of this technique.

Because the coefficient of x^2 is -1, factor -1 from the trinomial and then write the resulting trinomial in descending order.

$$15 - 2x - x^2 = -(x^2 + 2x - 15)$$

• $15 - 2x - x^2 = -1(-15 + 2x + x^2)$

$$= -(x^2 + 2x - 15)$$

$$= -(x + 5)(x - 3)$$

• Factor $x^2 + 2x - 15$. The two factors of -15 whose sum is 2 are 5 and -3.

Check: $-(x + 5)(x - 3) = -(x^2 + 2x - 15)$

$$= -x^2 - 2x + 15$$

$$= 15 - 2x - x^2$$

• The original polynomial

Example 3
Factor: $-3x^3 + 9x^2 + 12x$

Solution
The GCF is $-3x$.

$$-3x^3 + 9x^2 + 12x = -3x(x^2 - 3x - 4)$$

Factor the trinomial $x^2 - 3x - 4$. Find two factors of -4 whose sum is -3.

Factors	*Sum*
$-2, +2$	0
$+1, -4$	-3

$$-3x^3 + 9x^2 + 12x = -3x(x + 1)(x - 4)$$

You Try It 3
Factor: $-2x^3 + 14x^2 - 12x$

Your solution

Example 4
Factor: $4x^2 - 40xy + 84y^2$

Solution
The GCF is 4.

$$4x^2 - 40xy + 84y^2 = 4(x^2 - 10xy + 21y^2)$$

Factor the trinomial $x^2 - 10xy + 21y^2$. Find two negative factors of 21 whose sum is -10.

Factors	*Sum*
$-1, -21$	-22
$-3, -7$	-10

$$4x^2 - 40xy + 84y^2 = 4(x - 3y)(x - 7y)$$

You Try It 4
Factor: $3x^2 - 9xy - 12y^2$

Your solution

Solutions on p. A33

7.2 Exercises

Objective A

Factor.

1. $x^2 + 3x + 2$ **2.** $x^2 + 5x + 6$ **3.** $x^2 - x - 2$ **4.** $x^2 + x - 6$

5. $a^2 + a - 12$ **6.** $a^2 - 2a - 35$ **7.** $a^2 - 3a + 2$ **8.** $a^2 - 5a + 4$

9. $a^2 + a - 2$ **10.** $a^2 - 2a - 3$ **11.** $b^2 - 6b + 9$ **12.** $b^2 + 8b + 16$

13. $b^2 + 7b - 8$ **14.** $y^2 - y - 6$ **15.** $y^2 + 6y - 55$ **16.** $z^2 - 4z - 45$

17. $y^2 - 5y + 6$ **18.** $y^2 - 8y + 15$ **19.** $z^2 - 14z + 45$ **20.** $z^2 - 14z + 49$

21. $z^2 - 12z - 160$ **22.** $p^2 + 2p - 35$ **23.** $p^2 + 12p + 27$ **24.** $p^2 - 6p + 8$

25. $x^2 + 20x + 100$ **26.** $x^2 + 18x + 81$ **27.** $b^2 + 9b + 20$ **28.** $b^2 + 13b + 40$

29. $x^2 - 11x - 42$ **30.** $x^2 + 9x - 70$ **31.** $b^2 - b - 20$ **32.** $b^2 + 3b - 40$

33. $y^2 - 14y - 51$ **34.** $y^2 - y - 72$ **35.** $p^2 - 4p - 21$ **36.** $p^2 + 16p + 39$

37. $y^2 - 8y + 32$ **38.** $y^2 - 9y + 81$ **39.** $x^2 - 20x + 75$ **40.** $p^2 + 24p + 63$

Factor.

41. $x^2 - 15x + 56$ **42.** $x^2 + 21x + 38$ **43.** $x^2 + x - 56$ **44.** $x^2 + 5x - 36$

45. $a^2 - 21a - 72$ **46.** $a^2 - 7a - 44$ **47.** $a^2 - 15a + 36$ **48.** $a^2 - 21a + 54$

49. $z^2 - 9z - 136$ **50.** $z^2 + 14z - 147$ **51.** $c^2 - c - 90$ **52.** $c^2 - 3c - 180$

53. $z^2 + 15z + 44$ **54.** $p^2 + 24p + 135$ **55.** $c^2 + 19c + 34$ **56.** $c^2 + 11c + 18$

57. $x^2 - 4x - 96$ **58.** $x^2 + 10x - 75$ **59.** $x^2 - 22x + 112$ **60.** $x^2 + 21x - 100$

61. $b^2 + 8b - 105$ **62.** $b^2 - 22b + 72$ **63.** $a^2 - 9a - 36$ **64.** $a^2 + 42a - 135$

65. $b^2 - 23b + 102$ **66.** $b^2 - 25b + 126$ **67.** $a^2 + 27a + 72$ **68.** $z^2 + 24z + 144$

69. $x^2 + 25x + 156$ **70.** $x^2 - 29x + 100$ **71.** $x^2 - 10x - 96$ **72.** $x^2 + 9x - 112$

Objective B

Factor.

73. $2x^2 + 6x + 4$ **74.** $3x^2 + 15x + 18$ **75.** $18 + 7x - x^2$

76. $12 - 4x - x^2$ **77.** $ab^2 + 2ab - 15a$ **78.** $ab^2 + 7ab - 8a$

79. $xy^2 - 5xy + 6x$ **80.** $xy^2 + 8xy + 15x$ **81.** $z^3 - 7z^2 + 12z$

Factor.

82. $-2a^3 - 6a^2 - 4a$

83. $-3y^3 + 15y^2 - 18y$

84. $4y^3 + 12y^2 - 72y$

85. $3x^2 + 3x - 36$

86. $2x^3 - 2x^2 + 4x$

87. $5z^2 - 15z - 140$

88. $6z^2 + 12z - 90$

89. $2a^3 + 8a^2 - 64a$

90. $3a^3 - 9a^2 - 54a$

91. $x^2 - 5xy + 6y^2$

92. $x^2 + 4xy - 21y^2$

93. $a^2 - 9ab + 20b^2$

94. $a^2 - 15ab + 50b^2$

95. $x^2 - 3xy - 28y^2$

96. $s^2 + 2st - 48t^2$

97. $y^2 - 15yz - 41z^2$

98. $y^2 + 85yz + 36z^2$

99. $z^4 - 12z^3 + 35z^2$

100. $z^4 + 2z^3 - 80z^2$

101. $b^4 - 22b^3 + 120b^2$

102. $b^4 - 3b^3 - 10b^2$

103. $2y^4 - 26y^3 - 96y^2$

104. $3y^4 + 54y^3 + 135y^2$

105. $-x^4 - 7x^3 + 8x^2$

106. $-x^4 + 11x^3 + 12x^2$

107. $4x^2y + 20xy - 56y$

108. $3x^2y - 6xy - 45y$

109. $c^3 + 18c^2 - 40c$

110. $-3x^3 + 36x^2 - 81x$

111. $-4x^3 - 4x^2 + 24x$

112. $x^2 - 8xy + 15y^2$

113. $y^2 - 7xy - 8x^2$

114. $a^2 - 13ab + 42b^2$

115. $y^2 + 4yz - 21z^2$

116. $y^2 + 8yz + 7z^2$

117. $y^2 - 16yz + 15z^2$

Factor.

118. $3x^2y + 60xy - 63y$ **119.** $4x^2y - 68xy - 72y$ **120.** $3x^3 + 3x^2 - 36x$

121. $4x^3 + 12x^2 - 160x$ **122.** $4z^3 + 32z^2 - 132z$ **123.** $5z^3 - 50z^2 - 120z$

124. $4x^3 + 8x^2 - 12x$ **125.** $5x^3 + 30x^2 + 40x$ **126.** $5p^2 + 25p - 420$

127. $4p^2 - 28p - 480$ **128.** $p^4 + 9p^3 - 36p^2$ **129.** $p^4 + p^3 - 56p^2$

130. $t^2 - 12ts + 35s^2$ **131.** $a^2 - 10ab + 25b^2$ **132.** $a^2 - 8ab - 33b^2$

133. $x^2 + 4xy - 60y^2$ **134.** $5x^4 - 30x^3 + 40x^2$ **135.** $6x^3 - 6x^2 - 120x$

APPLYING THE CONCEPTS

Factor.

136. $2 + c^2 + 9c$ **137.** $x^2y - 54y - 3xy$ **138.** $45a^2 + a^2b^2 - 14a^2b$

Find all integers k such that the trinomial can be factored over the integers.

139. $x^2 + kx + 35$ **140.** $x^2 + kx + 18$ **141.** $x^2 + kx + 21$

Determine the positive integer values of k for which the following polynomials are factorable over the integers.

142. $y^2 + 4y + k$ **143.** $z^2 + 7z + k$ **144.** $a^2 - 6a + k$

145. $c^2 - 7c + k$ **146.** $x^2 - 3x + k$ **147.** $y^2 + 5y + k$

148. In Exercises 142–145, there was the stated requirement that $k > 0$. If k is allowed to be any integer, how many different values of k are possible for each polynomial?

7.3 Factoring Polynomials of the Form $ax^2 + bx + c$

Objective A   *To factor a trinomial of the form $ax^2 + bx + c$ by using trial factors* ....

Trinomials of the form $ax^2 + bx + c$, where a, b, and c are integers, are shown at the right.

$3x^2 - x + 4$; $a = 3$, $b = -1$, $c = 4$
$6x^2 + 2x - 3$; $a = 6$, $b = 2$, $c = -3$

These trinomials differ from those in the previous section in that the coefficient of x^2 is not 1. There are various methods of factoring these trinomials. The method described in this objective is factoring polynomials using trial factors.

To reduce the number of trial factors that must be considered, remember the following:

1. Use the signs of the constant term and the coefficient of x in the trinomial to determine the signs of the binomial factors. If the constant term is positive, the signs of the binomial factors will be the same as the sign of the coefficient of x in the trinomial. If the sign of the constant term is negative, the constant terms in the binomials have opposite signs.

2. If the terms of the trinomial do not have a common factor, then the terms of neither of the binomial factors will have a common factor.

➡ Factor: $2x^2 - 7x + 3$

The terms have no common factor. The constant term is positive. The coefficient of x is negative. The binomial constants will be negative.

Positive Factors of 2 (coefficient of x^2)	*Negative Factors of 3* (constant term)
1, 2	−1, −3

Write trial factors. Use the **O**uter and **I**nner products of FOIL to determine the middle term, $-7x$, of the trinomial.

Trial Factors	*Middle Term*
$(x - 1)(2x - 3)$	$-3x - 2x = -5x$
$(x - 3)(2x - 1)$	$-x - 6x = -7x$

Write the factors of the trinomial. $2x^2 - 7x + 3 = (x - 3)(2x - 1)$

➡ Factor: $3x^2 - 8x + 4$

The terms have no common factor. The constant term is positive. The coefficient of x is negative. The binomial constants will be negative.

Positive Factors of 3 (coefficient of x^2)	*Negative Factors of 4* (constant term)
1, 3	−1, −4
	−2, −2

Write trial factors. Use the **O**uter and **I**nner products of FOIL to determine the middle term, $-8x$, of the trinomial.

Trial Factors	*Middle Term*
$(x - 1)(3x - 4)$	$-4x - 3x = -7x$
$(x - 4)(3x - 1)$	$-x - 12x = -13x$
$(x - 2)(3x - 2)$	$-2x - 6x = -8x$

Write the factors of the trinomial. $3x^2 - 8x + 4 = (x - 2)(3x - 2)$

⇒ Factor: $6x^3 + 14x^2 - 12x$

Factor the GCF, $2x$, from the terms.

$6x^3 + 14x^2 - 12x = 2x(3x^2 + 7x - 6)$

Factor the trinomial $3x^2 + 7x - 6$. The constant term is negative. The binomial constants will have opposite signs.

Positive Factors of 3	Factors of -6
1, 3	$-1, \quad 6$
	$1, \quad -6$
	$-2, \quad 3$
	$2, \quad -3$

CONSIDER THIS

The binomial factor $3x + 6$ has a common factor of 3: $3x + 6 = 3(x + 2)$. Because $3x^2 + 7x - 6$ does not have a common factor, one of its binomial factors cannot have a common factor.

Write trial factors. Use the **O**uter and **I**nner products of FOIL to determine the middle term, $7x$, of the trinomial.

It is not necessary to test trial factors that have a common factor.

Trial Factors	Middle Term
$(x - 1)(3x + 6)$	Common factor
$(x + 6)(3x - 1)$	$-x + 18x = 17x$
$(x + 1)(3x - 6)$	Common factor
$(x - 6)(3x + 1)$	$x - 18x = -17x$
$(x - 2)(3x + 3)$	Common factor
$(x + 3)(3x - 2)$	$-2x + 9x = 7x$
$(x + 2)(3x - 3)$	Common factor
$(x - 3)(3x + 2)$	$2x - 9x = -7x$

Write the factors of the trinomial. $6x^3 + 14x^2 - 12x = 2x(x + 3)(3x - 2)$

For this example, all the trial factors were listed. Once the correct factors have been found, however, the remaining trial factors can be omitted. For the examples and solutions in this text, all trial factors except those that have a common factor will be listed.

Example 1
Factor: $3x^2 + x - 2$

Solution

Positive factors of 3: 1, 3	Factors of -2: $1, -2$
	$-1, \quad 2$

Trial Factors	Middle Term
$(1x + 1)(3x - 2)$	$-2x + 3x = x$
$(1x - 2)(3x + 1)$	$x - 6x = -5x$
$(1x - 1)(3x + 2)$	$2x - 3x = -x$
$(1x + 2)(3x - 1)$	$-x + 6x = 5x$

$3x^2 + x - 2 = (x + 1)(3x - 2)$

You Try It 1
Factor: $2x^2 - x - 3$

Your solution

Example 2
Factor: $-12x^3 - 32x^2 + 12x$

Solution
The GCF is $-4x$.
$-12x^3 - 32x^2 + 12x = -4x(3x^2 + 8x - 3)$
Factor the trinomial.

Positive factors of 3: 1, 3	Factors of -3: $1, -3$
	$-1, \quad 3$

Trial Factors	Middle Term
$(x - 3)(3x + 1)$	$x - 9x = -8x$
$(x + 3)(3x - 1)$	$-x + 9x = 8x$

$-12x^3 - 32x^2 + 12x = -4x(x + 3)(3x - 1)$

You Try It 2
Factor: $-45y^3 + 12y^2 + 12y$

Your solution

Solutions on p. A34

Objective B   *To factor a trinomial of the form $ax^2 + bx + c$ by grouping* ....

In the previous objective, trinomials of the form $ax^2 + bx + c$ were factored by using trial factors. In this objective, these trinomials will be factored by grouping.

To factor $ax^2 + bx + c$, first find two factors of $a \cdot c$ whose sum is b. Then use factoring by grouping to write the factorization of the trinomial.

➡ Factor: $2x^2 + 13x + 15$

Find two positive factors of 30 (2 · 15) whose sum is 13.

Positive Factors of 30	*Sum*
1, 30	31
2, 15	17
3, 10	13

• When the required sum has been found, the remaining factors need not be checked.

$$2x^2 + 13x + 15 = 2x^2 + 3x + 10x + 15$$

• Use the factors of 30 whose sum is 13 to write $13x$ as $3x + 10x$.

$$= (2x^2 + 3x) + (10x + 15)$$
$$= x(2x + 3) + 5(2x + 3)$$
$$= (2x + 3)(x + 5)$$

• Factor by grouping.

Check your answer. $(2x + 3)(x + 5) = 2x^2 + 10x + 3x + 15$
$$= 2x^2 + 13x + 15$$

➡ Factor: $6x^2 - 11x - 10$

Find two factors of -60 [6 · (−10)] whose sum is -11.

Factors of -60	*Sum*
1, −60	−59
−1, 60	59
2, −30	−28
−2, 30	28
3, −20	−17
−3, 20	17
4, −15	−11

$$6x^2 - 11x - 10 = 6x^2 + 4x - 15x - 10$$

• Use the factors of -60 whose sum is -11 to write $-11x$ as $4x - 15x$.

$$= (6x^2 + 4x) - (15x + 10)$$
$$= 2x(3x + 2) - 5(3x + 2)$$
$$= (3x + 2)(2x - 5)$$

• Factor by grouping. Recall that $-15x - 10 = -(15x + 10)$.

Check: $(3x + 2)(2x - 5) = 6x^2 - 15x + 4x - 10$
$$= 6x^2 - 11x - 10$$

➡️ Factor: $3x^2 - 2x - 4$

Find two factors of -12 $[3 \cdot (-4)]$ whose sum is -2.

Factors of -12	Sum
1, −12	−11
−1, 12	11
2, −6	−4
−2, 6	4
3, −4	−1
−3, 4	1

Because no integer factors of -12 have a sum of -2, $3x^2 - 2x - 4$ is nonfactorable over the integers. $3x^2 - 2x - 4$ is a **prime polynomial.**

Example 3
Factor: $2x^2 + 19x - 10$

Solution

Factors of -20 $[2(-10)]$	Sum
−1, 20	19

$$\begin{aligned} 2x^2 + 19x - 10 &= 2x^2 - x + 20x - 10 \\ &= (2x^2 - x) + (20x - 10) \\ &= x(2x - 1) + 10(2x - 1) \\ &= (2x - 1)(x + 10) \end{aligned}$$

You Try It 3
Factor: $2a^2 + 13a - 7$

Your solution

Example 4
Factor: $24x^2y - 76xy + 40y$

Solution
The GCF is $4y$.

$$24x^2y - 76xy + 40y = 4y(6x^2 - 19x + 10)$$

Negative Factors of 60 $[6(10)]$	Sum
−1, −60	−61
−2, −30	−32
−3, −20	−23
−4, −15	−19

$$\begin{aligned} 6x^2 - 19x + 10 &= 6x^2 - 4x - 15x + 10 \\ &= (6x^2 - 4x) - (15x - 10) \\ &= 2x(3x - 2) - 5(3x - 2) \\ &= (3x - 2)(2x - 5) \end{aligned}$$

$$\begin{aligned} 24x^2y - 76xy + 40y &= 4y(6x^2 - 19x + 10) \\ &= 4y(3x - 2)(2x - 5) \end{aligned}$$

You Try It 4
Factor: $15x^3 + 40x^2 - 80x$

Your solution

Solutions on p. A34

7.3 Exercises

Objective A

Factor by using trial factors.

1. $2x^2 + 3x + 1$

2. $5x^2 + 6x + 1$

3. $2y^2 + 7y + 3$

4. $3y^2 + 7y + 2$

5. $2a^2 - 3a + 1$

6. $3a^2 - 4a + 1$

7. $2b^2 - 11b + 5$

8. $3b^2 - 13b + 4$

9. $2x^2 + x - 1$

10. $4x^2 - 3x - 1$

11. $2x^2 - 5x - 3$

12. $3x^2 + 5x - 2$

13. $2t^2 - t - 10$

14. $2t^2 + 5t - 12$

15. $3p^2 - 16p + 5$

16. $6p^2 + 5p + 1$

17. $12y^2 - 7y + 1$

18. $6y^2 - 5y + 1$

19. $6z^2 - 7z + 3$

20. $9z^2 + 3z + 2$

21. $6t^2 - 11t + 4$

22. $10t^2 + 11t + 3$

23. $8x^2 + 33x + 4$

24. $7x^2 + 50x + 7$

25. $5x^2 - 62x - 7$

26. $9x^2 - 13x - 4$

27. $12y^2 + 19y + 5$

28. $5y^2 - 22y + 8$

29. $7a^2 + 47a - 14$

30. $11a^2 - 54a - 5$

31. $3b^2 - 16b + 16$

32. $6b^2 - 19b + 15$

33. $2z^2 - 27z - 14$

34. $4z^2 + 5z - 6$

35. $3p^2 + 22p - 16$

36. $7p^2 + 19p + 10$

Factor by using trial factors.

37. $4x^2 + 6x + 2$ **38.** $12x^2 + 33x - 9$ **39.** $15y^2 - 50y + 35$ **40.** $30y^2 + 10y - 20$

41. $2x^3 - 11x^2 + 5x$ **42.** $2x^3 - 3x^2 - 5x$ **43.** $3a^2b - 16ab + 16b$ **44.** $2a^2b - ab - 21b$

45. $3z^2 + 95z + 10$ **46.** $8z^2 - 36z + 1$ **47.** $36x - 3x^2 - 3x^3$ **48.** $-2x^3 + 2x^2 + 4x$

49. $80y^2 - 36y + 4$ **50.** $24y^2 - 24y - 18$ **51.** $8z^3 + 14z^2 + 3z$ **52.** $6z^3 - 23z^2 + 20z$

53. $6x^2y - 11xy - 10y$ **54.** $8x^2y - 27xy + 9y$ **55.** $10t^2 - 5t - 50$

56. $16t^2 + 40t - 96$ **57.** $3p^3 - 16p^2 + 5p$ **58.** $6p^3 + 5p^2 + p$

59. $26z^2 + 98z - 24$ **60.** $30z^2 - 87z + 30$ **61.** $10y^3 - 44y^2 + 16y$

62. $14y^3 + 94y^2 - 28y$ **63.** $4yz^3 + 5yz^2 - 6yz$ **64.** $12a^3 + 14a^2 - 48a$

65. $42a^3 + 45a^2 - 27a$ **66.** $36p^2 - 9p^3 - p^4$ **67.** $9x^2y - 30xy^2 + 25y^3$

68. $8x^2y - 38xy^2 + 35y^3$ **69.** $9x^3y - 24x^2y^2 + 16xy^3$ **70.** $9x^3y + 12x^2y + 4xy$

Objective B

Factor by grouping.

71. $6x^2 - 17x + 12$ **72.** $15x^2 - 19x + 6$ **73.** $5b^2 + 33b - 14$ **74.** $8x^2 - 30x + 25$

75. $6a^2 + 7a - 24$ **76.** $14a^2 + 15a - 9$ **77.** $4z^2 + 11z + 6$ **78.** $6z^2 - 25z + 14$

79. $22p^2 + 51p - 10$ **80.** $14p^2 - 41p + 15$ **81.** $8y^2 + 17y + 9$ **82.** $12y^2 - 145y + 12$

83. $18t^2 - 9t - 5$ **84.** $12t^2 + 28t - 5$ **85.** $6b^2 + 71b - 12$ **86.** $8b^2 + 65b + 8$

87. $9x^2 + 12x + 4$ **88.** $25x^2 - 30x + 9$ **89.** $6b^2 - 13b + 6$ **90.** $20b^2 + 37b + 15$

91. $33b^2 + 34b - 35$ **92.** $15b^2 - 43b + 22$ **93.** $18y^2 - 39y + 20$ **94.** $24y^2 + 41y + 12$

95. $15a^2 + 26a - 21$ **96.** $6a^2 + 23a + 21$ **97.** $8y^2 - 26y + 15$ **98.** $18y^2 - 27y + 4$

99. $8z^2 + 2z - 15$ **100.** $10z^2 + 3z - 4$ **101.** $15x^2 - 82x + 24$ **102.** $13z^2 + 49z - 8$

103. $10z^2 - 29z + 10$ **104.** $15z^2 - 44z + 32$ **105.** $36z^2 + 72z + 35$ **106.** $16z^2 + 8z - 35$

107. $3x^2 + xy - 2y^2$ **108.** $6x^2 + 10xy + 4y^2$ **109.** $3a^2 + 5ab - 2b^2$ **110.** $2a^2 - 9ab + 9b^2$

Factor by grouping.

111. $4y^2 - 11yz + 6z^2$ **112.** $2y^2 + 7yz + 5z^2$ **113.** $28 + 3z - z^2$ **114.** $15 - 2z - z^2$

115. $8 - 7x - x^2$ **116.** $12 + 11x - x^2$ **117.** $9x^2 + 33x - 60$ **118.** $16x^2 - 16x - 12$

119. $24x^2 - 52x + 24$ **120.** $60x^2 + 95x + 20$ **121.** $35a^4 + 9a^3 - 2a^2$

122. $15a^4 + 26a^3 + 7a^2$ **123.** $15b^2 - 115b + 70$ **124.** $25b^2 + 35b - 30$

125. $3x^2 - 26xy + 35y^2$ **126.** $4x^2 + 16xy + 15y^2$ **127.** $216y^2 - 3y - 3$

128. $360y^2 + 4y - 4$ **129.** $21 - 20x - x^2$ **130.** $18 + 17x - x^2$

131. $15a^2 + 11ab - 14b^2$ **132.** $15a^2 - 31ab + 10b^2$ **133.** $33z - 8z^2 - z^3$

APPLYING THE CONCEPTS

Factor.

134. $2(y + 2)^2 - (y + 2) - 3$ **135.** $3(a + 2)^2 - (a + 2) - 4$ **136.** $4(y - 1)^2 - 7(y - 1) - 2$

Find all integers k such that the trinomial can be factored over the integers.

137. $2x^2 + kx + 3$ **138.** $2x^2 + kx - 3$ **139.** $3x^2 + kx + 2$

140. Write the area of the shaded portion of each diagram in factored form.

a. 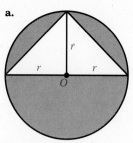 b. 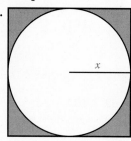 c. 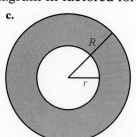 d.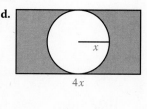

141. In your own words, explain how the signs of the last terms of the
[W] two binomial factors of a trinomial are determined.

7.4

Special Factoring

Objective A *To factor the difference of two perfect squares or a perfect-square trinomial*

The product of a term and itself is called a **perfect square.** The exponents on variables of perfect squares are always even numbers.

Term		Perfect Square
5	$5 \cdot 5 =$	25
x	$x \cdot x =$	x^2
$3y^4$	$3y^4 \cdot 3y^4 =$	$9y^8$
x^n	$x^n \cdot x^n =$	x^{2n}

The **square root** of a perfect square is one of the two equal factors of the perfect square. "$\sqrt{}$" is the symbol for square root. To find the exponent of the square root of a variable term, divide the exponent by 2.

$$\sqrt{25} = 5$$
$$\sqrt{x^2} = x$$
$$\sqrt{9y^8} = 3y^4$$
$$\sqrt{x^{2n}} = x^n$$

The difference of two perfect squares is the product of the sum and difference of two terms. The factors of the difference of two perfect squares are the sum and difference of the square roots of the perfect squares.

CONSIDER THIS

$(a + b)$ is the sum of the two terms a and b. $(a - b)$ is the difference of the two terms a and b. a and b are the square roots of a^2 and b^2.

> **Factors of the Difference of Two Perfect Squares**
>
> $a^2 - b^2 = (a + b)(a - b)$

The sum of two perfect squares, $a^2 + b^2$, is nonfactorable over the integers.

➡ Factor: $4x^2 - 81y^2$

$4x^2 - 81y^2 = (2x)^2 - (9y)^2$ • Write the binomial as the difference of two perfect squares.

$\qquad = (2x + 9y)(2x - 9y)$ • The factors are the sum and difference of the square roots of the perfect squares.

A perfect-square trinomial is the square of a binomial.

> **Factors of a Perfect-Square Trinomial**
>
> $a^2 + 2ab + b^2 = (a + b)^2$
> $a^2 - 2ab + b^2 = (a - b)^2$

In factoring a perfect-square trinomial, remember that the terms of the binomial are the square roots of the perfect squares of the trinomial. The sign in the binomial is the sign of the middle term of the trinomial.

➡ Factor: $4x^2 + 12x + 9$

Because $4x^2$ is a perfect square $[4x^2 = (2x)^2]$ and 9 is a perfect square $(9 = 3^2)$, try factoring $4x^2 + 12x + 9$ as the square of a binomial.

$$4x^2 + 12x + 9 \stackrel{?}{=} (2x + 3)^2$$

Check: $(2x + 3)^2 = (2x + 3)(2x + 3) = 4x^2 + 6x + 6x + 9 = 4x^2 + 12x + 9$

The check verifies that $4x^2 + 12x + 9 = (2x + 3)^2$.

It is important to check a proposed factorization as we did above. The next example illustrates the importance of this check.

➡ Factor: $x^2 + 13x + 36$

Because x^2 is a perfect square and 36 is a perfect square, try factoring $x^2 + 13x + 36$ as the square of a binomial.

$$x^2 + 13x + 36 \stackrel{?}{=} (x + 6)^2$$

Check: $(x + 6)^2 = (x + 6)(x + 6) = x^2 + 6x + 6x + 36 = x^2 + 12x + 36$

In this case, the proposed factorization of $x^2 + 13x + 36$ does *not* check. Try another factorization. The numbers 4 and 9 are factors of 36 whose sum is 13.

$$x^2 + 13x + 36 = (x + 4)(x + 9)$$

Example 1

Factor: $25x^2 - 1$

Solution

$$\begin{aligned} 25x^2 - 1 &= (5x)^2 - (1)^2 \\ &= (5x + 1)(5x - 1) \end{aligned}$$

You Try It 1

Factor: $x^2 - 36y^4$

Your solution

Example 2

Factor: $4x^2 - 20x + 25$

Solution

$$4x^2 - 20x + 25 = (2x - 5)^2$$

You Try It 2

Factor: $9x^2 + 12x + 4$

Your solution

Example 3

Factor: $(x + y)^2 - 4$

Solution

$$\begin{aligned} (x + y)^2 - 4 &= (x + y)^2 - (2)^2 \\ &= (x + y + 2)(x + y - 2) \end{aligned}$$

You Try It 3

Factor: $(a + b)^2 - (a - b)^2$

Your solution

Solutions on p. A35

Objective B   *To factor the sum or the difference of two cubes* ....

The product of the same three factors is called a **perfect cube.** The exponents on variables of perfect cubes are always divisible by 3.	*Term*		*Perfect Cube*
	2	$2 \cdot 2 \cdot 2 = 2^3 =$	8
	$3y$	$3y \cdot 3y \cdot 3y = (3y)^3 =$	$27y^3$
	y^2	$y^2 \cdot y^2 \cdot y^2 = (y^2)^3 =$	y^6

The **cube root** of a perfect cube is one of the three equal factors of the perfect cube. "$\sqrt[3]{}$" is the symbol for cube root. To find the exponent of the cube root of a variable term, divide the exponent by 3.

$$\sqrt[3]{8} = 2$$
$$\sqrt[3]{27y^3} = 3y$$
$$\sqrt[3]{y^6} = y^2$$

> **Factors of the Sum or Difference of Two Cubes**
>
> $a^3 + b^3 = (a + b)(a^2 - ab + b^2)$
> $a^3 - b^3 = (a - b)(a^2 + ab + b^2)$

➡ Factor: $8x^3 - 27$

Write the binomial as the difference of two perfect cubes.

$$8x^3 - 27 = (2x)^3 - 3^3$$

The terms of the binomial factor are the cube roots of the perfect cubes. The sign of the binomial factor is the same sign as in the given binomial. The trinomial factor is obtained from the binomial factor.

$$= (2x - 3)(4x^2 + 6x + 9)$$

Square of the first term ⬏

Opposite of the product of the two terms ⬏

Square of the last term ⬏

CONSIDER THIS

You can always check a proposed factorization by multiplying the factors.

Check:

$$
\begin{array}{r}
4x^2 + 6x + 9 \\
2x - 3 \\
\hline
-12x^2 - 18x - 27 \\
8x^3 + 12x^2 + 18x \\
\hline
8x^3 - 27
\end{array}
$$

  • **The original polynomial**

➡ Factor: $a^3 + 64y^3$

$a^3 + 64y^3 = a^3 + (4y)^3$ • **Write the binomial as the sum of two perfect cubes.**

$= (a + 4y)(a^2 - 4ay + 16y^2)$ • **Factor.**

➡ Factor: $64y^4 - 125y$

$64y^4 - 125y = y(64y^3 - 125)$ • **Factor out y, the GCF.**

$= y[(4y)^3 - 5^3]$ • **Write the binomial as the difference of two cubes.**

$= y(4y - 5)(16y^2 + 20y + 25)$ • **Factor.**

Example 4

Factor: $x^3y^3 - 1$

Solution

$x^3y^3 - 1 = (xy)^3 - 1^3$
$= (xy - 1)(x^2y^2 + xy + 1)$

You Try It 4

Factor: $8x^3 + y^3z^3$

Your solution

Example 5

Factor: $(x + y)^3 - x^3$

Solution

$(x + y)^3 - x^3$
$= [(x + y) - x][(x + y)^2 + x(x + y) + x^2]$
$= y(x^2 + 2xy + y^2 + x^2 + xy + x^2)$
$= y(3x^2 + 3xy + y^2)$

You Try It 5

Factor: $(x - y)^3 + (x + y)^3$

Your solution

Solutions on p. A35

Objective C *To factor a trinomial that is quadratic in form*

Certain trinomials that are not quadratic can be expressed as quadratic trinomials by making suitable variable substitutions. A trinomial is *quadratic in form* if it can be written as $au^2 + bu + c$.

Each of the trinomials shown below is quadratic in form.

$$x^4 + 5x^2 + 6 \qquad\qquad 2x^2y^2 + 3xy - 9$$
$$(x^2)^2 + 5(x^2) + 6 \qquad\qquad 2(xy)^2 + 3(xy) - 9$$

Let $u = x^2$. $\quad u^2 + 5u + 6 \qquad$ Let $u = xy$. $\quad 2u^2 + 3u - 9$

When we use this method to factor a trinomial that is quadratic in form, the variable part of the first term in each binomial will be u.

For example, $x^4 + 5x^2 + 6$ is quadratic in form when $u = x^2$.

$$x^4 + 5x^2 + 6 = (x^2)^2 + 5(x^2) + 6 = u^2 + 5u + 6$$

As shown below, the variable part of the first term in each binomial is x^2.

➡ Factor: $x^4 + 5x^2 + 6$

$x^4 + 5x^2 + 6 = u^2 + 5u + 6$ • Let $u = x^2$.

$\qquad\qquad\quad = (u + 3)(u + 2)$ • Factor.

$\qquad\qquad\quad = (x^2 + 3)(x^2 + 2)$ • Replace u by x^2.

Example 6
Factor: $6x^2y^2 - xy - 12$

Solution
Let $u = xy$.

$$6x^2y^2 - xy - 12 = 6u^2 - u - 12$$
$$= (3u + 4)(2u - 3)$$
$$= (3xy + 4)(2xy - 3)$$

You Try It 6
Factor: $6x^2y^2 - 19xy + 10$

Your solution

Example 7
Factor: $2x^4 + 5x^2 - 12$

Solution
Let $u = x^2$.

$$2x^4 + 5x^2 - 12 = 2u^2 + 5u - 12$$
$$= (2u - 3)(u + 4)$$
$$= (2x^2 - 3)(x^2 + 4)$$

You Try It 7
Factor: $3x^4 + 4x^2 - 4$

Your solution

Example 8
Factor: $x^4y^4 + 3x^2y^2 - 10$

Solution
Let $u = x^2y^2$.

$$x^4y^4 + 3x^2y^2 - 10 = u^2 + 3u - 10$$
$$= (u + 5)(u - 2)$$
$$= (x^2y^2 + 5)(x^2y^2 - 2)$$

You Try It 8
Factor: $a^4b^4 + 6a^2b^2 - 7$

Your solution

Solutions on p. A35

Objective D *To factor completely* ....

When factoring a polynomial completely, ask the following questions about the polynomial.

1. Is there a common factor? If so, factor out the GCF.

2. If the polynomial is a binomial, is it the difference of two perfect squares, the sum of two cubes, or the difference of two cubes? If so, factor.

3. If the polynomial is a trinomial, is it a perfect-square trinomial or the product of two binomials? If so, factor.

4. If the polynomial has four terms, can it be factored by grouping? If so, factor.

5. Is each factor nonfactorable over the integers? If not, factor.

Example 9
Factor: $6a^3 + 15a^2 - 36a$

Solution
$$6a^3 + 15a^2 - 36a = 3a(2a^2 + 5a - 12)$$
$$= 3a(2a - 3)(a + 4)$$

You Try It 9
Factor: $18x^3 - 6x^2 - 60x$

Your solution

Example 10
Factor: $x^2y + 2x^2 - y - 2$

Solution
$$x^2y + 2x^2 - y - 2 = (x^2y + 2x^2) - (y + 2)$$
$$= x^2(y + 2) - 1(y + 2)$$
$$= (y + 2)(x^2 - 1)$$
$$= (y + 2)(x + 1)(x - 1)$$

You Try It 10
Factor: $4x - 4y - x^3 + x^2y$

Your solution

Example 11
Factor: $x^{4n} - y^{4n}$

Solution
$$x^{4n} - y^{4n} = (x^{2n})^2 - (y^{2n})^2$$
$$= (x^{2n} + y^{2n})(x^{2n} - y^{2n})$$
$$= (x^{2n} + y^{2n})[(x^n)^2 - (y^n)^2]$$
$$= (x^{2n} + y^{2n})(x^n + y^n)(x^n - y^n)$$

You Try It 11
Factor: $x^{4n} - x^{2n}y^{2n}$

Your solution

Example 12
Factor: $x^{n+3} + x^ny^3$

Solution
$$x^{n+3} + x^ny^3 = x^n(x^3 + y^3)$$
$$= x^n(x + y)(x^2 - xy + y^2)$$

You Try It 12
Factor: $ax^5 - ax^2y^6$

Your solution

Solutions on p. A35

7.4 Exercises

Objective A

Factor.

1. $x^2 - 16$

2. $y^2 - 49$

3. $4x^2 - 1$

4. $81x^2 - 4$

5. $16x^2 - 121$

6. $49y^2 - 36$

7. $1 - 9a^2$

8. $16 - 81y^2$

9. $x^2y^2 - 100$

10. $a^2b^2 - 25$

11. $x^2 + 4$

12. $a^2 + 16$

13. $25 - a^2b^2$

14. $64 - x^2y^2$

15. $a^{2n} - 1$

16. $b^{2n} - 16$

17. $x^2 - 12x + 36$

18. $y^2 - 6y + 9$

19. $b^2 - 2b + 1$

20. $a^2 + 14a + 49$

21. $16x^2 - 40x + 25$

22. $49x^2 + 28x + 4$

23. $4a^2 + 4a - 1$

24. $9x^2 + 12x - 4$

25. $b^2 + 7b + 14$

26. $y^2 - 5y + 25$

27. $x^2 + 6xy + 9y^2$

28. $4x^2y^2 + 12xy + 9$

29. $25a^2 - 40ab + 16b^2$

30. $4a^2 - 36ab + 81b^2$

31. $x^{2n} + 6x^n + 9$

32. $y^{2n} - 16y^n + 64$

33. $(x - 4)^2 - 9$

34. $16 - (a - 3)^2$

35. $(x - y)^2 - (a + b)^2$

36. $(x - 2y)^2 - (x + y)^2$

Objective B

Factor.

37. $x^3 - 27$

38. $y^3 + 125$

39. $8x^3 - 1$

40. $64a^3 + 27$

41. $x^3 - y^3$

42. $x^3 - 8y^3$

43. $m^3 + n^3$

44. $27a^3 + b^3$

45. $64x^3 + 1$

46. $1 - 125b^3$

47. $27x^3 - 8y^3$

48. $64x^3 + 27y^3$

49. $x^3y^3 + 64$

50. $8x^3y^3 + 27$

51. $16x^3 - y^3$

52. $27x^3 - 8y^2$

53. $8x^3 - 9y^3$

54. $27a^3 - 16$

55. $(a - b)^3 - b^3$

56. $a^3 + (a + b)^3$

57. $x^{6n} + y^{3n}$

58. $x^{3n} + y^{3n}$

59. $x^{3n} + 8$

60. $a^{3n} + 64$

Objective C

Factor.

61. $x^2y^2 - 8xy + 15$

62. $x^2y^2 - 8xy - 33$

63. $x^2y^2 - 17xy + 60$

64. $a^2b^2 + 10ab + 24$

65. $x^4 - 9x^2 + 18$

66. $y^4 - 6y^2 - 16$

67. $b^4 - 13b^2 - 90$

68. $a^4 + 14a^2 + 45$

69. $x^4y^4 - 8x^2y^2 + 12$

Factor.

70. $a^4b^4 + 11a^2b^2 - 26$

71. $x^{2n} + 3x^n + 2$

72. $a^{2n} - a^n - 12$

73. $3x^2y^2 - 14xy + 15$

74. $5x^2y^2 - 59xy + 44$

75. $6a^2b^2 - 23ab + 21$

76. $10a^2b^2 + 3ab - 7$

77. $2x^4 - 13x^2 - 15$

78. $3x^4 + 20x^2 + 32$

79. $2x^{2n} - 7x^n + 3$

80. $4x^{2n} + 8x^n - 5$

81. $6a^{2n} + 19a^n + 10$

Objective D

Factor.

82. $5x^2 + 10x + 5$

83. $12x^2 - 36x + 27$

84. $3x^4 - 81x$

85. $27a^4 - a$

86. $7x^2 - 28$

87. $20x^2 - 5$

88. $y^4 - 10y^3 + 21y^2$

89. $y^5 + 6y^4 - 55y^3$

90. $x^4 - 16$

91. $16x^4 - 81$

92. $8x^5 - 98x^3$

93. $16a - 2a^4$

94. $x^3y^3 - x^3$

95. $a^3b^6 - b^3$

96. $x^6y^6 - x^3y^3$

97. $8x^4 - 40x^3 + 50x^2$

98. $6x^5 + 74x^4 + 24x^3$

99. $x^4 - y^4$

100. $16a^4 - b^4$

101. $x^6 + y^6$

102. $x^4 - 5x^2 - 4$

103. $a^4 - 25a^2 - 144$

104. $3b^5 - 24b^2$

105. $16a^4 - 2a$

Factor.

106. $x^4y^2 - 5x^3y^3 + 6x^2y^4$

107. $a^4b^2 - 8a^3b^3 - 48a^2b^4$

108. $16x^3y + 4x^2y^2 - 42xy^3$

109. $24a^2b^2 - 14ab^3 - 90b^4$

110. $x^3 - 2x^2 - x + 2$

111. $x^3 - 2x^2 - 4x + 8$

112. $4x^2y^2 - 4x^2 - 9y^2 + 9$

113. $4x^4 - x^2 - 4x^2y^2 + y^2$

114. $x^5 - 4x^3 - 8x^2 + 32$

115. $x^6y^3 + x^3 - x^3y^3 - 1$

116. $a^{2n+2} - 6a^{n+2} + 9a^2$

117. $x^{2n+1} + 2x^{n+1} + x$

118. $2x^{n+2} - 7x^{n+1} + 3x^n$

119. $3b^{n+2} + 4b^{n+1} - 4b^n$

APPLYING THE CONCEPTS

120. Find all integers k such that the trinomial is a perfect-square trinomial.
   a. $x^2 + kx + 36$ **b.** $4x^2 - kx + 25$ **c.** $49x^2 + kxy + 64y^2$
   d. $x^2 + 8x + k$ **e.** $x^2 - 12x + k$ **f.** $x^2 + 4xy + ky^2$

121. Factor: $x^2(x - 3) - 3x(x - 3) + 2(x - 3)$

122. Factor: $(a^2 + a - 6)^2 - (a^2 + 2a - 8)^2$

123. The area of a square is $(16x^2 + 24x + 9)$ m^2. Find the dimensions of the square in terms of the variable x. Can $x = 0$? What are the possible values of x?

$A = 16x^2 + 24x + 9$

124. Is $x^2 + 9$ a prime polynomial? Explain your answer.
[W]

125. Can a third-degree polynomial have factors $(x - 1)$, $(x + 1)$, $(x - 3)$,
[W] and $(x + 4)$? Why or why not?

126. Given that $(x - 3)$ and $(x + 4)$ are factors of $x^3 + 6x^2 - 7x - 60$,
[W] explain how you can find a third *first-degree* factor of $x^3 + 6x^2 - 7x - 60$. Then find the factor.

7.5 Solving Equations

Objective A   *To solve equations by factoring*

Recall that the Multiplication Property of Zero states that the product of a number and zero is zero. This property is restated below.

$$\text{If } a \text{ is a real number, then } a \cdot 0 = 0 \cdot a = 0.$$

Now consider $x \cdot y = 0$. For this to be a true equation, then either $x = 0$ or $y = 0$.

Principle of Zero Products

If the product of two factors is zero, then at least one of the factors must be zero.

If $a \cdot b = 0$, then $a = 0$ or $b = 0$.

The Principle of Zero Products is used to solve some equations.

➡ Solve: $(x - 2)(x - 3) = 0$

By the Principle of Zero Products, if $(x - 2)(x - 3) = 0$, then $x - 2 = 0$ or $x - 3 = 0$.

$(x - 2)(x - 3) = 0$
$x - 2 = 0 \quad\quad x - 3 = 0$ • Let each factor equal zero (the Principle of Zero Products).

$\quad\quad x = 2 \quad\quad\quad\quad x = 3$ • Solve each equation for *x*.

Check:

$$\begin{array}{c|c} (x - 2)(x - 3) = 0 & \\ \hline (2 - 2)(2 - 3) & 0 \\ 0(-1) & 0 \\ & 0 = 0 \ \text{True} \end{array}$$

$$\begin{array}{c|c} (x - 2)(x - 3) = 0 & \\ \hline (3 - 2)(3 - 3) & 0 \\ (1)(0) & 0 \\ & 0 = 0 \ \text{True} \end{array}$$

The solutions are 2 and 3.

An equation of the form $ax^2 + bx + c = 0, a \neq 0$, is a **quadratic equation**. A quadratic equation is in **standard form** when the polynomial is in descending order and equal to zero. The quadratic equations at the right are in standard form.

$3x^2 + 2x + 1 = 0$

$4x^2 - 3x + 2 = 0$

➡ Solve: $2x^2 + x = 6$

$$2x^2 + x = 6$$
$$2x^2 + x - 6 = 0$$ • Write the equation in standard form.
$$(2x - 3)(x + 2) = 0$$ • Factor.

$$2x - 3 = 0 \qquad x + 2 = 0$$ • Use the Principle of Zero Products.
$$2x = 3 \qquad\qquad x = -2$$ • Solve each equation for x.
$$x = \frac{3}{2}$$

$\frac{3}{2}$ and -2 check as solutions. The solutions are $\frac{3}{2}$ and -2.

Example 1
Solve: $x(x - 3) = 0$

Solution
$x(x - 3) = 0$

$x = 0 \qquad x - 3 = 0$
$\qquad\qquad x = 3$

The solutions are 0 and 3.

You Try It 1
Solve: $2x(x + 7) = 0$

Your solution

Example 2
Solve: $2x^2 - 50 = 0$

Solution
$$2x^2 - 50 = 0$$
$$2(x^2 - 25) = 0$$
$$2(x + 5)(x - 5) = 0$$

$x + 5 = 0 \qquad x - 5 = 0$
$\quad x = -5 \qquad\quad x = 5$

The solutions are -5 and 5.

You Try It 2
Solve: $4x^2 - 9 = 0$

Your solution

Example 3
Solve: $(x - 3)(x - 10) = -10$

Solution
$$(x - 3)(x - 10) = -10$$
$$x^2 - 13x + 30 = -10$$
$$x^2 - 13x + 40 = 0$$
$$(x - 8)(x - 5) = 0$$

• Multiply $(x - 3)(x - 10)$.
• Add 10 to each side of the equation. The equation is now in standard form.

$x - 8 = 0 \quad x - 5 = 0$
$\quad x = 8 \qquad x = 5$

The solutions are 8 and 5.

You Try It 3
Solve: $(x + 2)(x - 7) = 52$

Your solution

Solutions on p. A36

Objective B   *To solve application problems*

Recall that the integers are the numbers $\ldots, -3, -2, -1, 0, 1, 2, 3, \ldots$.

An **even integer** is an integer that is divisible by 2. Examples of even integers are -8, 0, and 22. An **odd integer** is an integer that is not divisible by 2. Examples of odd integers are -17, 1, and 39.

Consecutive integers are integers that follow one another in order. Examples of consecutive integers are shown at the right. (Assume the variable n represents an integer.)	11, 12, 13 $-8, -7, -6$ $n, n + 1, n + 2$
Examples of **consecutive even integers** are shown at the right. (Assume the variable n represents an even integer.)	24, 26, 28 $-10, -8, -6$ $n, n + 2, n + 4$
Examples of **consecutive odd integers** are shown at the right. (Assume the variable n represents an odd integer.)	19, 21, 23 $-1, 1, 3$ $n, n + 2, n + 4$

Example 4

The sum of the squares of two consecutive positive even integers is equal to 100. Find the two integers.

Strategy

First positive even integer: n
Second positive even integer: $n + 2$

The sum of the square of the first positive even integer and the square of the second positive even integer is 100.

Solution

$$n^2 + (n + 2)^2 = 100$$
$$n^2 + n^2 + 4n + 4 = 100$$
$$2n^2 + 4n + 4 = 100$$
$$2n^2 + 4n - 96 = 0$$
$$2(n^2 + 2n - 48) = 0$$
$$n^2 + 2n - 48 = 0$$

• Divide each side of the equation by 2.

$$(n - 6)(n + 8) = 0$$

$$n - 6 = 0 \qquad n + 8 = 0$$
$$n = 6 \qquad\quad n = -8$$

Because -8 is not a positive even integer, it is not a solution.

$$n = 6$$
$$n + 2 = 6 + 2 = 8$$

The two integers are 6 and 8.

You Try It 4

The sum of the squares of two positive consecutive integers is 61. Find the two integers.

Your strategy

Your solution

Solution on p. A36

Example 5

A stone is thrown into a well with an initial speed of 4 ft/s. The well is 420 ft deep. How many seconds later will the stone hit the bottom of the well? Use the equation $d = vt + 16t^2$, where d is the distance in feet, v is the initial speed, and t is the time in seconds.

Strategy

To find the time for the stone to drop to the bottom of the well, replace the variables d and v by their given values and solve for t.

You Try It 5

The length of a rectangle is 4 in. longer than twice the width. The area of the rectangle is 96 in². Find the length and width of the rectangle.

Your strategy

Solution

$$d = vt + 16t^2$$
$$420 = 4t + 16t^2$$
$$0 = -420 + 4t + 16t^2$$
$$16t^2 + 4t - 420 = 0$$
$$4(4t^2 + t - 105) = 0$$
$$4t^2 + t - 105 = 0 \qquad \bullet \text{ Divide each side of the equation by 4.}$$

$$(4t + 21)(t - 5) = 0$$

$$4t + 21 = 0 \qquad t - 5 = 0$$
$$4t = -21 \qquad t = 5$$
$$t = -\frac{21}{4}$$

Because the time cannot be a negative number, $-\frac{21}{4}$ is not a solution.

The time is 5 s.

Your solution

Solution on p. A36

7.5 Exercises

Objective A

Solve.

1. $(y + 3)(y + 2) = 0$ **2.** $(y - 3)(y - 5) = 0$ **3.** $(z - 7)(z - 3) = 0$ **4.** $(z + 8)(z - 9) = 0$

5. $x(x - 5) = 0$ **6.** $x(x + 2) = 0$ **7.** $a(a - 9) = 0$ **8.** $a(a + 12) = 0$

9. $y(2y + 3) = 0$ **10.** $t(4t - 7) = 0$ **11.** $2a(3a - 2) = 0$ **12.** $4b(2b + 5) = 0$

13. $(b + 2)(b - 5) = 0$ **14.** $(b - 8)(b + 3) = 0$ **15.** $x^2 - 81 = 0$ **16.** $x^2 - 121 = 0$

17. $4x^2 - 49 = 0$ **18.** $16x^2 - 1 = 0$ **19.** $9x^2 - 1 = 0$ **20.** $16x^2 - 49 = 0$

21. $x^2 + 6x + 8 = 0$ **22.** $x^2 - 8x + 15 = 0$ **23.** $z^2 + 5z - 14 = 0$ **24.** $z^2 + z - 72 = 0$

25. $x^2 - 5x + 6 = 0$ **26.** $x^2 - 3x - 10 = 0$ **27.** $y^2 + 4y - 21 = 0$ **28.** $2y^2 - y - 1 = 0$

29. $2a^2 - 9a - 5 = 0$ **30.** $3a^2 + 14a + 8 = 0$ **31.** $6z^2 + 5z + 1 = 0$ **32.** $6y^2 - 19y + 15 = 0$

33. $x^2 - 3x = 0$ **34.** $a^2 - 5a = 0$ **35.** $x^2 - 7x = 0$ **36.** $2a^2 - 8a = 0$

37. $a^2 + 5a = -4$ **38.** $a^2 - 5a = 24$ **39.** $y^2 - 5y = -6$ **40.** $y^2 - 7y = 8$

41. $2t^2 + 7t = 4$ **42.** $3t^2 + t = 10$ **43.** $3t^2 - 13t = -4$ **44.** $5t^2 - 16t = -12$

Solve.

45. $x(x - 12) = -27$ **46.** $x(x - 11) = 12$ **47.** $y(y - 7) = 18$ **48.** $y(y + 8) = -15$

49. $p(p + 3) = -2$ **50.** $p(p - 1) = 20$ **51.** $y(y + 4) = 45$ **52.** $y(y - 8) = -15$

53. $x(x + 3) = 28$ **54.** $p(p - 14) = 15$ **55.** $(x + 8)(x - 3) = -30$ **56.** $(x + 4)(x - 1) = 14$

57. $(z - 5)(z + 4) = 52$ **58.** $(z - 8)(z + 4) = -35$ **59.** $(z - 6)(z + 1) = -10$

60. $(a + 3)(a + 4) = 72$ **61.** $(a - 4)(a + 7) = -18$ **62.** $(2x + 5)(x + 1) = -1$

63. $(z + 3)(z - 10) = -42$ **64.** $(y + 3)(2y + 3) = 5$ **65.** $(y + 5)(3y - 2) = -14$

Objective B *Application Problems*

Solve.

66. The square of a positive number is six more than five times the positive number. Find the number.

67. The square of a negative number is sixteen more than six times the negative number. Find the number.

68. The sum of the squares of two consecutive positive integers is eighty-five. Find the two integers.

69. The sum of the squares of two consecutive positive even integers is one hundred thirty. Find the two integers.

70. The length of a rectangle is 5 in. more than twice the width. The area is 75 in². Find the length and width of the rectangle.

Solve.

71. The width of a rectangle is 5 ft less than the length. The area of the rectangle is 176 ft². Find the length and width of the rectangle.

72. The length of each side of a square is extended 2 in. The area of the resulting square is 144 in². Find the length of a side of the original square.

73. The length of each side of a square is extended 5 in. The area of the resulting square is 64 in². Find the length of a side of the original square.

74. The page of a book measures 6 in. by 9 in. A uniform border around the page leaves 28 in² for type. What are the dimensions of the type area?

75. A small garden measures 8 ft by 10 ft. A uniform border around the garden increases the total area to 168 ft². What is the width of the border?

76. The radius of a circle is increased by 3 in., increasing the area by 100 in². Find the radius of the original circle. Round to the nearest hundredth.

77. A circle has a radius of 10 in. Find the increase in area that occurs when the radius is increased by 2 in. Round to the nearest hundredth.

The formula $S = \dfrac{n^2 + n}{2}$ gives the sum S of the first n natural numbers. Use this formula for Exercises 78 and 79.

78. How many consecutive natural numbers beginning with 1 will give a sum of 78?

79. How many consecutive natural numbers beginning with 1 will give a sum of 120?

The formula $N = \dfrac{t^2 - t}{2}$ gives the number N of football games that must be scheduled in a league with t teams if each team is to play every other team once. Use this formula for Exercises 80 and 81.

80. How many teams are in a league that schedules 28 games in such a way that each team plays every other team once?

81. How many teams are in a league that schedules 45 games in such a way that each team plays every other team once?

The distance s that an object will fall (neglecting air resistance) in t seconds is given by $s = vt + 16t^2$, where v is the initial velocity of the object. Use this formula for Exercises 82 and 83.

82. An object is released from a plane at an altitude of 1600 ft. The initial velocity is 0 ft/s, and air resistance is neglected. How many seconds later will the object hit the ground?

83. An object is released from the top of a building 320 ft high. The initial velocity is 16 ft/s, and air resistance is neglected. How many seconds later will the object hit the ground?

The height h an object will attain (neglecting air resistance) in t seconds is given by $h = vt - 16t^2$, where v is the initial velocity of the object. Use this formula for Exercises 84 and 85.

84. A baseball player hits a "Baltimore chop," meaning the ball bounces off home plate after the batter hits it. The ball leaves home plate with an initial velocity of 64 ft/s. How many seconds after the ball hits home plate will the ball be 64 ft above the ground?

85. A golf ball is thrown onto a cement surface and rebounds straight up. The initial velocity of the rebound is 96 ft/s. How many seconds later will the golf ball return to the ground?

APPLYING THE CONCEPTS

Solve.

86. The length of a rectangle is 7 cm, and the width is 4 cm. If both the length and the width are increased by equal amounts, the area of the rectangle is increased by 42 cm². Find the length and width of the larger rectangle.

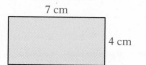

7 cm

4 cm

87. A rectangular piece of cardboard is 10 in. longer than it is wide. Squares 2 in. on a side are to be cut from each corner, and then the sides will be folded up to make an open box with a volume of 192 in³. Find the length and width of the piece of cardboard.

Solve.

88. $p^3 = 9p^2$

89. $(x + 3)(2x - 1) = (3 - x)(5 - 3x)$

90. Find $3n^2$ if $n(n + 5) = -4$.

91.
[W] Explain the error made in solving the equation at the right. Solve the equation correctly.

$(x + 2)(x - 3) = 6$
$x + 2 = 6 \quad x - 3 = 6$
$x = 4 \qquad x = 9$

92.
[W] In your own words, explain why it is possible to solve a quadratic equation using the Principle of Zero Products.

Projects in Mathematics

Inductive and Deductive Reasoning

Consider the following sums of odd positive integers.

$1 + 3 = 4 = 2^2$ Sum of first two odd numbers is 2^2.
$1 + 3 + 5 = 9 = 3^2$ Sum of first three odd numbers is 3^2.
$1 + 3 + 5 + 7 = 16 = 4^2$ Sum of first four odd numbers is 4^2.
$1 + 3 + 5 + 7 + 9 = 25 = 5^2$ Sum of first five odd numbers is 5^2.

1. Make a conjecture about the sum of $1 + 3 + 5 + 7 + 9 + 11 + 13 + 15$ without adding the numbers.

If the pattern continues, the sum of the first 8 odd numbers will be $8^2 = 64$. By adding the numbers, you can verify that this is correct. Inferring that the pattern established by a few cases is valid for all cases is an example of **inductive reasoning.** Just because a pattern seems to be true does not prove that it is true for all cases. The pattern of adding odd positive integers is, however, always true.

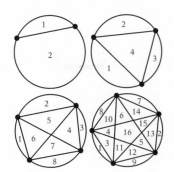

Now consider the problem of connecting the points on a circle with line segments (called chords). The number of different regions in the circle is then counted. A summary of the results in the figure at the left is given in the table.

Number of Points	Number of Regions
2	$2^{2-1} = 2^1 = 2$
3	$2^{3-1} = 2^2 = 4$
4	$2^{4-1} = 2^3 = 8$
5	$2^{5-1} = 2^4 = 16$

Fig. 1

2. Make a conjecture about the number of different regions when 6 points on a circle are all connected with chords.

Draw and label a circle to determine whether your conjecture is correct. You will find that there are 31 regions (or 30 if chords from opposite points meet at the center), not 32 as it appears from the table that it should be. This illustrates that inductive reasoning may lead to incorrect conclusions and that an inductive *proof* must be available to establish conjectures. There is such a proof, called **mathematical induction,** that you will study in future math courses.

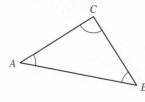

Now consider the true statement that the sum of the measures of the interior angles of a triangle is 180°. Figure 2 is a triangle. Therefore, the sum of the measures of the interior angles is 180°. This is an example of *deductive reasoning.*

Deductive reasoning uses a rule or statement of fact to reach a conclusion. The rule may be established fact, as in the case of the statement about triangles given above, or a conjectured rule that is apparently true.

Fig. 2

For instance, an instructor states that there will be a quiz every Monday. If today is Monday, you know that this instructor will give a quiz.

3. Use deductive reasoning to complete the following sentence. All even numbers are divisible by 2. Because 14,386 is an even number, ...

For each of the following, determine whether inductive or deductive reasoning is being used.

4. The tenth number in the list 1, 4, 9, 16, 25, ... is 100.

5. All quadrilaterals have four sides. A square is a quadrilateral. Therefore, a square has four sides.

6. Explain the difference between inductive and deductive reasoning.

Chapter Summary

Key Words   The *greatest common factor* (GCF) of two or more monomials is the product of the GCF of the coefficients and the common variable factors.

To *factor* a polynomial means to write the polynomial as a product of other polynomials.

To *factor* a trinomial of the form $ax^2 + bx + c$ means to express the trinomial as the product of two binomials.

A polynomial is *nonfactorable over the integers* if it does not factor using only integers. Such a polynomial is called a *prime polynomial*.

A product of a term and itself is a *perfect square*.

The product of the same three factors is a *perfect cube*.

An equation of the form $ax^2 + bx + c = 0$ is a *quadratic equation*.

A quadratic equation is in *standard form* when the polynomial is in descending order and equal to zero. The quadratic equation $ax^2 + bx + c = 0$ is in standard form.

Essential Rules

Factors of the Difference of Two Perfect Squares	$a^2 - b^2 = (a + b)(a - b)$
Factors of a Perfect-Square Trinomial	$a^2 + 2ab + b^2 = (a + b)^2$ $a^2 - 2ab + b^2 = (a - b)^2$
Factors of the Sum or Difference of Two Cubes	$a^3 + b^3 = (a + b)(a^2 - ab + b^2)$ $a^3 - b^3 = (a - b)(a^2 + ab + b^2)$

Principle of Zero Products

If the product of two factors is zero, then at least one of the factors must be zero.	If $a \cdot b = 0$, then $a = 0$ or $b = 0$.

General Factoring Strategy

1. Is there a common factor? If so, factor out the GCF.
2. If the polynomial is a binomial, is it the difference of two perfect squares, the sum of two cubes, or the difference of two cubes? If so, factor.
3. If the polynomial is a trinomial, is it a perfect-square trinomial or the product of two binomials? If so, factor.
4. If the polynomial has four terms, can it be factored by grouping? If so, factor.
5. Is each factor nonfactorable over the integers? If not, factor.

Chapter Review Exercises

. . . .

1. Factor: $5x^3 + 10x^2 + 35x$

2. Factor: $12a^2b + 3ab^2$

3. Factor: $14y^9 - 49y^6 + 7y^3$

4. Factor: $4x(x - 3) - 5(3 - x)$

5. Factor: $10x^2 + 25x + 4xy + 10y$

6. Factor: $21ax - 35bx - 10by + 6ay$

7. Factor: $b^2 - 13b + 30$

8. Factor: $c^2 + 8c + 12$

9. Factor: $y^2 + 5y - 36$

10. Factor: $3a^2 - 15a - 42$

11. Factor: $4x^3 - 20x^2 - 24x$

12. Factor: $n^4 - 2n^3 - 3n^2$

13. Factor $6x^2 - 29x + 28$ by using trial factors.

14. Factor $12y^2 + 16y - 3$ by using trial factors.

15. Factor $2x^2 - 5x + 6$ by using trial factors.

16. Factor $3x^2 - 17x + 10$ by grouping.

17. Factor $2a^2 - 19a - 60$ by grouping.

18. Factor $18a^2 - 3a - 10$ by grouping.

19. Factor: $x^2y^2 - 9$

20. Factor: $4x^2 + 12xy + 9y^2$

21. Factor: $x^{2n} - 12x^n + 36$

22. Factor: $64a^3 - 27b^3$

23. Factor: $15x^4 + x^2 - 6$

24. Factor: $21x^4y^4 + 23x^2y^2 + 6$

25. Factor: $3a^6 - 15a^4 - 18a^2$

26. Solve: $4x^2 + 27x = 7$

27. Solve: $(x + 1)(x - 5) = 16$

28. The length of a hockey field is 20 yd less than twice the width of the hockey field. The area of the hockey field is 6000 yd². Find the length and width of the hockey field.

29. A rectangular photograph has dimensions 15 in. by 12 in. A picture frame around the photograph increases the total area to 270 in². What is the width of the frame?

30. The length of each side of a square garden plot is extended 4 ft. The area of the resulting square is 576 ft². Find the length of a side of the original garden plot.

Cumulative Review Exercises

. . . .

1. Subtract: $-2 - (-3) - 5 - (-11)$

2. Simplify: $(3 - 7)^2 \div (-2) - 3(-4)$

3. Evaluate $-2a^2 \div (2b) - c$ when $a = -4$, $b = 2$, and $c = -1$.

4. Multiply: $-\dfrac{3}{4}(-20x^2)$

5. Simplify: $-2[4x - 2(3 - 2x) - 8x]$

6. Solve: $-\dfrac{5}{7}x = -\dfrac{10}{21}$

7. Solve: $3x - 2 = 12 - 5x$

8. Solve:
$-2 + 4[3x - 2(4 - x) - 3] = 4x + 2$

9. 120% of what number is 54?

10. Given $f(x) = -x^2 + 3x - 1$, find $f(2)$.

11. Graph $y = \dfrac{1}{4}x + 3$.

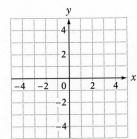

12. Graph $5x + 3y = 15$.

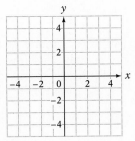

13. Find the equation of the line that contains the point $(-3, 4)$ and has slope $\dfrac{2}{3}$.

14. Solve by substitution: $8x - y = 2$
$y = 5x + 1$

15. Solve by the addition method:
$$5x + 2y = -9$$
$$12x - 7y = 2$$

16. Simplify: $(-3a^3b^2)^2$

17. Multiply: $(x + 2)(x^2 - 5x + 4)$

18. Divide: $(8x^2 + 4x - 3) \div (2x - 3)$

19. Simplify: $(x^{-4}y^3)^2$

20. Factor: $3a - 3b - ax + bx$

21. Factor: $15xy^2 - 20xy^4$

22. Factor: $x^2 - 5xy - 14y^2$

23. Solve: $3x^2 + 19x - 14 = 0$

24. Solve: $6x^2 + 60 = 39x$

25. A triangle has a 31° angle and a right angle. Find the measure of the third angle.

26. A rectangular flower garden has a perimeter of 86 ft. The length of the garden is 28 ft. What is the width of the garden?

27. A board 10 ft long is cut into two pieces. Four times the length of the shorter piece is 2 ft less than three times the length of the longer piece. Find the length of each piece.

28. An investment of $4000 was made at an annual simple interest rate of 8%. How much more money was invested at an annual simple interest rate of 11% if the total interest earned in one year was $1035?

29. A family drove to a resort at an average speed of 42 mph and later returned over the same road at an average speed of 56 mph. Find the distance to the resort if the total driving time was 7 h.

30. The length of the base of a triangle is three times the height. The area of the triangle is 24 in². Find the length of the base of the triangle.

Chapter

Rational Expressions

Objectives

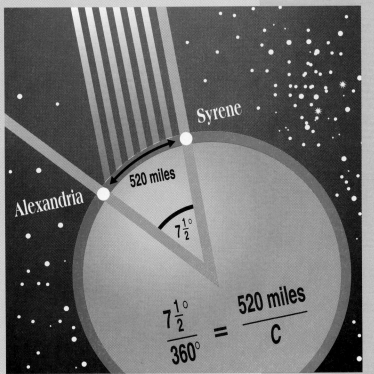

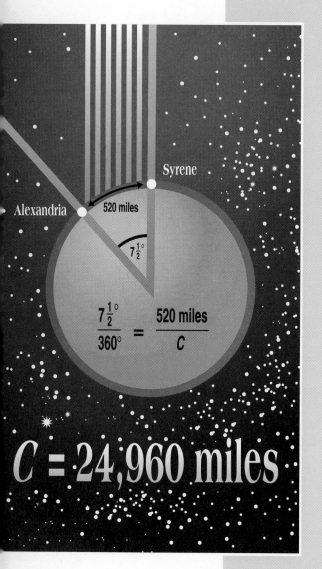

$$\frac{7\frac{1}{2}°}{360°} = \frac{520 \text{ miles}}{C}$$

$C = 24,960 \text{ miles}$

Measurement of the Circumference of the Earth

Distances on the earth, the circumference of the earth, and the distance to the moon and stars are known to great precision. Eratosthenes, the fifth librarian of Alexandria (230 B.C.), laid the foundation of scientific geography with his determination of the circumference of the earth.

Eratosthenes was familiar with certain astronomical data that enabled him to calculate the circumference of the earth by using a proportion statement.

Eratosthenes knew that on a mid-summer day, the sun was directly overhead at Syrene, as shown in the diagram. At the same time, at Alexandria the sun was at a $7\frac{1}{2}°$ angle from the zenith. The distance from Syrene to Alexandria was 5000 stadia (about 520 mi).

Knowing that the ratio of the $7\frac{1}{2}°$ angle to one revolution (360°) is equal to the ratio of the arc length (520 mi) to the circumference, Eratosthenes was able to write and solve a proportion.

This result, calculated over 2000 years ago, is very close to the accepted value of 24,800 miles.

8.1 Multiplication and Division of Rational Expressions

Objective A   *To simplify a rational expression*

A fraction in which the numerator or denominator is a polynomial is called a **rational expression.** Examples of rational expressions are shown at the right.

$$\frac{5}{z}, \quad \frac{x^2 + 1}{2x - 1}, \quad \frac{y^2 + y - 1}{4y^2 + 1}$$

Care must be exercised with a rational expression to ensure that when the variables are replaced with numbers, the resulting denominator is not zero.

Consider the rational expression at the right. The value of x cannot be 3, because the denominator would then be zero.

$$\frac{4x^2 - 9}{2x - 6}$$

$$\frac{4(3)^2 - 9}{2(3) - 6} = \frac{27}{0} \quad \text{Not a real number}$$

A rational expression is in simplest form when the numerator and denominator have no common factors. The Multiplication Property of One is used to write a rational expression in simplest form.

➡ Simplify: $\dfrac{x^2 - 4}{x^2 - 2x - 8}$

$$\frac{x^2 - 4}{x^2 - 2x - 8} = \frac{(x - 2)(x + 2)}{(x - 4)(x + 2)}$$

- Factor the numerator and denominator.

$$= \frac{x - 2}{x - 4} \cdot \boxed{\frac{x + 2}{x + 2}} = \frac{x - 2}{x - 4} \cdot 1$$

$$= \frac{x - 2}{x - 4}, \quad x \neq -2, 4$$

- The restrictions $x \neq -2, 4$ are necessary to prevent division by zero.

This simplification is usually shown with slashes through the common factors. The last simplification would be shown as follows:

$$\frac{x^2 - 4}{x^2 - 2x - 8} = \frac{(x - 2)\overset{1}{\cancel{(x + 2)}}}{(x - 4)\underset{1}{\cancel{(x + 2)}}}$$

- Factor the numerator and denominator.

$$= \frac{x - 2}{x - 4}, \quad x \neq -2, 4$$

- Divide by the common factors. The restrictions $x \neq -2, 4$ are necessary to prevent division by zero.

➡ Simplify: $\dfrac{10 + 3x - x^2}{x^2 - 4x - 5}$

$$\frac{10 + 3x - x^2}{x^2 - 4x - 5} = \frac{(5 - x)(2 + x)}{(x - 5)(x + 1)}$$

- Factor the numerator and denominator.

$$= \frac{\overset{-1}{\cancel{(5 - x)}}(2 + x)}{\underset{1}{\cancel{(x - 5)}}(x + 1)}$$

- Recall that $5 - x = -(x - 5)$. Therefore, $\dfrac{5 - x}{x - 5} = \dfrac{-(x - 5)}{x - 5} = \dfrac{-1}{1} = -1$.

$$= -\frac{x + 2}{x + 1}, \quad x \neq -1, 5$$

For the remaining examples, we will omit the restrictions on the variables that prevent division by zero and assume the values of the variables are such that division by zero is not possible.

Example 1

Simplify: $\dfrac{4x^3y^4}{6x^4y}$

Solution

$\dfrac{4x^3y^4}{6x^4y} = \dfrac{2y^3}{3x}$ • Use the rules of exponents.

You Try It 1

Simplify: $\dfrac{6x^5y}{12x^2y^3}$

Your solution

Example 2

Simplify: $\dfrac{9 - x^2}{x^2 + x - 12}$

Solution

$\dfrac{9 - x^2}{x^2 + x - 12} = \dfrac{\overset{-1}{\cancel{(3 - x)}}(3 + x)}{\underset{1}{\cancel{(x - 3)}}(x + 4)} = -\dfrac{x + 3}{x + 4}$

You Try It 2

Simplify: $\dfrac{x^2 + 2x - 24}{16 - x^2}$

Your solution

Example 3

Simplify: $\dfrac{x^2 + 2x - 15}{x^2 - 7x + 12}$

Solution

$\dfrac{x^2 + 2x - 15}{x^2 - 7x + 12} = \dfrac{(x + 5)\overset{1}{\cancel{(x - 3)}}}{\underset{1}{\cancel{(x - 3)}}(x - 4)} = \dfrac{x + 5}{x - 4}$

You Try It 3

Simplify: $\dfrac{x^2 + 4x - 12}{x^2 - 3x + 2}$

Your solution

Solutions on p. A37

Objective B *To multiply rational expressions* ....

The product of two fractions is a fraction whose numerator is the product of the numerators of the two fractions and whose denominator is the product of the denominators of the two fractions.

If $\dfrac{a}{b}$ and $\dfrac{c}{d}$ are rational numbers, then $\dfrac{a}{b} \cdot \dfrac{c}{d} = \dfrac{ac}{bd}$.

$\dfrac{2}{3} \cdot \dfrac{4}{5} = \dfrac{8}{15}$ $\qquad\qquad$ $\dfrac{3x}{y} \cdot \dfrac{2}{z} = \dfrac{6x}{yz}$ $\qquad\qquad$ $\dfrac{x + 2}{x} \cdot \dfrac{3}{x - 2} = \dfrac{3x + 6}{x^2 - 2x}$

➡ Multiply: $\dfrac{x^2 + 3x}{x^2 - 3x - 4} \cdot \dfrac{x^2 - 5x + 4}{x^2 + 2x - 3}$

$$\dfrac{x^2 + 3x}{x^2 - 3x - 4} \cdot \dfrac{x^2 - 5x + 4}{x^2 + 2x - 3}$$

$$= \dfrac{x(x + 3)}{(x - 4)(x + 1)} \cdot \dfrac{(x - 4)(x - 1)}{(x + 3)(x - 1)}$$

- Factor the numerator and denominator of each fraction.

$$= \dfrac{x(\overset{1}{\cancel{x + 3}})(\overset{1}{\cancel{x - 4}})(\overset{1}{\cancel{x - 1}})}{(\underset{1}{\cancel{x - 4}})(x + 1)(\underset{1}{\cancel{x + 3}})(\underset{1}{\cancel{x - 1}})}$$

- Multiply the fractions. Then divide the numerator and denominator by their common factors.

$$= \dfrac{x}{x + 1}$$

- Write the answer in simplest form.

Example 4

Multiply: $\dfrac{10x^2 - 15x}{12x - 8} \cdot \dfrac{3x - 2}{20x - 25}$

Solution

$$\dfrac{10x^2 - 15x}{12x - 8} \cdot \dfrac{3x - 2}{20x - 25}$$

$$= \dfrac{5x(2x - 3)}{4(3x - 2)} \cdot \dfrac{(3x - 2)}{5(4x - 5)}$$

$$= \dfrac{\overset{1}{\cancel{5}}x(2x - 3)(\overset{1}{\cancel{3x - 2}})}{4(\underset{1}{\cancel{3x - 2}})\underset{1}{\cancel{5}}(4x - 5)} = \dfrac{x(2x - 3)}{4(4x - 5)}$$

You Try It 4

Multiply: $\dfrac{12x^2 + 3x}{10x - 15} \cdot \dfrac{8x - 12}{9x + 18}$

Your solution

Example 5

Multiply: $\dfrac{x^2 + x - 6}{x^2 + 7x + 12} \cdot \dfrac{x^2 + 3x - 4}{4 - x^2}$

Solution

$$\dfrac{x^2 + x - 6}{x^2 + 7x + 12} \cdot \dfrac{x^2 + 3x - 4}{4 - x^2}$$

$$= \dfrac{(x + 3)(x - 2)}{(x + 3)(x + 4)} \cdot \dfrac{(x + 4)(x - 1)}{(2 - x)(2 + x)}$$

$$= \dfrac{(\overset{1}{\cancel{x + 3}})(\overset{-1}{\cancel{x - 2}})(\overset{1}{\cancel{x + 4}})(x - 1)}{(\underset{1}{\cancel{x + 3}})(\underset{1}{\cancel{x + 4}})(\underset{1}{\cancel{2 - x}})(2 + x)} = -\dfrac{x - 1}{x + 2}$$

You Try It 5

Multiply: $\dfrac{x^2 + 2x - 15}{9 - x^2} \cdot \dfrac{x^2 - 3x - 18}{x^2 - 7x + 6}$

Your solution

Solutions on p. A37

Content and Format © 1996 HMCo.

Objective C *To divide rational expressions* ....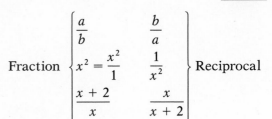

The **reciprocal** of a fraction is a fraction with the numerator and denominator interchanged.

$$\text{Fraction} \left\{ \begin{array}{ll} \dfrac{a}{b} & \dfrac{b}{a} \\ x^2 = \dfrac{x^2}{1} & \dfrac{1}{x^2} \\ \dfrac{x+2}{x} & \dfrac{x}{x+2} \end{array} \right\} \text{Reciprocal}$$

> To divide two fractions, multiply by the reciprocal of the divisor.
>
> $$\frac{a}{b} \div \frac{c}{d} = \frac{a}{b} \cdot \frac{d}{c} = \frac{ad}{bc}$$

$$\frac{4}{x} \div \frac{y}{5} = \frac{4}{x} \cdot \frac{5}{y} = \frac{20}{xy} \qquad\qquad \frac{x+4}{x} \div \frac{x-2}{4} = \frac{x+4}{x} \cdot \frac{4}{x-2} = \frac{4x+16}{x^2-2x}$$

The basis for the division rule is shown at the right.

$$\frac{a}{b} \div \frac{c}{d} = \frac{\dfrac{a}{b}}{\dfrac{c}{d}} \cdot \frac{\dfrac{d}{c}}{\dfrac{d}{c}} = \frac{\dfrac{a}{b} \cdot \dfrac{d}{c}}{\dfrac{c}{d} \cdot \dfrac{d}{c}} = \frac{\dfrac{a}{b} \cdot \dfrac{d}{c}}{1} = \frac{a}{b} \cdot \frac{d}{c}$$

Example 6

Divide: $\dfrac{xy^2 - 3x^2y}{z^2} \div \dfrac{6x^2 - 2xy}{z^3}$

Solution

$$\frac{xy^2 - 3x^2y}{z^2} \div \frac{6x^2 - 2xy}{z^3}$$

$$= \frac{xy^2 - 3x^2y}{z^2} \cdot \frac{z^3}{6x^2 - 2xy}$$

$$= \frac{xy(y \overset{-1}{\overline{- 3x}}) \cdot z^3}{z^2 \cdot 2x(3x \overline{- y})} = -\frac{yz}{2}$$

You Try It 6

Divide: $\dfrac{a^2}{4bc^2 - 2b^2c} \div \dfrac{a}{6bc - 3b^2}$

Your solution

Example 7

Divide: $\dfrac{2x^2 + 5x + 2}{2x^2 + 3x - 2} \div \dfrac{3x^2 + 13x + 4}{2x^2 + 7x - 4}$

Solution

$$\frac{2x^2 + 5x + 2}{2x^2 + 3x - 2} \div \frac{3x^2 + 13x + 4}{2x^2 + 7x - 4}$$

$$= \frac{2x^2 + 5x + 2}{2x^2 + 3x - 2} \cdot \frac{2x^2 + 7x - 4}{3x^2 + 13x + 4}$$

$$= \frac{(2x+1)(x+2) \cdot (2x-1)(x+4)}{(2x-1)(x+2) \cdot (3x+1)(x+4)} = \frac{2x+1}{3x+1}$$

You Try It 7

Divide: $\dfrac{3x^2 + 26x + 16}{3x^2 - 7x - 6} \div \dfrac{2x^2 + 9x - 5}{x^2 + 2x - 15}$

Your solution

Solutions on p. A37

8.1 Exercises

· ·

Objective A

Simplify.

1. $\dfrac{9x^3}{12x^4}$

2. $\dfrac{16x^2y}{24xy^3}$

3. $\dfrac{(x+3)^2}{(x+3)^3}$

4. $\dfrac{(2x-1)^5}{(2x-1)^4}$

5. $\dfrac{3n-4}{4-3n}$

6. $\dfrac{5-2x}{2x-5}$

7. $\dfrac{6y(y+2)}{9y^2(y+2)}$

8. $\dfrac{12x^2(3-x)}{18x(3-x)}$

9. $\dfrac{6x(x-5)}{8x^2(5-x)}$

10. $\dfrac{14x^3(7-3x)}{21x(3x-7)}$

11. $\dfrac{a^2+4a}{ab+4b}$

12. $\dfrac{x^2-3x}{2x-6}$

13. $\dfrac{4-6x}{3x^2-2x}$

14. $\dfrac{5xy-3y}{9-15x}$

15. $\dfrac{y^2-3y+2}{y^2-4y+3}$

16. $\dfrac{x^2+5x+6}{x^2+8x+15}$

17. $\dfrac{x^2+3x-10}{x^2+2x-8}$

18. $\dfrac{a^2+7a-8}{a^2+6a-7}$

19. $\dfrac{x^2+x-12}{x^2-6x+9}$

20. $\dfrac{x^2+8x+16}{x^2-2x-24}$

21. $\dfrac{x^2-3x-10}{25-x^2}$

22. $\dfrac{4-y^2}{y^2-3y-10}$

23. $\dfrac{2x^3+2x^2-4x}{x^3+2x^2-3x}$

24. $\dfrac{3x^3-12x}{6x^3-24x^2+24x}$

25. $\dfrac{6x^2-7x+2}{6x^2+5x-6}$

26. $\dfrac{2n^2-9n+4}{2n^2-5n-12}$

27. $\dfrac{x^2+3x-28}{24-2x-x^2}$

Objective B

Multiply.

28. $\dfrac{8x^2}{9y^3} \cdot \dfrac{3y^2}{4x^3}$

29. $\dfrac{14a^2b^3}{15x^5y^2} \cdot \dfrac{25x^3y}{16ab}$

30. $\dfrac{12x^3y^4}{7a^2b^3} \cdot \dfrac{14a^3b^4}{9x^2y^2}$

31. $\dfrac{18a^4b^2}{25x^2y^3} \cdot \dfrac{50x^5y^6}{27a^6b^2}$

32. $\dfrac{3x - 6}{5x - 20} \cdot \dfrac{10x - 40}{27x - 54}$

33. $\dfrac{8x - 12}{14x + 7} \cdot \dfrac{42x + 21}{32x - 48}$

34. $\dfrac{3x^2 + 2x}{2xy - 3y} \cdot \dfrac{2xy^3 - 3y^3}{3x^3 + 2x^2}$

35. $\dfrac{4a^2x - 3a^2}{2by + 5b} \cdot \dfrac{2b^3y + 5b^3}{4ax - 3a}$

36. $\dfrac{x^2 + 5x + 4}{x^3y^2} \cdot \dfrac{x^2y^3}{x^2 + 2x + 1}$

37. $\dfrac{x^2 + x - 2}{xy^2} \cdot \dfrac{x^3y}{x^2 + 5x + 6}$

38. $\dfrac{x^4y^2}{x^2 + 3x - 28} \cdot \dfrac{x^2 - 49}{xy^4}$

39. $\dfrac{x^5y^3}{x^2 + 13x + 30} \cdot \dfrac{x^2 + 2x - 3}{x^7y^2}$

40. $\dfrac{2x^2 - 5x}{2xy + y} \cdot \dfrac{2xy^2 + y^2}{5x^2 - 2x^3}$

41. $\dfrac{3a^3 + 4a^2}{5ab - 3b} \cdot \dfrac{3b^3 - 5ab^3}{3a^2 + 4a}$

42. $\dfrac{x^2 - 2x - 24}{x^2 - 5x - 6} \cdot \dfrac{x^2 + 5x + 6}{x^2 + 6x + 8}$

43. $\dfrac{x^2 - 8x + 7}{x^2 + 3x - 4} \cdot \dfrac{x^2 + 3x - 10}{x^2 - 9x + 14}$

44. $\dfrac{x^2 + 2x - 35}{x^2 + 4x - 21} \cdot \dfrac{x^2 + 3x - 18}{x^2 + 9x + 18}$

45. $\dfrac{y^2 + y - 20}{y^2 + 2y - 15} \cdot \dfrac{y^2 + 4y - 21}{y^2 + 3y - 28}$

46. $\dfrac{x^2 - 3x - 4}{x^2 + 6x + 5} \cdot \dfrac{x^2 + 5x + 6}{8 + 2x - x^2}$

47. $\dfrac{25 - n^2}{n^2 - 2n - 35} \cdot \dfrac{n^2 - 8n - 20}{n^2 - 3n - 10}$

Multiply.

48. $\dfrac{12x^2 - 6x}{x^2 + 6x + 5} \cdot \dfrac{2x^4 + 10x^3}{4x^2 - 1}$

49. $\dfrac{8x^3 + 4x^2}{x^2 - 3x + 2} \cdot \dfrac{x^2 - 4}{16x^2 + 8x}$

50. $\dfrac{16 + 6x - x^2}{x^2 - 10x - 24} \cdot \dfrac{x^2 - 6x - 27}{x^2 - 17x + 72}$

51. $\dfrac{x^2 - 11x + 28}{x^2 - 13x + 42} \cdot \dfrac{x^2 + 7x + 10}{20 - x - x^2}$

52. $\dfrac{2x^2 + 5x + 2}{2x^2 + 7x + 3} \cdot \dfrac{x^2 - 7x - 30}{x^2 - 6x - 40}$

53. $\dfrac{x^2 - 4x - 32}{x^2 - 8x - 48} \cdot \dfrac{3x^2 + 17x + 10}{3x^2 - 22x - 16}$

54. $\dfrac{2x^2 + x - 3}{2x^2 - x - 6} \cdot \dfrac{2x^2 - 9x + 10}{2x^2 - 3x + 1}$

55. $\dfrac{3y^2 + 14y + 8}{2y^2 + 7y - 4} \cdot \dfrac{2y^2 + 9y - 5}{3y^2 + 16y + 5}$

Objective C

Divide.

56. $\dfrac{4x^2y^3}{15a^2b^3} \div \dfrac{6xy}{5a^3b^5}$

57. $\dfrac{9x^3y^4}{16a^4b^2} \div \dfrac{45x^4y^2}{14a^7b}$

58. $\dfrac{6x - 12}{8x + 32} \div \dfrac{18x - 36}{10x + 40}$

59. $\dfrac{28x + 14}{45x - 30} \div \dfrac{14x + 7}{30x - 20}$

60. $\dfrac{6x^3 + 7x^2}{12x - 3} \div \dfrac{6x^2 + 7x}{36x - 9}$

61. $\dfrac{5a^2y + 3a^2}{2x^3 + 5x^2} \div \dfrac{10ay + 6a}{6x^3 + 15x^2}$

62. $\dfrac{x^2 + 4x + 3}{x^2y} \div \dfrac{x^2 + 2x + 1}{xy^2}$

63. $\dfrac{x^3y^2}{x^2 - 3x - 10} \div \dfrac{xy^4}{x^2 - x - 20}$

64. $\dfrac{x^2 - 49}{x^4y^3} \div \dfrac{x^2 - 14x + 49}{x^4y^3}$

65. $\dfrac{x^2y^5}{x^2 - 11x + 30} \div \dfrac{xy^6}{x^2 - 7x + 10}$

66. $\dfrac{4ax - 8a}{c^2} \div \dfrac{2y - xy}{c^3}$

67. $\dfrac{3x^2y - 9xy}{a^2b} \div \dfrac{3x^2 - x^3}{ab^2}$

Divide.

68. $\dfrac{x^2 - 5x + 6}{x^2 - 9x + 18} \div \dfrac{x^2 - 6x + 8}{x^2 - 9x + 20}$

69. $\dfrac{x^2 + 3x - 40}{x^2 + 2x - 35} \div \dfrac{x^2 + 2x - 48}{x^2 + 3x - 18}$

70. $\dfrac{x^2 + 2x - 15}{x^2 - 4x - 45} \div \dfrac{x^2 + x - 12}{x^2 - 5x - 36}$

71. $\dfrac{y^2 - y - 56}{y^2 + 8y + 7} \div \dfrac{y^2 - 13y + 40}{y^2 - 4y - 5}$

72. $\dfrac{8 + 2x - x^2}{x^2 + 7x + 10} \div \dfrac{x^2 - 11x + 28}{x^2 - x - 42}$

73. $\dfrac{x^2 - x - 2}{x^2 - 7x + 10} \div \dfrac{x^2 - 3x - 4}{40 - 3x - x^2}$

74. $\dfrac{2x^2 - 3x - 20}{2x^2 - 7x - 30} \div \dfrac{2x^2 - 5x - 12}{4x^2 + 12x + 9}$

75. $\dfrac{6n^2 + 13n + 6}{4n^2 - 9} \div \dfrac{6n^2 + n - 2}{4n^2 - 1}$

76. $\dfrac{9x^2 - 16}{6x^2 - 11x + 4} \div \dfrac{6x^2 + 11x + 4}{8x^2 + 10x + 3}$

77. $\dfrac{15 - 14x - 8x^2}{4x^2 + 4x - 15} \div \dfrac{4x^2 + 13x - 12}{3x^2 + 13x + 4}$

APPLYING THE CONCEPTS

78. Given the expression $\dfrac{9}{x^2 + 1}$, choose some values of x and evaluate the expression for those values. Is it possible to choose a value of x for which the value of the expression is greater than 10? If so, what is that value of x? If not, explain why it is not possible.

79. Given the expression $\dfrac{1}{y - 3}$, choose some values of y and evaluate the expression for those values. Is it possible to choose a value of y for which the value of the expression is greater than 10,000,000? If so, what is that value of y? If not, explain why it is not possible.

For what values of x is the algebraic fraction undefined?

80. $\dfrac{x}{(x - 2)(x + 5)}$

81. $\dfrac{7}{x^2 - 25}$

82. $\dfrac{3x - 8}{3x^2 - 10x - 8}$

Simplify.

83. $\dfrac{xy}{3} \cdot \dfrac{x}{y^2} \div \dfrac{x}{4}$

84. $\left(\dfrac{y}{3}\right) \div \left(\dfrac{y}{2} \cdot \dfrac{y}{4}\right)$

85. $\left(\dfrac{x - 4}{y^2}\right)^3 \cdot \left(\dfrac{y}{4 - x}\right)^3$

86. $\dfrac{x - 2}{x + 5} \div \dfrac{x - 3}{x + 5} \cdot \dfrac{x - 3}{x - 2}$

Addition and Subtraction of Rational Expressions

Objective A   *To find the least common multiple (LCM) of two or more polynomials* ....

The **least common multiple (LCM)** of two or more numbers is the smallest number that contains the prime factorization of each number.

CONSIDER THIS
The LCM of 12 and 18 is 36 because 36 is the smallest number that both 12 and 18 divide evenly into.

The LCM of 12 and 18 is 36 because 36 contains the prime factors of 12 and the prime factors of 18.

$12 = 2 \cdot 2 \cdot 3$
$18 = 2 \cdot 3 \cdot 3$

Factors of 12

$$\text{LCM} = 36 = \overbrace{2 \cdot \underbrace{2 \cdot 3 \cdot 3}}$$

Factors of 18

The least common multiple of two or more polynomials is the polynomial of least degree that contains the factors of each polynomial.

To find the LCM of two or more polynomials, first factor each polynomial completely. The LCM is the product of each factor the greatest number of times it occurs in any one factorization.

➡ Find the LCM of $4x^2 + 4x$ and $x^2 + 2x + 1$.

CONSIDER THIS
The LCM must contain the factors of each polynomial. As shown with the braces at the right, the LCM contains the factors of $4x^2 + 4x$ and the factors of $x^2 + 2x + 1$.

The LCM of the polynomials is the product of the LCM of the numerical coefficients and each variable factor the greatest number of times it occurs in any one factorization.

$4x^2 + 4x = 4x(x + 1) = 2 \cdot 2 \cdot x(x + 1)$
$x^2 + 2x + 1 = (x + 1)(x + 1)$

Factors of $4x^2 + 4x$

$$\text{LCM} = 2 \cdot 2 \cdot \underbrace{x(x + 1)(x + 1)} = 4x(x + 1)(x + 1)$$

Factors of $x^2 + 2x + 1$

Example 1
Find the LCM of $4x^2y$ and $6xy^2$.

Solution
$4x^2y = 2 \cdot 2 \cdot x \cdot x \cdot y$
$6xy^2 = 2 \cdot 3 \cdot x \cdot y \cdot y$
$\text{LCM} = 2 \cdot 2 \cdot 3 \cdot x \cdot x \cdot y \cdot y = 12x^2y^2$

You Try It 1
Find the LCM of $8uv^2$ and $12uw$.

Your solution

Example 2
Find the LCM of $x^2 - x - 6$ and $9 - x^2$.

Solution
$x^2 - x - 6 = (x - 3)(x + 2)$
$9 - x^2 = -(x^2 - 9) = -(x + 3)(x - 3)$
$\text{LCM} = (x - 3)(x + 2)(x + 3)$

You Try It 2
Find the LCM of $m^2 - 6m + 9$ and $m^2 - 2m - 3$.

Your solution

Solutions on p. A37

Objective B   *To express two fractions in terms of the LCM of their denominators* ....

When adding and subtracting fractions, it is frequently necessary to express two or more fractions in terms of a common denominator. This common denominator is the LCM of the denominators of the fractions.

CONSIDER THIS

$\dfrac{3(x-2)}{3(x-2)} = 1$ and $\dfrac{2x}{2x} = 1$. We are multiplying each fraction by 1, so we are not changing the value of either fraction.

➡ Write the fractions $\dfrac{x+1}{4x^2}$ and $\dfrac{x-3}{6x^2 - 12x}$ in terms of the LCM of the denominators.

Find the LCM of the denominators.

$6x^2 - 12x = 6x(x-2)$
The LCM is $12x^2(x-2)$.

For each fraction, multiply the numerator and denominator by the factors whose product with the denominator is the LCM.

$$\dfrac{x+1}{4x^2} = \dfrac{x+1}{4x^2} \cdot \dfrac{3(x-2)}{3(x-2)} = \dfrac{3x^2 - 3x - 6}{12x^2(x-2)} \;\longleftarrow$$

$$\dfrac{x-3}{6x^2 - 12x} = \dfrac{x-3}{6x(x-2)} \cdot \dfrac{2x}{2x} = \dfrac{2x^2 - 6x}{12x^2(x-2)} \;\longleftarrow$$

LCM

Example 3

Write the fractions $\dfrac{x+2}{3x^2}$ and $\dfrac{x-1}{8xy}$ in terms of the LCM of the denominators.

Solution
The LCM is $24x^2y$.

$$\dfrac{x+2}{3x^2} = \dfrac{x+2}{3x^2} \cdot \dfrac{8y}{8y} = \dfrac{8xy + 16y}{24x^2y}$$

$$\dfrac{x-1}{8xy} = \dfrac{x-1}{8xy} \cdot \dfrac{3x}{3x} = \dfrac{3x^2 - 3x}{24x^2y}$$

You Try It 3

Write the fractions $\dfrac{x-3}{4xy^2}$ and $\dfrac{2x+1}{9y^2z}$ in terms of the LCM of the denominators.

Your solution

Example 4

Write the fractions $\dfrac{2x-1}{2x - x^2}$ and $\dfrac{x}{x^2 + x - 6}$ in terms of the LCM of the denominators.

Solution

$$\dfrac{2x-1}{2x - x^2} = \dfrac{2x-1}{-(x^2 - 2x)} = -\dfrac{2x-1}{x^2 - 2x}$$

The LCM is $x(x-2)(x+3)$.

$$\dfrac{2x-1}{2x - x^2} = -\dfrac{2x-1}{x(x-2)} \cdot \dfrac{x+3}{x+3} = -\dfrac{2x^2 + 5x - 3}{x(x-2)(x+3)}$$

$$\dfrac{x}{x^2 + x - 6} = \dfrac{x}{(x-2)(x+3)} \cdot \dfrac{x}{x} = \dfrac{x^2}{x(x-2)(x+3)}$$

You Try It 4

Write the fractions $\dfrac{x+4}{x^2 - 3x - 10}$ and $\dfrac{2x}{25 - x^2}$ in terms of the LCM of the denominators.

Your solution

Solutions on p. A37

Objective C **To add or subtract rational expressions with the same denominator** ....

When adding rational expressions in which the denominators are the same, add the numerators. The denominator of the sum is the common denominator.

$$\frac{a}{b} + \frac{c}{b} = \frac{a + c}{b}$$

$$\frac{5x}{18} + \frac{7x}{18} = \frac{5x + 7x}{18} = \frac{12x}{18} = \frac{2x}{3}$$

$$\frac{x}{x^2 - 1} + \frac{1}{x^2 - 1} = \frac{x + 1}{x^2 - 1} = \frac{\overset{1}{\cancel{(x + 1)}}}{(x - 1)\underset{1}{\cancel{(x + 1)}}} = \frac{1}{x - 1}$$

Note that the sum is written in simplest form.

When subtracting rational expressions with like denominators, subtract the numerators. The denominator of the difference is the common denominator. Write the answer in simplest form.

CONSIDER THIS

Be careful with signs when subtracting algebraic fractions. Note that we must subtract the *entire* numerator $2x + 3$.
$(3x - 1) - (2x + 3) =$
$3x - 1 - 2x - 3.$

$$\frac{2x}{x - 2} - \frac{4}{x - 2} = \frac{2x - 4}{x - 2} = \frac{2\overset{1}{\cancel{(x - 2)}}}{\underset{1}{\cancel{x - 2}}} = 2$$

$$\frac{3x - 1}{x^2 - 5x + 4} - \frac{2x + 3}{x^2 - 5x + 4} = \frac{(3x - 1) - (2x + 3)}{x^2 - 5x + 4} = \frac{3x - 1 - 2x - 3}{x^2 - 5x + 4}$$

$$= \frac{x - 4}{x^2 - 5x + 4} = \frac{\overset{1}{\cancel{(x - 4)}}}{\underset{1}{\cancel{(x - 4)}}(x - 1)} = \frac{1}{x - 1}$$

Example 5

Add: $\dfrac{7}{x^2} + \dfrac{9}{x^2}$

Solution

$$\frac{7}{x^2} + \frac{9}{x^2} = \frac{7 + 9}{x^2} = \frac{16}{x^2}$$

You Try It 5

Add: $\dfrac{3}{xy} + \dfrac{12}{xy}$

Your solution

Example 6

Subtract: $\dfrac{3x^2}{x^2 - 1} - \dfrac{x + 4}{x^2 - 1}$

Solution

$$\frac{3x^2}{x^2 - 1} - \frac{x + 4}{x^2 - 1} = \frac{3x^2 - (x + 4)}{x^2 - 1}$$

$$= \frac{3x^2 - x - 4}{x^2 - 1}$$

$$= \frac{(3x - 4)\overset{1}{\cancel{(x + 1)}}}{(x - 1)\underset{1}{\cancel{(x + 1)}}} = \frac{3x - 4}{x - 1}$$

You Try It 6

Subtract: $\dfrac{2x^2}{x^2 - x - 12} - \dfrac{7x + 4}{x^2 - x - 12}$

Your solution

Solutions on p. A38

Example 7

Simplify:

$$\frac{2x^2 + 5}{x^2 + 2x - 3} - \frac{x^2 - 3x}{x^2 + 2x - 3} + \frac{x - 2}{x^2 + 2x - 3}$$

Solution

$$\frac{2x^2 + 5}{x^2 + 2x - 3} - \frac{x^2 - 3x}{x^2 + 2x - 3} + \frac{x - 2}{x^2 + 2x - 3}$$

$$= \frac{(2x^2 + 5) - (x^2 - 3x) + (x - 2)}{x^2 + 2x - 3}$$

$$= \frac{2x^2 + 5 - x^2 + 3x + x - 2}{x^2 + 2x - 3}$$

$$= \frac{x^2 + 4x + 3}{x^2 + 2x - 3} = \frac{\overset{1}{\cancel{(x + 3)}}(x + 1)}{\underset{1}{\cancel{(x + 3)}}(x - 1)} = \frac{x + 1}{x - 1}$$

You Try It 7

Simplify:

$$\frac{x^2 - 1}{x^2 - 8x + 12} - \frac{2x + 1}{x^2 - 8x + 12} + \frac{x}{x^2 - 8x + 12}$$

Your solution

Solution on p. A38

Objective D *To add or subtract rational expressions with*
different denominators ....

Before two fractions with unlike denominators can be added or subtracted, each fraction must be expressed in terms of a common denominator. This common denominator is the LCM of the denominators of the fractions.

➡ Add: $\dfrac{x - 3}{x^2 - 2x} + \dfrac{6}{x^2 - 4}$

Find the LCM of the denominators.   The LCM is $x(x - 2)(x + 2)$.

$$\frac{x - 3}{x^2 - 2x} + \frac{6}{x^2 - 4} = \frac{x - 3}{x(x - 2)} \cdot \frac{x + 2}{x + 2} + \frac{6}{(x - 2)(x + 2)} \cdot \frac{x}{x}$$

• Write each fraction in terms of the LCM.

$$= \frac{x^2 - x - 6}{x(x - 2)(x + 2)} + \frac{6x}{x(x - 2)(x + 2)}$$

• Multiply the factors in the numerator.

$$= \frac{(x^2 - x - 6) + 6x}{x(x - 2)(x + 2)}$$

• Add the fractions.

$$= \frac{x^2 + 5x - 6}{x(x - 2)(x + 2)}$$

$$= \frac{(x + 6)(x - 1)}{x(x - 2)(x + 2)}$$

The last step is to factor the numerator to determine whether there are common factors in the numerator and denominator. For this example there are no common factors, so the answer is in simplest form.

Example 8

Simplify: $\dfrac{y}{x} - \dfrac{4y}{3x} + \dfrac{3y}{4x}$

Solution

The LCM of the denominators is $12x$.

$$\dfrac{y}{x} - \dfrac{4y}{3x} + \dfrac{3y}{4x} = \dfrac{y}{x} \cdot \dfrac{12}{12} - \dfrac{4y}{3x} \cdot \dfrac{4}{4} + \dfrac{3y}{4x} \cdot \dfrac{3}{3}$$

$$= \dfrac{12y}{12x} - \dfrac{16y}{12x} + \dfrac{9y}{12x}$$

$$= \dfrac{12y - 16y + 9y}{12x} = \dfrac{5y}{12x}$$

You Try It 8

Simplify: $\dfrac{z}{8y} - \dfrac{4z}{3y} + \dfrac{5z}{4y}$

Your solution

Example 9

Subtract: $\dfrac{2x}{x - 3} - \dfrac{5}{3 - x}$

Solution

Remember: $3 - x = -(x - 3)$.

Therefore, $\dfrac{5}{3 - x} = \dfrac{5}{-(x - 3)} = \dfrac{-5}{x - 3}$.

$$\dfrac{2x}{x - 3} - \dfrac{5}{3 - x} = \dfrac{2x}{x - 3} - \dfrac{-5}{x - 3}$$

$$= \dfrac{2x - (-5)}{x - 3} = \dfrac{2x + 5}{x - 3}$$

You Try It 9

Subtract: $\dfrac{5x}{x - 2} - \dfrac{3}{2 - x}$

Your solution

Example 10

Subtract: $\dfrac{2x}{2x - 3} - \dfrac{1}{x + 1}$

Solution

The LCM is $(2x - 3)(x + 1)$.

$$\dfrac{2x}{2x - 3} - \dfrac{1}{x + 1}$$

$$= \dfrac{2x}{2x - 3} \cdot \dfrac{x + 1}{x + 1} - \dfrac{1}{x + 1} \cdot \dfrac{2x - 3}{2x - 3}$$

$$= \dfrac{2x^2 + 2x}{(2x - 3)(x + 1)} - \dfrac{2x - 3}{(2x - 3)(x + 1)}$$

$$= \dfrac{(2x^2 + 2x) - (2x - 3)}{(2x - 3)(x + 1)} = \dfrac{2x^2 + 3}{(2x - 3)(x + 1)}$$

You Try It 10

Subtract: $\dfrac{4x}{3x - 1} - \dfrac{9}{x + 4}$

Your solution

Solutions on p. A38

Example 11

Add: $1 + \dfrac{3}{x^2}$

Solution

The LCM is x^2.

$$1 + \frac{3}{x^2} = 1 \cdot \frac{x^2}{x^2} + \frac{3}{x^2}$$

$$= \frac{x^2}{x^2} + \frac{3}{x^2} = \frac{x^2 + 3}{x^2}$$

You Try It 11

Subtract: $2 - \dfrac{1}{x - 3}$

Your solution

Example 12

Subtract: $\dfrac{x}{2x - 4} - \dfrac{4 - x}{x^2 - 2x}$

Solution

$2x - 4 = 2(x - 2); \quad x^2 - 2x = x(x - 2)$

The LCM is $2x(x - 2)$.

$$\frac{x}{2x - 4} - \frac{4 - x}{x^2 - 2x} = \frac{x}{2(x - 2)} \cdot \frac{x}{x} - \frac{4 - x}{x(x - 2)} \cdot \frac{2}{2}$$

$$= \frac{x^2 - (4 - x)2}{2x(x - 2)}$$

$$= \frac{x^2 - (8 - 2x)}{2x(x - 2)} = \frac{x^2 + 2x - 8}{2x(x - 2)}$$

$$= \frac{(x + 4)\overset{1}{\cancel{(x - 2)}}}{2x\underset{1}{\cancel{(x - 2)}}} = \frac{x + 4}{2x}$$

You Try It 12

Add: $\dfrac{a - 3}{a^2 - 5a} + \dfrac{a - 9}{a^2 - 25}$

Your solution

Example 13

Simplify: $\dfrac{3x + 2}{2x^2 - x - 1} - \dfrac{3}{2x + 1} + \dfrac{4}{x - 1}$

Solution

The LCM is $(2x + 1)(x - 1)$.

$$\frac{3x + 2}{2x^2 - x - 1} - \frac{3}{2x + 1} + \frac{4}{x - 1}$$

$$= \frac{3x + 2}{(2x + 1)(x - 1)} - \frac{3}{2x + 1} \cdot \frac{x - 1}{x - 1} + \frac{4}{x - 1} \cdot \frac{2x + 1}{2x + 1}$$

$$= \frac{3x + 2}{(2x + 1)(x - 1)} - \frac{3x - 3}{(2x + 1)(x - 1)} + \frac{8x + 4}{(2x + 1)(x - 1)}$$

$$= \frac{(3x + 2) - (3x - 3) + (8x + 4)}{(2x + 1)(x - 1)}$$

$$= \frac{3x + 2 - 3x + 3 + 8x + 4}{(2x + 1)(x - 1)} = \frac{8x + 9}{(2x + 1)(x - 1)}$$

You Try It 13

Simplify: $\dfrac{2x - 3}{3x^2 - x - 2} + \dfrac{5}{3x + 2} - \dfrac{1}{x - 1}$

Your solution

Solutions on pp. A38–A39

8.2 Exercises

· ·

Objective A

Find the LCM of the expressions.

1. $8x^3y$
$12xy^2$

2. $6ab^2$
$18ab^3$

3. $10x^4y^2$
$15x^3y$

4. $12a^2b$
$18ab^3$

5. $8x^2$
$4x^2 + 8x$

6. $6y^2$
$4y + 12$

7. $2x^2y$
$3x^2 + 12x$

8. $4xy^2$
$6xy^2 + 12y^2$

9. $9x(x + 2)$
$12(x + 2)^2$

10. $8x^2(x - 1)^2$
$10x^3(x - 1)$

11. $3x + 3$
$2x^2 + 4x + 2$

12. $4x - 12$
$2x^2 - 12x + 18$

13. $(x - 1)(x + 2)$
$(x - 1)(x + 3)$

14. $(2x - 1)(x + 4)$
$(2x + 1)(x + 4)$

15. $(2x + 3)^2$
$(2x + 3)(x - 5)$

16. $(x - 7)(x + 2)$
$(x - 7)^2$

17. $(x - 1)$
$(x - 2)$
$(x - 1)(x - 2)$

18. $(x + 4)(x - 3)$
$x + 4$
$x - 3$

19. $x^2 - x - 6$
$x^2 + x - 12$

20. $x^2 + 3x - 10$
$x^2 + 5x - 14$

21. $x^2 + 5x + 4$
$x^2 - 3x - 28$

22. $x^2 - 10x + 21$
$x^2 - 8x + 15$

23. $x^2 - 2x - 24$
$x^2 - 36$

24. $x^2 + 7x + 10$
$x^2 - 25$

25. $x^2 - 7x - 30$
$x^2 - 5x - 24$

26. $2x^2 - 7x + 3$
$2x^2 + x - 1$

27. $3x^2 - 11x + 6$
$3x^2 + 4x - 4$

28. $2x^2 - 9x + 10$
$2x^2 + x - 15$

29. $6 + x - x^2$
$x + 2$
$x - 3$

30. $15 + 2x - x^2$
$x - 5$
$x + 3$

31. $5 + 4x - x^2$
$x - 5$
$x + 1$

32. $x^2 + 3x - 18$
$3 - x$
$x + 6$

33. $x^2 - 5x + 6$
$1 - x$
$x - 6$

Objective B

Write each fraction in terms of the LCM of the denominators.

34. $\dfrac{4}{x}, \dfrac{3}{x^2}$

35. $\dfrac{5}{ab^2}, \dfrac{6}{ab}$

36. $\dfrac{x}{3y^2}, \dfrac{z}{4y}$

37. $\dfrac{5y}{6x^2}, \dfrac{7}{9xy}$

38. $\dfrac{y}{x(x-3)}, \dfrac{6}{x^2}$

39. $\dfrac{a}{y^2}, \dfrac{6}{y(y+5)}$

40. $\dfrac{9}{(x-1)^2}, \dfrac{6}{x(x-1)}$

41. $\dfrac{a^2}{y(y+7)}, \dfrac{a}{(y+7)^2}$

42. $\dfrac{3}{x-3}, \dfrac{5}{x(3-x)}$

43. $\dfrac{b}{y(y-4)}, \dfrac{b^2}{4-y}$

44. $\dfrac{3}{(x-5)^2}, \dfrac{2}{5-x}$

45. $\dfrac{3}{7-y}, \dfrac{2}{(y-7)^2}$

46. $\dfrac{3}{x^2+2x}, \dfrac{4}{x^2}$

47. $\dfrac{2}{y-3}, \dfrac{3}{y^3-3y^2}$

48. $\dfrac{x-2}{x+3}, \dfrac{x}{x-4}$

49. $\dfrac{x^2}{2x-1}, \dfrac{x+1}{x+4}$

50. $\dfrac{3}{x^2+x-2}, \dfrac{x}{x+2}$

51. $\dfrac{3x}{x-5}, \dfrac{4}{x^2-25}$

52. $\dfrac{5}{2x^2-9x+10}, \dfrac{x-1}{2x-5}$

53. $\dfrac{x-3}{3x^2+4x-4}, \dfrac{2}{x+2}$

54. $\dfrac{x}{x^2+x-6}, \dfrac{2x}{x^2-9}$

55. $\dfrac{x-1}{x^2+2x-15}, \dfrac{x}{x^2+6x+5}$

Objective C

Add or subtract.

56. $\dfrac{3}{y^2}+\dfrac{8}{y^2}$

57. $\dfrac{6}{ab}-\dfrac{2}{ab}$

58. $\dfrac{3}{x+4}-\dfrac{10}{x+4}$

59. $\dfrac{x}{x+6}-\dfrac{2}{x+6}$

60. $\dfrac{3x}{2x+3}+\dfrac{5x}{2x+3}$

61. $\dfrac{6y}{4y+1}-\dfrac{11y}{4y+1}$

Add or subtract.

62. $\dfrac{2x + 1}{x - 3} + \dfrac{3x + 6}{x - 3}$

63. $\dfrac{4x + 3}{2x - 7} + \dfrac{3x - 8}{2x - 7}$

64. $\dfrac{5x - 1}{x + 9} - \dfrac{3x + 4}{x + 9}$

65. $\dfrac{6x - 5}{x - 10} - \dfrac{3x - 4}{x - 10}$

66. $\dfrac{x - 7}{2x + 7} - \dfrac{4x - 3}{2x + 7}$

67. $\dfrac{2n}{3n + 4} - \dfrac{5n - 3}{3n + 4}$

68. $\dfrac{x}{x^2 + 2x - 15} - \dfrac{3}{x^2 + 2x - 15}$

69. $\dfrac{3x}{x^2 + 3x - 10} - \dfrac{6}{x^2 + 3x - 10}$

70. $\dfrac{2x + 3}{x^2 - x - 30} - \dfrac{x - 2}{x^2 - x - 30}$

71. $\dfrac{3x - 1}{x^2 + 5x - 6} - \dfrac{2x - 7}{x^2 + 5x - 6}$

72. $\dfrac{4y + 7}{2y^2 + 7y - 4} - \dfrac{y - 5}{2y^2 + 7y - 4}$

73. $\dfrac{x + 1}{2x^2 - 5x - 12} + \dfrac{x + 2}{2x^2 - 5x - 12}$

74. $\dfrac{2x^2 + 3x}{x^2 - 9x + 20} + \dfrac{2x^2 - 3}{x^2 - 9x + 20} - \dfrac{4x^2 + 2x + 1}{x^2 - 9x + 20}$

75. $\dfrac{2x^2 + 3x}{x^2 - 2x - 63} - \dfrac{x^2 - 3x + 21}{x^2 - 2x - 63} - \dfrac{x - 7}{x^2 - 2x - 63}$

Objective D

Add or subtract.

76. $\dfrac{4}{x} + \dfrac{5}{y}$

77. $\dfrac{7}{a} + \dfrac{5}{b}$

78. $\dfrac{12}{x} - \dfrac{5}{2x}$

79. $\dfrac{5}{3a} - \dfrac{3}{4a}$

80. $\dfrac{1}{2x} - \dfrac{5}{4x} + \dfrac{7}{6x}$

81. $\dfrac{7}{4y} + \dfrac{11}{6y} - \dfrac{8}{3y}$

Add or subtract.

82. $\dfrac{5}{3x} - \dfrac{2}{x^2} + \dfrac{3}{2x}$

83. $\dfrac{6}{y^2} + \dfrac{3}{4y} - \dfrac{2}{5y}$

84. $\dfrac{2}{x} - \dfrac{3}{2y} + \dfrac{3}{5x} - \dfrac{1}{4y}$

85. $\dfrac{5}{2a} + \dfrac{7}{3b} - \dfrac{2}{b} - \dfrac{3}{4a}$

86. $\dfrac{2x+1}{3x} + \dfrac{x-1}{5x}$

87. $\dfrac{4x-3}{6x} + \dfrac{2x+3}{4x}$

88. $\dfrac{x-3}{6x} + \dfrac{x+4}{8x}$

89. $\dfrac{2x-3}{2x} + \dfrac{x+3}{3x}$

90. $\dfrac{2x+9}{9x} - \dfrac{x-5}{5x}$

91. $\dfrac{3y-2}{12y} - \dfrac{y-3}{18y}$

92. $\dfrac{x+4}{2x} - \dfrac{x-1}{x^2}$

93. $\dfrac{x-2}{3x^2} - \dfrac{x+4}{x}$

94. $\dfrac{x-10}{4x^2} + \dfrac{x+1}{2x}$

95. $\dfrac{x+5}{3x^2} + \dfrac{2x+1}{2x}$

96. $\dfrac{4}{x+4} - x$

97. $2x + \dfrac{1}{x}$

98. $5 - \dfrac{x-2}{x+1}$

99. $3 + \dfrac{x-1}{x+1}$

100. $\dfrac{x+3}{6x} - \dfrac{x-3}{8x^2}$

101. $\dfrac{x+2}{xy} - \dfrac{3x-2}{x^2y}$

102. $\dfrac{3x-1}{xy^2} - \dfrac{2x+3}{xy}$

103. $\dfrac{4x-3}{3x^2y} + \dfrac{2x+1}{4xy^2}$

104. $\dfrac{5x+7}{6xy^2} - \dfrac{4x-3}{8x^2y}$

105. $\dfrac{x-2}{8x^2} - \dfrac{x+7}{12xy}$

Add or subtract.

106. $\dfrac{3x - 1}{6y^2} - \dfrac{x + 5}{9xy}$

107. $\dfrac{4}{x - 2} + \dfrac{5}{x + 3}$

108. $\dfrac{2}{x - 3} + \dfrac{5}{x - 4}$

109. $\dfrac{6}{x - 7} - \dfrac{4}{x + 3}$

110. $\dfrac{3}{y + 6} - \dfrac{4}{y - 3}$

111. $\dfrac{2x}{x + 1} + \dfrac{1}{x - 3}$

112. $\dfrac{3x}{x - 4} + \dfrac{2}{x + 6}$

113. $\dfrac{4x}{2x - 1} - \dfrac{5}{x - 6}$

114. $\dfrac{6x}{x + 5} - \dfrac{3}{2x + 3}$

115. $\dfrac{2a}{a - 7} + \dfrac{5}{7 - a}$

116. $\dfrac{4x}{6 - x} + \dfrac{5}{x - 6}$

117. $\dfrac{x}{x^2 - 9} + \dfrac{3}{x - 3}$

118. $\dfrac{y}{y^2 - 16} + \dfrac{1}{y - 4}$

119. $\dfrac{2x}{x^2 - x - 6} - \dfrac{3}{x + 2}$

120. $\dfrac{(x - 1)^2}{(x + 1)^2} - 1$

121. $1 - \dfrac{(y - 2)^2}{(y + 2)^2}$

122. $\dfrac{x}{1 - x^2} - 1 + \dfrac{x}{1 + x}$

123. $\dfrac{y}{x - y} + 2 - \dfrac{x}{y - x}$

124. $\dfrac{3x - 1}{x^2 - 10x + 25} - \dfrac{3}{x - 5}$

125. $\dfrac{2a + 3}{a^2 - 7a + 12} - \dfrac{2}{a - 3}$

126. $\dfrac{x + 4}{x^2 - x - 42} + \dfrac{3}{7 - x}$

127. $\dfrac{x + 3}{x^2 - 3x - 10} + \dfrac{2}{5 - x}$

Add or subtract.

128. $\dfrac{1}{x+1} + \dfrac{x}{x-6} - \dfrac{5x-2}{x^2-5x-6}$

129. $\dfrac{x}{x-4} + \dfrac{5}{x+5} - \dfrac{11x-8}{x^2+x-20}$

130. $\dfrac{3x+1}{x-1} - \dfrac{x-1}{x-3} + \dfrac{x+1}{x^2-4x+3}$

131. $\dfrac{4x+1}{x-8} - \dfrac{3x+2}{x+4} - \dfrac{49x+4}{x^2-4x-32}$

132. $\dfrac{2x+9}{3-x} + \dfrac{x+5}{x+7} - \dfrac{2x^2+3x-3}{x^2+4x-21}$

133. $\dfrac{3x+5}{x+5} - \dfrac{x+1}{2-x} - \dfrac{4x^2-3x-1}{x^2+3x-10}$

APPLYING THE CONCEPTS

134. Simplify.

a. $\left(\dfrac{y+1}{y-1}\right)^2 - 1$

b. $\left(\dfrac{1}{3} - \dfrac{2}{a}\right) \div \left(\dfrac{3}{a} - 2 + \dfrac{a}{4}\right)$

135. Rewrite as the sum of two fractions in simplest form.

a. $\dfrac{3x+6y}{xy}$

b. $\dfrac{4a^2+3ab}{a^2b^2}$

c. $\dfrac{3m^2n+2mn^2}{12m^3n^2}$

136. Let $f(x) = \dfrac{x}{x+2}$, $g(x) = \dfrac{4}{x-3}$, and $S(x) = \dfrac{x^2+x+8}{x^2-x-6}$. Evaluate $f(4)$, $g(4)$, and $S(4)$. Does $f(4) + g(4) = S(4)$? Let a be a real number ($a \neq -2$, $a \neq 3$). Express $S(a)$ in terms of $f(a)$ and $g(a)$.

137. Find the sum of the following: $\dfrac{1}{1 \cdot 2} + \dfrac{1}{2 \cdot 3}$

$$\dfrac{1}{1 \cdot 2} + \dfrac{1}{2 \cdot 3} + \dfrac{1}{3 \cdot 4}$$

$$\dfrac{1}{1 \cdot 2} + \dfrac{1}{2 \cdot 3} + \dfrac{1}{3 \cdot 4} + \dfrac{1}{4 \cdot 5}$$

Note the pattern in these sums, and find the sum of 50 terms, of 100 terms, and of 1000 terms.

138. In your own words, explain the procedure for adding rational ex-
[W]   pressions with different denominators.

139. When is the LCM of two expressions equal to their product?
[W]

Complex Fractions

Objective A   *To simplify a complex fraction*

POINT OF INTEREST

There are many instances of complex fractions in application problems. The

fraction $\dfrac{1}{\dfrac{1}{R_1} + \dfrac{1}{R_2}}$ is used to

determine the total resistance in certain electric circuits.

A **complex fraction** is a fraction whose numerator or denominator contains one or more fractions. Examples of complex fractions are shown at the right.

$$\frac{3}{2 - \dfrac{1}{2}}, \quad \frac{4 + \dfrac{1}{x}}{3 + \dfrac{2}{x}}, \quad \frac{\dfrac{1}{x - 1} + x + 3}{x - 3 + \dfrac{1}{x + 4}}$$

➡ Simplify: $\dfrac{1 - \dfrac{4}{x^2}}{1 + \dfrac{2}{x}}$

CONSIDER THIS

First of all, we are multiplying

the complex fraction by $\dfrac{x^2}{x^2}$,

which equals 1, so we are not changing the value of the fraction.

Second, we are using the Distributive Property to

multiply $\left(1 - \dfrac{4}{x^2}\right)x^2$ and

$\left(1 + \dfrac{2}{x}\right)x^2$.

Find the LCM of the denominators of the fractions in the numerator and denominator. The LCM of x and x^2 is x^2.

$$\frac{1 - \dfrac{4}{x^2}}{1 + \dfrac{2}{x}} = \frac{1 - \dfrac{4}{x^2}}{1 + \dfrac{2}{x}} \cdot \frac{x^2}{x^2}$$

- **Multiply the numerator and denominator by the LCM.**

$$= \frac{1 \cdot x^2 - \dfrac{4}{x^2} \cdot x^2}{1 \cdot x^2 + \dfrac{2}{x} \cdot x^2}$$

- **Simplify.**

$$= \frac{x^2 - 4}{x^2 + 2x} = \frac{(x - 2)\overset{1}{\cancel{(x + 2)}}}{x\underset{1}{\cancel{(x + 2)}}}$$

$$= \frac{x - 2}{x}$$

Example 1

Simplify: $\dfrac{\dfrac{1}{x} + \dfrac{1}{2}}{\dfrac{1}{x^2} - \dfrac{1}{4}}$

Solution

The LCM of x, 2, x^2, and 4 is $4x^2$.

$$\frac{\dfrac{1}{x} + \dfrac{1}{2}}{\dfrac{1}{x^2} - \dfrac{1}{4}} = \frac{\dfrac{1}{x} + \dfrac{1}{2}}{\dfrac{1}{x^2} - \dfrac{1}{4}} \cdot \frac{4x^2}{4x^2} = \frac{\dfrac{1}{x} \cdot 4x^2 + \dfrac{1}{2} \cdot 4x^2}{\dfrac{1}{x^2} \cdot 4x^2 - \dfrac{1}{4} \cdot 4x^2}$$

$$= \frac{4x + 2x^2}{4 - x^2} = \frac{2x\overset{1}{\cancel{(2 + x)}}}{(2 - x)\underset{1}{\cancel{(2 + x)}}} = \frac{2x}{2 - x}$$

You Try It 1

Simplify: $\dfrac{\dfrac{1}{3} - \dfrac{1}{x}}{\dfrac{1}{9} - \dfrac{1}{x^2}}$

Your solution

Solution on p. A39

Example 2

Simplify: $\dfrac{1 - \dfrac{2}{x} - \dfrac{15}{x^2}}{1 - \dfrac{11}{x} + \dfrac{30}{x^2}}$

Solution

The LCM of x and x^2 is x^2.

$$\dfrac{1 - \dfrac{2}{x} - \dfrac{15}{x^2}}{1 - \dfrac{11}{x} + \dfrac{30}{x^2}} = \dfrac{1 - \dfrac{2}{x} - \dfrac{15}{x^2}}{1 - \dfrac{11}{x} + \dfrac{30}{x^2}} \cdot \dfrac{x^2}{x^2}$$

$$= \dfrac{1 \cdot x^2 - \dfrac{2}{x} \cdot x^2 - \dfrac{15}{x^2} \cdot x^2}{1 \cdot x^2 - \dfrac{11}{x} \cdot x^2 + \dfrac{30}{x^2} \cdot x^2}$$

$$= \dfrac{x^2 - 2x - 15}{x^2 - 11x + 30}$$

$$= \dfrac{\overset{1}{\cancel{(x - 5)}}(x + 3)}{\underset{1}{\cancel{(x - 5)}}(x - 6)} = \dfrac{x + 3}{x - 6}$$

Example 3

Simplify: $\dfrac{x - 8 + \dfrac{20}{x + 4}}{x - 10 + \dfrac{24}{x + 4}}$

Solution

The LCM is $x + 4$.

$$\dfrac{x - 8 + \dfrac{20}{x + 4}}{x - 10 + \dfrac{24}{x + 4}}$$

$$= \dfrac{x - 8 + \dfrac{20}{x + 4}}{x - 10 + \dfrac{24}{x + 4}} \cdot \dfrac{x + 4}{x + 4}$$

$$= \dfrac{x(x + 4) - 8(x + 4) + \dfrac{20}{x + 4} \cdot (x + 4)}{x(x + 4) - 10(x + 4) + \dfrac{24}{x + 4} \cdot (x + 4)}$$

$$= \dfrac{x^2 + 4x - 8x - 32 + 20}{x^2 + 4x - 10x - 40 + 24} = \dfrac{x^2 - 4x - 12}{x^2 - 6x - 16}$$

$$= \dfrac{(x - 6)\overset{1}{\cancel{(x + 2)}}}{(x - 8)\underset{1}{\cancel{(x + 2)}}} = \dfrac{x - 6}{x - 8}$$

You Try It 2

Simplify: $\dfrac{1 + \dfrac{4}{x} + \dfrac{3}{x^2}}{1 + \dfrac{10}{x} + \dfrac{21}{x^2}}$

Your solution

You Try It 3

Simplify: $\dfrac{x + 3 - \dfrac{20}{x - 5}}{x + 8 + \dfrac{30}{x - 5}}$

Your solution

Solutions on p. A39

8.3 Exercises

Objective A

Simplify.

1. $\dfrac{1 + \dfrac{3}{x}}{1 - \dfrac{9}{x^2}}$

2. $\dfrac{1 + \dfrac{4}{x}}{1 - \dfrac{16}{x^2}}$

3. $\dfrac{2 - \dfrac{8}{x + 4}}{3 - \dfrac{12}{x + 4}}$

4. $\dfrac{5 - \dfrac{25}{x + 5}}{1 - \dfrac{3}{x + 5}}$

5. $\dfrac{1 + \dfrac{5}{y - 2}}{1 - \dfrac{2}{y - 2}}$

6. $\dfrac{2 - \dfrac{11}{2x - 1}}{3 - \dfrac{17}{2x - 1}}$

7. $\dfrac{4 - \dfrac{2}{x + 7}}{5 + \dfrac{1}{x + 7}}$

8. $\dfrac{5 + \dfrac{3}{x - 8}}{2 - \dfrac{1}{x - 8}}$

9. $\dfrac{1 - \dfrac{1}{x} - \dfrac{6}{x^2}}{1 - \dfrac{9}{x^2}}$

10. $\dfrac{1 + \dfrac{4}{x} + \dfrac{4}{x^2}}{1 - \dfrac{2}{x} - \dfrac{8}{x^2}}$

11. $\dfrac{1 - \dfrac{5}{x} - \dfrac{6}{x^2}}{1 + \dfrac{6}{x} + \dfrac{5}{x^2}}$

12. $\dfrac{1 - \dfrac{7}{a} + \dfrac{12}{a^2}}{1 + \dfrac{1}{a} - \dfrac{20}{a^2}}$

13. $\dfrac{1 - \dfrac{6}{x} + \dfrac{8}{x^2}}{\dfrac{4}{x^2} + \dfrac{3}{x} - 1}$

14. $\dfrac{1 + \dfrac{3}{x} - \dfrac{18}{x^2}}{\dfrac{21}{x^2} - \dfrac{4}{x} - 1}$

15. $\dfrac{x - \dfrac{4}{x + 3}}{1 + \dfrac{1}{x + 3}}$

16. $\dfrac{y + \dfrac{1}{y - 2}}{1 + \dfrac{1}{y - 2}}$

17. $\dfrac{1 - \dfrac{x}{2x + 1}}{x - \dfrac{1}{2x + 1}}$

18. $\dfrac{1 - \dfrac{2x - 2}{3x - 1}}{x - \dfrac{4}{3x - 1}}$

Simplify.

19. $\dfrac{x - 5 + \dfrac{14}{x + 4}}{x + 3 - \dfrac{2}{x + 4}}$

20. $\dfrac{a + 4 + \dfrac{5}{a - 2}}{a + 6 + \dfrac{15}{a - 2}}$

21. $\dfrac{x + 3 - \dfrac{10}{x - 6}}{x + 2 - \dfrac{20}{x - 6}}$

22. $\dfrac{x - 7 + \dfrac{5}{x - 1}}{x - 3 + \dfrac{1}{x - 1}}$

23. $\dfrac{y - 6 + \dfrac{22}{2y + 3}}{y - 5 + \dfrac{11}{2y + 3}}$

24. $\dfrac{x + 2 - \dfrac{12}{2x - 1}}{x + 1 - \dfrac{9}{2x - 1}}$

25. $\dfrac{x - \dfrac{2}{2x - 3}}{2x - 1 - \dfrac{8}{2x - 3}}$

26. $\dfrac{x + 3 - \dfrac{18}{2x + 1}}{x - \dfrac{6}{2x + 1}}$

27. $\dfrac{\dfrac{1}{x} - \dfrac{2}{x - 1}}{\dfrac{3}{x} + \dfrac{1}{x - 1}}$

28. $\dfrac{\dfrac{3}{n + 1} + \dfrac{1}{n}}{\dfrac{2}{n + 1} + \dfrac{3}{n}}$

29. $\dfrac{\dfrac{3}{2x - 1} - \dfrac{1}{x}}{\dfrac{4}{x} + \dfrac{2}{2x - 1}}$

30. $\dfrac{\dfrac{4}{3x + 1} + \dfrac{3}{x}}{\dfrac{6}{x} - \dfrac{2}{3x + 1}}$

APPLYING THE CONCEPTS

Simplify.

31. $1 + \dfrac{1}{1 + \dfrac{1}{2}}$

32. $1 + \dfrac{1}{1 + \dfrac{1}{1 + \dfrac{1}{2}}}$

33. $1 - \dfrac{1}{1 - \dfrac{1}{x}}$

34. $\dfrac{a^{-1} - b^{-1}}{a^{-2} - b^{-2}}$

35. $\left(\dfrac{y}{4} - \dfrac{4}{y}\right) \div \left(\dfrac{4}{y} - 3 + \dfrac{y}{2}\right)$

36. $\dfrac{1 + x^{-1}}{1 - x^{-1}}$

37.
[W] How would you explain to a classmate why we multiply the numerator and denominator of a complex fraction by the LCM of the denominators of the fractions in the numerator and denominator?

Rational Equations

Objective A *To solve rational equations* ....

To solve an equation containing fractions, **clear denominators** by multiplying each side of the equation by the LCM of the denominators. Then solve for the variable.

➡ Solve: $\dfrac{3x-1}{4x} + \dfrac{2}{3x} = \dfrac{7}{6x}$

$$\dfrac{3x-1}{4x} + \dfrac{2}{3x} = \dfrac{7}{6x}$$

- The LCM of $4x$, $3x$, and $6x$ is $12x$.

$$12x\left(\dfrac{3x-1}{4x} + \dfrac{2}{3x}\right) = 12x\left(\dfrac{7}{6x}\right)$$

- Multiply each side of the equation by the LCM of the denominators.

$$12x\left(\dfrac{3x-1}{4x}\right) + 12x\left(\dfrac{2}{3x}\right) = 12x\left(\dfrac{7}{6x}\right)$$

- Simplify using the Distributive Property.

$$\dfrac{12x}{1}\left(\dfrac{3x-1}{4x}\right) + \dfrac{12x}{1}\left(\dfrac{2}{3x}\right) = \dfrac{12x}{1}\left(\dfrac{7}{6x}\right)$$

$$3(3x-1) + 4(2) = 2(7)$$

- Solve for x.

$$9x - 3 + 8 = 14$$
$$9x + 5 = 14$$
$$9x = 9$$
$$x = 1$$

1 checks as a solution. The solution is 1.

Occasionally, a value of the variable that appears to be a solution of an equation will make one of the denominators zero. In this case, the equation has no solution for that value of the variable.

➡ Solve: $\dfrac{2x}{x-2} = 1 + \dfrac{4}{x-2}$

$$\dfrac{2x}{x-2} = 1 + \dfrac{4}{x-2}$$

$$(x-2)\dfrac{2x}{x-2} = (x-2)\left(1 + \dfrac{4}{x-2}\right)$$

- The LCM is $x-2$. Multiply each side of the equation by the LCM.

$$(x-2)\dfrac{2x}{x-2} = (x-2)\cdot 1 + (x-2)\dfrac{4}{x-2}$$

- Simplify using the Distributive Property and the Properties of Fractions.

$$\dfrac{(x-2)}{1}\cdot\dfrac{2x}{x-2} = (x-2)\cdot 1 + \dfrac{(x-2)}{1}\cdot\dfrac{4}{x-2}$$

$$2x = x - 2 + 4$$

- Solve for x.

$$2x = x + 2$$
$$x = 2$$

When x is replaced by 2, the denominators of $\dfrac{2x}{x-2}$ and $\dfrac{4}{x-2}$ are zero. Therefore, the equation has no solution.

Example 1

Solve: $\dfrac{x}{x+4} = \dfrac{2}{x}$

Solution

The LCM is $x(x+4)$.

$$\frac{x}{x+4} = \frac{2}{x}$$

$$x(x+4)\left(\frac{x}{x+4}\right) = x(x+4)\left(\frac{2}{x}\right)$$

$$\frac{\overset{1}{x(x+4)}}{1} \cdot \frac{x}{\underset{1}{x+4}} = \frac{\overset{1}{x}(x+4)}{1} \cdot \frac{2}{\underset{1}{x}}$$

$$x^2 = (x+4)2$$

$$x^2 = 2x + 8$$

Solve the quadratic equation by factoring.

$$x^2 - 2x - 8 = 0$$

$$(x-4)(x+2) = 0$$

$$x - 4 = 0 \qquad x + 2 = 0$$

$$x = 4 \qquad\quad x = -2$$

Both 4 and −2 check as solutions.

The solutions are 4 and −2.

Example 2

Solve: $\dfrac{3x}{x-4} = 5 + \dfrac{12}{x-4}$

Solution

The LCM is $x - 4$.

$$\frac{3x}{x-4} = 5 + \frac{12}{x-4}$$

$$(x-4)\left(\frac{3x}{x-4}\right) = (x-4)\left(5 + \frac{12}{x-4}\right)$$

$$\frac{\overset{1}{(x-4)}}{1} \cdot \frac{3x}{\underset{1}{x-4}} = (x-4)5 + \frac{\overset{1}{(x-4)}}{1} \cdot \frac{12}{\underset{1}{x-4}}$$

$$3x = (x-4)5 + 12$$

$$3x = 5x - 20 + 12$$

$$3x = 5x - 8$$

$$-2x = -8$$

$$x = 4$$

4 does not check as a solution.

The equation has no solution.

You Try It 1

Solve: $\dfrac{x}{x+6} = \dfrac{3}{x}$

Your solution

You Try It 2

Solve: $\dfrac{5x}{x+2} = 3 - \dfrac{10}{x+2}$

Your solution

Solutions on p. A40

Objective B   *To solve proportions* ....

Quantities such as 4 meters, 15 seconds, and 8 gallons are number quantities written with units. In these examples the units are meters, seconds, and gallons.

A **ratio** is the quotient of two quantities that have the same unit.

The length of a living room is 16 ft and the width is 12 ft. The ratio of the length to the width is written

$$\frac{16 \text{ ft}}{12 \text{ ft}} = \frac{16}{12} = \frac{4}{3}$$ A ratio is in simplest form when the two numbers do not have a common factor. Note that the units are not written.

A **rate** is the quotient of two quantities that have different units.

There are 2 lb of salt in 8 gal of water. The salt-to-water rate is

$$\frac{2 \text{ lb}}{8 \text{ gal}} = \frac{1 \text{ lb}}{4 \text{ gal}}$$ A rate is in simplest form when the two numbers do not have a common factor. The units are written as part of the rate.

A **proportion** is an equation that states the equality of two ratios or rates. Examples of proportions are shown at the right.

$$\frac{30 \text{ mi}}{4 \text{ h}} = \frac{15 \text{ mi}}{2 \text{ h}}$$

$$\frac{4}{6} = \frac{8}{12}$$

$$\frac{3}{4} = \frac{x}{8}$$

➡ Solve: $\dfrac{4}{x} = \dfrac{2}{3}$

$$\frac{4}{x} = \frac{2}{3}$$ • The LCM of x and 3 is $3x$.

$$3x\left(\frac{4}{x}\right) = 3x\left(\frac{2}{3}\right)$$ • Multiply each side of the proportion by $3x$.

$$12 = 2x$$ • Solve the equation.

$$6 = x$$

The solution is 6.

Example 3

Solve: $\dfrac{8}{x + 3} = \dfrac{4}{x}$

Solution

$$\frac{8}{x + 3} = \frac{4}{x}$$

$$x(x + 3)\frac{8}{x + 3} = x(x + 3)\frac{4}{x}$$

$$8x = 4(x + 3)$$
$$8x = 4x + 12$$
$$4x = 12$$
$$x = 3$$

The solution is 3.

You Try It 3

Solve: $\dfrac{2}{x + 3} = \dfrac{6}{5x + 5}$

Your solution

Solution on p. A40

Objective C *To solve similar triangles* ....

Similar objects have the same shape but not necessarily the same size. A tennis ball is similar to a basketball. A model ship is similar to an actual ship.

Similar objects have corresponding parts; for example, the rudder on the model ship corresponds to the rudder on the actual ship. The relationship between the sizes of the corresponding parts can be written as a ratio, and each ratio will be the same. If the rudder on the model ship is $\frac{1}{100}$ the size of the rudder on the actual ship, then the model wheelhouse is $\frac{1}{100}$ the size of the actual wheelhouse, the width of the model is $\frac{1}{100}$ the width of the actual ship, and so on.

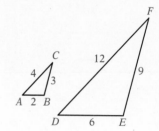

The two triangles *ABC* and *DEF* shown at the right are similar. Side *AB* corresponds to side *DE*, side *BC* corresponds to side *EF*, and side *AC* corresponds to side *DF*. The ratios of corresponding sides are equal.

$$\frac{AB}{DE} = \frac{2}{6} = \frac{1}{3}, \frac{BC}{EF} = \frac{3}{9} = \frac{1}{3}, \text{ and } \frac{AC}{DF} = \frac{4}{12} = \frac{1}{3}.$$

Because the ratio of corresponding sides is equal, three proportions can be formed:

$$\frac{AB}{DE} = \frac{BC}{EF}, \frac{AB}{DE} = \frac{AC}{DF}, \text{ and } \frac{BC}{EF} = \frac{AC}{DF}$$

The corresponding angles in similar triangles are equal. Therefore,

$$\angle A = \angle D, \angle B = \angle E, \text{ and } \angle C = \angle F$$

Triangles *ABC* and *DEF* at the right are similar triangles. *AH* and *DK* are the heights of the triangles. The ratio of heights of similar triangles equals the ratio of corresponding sides.

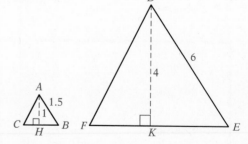

Ratio of corresponding sides = $\frac{1.5}{6} = \frac{1}{4}$

Ratio of heights = $\frac{1}{4}$

> **Properties of Similar Triangles**
>
> For similar triangles, the ratios of corresponding sides are equal. The ratio of corresponding heights is equal to the ratio of corresponding sides.

➡ The two triangles at the right are similar triangles. Find the length of side *EF*. Round to the nearest tenth.

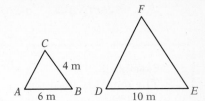

$$\frac{EF}{BC} = \frac{DE}{AB}$$

$$\frac{EF}{4} = \frac{10}{6}$$

● The triangles are similar, so the ratios of corresponding sides are equal.

$$12\left(\frac{EF}{4}\right) = 12\left(\frac{10}{6}\right)$$

$$3(EF) = 2(10)$$

$$3(EF) = 20$$

$$EF \approx 6.7$$

The length of side *EF* is approximately 6.7 m.

Example 4

Triangles *ABC* and *DEF* are similar. Find *FG*, the height of triangle *DEF*.

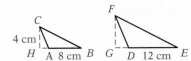

Strategy

To find *FG*, write a proportion using the fact that, in similar triangles, the ratio of corresponding sides equals the ratio of corresponding heights. Solve the proportion for *FG*.

Solution

$$\frac{AB}{DE} = \frac{CH}{FG}$$

$$\frac{8}{12} = \frac{4}{FG}$$

$$12(FG)\left(\frac{8}{12}\right) = 12(FG)\left(\frac{4}{FG}\right)$$

$$8(FG) = 12(4)$$

$$8(FG) = 48$$

$$FG = 6$$

The height *FG* of triangle *DEF* is 6 cm.

You Try It 4

Triangles *ABC* and *DEF* are similar. Find *FG*, the height of triangle *DEF*.

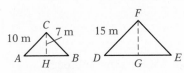

Your strategy

Your solution

Solution on p. A40

Objective D   To solve application problems ....

Example 5

The monthly loan payment for a car is
$28.35 for each $1000 borrowed. At this
rate, find the monthly payment for a
$6000 car loan.

Strategy

To find the monthly payment, write and
solve a proportion, using P to represent
the monthly car payment.

Solution

$$\frac{28.35}{1000} = \frac{P}{6000}$$

$$6000\left(\frac{28.35}{1000}\right) = 6000\left(\frac{P}{6000}\right)$$

$$170.10 = P$$

The monthly payment is $170.10.

You Try It 5

Sixteen ceramic tiles are needed to tile an
area of 9 ft². At this rate, how many square
feet can be tiled using 256 ceramic tiles?

Your strategy

Your solution

Example 6

An investment of $500 earns $60 each
year. At the same rate, how much
additional money must be invested to
earn $90 each year?

Strategy

To find the additional amount of money
that must be invested, write and solve a
proportion, using x to represent the
additional money. Then $500 + x$ is the
total amount invested.

Solution

$$\frac{60}{500} = \frac{90}{500 + x}$$

$$\frac{3}{25} = \frac{90}{500 + x}$$

$$25(500 + x)\left(\frac{3}{25}\right) = 25(500 + x)\left(\frac{90}{500 + x}\right)$$

$$(500 + x)3 = 25(90)$$

$$1500 + 3x = 2250$$

$$3x = 750$$

$$x = 250$$

An additional $250 must be invested.

You Try It 6

Three ounces of a certain medication are
required for a 150-pound adult. At the
same rate, how many additional ounces of
this medication are required for a
200-pound adult?

Your strategy

Your solution

Solutions on p. A41

8.4 Exercises

· ·

Objective A

Solve.

1. $\dfrac{2x}{3} - \dfrac{5}{2} = -\dfrac{1}{2}$

2. $\dfrac{x}{3} - \dfrac{1}{4} = \dfrac{1}{12}$

3. $\dfrac{x}{3} - \dfrac{1}{4} = \dfrac{x}{4} - \dfrac{1}{6}$

4. $\dfrac{2y}{9} - \dfrac{1}{6} = \dfrac{y}{9} + \dfrac{1}{6}$

5. $\dfrac{2x-5}{8} + \dfrac{1}{4} = \dfrac{x}{8} + \dfrac{3}{4}$

6. $\dfrac{3x+4}{12} - \dfrac{1}{3} = \dfrac{5x+2}{12} - \dfrac{1}{2}$

7. $\dfrac{6}{2a+1} = 2$

8. $\dfrac{12}{3x-2} = 3$

9. $\dfrac{9}{2x-5} = -2$

10. $\dfrac{6}{4-3x} = 3$

11. $2 + \dfrac{5}{x} = 7$

12. $3 + \dfrac{8}{n} = 5$

13. $1 - \dfrac{9}{x} = 4$

14. $3 - \dfrac{12}{x} = 7$

15. $\dfrac{2}{y} + 5 = 9$

16. $\dfrac{6}{x} + 3 = 11$

17. $\dfrac{3}{x-2} = \dfrac{4}{x}$

18. $\dfrac{5}{x+3} = \dfrac{3}{x-1}$

19. $\dfrac{2}{3x-1} = \dfrac{3}{4x+1}$

20. $\dfrac{5}{3x-4} = \dfrac{-3}{1-2x}$

21. $\dfrac{-3}{2x+5} = \dfrac{2}{x-1}$

22. $\dfrac{4}{5y-1} - \dfrac{2}{2y-1}$

23. $\dfrac{4x}{x-4} + 5 = \dfrac{5x}{x-4}$

24. $\dfrac{2x}{x+2} - 5 = \dfrac{7x}{x+2}$

Solve.

25. $2 + \dfrac{3}{a-3} = \dfrac{a}{a-3}$

26. $\dfrac{x}{x+4} = 3 - \dfrac{4}{x+4}$

27. $\dfrac{x}{x-1} = \dfrac{8}{x+2}$

28. $\dfrac{x}{x+12} = \dfrac{1}{x+5}$

29. $\dfrac{2x}{x+4} = \dfrac{3}{x-1}$

30. $\dfrac{5}{3n-8} = \dfrac{n}{n+2}$

31. $x + \dfrac{6}{x-2} = \dfrac{3x}{x-2}$

32. $x - \dfrac{6}{x-3} = \dfrac{2x}{x-3}$

33. $\dfrac{8}{y} = \dfrac{2}{y-2} + 1$

Objective B

Solve.

34. $\dfrac{x}{12} = \dfrac{3}{4}$

35. $\dfrac{6}{x} = \dfrac{2}{3}$

36. $\dfrac{4}{9} = \dfrac{x}{27}$

37. $\dfrac{16}{9} = \dfrac{64}{x}$

38. $\dfrac{x+3}{12} = \dfrac{5}{6}$

39. $\dfrac{3}{5} = \dfrac{x-4}{10}$

40. $\dfrac{18}{x+4} = \dfrac{9}{5}$

41. $\dfrac{2}{11} = \dfrac{20}{x-3}$

42. $\dfrac{2}{x} = \dfrac{4}{x+1}$

43. $\dfrac{16}{x-2} = \dfrac{8}{x}$

44. $\dfrac{x+3}{4} = \dfrac{x}{8}$

45. $\dfrac{x-6}{3} = \dfrac{x}{5}$

46. $\dfrac{2}{x-1} = \dfrac{6}{2x+1}$

47. $\dfrac{9}{x+2} = \dfrac{3}{x-2}$

48. $\dfrac{2x}{7} = \dfrac{x-2}{14}$

Objective C

Find the ratio of corresponding sides for the similar triangles.

49.

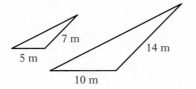

50.

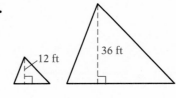

51.

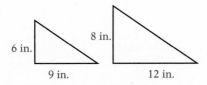

52.

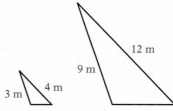

In Exercises 53–60, triangles *ABC* and *DEF* are similar triangles. Solve and round to the nearest tenth.

53. Find side *DE*.

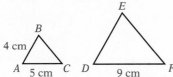

54. Find side *DE*.

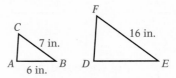

55. Find the height of triangle *DEF*.

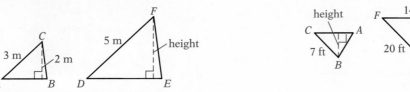

56. Find the height of triangle *ABC*.

57. Find the perimeter of triangle *ABC*.

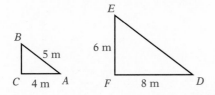

58. Find the perimeter of triangle *DEF*.

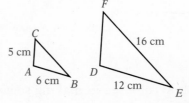

59. Find the perimeter of triangle *ABC*.

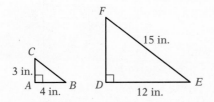

60. Find the area of triangle *DEF*.

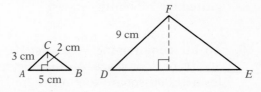

In Exercises 61–62, triangles *ABC* and *DEF* are similar triangles. Solve and round to the nearest tenth.

61. Find the area of triangle *ABC*.

62. Find the area of triangle *DEF*.

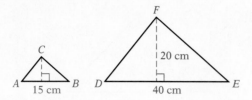 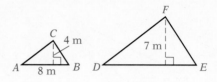

The sun's rays, objects on Earth, and the shadows cast by them form similar triangles. Use this fact to solve Exercises 63–66.

63. Find the height of the flagpole.

64. Find the height of the flagpole.

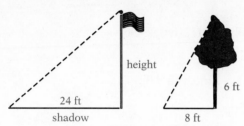 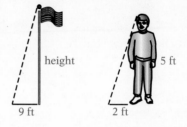

65. Find the height of the building.

66. Find the height of the building.

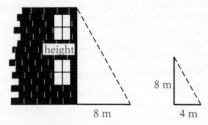 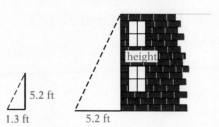

Objective D *Application Problems*

Solve.

67. Simple syrup used in making some desserts requires 2 c sugar for every $\frac{2}{3}$ c boiling water. At this rate, how many cups of sugar are required for 2 c boiling water?

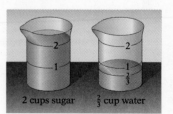

2 cups sugar $\frac{2}{3}$ cup water

68. An air conditioning specialist recommends 2 air vents for each 300 ft² of floor space. At this rate, how many air vents are required for a 21,000-square-foot office building?

Solve.

69. An exit poll survey showed that 4 out of every 7 voters cast a ballot in favor of an amendment to a city charter. At this rate, how many voters voted in favor of the amendment if 35,000 people voted?

70. A company decides to accept a large shipment of 10,000 computer disks if there are 2 or fewer defects in a sample of 100 randomly chosen disks. Assuming that there are 300 defective disks in the shipment and that the rate of defective disks in the sample is the same as the rate in the shipment, will the shipment be accepted?

71. A company decides to accept a large shipment of 20,000 precision bearings if there are 3 or fewer defects in a sample of 100 randomly chosen bearings. Assuming that there are 400 defective bearings in the shipment and that the rate of defective bearings in the sample is the same as the rate in the shipment, will the shipment be accepted?

72. The engine of a small rocket burns 170,000 lb of fuel in 1 min. At this rate, how many pounds of fuel does the rocket burn in 45 s?

73. The lighting for some billboards is provided by using solar energy. If 3 small solar energy panels can generate 10 watts of power, how many panels are necessary to provide 600 watts of power?

74. A laser printer is rated by the number of pages per minute it can print. An inexpensive laser printer can print 5 pages every 2 min. At this rate, how long would it take to print a document 45 pages long?

75. On a map, two cities are $5\frac{5}{8}$ in. apart. If $\frac{3}{4}$ in. on the map represents 100 mi, find the number of miles in the distance between the two cities.

76. In a wildlife preserve, 10 elk are captured, tagged, and then released. Later 15 elk are captured and 2 are found to have tags. Estimate the number of elk in the preserve.

77. As part of a conservation effort for a lake, 40 fish are caught, tagged, and then released. Later 80 fish are caught. Four of the 80 fish are found to have tags. Estimate the number of fish in the lake.

Solve.

78. To conserve energy and still allow for as much natural lighting as possible, an architect suggests that the ratio of the area of a window to the area of the total wall surface be 5 to 12. Using this ratio, determine the recommended area of a window to be installed in a wall that measures 8 ft by 12 ft.

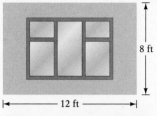

79. A painter estimates that 5 gal of paint will cover 1200 ft² of wall space. At this rate, how many additional gallons will be necessary to cover 1680 ft²?

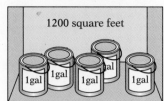

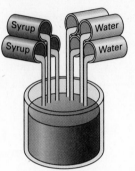

80. A soft drink is made by mixing 4 parts carbonated water with every 3 parts syrup. How many milliliters of water are in 280 ml of soft drink?

APPLYING THE CONCEPTS

Solve.

81. $\dfrac{b+2}{5} = \dfrac{1}{4}b - \dfrac{3}{10}(b-1)$

82. $\dfrac{x}{2x^2 - x - 1} = \dfrac{3}{x^2 - 1} + \dfrac{3}{2x + 1}$

83. Three people put their money together to buy lottery tickets. The first person put in $25, the second person put in $30, and the third person put in $35. One of their tickets was a winning ticket. If they won $4.5 million, what was the first person's share of the winnings?

84. No one belongs to both the Math Club and the Photography Club, but the two clubs join to hold a car wash. Ten members of the Math Club and 6 members of the Photography Club participate. The profits from the car wash are $120. If each club's profits are proportional to the number of members participating, what share of the profits does the Math Club receive?

85. A basketball player has made 5 out of every 6 foul shots attempted in one year of play. If 42 foul shots were missed that year, how many shots did the basketball player make?

86. Explain the procedure for solving an equation that contains fractions. Include in your discussion how the LCM is used to eliminate fractions in the equation.
[W]

8.5 Literal Equations

Objective A   *To solve a literal equation for one of the variables* ....

A **literal equation** is an equation that contains more than one variable. Examples of literal equations are shown at the right.

$$2x + 3y = 6$$
$$4w - 2x + z = 0$$

Formulas are used to express a relationship among physical quantities. A **formula** is a literal equation that states rules about measurements. Examples of formulas are shown at the right.

$$\frac{1}{R_1} + \frac{1}{R_2} = \frac{1}{R} \quad \text{(Physics)}$$
$$s = a + (n - 1)d \quad \text{(Mathematics)}$$
$$A = P + Prt \quad \text{(Business)}$$

The Addition and Multiplication Properties can be used to solve a literal equation for one of the variables. The goal is to rewrite the equation so that the variable being solved for is alone on one side of the equation and all the other numbers and variables are on the other side.

➡ Solve $A = P(1 + i)$ for i.

The goal is to rewrite the equation so that i is on one side of the equation and all other variables are on the other side.

$$A = P(1 + i)$$
$$A = P + Pi \qquad \bullet \text{ Use the Distributive Property to remove parentheses.}$$
$$A - P = P - P + Pi \qquad \bullet \text{ Subtract } P \text{ from each side of the equation.}$$
$$A - P = Pi$$
$$\frac{A - P}{P} = \frac{Pi}{P} \qquad \bullet \text{ Divide each side of the equation by } P.$$
$$\frac{A - P}{P} = i$$

Example 1
Solve $s = a + (n - 1)d$ for d.

Solution
$$s = a + (n - 1)d$$
$$s - a = (n - 1)d$$
$$\frac{s - a}{n - 1} = \frac{(n - 1)d}{n - 1}$$
$$\frac{s - a}{n - 1} = d$$

You Try It 1
Solve $A = P + Prt$ for r.

Your solution

Solution on p. A41

Example 2

Solve $I = \dfrac{E}{R + r}$ for R.

Solution

$$I = \dfrac{E}{R + r}$$

$$(R + r)I = (R + r)\dfrac{E}{R + r}$$

$$RI + rI = E$$

$$RI + rI - rI = E - rI$$

$$RI = E - rI$$

$$\dfrac{RI}{I} = \dfrac{E - rI}{I}$$

$$R = \dfrac{E - rI}{I}$$

You Try It 2

Solve $s = \dfrac{A + L}{2}$ for L.

Your solution

Example 3

Solve $L = a(1 + ct)$ for c.

Solution

$$L = a(1 + ct)$$

$$L = a + act$$

$$L - a = a - a + act$$

$$L - a = act$$

$$\dfrac{L - a}{at} = \dfrac{act}{at}$$

$$\dfrac{L - a}{at} = c$$

You Try It 3

Solve $S = a + (n - 1)d$ for n.

Your solution

Example 4

Solve $S = C - rC$ for C.

Solution

$$S = C - rC$$

$$S = (1 - r)C$$

$$\dfrac{S}{1 - r} = \dfrac{(1 - r)C}{1 - r}$$

$$\dfrac{S}{1 - r} = C$$

You Try It 4

Solve $S = C + rC$ for C.

Your solution

Solutions on p. A42

Content and Format © 1996 HMCo.

8.5 Exercises

. . . .

Objective A

Solve the formula for the given variable.

1. $A = \dfrac{1}{2}bh$; h (Geometry)

2. $P = a + b + c$; b (Geometry)

3. $d = rt$; t (Physics)

4. $E = IR$; R (Physics)

5. $PV = nRT$; T (Chemistry)

6. $A = bh$; h (Geometry)

7. $P = 2l + 2w$; l (Geometry)

8. $F = \dfrac{9}{5}C + 32$; C (Temperature conversion)

9. $A = \dfrac{1}{2}h(b_1 + b_2)$; b_1 (Geometry)

10. $C = \dfrac{5}{9}(F - 32)$; F (Temperature conversion)

11. $V = \dfrac{1}{3}Ah$; h (Geometry)

12. $P = R - C$; C (Business)

13. $R = \dfrac{C - S}{t}$; S (Business)

14. $P = \dfrac{R - C}{n}$; R (Business)

15. $A = P + Prt$; P (Business)

16. $T = fm - gm$; m (Engineering)

17. $A = Sw + w$; w (Physics)

18. $a = S - Sr$; S (Mathematics)

APPLYING THE CONCEPTS

The surface area of a right circular cylinder is given by the formula $S = 2\pi rh + 2\pi r^2$, where r is the radius of the base, and h is the height of the cylinder.

19. **a.**   Solve the formula $S = 2\pi rh + 2\pi r^2$ for h.
    b.   Use your answer to part **a** to find the height of a right circular cylinder when the surface area is 12π in² and the radius is 1 in.
    c.   Use your answer to part **a** to find the height of a right circular cylinder when the surface area is 24π in² and the radius is 2 in.

1 in.

$S = 12\pi$ in²

When markup is based on selling price, the selling price of a product is given by the formula $S = \dfrac{C}{1-r}$, where C is the cost of the product, and r is the markup rate.

20. **a.** Solve the formula $S = \dfrac{C}{1-r}$ for r.

      b. Use your answer to part **a** to find the markup rate on a tennis racket when the cost is \$112 and the selling price is \$140.

      c. Use your answer to part **a** to find the markup rate on a radio when the cost is \$50.40 and the selling price is \$72.

Break-even analysis is a method used to determine the sales volume required for a company to break even, or experience neither a profit nor a loss on the sale of a product. The break-even point represents the number of units that must be made and sold for income from sales to equal the cost of the product. The break-even point can be calculated using the formula $B = \dfrac{F}{S-V}$, where F is the fixed costs, S is the selling price per unit, and V is the variable costs per unit.

21. **a.** Solve the formula $B = \dfrac{F}{S-V}$ for S.

      b. Use your answer to part **a** to find the required selling price per desk for a company to break even. The fixed costs are \$20,000, the variable costs per desk are \$80, and the company plans to make and sell 200 desks.

      c. Use your answer to part **a** to find the required selling price per camera for a company to break even. The fixed costs are \$15,000, the variable costs per camera are \$50, and the company plans to make and sell 600 cameras.

Resistors are used to control the flow of current. The total resistance of two resistors in a circuit can be given by the formula $R = \dfrac{1}{\dfrac{1}{R_1} + \dfrac{1}{R_2}}$, where R_1 and R_2 are the two resistors in the circuit. Resistance is measured in ohms.

22. **a.** Solve the formula $R = \dfrac{1}{\dfrac{1}{R_1} + \dfrac{1}{R_2}}$ for R_1.

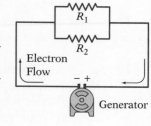

      b. Use your answer to part **a** to find the resistance in R_1 if the resistance in R_2 is 30 ohms and the total resistance is 12 ohms.

      c. Use your answer to part **a** to find the resistance in R_1 if the resistance in R_2 is 15 ohms and the total resistance is 6 ohms.

Work and Uniform Motion Problems

Objective A   *To solve work problems* ....

If a painter can paint a room in 4 h, then in 1 h the painter can paint $\frac{1}{4}$ of the room. The painter's rate of work is $\frac{1}{4}$ of the room each hour. The **rate of work** is the part of a task that is completed in one unit of time.

A pipe can fill a tank in 30 min. This pipe can fill $\frac{1}{30}$ of the tank in 1 min. The rate of work is $\frac{1}{30}$ of the tank each minute. If a second pipe can fill the tank in x min, the rate of work for the second pipe is $\frac{1}{x}$ of the tank each minute.

In solving a work problem, the goal is to determine the time it takes to complete a task. The basic equation that is used to solve work problems is

Rate of work × time worked = part of task completed

For example, if a faucet can fill a sink in 6 min, then in 5 min the faucet can fill $\frac{1}{6} \times 5 = \frac{5}{6}$ of the sink. In 5 min the faucet completes $\frac{5}{6}$ of the task.

➡ A painter can paint a wall in 20 min. The painter's apprentice can paint the same wall in 30 min. How long will it take them to paint the wall when they work together?

Strategy for Solving a Work Problem

1. For each person or machine, write a numerical or variable expression for the rate of work, the time worked, and the part of the task completed. The results can be recorded in a table.

Unknown time to paint the wall working together: t

	Rate of Work	·	*Time Worked*	=	*Part of Task Completed*
Painter	$\frac{1}{20}$	·	t	=	$\frac{t}{20}$
Apprentice	$\frac{1}{30}$	·	t	=	$\frac{t}{30}$

---

2. Determine how the parts of the task completed are related. Use the fact that the sum of the parts of the task completed must equal 1; the complete task.

$$\frac{t}{20} + \frac{t}{30} = 1$$

● The sum of the part of the task completed by the painter and the part of the task completed by the apprentice is 1.

$$60\left(\frac{t}{20} + \frac{t}{30}\right) = 60 \cdot 1$$

● Multiply by the LCM of 20 and 30.

$$3t + 2t = 60$$
$$5t = 60$$
$$t = 12$$

Working together, they will paint the wall in 12 min.

Example 1

A small water pipe takes three times longer to fill a tank than does a large water pipe. With both pipes open, it takes 4 h to fill the tank. Find the time it would take the small pipe, working alone, to fill the tank.

You Try It 1

Two computer printers that work at the same rate are working together to print the payroll checks for a large corporation. After they work together for 2 h, one of the printers quits. The second requires 3 more hours to complete the payroll checks. Find the time it would take one printer, working alone, to print the payroll.

Strategy

• Time for large pipe to fill the tank: t
  Time for small pipe to fill the tank: $3t$

	Rate	Time	Part
Small pipe	$\dfrac{1}{3t}$	4	$\dfrac{4}{3t}$
Large pipe	$\dfrac{1}{t}$	4	$\dfrac{4}{t}$

• The sum of the parts of the task completed by each pipe must equal 1.

Your strategy

Solution

$$\frac{4}{3t} + \frac{4}{t} = 1$$

$$3t\left(\frac{4}{3t} + \frac{4}{t}\right) = 3t \cdot 1$$

$$4 + 12 = 3t$$

$$16 = 3t$$

$$\frac{16}{3} = t$$

$$3t = 3\left(\frac{16}{3}\right) = 16$$

The small pipe, working alone, takes 16 h to fill the tank.

Your solution

Solution on p. A42

Objective B   *To solve uniform motion problems* ....

A car that travels constantly in a straight line at 30 mph is in uniform motion. **Uniform motion** means that the speed or direction of an object does not change.

The basic equation used to solve uniform motion problems is

Distance = rate × time

An alternative form of this equation can be written by solving the equation for time.

$$\frac{\textbf{Distance}}{\textbf{Rate}} = \textbf{time}$$

This form of the equation is useful when the total time of travel for two objects or the time of travel between two points is known.

➡ The speed of a boat in still water is 20 mph. The boat traveled 75 mi down a river in the same amount of time it took to travel 45 mi up the river. Find the rate of the river's current.

Strategy for Solving a Uniform Motion Problem
1. For each object, write a numerical or variable expression for the distance, rate, and time. The results can be recorded in a table.

CONSIDER THIS

Use the information given in the problem to fill in the "Distance" and "Rate" columns of the table. Fill in the "Time" column by dividing the two expressions you wrote in each row.

The unknown rate of the river's current: r

	Distance	÷	*Rate*	=	*Time*
Down river	75	÷	$20 + r$	=	$\frac{75}{20 + r}$
Up river	45	÷	$20 - r$	=	$\frac{45}{20 - r}$

2. Determine how the times traveled by each object are related. For example, it may be known that the times are equal, or the total time may be known.

$$\frac{75}{20 + r} = \frac{45}{20 - r}$$ • The time down the river is equal to the time up the river.

$$(20 + r)(20 - r)\frac{75}{20 + r} = (20 + r)(20 - r)\frac{45}{20 - r}$$ • Multiply by the LCM of the denominators.

$$(20 - r)75 = (20 + r)45$$
$$1500 - 75r = 900 + 45r$$
$$-120r = -600$$
$$r = 5$$

The rate of the river's current is 5 mph.

Example 2

A cyclist rode the first 20 mi of a trip at a constant rate. For the next 16 mi, the cyclist reduced the speed by 2 mph. The total time for the 36 mi was 4 h. Find the rate of the cyclist for each leg of the trip.

Strategy

- Rate for the first 20 mi: r
  Rate for the next 16 mi: $r - 2$

	Distance	Rate	Time
First 20 mi	20	r	$\dfrac{20}{r}$
Next 16 mi	16	$r - 2$	$\dfrac{16}{r - 2}$

- The total time for the trip was 4 h.

Solution

$$\frac{20}{r} + \frac{16}{r - 2} = 4$$

$$r(r - 2)\left[\frac{20}{r} + \frac{16}{r - 2}\right] = r(r - 2) \cdot 4$$

$$(r - 2)20 + 16r = 4r^2 - 8r$$

$$20r - 40 + 16r = 4r^2 - 8r$$

$$36r - 40 = 4r^2 - 8r$$

Solve the quadratic equation by factoring.

$$0 = 4r^2 - 44r + 40$$

$$0 = 4(r^2 - 11r + 10)$$

$$0 = r^2 - 11r + 10$$

$$0 = (r - 10)(r - 1)$$

$$r - 10 = 0 \qquad r - 1 = 0$$
$$r = 10 \qquad\quad r = 1$$

The solution $r = 1$ mph is not possible, because the rate on the last 16 mi would then be -1 mph.

$$r - 2 = 10 - 2 = 8$$

10 mph was the rate for the first 20 mi.
8 mph was the rate for the next 16 mi.

You Try It 2

The total time it took for a sailboat to sail back and forth across a lake 6 km wide was 2 h. The rate sailing back was three times the rate sailing across. Find the rate sailing out across the lake.

Your strategy

Your solution

Solution on p. A42

8.6 Exercises

Objective A   *Application Problems*

Solve.

1. An experienced painter can paint a garage twice as fast as an apprentice. Working together, the painters require 4 h to paint the garage. How long would it take the experienced painter, working alone, to paint the garage?

2. One grocery clerk can stock a shelf in 20 min, whereas a second clerk requires 30 min to stock the same shelf. How long would it take to stock the shelf if the two clerks worked together?

3. One person with a skiploader requires 12 h to remove a large quantity of earth. A second, larger skiploader can remove the same amount of earth in 4 h. How long would it take to remove the earth with both skiploaders working together?

4. One worker can dig the trenches for a sprinkler system in 3 h, whereas a second worker requires 6 h to do the same task. How long would it take to dig the trenches with both people working together?

5. One computer can solve a complex prime factorization problem in 75 h. A second computer can solve the same problem in 50 h. How long would it take both computers, working together, to solve the problem?

6. A new machine can make 10,000 aluminum cans three times faster than an older machine. With both machines working, 10,000 cans can be made in 9 h. How long would it take the new machine, working alone, to make 10,000 cans?

7. A small air conditioner will cool a room 2° in 15 min. A larger air conditioner will cool the room 2° in 10 min. How long would it take to cool the room 2° with both air conditioners operating?

8. One printing press can print the copies of the first edition of a book in 55 min, whereas a second printing press requires 66 min to print the same number of copies. How long would it take to print the first edition with both presses operating?

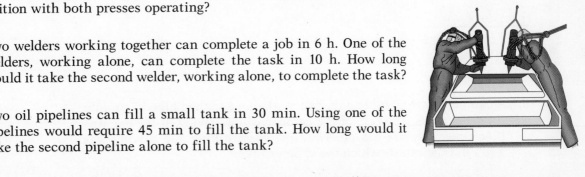

9. Two welders working together can complete a job in 6 h. One of the welders, working alone, can complete the task in 10 h. How long would it take the second welder, working alone, to complete the task?

10. Two oil pipelines can fill a small tank in 30 min. Using one of the pipelines would require 45 min to fill the tank. How long would it take the second pipeline alone to fill the tank?

Solve.

11. With two harvesters, a plot of land can be harvested in 1 h. One harvester, working alone, requires 1.5 h to harvest the field. How long would it take the second harvester, working alone, to harvest the field?

12. A wallpaper hanger requires 2 h to hang the wallpaper on one wall of a room. A second wallpaper hanger requires 4 h to hang the same amount of paper. The first wallpaper hanger worked alone for 1 h and then quit. How long will it take the second wallpaper hanger, working alone, to complete the wall?

13. A cement mason can build a barbeque in 8 h, whereas it takes a second mason 12 h to do the same task. After working alone for 4 h, the first mason quit. How long will it take the second mason to complete the task?

14. A mechanic requires 2 h to repair a transmission, whereas an apprentice requires 6 h to make the same repairs. The mechanic worked alone for 1 h and then stopped. How long will it take the apprentice, working alone, to complete the repairs?

15. One computer technician can wire a modem in 4 h, whereas it takes 6 h for a second technician to do the same job. After working alone for 2 h, the first technician quit. How long will it take the second technician to complete the wiring?

16. Two welders who work at the same rate are welding the girders of a building. After they work together for 10 h, one of the welders quits. The second welder requires 20 more hours to complete the welds. Find the time it would have taken one of the welders, working alone, to complete the welds.

17. A large and a small heating unit are being used to heat the water of a pool. The larger unit, working alone, requires 8 h to heat the pool. After both units have been operating 2 h, the larger unit is turned off. The small unit requires 9 more hours to heat the pool. How long would it take the small unit, working alone, to heat the pool?

18. Two machines that fill cereal boxes work at the same rate. After they work together for 7 h, one machine breaks down. The second machine requires 14 more hours to finish filling the boxes. How long would it have taken one of the machines, working alone, to fill the boxes?

Objective B   *Application Problems*

Solve.

19. A camper drove 90 mi to a recreational area and then hiked 5 mi into the wilderness. The rate of the camper while driving in the car was nine times the rate hiking. The time spent hiking and driving was 3 h. Find the rate at which the camper hiked.

Solve.

20. The president of a company traveled 1800 mi by jet and 300 mi on a prop plane. The rate of the jet was four times the rate of the prop plane. The entire trip took a total of 5 h. Find the rate of the jet plane.

21. As part of a conditioning program, a jogger ran 8 mi in the same time it took a cyclist to ride 20 mi. The rate of the cyclist was 12 mph faster than the rate of the jogger. Find the rate of the jogger and that of the cyclist.

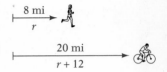

22. An express train travels 600 mi in the same amount of time it takes a freight train to travel 360 mi. The rate of the express train is 20 mph faster than that of the freight train. Find the rate of each train.

23. To assess the damage done by a fire, a forest ranger traveled 1080 mi by jet and then an additional 180 mi by helicopter. The rate of the jet was four times the rate of the helicopter. The entire trip took a total of 5 h. Find the rate of the jet.

24. A twin-engine plane can fly 800 mi in the same time that it takes a single-engine plane to fly 600 mi. The rate of the twin-engine plane is 50 mph faster than that of the single-engine plane. Find the rate of the twin-engine plane.

25. Two planes leave an airport and head for another airport 900 mi away. The rate of the first plane is twice that of the second plane. The second plane arrives at the airport 3 h after the first plane. Find the rate of the second plane.

26. A car and a bus leave a town at 1 P.M. and head for a town 300 mi away. The rate of the car is twice the rate of the bus. The car arrives 5 h ahead of the bus. Find the rate of the car.

27. A car is traveling at a rate that is 36 mph faster than the rate of a cyclist. The car travels 384 mi in the same time it takes the cyclist to travel 96 mi. Find the rate of the car.

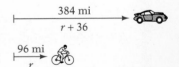

28. An engineer traveled 165 mi by car and then an additional 660 mi by plane. The rate of the plane was four times the rate of the car, and the total trip took 6 h. Find the rate of the car.

29. A backpacker hiking into a wilderness area walked 9 mi at a constant rate and then reduced this rate by 1 mph. Another 4 mi was hiked at this reduced rate. The time required to hike the 4 mi was 1 h less than the time required to walk the 9 mi. Find the rate at which the hiker walked the first 9 mi.

Solve.

30. A sailboat sailed 15 mi on the first leg of a trip before changing direction and sailing an additional 7 mi. Because of the wind, the change caused the sailboat to increase its speed by 2 mph. The total sailing time was 4 h. Find the rate of sailing for the first leg of the trip.

31. A small motor on a fishing boat can move the boat 6 mph in calm water. When trolling in a river, the amount of time it takes the boat to travel 12 mi against the river's current is the same as the time it takes to travel 24 mi with the current. Find the rate of the current.

32. A commercial jet can fly 550 mph in calm air. Traveling with the jet stream, the plane flew 2400 mi in the same amount of time it takes to fly 2000 mi against the jet stream. Find the rate of the jet stream.

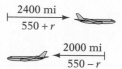

33. A cruise ship can sail at 28 mph in calm water. Sailing with the gulf current, the ship can sail 170 mi in the same amount of time that it can sail 110 mi against the gulf current. Find the rate of the gulf current.

34. Paddling in calm water, a canoeist can paddle at a rate of 8 mph. Traveling with the current, the canoeist went 30 mi in the same amount of time it took to travel 18 mi against the current. Find the rate of the current.

35. On a recent trip, a trucker traveled 330 mi at a constant rate. Because of road construction, the trucker then had to reduce the speed by 25 mph. An additional 30 mi was traveled at the reduced rate. The total time for the entire trip was 7 h. Find the rate of the trucker for the first 330 mi.

36. Commuting from work to home, a lab technician traveled 10 mi at a constant rate through congested traffic. Upon reaching the expressway, the technician increased the speed by 20 mph. An additional 20 mi was traveled at the increased speed. The total time for the trip was 1 h. Find the rate of travel through the congested traffic.

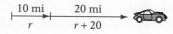

APPLYING THE CONCEPTS

37. One pipe can fill a tank in 2 h, a second pipe can fill the tank in 4 h, and a third pipe can fill the tank in 5 h. How long will it take to fill the tank with all three pipes working?

38. A surveyor traveled 32 mi by canoe and then hiked 4 mi. The rate of speed by boat was four times the rate on foot. If the time spent walking was 1 h less than the time spent canoeing, find the amount of time spent traveling by canoe.

39. Because of bad weather, a bus driver reduced the usual speed along a 150-mi bus route by 10 mph. The bus arrived only 30 min later than its usual arrival time. How fast does the bus usually travel?

8.7 Variation

Objective A *To solve variation problems* ....

Direct variation is a special function that can be expressed as the equation $y = kx$, where k is a constant. The equation $y = kx$ is read "y varies directly as x" or "y is proportional to x." The constant k is called the **constant of variation** or the **constant of proportionality.**

The circumference (C) of a circle varies directly as the diameter (d). The direct variation equation is written $C = \pi d$. The constant of variation is π.

A nurse makes $20 per hour. The total wage (w) of the nurse is directly proportional to the number of hours (h) worked. The equation of variation is $w = 20h$. The constant of proportionality is 20.

A direct variation equation can be written in the form $y = kx^n$, where n is a positive number. For example, the equation $y = kx^2$ is read "y varies directly as the square of x."

The area (A) of a circle varies directly as the square of the radius (r) of the circle. The direct variation equation is $A = \pi r^2$. The constant of variation is π.

➡ Given that V varies directly as r and that $V = 20$ when $r = 4$, find the constant of variation and the equation of variation.

$V = kr$	• Write the basic direct variation equation.
$20 = k \cdot 4$	• Replace V and r by the given values. Then solve for k.
$5 = k$	• This is the constant of variation.
$V = 5r$	• Write the direct variation equation by substituting the value of k into the basic direct variation equation.

➡ The tension (T) in a spring varies directly as the distance (x) it is stretched. If $T = 8$ lb when $x = 2$ in., find T when $x = 4$ in.

$T = kx$	• Write the basic direct variation equation.
$8 = k \cdot 2$	• Replace T and x by the given values.
$4 = k$	• Solve for the constant of variation.
$T = 4x$	• Write the direct variation equation.
$T = 4x = 4 \cdot 4 = 16$	• To find T when $x = 4$, substitute 4 for x in the equation and solve for T.

The tension is 16 lb.

Inverse variation is a function that can be expressed as the equation $y = \dfrac{k}{x}$, where k is a constant. The equation $y = \dfrac{k}{x}$ is read "y varies inversely as x" or "y is inversely proportional to x."

In general, an inverse variation equation can be written $y = \dfrac{k}{x^n}$, where n is a positive number. For example, the equation $y = \dfrac{k}{x^2}$ is read "y varies inversely as the square of x."

➡ Given that P varies inversely as the square of x and that $P = 5$ when $x = 2$, find the variation constant and the equation of variation.

$P = \dfrac{k}{x^2}$ • Write the basic inverse variation equation.

$5 = \dfrac{k}{2^2}$ • Replace P and x by the given values. Then solve for k.

$5 = \dfrac{k}{4}$

$20 = k$ • This is the constant of variation.

$P = \dfrac{20}{x^2}$ • Write the inverse variation equation by substituting the value of k into the basic inverse variation equation.

➡ The length (L) of a rectangle with fixed area is inversely proportional to the width (W). If $L = 6$ ft when $W = 2$ ft, find L when $W = 3$ ft.

$L = \dfrac{k}{W}$ • Write the basic inverse variation equation.

$6 = \dfrac{k}{2}$ • Replace L and W by the given values.

$12 = k$ • Solve for the constant of variation.

$L = \dfrac{12}{W}$ • Write the inverse variation equation.

$L = \dfrac{12}{W} = \dfrac{12}{3} = 4$ • To find L when $W = 3$ ft, substitute 3 for W in the equation and solve for L.

The length is 4 ft.

Joint variation is a variation in which a variable varies directly as the product of two or more other variables. A joint variation can be expressed as the equation $z = kxy$, where k is a constant. The equation $z = kxy$ is read "z varies jointly as x and y."

The area (A) of a triangle varies jointly as the base (b) and the height (h). The joint variation equation is written $A = \dfrac{1}{2}bh$. The constant of variation is $\dfrac{1}{2}$.

A **combined variation** is a variation in which two or more types of variation occur at the same time. For example, in physics, the volume (V) of a gas varies directly as the temperature (T) and inversely as the pressure (P). This combined variation is written $V = \dfrac{kT}{P}$.

⇒ A ball is being twirled on the end of a string. The tension (T) in the string is directly proportional to the square of the speed (v) of the ball and inversely proportional to the length (r) of the string. The tension is 96 lb when the length of the string is 0.5 ft and the speed is 4 ft/s. Find the tension when the length of the string is 1 ft and the speed is 5 ft/s.

$$T = \frac{kv^2}{r}$$ • Write the basic combined variation equation.

$$96 = \frac{k \cdot 4^2}{0.5}$$ • Replace T, v, and r by the given values.

$$96 = \frac{k \cdot 16}{0.5}$$

$$96 = k \cdot 32$$ • Solve for the constant of variation.
$$3 = k$$

$$T = \frac{3v^2}{r}$$ • Write the combined variation equation.

$$T = \frac{3v^2}{r} = \frac{3 \cdot 5^2}{1} = 3 \cdot 25 = 75$$ • To find T when $r = 1$ ft and $v = 5$ ft/s, substitute 1 for r and 5 for v and solve for T.

The tension is 75 lb.

Example 1

The amount (A) of medication prescribed for a person is directly related to the person's weight (W). For a 50-kilogram person, 2 ml of medication are prescribed. How many milliliters of medication are required for a person who weighs 75 kg?

Strategy

To find the required amount of medication:

- Write the basic direct variation equation, replace the variables by the given values, and solve for k.
- Write the direct variation equation, replacing k by its value. Substitute 75 for W and solve for A.

Solution

$A = kW$

$2 = k \cdot 50$

$\dfrac{1}{25} = k$

$A = \dfrac{1}{25}W = \dfrac{1}{25} \cdot 75 = 3$

The required amount of medication is 3 ml.

You Try It 1

The distance (s) a body falls from rest varies directly as the square of the time (t) of the fall. An object falls 64 ft in 2 s. How far will it fall in 5 s?

Your strategy

Your solution

Solution on p. A43

Example 2

A company that produces personal computers has determined that the number of computers it can sell (s) is inversely proportional to the price (P) of the computer. Two thousand computers can be sold when the price is $2500. How many computers can be sold when the price of a computer is $2000?

Strategy

To find the number of computers:

- Write the basic inverse variation equation, replace the variables by the given values, and solve for k.
- Write the inverse variation equation, replacing k by its value. Substitute 2000 for P and solve for s.

Solution $s = \dfrac{k}{P}$

$$2000 = \dfrac{k}{2500}$$
$$5,000,000 = k$$
$$s = \dfrac{5,000,000}{P} = \dfrac{5,000,000}{2000} = 2500$$

At $2000 each, 2500 computers can be sold.

You Try It 2

The resistance (R) to the flow of electric current in a wire of fixed length is inversely proportional to the square of the diameter (d) of a wire. If a wire of diameter 0.01 cm has a resistance of 0.5 ohm, what is the resistance in a wire that is 0.02 cm in diameter?

Your strategy

Your solution

Example 3

The pressure (P) of a gas varies directly as the temperature (T) and inversely as the volume (V). When $T = 50°$ and $V = 275$ in^3, $P = 20$ lb/in^2. Find the pressure of a gas when $T = 60°$ and $V = 250$ in^3.

Strategy

To find the pressure:

- Write the basic combined variation equation, replace the variables by the given values, and solve for k.
- Write the combined variation equation, replacing k by its value. Substitute 60 for T and 250 for V, and solve for P.

Solution $P = \dfrac{kT}{V}$

$$20 = \dfrac{k \cdot 50}{275}$$
$$110 = k$$
$$P = \dfrac{110T}{V} = \dfrac{110 \cdot 60}{250} = 26.4$$

The pressure is 26.4 lb/in^2.

You Try It 3

The strength (s) of a rectangular beam varies jointly as its width (w) and the square of its depth (d) and inversely as its length (L). If the strength of a beam 2 in. wide, 12 in. deep, and 12 ft long is 1200 lb, find the strength of a beam that is 4 in. wide, 8 in. deep, and 16 ft long.

Your strategy

Your solution

Solutions on p. A43

8.7 Exercises

· ·

Objective A   *Application Problems*

Solve.

1. The profit (P) realized by a company varies directly as the number of products it sells (s). If a company makes a profit of $4000 on the sale of 250 products, what is the profit when the company sells 5000 products?

2. The number of bushels of wheat (b) produced by a farm is directly proportional to the number of acres (A) planted in wheat. If a 20-acre farm yields 450 bushels of wheat, what is the yield of a farm that has 30 acres of wheat?

3. The pressure (p) on a diver in the water varies directly as the depth (d). If the pressure is 4.5 lb/in^2 when the depth is 10 ft, what is the pressure when the depth is 15 ft?

4. The distance (d) a spring will stretch varies directly as the force (f) applied to the spring. If a force of 6 lb is required to stretch a spring 3 in., what force is required to stretch the spring 4 in.?

5. The distance (d) a person can see to the horizon from a point above the surface of the earth varies directly as the square root of the height (H). If, for a height of 500 ft, the horizon is 19 mi away, how far is the horizon from a point that is 800 ft high? Round to the nearest hundredth.

6. The period (p) of a pendulum, or the time it takes the pendulum to make one complete swing, varies directly as the square root of the length (L) of the pendulum. If the period of a pendulum is 1.5 s when the length is 2 ft, find the period when the length is 5 ft. Round to the nearest hundredth.

7. The distance (s) a ball will roll down an inclined plane is directly proportional to the square of the time (t). If the ball rolls 6 ft in 1 s, how far will it roll in 3 s?

8. The stopping distance (s) of a car varies directly as the square of its speed (v). If a car traveling 50 mph requires 170 ft to stop, find the stopping distance for a car traveling 60 mph.

9. The length (L) of a rectangle of fixed area varies inversely as the width (W). If the length of a rectangle is 8 ft when the width is 5 ft, find the length of the rectangle when the width is 4 ft.

10. The number of items (n) that can be purchased for a given amount of money is inversely proportional to the cost (C) of an item. If 60 items can be purchased when the cost per item is $.25, how many items can be purchased when the cost per item is $.20?

Solve.

11. For a constant temperature, the pressure (P) of a gas varies inversely as the volume (V). The pressure is 30 lb/in^2 when the volume is 500 ft^3. Find the pressure when the volume is 200 ft^3.

12. The speed (v) of a gear varies inversely as the number of teeth (t). If a gear that has 45 teeth makes 24 revolutions per minute, how many revolutions per minute will a gear that has 36 teeth make?

13. The pressure (p) in a liquid varies directly as the product of the depth (d) and the density (D) of the liquid. The pressure is 150 lb/in^2 when the depth is 100 in. and the density is 1.2. Find the pressure when the density remains the same and the depth is 75 in.

14. The current (l) in a wire varies directly as the voltage (v) and inversely as the resistance (r). The current is 22 amps when the voltage is 110 volts and the resistance is 5 ohms. Find the current when the voltage is 195 volts and the resistance is 15 ohms.

15. The repulsive force (f) between the north poles of two magnets is inversely proportional to the square of the distance (d) between them. The repulsive force is 20 lb when the distance is 4 in. Find the repulsive force when the distance is 2 in.

16. The intensity (l) of a light source is inversely proportional to the square of the distance (d) from the source. The intensity is 12 lumens at a distance of 10 ft. What is the intensity when the distance is 5 ft?

17. The resistance (R) of a wire varies directly as the length (L) of the wire and inversely as the square of the diameter (d). The resistance is 9 ohms in 50 ft of wire that has a diameter of 0.05 in. Find the resistance in 50 ft of a similar wire that has a diameter of 0.02 in.

APPLYING THE CONCEPTS

18. In the inverse variation equation $y = \dfrac{k}{x}$, what is the effect on x when y doubles?

19. In the direct variation equation $y = kx$, what is the effect on y when x doubles?

Complete using the word *directly* or *inversely*.

20. If a varies directly as b and inversely as c, then c varies _____ as b and _____ as a.

21. If a varies _____ as b and c, then abc is constant.

22. If the length of a rectangle is held constant, the area of the rectangle varies _____ as the width.

23. If the area of a rectangle is held constant, the length of the rectangle varies _____ as the width.

Projects in Mathematics

Intensity of Illumination

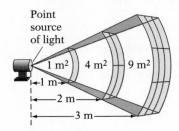

The rate at which light falls upon a one-square-unit area of surface is called the **intensity of illumination.** Intensity of illumination is measured in **lumens** (lm). A lumen is defined in the following illustration.

Picture a source of light equal to 1 candela positioned at the center of a hollow sphere that has a radius of 1 m. The rate at which light falls upon 1 m² of the inner surface of the sphere is equal to one lumen (1 lm). If a light source equal to 4 candelas is positioned at the center of the sphere, each square meter of the inner surface receives four times as much illumination, or 4 lm.

Light rays diverge as they leave a light source. The light that falls upon an area of 1 m² at a distance of 1 m from the source of light spreads out over an area of 4 m² when it is 2 m from the source. The same light spreads out over an area of 9 m² when it is 3 m from the light source and over an area of 16 m² when it is 4 m from the light source. Therefore, as a surface moves farther away from the source of light, the intensity of illumination on the surface decreases from its value at 1 m to $\left(\frac{1}{2}\right)^2$, or $\frac{1}{4}$, that value at 2 m; to $\left(\frac{1}{3}\right)^2$, or $\frac{1}{9}$, that value at 3 m; and to $\left(\frac{1}{4}\right)^2$, or $\frac{1}{16}$, that value at 4 m. The formula for the intensity of illumination is

$$I = \frac{s}{r^2}$$

where I is the intensity of illumination, s is the strength of the light source in candelas, and r is the distance in meters between the light source and the illuminated surface.

A 30-candela lamp is 0.5 m above a desk. Find the illumination on the desk.

$$I = \frac{s}{r^2} = \frac{30}{(0.5)^2} = 120$$

The illumination on the desk is 120 lm.

Solve.

1. A 100-candela light is hanging 5 m above a floor. What is the intensity of illumination on the floor beneath it?
2. A 25-candela source of light is 2 m above a desk. Find the intensity of illumination on the desk.
3. How strong a light source is needed to cast 80 lm of light on a surface 5 m from the source?
4. Two lights cast the same intensity of illumination on a wall. One light is 6 m from the wall and has a rating of 36 candelas. The second light is 8 m from the wall. Find the candela rating of the second light.

Chapter Summary

Key Words
A *rational expression* is a fraction in which the numerator or denominator is a polynomial. A rational expression is in *simplest form* when the numerator and denominator have no common factors.

The *least common multiple* (LCM) of two or more polynomials is the simplest polynomial that contains the factors of each polynomial.

The *reciprocal* of a rational expression is the rational expression with the numerator and denominator interchanged.

A *complex fraction* is a fraction whose numerator or denominator contains one or more fractions.

A *ratio* is the quotient of two quantities that have the same unit.

A *rate* is the quotient of two quantities that have different units.

A *proportion* is an equation that states the equality of two ratios or rates.

A *literal equation* is an equation that contains more than one variable.

Direct variation is a special function that can be expressed as the equation $y = kx$, where k is a constant called the *constant of variation* or the *constant of proportionality*.

Inverse variation is a function that can be expressed as the equation $y = \dfrac{k}{x}$, where k is a constant.

Joint variation is a variation in which a variable varies directly as the product of two or more variables. A joint variation can be expressed as the equation $z = kxy$, where k is a constant.

Combined variation is a variation in which two or more types of variation occur at the same time.

Essential Rules

To Multiply Fractions
$$\frac{a}{b} \cdot \frac{c}{d} = \frac{ac}{bd}$$

To Divide Fractions
$$\frac{a}{b} \div \frac{c}{d} = \frac{a}{b} \cdot \frac{d}{c} = \frac{ad}{bc}$$

To Add Fractions
$$\frac{a}{c} + \frac{b}{c} = \frac{a + b}{c}$$

To Subtract Fractions
$$\frac{a}{c} - \frac{b}{c} = \frac{a - b}{c}$$

Equation for Work Problems
$$\text{Rate of work} \times \text{time worked} = \text{part of task completed}$$

Uniform Motion Equation
$$\text{Distance} = \text{rate} \times \text{time}$$

Chapter Review Exercises

. . . .

1. Simplify: $\dfrac{16x^5y}{24x^2y^4}$

2. Simplify: $\dfrac{x^2 + 4x - 5}{1 - x^2}$

3. Multiply: $\dfrac{x^3y^4}{x^2 - 4x + 4} \cdot \dfrac{x^2 - x - 2}{x^6y^4}$

4. Multiply: $\dfrac{x^2 + 2x - 3}{x^2 + 6x + 9} \cdot \dfrac{2x^2 - 11x + 5}{2x^2 + 3x - 5}$

5. Divide: $\dfrac{x^2 + 3x + 2}{x^2 + 5x + 4} \div \dfrac{x^2 - x - 6}{x^2 + 2x - 15}$

6. Find the LCM of $6x - 3$ and $2x^2 + x - 1$.

7. Write each fraction in terms of the LCM of the denominators.
$$\dfrac{3}{x^2 - 2x}, \dfrac{x}{x^2 - 4}$$

8. Subtract: $\dfrac{2x}{x^2 + 3x - 10} - \dfrac{4}{x^2 + 3x - 10}$

9. Subtract: $\dfrac{2}{2x - 1} - \dfrac{3}{3x + 1}$

10. Subtract: $\dfrac{x}{x + 3} - \dfrac{2x - 5}{x^2 + x - 6}$

11. Simplify: $\dfrac{1 + \dfrac{1}{x} - \dfrac{12}{x^2}}{1 + \dfrac{2}{x} - \dfrac{8}{x^2}}$

12. Solve: $\dfrac{6}{x} - 2 = 1$

13. Solve: $\dfrac{2x}{x+1} - 3 = \dfrac{-2}{x+1}$

14. Solve: $\dfrac{3}{x+4} = \dfrac{5}{x+6}$

15. Triangles *ABC* and *DEF* are similar. Find the area of triangle *DEF*.

 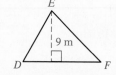

16. Solve $d = s + rt$ for t.

17. An interior designer uses 2 rolls of wallpaper for every 45 ft² of wall space in an office. At this rate, how many rolls of wallpaper are needed for an office that has 315 ft² of wall space?

18. One landscaper can till the soil for a lawn in 30 min, whereas it takes a second landscaper 15 min to do the same job. How long would it take to till the soil for the lawn with both landscapers working together?

19. A cyclist travels 20 mi in the same amount of time as it takes a hiker to walk 6 mi. The rate of the cyclist is 7 mph faster than the rate of the hiker. Find the rate of the cyclist.

20. The electrical resistance (r) of a cable varies directly as its length (l) and inversely as the square of its diameter (d). If a cable 16,000 ft long and $\dfrac{1}{4}$ in. in diameter has a resistance of 3.2 ohms, what is the resistance of a cable that is 8000 ft long and $\dfrac{1}{2}$ in. in diameter?

Cumulative Review Exercises

1. Simplify: $\left(\frac{2}{3}\right)^2 \div \left(\frac{3}{2} - \frac{2}{3}\right) + \frac{1}{2}$

2. Evaluate $-a^2 + (a - b)^2$ when $a = -2$ and $b = 3$.

3. Simplify: $-2x - (-3y) + 7x - 5y$

4. Simplify: $2[3x - 7(x - 3) - 8]$

5. Solve: $4 - \frac{2}{3}x = 7$

6. Solve: $3[x - 2(x - 3)] = 2(3 - 2x)$

7. Find $16\frac{2}{3}\%$ of 60.

8. Solve: $x - 3(1 - 2x) \geq 1 - 4(3 - 2x)$

9. Find the volume of the rectangular solid shown in the figure.

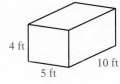

4 ft
5 ft
10 ft

10. Graph $-3x + 5y = -15$.

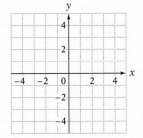

11. Given $P(x) = \frac{x - 1}{2x - 3}$, find $P(-2)$.

12. Find the equation of the line that contains the point $(-2, -1)$ and is parallel to the line $3x - 2y = 6$.

13. Evaluate the determinant:

$$\begin{vmatrix} 6 & 5 \\ 2 & -3 \end{vmatrix}$$

14. Multiply: $(a^2b^5)(ab^2)$

15. Simplify: $\dfrac{(2a^{-2}b^3)^{-2}}{(4a)^{-1}}$

16. Write 0.000000035 in scientific notation.

17. Multiply: $(2a^2 - 3a + 1)(-2a^2)$

18. Multiply: $(a - 3b)(a + 4b)$

19. Divide: $(x^3 - 8) \div (x - 2)$

20. Factor: $y^2 - 7y + 6$

21. Factor: $12x^2 - x - 1$

22. Factor: $2a^3 + 7a^2 - 15a$

23. Factor: $4b^2 - 100$

24. Solve: $(x + 3)(2x - 5) = 0$

25. Simplify: $\dfrac{12x^4y^2}{18xy^7}$

26. Simplify: $\dfrac{x^2 - 7x + 10}{25 - x^2}$

27. Divide: $\dfrac{x^2 - x - 56}{x^2 + 8x + 7} \div \dfrac{x^2 - 13x + 40}{x^2 - 4x - 5}$

28. Subtract: $\dfrac{2}{2x - 1} - \dfrac{1}{x + 1}$

29. Simplify: $\dfrac{1 - \dfrac{2}{x} - \dfrac{15}{x^2}}{1 - \dfrac{25}{x^2}}$

30. Solve: $\dfrac{3x}{x - 3} - 2 = \dfrac{10}{x - 3}$

31. A silversmith mixes 60 g of an alloy that is 40% silver with 120 g of another silver alloy. The resulting alloy is 60% silver. Find the percent of silver in the 120 g of alloy.

32. A life insurance policy costs $16 for every $1000 of coverage. At this rate, how much money would a policy of $5000 cost?

33. One water pipe can fill a tank in 9 min, whereas a second pipe requires 18 min to fill the tank. How long would it take both pipes, working together, to fill the tank?

Chapter

Rational Exponents and Radicals

Objectives

Golden Rectangle

The golden rectangle fascinated the early Greeks and appeared in much of their architecture. They considered this particular rectangle the most pleasing to the eye and believed that when it was used in the design of a building, it would make the structure pleasant to see.

The golden rectangle is constructed from a square by drawing a line from the midpoint of the base of the square to the opposite vertex. Now extend the base of the square, starting from the midpoint, the length of that line. The resulting rectangle is called the golden rectangle.

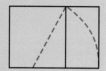

The Parthenon in Athens, Greece, is the classic example of the use of the golden rectangle in Greek architecture. A rendering of the Parthenon is shown here.

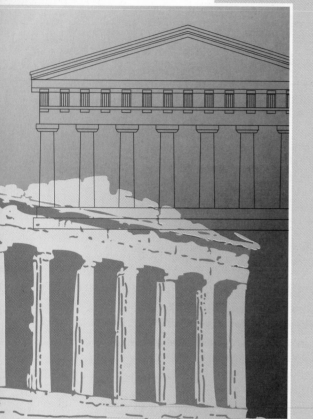

Rational Exponents and Radical Expressions

. . . .

Objective A   ***To simplify expressions with rational exponents*** . . . .

POINT OF INTEREST

Nicolas Chuquet (c. 1475), a French physician, wrote an algebra text in which he used a notation for expressions with fractional exponents. He wrote $R^2 6$ to mean $6^{1/2}$ and $R^3 15$ to mean $15^{1/3}$. This was an improvement over earlier notations that used words for these expressions.

In this section, the definition of an exponent is extended beyond integers so that any rational number can be used as an exponent. The definition is expressed in such a way that the Rules of Exponents hold true for rational exponents.

Consider the expression $(a^{1/n})^n$ for $a > 0$ and n a positive integer. Now simplify, assuming that the Rule for Simplifying Powers of Exponential Expressions is true.

$$(a^{1/n})^n = a^{\frac{1}{n} \cdot n} = a^1 = a$$

Because $(a^{1/n})^n = a$, the number $a^{1/n}$ is the number whose nth power is a.

> If $a > 0$ and n is a positive number, then $a^{1/n}$ is called the nth root of a.

$25^{1/2} = 5$ because $(5)^2 = 25$. $8^{1/3} = 2$ because $(2)^3 = 8$.

In the expression $a^{1/n}$, if a is a negative number and n is a positive even integer, then $a^{1/n}$ is not a real number.

$(-4)^{1/2}$ is not a real number, because there is no real number whose second power is -4.

When n is a positive odd integer, a can be a positive or a negative number.

$(-27)^{1/3} = -3$ because $(-3)^3 = -27$.

Using the definition of $a^{1/n}$ and the Rules of Exponents, it is possible to define any exponential expression that contains a rational exponent.

> If m and n are positive integers and $a^{1/n}$ is a real number, then
> $$a^{m/n} = (a^{1/n})^m$$

The expression $a^{m/n}$ can also be written $a^{m/n} = a^{m \cdot \frac{1}{n}} = (a^m)^{1/n}$.

As shown above, expressions that contain rational exponents do not always represent real numbers when the base of the exponential expression is a negative number. For this reason, all variables in this chapter represent positive numbers unless otherwise stated.

➡ Simplify: $27^{2/3}$

$27^{2/3} = (3^3)^{2/3}$ • **Rewrite 27 as 3^3.**

$\qquad\quad = 3^{3(2/3)}$ • **Use the Rule for Simplifying Powers of Exponential Expressions.**

$\qquad\quad = 3^2$ • **Simplify.**

$\qquad\quad = 9$

➡ Simplify: $32^{-2/5}$

$32^{-2/5} = (2^5)^{-2/5}$ • **Rewrite 32 as 2^5.**

$\qquad\quad = 2^{-2}$ • **Use the Rule for Simplifying Powers of Exponential Expressions.**

CONSIDER THIS

Note that $32^{-2/5} = \dfrac{1}{4}$, a positive number. The negative exponent does not affect the sign of a number.

$\qquad\quad = \dfrac{1}{2^2}$ • **Use the Definition of a Negative Exponent.**

$\qquad\quad = \dfrac{1}{4}$ • **Simplify.**

➡ Simplify: $a^{1/2} \cdot a^{2/3} \cdot a^{-1/4}$

$a^{1/2} \cdot a^{2/3} \cdot a^{-1/4} = a^{1/2+2/3-1/4}$ • **Use the Rule for Multiplying Exponential Expressions.**

$\qquad\qquad\qquad\quad = a^{6/12+8/12-3/12}$

$\qquad\qquad\qquad\quad = a^{11/12}$ • **Simplify.**

➡ Simplify: $(x^6y^4)^{3/2}$

$(x^6y^4)^{3/2} = x^{6(3/2)}y^{4(3/2)}$ • **Use the Rule for Simplifying Powers of Products.**

$\qquad\qquad = x^9y^6$ • **Simplify.**

➡ Simplify: $\left(\dfrac{8a^3b^{-4}}{64a^{-9}b^2}\right)^{2/3}$

$\left(\dfrac{8a^3b^{-4}}{64a^{-9}b^2}\right)^{2/3} = \left(\dfrac{2^3a^3b^{-4}}{2^6a^{-9}b^2}\right)^{2/3}$ • **Rewrite 8 as 2^3 and 64 as 2^6.**

$\qquad\qquad\qquad = (2^{-3}a^{12}b^{-6})^{2/3}$ • **Use the Rule for Dividing Exponential Expressions.**

$\qquad\qquad\qquad = 2^{-2}a^8b^{-4}$ • **Use the Rule for Simplifying Powers of Products.**

$\qquad\qquad\qquad = \dfrac{a^8}{2^2b^4} = \dfrac{a^8}{4b^4}$ • **Use the Definition of Negative Exponents and simplify.**

Example 1   Simplify: $64^{-2/3}$

Solution $64^{-2/3} = (2^6)^{-2/3} = 2^{-4}$

$$= \frac{1}{2^4} = \frac{1}{16}$$

You Try It 1   Simplify: $16^{-3/4}$

Your solution

Example 2   Simplify: $(-49)^{3/2}$

Solution   The base of the exponential expression is a negative number, while the denominator of the exponent is a positive even number.

Therefore, $(-49)^{3/2}$ is not a real number.

You Try It 2   Simplify: $(-81)^{3/4}$

Your solution

Example 3   Simplify: $(x^{1/2}y^{-3/2}z^{1/4})^{-3/2}$

Solution $(x^{1/2}y^{-3/2}z^{1/4})^{-3/2}$
$= x^{-3/4}y^{9/4}z^{-3/8}$

$$= \frac{y^{9/4}}{x^{3/4}z^{3/8}}$$

You Try It 3   Simplify: $(x^{3/4}y^{1/2}z^{-2/3})^{-4/3}$

Your solution

Example 4   Simplify: $\dfrac{x^{1/2}y^{-5/4}}{x^{-4/3}y^{1/3}}$

Solution $\dfrac{x^{1/2}y^{-5/4}}{x^{-4/3}y^{1/3}}$

$$= x^{3/6-(-8/6)}y^{-15/12-4/12}$$

$$= x^{11/6}y^{-19/12} = \frac{x^{11/6}}{y^{19/12}}$$

You Try It 4   Simplify: $\left(\dfrac{16a^{-2}b^{4/3}}{9a^4b^{-2/3}}\right)^{-1/2}$

Your solution

Solutions on p. A44

Objective B   ***To write exponential expressions as radical expressions and to write radical expressions as exponential expressions***

Content and Format © 1996 HMCo.

POINT OF INTEREST

The radical sign was introduced in 1525 in a book by Christoff Rudolff called *Coss*. He modified the symbol to indicate square roots, cube roots, and fourth roots. The idea of using an index, as we use it in our modern notation, did not occur until some years later.

Recall that $a^{1/n}$ is the nth root of a. The expression $\sqrt[n]{a}$ is another symbol for the nth root of a.

> If a is a real number, then $a^{1/n} = \sqrt[n]{a}$.

In the expression $\sqrt[n]{a}$, the symbol $\sqrt{}$ is called a **radical sign,** n is the **index,** and a is the **radicand.** When $n = 2$, the radical expression represents a square root and the index 2 is usually not written.

An exponential expression with a rational exponent can be written as a radical expression.

> If $a^{1/n}$ is a real number, then $a^{m/n} = a^{m \cdot 1/n} = (a^m)^{1/n} = \sqrt[n]{a^m}$.

The expression $a^{m/n}$ can also be written $a^{m/n} = a^{1/n \cdot m} = (\sqrt[n]{a})^m$.

The exponential expression at the right has been written as a radical expression.

$$y^{2/3} = (y^2)^{1/3}$$
$$= \sqrt[3]{y^2}$$

The radical expressions at the right have been written as exponential expressions.

$$\sqrt[5]{x^6} = (x^6)^{1/5} = x^{6/5}$$
$$\sqrt{17} = (17)^{1/2} = 17^{1/2}$$

⇒ Write $(5x)^{2/5}$ as a radical expression.

The denominator of the rational exponent is the index of the radical. The numerator is the power of the radicand.

$$(5x)^{2/5} = \sqrt[5]{(5x)^2}$$
$$= \sqrt[5]{25x^2}$$

⇒ Write $\sqrt[3]{x^4}$ as an exponential expression with a rational exponent.

The index of the radical is the denominator of the rational exponent. The power of the radicand is the numerator of the rational exponent.

$$\sqrt[3]{x^4} = (x^4)^{1/3} = x^{4/3}$$

⇒ Write $\sqrt[3]{a^3 + b^3}$ as an exponential expression with a rational exponent.

$$\sqrt[3]{a^3 + b^3} = (a^3 + b^3)^{1/3}$$

Note that $(a^3 + b^3)^{1/3} \neq a + b$.

Example 5   Write $(3x)^{5/4}$ as a radical expression.

Solution $(3x)^{5/4} = \sqrt[4]{(3x)^5} = \sqrt[4]{243x^5}$

You Try It 5   Write $(2x^3)^{3/4}$ as a radical expression.

Your solution

Example 6   Write $-2x^{2/3}$ as a radical expression.

Solution $-2x^{2/3} = -2(x^2)^{1/3} = -2\sqrt[3]{x^2}$

You Try It 6   Write $-5a^{5/6}$ as a radical expression.

Your solution

Example 7   Write $\sqrt[4]{3a}$ as an exponential expression.

Solution $\sqrt[4]{3a} = (3a)^{1/4}$

You Try It 7   Write $\sqrt[3]{3ab}$ as an exponential expression.

Your solution

Example 8   Write $\sqrt{a^2 - b^2}$ as an exponential expression.

Solution $\sqrt{a^2 - b^2} = (a^2 - b^2)^{1/2}$

You Try It 8   Write $\sqrt[4]{x^4 + y^4}$ as an exponential expression.

Your solution

Solutions on p. A44

Objective C   ***To simplify radical expressions that are roots of perfect powers*** ....

Every positive number has two square roots, one a positive and one a negative number. For example, because $(5)^2 = 25$ and $(-5)^2 = 25$, there are two square roots of 25: 5 and -5.

The symbol $\sqrt{}$ is used to indicate the positive or **principal square root.** To indicate the negative square root of a number, a negative sign is placed in front of the radical.

$$\sqrt{25} = 5$$

$$-\sqrt{25} = -5$$

The square root of zero is zero.

$$\sqrt{0} = 0$$

The square root of a negative number is not a real number, because the square of a real number must be positive.

$\sqrt{-25}$ is not a real number.

Note that

$$\sqrt{(-5)^2} = \sqrt{25} = 5 \quad \text{and} \quad \sqrt{5^2} = \sqrt{25} = 5$$

This is true for all real numbers and is stated as the following result.

> For any real number a, $\sqrt{a^2} = |a|$ and $-\sqrt{a^2} = -|a|$. If a is a positive real number, then $\sqrt{a^2} = a$ and $(\sqrt{a})^2 = a$.

Besides square roots, we can also determine cube roots, fourth roots, and so on.

$\sqrt[3]{8} = 2$, because $2^3 = 8$.   • **The cube root of a positive number is positive.**

$\sqrt[3]{-8} = -2$, because $(-2)^3 = -8$.   • **The cube root of a negative number is negative.**

$\sqrt[4]{625} = 5$, because $5^4 = 625$.

$\sqrt[5]{243} = 3$, because $3^5 = 243$.

The following properties hold true for finding the nth root of a real number.

CONSIDER THIS

Note that when the index is an even natural number, the nth root requires absolute value symbols.

$\sqrt[6]{y^6} = |y|$ but $\sqrt[5]{y^5} = y$

Because we stated that variables within radicals represent *positive* numbers, we will omit the absolute value symbols when writing an answer.

> If n is an even integer, then $\sqrt[n]{a^n} = |a|$ and $-\sqrt[n]{a^n} = -|a|$. If n is an odd integer, then $\sqrt[n]{a^n} = a$.

For example,

$$\sqrt[6]{y^6} = |y| \qquad -\sqrt[12]{x^{12}} = -|x| \qquad \sqrt[5]{b^5} = b$$

We have agreed that all variables in this chapter represent positive numbers unless otherwise stated, so it is not necessary to use the absolute value signs.

➡ Simplify: $\sqrt[3]{64}$

$\sqrt[3]{64} = \sqrt[3]{2^6}$ • Write the prime factorization of the radicand in simplest form.

$\phantom{\sqrt[3]{64}} = (2^6)^{1/3}$ • Write the radical expression as an exponential expression.

$\phantom{\sqrt[3]{64}} = 2^2 = 4$ • Use the Rule for Simplifying Powers of Exponential Expressions.

➡ Simplify: $\sqrt[4]{x^4 y^8}$

The radicand is a perfect fourth power because the exponents on the variables are divisible by 4.

CONSIDER THIS

To simplify the root of a perfect power, remove the radical sign and divide each exponent by the index.

$\sqrt[4]{x^4 y^8} = (x^4 y^8)^{1/4}$ • Write the radical expression as an exponential expression.

$\phantom{\sqrt[4]{x^4 y^8}} = xy^2$ • Use the Rule for Simplifying Powers of Products.

➡ Simplify: $\sqrt[5]{32x^{15}}$

$\sqrt[5]{32x^{15}} = (2^5 x^{15})^{1/5}$ • Write the prime factorization of 32. Write the radical expression as an exponential expression.

$\phantom{\sqrt[5]{32x^{15}}} = 2x^3$ • Use the Rule for Simplifying Powers of Products.

Example 9   Simplify: $\sqrt{49x^2 y^{12}}$

Solution $\sqrt{49x^2 y^{12}} = \sqrt{7^2 x^2 y^{12}}$

$\phantom{\sqrt{49x^2 y^{12}}} = (7^2 x^2 y^{12})^{1/2}$

$\phantom{\sqrt{49x^2 y^{12}}} = 7xy^6$

You Try It 9   Simplify: $\sqrt{121x^{10} y^4}$

Your solution

Example 10   Simplify: $\sqrt[3]{-125a^6 b^9}$

Solution $\sqrt[3]{-125a^6 b^9} = \sqrt[3]{(-5)^3 a^6 b^9}$

$\phantom{\sqrt[3]{-125a^6 b^9}} = [(-5)^3 a^6 b^9]^{1/3}$

$\phantom{\sqrt[3]{-125a^6 b^9}} = -5a^2 b^3$

You Try It 10   Simplify: $\sqrt[3]{-8x^{12} y^3}$

Your solution

Example 11   Simplify: $- \sqrt[4]{16a^4 b^8}$

Solution $- \sqrt[4]{16a^4 b^8} = - \sqrt[4]{2^4 a^4 b^8}$

$\phantom{- \sqrt[4]{16a^4 b^8}} = -(2^4 a^4 b^8)^{1/4}$

$\phantom{- \sqrt[4]{16a^4 b^8}} = -2ab^2$

You Try It 11   Simplify: $- \sqrt[4]{81x^{12} y^8}$

Your solution

Solutions on p. A44

9.1 Exercises

Objective A

Simplify.

1. $8^{1/3}$ **2.** $16^{1/2}$ **3.** $9^{3/2}$ **4.** $25^{3/2}$ **5.** $27^{-2/3}$

6. $64^{-1/3}$ **7.** $32^{2/5}$ **8.** $16^{3/4}$ **9.** $(-25)^{5/2}$ **10.** $(-36)^{1/4}$

11. $\left(\dfrac{25}{49}\right)^{-3/2}$ **12.** $\left(\dfrac{8}{27}\right)^{-2/3}$ **13.** $x^{1/2}x^{1/2}$ **14.** $a^{1/3}a^{5/3}$ **15.** $y^{-1/4}y^{3/4}$

16. $x^{2/5} \cdot x^{-4/5}$ **17.** $x^{-2/3} \cdot x^{3/4}$ **18.** $x \cdot x^{-1/2}$ **19.** $a^{1/3} \cdot a^{3/4} \cdot a^{-1/2}$ **20.** $y^{-1/6} \cdot y^{2/3} \cdot y^{1/2}$

21. $\dfrac{a^{1/2}}{a^{3/2}}$ **22.** $\dfrac{b^{1/3}}{b^{4/3}}$ **23.** $\dfrac{y^{-3/4}}{y^{1/4}}$ **24.** $\dfrac{x^{-3/5}}{x^{1/5}}$ **25.** $\dfrac{y^{2/3}}{y^{-5/6}}$

26. $\dfrac{b^{3/4}}{b^{-3/2}}$ **27.** $(x^2)^{-1/2}$ **28.** $(a^8)^{-3/4}$ **29.** $(x^{-2/3})^6$ **30.** $(y^{-5/6})^{12}$

31. $(a^{-1/2})^{-2}$ **32.** $(b^{-2/3})^{-6}$ **33.** $(x^{-3/8})^{-4/5}$ **34.** $(y^{-3/2})^{-2/9}$

35. $(a^{1/2} \cdot a)^2$ **36.** $(b^{2/3} \cdot b^{1/6})^6$ **37.** $(x^{-1/2} \cdot x^{3/4})^{-2}$ **38.** $(a^{1/2} \cdot a^{-2})^3$

39. $(y^{-1/2} \cdot y^{2/3})^{2/3}$ **40.** $(b^{-2/3} \cdot b^{1/4})^{-4/3}$ **41.** $(x^8 y^2)^{1/2}$ **42.** $(a^3 b^9)^{2/3}$

Simplify.

43. $(x^4 y^2 z^6)^{3/2}$ **44.** $(a^8 b^4 c^4)^{3/4}$ **45.** $(x^{-3} y^6)^{-1/3}$ **46.** $(a^2 b^{-6})^{-1/2}$

47. $(x^{-2} y^{1/3})^{-3/4}$ **48.** $(a^{-2/3} b^{2/3})^{3/2}$ **49.** $\left(\dfrac{x^{1/2}}{y^2}\right)^4$ **50.** $\left(\dfrac{b^{-3/4}}{a^{-1/2}}\right)^8$

51. $\dfrac{x^{1/4} \cdot x^{-1/2}}{x^{2/3}}$ **52.** $\dfrac{b^{1/2} \cdot b^{-3/4}}{b^{1/4}}$ **53.** $\left(\dfrac{y^{2/3} \cdot y^{-5/6}}{y^{1/9}}\right)^9$ **54.** $\left(\dfrac{a^{1/3} \cdot a^{-2/3}}{a^{1/2}}\right)^4$

55. $\left(\dfrac{b^2 \cdot b^{-3/4}}{b^{-1/2}}\right)^{-1/2}$ **56.** $\dfrac{(x^{-5/6} \cdot x^3)^{-2/3}}{x^{4/3}}$ **57.** $(a^{2/3} b^2)^6 (a^3 b^3)^{1/3}$ **58.** $(x^3 y^{-1/2})^{-2} (x^{-3} y^2)^{1/6}$

59. $(16 m^{-2} n^4)^{-1/2} (m n^{1/2})$ **60.** $(27 m^3 n^{-6})^{1/3} (m^{-1/3} n^{5/6})^6$

61. $\left(\dfrac{x^{1/2} y^{-3/4}}{y^{2/3}}\right)^{-6}$ **62.** $\left(\dfrac{x^{1/2} y^{-5/4}}{y^{-3/4}}\right)^{-4}$ **63.** $\left(\dfrac{2^{-6} b^{-3}}{a^{-1/2}}\right)^{-2/3}$ **64.** $\left(\dfrac{49 c^{5/3}}{a^{-1/4} b^{5/6}}\right)^{-3/2}$

65. $y^{3/2} (y^{1/2} - y^{-1/2})$ **66.** $y^{3/5} (y^{2/5} + y^{-3/5})$ **67.** $a^{-1/4} (a^{5/4} - a^{9/4})$ **68.** $x^{4/3} (x^{2/3} + x^{-1/3})$

69. $x^n \cdot x^{3n}$ **70.** $a^{2n} \cdot a^{-5n}$ **71.** $x^n \cdot x^{n/2}$ **72.** $a^{n/2} \cdot a^{-n/3}$

73. $\dfrac{y^{n/2}}{y^{-n}}$ **74.** $\dfrac{b^{m/3}}{b^m}$ **75.** $(x^{2n})^n$ **76.** $(x^{5n})^{2n}$

77. $(x^{n/4} y^{n/8})^8$ **78.** $(x^{n/2} y^{n/3})^6$ **79.** $(x^{n/5} y^{n/10})^{20}$ **80.** $(x^{n/2} y^{n/5})^{10}$

___

Objective B

Rewrite the exponential expressions as a radical expression.

81. $3^{1/4}$

82. $5^{1/2}$

83. $a^{3/2}$

84. $b^{4/3}$

85. $(2t)^{5/2}$

86. $(3x)^{2/3}$

87. $-2x^{2/3}$

88. $-3a^{2/5}$

89. $(a^2b)^{2/3}$

90. $(x^2y^3)^{3/4}$

91. $(a^2b^4)^{3/5}$

92. $(a^3b^7)^{3/2}$

93. $(4x - 3)^{3/4}$

94. $(3x - 2)^{1/3}$

95. $x^{-2/3}$

96. $b^{-3/4}$

Rewrite the radical expression as an exponential expression.

97. $\sqrt{14}$

98. $\sqrt{7}$

99. $\sqrt[3]{x}$

100. $\sqrt[4]{y}$

101. $\sqrt[3]{x^4}$

102. $\sqrt[4]{a^3}$

103. $\sqrt[5]{b^3}$

104. $\sqrt[4]{b^5}$

105. $\sqrt[3]{2x^2}$

106. $\sqrt[5]{4y^7}$

107. $-\sqrt{3x^5}$

108. $-\sqrt[4]{4x^5}$

109. $3x\sqrt[3]{y^2}$

110. $2y\sqrt{x^3}$

111. $\sqrt{a^2 - 2}$

112. $\sqrt{3 - y^2}$

___

Objective C

Simplify.

113. $\sqrt{x^{16}}$

114. $\sqrt{y^{14}}$

115. $-\sqrt{x^8}$

116. $-\sqrt{a^6}$

Simplify.

117. $\sqrt[3]{x^3y^9}$ **118.** $\sqrt[3]{a^6b^{12}}$ **119.** $-\sqrt[3]{x^{15}y^3}$ **120.** $-\sqrt[3]{a^9b^9}$

121. $\sqrt{16a^4b^{12}}$ **122.** $\sqrt{25x^8y^2}$ **123.** $\sqrt{-16x^4y^2}$ **124.** $\sqrt{-9a^4b^8}$

125. $\sqrt[3]{27x^9}$ **126.** $\sqrt[3]{8a^{21}b^6}$ **127.** $\sqrt[3]{-64x^9y^{12}}$ **128.** $\sqrt[3]{-27a^3b^{15}}$

129. $-\sqrt[4]{x^8y^{12}}$ **130.** $-\sqrt[4]{a^{16}b^4}$ **131.** $\sqrt[5]{x^{20}y^{10}}$ **132.** $\sqrt[5]{a^5b^{25}}$

133. $\sqrt[4]{81x^4y^{20}}$ **134.** $\sqrt[4]{16a^8b^{20}}$ **135.** $\sqrt[5]{32a^5b^{10}}$ **136.** $\sqrt[5]{-32x^{15}y^{20}}$

APPLYING THE CONCEPTS

137. Determine whether the following statements are true or false. If the statement is false, correct the right side of the equation.

   a. $\sqrt{(-2)^2} = -2$ **b.** $\sqrt[3]{(-3)^3} = -3$ **c.** $\sqrt[n]{a} = a^{1/n}$

   d. $\sqrt[n]{a^n + b^n} = a + b$ **e.** $(a^{1/2} + b^{1/2})^2 = a + b$ **f.** $\sqrt[m]{a^n} = a^{mn}$

138. Simplify.

   a. $\sqrt[3]{\sqrt{x^6}}$ **b.** $\sqrt[4]{\sqrt{a^8}}$ **c.** $\sqrt{\sqrt{81y^8}}$

   d. $\sqrt{\sqrt[n]{a^{4n}}}$ **e.** $\sqrt[n]{\sqrt{b^{6n}}}$ **f.** $\sqrt{\sqrt[3]{x^{12}y^{24}}}$

139. Show how to locate $\sqrt{2}$ on the number line.
[W]

140. Is $\sqrt{x^2} = x$ always true? Show why or why not.
[W]

9.2 Operations on Radical Expressions

Objective A   *To simplify radical expressions*

If a number is not a perfect power, its root can only be approximated; examples include $\sqrt{5}$ and $\sqrt[3]{3}$. These numbers are **irrational numbers.** Their decimal representations never terminate or repeat.

$$\sqrt{5} = 2.2360679\ldots \qquad \sqrt[3]{3} = 1.4422495\ldots$$

A radical expression is in simplest form when the radicand contains no factor that is a perfect power. The Product Property of Radicals is used to simplify radical expressions whose radicands are not perfect powers.

The Product Property of Radicals

If $\sqrt[n]{a}$ and $\sqrt[n]{b}$ are positive real numbers, then $\sqrt[n]{ab} = \sqrt[n]{a} \cdot \sqrt[n]{b}$ and $\sqrt[n]{a} \cdot \sqrt[n]{b} = \sqrt[n]{ab}$.

➡ Simplify: $\sqrt[3]{x^7}$

$\sqrt[3]{x^7} = \sqrt[3]{x^6 \cdot x}$
- Write the radicand as the product of a perfect cube and a factor that does not contain a perfect cube.

$= \sqrt[3]{x^6}\sqrt[3]{x}$
- Use the Product Property of Radicals to write the expression as a product.

$= x^2\sqrt[3]{x}$
- Simplify.

➡ Simplify: $\sqrt[4]{32x^7}$

$\sqrt[4]{32x^7} = \sqrt[4]{2^5 x^7}$
- Write the prime factorization of the coefficient of the radicand in exponential form.

$= \sqrt[4]{2^4 x^4 (2x^3)}$
- Write the radicand as the product of a perfect fourth power and factors that do not contain a perfect fourth power.

$= \sqrt[4]{2^4 x^4}\sqrt[4]{2x^3}$
- Use the Product Property of Radicals to write the expression as a product.

$= 2x\sqrt[4]{2x^3}$
- Simplify.

Example 1   Simplify: $\sqrt[4]{x^9}$

Solution $\sqrt[4]{x^9} = \sqrt[4]{x^8 \cdot x} = \sqrt[4]{x^8}\sqrt[4]{x}$
$= x^2\sqrt[4]{x}$

You Try It 1   Simplify: $\sqrt[5]{x^7}$

Your solution

Solution on p. A44

Example 2

Simplify: $\sqrt[3]{-27a^5b^{12}}$

Solution

$$\sqrt[3]{-27a^5b^{12}} = \sqrt[3]{(-3)^3a^5b^{12}}$$
$$= \sqrt[3]{(-3)^3a^3b^{12}(a^2)}$$
$$= \sqrt[3]{(-3)^3a^3b^{12}}\sqrt[3]{a^2}$$
$$= -3ab^4\sqrt[3]{a^2}$$

You Try It 2

Simplify: $\sqrt[3]{-64x^8y^{18}}$

Your solution

Solution on p. A44

Objective B   *To add or subtract radical expressions* ....

The Distributive Property is used to simplify the sum or difference of radical expressions that have the same radicand and the same index. For example,

$$3\sqrt{5} + 8\sqrt{5} = (3 + 8)\sqrt{5} = 11\sqrt{5}$$
$$2\sqrt[3]{3x} - 9\sqrt[3]{3x} = (2 - 9)\sqrt[3]{3x} = -7\sqrt[3]{3x}$$

CONSIDER THIS

Adding and subtracting radicals is similar to combining like terms.

Radical expressions that are in simplest form and have unlike radicands or different indices cannot be simplified by the Distributive Property. The expressions below cannot be simplified by the Distributive Property.

$$3\sqrt[4]{2} - 6\sqrt[4]{3} \qquad\qquad 2\sqrt[4]{4x} + 3\sqrt[3]{4x}$$

➡ Simplify: $3\sqrt{32x^2} - 2x\sqrt{2} + \sqrt{128x^2}$

First simplify each term. Then combine like terms by using the Distributive Property.

$$3\sqrt{32x^2} - 2x\sqrt{2} + \sqrt{128x^2} = 3\sqrt{2^5x^2} - 2x\sqrt{2} + \sqrt{2^7x^2}$$
$$= 3\sqrt{2^4x^2}\sqrt{2} - 2x\sqrt{2} + \sqrt{2^6x^2}\sqrt{2}$$
$$= 3 \cdot 2^2x\sqrt{2} - 2x\sqrt{2} + 2^3x\sqrt{2}$$
$$= 12x\sqrt{2} - 2x\sqrt{2} + 8x\sqrt{2}$$
$$= 18x\sqrt{2}$$

Example 3

Subtract: $5b\sqrt[4]{32a^7b^5} - 2a\sqrt[4]{162a^3b^9}$

Solution

$5b\sqrt[4]{32a^7b^5} - 2a\sqrt[4]{162a^3b^9}$
$$= 5b\sqrt[4]{2^5a^7b^5} - 2a\sqrt[4]{3^4 \cdot 2a^3b^9}$$
$$= 5b\sqrt[4]{2^4a^4b^4}\sqrt[4]{2a^3b} - 2a\sqrt[4]{3^4b^8}\sqrt[4]{2a^3b}$$
$$= 5b \cdot 2ab\sqrt[4]{2a^3b} - 2a \cdot 3b^2\sqrt[4]{2a^3b}$$
$$= 10ab^2\sqrt[4]{2a^3b} - 6ab^2\sqrt[4]{2a^3b}$$
$$= 4ab^2\sqrt[4]{2a^3b}$$

You Try It 3

Subtract: $3xy\sqrt[3]{81x^5y} - \sqrt[3]{192x^8y^4}$

Your solution

Solution on p. A44

Objective C   *To multiply radical expressions* ....

The Product Property of Radicals is used to multiply radical expressions with the same index.

$$\sqrt{3x} \cdot \sqrt{5y} = \sqrt{3x \cdot 5y}$$
$$= \sqrt{15xy}$$

When the expression $(\sqrt{x})^2$ is simplified using the Product Property of Radicals, the result is x.

$$(\sqrt{x})^2 = \sqrt{x} \cdot \sqrt{x} = \sqrt{x \cdot x}$$
$$= \sqrt{x^2} = x$$

➡ Multiply: $\sqrt[3]{2a^5b}\ \sqrt[3]{16a^2b^2}$

$$\sqrt[3]{2a^5b}\ \sqrt[3]{16a^2b^2} = \sqrt[3]{32a^7b^3}$$
• Use the Product Property of Radicals to multiply the radicands.

$$= \sqrt[3]{2^5a^7b^3}$$
• Simplify.

$$= \sqrt[3]{2^3a^6b^3}\ \sqrt[3]{2^2a}$$

$$= 2a^2b\ \sqrt[3]{4a}$$

➡ Multiply: $\sqrt{2x}\,(\sqrt{8x} - \sqrt{3}\,)$

$$\sqrt{2x}\,(\sqrt{8x} - \sqrt{3}\,) = \sqrt{2x}\,(\sqrt{8x}\,) - \sqrt{2x}\,(\sqrt{3}\,)$$
• Use the Distributive Property.

$$= \sqrt{16x^2} - \sqrt{6x}$$
• Simplify.

$$= \sqrt{2^4x^2} - \sqrt{6x}$$

$$= 2^2x - \sqrt{6x}$$

$$= 4x - \sqrt{6x}$$

➡ Multiply: $(2\sqrt{5} - 3)(3\sqrt{5} + 4)$

$$(2\sqrt{5} - 3)(3\sqrt{5} + 4) = 6(\sqrt{5}\,)^2 + 8\sqrt{5} - 9\sqrt{5} - 12$$
• Use the FOIL method to multiply the numbers.

$$= 30 + 8\sqrt{5} - 9\sqrt{5} - 12$$

$$= 18 - \sqrt{5}$$
• Combine like terms.

➡ Multiply: $(4\sqrt{a} - \sqrt{b}\,)(2\sqrt{a} + 5\sqrt{b}\,)$

$$(4\sqrt{a} - \sqrt{b}\,)(2\sqrt{a} + 5\sqrt{b}\,)$$
$$= 8(\sqrt{a}\,)^2 + 20\sqrt{ab} - 2\sqrt{ab} - 5(\sqrt{b}\,)^2$$
• Use the FOIL method.
$$= 8a + 18\sqrt{ab} - 5b$$

CONSIDER THIS

The concept of conjugate is used in a number of different instances. Make sure you understand this idea.

The conjugate of $\sqrt{3} - 4$ is $\sqrt{3} + 4$.

The conjugate of $\sqrt{3} + 4$ is $\sqrt{3} - 4$.

The conjugate of $\sqrt{5a} + \sqrt{b}$ is $\sqrt{5a} - \sqrt{b}$.

The expressions $a + b$ and $a - b$ are **conjugates** of each other. Recall that $(a + b)(a - b) = a^2 - b^2$.

➡ Multiply: $(\sqrt{11} - 3)(\sqrt{11} + 3)$

$$(\sqrt{11} - 3)(\sqrt{11} + 3) = (\sqrt{11}\,)^2 - 3^2$$
• The expressions are conjugates.

$$= 11 - 9 = 2$$

Example 4

Multiply: $\sqrt{3x}\left(\sqrt{27x^2} - \sqrt{3x}\right)$

Solution

$$\sqrt{3x}\left(\sqrt{27x^2} - \sqrt{3x}\right) = \sqrt{81x^3} - \left(\sqrt{3x}\right)^2$$

$$= \sqrt{3^4x^3} - 3x$$

$$= \sqrt{3^4x^2}\sqrt{x} - 3x$$

$$= 3^2x\sqrt{x} - 3x$$

$$= 9x\sqrt{x} - 3x$$

You Try It 4

Multiply: $\sqrt{5b}\left(\sqrt{3b} - \sqrt{10}\right)$

Your solution

Example 5

Multiply: $\left(2\sqrt[3]{x} - 3\right)\left(3\sqrt[3]{x} - 4\right)$

Solution

$$\left(2\sqrt[3]{x} - 3\right)\left(3\sqrt[3]{x} - 4\right)$$

$$= 6\sqrt[3]{x^2} - 8\sqrt[3]{x} - 9\sqrt[3]{x} + 12$$

$$= 6\sqrt[3]{x^2} - 17\sqrt[3]{x} + 12$$

You Try It 5

Multiply: $\left(2\sqrt[3]{2x} - 3\right)\left(\sqrt[3]{2x} - 5\right)$

Your solution

Example 6

Multiply: $\left(2\sqrt{x} - \sqrt{2y}\right)\left(2\sqrt{x} + \sqrt{2y}\right)$

Solution

$$\left(2\sqrt{x} - \sqrt{2y}\right)\left(2\sqrt{x} + \sqrt{2y}\right)$$

$$= \left(2\sqrt{x}\right)^2 - \left(\sqrt{2y}\right)^2$$

$$= 4x - 2y$$

You Try It 6

Multiply: $\left(\sqrt{a} - 3\sqrt{y}\right)\left(\sqrt{a} + 3\sqrt{y}\right)$

Your solution

Solutions on pp. A44–A45

Objective D   *To divide radical expressions* ....

The Quotient Property of Radicals is used to divide radical expressions with the same index.

The Quotient Property of Radicals

If $\sqrt[n]{a}$ and $\sqrt[n]{b}$ are real numbers, and $b \neq 0$, then

$$\sqrt[n]{\frac{a}{b}} = \frac{\sqrt[n]{a}}{\sqrt[n]{b}} \quad \text{and} \quad \frac{\sqrt[n]{a}}{\sqrt[n]{b}} = \sqrt[n]{\frac{a}{b}}$$

POINT OF INTEREST

A radical expression that occurs in Einstein's Theory of Relativity is

$$\frac{1}{\sqrt{1 - \dfrac{v^2}{c^2}}}$$

where v is the velocity of an object and c is the speed of light.

➡ Simplify: $\sqrt[3]{\dfrac{81x^5}{y^6}}$

$$\sqrt[3]{\frac{81x^5}{y^6}} = \frac{\sqrt[3]{81x^5}}{\sqrt[3]{y^6}}$$

$$= \frac{\sqrt[3]{3^4x^5}}{\sqrt[3]{y^6}}$$

$$= \frac{\sqrt[3]{3^3x^3}\,\sqrt[3]{3x^2}}{\sqrt[3]{y^6}}$$

$$= \frac{3x\,\sqrt[3]{3x^2}}{y^2}$$

- Use the Quotient Property of Radicals.

- Simplify each radical expression.

➡ Simplify: $\dfrac{\sqrt{5a^4b^7c^2}}{\sqrt{ab^3c}}$

$$\frac{\sqrt{5a^4b^7c^2}}{\sqrt{ab^3c}} = \sqrt{\frac{5a^4b^7c^2}{ab^3c}}$$

$$= \sqrt{5a^3b^4c}$$

$$= \sqrt{a^2b^4}\,\sqrt{5ac}$$

$$= ab^2\,\sqrt{5ac}$$

- Use the Quotient Property of Radicals.

- Simplify the radicand.

A radical expression is in simplest form when no radical remains in the denominator of the radical expression. The procedure used to remove a radical from the denominator is called **rationalizing the denominator.**

➡ Simplify: $\dfrac{5}{\sqrt{2}}$

Multiply by $\dfrac{\sqrt{2}}{\sqrt{2}}$, which equals 1.

$$\frac{5}{\sqrt{2}} = \frac{5}{\sqrt{2}} \cdot 1 = \frac{5}{\sqrt{2}} \cdot \frac{\sqrt{2}}{\sqrt{2}}$$

$$= \frac{5\sqrt{2}}{2}$$

- $\dfrac{\sqrt{2}}{\sqrt{2}} = 1$

- $\sqrt{2} \cdot \sqrt{2} = (\sqrt{2})^2 = 2$

CONSIDER THIS

Note that multiplying by $\dfrac{\sqrt[3]{4x}}{\sqrt[3]{4x}}$ will not rationalize the denominator of $\dfrac{3x}{\sqrt[3]{4x}}$.

$$\frac{3x}{\sqrt[3]{4x}} \cdot \frac{\sqrt[3]{4x}}{\sqrt[3]{4x}} = \frac{3x\sqrt[3]{4x}}{\sqrt[3]{16x^2}}$$

Because the radicand of $\sqrt[3]{16x^2}$ is not a perfect cube, the denominator still contains a radical expression.

➡ Simplify: $\dfrac{3x}{\sqrt[3]{4x}}$

$$\frac{3x}{\sqrt[3]{4x}} = \frac{3x}{\sqrt[3]{2^2x}} \cdot \frac{\sqrt[3]{2x^2}}{\sqrt[3]{2x^2}}$$

$$= \frac{3x\,\sqrt[3]{2x^2}}{\sqrt[3]{2^3x^3}} = \frac{3x\,\sqrt[3]{2x^2}}{2x}$$

$$= \frac{3\,\sqrt[3]{2x^2}}{2}$$

- $\sqrt[3]{2^2x} \cdot \sqrt[3]{2x^2} = \sqrt[3]{2^3x^3}$, the cube root of a perfect cube. Multiply the numerator and denominator by $\dfrac{\sqrt[3]{2x^2}}{\sqrt[3]{2x^2}}$, which equals 1.

- Simplify.

CONSIDER THIS

Here is an example of using a conjugate to simplify a radical expression.

➡ Simplify: $\dfrac{\sqrt{x} - \sqrt{y}}{\sqrt{x} + \sqrt{y}}$

To simplify a fraction that has a binomial radical expression in the denominator, multiply the numerator and denominator by the conjugate of the denominator. Then simplify.

$$\frac{\sqrt{x} - \sqrt{y}}{\sqrt{x} + \sqrt{y}} = \frac{\sqrt{x} - \sqrt{y}}{\sqrt{x} + \sqrt{y}} \cdot \frac{\sqrt{x} - \sqrt{y}}{\sqrt{x} - \sqrt{y}}$$

$$= \frac{(\sqrt{x})^2 - \sqrt{xy} - \sqrt{xy} + (\sqrt{y})^2}{(\sqrt{x})^2 - (\sqrt{y})^2}$$

$$= \frac{x - 2\sqrt{xy} + y}{x - y}$$

Example 7 Simplify: $\dfrac{5}{\sqrt{5x}}$

Solution $\dfrac{5}{\sqrt{5x}} = \dfrac{5}{\sqrt{5x}} \cdot \dfrac{\sqrt{5x}}{\sqrt{5x}}$

$$= \frac{5\sqrt{5x}}{(\sqrt{5^2 x^2})}$$

$$= \frac{5\sqrt{5x}}{5x}$$

$$= \frac{\sqrt{5x}}{x}$$

You Try It 7 Simplify: $\dfrac{y}{\sqrt{3y}}$

Your solution

Example 8 Simplify: $\dfrac{3}{\sqrt[4]{2x}}$

Solution $\dfrac{3}{\sqrt[4]{2x}} = \dfrac{3}{\sqrt[4]{2x}} \cdot \dfrac{\sqrt[4]{2^3 x^3}}{\sqrt[4]{2^3 x^3}}$

$$= \frac{3\sqrt[4]{8x^3}}{\sqrt[4]{2^4 x^4}}$$

$$= \frac{3\sqrt[4]{8x^3}}{2x}$$

You Try It 8 Simplify: $\dfrac{3}{\sqrt[3]{3x^2}}$

Your solution

Example 9 Simplify: $\dfrac{3}{5 - 2\sqrt{3}}$

Solution $\dfrac{3}{5 - 2\sqrt{3}} \cdot \dfrac{5 + 2\sqrt{3}}{5 + 2\sqrt{3}} = \dfrac{15 + 6\sqrt{3}}{5^2 - (2\sqrt{3})^2}$

$$= \frac{15 + 6\sqrt{3}}{25 - 12}$$

$$= \frac{15 + 6\sqrt{3}}{13}$$

You Try It 9 Simplify: $\dfrac{3 + \sqrt{6}}{2 - \sqrt{6}}$

Your solution

Solutions on p. A45

9.2 Exercises

Objective A

Simplify.

1. $\sqrt{x^4 y^3 z^5}$

2. $\sqrt{x^3 y^6 z^9}$

3. $\sqrt{8a^3 b^8}$

4. $\sqrt{24a^9 b^6}$

5. $\sqrt{45x^2 y^3 z^5}$

6. $\sqrt{60xy^7 z^{12}}$

7. $\sqrt{-9x^3}$

8. $\sqrt{-x^2 y^5}$

9. $\sqrt[3]{a^{16} b^8}$

10. $\sqrt[3]{a^5 b^8}$

11. $\sqrt[3]{-125x^2 y^4}$

12. $\sqrt[3]{-216x^5 y^9}$

13. $\sqrt[3]{a^4 b^5 c^6}$

14. $\sqrt[3]{a^8 b^{11} c^{15}}$

15. $\sqrt[4]{16x^9 y^5}$

16. $\sqrt[4]{64x^8 y^{10}}$

Objective B

Add or subtract.

17. $2\sqrt{x} - 8\sqrt{x}$

18. $3\sqrt{y} + 12\sqrt{y}$

19. $\sqrt{8} - \sqrt{32}$

20. $\sqrt{27a} - \sqrt{8a}$

21. $\sqrt{18b} + \sqrt{75b}$

22. $2\sqrt{2x^3} + 4x\sqrt{8x}$

23. $3\sqrt{8x^2 y^3} - 2x\sqrt{32y^3}$

24. $2\sqrt{32x^2 y^3} - xy\sqrt{98y}$

25. $2a\sqrt{27ab^5} + 3b\sqrt{3a^3 b}$

26. $\sqrt[3]{128} + \sqrt[3]{250}$

27. $\sqrt[3]{16} - \sqrt[3]{54}$

28. $2\sqrt[3]{3a^4} - 3a\sqrt[3]{81a}$

29. $2b\sqrt[3]{16b^2} + \sqrt[3]{128b^5}$

30. $3\sqrt[3]{x^5 y^7} - 8xy\sqrt[3]{x^2 y^4}$

Chapter 9 / Rational Exponents and Radicals

Add or subtract.

31. $3\sqrt[4]{32a^5} - a\sqrt[4]{162a}$

32. $2a\sqrt[4]{16ab^5} + 3b\sqrt[4]{256a^5b}$

33. $2\sqrt{50} - 3\sqrt{125} + \sqrt{98}$

34. $3\sqrt{108} - 2\sqrt{18} - 3\sqrt{48}$

35. $\sqrt{9b^3} - \sqrt{25b^3} + \sqrt{49b^3}$

36. $\sqrt{4x^7y^5} + 9x^2\sqrt{x^3y^5} - 5xy\sqrt{x^5y^3}$

37. $2x\sqrt{8xy^2} - 3y\sqrt{32x^3} + \sqrt{4x^3y^3}$

38. $5a\sqrt{3a^3b} + 2a^2\sqrt{27ab} - 4\sqrt{75a^5b}$

39. $\sqrt[3]{54xy^3} - 5\sqrt[3]{2xy^3} + y\sqrt[3]{128x}$

40. $2\sqrt[3]{24x^3y^4} + 4x\sqrt[3]{81y^4} - 3y\sqrt[3]{24x^3y}$

41. $2a\sqrt[4]{32b^5} - 3b\sqrt[4]{162a^4b} + \sqrt[4]{2a^4b^5}$

42. $6y\sqrt[4]{48x^5} - 2x\sqrt[4]{243xy^4} - 4\sqrt[4]{3x^5y^4}$

Objective C

Multiply.

43. $\sqrt{8}\sqrt{32}$

44. $\sqrt{14}\sqrt{35}$

45. $\sqrt[3]{4}\sqrt[3]{8}$

46. $\sqrt[3]{6}\sqrt[3]{36}$

47. $\sqrt{x^2y^5}\sqrt{xy}$

48. $\sqrt{a^3b}\sqrt{ab^4}$

49. $\sqrt{2x^2y}\sqrt{32xy}$

50. $\sqrt{5x^3y}\sqrt{10x^3y^4}$

51. $\sqrt[3]{x^2y}\sqrt[3]{16x^4y^2}$

52. $\sqrt[3]{4a^2b^3}\sqrt[3]{8ab^5}$

53. $\sqrt[4]{12ab^3}\sqrt[4]{4a^5b^2}$

54. $\sqrt[4]{36a^2b^4}\sqrt[4]{12a^5b^3}$

55. $\sqrt{3}(\sqrt{27} - \sqrt{3})$

56. $\sqrt{10}(\sqrt{10} - \sqrt{5})$

57. $\sqrt{x}(\sqrt{x} - \sqrt{2})$

58. $\sqrt{y}(\sqrt{y} - \sqrt{5})$

59. $\sqrt{2x}(\sqrt{8x} - \sqrt{32})$

60. $\sqrt{3a}(\sqrt{27a^2} - \sqrt{a})$

Content and Format © 1996 HMCo.

Multiply.

61. $(\sqrt{x} - 3)^2$

62. $(\sqrt{2x} + 4)^2$

63. $(4\sqrt{5} + 2)^2$

64. $2\sqrt{3x^2} \cdot 3\sqrt{12xy^3} \cdot \sqrt{6x^3y}$

65. $2\sqrt{14xy} \cdot 4\sqrt{7x^2y} \cdot 3\sqrt{8xy^2}$

66. $\sqrt[3]{8ab}\,\sqrt[3]{4a^2b^3}\,\sqrt[3]{9ab^4}$

67. $\sqrt[3]{2a^2b}\,\sqrt[3]{4a^3b^2}\,\sqrt[3]{8a^5b^6}$

68. $(\sqrt{2} - 3)(\sqrt{2} + 4)$

69. $(\sqrt{5} - 5)(2\sqrt{5} + 2)$

70. $(\sqrt{y} - 2)(\sqrt{y} + 2)$

71. $(\sqrt{x} - y)(\sqrt{x} + y)$

72. $(\sqrt{2x} - 3\sqrt{y})(\sqrt{2x} + 3\sqrt{y})$

73. $(2\sqrt{3x} - \sqrt{y})(2\sqrt{3x} + \sqrt{y})$

Objective D

Simplify.

74. $\dfrac{\sqrt{32x^2}}{\sqrt{2x}}$

75. $\dfrac{\sqrt{60y^4}}{\sqrt{12y}}$

76. $\dfrac{\sqrt{42a^3b^5}}{\sqrt{14a^2b}}$

77. $\dfrac{\sqrt{65ab^4}}{\sqrt{5ab}}$

78. $\dfrac{1}{\sqrt{5}}$

79. $\dfrac{1}{\sqrt{2}}$

80. $\dfrac{1}{\sqrt{2x}}$

81. $\dfrac{2}{\sqrt{3y}}$

82. $\dfrac{5}{\sqrt{5x}}$

83. $\dfrac{9}{\sqrt{3a}}$

84. $\sqrt{\dfrac{x}{5}}$

85. $\sqrt{\dfrac{y}{2}}$

86. $\dfrac{3}{\sqrt[3]{2}}$

87. $\dfrac{5}{\sqrt[3]{9}}$

88. $\dfrac{3}{\sqrt[3]{4x^2}}$

89. $\dfrac{5}{\sqrt[3]{3y}}$

90. $\dfrac{\sqrt{40x^3y^2}}{\sqrt{80x^2y^3}}$

91. $\dfrac{\sqrt{15a^2b^5}}{\sqrt{30a^5b^3}}$

92. $\dfrac{\sqrt{24a^2b}}{\sqrt{18ab^4}}$

93. $\dfrac{\sqrt{12x^3y}}{\sqrt{20x^4y}}$

Simplify.

94. $\dfrac{5}{\sqrt{3} - 2}$

95. $\dfrac{-2}{1 - \sqrt{2}}$

96. $\dfrac{-3}{2 - \sqrt{3}}$

97. $\dfrac{-4}{3 - \sqrt{2}}$

98. $\dfrac{2}{\sqrt{5} + 2}$

99. $\dfrac{5}{2 - \sqrt{7}}$

100. $\dfrac{3}{\sqrt{y} - 2}$

101. $\dfrac{-7}{\sqrt{x} - 3}$

102. $\dfrac{\sqrt{2} - \sqrt{3}}{\sqrt{2} + \sqrt{3}}$

103. $\dfrac{\sqrt{3} + \sqrt{4}}{\sqrt{2} + \sqrt{3}}$

104. $\dfrac{2 + 3\sqrt{7}}{5 - 2\sqrt{7}}$

105. $\dfrac{2 + 3\sqrt{5}}{1 - \sqrt{5}}$

106. $\dfrac{2\sqrt{3} - 1}{3\sqrt{3} + 2}$

107. $\dfrac{2\sqrt{a} - \sqrt{b}}{4\sqrt{a} + 3\sqrt{b}}$

108. $\dfrac{2\sqrt{x} - 4}{\sqrt{x} + 2}$

109. $\dfrac{3\sqrt{y} - y}{\sqrt{y} + 2y}$

110. $\dfrac{3\sqrt{x} - 4\sqrt{y}}{3\sqrt{x} - 2\sqrt{y}}$

APPLYING THE CONCEPTS

111. Determine whether the following statements are true or false. If the statement is false, correct the right side of the equation.

    a. $\sqrt[2]{3} \cdot \sqrt[3]{4} = \sqrt[5]{12}$
    b. $\sqrt{3} \cdot \sqrt{3} = 3$
    c. $\sqrt[3]{x} \cdot \sqrt[3]{x} = x$

    d. $\sqrt{x} + \sqrt{y} = \sqrt{x + y}$
    e. $\sqrt[2]{2} + \sqrt[3]{3} = \sqrt[5]{2 + 3}$
    f. $8\sqrt[5]{a} - 2\sqrt[5]{a} = 6\sqrt[5]{a}$

112. Simplify.

    a. $(\sqrt{2} - 2)^3$
    b. $\dfrac{2}{\sqrt{x + 4} + 2}$
    c. $\dfrac{\sqrt{b + 9} - 3}{\sqrt{b + 9} + 3}$

113. Multiply: $(\sqrt[3]{a} + \sqrt[3]{b})(\sqrt[3]{a^2} - \sqrt[3]{ab} + \sqrt[3]{b^2})$

114. Rewrite $\dfrac{\sqrt[4]{(a + b)^3}}{\sqrt{a + b}}$ as an expression with a single radical.

115.
[W] Describe in your own words what it means to rationalize the denominator of a radical expression, and explain how to do so.

Complex Numbers

Objective A   *To simplify a complex number* ....

The radical expression $\sqrt{-4}$ is not a real number because there is no real number whose square is -4. However, the solution of an algebraic equation is sometimes the square root of a negative number.

For example, the equation $x^2 + 1 = 0$ does not have a real number solution because there is no real number whose square is a negative number.

$$x^2 + 1 = 0$$
$$x^2 = -1$$

Around the 17th century, a new number, called an **imaginary number,** was defined so that a negative number would have a square root. The letter i was chosen to represent the number whose square is -1.

$$i^2 = -1$$

An imaginary number is defined in terms of i.

POINT OF INTEREST
The first written occurrence of an imaginary number was in a book published in 1545 by Hieronimo Cardan, where he wrote (in our modern notation) $5 + \sqrt{-15}$. He went on to say that the number "is as refined as it is useless." It was not until the 20th century that applications of complex numbers were found.

> If a is a positive real number, then the principal square root of negative a is the imaginary number $i\sqrt{a}$.
> $$\sqrt{-a} = i\sqrt{a}$$

Here are some examples.

$$\sqrt{-16} = i\sqrt{16} = 4i$$
$$\sqrt{-12} = i\sqrt{12} = 2i\sqrt{3}$$
$$\sqrt{-21} = i\sqrt{21}$$
$$\sqrt{-1} = i\sqrt{1} = i$$

It is customary to write i in front of a radical to avoid confusing $\sqrt{a}\,i$ with $\sqrt{ai}$.

The real numbers and imaginary numbers make up the complex numbers.

> **Complex Number**
>
> A **complex number** is a number of the form $a + bi$, where a and b are real numbers and $i = \sqrt{-1}$. The number a is the **real part** of $a + bi$, and b is the **imaginary part.**

CONSIDER THIS
The *imaginary part* of a complex number is a real number. As another example, the imaginary part of $6 - 8i$ is -8.

Examples of complex numbers are shown at the right.

Real Part	*Imaginary Part*
$a + bi$	
$3 + 2i$	
$8 - 10i$	

Complex Numbers $a + bi$

┌ Real Numbers $a + 0i$

└ Imaginary Numbers $0 + bi$

A *real number* is a complex number in which $b = 0$.

An *imaginary number* is a complex number in which $a = 0$.

➡ Simplify: $\sqrt{20} - \sqrt{-50}$

$\sqrt{20} - \sqrt{-50} = \sqrt{20} - i\sqrt{50}$

$= \sqrt{2^2 \cdot 5} - i\sqrt{5^2 \cdot 2}$

$= 2\sqrt{5} - 5i\sqrt{2}$

- Write the complex number in the form $a + bi$.
- Use the Product Property of Radicals to simplify each radical.

Example 1
Simplify: $\sqrt{-80}$

Solution
$\sqrt{-80} = i\sqrt{80} = i\sqrt{2^4 \cdot 5} = 4i\sqrt{5}$

You Try It 1
Simplify: $\sqrt{-45}$

Your solution

Example 2
Simplify: $\sqrt{25} + \sqrt{-40}$

Solution
$\sqrt{25} + \sqrt{-40} = \sqrt{25} + i\sqrt{40}$

$= \sqrt{5^2} + i\sqrt{2^2 \cdot 2 \cdot 5}$

$= 5 + 2i\sqrt{10}$

You Try It 2
Simplify: $\sqrt{98} - \sqrt{-60}$

Your solution

Solutions on p. A45

Objective B   To add or subtract complex numbers ....

To add two complex numbers, add the real parts and add the imaginary parts. To subtract two complex numbers, subtract the real parts and subtract the imaginary parts.

$$(a + bi) + (c + di) = (a + c) + (b + d)i$$
$$(a + bi) - (c + di) = (a - c) + (b - d)i$$

➡ Subtract: $(3 - 7i) - (4 - 2i)$

Subtract the real parts and subtract the imaginary parts of the complex numbers.

$(3 - 7i) - (4 - 2i) = \overline{(3 - 4) + [-7 - (-2)]i}$

$= -1 - 5i$

- Do this step mentally.

➡ Add: $(3 + \sqrt{-12}) + (7 - \sqrt{-27})$

$(3 + \sqrt{-12}) + (7 - \sqrt{-27})$
$= (3 + i\sqrt{12}) + (7 - i\sqrt{27})$ • Write each complex number in the form $a + bi$.

$= (3 + i\sqrt{2^2 \cdot 3}) + (7 - i\sqrt{3^2 \cdot 3})$ • Use the Product Property of Radicals to simplify each radical.

$= (3 + 2i\sqrt{3}) + (7 - 3i\sqrt{3})$
$= 10 - i\sqrt{3}$ • Add the complex numbers.

Example 3
Add: $(3 + 2i) + (6 - 5i)$

Solution
$(3 + 2i) + (6 - 5i) = 9 - 3i$

You Try It 3
Subtract: $(-4 + 2i) - (6 - 8i)$

Your solution

Example 4
Subtract: $(9 - \sqrt{-8}) - (5 + \sqrt{-32})$

Solution
$(9 - \sqrt{-8}) - (5 + \sqrt{-32})$
$= (9 - i\sqrt{8}) - (5 + i\sqrt{32})$
$= (9 - i\sqrt{2^2 \cdot 2}) - (5 + i\sqrt{2^4 \cdot 2})$
$= (9 - 2i\sqrt{2}) - (5 + 4i\sqrt{2})$
$= 4 - 6i\sqrt{2}$

You Try It 4
Subtract: $(16 - \sqrt{-45}) - (3 + \sqrt{-20})$

Your solution

Example 5
Add: $(6 + 4i) + (-6 - 4i)$

Solution
$(6 + 4i) + (-6 - 4i) = 0 + 0i = 0$

This illustrates that the additive inverse of $a + bi$ is $-a - bi$.

You Try It 5
Add: $(3 - 2i) + (-3 + 2i)$

Your solution

Solutions on p. A45

Objective C   *To multiply complex numbers* ....

When multiplying complex numbers, we often find that i^2 is a part of the product. Recall that $i^2 = -1$.

➡ Multiply: $2i \cdot 3i$

$2i \cdot 3i = 6i^2$ • Multiply the imaginary numbers.
$= 6(-1)$ • Replace i^2 by -1.
$= -6$ • Simplify.

CONSIDER THIS

This example illustrates an important point. When working with a number that has a square root of a negative number, always rewrite the number with i before continuing.

➡ Multiply: $\sqrt{-6} \cdot \sqrt{-24}$

$$\sqrt{-6} \cdot \sqrt{-24} = i\sqrt{6} \cdot i\sqrt{24}$$

- Write each radical as the product of a real number and i.

$$= i^2\sqrt{144}$$
- Multiply the imaginary numbers.

$$= -\sqrt{144}$$
- Replace i^2 by -1.

$$= -12$$
- Simplify the radical expression.

Note from the last example that it would have been incorrect to multiply the radicands of the two radical expressions. To illustrate,

$$\sqrt{-6} \cdot \sqrt{-24} = \sqrt{(-6)(-24)} = \sqrt{144} = 12,\ not\ -12$$

➡ Multiply: $4i(3 - 2i)$

$$4i(3 - 2i) = 12i - 8i^2$$
- Use the Distributive Property to remove parentheses.

$$= 12i - 8(-1)$$
- Replace i^2 by -1.

$$= 8 + 12i$$
- Write the answer in the form $a + bi$.

The product of two complex numbers is defined as follows.

Product of Complex Numbers

$(a + bi)(c + di) = (ac - bd) + (ad + bc)i$

One way to remember this rule is to use the FOIL method.

➡ Multiply: $(2 + 4i)(3 - 5i)$

$$(2 + 4i)(3 - 5i) = 6 - 10i + 12i - 20i^2$$
- Use the FOIL method to find the product.

$$= 6 + 2i - 20i^2$$

$$= 6 + 2i - 20(-1)$$
- Replace i^2 by -1.

$$= 26 + 2i$$
- Write the answer in the form $a + bi$.

The conjugate of $a + bi$ is $a - bi$. The product of conjugates, $(a + bi)(a - bi)$, is the real number $a^2 + b^2$.

$$(a + bi)(a - bi) = a^2 - b^2i^2$$
$$= a^2 - b^2(-1)$$
$$= a^2 + b^2$$

➡ Multiply: $(2 + 3i)(2 - 3i)$

$$(2 + 3i)(2 - 3i) = 2^2 + 3^2$$
- The product of conjugates is $a^2 + b^2$.

$$= 4 + 9$$

$$= 13$$

Note that the product of a complex number and its conjugate is a real number.

Example 6
Multiply: $(2i)(-5i)$

Solution
$(2i)(-5i) = -10i^2 = (-10)(-1) = 10$

You Try It 6
Multiply: $(-3i)(-10i)$

Your solution

Example 7
Multiply: $\sqrt{-10} \cdot \sqrt{-5}$

Solution
$$\sqrt{-10} \cdot \sqrt{-5} = i\sqrt{10} \cdot i\sqrt{5}$$
$$= i^2\sqrt{50} = -\sqrt{5^2 \cdot 2} = -5\sqrt{2}$$

You Try It 7
Multiply: $-\sqrt{-8} \cdot \sqrt{-5}$

Your solution

Example 8
Multiply: $3i(2 - 4i)$

Solution
$$3i(2 - 4i) = 6i - 12i^2 = 6i - 12(-1)$$
$$= 12 + 6i$$

You Try It 8
Multiply: $-6i(3 + 4i)$

Your solution

Example 9
Multiply: $\sqrt{-8}(\sqrt{6} - \sqrt{-2})$

Solution
$$\sqrt{-8}(\sqrt{6} - \sqrt{-2}) = i\sqrt{8}(\sqrt{6} - i\sqrt{2})$$
$$= i\sqrt{48} - i^2\sqrt{16}$$
$$= i\sqrt{2^4 \cdot 3} - (-1)\sqrt{2^4}$$
$$= 4i\sqrt{3} + 4 = 4 + 4i\sqrt{3}$$

You Try It 9
Multiply: $\sqrt{-3}(\sqrt{27} - \sqrt{-6})$

Your solution

Example 10
Multiply: $(3 - 4i)(2 + 5i)$

Solution
$$(3 - 4i)(2 + 5i) = 6 + 15i - 8i - 20i^2$$
$$= 6 + 7i - 20i^2$$
$$= 6 + 7i - 20(-1) = 26 + 7i$$

You Try It 10
Multiply: $(4 - 3i)(2 - i)$

Your solution

Example 11
Multiply: $(4 + 5i)(4 - 5i)$

Solution
$(4 + 5i)(4 - 5i) = 4^2 + 5^2 = 16 + 25 = 41$

You Try It 11
Multiply: $(3 + 6i)(3 - 6i)$

Your solution

Example 12
Multiply: $\left(\dfrac{9}{10} + \dfrac{3}{10}i\right)\left(1 - \dfrac{1}{3}i\right)$

Solution
$$\left(\frac{9}{10} + \frac{3}{10}i\right)\left(1 - \frac{1}{3}i\right) = \frac{9}{10} - \frac{3}{10}i + \frac{3}{10}i - \frac{1}{10}i^2$$
$$= \frac{9}{10} - \frac{1}{10}i^2 = \frac{9}{10} - \frac{1}{10}(-1)$$
$$= \frac{9}{10} + \frac{1}{10} = 1$$

You Try It 12
Multiply: $(3 - i)\left(\dfrac{3}{10} + \dfrac{1}{10}i\right)$

Your solution

Solutions on pp. A45–A46

Objective D   *To divide complex numbers* ....

A rational expression containing one or more complex numbers is in simplest form when no imaginary number remains in the denominator.

➡ Simplify: $\dfrac{2 - 3i}{2i}$

$$\dfrac{2 - 3i}{2i} = \dfrac{2 - 3i}{2i} \cdot \dfrac{i}{i}$$

• Multiply the expression by $\dfrac{i}{i}$.

$$= \dfrac{2i - 3i^2}{2i^2}$$

$$= \dfrac{2i - 3(-1)}{2(-1)}$$

• Replace i^2 by -1.

$$= \dfrac{3 + 2i}{-2}$$

• Simplify.

$$= -\dfrac{3}{2} - i$$

• Write the answer in the form $a + bi$.

➡ Simplify: $\dfrac{3 + 2i}{1 + i}$

$$\dfrac{3 + 2i}{1 + i} = \dfrac{3 + 2i}{1 + i} \cdot \dfrac{1 - i}{1 - i}$$

• Multiply the numerator and denominator by the conjugate of $1 + i$.

$$= \dfrac{3 - 3i + 2i - 2i^2}{1^2 + 1^2}$$

• In $1 + i$, $a = 1$ and $b = 1$.
  $a^2 + b^2 = 1^2 + 1^2$.

$$= \dfrac{3 - i - 2(-1)}{2}$$

• Replace i^2 by -1 and simplify.

$$= \dfrac{5 - i}{2} = \dfrac{5}{2} - \dfrac{1}{2}i$$

• Write the answer in the form $a + bi$.

Example 13

Simplify: $\dfrac{5 + 4i}{3i}$

Solution

$$\dfrac{5 + 4i}{3i} = \dfrac{5 + 4i}{3i} \cdot \dfrac{i}{i} = \dfrac{5i + 4i^2}{3i^2}$$

$$= \dfrac{5i + 4(-1)}{3(-1)} = \dfrac{-4 + 5i}{-3} = \dfrac{4}{3} - \dfrac{5}{3}i$$

You Try It 13

Simplify: $\dfrac{2 - 3i}{4i}$

Your solution

Example 14

Simplify: $\dfrac{5 - 3i}{4 + 2i}$

Solution

$$\dfrac{5 - 3i}{4 + 2i} = \dfrac{5 - 3i}{4 + 2i} \cdot \dfrac{4 - 2i}{4 - 2i}$$

$$= \dfrac{20 - 10i - 12i + 6i^2}{4^2 + 2^2}$$

$$= \dfrac{20 - 22i + 6(-1)}{20}$$

$$= \dfrac{14 - 22i}{20} = \dfrac{14}{20} - \dfrac{22}{20}i = \dfrac{7}{10} - \dfrac{11}{10}i$$

You Try It 14

Simplify: $\dfrac{2 + 5i}{3 - 2i}$

Your solution

Solutions on p. A46

9.3 Exercises

Objective A

Simplify.

1. $\sqrt{-4}$

2. $\sqrt{-64}$

3. $\sqrt{-98}$

4. $\sqrt{-72}$

5. $\sqrt{-27}$

6. $\sqrt{-75}$

7. $\sqrt{16} + \sqrt{-4}$

8. $\sqrt{25} + \sqrt{-9}$

9. $\sqrt{12} - \sqrt{-18}$

10. $\sqrt{60} - \sqrt{-48}$

11. $\sqrt{160} - \sqrt{-147}$

12. $\sqrt{96} - \sqrt{-125}$

Objective B

Add or subtract.

13. $(2 + 4i) + (6 - 5i)$

14. $(6 - 9i) + (4 + 2i)$

15. $(-2 - 4i) - (6 - 8i)$

16. $(3 - 5i) + (8 - 2i)$

17. $(8 - \sqrt{-4}) - (2 + \sqrt{-16})$

18. $(5 - \sqrt{-25}) - (11 - \sqrt{-36})$

19. $(12 - \sqrt{-50}) + (7 - \sqrt{-8})$

20. $(5 - \sqrt{-12}) - (9 + \sqrt{-108})$

21. $(\sqrt{8} + \sqrt{-18}) + (\sqrt{32} - \sqrt{-72})$

22. $(\sqrt{40} - \sqrt{-98}) - (\sqrt{90} + \sqrt{-32})$

Objective C

Multiply.

23. $(7i)(-9i)$

24. $(-6i)(-4i)$

25. $\sqrt{-2}\sqrt{-8}$

26. $\sqrt{-5}\sqrt{-45}$

27. $\sqrt{-3}\sqrt{-6}$

28. $\sqrt{-5}\sqrt{-10}$

Multiply.

29. $2i(6 + 2i)$

30. $-3i(4 - 5i)$

31. $\sqrt{-2}(\sqrt{8} + \sqrt{-2})$

32. $\sqrt{-3}(\sqrt{12} - \sqrt{-6})$

33. $(5 - 2i)(3 + i)$

34. $(2 - 4i)(2 - i)$

35. $(6 + 5i)(3 + 2i)$

36. $(4 - 7i)(2 + 3i)$

37. $(1 - i)\left(\dfrac{1}{2} + \dfrac{1}{2}i\right)$

38. $\left(\dfrac{4}{5} - \dfrac{2}{5}i\right)\left(1 + \dfrac{1}{2}i\right)$

39. $\left(\dfrac{6}{5} + \dfrac{3}{5}i\right)\left(\dfrac{2}{3} - \dfrac{1}{3}i\right)$

40. $(2 - i)\left(\dfrac{2}{5} + \dfrac{1}{5}i\right)$

Objective D

Simplify.

41. $\dfrac{3}{i}$

42. $\dfrac{4}{5i}$

43. $\dfrac{2 - 3i}{-4i}$

44. $\dfrac{16 + 5i}{-3i}$

45. $\dfrac{4}{5 + i}$

46. $\dfrac{6}{5 + 2i}$

47. $\dfrac{2}{2 - i}$

48. $\dfrac{5}{4 - i}$

49. $\dfrac{1 - 3i}{3 + i}$

50. $\dfrac{2 + 12i}{5 + i}$

51. $\dfrac{\sqrt{-10}}{\sqrt{8} - \sqrt{-2}}$

52. $\dfrac{\sqrt{-2}}{\sqrt{12} - \sqrt{-8}}$

53. $\dfrac{2 - 3i}{3 + i}$

54. $\dfrac{3 + 5i}{1 - i}$

55. $\dfrac{5 + 3i}{3 - i}$

56. $\dfrac{3 - 2i}{2i + 3}$

APPLYING THE CONCEPTS

57. The property that the product $(a + bi)(a - bi)$ is equal to $a^2 + b^2$ can be used to factor the sum of two perfect squares over the set of complex numbers. For instance, $x^2 + 4 = (x + 2i)(x - 2i)$. Factor the following expressions over the set of complex numbers.
a. $y^2 + 1$
b. $49x^2 + 16$
c. $9a^2 + 64$

58. a. Is $3i$ a solution of $2x^2 + 18 = 0$?
b. Is $3 + i$ a solution of $x^2 - 6x + 10 = 0$?

59. Evaluate i^n for $n = 0, 1, 2, 3, 4, 5, 6,$ and 7. Make a conjecture for the value of i^n for any natural number. Using your conjecture, evaluate i^{76}.

9.4 Solving Equations Containing Radical Expressions

Objective A *To solve a radical equation*

An equation that contains a variable expression in a radicand is a **radical equation.**

$$\left.\begin{array}{l}\sqrt[3]{2x-5}+x=7\\\sqrt{x+1}-\sqrt{x}=4\end{array}\right\}\begin{array}{l}\text{Radical}\\\text{Equations}\end{array}$$

The following property is used to solve a radical equation.

> **The Property of Raising Each Side of an Equation to a Power**
>
> If two numbers are equal, then the same powers of the numbers are equal.
>
> If $a = b$, then $a^n = b^n$.

Solve: $\sqrt{x-2}-6=0$

$\sqrt{x-2}-6=0$

$\sqrt{x-2}=6$ • Isolate the radical by adding 6 to each side of the equation.

$(\sqrt{x-2})^2=6^2$ • Square each side of the equation.

$x-2=36$ • Simplify and solve for x.

$x=38$

Check:
$$\begin{array}{c|c}\sqrt{x-2}-6=0\\\hline\sqrt{38-2}-6 & 0\\\sqrt{36}-6 & 0\\6-6 & 0\\0=0\end{array}$$

38 checks as a solution. The solution is 38.

Solve: $\sqrt[3]{x+2}=-3$

$\sqrt[3]{x+2}=-3$

$(\sqrt[3]{x+2})^3=(-3)^3$ • Cube each side of the equation.

$x+2=-27$ • Solve the resulting equation.

$x=-29$

Check:
$$\begin{array}{c|c}\sqrt[3]{x+2}=-3\\\hline\sqrt[3]{-29+2} & -3\\\sqrt[3]{-27} & -3\\-3=-3\end{array}$$

−29 checks as a solution. The solution is −29.

Raising each side of an equation to an even power may result in an equation that has a solution that is not a solution of the original equation. This is called an **extraneous solution.** Here is an example:

→ Solve: $\sqrt{2x - 1} + \sqrt{x} = 2$

$$\sqrt{2x - 1} + \sqrt{x} = 2$$

$$\sqrt{2x - 1} = 2 - \sqrt{x}$$

- Solve for one of the radical expressions.

$$(\sqrt{2x - 1})^2 = (2 - \sqrt{x})^2$$

- Square each side. Recall that $(a - b)^2 = a^2 - 2ab + b^2$.

$$2x - 1 = 4 - 4\sqrt{x} + x$$

$$x - 5 = -4\sqrt{x}$$

$$(x - 5)^2 = (-4\sqrt{x})^2$$

- Square each side.

$$x^2 - 10x + 25 = 16x$$

$$x^2 - 26x + 25 = 0$$

$$(x - 25)(x - 1) = 0$$

$$x = 25 \quad \text{or} \quad x = 1$$

- Solve the quadratic equation by factoring.

CONSIDER THIS

Note that

$(2 - \sqrt{x})^2 =$
$(2 - \sqrt{x})(2 - \sqrt{x})$
$= 4 - 4\sqrt{x} + x$

CONSIDER THIS

You must check the proposed solutions to radical equations. The proposed solutions of the equation on the right were 1 and 25. However, 25 did not check as a solution. Here 25 is an extraneous solution.

Check:

$\sqrt{2x - 1} + \sqrt{x} = 2$		$\sqrt{2x - 1} + \sqrt{x} = 2$	
$\sqrt{2(25) - 1} + \sqrt{25}$	2	$\sqrt{2(1) - 1} + \sqrt{1}$	2
$7 + 5$	2	$1 + 1$	2
$12 \neq 2$		$2 = 2$	

25 does not check as a solution. 1 checks as a solution. The solution is 1.

Example 1

Solve: $\sqrt[3]{3x - 1} = -4$

Solution

$$\sqrt[3]{3x - 1} = -4$$

$$(\sqrt[3]{3x - 1})^3 = (-4)^3$$

$$3x - 1 = -64$$

$$3x = -63$$

$$x = -21$$

Check:

$\sqrt[3]{3x - 1} = -4$	
$\sqrt[3]{3(-21) - 1}$	-4
$\sqrt[3]{-64}$	-4
$-4 = -4$	

The solution is -21.

You Try It 1

Solve: $\sqrt[4]{x - 8} = 3$

Your solution

Example 2

Solve: $\sqrt{x - 1} + \sqrt{x + 4} = 5$

Solution

$$\sqrt{x - 1} + \sqrt{x + 4} = 5$$

$$\sqrt{x + 4} = 5 - \sqrt{x - 1}$$

$$(\sqrt{x + 4})^2 = (5 - \sqrt{x - 1})^2$$

$$x + 4 = 25 - 10\sqrt{x - 1} + x - 1$$

$$2 = \sqrt{x - 1}$$

$$2^2 = (\sqrt{x - 1})^2$$

$$4 = x - 1$$

$$5 = x$$

5 checks as a solution. The solution is 5.

You Try It 2

Solve: $\sqrt{x} - \sqrt{x + 5} = 1$

Your solution

Solutions on p. A46

Objective B   To solve application problems ....

A right triangle contains one 90° angle. The side opposite the 90° angle is called the **hypotenuse.** The other two sides are called **legs.**

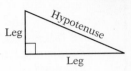

POINT OF INTEREST
The first known proof of this theorem occurs in a Chinese text, *Arithmetic Classic*, which was first written around 600 B.C.

Pythagoras, a Greek mathematician, discovered that the square of the hypotenuse of a right triangle is equal to the sum of the squares of the two legs. This is called the **Pythagorean Theorem.**

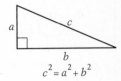

$$c^2 = a^2 + b^2$$

Example 3

A ladder 20 ft long is leaning against a building. How high on the building will the ladder reach when the bottom of the ladder is 8 ft from the building? Round to the nearest tenth.

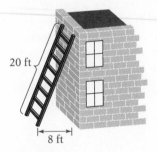

20 ft

8 ft

Strategy

To find the distance, use the Pythagorean Theorem. The hypotenuse is the length of the ladder. One leg is the distance from the bottom of the ladder to the base of the building. The distance along the building from the ground to the top of the ladder is the unknown leg.

Solution

$$c^2 = a^2 + b^2$$
$$20^2 = 8^2 + b^2$$
$$400 = 64 + b^2$$
$$336 = b^2$$
$$(336)^{1/2} = (b^2)^{1/2}$$
$$\sqrt{336} = b$$
$$18.3 \approx b$$

The distance is 18.3 ft.

You Try It 3

Find the diagonal of a rectangle that is 6 cm long and 3 cm wide. Round to the nearest tenth.

Your strategy

Your solution

Solution on p. A47

Example 4

An object is dropped from a high building. Find the distance the object has fallen when the speed reaches 96 ft/s. Use the equation $v = \sqrt{64d}$, where v is the speed of the object and d is the distance.

Strategy

To find the distance the object has fallen, replace v in the equation with the given value and solve for d.

Solution

$$v = \sqrt{64d}$$
$$96 = \sqrt{64d}$$
$$(96)^2 = (\sqrt{64d})^2$$
$$9216 = 64d$$
$$144 = d$$

The object has fallen 144 ft.

Example 5

Find the length of a pendulum that makes one swing in 1.5 s. The equation for the time of one swing is given by $T = 2\pi\sqrt{\dfrac{L}{32}}$, where T is the time in seconds and L is the length in feet. Round to the nearest hundredth.

Strategy

To find the length of the pendulum, replace T in the equation with the given value and solve for L.

Solution

$$T = 2\pi\sqrt{\dfrac{L}{32}}$$

$$1.5 = 2\pi\sqrt{\dfrac{L}{32}}$$

$$\dfrac{1.5}{2\pi} = \sqrt{\dfrac{L}{32}} \qquad \bullet \text{ Divide each side by } 2\pi.$$

$$\left(\dfrac{1.5}{2\pi}\right)^2 = \dfrac{L}{32} \qquad \bullet \text{ Square each side.}$$

$$32\left(\dfrac{1.5}{2\pi}\right)^2 = L \qquad \bullet \text{ Multiply each side by 32.}$$

$$1.82 \approx L \qquad \bullet \text{ Use the } \pi \text{ key on your calculator.}$$

The length of the pendulum is 1.82 ft.

You Try It 4

How far would a submarine periscope have to be above the water for the lookout to locate a ship 5.5 mi away? The equation for the distance in miles that the lookout can see is $d = \sqrt{1.5h}$, where h is the height in feet above the surface of the water. Round to the nearest hundredth.

Your strategy

Your solution

You Try It 5

Find the distance required for a car to reach a velocity of 88 ft/s when the acceleration is 22 ft/s². Use the equation $v = \sqrt{2as}$, where v is the velocity, a is the acceleration, and s is the distance.

Your strategy

Your solution

Solutions on p. A47

9.4 Exercises

Objective A

Solve.

1. $\sqrt{x} = 5$

2. $\sqrt{y} = 2$

3. $\sqrt[3]{a} = 3$

4. $\sqrt[3]{y} = 5$

5. $\sqrt{3x} = 12$

6. $\sqrt{5x} = 10$

7. $\sqrt[3]{4x} = -2$

8. $\sqrt[3]{6x} = -3$

9. $\sqrt{2x} = -4$

10. $\sqrt{5x} = -5$

11. $\sqrt{3x - 2} = 5$

12. $\sqrt{5x - 4} = 9$

13. $\sqrt{3 - 2x} = 7$

14. $\sqrt{9 - 4x} = 4$

15. $7 = \sqrt{1 - 3x}$

16. $6 = \sqrt{8 - 7x}$

17. $\sqrt[3]{4x - 1} = 2$

18. $\sqrt[3]{5x + 2} = 3$

19. $\sqrt[3]{1 - 2x} = -3$

20. $\sqrt[3]{3 - 2x} = -2$

21. $\sqrt[3]{9x + 1} = 4$

22. $\sqrt{3x + 9} - 12 = 0$

23. $\sqrt{4x - 3} - 5 = 0$

24. $\sqrt{x - 2} = 4$

25. $\sqrt[3]{x - 3} + 5 = 0$

26. $\sqrt[3]{x - 2} = 3$

27. $\sqrt[3]{2x - 6} = 4$

28. $\sqrt{x^2 - 8x} = 3$

29. $\sqrt{x^2 + 7x + 11} = 1$

30. $\sqrt[4]{4x + 1} = 2$

31. $\sqrt[4]{2x - 9} = 3$

32. $\sqrt{2x - 3} - 2 = 1$

33. $\sqrt{3x - 5} - 5 = 3$

Solve.

34. $\sqrt[3]{2x - 3} + 5 = 2$

35. $\sqrt[3]{x - 4} + 7 = 5$

36. $\sqrt{5x - 16} + 1 = 4$

37. $\sqrt{3x - 5} - 2 = 3$

38. $\sqrt{2x - 1} - 8 = -5$

39. $\sqrt{7x + 2} - 10 = -7$

40. $\sqrt[3]{4x - 3} - 2 = 3$

41. $\sqrt[3]{1 - 3x} + 5 = 3$

42. $1 - \sqrt{4x + 3} = -5$

43. $7 - \sqrt{3x + 1} = -1$

44. $\sqrt{x + 1} = 2 - \sqrt{x}$

45. $\sqrt{2x + 4} = 3 - \sqrt{2x}$

46. $\sqrt{x^2 + 3x - 2} - x = 1$

47. $\sqrt{x^2 - 4x - 1} + 3 = x$

48. $\sqrt{x^2 - 3x - 1} = 3$

49. $\sqrt{x^2 - 2x + 1} = 3$

50. $\sqrt{2x + 5} - \sqrt{3x - 2} = 1$

51. $\sqrt{4x + 1} - \sqrt{2x + 4} = 1$

52. $\sqrt{5x - 1} - \sqrt{3x - 2} = 1$

53. $\sqrt{5x + 4} - \sqrt{3x + 1} = 1$

54. $\sqrt[3]{x^2 + 2} - 3 = 0$

55. $\sqrt[3]{x^2 + 4} - 2 = 0$

56. $\sqrt[4]{x^2 + 2x + 8} - 2 = 0$

57. $\sqrt[4]{x^2 + x - 1} - 1 = 0$

58. $4\sqrt{x + 1} - x = 1$

59. $3\sqrt{x - 2} + 2 = x$

60. $x + 3\sqrt{x - 2} = 12$

Objective B   *Application Problems*

Solve.

61. Find the width of a rectangle that has a diagonal of 10 ft and a length of 8 ft.

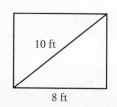

10 ft

8 ft

Solve.

62. Find the length of a rectangle that has a diagonal of 15 m and a width of 9 m.

63. A 26-foot ladder is leaning against a building. How far is the bottom of the ladder from the building when the ladder reaches a height of 24 ft on the building?

64. A 16-foot ladder is leaning against a building. How high on the building will the ladder reach when the bottom of the ladder is 5 ft from the building? Round to the nearest tenth.

16 ft

5 ft

65. An object is dropped from an airplane. Find the distance the object has fallen when the speed reaches 400 ft/s. Use the equation $v = \sqrt{64d}$, where v is the speed of the object and d is the distance.

66. An object is dropped from a bridge. Find the distance the object has fallen when the speed reaches 100 ft/s. Use the equation $v = \sqrt{64d}$, where v is the speed of the object and d is the distance.

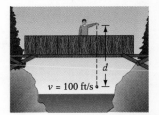

$v = 100$ ft/s

67. How far would a submarine periscope have to be above the water for the lookout to locate a ship 3.6 mi away? The equation for the distance in miles that the lookout can see is $d = \sqrt{1.5h}$, where h is the height in feet above the surface of the water.

68. How far would a submarine periscope have to be above the water for the lookout to locate a ship 4.2 mi away? The equation for the distance in miles that the lookout can see is $d = \sqrt{1.5h}$, where h is the height in feet above the surface of the water.

69. Find the length of a pendulum on a clock that makes one swing in 2.4 s. The equation for the time of one swing of a pendulum is given by $T = 2\pi\sqrt{\dfrac{L}{32}}$, where T is the time in seconds and L is the length in feet. Round to the nearest hundredth.

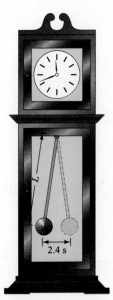

L

2.4 s

70. Find the length of a pendulum that makes one swing in 3 s. The equation for the time of one swing of a pendulum is given by $T = 2\pi\sqrt{\dfrac{L}{32}}$, where T is the time in seconds and L is the length in feet. Round to the nearest hundredth.

Solve.

71. Find the distance required for a car to reach a velocity of 60 m/s when the acceleration is 10 m/s². Use the equation $v = \sqrt{2as}$, where v is the velocity, a is the acceleration, and s is the distance.

72. Find the distance required for a car to reach a velocity of 48 ft/s when the acceleration is 12 ft/s². Use the equation $v = \sqrt{2as}$, where v is the velocity, a is the acceleration, and s is the distance.

APPLYING THE CONCEPTS

73. Solve: $\sqrt{3x - 2} = \sqrt{2x - 3} + \sqrt{x - 1}$

74. Solve each equation. Describe the solution by using the following terms: integer, rational number, irrational number, real number, and imaginary number. Note that more than one term may be used to describe the answer.

   a. $x^2 + 3 = 7$ **b.** $x^2 + 1 = 0$ **c.** $\dfrac{5}{8}x = \dfrac{2}{3}$

   d. $x^2 + 1 = 9$ **e.** $x^{3/4} = 8$ **f.** $\sqrt[3]{x} = -27$

75. Solve $a^2 + b^2 = c^2$ for a.

76. Solve $V = \dfrac{4}{3}\pi r^3$ for r.

77. Find the length of the side labeled x.

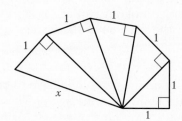

78. Beginning with a square with each side of length s, draw a line segment from the midpoint of the base of the square to a vertex as shown. Call this distance a. Make a rectangle whose width is that of the square and whose length is $\dfrac{s}{2} + a$. Find:

   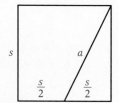

   a. The area of the rectangle in terms of s.
   b. The ratio of the length to the width.

   The rectangle formed in this way is called a golden rectangle. It was described at the beginning of this chapter.

79. Two cyclists left an intersection at the same time. The first cyclist headed due south. The second cyclist headed due east. When the first cyclist had traveled 10 mi farther than the second cyclist, the cyclists were 50 mi apart. How far had each cyclist traveled?

Project in Mathematics

Pythagorean Triples   The Pythagorean Theorem states that if a and b are the legs of a right triangle and c is the length of the hypotenuse, then $a^2 + b^2 = c^2$.

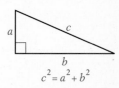

$$c^2 = a^2 + b^2$$

For instance, the triangle with legs 3 and 4 and hypotenuse 5 is a right triangle because $3^2 + 4^2 = 5^2$. The numbers 3, 4, and 5 are called a **Pythagorean triple** because they are natural numbers that satisfy the equation of the Pythagorean Theorem.

1. Determine if the numbers are a Pythagorean triple.
   a. 5, 7, and 9   **b.** 8, 15, and 17
   c. 11, 60, and 61   **d.** 28, 45, and 53

Mathematicians have investigated Pythagorean triples and have found formulas that will generate these triples. One such formula is

$$a = m^2 - n^2 \qquad b = 2mn \qquad c = m^2 + n^2, \text{ where } m > n$$

For instance, let $m = 2$ and $n = 1$. Then $a = 2^2 - 1^2 = 3$, $b = 2(2)(1) = 4$, and $c = 2^2 + 1^2 = 5$. This is the Pythagorean triple given above.

2. Find the Pythagorean triple produced by each of the following.
   a. $m = 3$ and $n = 1$ **b.** $m = 5$ and $n = 2$
   c. $m = 4$ and $n = 2$ **d.** $m = 6$ and $n = 1$

3. Find values of m and n that yield the Pythagorean triple 11, 60, 61.

4. Verify that $a^2 + b^2 = c^2$ when $a = m^2 - n^2$, $b = 2mn$, and $c = m^2 + n^2$.

5. The early Greek builders used a rope with 12 equally spaced knots to make right-angle corners for buildings. Explain how they used the rope.

6. Find three odd integers, a, b, c, such that $a^2 + b^2 = c^2$.

Chapter Summary

Key Words   The *nth root of a* is $a^{\frac{1}{n}}$. The expression $\sqrt[n]{a}$ is another symbol for the *nth* root of a. In the expression $\sqrt[n]{a}$, the symbol $\sqrt{}$ is called a *radical sign*, n is the *index*, and a is the *radicand*.

If $a^{\frac{1}{n}}$ is a real number, then $a^{\frac{m}{n}} = \sqrt[n]{a^m} = (\sqrt[n]{a})^m$.

The symbol $\sqrt{}$ is used to indicate the positive or *principal square root* of a number.

The expressions $a + b$ and $a - b$ are called *conjugates* of each other. The product of conjugates of the form $(a + b)(a - b)$ is $a^2 - b^2$.

The procedure used to remove a radical from the denominator of a radical expression is called *rationalizing the denominator*.

A *complex number* is a number of the form $a + bi$, where a and b are real numbers and $i = \sqrt{-1}$. For the complex number $a + bi$, a is the *real part* of the complex number, and b is the *imaginary part* of the complex number.

A *radical equation* is an equation that contains a variable expression in a radicand.

Essential Rules

The Product Property of Radicals

If a and b are positive real numbers, then $\sqrt[n]{ab} = \sqrt[n]{a}\sqrt[n]{b}$.

The Quotient Property of Radicals

If a and b are positive real numbers, then $\sqrt[n]{\dfrac{a}{b}} = \dfrac{\sqrt[n]{a}}{\sqrt[n]{b}}$.

Addition of Complex Numbers

If $a + bi$ and $c + di$ are complex numbers, then
$(a + bi) + (c + di) = (a + c) + (b + d)i$.

Subtraction of Complex Numbers

If $a + bi$ and $c + di$ are complex numbers, then
$(a + bi) - (c + di) = (a - c) + (b - d)i$.

The Property of Raising Each Side of an Equation to a Power

If a and b are real numbers and $a = b$, then $a^n = b^n$.

The Pythagorean Theorem

The square of the hypotenuse of a right triangle is equal to the sum of the squares of the two legs.
$c^2 = a^2 + b^2$

Chapter Review Exercises

1. Simplify: $\dfrac{r^{2/3} r^{-1}}{r^{-1/2}}$

2. Simplify: $\dfrac{(2x^{1/3} y^{-2/3})^6}{(x^{-4} y^8)^{1/4}}$

3. Simplify: $\left(\dfrac{4a^4}{b^2}\right)^{-3/2}$

4. Write $3y^{2/5}$ as a radical expression.

5. Write $\dfrac{1}{2}\sqrt[4]{x^3}$ as an exponential expression.

6. Simplify: $\sqrt[3]{8x^3 y^6}$

7. Simplify: $\sqrt{32x^4 y^7}$

8. Simplify: $\sqrt[3]{27a^4 b^3 c^7}$

9. Add: $\sqrt{18a^3} + a\sqrt{50a}$

10. Subtract:
    $\sqrt[3]{54x^7 y^3} - x\sqrt[3]{128x^4 y^3} - x^2\sqrt[3]{2xy^3}$

11. Multiply: $\sqrt{3x}\,(\sqrt{x} - \sqrt{25x}\,)$

12. Multiply: $(2\sqrt{3} + 4)(3\sqrt{3} - 1)$

13. Multiply: $(\sqrt{a} - 3\sqrt{b}\,)(2\sqrt{a} + 5\sqrt{b}\,)$

14. Simplify: $(2\sqrt{x} + \sqrt{y}\,)^2$

15. Simplify: $\dfrac{\sqrt{32x^5y}}{\sqrt{2xy^3}}$

16. Simplify: $\dfrac{4 - 2\sqrt{5}}{2 - \sqrt{5}}$

17. Simplify: $\dfrac{\sqrt{x}}{\sqrt{x} - \sqrt{y}}$

18. Simplify: $\sqrt{-50}$

19. Subtract: $(5 - 2i) - (8 - 4i)$

20. Multiply: $(2 + 5i)(4 - 2i)$

21. Simplify: $\dfrac{2 + 3i}{1 - 2i}$

22. Multiply: $(\sqrt{-8})(\sqrt{-2})$

23. Solve: $\sqrt{x + 12} - \sqrt{x} = 2$

24. Solve: $\sqrt[3]{2x - 2} + 4 = 2$

25. An object is dropped from a high building. Find the distance the object has fallen when the speed reaches 192 ft/s. Use the equation $v = \sqrt{64d}$, where v is the speed of the object and d is the distance.

Cumulative Review Exercises

1. Simplify: $2^3 \cdot 3 - 4(3 - 4 \cdot 5)$

2. Evaluate $4a^2b - a^3$ when $a = -2$ and $b = 3$.

3. Simplify: $-3(4x - 1) - 2(1 - x)$

4. Solve: $5 - \dfrac{2}{3}x = 4$

5. Solve: $2[4 - 2(3 - 2x)] = 4(1 - x)$

6. Solve: $6x - 3(2x + 2) > 3 - 3(x + 2)$

7. Solve: $2 + |4 - 3x| = 5$

8. Solve: $|2x + 3| \le 9$

9. Find the area of the triangle shown in the figure below.

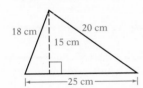

10. Find the value of the determinant:

$$\begin{vmatrix} 1 & 2 & -3 \\ 0 & -1 & 2 \\ 3 & 1 & -2 \end{vmatrix}$$

11. Find the slope and y-intercept and graph $3x - 2y = -6$.

12. Graph the solution set of $3x + 2y \le 4$.

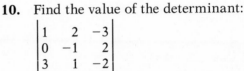

13. Find the equation of the line that passes through the points $(2, 3)$ and $(-1, 2)$.

14. Solve by using Cramer's Rule:
$$2x - y = 4$$
$$-2x + 3y = 5$$

15. Simplify: $(2^{-1}x^2y^{-6})(2^{-1}y^{-4})^{-2}$

16. Factor: $81x^2 - y^2$

17. Factor: $x^5 + 2x^3 - 3x$

18. Solve $P = \dfrac{R - C}{n}$ for C.

19. Simplify: $\left(\dfrac{x^{-2/3}y^{1/2}}{y^{-1/3}}\right)^6$

20. Subtract: $\sqrt{40x^3} - x\sqrt{90x}$

21. Multiply: $(\sqrt{3} - 2)(\sqrt{3} - 5)$

22. Simplify: $\dfrac{4}{\sqrt{6} - \sqrt{2}}$

23. Simplify: $\dfrac{2i}{3 - i}$

24. Solve: $\sqrt[3]{3x - 4} + 5 = 1$

25. The two triangles are similar triangles. Find the length of side *DE*.

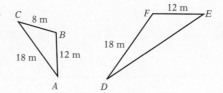

26. An investment of $2500 is made at an annual simple interest rate of 7.2%. How much additional money must be invested at an annual simple interest rate of 8.4% so that the total interest earned is $516?

27. A sales executive traveled 25 mi by car and then an additional 625 mi by plane. The rate of the plane was five times faster than the rate of the car. The total time of the trip was 3 h. Find the rate of the plane.

28. How long does it take light to travel to the earth from the moon when the moon is 232,500 mi from the earth? Light travels 1.86×10^5 mi/s.

29. The graph shows the amount invested and the annual income from an investment. Find the slope of the line between the two points shown on the graph. Then write a sentence that states the meaning of the slope.

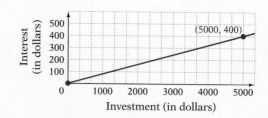

30. How far would a submarine periscope have to be above the water for the lookout to locate a ship 7 mi away? The equation for the distance in miles that the lookout can see is $d = \sqrt{1.5h}$, where h is the height in feet above the surface of the water. Round to the nearest tenth of a foot.

Quadratic Equations

Objectives

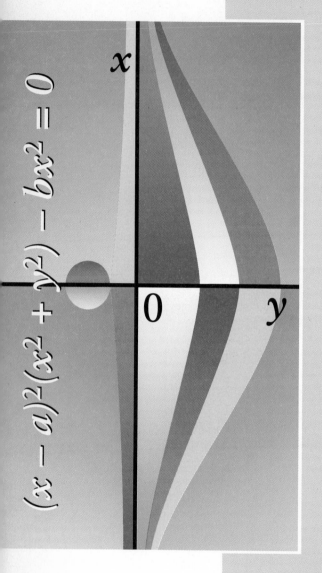

$(x - a)^2 (x^2 + y^2) - bx^2 = 0$

Algebraic Symbolism

The way in which an algebraic expression or equation is written has gone through several stages of development. First there was the *rhetoric*, which was in vogue until the late 13th century. In this method, expressions were written out in sentences. The word *res* was used to represent an unknown.

Rhetoric: From the additive *res* in the additive *res* results in a square *res*. From the three in an additive *x* comes three additive *res* and from the subtractive four in the additive *res* comes subtractive four *res*. From three in subtractive four comes subtractive twelve.

Modern: $(x + 3)(x - 4) = x^2 - x - 12$

The second stage was *syncoptic*, which was a shorthand in which abbreviations were used for words.

Syncoptic: *a* 6 in *b* quad − *c* plano 4 in *b* + *b* cub

Modern: $6ab^2 - 4cb + b^3$

The current modern stage, called the *symbolic* stage, began with the use of exponents rather than words to symbolize exponential expressions. This occurred near the beginning of the 17th century with the publication of the book *La Geometrie* by René Descartes. Modern notation is still evolving as mathematicians continue to search for convenient methods to symbolize concepts.

10.1 Solving Quadratic Equations by Factoring or by Taking Square Roots

Objective A   *To solve a quadratic equation by factoring* ································

Recall that a **quadratic equation** is an equation of the form $ax^2 + bx + c = 0$, $a \neq 0$.

A quadratic equation is in **standard form** when the polynomial is in descending order and equal to zero. Because the degree of the polynomial $ax^2 + bx + c$ is 2, a quadratic equation is also called a **second-degree equation.**

As we discussed earlier, quadratic equations sometimes can be solved by using the Principle of Zero Products. This method is reviewed here.

> **The Principle of Zero Products**
>
> If a and b are real numbers and $ab = 0$, then $a = 0$ or $b = 0$.

The Principle of Zero Products states that if the product of two factors is zero, then at least one of the factors must be zero.

➡ Solve by factoring: $3x^2 = 2 - 5x$

$$3x^2 = 2 - 5x$$

$3x^2 + 5x - 2 = 0$		• Write the equation in standard form.
$(3x - 1)(x + 2) = 0$		• Factor.
$3x - 1 = 0 \qquad\quad x + 2 = 0$		• Use the Principle of Zero Products to write two equations.
$3x = 1 \qquad\qquad\ \ x = -2$		• Solve each equation for x.
$x = \dfrac{1}{3}$		

$\dfrac{1}{3}$ and -2 check as solutions. The solutions are $\dfrac{1}{3}$ and -2.

➡ Solve by factoring: $x^2 - 6x = -9$

$$x^2 - 6x = -9$$

$x^2 - 6x + 9 = 0$		• Write the equation in standard form.
$(x - 3)(x - 3) = 0$		• Factor.
$x - 3 = 0 \qquad\quad x - 3 = 0$		• Use the Principle of Zero Products.
$x = 3 \qquad\qquad x = 3$		• Solve each equation for x.

3 checks as a solution. The solution is 3.

When a quadratic equation has two solutions that are the same number, the solution is called a **double root** of the equation. The solution 3 is a double root of the equation $x^2 - 6x = -9$.

Example 1

Solve by factoring: $2x(x - 3) = x + 4$

Solution

$$2x(x - 3) = x + 4$$
$$2x^2 - 6x = x + 4$$
$$2x^2 - 7x - 4 = 0$$
$$(2x + 1)(x - 4) = 0$$

$$2x + 1 = 0 \qquad x - 4 = 0$$
$$2x = -1 \qquad\quad x = 4$$
$$x = -\frac{1}{2}$$

The solutions are $-\frac{1}{2}$ and 4.

You Try It 1

Solve by factoring: $2x^2 = 7x - 3$

Your solution

Example 2

Solve for x by factoring:
$x^2 - 4ax - 5a^2 = 0$

Solution

This is a literal equation. Solve for x in terms of a.

$$x^2 - 4ax - 5a^2 = 0$$
$$(x + a)(x - 5a) = 0$$

$$x + a = 0 \qquad x - 5a = 0$$
$$x = -a \qquad\quad x = 5a$$

The solutions are $-a$ and $5a$.

You Try It 2

Solve for x by factoring:
$x^2 - 3ax - 4a^2 = 0$

Your solution

Solutions on p. A48

Objective B *To write a quadratic equation given its solutions*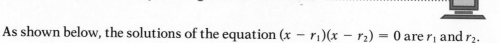

CONSIDER THIS

Here is an extension of these ideas.
$(x - r_1)(x - r_2) = x^2 - (r_1 + r_2)x + r_1 r_2$
Thus the coefficient of x is the opposite of the sum of the roots, and the constant term is the product of the roots. This can be used as a quick way of checking the solution of a quadratic equation in the next sections. For instance, $1 + \sqrt{2}$ and $1 - \sqrt{2}$ are solutions of the equation $x^2 - 2x - 1 = 0$ because
$-[(1 + \sqrt{2}) + (1 - \sqrt{2})]$
$= -2$, the coefficient of x, and
$(1 + \sqrt{2})(1 - \sqrt{2}) = -1$,
the constant term.

As shown below, the solutions of the equation $(x - r_1)(x - r_2) = 0$ are r_1 and r_2.

$$(x - r_1)(x - r_2) = 0$$

$$x - r_1 = 0 \qquad\qquad x - r_2 = 0$$
$$x = r_1 \qquad\qquad\quad x = r_2$$

Check:
$$\frac{(x - r_1)(x - r_2) = 0}{(r_1 - r_1)(r_1 - r_2)} \Big| \qquad \frac{(x - r_1)(x - r_2) = 0}{(r_2 - r_1)(r_2 - r_2)} \Big|$$
$$0 \cdot (r_1 - r_2) \Big| \qquad\qquad (r_2 - r_1) \cdot 0 \Big|$$
$$0 = 0 \qquad\qquad\qquad\qquad 0 = 0$$

Using the equation $(x - r_1)(x - r_2) = 0$ and the fact that r_1 and r_2 are solutions of this equation, it is possible to write a quadratic equation given its solutions.

➡ Write a quadratic equation that has solutions 4 and -5.

$$(x - r_1)(x - r_2) = 0$$
$$(x - 4)[x - (-5)] = 0$$
$$(x - 4)(x + 5) = 0$$
$$x^2 + x - 20 = 0$$

- Replace r_1 by 4 and r_2 by -5.
- Simplify.
- Multiply.

Content and Format © 1996 HMCo.

➡ Write a quadratic equation with integer coefficients and solutions $\frac{2}{3}$ and $\frac{1}{2}$.

$$(x - r_1)(x - r_2) = 0$$

$$\left(x - \frac{2}{3}\right)\left(x - \frac{1}{2}\right) = 0$$ • **Replace r_1 by $\frac{2}{3}$ and r_2 by $\frac{1}{2}$.**

$$x^2 - \frac{7}{6}x + \frac{1}{3} = 0$$ • **Multiply.**

$$6\left(x^2 - \frac{7}{6}x + \frac{1}{3}\right) = 6 \cdot 0$$ • **Multiply each side of the equation by the LCM of the denominators.**

$$6x^2 - 7x + 2 = 0$$

Example 3

Write a quadratic equation with integer coefficients and solutions $\frac{1}{2}$ and -4.

Solution

$$(x - r_1)(x - r_2) = 0$$

$$\left(x - \frac{1}{2}\right)[x - (-4)] = 0$$

$$\left(x - \frac{1}{2}\right)(x + 4) = 0$$

$$x^2 + \frac{7}{2}x - 2 = 0$$

$$2\left(x^2 + \frac{7}{2}x - 2\right) = 2 \cdot 0$$

$$2x^2 + 7x - 4 = 0$$

You Try It 3

Write a quadratic equation with integer coefficients and solutions 3 and $-\frac{1}{2}$.

Your solution

Solution on p. A48

Objective C   To solve a quadratic equation by taking square roots ....

The solution of the quadratic equation $x^2 = 16$ is shown at the right.

$$x^2 = 16$$
$$x^2 - 16 = 0$$
$$(x - 4)(x + 4) = 0$$
$$x - 4 = 0 \quad x + 4 = 0$$
$$x = 4 \qquad x = -4$$

As shown below, the solutions can also be found by taking the square root of each side of the equation and writing the positive and negative square roots of the number. The notation $x = \pm 4$ means $x = 4$ or $x = -4$.

$$x^2 = 16$$
$$\sqrt{x^2} = \pm\sqrt{16}$$
$$x = \pm 4$$

The solutions are 4 and -4.

➡ Solve by taking square roots: $3x^2 = 54$

$$3x^2 = 54$$
$$x^2 = 18$$ • **Solve for x^2.**
$$\sqrt{x^2} = \pm\sqrt{18}$$ • **Take the square root of each side of the equation.**
$$x = \pm 3\sqrt{2}$$ • **Simplify.**

$3\sqrt{2}$ and $-3\sqrt{2}$ check as solutions. The solutions are $3\sqrt{2}$ and $-3\sqrt{2}$.

Solving a quadratic equation by taking the square root of each side of the equation can lead to solutions that are complex numbers.

➡ Solve by taking square roots: $2x^2 + 18 = 0$

$$2x^2 + 18 = 0$$

$$2x^2 = -18 \qquad \bullet \text{ Solve for } x^2.$$

$$x^2 = -9$$

$$\sqrt{x^2} = \pm\sqrt{-9} \qquad \bullet \text{ Take the square root of each side of the equation.}$$

$$x = \pm 3i \qquad \bullet \text{ Simplify.}$$

Check:

$2x^2 + 18 = 0$		$2x^2 + 18 = 0$	
$2(3i)^2 + 18$	0	$2(-3i)^2 + 18$	0
$2(-9) + 18$	0	$2(-9) + 18$	0
$-18 + 18$	0	$-18 + 18$	0
	$0 = 0$		$0 = 0$

The solutions are $3i$ and $-3i$.

An equation that contains the square of a binomial can be solved by taking square roots.

➡ Solve by taking square roots: $(x + 2)^2 - 24 = 0$

$$(x + 2)^2 - 24 = 0$$

$$(x + 2)^2 = 24 \qquad \bullet \text{ Solve for } (x + 2)^2.$$

$$\sqrt{(x + 2)^2} = \pm\sqrt{24} \qquad \bullet \text{ Take the square root of each side of the equation.}$$

$$x + 2 = \pm 2\sqrt{6} \qquad \bullet \text{ Simplify.}$$

$$x + 2 = 2\sqrt{6} \qquad x + 2 = -2\sqrt{6} \qquad \bullet \text{ Solve for } x.$$

$$x = -2 + 2\sqrt{6} \qquad x = -2 - 2\sqrt{6}$$

The solutions are $-2 + 2\sqrt{6}$ and $-2 - 2\sqrt{6}$.

Example 4
Solve by taking square roots:
$3(x - 2)^2 + 12 = 0$

Solution
$$3(x - 2)^2 + 12 = 0$$
$$3(x - 2)^2 = -12$$
$$(x - 2)^2 = -4$$
$$\sqrt{(x - 2)^2} = \pm\sqrt{-4}$$
$$x - 2 = \pm 2i$$

$$x - 2 = 2i \qquad x - 2 = -2i$$
$$x = 2 + 2i \qquad x = 2 - 2i$$

The solutions are $2 + 2i$ and $2 - 2i$.

You Try It 4
Solve by taking square roots:
$2(x + 1)^2 - 24 = 0$

Your solution

Solution on p. A48

10.1 Exercises

Objective A

Solve by factoring.

1. $x^2 - 4x = 0$

2. $y^2 + 6y = 0$

3. $t^2 - 25 = 0$

4. $p^2 - 81 = 0$

5. $s^2 - s - 6 = 0$

6. $v^2 + 4v - 5 = 0$

7. $y^2 - 6y + 9 = 0$

8. $x^2 + 10x + 25 = 0$

9. $9z^2 - 18z = 0$

10. $4y^2 + 20y = 0$

11. $r^2 - 3r = 10$

12. $p^2 + 5p = 6$

13. $v^2 + 10 = 7v$

14. $t^2 - 16 = 15t$

15. $2x^2 - 9x - 18 = 0$

16. $3y^2 - 4y - 4 = 0$

17. $4z^2 - 9z + 2 = 0$

18. $2s^2 - 9s + 9 = 0$

19. $3w^2 + 11w = 4$

20. $2r^2 + r = 6$

21. $6x^2 = 23x + 18$

22. $6x^2 = 7x - 2$

23. $4 - 15u - 4u^2 = 0$

24. $3 - 2y - 8y^2 = 0$

25. $x + 18 = x(x - 6)$

26. $t + 24 = t(t + 6)$

27. $4s(s + 3) = s - 6$

28. $3v(v - 2) = 11v + 6$

29. $u^2 - 2u + 4 = (2u - 3)(u + 2)$

30. $(3v - 2)(2v + 1) = 3v^2 - 11v - 10$

31. $(3x - 4)(x + 4) = x^2 - 3x - 28$

Solve for x by factoring.

32. $x^2 + 14ax + 48a^2 = 0$

33. $x^2 - 9bx + 14b^2 = 0$

34. $x^2 + 9xy - 36y^2 = 0$

35. $x^2 - 6cx - 7c^2 = 0$

36. $x^2 - ax - 20a^2 = 0$

37. $2x^2 + 3bx + b^2 = 0$

38. $3x^2 - 4cx + c^2 = 0$

39. $3x^2 - 14ax + 8a^2 = 0$

40. $3x^2 - 11xy + 6y^2 = 0$

41. $3x^2 - 8ax - 3a^2 = 0$

42. $3x^2 - 4bx - 4b^2 = 0$

43. $4x^2 + 8xy + 3y^2 = 0$

44. $6x^2 - 11cx + 3c^2 = 0$

45. $6x^2 + 11ax + 4a^2 = 0$

46. $12x^2 - 5xy - 2y^2 = 0$

Objective B

Write a quadratic equation that has integer coefficients and has as solutions the given pair of numbers.

47. 2 and 5

48. 3 and 1

49. -2 and -4

50. -1 and -3

51. 6 and -1

52. -2 and 5

53. 3 and -3

54. 5 and -5

55. 4 and 4

56. 2 and 2

57. 0 and 5

58. 0 and -2

59. 0 and 3

60. 0 and -1

61. 3 and $\frac{1}{2}$

Write a quadratic equation that has integer coefficients and has as solutions the given pair of numbers.

62. 2 and $\frac{2}{3}$

63. $-\frac{3}{4}$ and 2

64. $-\frac{1}{2}$ and 5

65. $-\frac{5}{3}$ and -2

66. $-\frac{3}{2}$ and -1

67. $-\frac{2}{3}$ and $\frac{2}{3}$

68. $-\frac{1}{2}$ and $\frac{1}{2}$

69. $\frac{1}{2}$ and $\frac{1}{3}$

70. $\frac{3}{4}$ and $\frac{2}{3}$

71. $\frac{6}{5}$ and $-\frac{1}{2}$

72. $\frac{3}{4}$ and $-\frac{3}{2}$

73. $-\frac{1}{4}$ and $-\frac{1}{2}$

74. $-\frac{5}{6}$ and $-\frac{2}{3}$

75. $\frac{3}{5}$ and $-\frac{1}{10}$

76. $\frac{7}{2}$ and $-\frac{1}{4}$

Objective C

Solve by taking square roots.

77. $y^2 = 49$

78. $x^2 = 64$

79. $z^2 = -4$

80. $v^2 = -16$

81. $s^2 - 4 = 0$

82. $r^2 - 36 = 0$

83. $4x^2 - 81 = 0$

84. $9x^2 - 16 = 0$

85. $y^2 + 49 = 0$

86. $z^2 + 16 = 0$

87. $v^2 - 48 = 0$

88. $s^2 - 32 = 0$

89. $r^2 - 75 = 0$

90. $u^2 - 54 = 0$

91. $z^2 + 18 = 0$

Solve by taking square roots.

92. $t^2 + 27 = 0$

93. $(x - 1)^2 = 36$

94. $(x + 2)^2 = 25$

95. $3(y + 3)^2 = 27$

96. $4(s - 2)^2 = 36$

97. $5(z + 2)^2 = 125$

98. $2(y - 3)^2 = 18$

99. $\left(v - \dfrac{1}{2}\right)^2 = \dfrac{1}{4}$

100. $\left(r + \dfrac{2}{3}\right)^2 = \dfrac{1}{9}$

101. $(x + 5)^2 - 6 = 0$

102. $(t - 1)^2 - 15 = 0$

103. $(v - 3)^2 + 45 = 0$

104. $(x + 5)^2 + 32 = 0$

105. $\left(u + \dfrac{2}{3}\right)^2 - 18 = 0$

106. $\left(z - \dfrac{1}{2}\right)^2 - 20 = 0$

APPLYING THE CONCEPTS

Write a quadratic equation that has as solutions the given pair of numbers.

107. $\sqrt{2}$ and $-\sqrt{2}$

108. $2i$ and $-2i$

109. $3\sqrt{2}$ and $-3\sqrt{2}$

110. $2i\sqrt{3}$ and $-2i\sqrt{3}$

Solve for x.

111. $4a^2x^2 = 36b^2, a > 0, b > 0$

112. $(x + a)^2 - 4 = 0$

113. $(2x - 1)^2 = (2x + 3)^2$

114. [W] Show that the solutions of the equation $ax^2 + bx = 0, a > 0, b > 0,$ are 0 and $-\dfrac{b}{a}$.

115. [W] Show that the solutions of the equation $ax^2 + c = 0, a > 0, c > 0,$ are $\dfrac{\sqrt{ca}}{a}i$ and $-\dfrac{\sqrt{ca}}{a}i$.

116. [W] Explain why the restriction that $a \neq 0$ is given in the definition of a quadratic equation.

10.2 Solving Quadratic Equations by Completing the Square

Objective A *To solve a quadratic equation by completing the square*

Recall that a perfect-square trinomial is the square of a binomial.

Perfect-Square Trinomial		*Square of a Binomial*
$x^2 + 8x + 16$	$=$	$(x + 4)^2$
$x^2 - 10x + 25$	$=$	$(x - 5)^2$
$x^2 + 2ax + a^2$	$=$	$(x + a)^2$

For each perfect-square trinomial, the square of $\frac{1}{2}$ of the coefficient of x equals the constant term.

$$x^2 + 8x + 16, \quad \left(\frac{1}{2} \cdot 8\right)^2 = 16$$

$$x^2 - 10x + 25, \quad \left[\frac{1}{2}(-10)\right]^2 = 25$$

$$x^2 + 2ax + a^2, \quad \left(\frac{1}{2} \cdot 2a\right)^2 = a^2$$

$$\left(\frac{1}{2} \text{ coefficient of } x\right)^2 = \textbf{constant term}$$

POINT OF INTEREST

Early attempts to solve quadratic equations were primarily geometric. The Persian mathematician al-Khowarizmi (c. A.D. 800) essentially completed a square of $x^2 + 12x$ as follows.

This relationship can be used to write the constant term for a perfect-square trinomial. Adding to a binomial the constant term that makes it a perfect-square trinomial is called **completing the square.**

➡ Complete the square on $x^2 + 12x$. Write the resulting perfect-square trinomial as the square of a binomial.

$$\left[\frac{1}{2}(12)\right]^2 = (6)^2 = 36$$
$$x^2 + 12x + 36$$

$$x^2 + 12x + 36 = (x + 6)^2$$

- Find the constant term.
- Complete the square on $x^2 + 12x$ by adding the constant term.
- Write the resulting perfect-square trinomial as the square of a binomial.

➡ Complete the square on $z^2 - 3z$. Write the resulting perfect-square trinomial as the square of a binomial.

$$\left[\frac{1}{2} \cdot (-3)\right]^2 = \left(-\frac{3}{2}\right)^2 = \frac{9}{4}$$

$$z^2 - 3z + \frac{9}{4}$$

$$z^2 - 3z + \frac{9}{4} = \left(z - \frac{3}{2}\right)^2$$

- Find the constant term.
- Complete the square on $z^2 - 3z$ by adding the constant term.
- Write the resulting perfect-square trinomial as the square of a binomial.

Any quadratic equation can be solved by completing the square. Add to each side of the equation the term that completes the square. Rewrite the equation in the form $(x + a)^2 = b$. Take the square root of each side of the equation.

➡ Solve by completing the square: $x^2 - 6x - 15 = 0$

$$x^2 - 6x - 15 = 0$$
$$x^2 - 6x = 15$$
$$x^2 - 6x + 9 = 15 + 9$$

- Add 15 to each side of the equation.
- Complete the square. Add
  $\left[\dfrac{1}{2}(-6)\right]^2 = (-3)^2 = 9$ to each side of the equation.

$$(x - 3)^2 = 24$$

- Factor the trinomial.

$$\sqrt{(x - 3)^2} = \pm\sqrt{24}$$

- Take the square root of each side of the equation.

$$x - 3 = \pm 2\sqrt{6}$$

- Solve for x.

$$x - 3 = 2\sqrt{6} \qquad x - 3 = -2\sqrt{6}$$
$$x = 3 + 2\sqrt{6} \qquad x = 3 - 2\sqrt{6}$$

Check:

$$
\begin{array}{c|c}
x^2 - 6x - 15 = 0 & \\
\hline
(3 + 2\sqrt{6})^2 - 6(3 + 2\sqrt{6}) - 15 & 0 \\
9 + 12\sqrt{6} + 24 - 18 - 12\sqrt{6} - 15 & 0 \\
0 = 0
\end{array}
$$

$$
\begin{array}{c|c}
x^2 - 6x - 15 = 0 & \\
\hline
(3 - 2\sqrt{6})^2 - 6(3 - 2\sqrt{6}) - 15 & 0 \\
9 - 12\sqrt{6} + 24 - 18 + 12\sqrt{6} - 15 & 0 \\
0 = 0
\end{array}
$$

The solutions are $3 + 2\sqrt{6}$ and $3 - 2\sqrt{6}$.

➡ Solve by completing the square: $2x^2 - x - 2 = 0$

In order for us to complete the square on an expression, the coefficient of x^2 must be 1. After adding the constant term to each side of the equation, multiply each side of the equation by the reciprocal of the coefficient of x^2.

$$2x^2 - x - 2 = 0$$
$$2x^2 - x = 2$$

- Add 2 to each side of the equation.

$$\frac{1}{2}(2x^2 - x) = \frac{1}{2} \cdot 2$$

- Multiply each side of the equation by $\dfrac{1}{2}$.

$$x^2 - \frac{1}{2}x = 1$$

- The coefficient of x^2 is now 1.

$$x^2 - \frac{1}{2}x + \frac{1}{16} = 1 + \frac{1}{16}$$

- Complete the square. Add
  $\left[\dfrac{1}{2}\left(-\dfrac{1}{2}\right)\right]^2 = \left(-\dfrac{1}{4}\right)^2 = \dfrac{1}{16}$ to each side of the equation.

$$\left(x - \frac{1}{4}\right)^2 = \frac{17}{16}$$

- Factor the perfect-square trinomial.

$$\sqrt{\left(x - \frac{1}{4}\right)^2} = \pm\sqrt{\frac{17}{16}}$$

- Take the square root of each side of the equation.

$$x - \frac{1}{4} = \pm\frac{\sqrt{17}}{4}$$

$$x - \frac{1}{4} = \frac{\sqrt{17}}{4} \qquad x - \frac{1}{4} = -\frac{\sqrt{17}}{4}$$

- Solve for x.

$$x = \frac{1}{4} + \frac{\sqrt{17}}{4} \qquad x = \frac{1}{4} - \frac{\sqrt{17}}{4}$$

The solutions are $\dfrac{1 + \sqrt{17}}{4}$ and $\dfrac{1 - \sqrt{17}}{4}$.

The solutions of the quadratic equation in the example below are complex numbers.

POINT OF INTEREST

Mathematicians have studied quadratic equations for centuries. Many of the initial equations were a result of trying to solve a geometry problem. One of the most famous, which dates from around 500 B.C., is "squaring the circle." The question was "Is it possible to construct a square whose area is that of a given circle?" For these early mathematicians, to *construct* meant to draw with only a straightedge and a compass. It was approximately 2300 years later that mathematicians were able to prove that such a construction is impossible.

➡ Solve by completing the square: $x^2 - 3x + 5 = 0$

$$x^2 - 3x + 5 = 0$$
$$x^2 - 3x = -5$$

- Add the opposite of the constant term to each side of the equation.

$$x^2 - 3x + \frac{9}{4} = -5 + \frac{9}{4}$$

- Add to each side of the equation the term that completes the square on $x^2 - 3x$.

$$\left(x - \frac{3}{2}\right)^2 = -\frac{11}{4}$$

- Factor the perfect-square trinomial.

$$\sqrt{\left(x - \frac{3}{2}\right)^2} = \pm\sqrt{-\frac{11}{4}}$$

- Take the square root of each side of the equation.

$$x - \frac{3}{2} = \pm\frac{i\sqrt{11}}{2}$$

- Simplify.

$$x - \frac{3}{2} = \frac{i\sqrt{11}}{2} \qquad x - \frac{3}{2} = -\frac{i\sqrt{11}}{2}$$

- Solve for x.

$$x = \frac{3}{2} + \frac{i\sqrt{11}}{2} \qquad x = \frac{3}{2} - \frac{i\sqrt{11}}{2}$$

$\dfrac{3}{2} + \dfrac{\sqrt{11}}{2}i$ and $\dfrac{3}{2} - \dfrac{\sqrt{11}}{2}i$ check as solutions.

The solutions are $\dfrac{3}{2} + \dfrac{\sqrt{11}}{2}i$ and $\dfrac{3}{2} - \dfrac{\sqrt{11}}{2}i$.

Example 1

Solve by completing the square:
$4x^2 - 8x + 1 = 0$

Solution

$4x^2 - 8x + 1 = 0$

$\quad\ 4x^2 - 8x = -1$

$\frac{1}{4}(4x^2 - 8x) = \frac{1}{4}(-1)$

$\quad\quad x^2 - 2x = -\frac{1}{4}$

$x^2 - 2x + 1 = -\frac{1}{4} + 1$ • Complete the square.

$\quad\quad (x - 1)^2 = \frac{3}{4}$

$\quad\ \sqrt{(x - 1)^2} = \pm\sqrt{\frac{3}{4}}$

$\quad\quad\ x - 1 = \pm\frac{\sqrt{3}}{2}$

$x - 1 = \frac{\sqrt{3}}{2} \qquad\quad x - 1 = -\frac{\sqrt{3}}{2}$

$\quad x = 1 + \frac{\sqrt{3}}{2} \qquad\quad x = 1 - \frac{\sqrt{3}}{2}$

$\quad\quad = \frac{2 + \sqrt{3}}{2} \qquad\quad\quad = \frac{2 - \sqrt{3}}{2}$

The solutions are $\frac{2 + \sqrt{3}}{2}$ and $\frac{2 - \sqrt{3}}{2}$.

You Try It 1

Solve by completing the square:
$4x^2 - 4x - 1 = 0$

Your solution

Example 2

Solve by completing the square:
$x^2 + 4x + 5 = 0$

Solution

$x^2 + 4x + 5 = 0$

$\quad\ x^2 + 4x = -5$

$x^2 + 4x + 4 = -5 + 4$ • Complete the square.

$\quad\ (x + 2)^2 = -1$

$\sqrt{(x + 2)^2} = \pm\sqrt{-1}$

$\quad\quad x + 2 = \pm i$

$x + 2 = i \qquad\quad x + 2 = -i$

$\quad x = -2 + i \qquad\quad x = -2 - i$

The solutions are $-2 + i$ and $-2 - i$.

You Try It 2

Solve by completing the square:
$2x^2 + x - 5 = 0$

Your solution

Solutions on p. A48

10.2 Exercises

Objective A

Solve by completing the square.

1. $x^2 - 4x - 5 = 0$

2. $y^2 + 6y + 5 = 0$

3. $v^2 + 8v - 9 = 0$

4. $w^2 - 2w - 24 = 0$

5. $z^2 - 6z + 9 = 0$

6. $u^2 + 10u + 25 = 0$

7. $r^2 + 4r - 7 = 0$

8. $s^2 + 6s - 1 = 0$

9. $x^2 - 6x + 7 = 0$

10. $y^2 + 8y + 13 = 0$

11. $z^2 - 2z + 2 = 0$

12. $t^2 - 4t + 8 = 0$

13. $s^2 - 5s - 24 = 0$

14. $v^2 + 7v - 44 = 0$

15. $x^2 + 5x - 36 = 0$

16. $y^2 - 9y + 20 = 0$

17. $p^2 - 3p + 1 = 0$

18. $r^2 - 5r - 2 = 0$

19. $t^2 - t - 1 = 0$

20. $u^2 - u - 7 = 0$

21. $y^2 - 6y = 4$

22. $w^2 + 4w = 2$

23. $x^2 = 8x - 15$

24. $z^2 = 4z - 3$

25. $v^2 = 4v - 13$

26. $x^2 = 2x - 17$

27. $p^2 + 6p = -13$

28. $x^2 + 4x = -20$

29. $y^2 - 2y = 17$

30. $x^2 + 10x = 7$

31. $z^2 = z + 4$

32. $r^2 = 3r - 1$

33. $x^2 + 13 = 2x$

Solve by completing the square.

34. $6v^2 - 7v = 3$

35. $4x^2 - 4x + 5 = 0$

36. $4t^2 - 4t + 17 = 0$

37. $9x^2 - 6x + 2 = 0$

38. $9y^2 - 12y + 13 = 0$

39. $2s^2 = 4s + 5$

40. $3u^2 = 6u + 1$

41. $2r^2 = 3 - r$

42. $2x^2 = 12 - 5x$

43. $y - 2 = (y - 3)(y + 2)$

44. $8s - 11 = (s - 4)(s - 2)$

45. $6t - 2 = (2t - 3)(t - 1)$

46. $2z + 9 = (2z + 3)(z + 2)$

47. $(x - 4)(x + 1) = x - 3$

48. $(y - 3)^2 = 2y + 10$

APPLYING THE CONCEPTS

Solve for x by completing the square.

49. $x^2 - ax - 2a^2 = 0$

50. $x^2 + 3ax - 4a^2 = 0$

51. $x^2 + 3ax - 10a^2 = 0$

52. After a baseball is hit, the height h (in feet) of the ball above the ground t seconds after it is hit can be approximated by the equation $h = -16t^2 + 70t + 4$. Using this equation, determine when the ball will hit the ground. (*Hint:* The ball hits the ground when $h = 0$.)

53. After a baseball is hit, there are two equations that can be considered. One gives the height h (in feet) the ball is above the ground t seconds after it is hit. The second is the horizontal distance s (in feet) the ball is from home plate t seconds after it is hit. A model of this situation is given by $h = -16t^2 + 70t + 4$ and $s = 44.5t$. Using this model, determine whether the ball will clear a 6 ft fence 325 ft from home plate.

54.
[W]
Explain how to complete the square of $x^2 + bx$.

55.
[W]
Write out the steps for solving a quadratic equation by completing the square.

10.3 Solving Quadratic Equations by Using the Quadratic Formula

Objective A   *To solve a quadratic equation by using the quadratic formula*

A general formula known as the **quadratic formula** can be derived by applying the method of completing the square to the standard form of a quadratic equation. This formula can be used to solve any quadratic equation.

Solve $ax^2 + bx + c = 0$ by completing the square.

$$ax^2 + bx + c = 0$$

Subtract the constant term from each side of the equation.

$$ax^2 + bx + c - c = 0 - c$$

$$ax^2 + bx = -c$$

Multiply each side of the equation by the reciprocal of a, the coefficient of x^2.

$$\frac{1}{a}(ax^2 + bx) = \frac{1}{a}(-c)$$

$$x^2 + \frac{b}{a}x = -\frac{c}{a}$$

Complete the square by adding $\left(\frac{1}{2} \cdot \frac{b}{a}\right)^2 = \frac{b^2}{4a^2}$ to each side of the equation.

$$x^2 + \frac{b}{a}x + \frac{b^2}{4a^2} = \frac{b^2}{4a^2} - \frac{c}{a}$$

Simplify the right side of the equation.

$$x^2 + \frac{b}{a}x + \frac{b^2}{4a^2} = \frac{b^2}{4a^2} - \left(\frac{c}{a} \cdot \frac{4a}{4a}\right)$$

$$x^2 + \frac{b}{a}x + \frac{b^2}{4a^2} = \frac{b^2}{4a^2} - \frac{4ac}{4a^2}$$

$$x^2 + \frac{b}{a}x + \frac{b^2}{4a^2} = \frac{b^2 - 4ac}{4a^2}$$

Factor the perfect-square trinomial on the left side of the equation.

$$\left(x + \frac{b}{2a}\right)^2 = \frac{b^2 - 4ac}{4a^2}$$

Take the square root of each side of the equation.

$$\sqrt{\left(x + \frac{b}{2a}\right)^2} = \pm\sqrt{\frac{b^2 - 4ac}{4a^2}}$$

$$x + \frac{b}{2a} = \pm\frac{\sqrt{b^2 - 4ac}}{2a}$$

Solve for x.

$$x + \frac{b}{2a} = \frac{\sqrt{b^2 - 4ac}}{2a}$$

$$x = -\frac{b}{2a} + \frac{\sqrt{b^2 - 4ac}}{2a}$$

$$= \frac{-b + \sqrt{b^2 - 4ac}}{2a}$$

$$x + \frac{b}{2a} = -\frac{\sqrt{b^2 - 4ac}}{2a}$$

$$x = -\frac{b}{2a} - \frac{\sqrt{b^2 - 4ac}}{2a}$$

$$= \frac{-b - \sqrt{b^2 - 4ac}}{2a}$$

POINT OF INTEREST

Although mathematicians have studied quadratic equations since around 500 B.C., it was not until the 18th century that the formula was written as it is today. Of further note, the word *quadratic* has the same Latin root as does the word *square*.

The Quadratic Formula

The solutions of $ax^2 + bx + c = 0$, $a \neq 0$, are

$$\frac{-b + \sqrt{b^2 - 4ac}}{2a} \quad \text{and} \quad \frac{-b - \sqrt{b^2 - 4ac}}{2a}$$

The quadratic formula is frequently written as $x = \dfrac{-b \pm \sqrt{b^2 - 4ac}}{2a}$.

➡ Solve by using the quadratic formula: $2x^2 + 5x + 3 = 0$

From the standard form of the equation, $a = 2$, $b = 5$, and $c = 3$. Replace a, b, and c in the quadratic formula with these values.

CONSIDER THIS

The solutions of this quadratic equation are rational numbers. When this happens, the equation could have been solved by factoring and using the Principle of Zero Products. This may be easier than applying the quadratic formula.

$$x = \frac{-b \pm \sqrt{b^2 - 4ac}}{2a} = \frac{-(5) \pm \sqrt{(5)^2 - 4(2)(3)}}{2(2)}$$

$$= \frac{-5 \pm \sqrt{25 - 24}}{4}$$

$$= \frac{-5 \pm \sqrt{1}}{4} = \frac{-5 \pm 1}{4}$$

$$x = \frac{-5 + 1}{4} = \frac{-4}{4} = -1 \qquad x = \frac{-5 - 1}{4} = \frac{-6}{4} = -\frac{3}{2}$$

The solutions are -1 and $-\dfrac{3}{2}$.

➡ Solve by using the quadratic formula: $3x^2 = 4x + 6$

$$3x^2 = 4x + 6$$
$$3x^2 - 4x - 6 = 0$$

- Write the equation in standard form.
- Subtract $4x$ and 6 from each side of the equation.

CONSIDER THIS

Remember that the sign that precedes a number is the sign of the number. For the equation at the right, we have

$$\overset{a}{\textcircled{3}}\,x^2 \overset{b}{\textcircled{- 4}}\,x \overset{c}{\textcircled{- 6}} = 0$$

From the standard form of the equation, $a = 3$, $b = -4$, and $c = -6$. Replace a, b, and c in the quadratic formula with these values.

$$x = \frac{-b \pm \sqrt{b^2 - 4ac}}{2a} = \frac{-(-4) \pm \sqrt{(-4)^2 - 4(3)(-6)}}{2(3)}$$

$$= \frac{4 \pm \sqrt{16 - (-72)}}{6}$$

$$= \frac{4 \pm \sqrt{88}}{6} = \frac{4 \pm 2\sqrt{22}}{6}$$

$$= \frac{2(2 \pm \sqrt{22})}{2 \cdot 3} = \frac{2 \pm \sqrt{22}}{3}$$

Check:

$3x^2 = 4x + 6$		$3x^2 = 4x + 6$	
$3\left(\dfrac{2 + \sqrt{22}}{3}\right)^2$	$4\left(\dfrac{2 + \sqrt{22}}{3}\right) + 6$	$3\left(\dfrac{2 - \sqrt{22}}{3}\right)^2$	$4\left(\dfrac{2 - \sqrt{22}}{3}\right) + 6$
$3\left(\dfrac{4 + 4\sqrt{22} + 22}{9}\right)$	$\dfrac{8}{3} + \dfrac{4\sqrt{22}}{3} + \dfrac{18}{3}$	$3\left(\dfrac{4 - 4\sqrt{22} + 22}{9}\right)$	$\dfrac{8}{3} - \dfrac{4\sqrt{22}}{3} + \dfrac{18}{3}$
$3\left(\dfrac{26 + 4\sqrt{22}}{9}\right)$	$\dfrac{26}{3} + \dfrac{4\sqrt{22}}{3}$	$3\left(\dfrac{26 - 4\sqrt{22}}{9}\right)$	$\dfrac{26}{3} - \dfrac{4\sqrt{22}}{3}$
$\dfrac{26 + 4\sqrt{22}}{3} = \dfrac{26 + 4\sqrt{22}}{3}$		$\dfrac{26 - 4\sqrt{22}}{3} = \dfrac{26 - 4\sqrt{22}}{3}$	

The solutions are $\dfrac{2 + \sqrt{22}}{3}$ and $\dfrac{2 - \sqrt{22}}{3}$.

➡ Solve by using the quadratic formula: $4x^2 - 8x + 13 = 0$

$4x^2 - 8x + 13 = 0$

$x = \dfrac{-b \pm \sqrt{b^2 - 4ac}}{2a}$

• Use the quadratic formula.

$= \dfrac{-(-8) \pm \sqrt{(-8)^2 - 4 \cdot 4 \cdot 13}}{2 \cdot 4}$

• $a = 4$, $b = -8$, $c = 13$

$= \dfrac{8 \pm \sqrt{64 - 208}}{8} = \dfrac{8 \pm \sqrt{-144}}{8}$

$= \dfrac{8 \pm 12i}{8} = \dfrac{4(2 \pm 3i)}{4 \cdot 2} = \dfrac{2 \pm 3i}{2}$

The solutions are $1 + \dfrac{3}{2}i$ and $1 - \dfrac{3}{2}i$.

Of the three examples above, the first two had real number solutions; the last one had complex number solutions.

In the quadratic formula, the quantity $b^2 - 4ac$ is called the **discriminant.** When a, b, and c are real numbers, the discriminant determines whether a quadratic equation will have a double root, two real number solutions that are not equal, or two complex number solutions.

THE EFFECT OF THE DISCRIMINANT ON THE SOLUTIONS OF A QUADRATIC EQUATION

1. If $b^2 - 4ac = 0$, the equation has one real number solution, a double root.
2. If $b^2 - 4ac > 0$, the equation has two unequal real number solutions.
3. If $b^2 - 4ac < 0$, the equation has two complex number solutions.

➡ Use the discriminant to determine whether $x^2 - 4x - 5 = 0$ has one real number solution, two real number solutions, or two complex number solutions.

$b^2 - 4ac$
$(-4)^2 - 4(1)(-5) = 16 + 20 = 36$
$36 > 0$

• Evaluate the discriminant.
• $a = 1$, $b = -4$, $c = -5$

Because $b^2 - 4ac > 0$, the equation has two real number solutions.

Example 1 Solve by using the quadratic formula: $2x^2 - x + 5 = 0$

You Try It 1 Solve by using the quadratic formula: $x^2 - 2x + 10 = 0$

Solution $2x^2 - x + 5 = 0$
$a = 2$, $b = -1$, $c = 5$
$x = \dfrac{-b \pm \sqrt{b^2 - 4ac}}{2a}$

$= \dfrac{-(-1) \pm \sqrt{(-1)^2 - 4(2)(5)}}{2 \cdot 2}$

$= \dfrac{1 \pm \sqrt{1 - 40}}{4} = \dfrac{1 \pm \sqrt{-39}}{4}$

$= \dfrac{1 \pm i\sqrt{39}}{4}$

Your solution

The solutions are $\dfrac{1}{4} + \dfrac{\sqrt{39}}{4}i$ and $\dfrac{1}{4} - \dfrac{\sqrt{39}}{4}i$.

Solution on p. A49

Example 2

Solve by using the quadratic formula:
$2x^2 = (x - 2)(x - 3)$

Solution

$$2x^2 = (x - 2)(x - 3)$$
$$2x^2 = x^2 - 5x + 6$$

$$x^2 + 5x - 6 = 0$$

$$a = 1, b = 5, c = -6$$

$$x = \frac{-b \pm \sqrt{b^2 - 4ac}}{2a}$$

$$= \frac{-5 \pm \sqrt{5^2 - 4(1)(-6)}}{2 \cdot 1}$$

$$= \frac{-5 \pm \sqrt{25 + 24}}{2} = \frac{-5 \pm \sqrt{49}}{2}$$

$$= \frac{-5 \pm 7}{2}$$

$$x = \frac{-5 + 7}{2} \qquad x = \frac{-5 - 7}{2}$$

$$= \frac{2}{2} = 1 \qquad = \frac{-12}{2} = -6$$

The solutions are 1 and −6.

You Try It 2

Solve by using the quadratic formula:
$4x^2 = 4x - 1$

Your solution

Example 3

Use the discriminant to determine whether $4x^2 - 2x + 5 = 0$ has one real number solution, two real number solutions, or two complex number solutions.

Solution

$$a = 4, b = -2, c = 5$$
$$b^2 - 4ac = (-2)^2 - 4(4)(5)$$
$$= 4 - 80$$
$$= -76$$

$$-76 < 0$$

Because the discriminant is less than zero, the equation has two complex number solutions.

You Try It 3

Use the discriminant to determine whether $3x^2 - x - 1 = 0$ has one real number solution, two real number solutions, or two complex number solutions.

Your solution

Solutions on p. A49

10.3 Exercises

Objective A

Solve by using the quadratic formula.

1. $x^2 - 3x - 10 = 0$

2. $z^2 - 4z - 8 = 0$

3. $y^2 + 5y - 36 = 0$

4. $z^2 - 3z - 40 = 0$

5. $w^2 = 8w + 72$

6. $t^2 = 2t + 35$

7. $v^2 = 24 - 5v$

8. $x^2 = 18 - 7x$

9. $2y^2 + 5y - 3 = 0$

10. $4p^2 - 7p + 3 = 0$

11. $8s^2 = 10s + 3$

12. $12t^2 = 5t + 2$

13. $x^2 = 14x - 24$

14. $v^2 = 12v - 24$

15. $2z^2 - 2z - 1 = 0$

16. $6w^2 = 19w - 10$

17. $z^2 + 2z + 2 = 0$

18. $p^2 - 4p + 5 = 0$

19. $y^2 - 2y + 5 = 0$

20. $x^2 + 6x + 13 = 0$

21. $s^2 - 4s + 13 = 0$

22. $t^2 - 6t + 10 = 0$

23. $2w^2 - 2w + 5 = 0$

24. $4v^2 + 8v + 3 = 0$

25. $2x^2 + 6x + 5 = 0$

26. $2y^2 + 2y + 13 = 0$

27. $4t^2 - 6t + 9 = 0$

Use the discriminant to determine whether the quadratic equation has one real number solution, two real number solutions, or two complex number solutions.

28. $2z^2 - z + 5 = 0$

29. $3y^2 + y + 1 = 0$

30. $9x^2 - 12x + 4 = 0$

31. $4x^2 + 20x + 25 = 0$

32. $2v^2 - 3v - 1 = 0$

33. $3w^2 + 3w - 2 = 0$

APPLYING THE CONCEPTS

For what values of p does the quadratic equation have two real number solutions that are not equal? Write the answer in set-builder notation.

34. $x^2 - 6x + p = 0$

35. $x^2 + 10x + p = 0$

For what values of p does the quadratic equation have two complex number solutions? Write the answer in set-builder notation.

36. $x^2 - 2x + p = 0$

37. $x^2 + 4x + p = 0$

38. Show that the equation $x^2 + bx - 1 = 0$ always has real number solutions regardless of the value of b.

39. The National Forest Management Act of 1976 specifies that harvesting timber on national forests must be accomplished in conjunction with environmental considerations. One such consideration is providing a habitat for the spotted owl. One model of the survival of the spotted owl requires the solution of the equation $x^2 - s_a x - s_j s_s f = 0$ for x. Different values of s_a, s_j, s_s, and f are given in the table at the right. The values are particularly important because they are related to the survival of the owl. If $x > 1$, then the model predicts a growth in the population; if $x = 1$, the population remains steady; for $x < 1$, the population decreases. The important solution of the equation is the larger of the two roots of the equation.

	U.S. Forest Service	Lande
s_j	0.34	0.11
s_s	0.97	0.71
s_a	0.97	0.94
f	0.24	0.24

Source: Biles, Charles and Barry Noon. "The Spotted Owl." *The Journal of Undergraduate Mathematics and Its Application*, vol. 11, no. 2, 1990.

  a. Determine the larger root of this equation for values provided by the U.S. Forest Service. Round it to the nearest hundredth. Does it predict that the population will increase, remain steady, or decrease?

  b. Determine the larger root of this equation for the values provided by R. Lande in *Oecologia* (vol. 75, 1988). Round it to the nearest hundredth. Does it predict that the population will increase, remain steady, or decrease?

10.4 Solving Equations That Are Reducible to Quadratic Equations

Objective A   *To solve an equation that is quadratic in form* ....

Certain equations that are not quadratic can be expressed in quadratic form by making suitable substitutions. An equation is quadratic in form if it can be written as $au^2 + bu + c = 0$.

The equation $x^4 - 4x^2 - 5 = 0$ is quadratic in form.

$$x^4 - 4x^2 - 5 = 0$$
$$(x^2)^2 - 4(x^2) - 5 = 0 \qquad \bullet \text{ Let } x^2 = u.$$
$$u^2 - 4u - 5 = 0$$

The equation $y - y^{1/2} - 6 = 0$ is quadratic in form.

$$y - y^{1/2} - 6 = 0$$
$$(y^{1/2})^2 - (y^{1/2}) - 6 = 0 \qquad \bullet \text{ Let } y^{1/2} = u.$$
$$u^2 - u - 6 = 0$$

Here is the key to recognizing equations that are quadratic in form: When the equation is written in standard form, the exponent on one variable term is $\frac{1}{2}$ the exponent on the other variable term.

➡ Solve: $z + 7z^{1/2} - 18 = 0$

$$z + 7z^{1/2} - 18 = 0$$
$$(z^{1/2})^2 + 7(z^{1/2}) - 18 = 0 \qquad \bullet \text{ The equation is quadratic in form.}$$
$$u^2 + 7u - 18 = 0 \qquad \bullet \text{ Let } z^{1/2} = u.$$
$$(u - 2)(u + 9) = 0 \qquad \bullet \text{ Solve by factoring.}$$

$$\begin{array}{ll} u - 2 = 0 & u + 9 = 0 \\ u = 2 & u = -9 \end{array}$$

$$\begin{array}{ll} z^{1/2} = 2 & z^{1/2} = -9 \\ \sqrt{z} = 2 & \sqrt{z} = -9 \end{array} \qquad \bullet \text{ Replace } u \text{ by } z^{1/2}.$$

$$\begin{array}{ll} (\sqrt{z})^2 = 2^2 & (\sqrt{z})^2 = (-9)^2 \\ z = 4 & z = 81 \end{array} \qquad \bullet \text{ Solve for } z.$$

CONSIDER THIS
When each side of an equation is squared, the resulting equation may have a solution that is not a solution of the original equation.

Check each solution.

Check:
$$\begin{array}{c|c} z + 7z^{1/2} - 18 = 0 & \\ \hline 4 + 7(4)^{1/2} - 18 & 0 \\ 4 + 7 \cdot 2 - 18 & \\ 4 + 14 - 18 & \\ & 0 = 0 \end{array}$$

$$\begin{array}{c|c} z + 7z^{1/2} - 18 = 0 & \\ \hline 81 + 7(81)^{1/2} - 18 & 0 \\ 81 + 7 \cdot 9 - 18 & \\ 81 + 63 - 18 & \\ & 126 \neq 0 \end{array}$$

4 checks as a solution, but 81 does not check as a solution.
The solution is 4.

Example 1   Solve: $x^4 + x^2 - 12 = 0$

Solution
$$x^4 + x^2 - 12 = 0$$
$$(x^2)^2 + (x^2) - 12 = 0$$
$$u^2 + u - 12 = 0$$
$$(u - 3)(u + 4) = 0$$

$$u - 3 = 0 \qquad u + 4 = 0$$
$$u = 3 \qquad\quad u = -4$$

Replace u by x^2.

$$\begin{array}{ll} x^2 = 3 & x^2 = -4 \\ \sqrt{x^2} = \pm\sqrt{3} & \sqrt{x^2} = \pm\sqrt{-4} \\ x = \pm\sqrt{3} & x = \pm 2i \end{array}$$

The solutions are $\sqrt{3}$, $-\sqrt{3}$, $2i$, and $-2i$.

You Try It 1   Solve: $x - 5x^{1/2} + 6 = 0$

Your solution

Solution on p. A49

Objective B   *To solve a radical equation that is reducible to a quadratic equation* ....

Certain equations that contain radicals can be expressed as quadratic equations.

➡ Solve: $\sqrt{x + 2} + 4 = x$

$$\sqrt{x + 2} + 4 = x$$

$$\sqrt{x + 2} = x - 4$$ • Solve for the radical expression.

$$(\sqrt{x + 2})^2 = (x - 4)^2$$ • Square each side of the equation.

$$x + 2 = x^2 - 8x + 16$$ • Simplify.

$$0 = x^2 - 9x + 14$$ • Write the equation in standard form.

$$0 = (x - 7)(x - 2)$$ • Solve for x.

$$\begin{array}{ll} x - 7 = 0 & x - 2 = 0 \\ x = 7 & x = 2 \end{array}$$

CONSIDER THIS
You should always check your solutions by substituting the proposed solutions back into the *original* equation.

Check:
$$\begin{array}{c|c} \sqrt{x + 2} + 4 = x & \\ \hline \sqrt{7 + 2} + 4 & 7 \\ \sqrt{9} + 4 & \\ 3 + 4 & \\ 7 = 7 & \end{array}$$

$$\begin{array}{c|c} \sqrt{x + 2} + 4 = x & \\ \hline \sqrt{2 + 2} + 4 & 2 \\ \sqrt{4} + 4 & \\ 2 + 4 & \\ 6 \neq 2 & \end{array}$$

7 checks as a solution, but 2 does not check as a solution. The solution is 7.

Example 2

Solve: $\sqrt{7y - 3} + 3 = 2y$

Solution

$$\sqrt{7y - 3} + 3 = 2y$$
$$\sqrt{7y - 3} = 2y - 3$$
$$(\sqrt{7y - 3})^2 = (2y - 3)^2$$
$$7y - 3 = 4y^2 - 12y + 9$$
$$0 = 4y^2 - 19y + 12$$
$$0 = (4y - 3)(y - 4)$$

$$4y - 3 = 0 \qquad y - 4 = 0$$
$$4y = 3 \qquad\quad y = 4$$
$$y = \frac{3}{4}$$

4 checks as a solution.

$\frac{3}{4}$ does not check as a solution.

The solution is 4.

You Try It 2

Solve: $\sqrt{2x + 1} + x = 7$

Your solution

Example 3

Solve: $\sqrt{2y + 1} - \sqrt{y} = 1$

Solution

$$\sqrt{2y + 1} - \sqrt{y} = 1$$

Solve for one of the radical expressions.

$$\sqrt{2y + 1} = \sqrt{y} + 1$$
$$(\sqrt{2y + 1})^2 = (\sqrt{y} + 1)^2$$
$$2y + 1 = y + 2\sqrt{y} + 1$$
$$y = 2\sqrt{y}$$

Square each side of the equation.

$$y^2 = (2\sqrt{y})^2$$
$$y^2 = 4y$$
$$y^2 - 4y = 0$$
$$y(y - 4) = 0$$
$$y = 0 \qquad y - 4 = 0$$
$$y = 4$$

0 and 4 check as solutions.

The solutions are 0 and 4.

You Try It 3

Solve: $\sqrt{2x - 1} + \sqrt{x} = 2$

Your solution

Solutions on pp. A49–A50

Objective C   *To solve a fractional equation that is reducible to a quadratic equation* ....

After each side of a fractional equation has been multiplied by the LCM of the denominators, the resulting equation may be a quadratic equation.

➡ Solve: $\dfrac{1}{r} + \dfrac{1}{r+1} = \dfrac{3}{2}$

$$\frac{1}{r} + \frac{1}{r+1} = \frac{3}{2}$$

$$2r(r+1)\left(\frac{1}{r} + \frac{1}{r+1}\right) = 2r(r+1) \cdot \frac{3}{2}$$

   • Multiply each side of the equation by the LCM of the denominators.

$$2(r+1) + 2r = r(r+1) \cdot 3$$
$$2r + 2 + 2r = 3r(r+1)$$
$$4r + 2 = 3r^2 + 3r$$
$$0 = 3r^2 - r - 2$$

   • Write the equation in standard form.

$$0 = (3r+2)(r-1)$$

   • Solve for r by factoring.

$$
\begin{array}{ll}
3r + 2 = 0 & r - 1 = 0 \\
3r = -2 & r = 1 \\
r = -\dfrac{2}{3} &
\end{array}
$$

$-\dfrac{2}{3}$ and 1 check as solutions. The solutions are $-\dfrac{2}{3}$ and 1.

Example 4

Solve: $\dfrac{9}{x-3} = 2x + 1$

Solution

$$\frac{9}{x-3} = 2x + 1$$

$$(x-3)\frac{9}{x-3} = (x-3)(2x+1)$$

$$9 = 2x^2 - 5x - 3$$
$$0 = 2x^2 - 5x - 12$$
$$0 = (2x+3)(x-4)$$

$$
\begin{array}{ll}
2x + 3 = 0 & x - 4 = 0 \\
2x = -3 & x = 4 \\
x = -\dfrac{3}{2} &
\end{array}
$$

$-\dfrac{3}{2}$ and 4 check as solutions.

The solutions are $-\dfrac{3}{2}$ and 4.

You Try It 4

Solve: $3y + \dfrac{25}{3y-2} = -8$

Your solution

Solution on p. A50

10.4 Exercises

. . . .

Objective A

Solve.

1. $x^4 - 13x^2 + 36 = 0$

2. $y^4 - 5y^2 + 4 = 0$

3. $z^4 - 6z^2 + 8 = 0$

4. $t^4 - 12t^2 + 27 = 0$

5. $p - 3p^{1/2} + 2 = 0$

6. $v - 7v^{1/2} + 12 = 0$

7. $x - x^{1/2} - 12 = 0$

8. $w - 2w^{1/2} - 15 = 0$

9. $z^4 + 3z^2 - 4 = 0$

10. $y^4 + 5y^2 - 36 = 0$

11. $x^4 + 12x^2 - 64 = 0$

12. $x^4 - 81 = 0$

13. $p + 2p^{1/2} - 24 = 0$

14. $v + 3v^{1/2} - 4 = 0$

15. $y^{2/3} - 9y^{1/3} + 8 = 0$

16. $z^{2/3} - z^{1/3} - 6 = 0$

17. $9w^4 - 13w^2 + 4 = 0$

18. $4y^4 - 7y^2 - 36 = 0$

Objective B

Solve.

19. $\sqrt{x + 1} + x = 5$

20. $\sqrt{x - 4} + x = 6$

21. $x = \sqrt{x} + 6$

22. $\sqrt{2y - 1} = y - 2$

23. $\sqrt{3w + 3} = w + 1$

24. $\sqrt{2s + 1} = s - 1$

25. $\sqrt{4y + 1} - y = 1$

26. $\sqrt{3s + 4} + 2s = 12$

27. $\sqrt{10x + 5} - 2x = 1$

Solve.

28. $\sqrt{t + 8} = 2t + 1$ $\qquad$ **29.** $\sqrt{p + 11} = 1 - p$ $\qquad$ **30.** $x - 7 = \sqrt{x - 5}$

31. $\sqrt{x - 1} - \sqrt{x} = -1$ $\qquad$ **32.** $\sqrt{y + 1} = \sqrt{y + 5}$ $\qquad$ **33.** $\sqrt{2x - 1} = 1 - \sqrt{x - 1}$

34. $\sqrt{x + 6} + \sqrt{x + 2} = 2$ $\qquad$ **35.** $\sqrt{t + 3} + \sqrt{2t + 7} = 1$ $\qquad$ **36.** $\sqrt{5 - 2x} = \sqrt{2 - x} + 1$

Objective C

Solve.

37. $x = \dfrac{10}{x - 9}$ $\qquad$ **38.** $z = \dfrac{5}{z - 4}$ $\qquad$ **39.** $\dfrac{t}{t + 1} = \dfrac{-2}{t - 1}$

40. $\dfrac{2v}{v - 1} = \dfrac{5}{v + 2}$ $\qquad$ **41.** $\dfrac{y - 1}{y + 2} + y = 1$ $\qquad$ **42.** $\dfrac{2p - 1}{p - 2} + p = 8$

43. $\dfrac{3r + 2}{r + 2} - 2r = 1$ $\qquad$ **44.** $\dfrac{2v + 3}{v + 4} + 3v = 4$ $\qquad$ **45.** $\dfrac{2}{2x + 1} + \dfrac{1}{x} = 3$

46. $\dfrac{3}{s} - \dfrac{2}{2s - 1} = 1$ $\qquad$ **47.** $\dfrac{16}{z - 2} + \dfrac{16}{z + 2} = 6$ $\qquad$ **48.** $\dfrac{2}{y + 1} + \dfrac{1}{y - 1} = 1$

49. $\dfrac{t}{t - 2} + \dfrac{2}{t - 1} = 4$ $\qquad$ **50.** $\dfrac{4t + 1}{t + 4} + \dfrac{3t - 1}{t + 1} = 2$ $\qquad$ **51.** $\dfrac{5}{2p - 1} + \dfrac{4}{p + 1} = 2$

APPLYING THE CONCEPTS

Solve.

52. $(\sqrt{x} - 2)^2 - 5\sqrt{x} + 14 = 0$
    Hint: Let $u = \sqrt{x} - 2$.

53. $(\sqrt{x} + 3)^2 - 4\sqrt{x} - 17 = 0$
    Hint: Let $u = \sqrt{x} + 3$.

10.5 Applications of Quadratic Equations

Objective A   *To solve application problems* ....

The application problems in this section are similar to problems solved earlier in the text. The strategies for these problems will result in quadratic equations.

➡ A small pipe takes 16 min longer to empty a tank than does a larger pipe. Working together, the pipes can empty the tank in 6 min. How long would it take the larger pipe, working alone, to empty the tank?

Strategy for Solving an Application Problem

1. Determine the type of problem. Is it a uniform motion problem, a geometry problem, an integer problem, or a work problem?

The problem is a work problem.

---

2. Choose a variable to represent the unknown quantity. Write numerical or variable expressions for all the remaining quantities. These results can be recorded in a table.

The unknown time of the larger pipe: t
The unknown time of the smaller pipe: $t + 16$

	Rate of Work	·	*Time Worked*	=	*Part of Task Completed*
Larger pipe	$\dfrac{1}{t}$	·	6	=	$\dfrac{6}{t}$
Smaller pipe	$\dfrac{1}{t + 16}$	·	6	=	$\dfrac{6}{t + 16}$

---

3. Determine how the quantities are related.

$$\frac{6}{t} + \frac{6}{t + 16} = 1$$

• The sum of the parts of the task completed must equal 1.

$$t(t + 16)\left(\frac{6}{t} + \frac{6}{t + 16}\right) = t(t + 16) \cdot 1$$
$$(t + 16)6 + 6t = t^2 + 16t$$
$$6t + 96 + 6t = t^2 + 16t$$
$$0 = t^2 + 4t - 96$$
$$0 = (t + 12)(t - 8)$$

$$t + 12 = 0 \qquad t - 8 = 0$$
$$t = -12 \qquad\quad t = 8$$

Because time cannot be negative, the solution $t = -12$ is not possible.

The larger pipe requires 8 min to empty the tank.

Example 1

In 8 h, two canoers paddled 15 mi down a river and then paddled back to their campsite. The rate of the river's current was 1 mph. Find the rate at which the canoers paddled.

Strategy

* This is a uniform motion problem.
* Unknown paddling rate of the canoers: r

	Distance	Rate	Time
Down river	15	$r + 1$	$\dfrac{15}{r + 1}$
Up river	15	$r - 1$	$\dfrac{15}{r - 1}$

* The total time of the trip was 8 h.

Solution

$$\frac{15}{r + 1} + \frac{15}{r - 1} = 8$$
$$(r + 1)(r - 1)\left(\frac{15}{r + 1} + \frac{15}{r - 1}\right) = (r + 1)(r - 1)8$$
$$(r - 1)15 + (r + 1)15 = (r^2 - 1)8$$
$$15r - 15 + 15r + 15 = 8r^2 - 8$$
$$30r = 8r^2 - 8$$
$$0 = 8r^2 - 30r - 8$$
$$0 = 2(4r^2 - 15r - 4)$$
$$0 = 4r^2 - 15r - 4$$
$$0 = (4r + 1)(r - 4)$$

$$4r + 1 = 0 \qquad r - 4 = 0$$
$$4r = -1 \qquad\quad r = 4$$
$$r = -\frac{1}{4}$$

The solution $r = -\dfrac{1}{4}$ is not possible, because the rate cannot be a negative number.

The paddling rate was 4 mph.

You Try It 1

The length of a rectangle is 3 m more than the width. The area is 54 m². Find the length of the rectangle.

Your strategy

Your solution

Solution on p. A50

10.5 Exercises

· ·

Objective A   *Application Problems*

1. The base of a triangle is one less than five times the height of the triangle. The area of the triangle is 21 cm². Find the height and the length of the base of the triangle.

2. The height of a triangle is 3 in. less than the base of the triangle. The area of the triangle is 90 in². Find the height and the length of the base of the triangle.

3. The length of a rectangle is 2 ft less than three times the width of the rectangle. The area of the rectangle is 65 ft². Find the length and width of the rectangle.

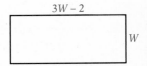

4. The length of a rectangle is 2 cm less than twice the width. The area of the rectangle is 180 cm². Find the length and width of the rectangle.

5. The sum of the squares of two consecutive odd integers is thirty-four. Find the two integers.

6. The sum of the squares of three consecutive odd integers is eighty-three. Find the three integers.

7. Five times an integer plus the square of the integer is twenty-four. Find the integer.

8. The sum of five times an integer and twice the square of the integer is three. Find the integer.

9. The height of a projectile fired upward is given by the formula $s = v_0t - 16t^2$, where s is the height, v_0 is the initial velocity, and t is the time. Find the time for a projectile to return to earth if it has an initial velocity of 200 ft/s.

10. The height of a projectile fired upward is given by the formula $s = v_0t - 16t^2$, where s is the height, v_0 is the initial velocity, and t is the time. Find the time for a projectile to reach a height of 64 ft if it has an initial velocity of 128 ft/s. Round to the nearest hundredth of a second.

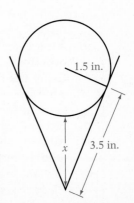

11. A perfectly spherical scoop of mint chocolate chip ice cream is placed in a cone as shown at the right. How far is the bottom of the scoop of ice cream from the bottom of the cone? *Hint:* A line segment from the center of the scoop of ice cream to the point at which the ice cream touches the cone is perpendicular to the edge of the cone.

Solve.

12. A small pipe can fill a tank in 6 min more time than it takes a larger pipe to fill the same tank. Working together, both pipes can fill the tank in 4 min. How long would it take each pipe, working alone, to fill the tank?

13. A cruise ship made a trip of 100 mi in 8 h. The ship traveled the first 40 mi at a constant rate before increasing its speed by 5 mph. Then it traveled another 60 mi at the increased speed. Find the rate of the cruise ship for the first 40 mi.

14. The rate of a single-engine plane in calm air is 100 mph. Flying with the wind, the plane can fly 240 mi in one hour less time than it requires to make the return trip of 240 mi. Find the rate of the wind.

15. A car travels 120 mi. A second car, traveling 10 mph faster than the first car, makes the same trip in 1 h less time. Find the speed of each car.

16. The rate of a river's current is 2 mph. A rowing crew can row 16 mi down this river and back in 6 h. Find the rowing rate of the crew in calm water.

17. The height of an arch is given by the equation

$$h(x) = -\frac{3}{64}x^2 + 27, \quad -24 \le x \le 24$$

where $|x|$ is the distance in feet from the center of the arch.
a. What is the maximum height of the arch?
b. What is the height of the arch 8 ft to the right of center?
c. How far from the center is the arch 8 ft tall? Round to the nearest hundredth.

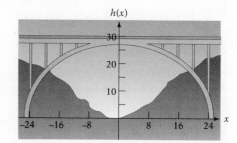

APPLYING THE CONCEPTS

18. The percent increases in average major league baseball salaries (using 1980 as the base year) are shown in the table at the right. A quadratic function that approximately models this data is given by $y = 1.31364x^2 - 15.8409x + 53.8091$, where x is the number of years after 1980 and y is the percent increase in salary for year x.
a. According to the table, baseball players received an approximately 13% increase in salaries in 1985 and 1989. Using the quadratic formula and the model, determine in what years, to the nearest year, the model predicts that the percent increase in salaries was 13%?
b. Explain why there are two answers to part **a**.
c. In 1991, the percent increase was 49.1%. What does the model predict for 1991?
d. In what year does the model predict that players received a 0% increase in pay?

Year	Percent Increase
1981	29.1
1982	35.5
1983	19.8
1984	13.9
1985	12.8
1986	11.0
1987	0.0
1988	6.4
1989	13.3
1990	20.2
1991	49.1
1992	13.6

Source: *USA Today*, May 11, 1993

10.6 Quadratic Inequalities and Rational Inequalities

Objective A   *To solve a nonlinear inequality*

A **quadratic inequality** is one that can be written in the form $ax^2 + bx + c < 0$ or $ax^2 + bx + c > 0$, where $a \neq 0$. The symbols $\leq$ and $\geq$ can also be used. The solution set of a quadratic inequality can be found by solving a compound inequality.

To solve $x^2 - 3x - 10 > 0$, first factor the trinomial.

$$x^2 - 3x - 10 > 0$$
$$(x + 2)(x - 5) > 0$$

There are two cases for which the product of the factors $x + 2$ and $x - 5$ will be positive:
(1) Both factors are positive, or
(2) Both factors are negative.

(1) $x + 2 > 0$ and $x - 5 > 0$
(2) $x + 2 < 0$ and $x - 5 < 0$

Solve each pair of compound inequalities.

(1) $x + 2 > 0$ and $x - 5 > 0$
     $x > -2$ $x > 5$
$\{x \mid x > -2\} \cap \{x \mid x > 5\} = \{x \mid x > 5\}$

(2) $x + 2 < 0$ and $x - 5 < 0$
     $x < -2$ $x < 5$
$\{x \mid x < -2\} \cap \{x \mid x < 5\} = \{x \mid x < -2\}$

Because the two cases for which the product will be positive are connected by *or*, the solution set is the union of the solution sets of the individual inequalities.

$\{x \mid x > 5\} \cup \{x \mid x < -2\}$
     $= \{x \mid x > 5 \;\; \text{or} \;\; x < -2\}$

Although the solution set of any quadratic inequality can be found by using the method outlined above, frequently a graphical method is easier to use.

➡ Solve and graph the solution set of $x^2 - x - 6 < 0$.

Factor the trinomial.

$$x^2 - x - 6 < 0$$
$$(x - 3)(x + 2) < 0$$

CONSIDER THIS
$x - 3 = 0$ when $x = 3$. If x is any number less than 3, $x - 3$ is negative. If x is any number greater than 3, $x - 3$ is positive.
$x + 2 = 0$ when $x = -2$. If x is any number less than -2, $x + 2$ is negative. If x is any number greater than -2, $x + 2$ is positive.

On a number line, draw lines indicating the numbers that make each factor equal to zero.

$x - 3 = 0$ $x + 2 = 0$
    $x = 3$ $x = -2$

For each factor, place plus signs above the number line for those regions where the factor is positive and negative signs where the factor is negative.

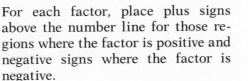

Because $x^2 - x - 6 < 0$, the solution set will be the regions where one factor is positive and the other factor is negative.

Write the solution set.

$\{x \mid -2 < x < 3\}$

Graph the solution set.

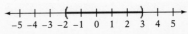

➡ Solve and graph the solution set of $(x - 2)(x + 1)(x - 4) > 0$.

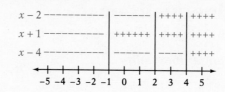

- On a number line, identify for each factor the regions where the factor is positive and those where the factor is negative.
  There are two regions where the product of the three factors is positive.

$\{x \mid -1 < x < 2 \text{ or } x > 4\}$

- Write the solution set.

- Graph the solution set.

➡ Solve: $\dfrac{2x - 5}{x - 4} \le 1$

$$\frac{2x - 5}{x - 4} \le 1$$

$$\frac{2x - 5}{x - 4} - 1 \le 0$$

- Rewrite the inequality so that 0 appears on the right side of the inequality.

$$\frac{2x - 5}{x - 4} - \frac{x - 4}{x - 4} \le 0$$

$$\frac{x - 1}{x - 4} \le 0$$

- Simplify.

$x - 1$ ----------------- $|$ ++++++ $|$ +++
$x - 4$ ----------------- $|$ ------ $|$ +++

$$\begin{array}{cccccccccccc} + & + & + & + & + & + & + & + & + & + & + \\ -5 & -4 & -3 & -2 & -1 & 0 & 1 & 2 & 3 & 4 & 5 \end{array}$$

- On a number line, identify for each factor of the numerator and each factor of the denominator the regions where the factor is positive and those where the factor is negative.
  The region where the quotient of the two factors is negative is between 1 and 4.

$\{x \mid 1 \le x < 4\}$

- Write the solution set.

Note that 1 is part of the solution set but that 4 is not because the denominator of the rational expression is zero when $x = 4$.

Example 1
Solve and graph the solution set of $2x^2 - x - 3 \ge 0$.

Solution
$$2x^2 - x - 3 \ge 0$$
$$(2x - 3)(x + 1) \ge 0$$

$2x - 3$ --- $|$ ------ $|$ +++
$x + 1$ --- $|$ ++++++ $|$ +++

$$\begin{array}{ccccc} + & + & + & + & + \\ -2 & -1 & 0 & 1 & 2 \end{array}$$

$$\left\{ x \mid x \le -1 \text{ or } x \ge \frac{3}{2} \right\}$$

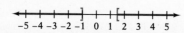

You Try It 1
Solve and graph the solution set of $2x^2 - x - 10 \le 0$.

Your solution

Solution on p. A51

10.6 Exercises

Objective A

Solve and graph the solution set.

1. $(x - 4)(x + 2) > 0$

$$\xleftarrow{\;\;}\!\!\!+\!\!+\!\!+\!\!+\!\!+\!\!+\!\!+\!\!+\!\!+\!\!+\!\!+\!\xrightarrow{\;\;}$$
$$-5\;-4\;-3\;-2\;-1\;\;0\;\;1\;\;2\;\;3\;\;4\;\;5$$

2. $(x + 1)(x - 3) > 0$

$$\xleftarrow{\;\;}\!\!\!+\!\!+\!\!+\!\!+\!\!+\!\!+\!\!+\!\!+\!\!+\!\!+\!\!+\!\xrightarrow{\;\;}$$
$$-5\;-4\;-3\;-2\;-1\;\;0\;\;1\;\;2\;\;3\;\;4\;\;5$$

3. $x^2 - 3x + 2 \geq 0$

$$\xleftarrow{\;\;}\!\!\!+\!\!+\!\!+\!\!+\!\!+\!\!+\!\!+\!\!+\!\!+\!\!+\!\!+\!\xrightarrow{\;\;}$$
$$-5\;-4\;-3\;-2\;-1\;\;0\;\;1\;\;2\;\;3\;\;4\;\;5$$

4. $x^2 + 5x + 6 > 0$

$$\xleftarrow{\;\;}\!\!\!+\!\!+\!\!+\!\!+\!\!+\!\!+\!\!+\!\!+\!\!+\!\!+\!\!+\!\xrightarrow{\;\;}$$
$$-5\;-4\;-3\;-2\;-1\;\;0\;\;1\;\;2\;\;3\;\;4\;\;5$$

5. $x^2 - x - 12 < 0$

$$\xleftarrow{\;\;}\!\!\!+\!\!+\!\!+\!\!+\!\!+\!\!+\!\!+\!\!+\!\!+\!\!+\!\!+\!\xrightarrow{\;\;}$$
$$-5\;-4\;-3\;-2\;-1\;\;0\;\;1\;\;2\;\;3\;\;4\;\;5$$

6. $x^2 + x - 20 < 0$

$$\xleftarrow{\;\;}\!\!\!+\!\!+\!\!+\!\!+\!\!+\!\!+\!\!+\!\!+\!\!+\!\!+\!\!+\!\xrightarrow{\;\;}$$
$$-5\;-4\;-3\;-2\;-1\;\;0\;\;1\;\;2\;\;3\;\;4\;\;5$$

7. $(x - 1)(x + 2)(x - 3) < 0$

$$\xleftarrow{\;\;}\!\!\!+\!\!+\!\!+\!\!+\!\!+\!\!+\!\!+\!\!+\!\!+\!\!+\!\!+\!\xrightarrow{\;\;}$$
$$-5\;-4\;-3\;-2\;-1\;\;0\;\;1\;\;2\;\;3\;\;4\;\;5$$

8. $(x + 4)(x - 2)(x + 1) > 0$

$$\xleftarrow{\;\;}\!\!\!+\!\!+\!\!+\!\!+\!\!+\!\!+\!\!+\!\!+\!\!+\!\!+\!\!+\!\xrightarrow{\;\;}$$
$$-5\;-4\;-3\;-2\;-1\;\;0\;\;1\;\;2\;\;3\;\;4\;\;5$$

9. $(x + 4)(x - 2)(x - 1) \geq 0$

$$\xleftarrow{\;\;}\!\!\!+\!\!+\!\!+\!\!+\!\!+\!\!+\!\!+\!\!+\!\!+\!\!+\!\!+\!\xrightarrow{\;\;}$$
$$-5\;-4\;-3\;-2\;-1\;\;0\;\;1\;\;2\;\;3\;\;4\;\;5$$

10. $(x - 1)(x + 5)(x - 2) \leq 0$

$$\xleftarrow{\;\;}\!\!\!+\!\!+\!\!+\!\!+\!\!+\!\!+\!\!+\!\!+\!\!+\!\!+\!\!+\!\xrightarrow{\;\;}$$
$$-5\;-4\;-3\;-2\;-1\;\;0\;\;1\;\;2\;\;3\;\;4\;\;5$$

11. $\dfrac{x - 4}{x + 2} > 0$

$$\xleftarrow{\;\;}\!\!\!+\!\!+\!\!+\!\!+\!\!+\!\!+\!\!+\!\!+\!\!+\!\!+\!\!+\!\xrightarrow{\;\;}$$
$$-5\;-4\;-3\;-2\;-1\;\;0\;\;1\;\;2\;\;3\;\;4\;\;5$$

12. $\dfrac{x + 2}{x - 3} > 0$

$$\xleftarrow{\;\;}\!\!\!+\!\!+\!\!+\!\!+\!\!+\!\!+\!\!+\!\!+\!\!+\!\!+\!\!+\!\xrightarrow{\;\;}$$
$$-5\;-4\;-3\;-2\;-1\;\;0\;\;1\;\;2\;\;3\;\;4\;\;5$$

13. $\dfrac{x - 3}{x + 1} \leq 0$

$$\xleftarrow{\;\;}\!\!\!+\!\!+\!\!+\!\!+\!\!+\!\!+\!\!+\!\!+\!\!+\!\!+\!\!+\!\xrightarrow{\;\;}$$
$$-5\;-4\;-3\;-2\;-1\;\;0\;\;1\;\;2\;\;3\;\;4\;\;5$$

14. $\dfrac{x - 1}{x} > 0$

$$\xleftarrow{\;\;}\!\!\!+\!\!+\!\!+\!\!+\!\!+\!\!+\!\!+\!\!+\!\!+\!\!+\!\!+\!\xrightarrow{\;\;}$$
$$-5\;-4\;-3\;-2\;-1\;\;0\;\;1\;\;2\;\;3\;\;4\;\;5$$

15. $\dfrac{(x - 1)(x + 2)}{x - 3} \leq 0$

$$\xleftarrow{\;\;}\!\!\!+\!\!+\!\!+\!\!+\!\!+\!\!+\!\!+\!\!+\!\!+\!\!+\!\!+\!\xrightarrow{\;\;}$$
$$-5\;-4\;-3\;-2\;-1\;\;0\;\;1\;\;2\;\;3\;\;4\;\;5$$

16. $\dfrac{(x + 3)(x - 1)}{x - 2} \geq 0$

$$\xleftarrow{\;\;}\!\!\!+\!\!+\!\!+\!\!+\!\!+\!\!+\!\!+\!\!+\!\!+\!\!+\!\!+\!\xrightarrow{\;\;}$$
$$-5\;-4\;-3\;-2\;-1\;\;0\;\;1\;\;2\;\;3\;\;4\;\;5$$

Solve.

17. $x^2 - 16 > 0$

18. $x^2 - 4 \geq 0$

19. $x^2 - 9x \leq 36$

20. $x^2 + 4x > 21$

21. $4x^2 - 8x + 3 < 0$

22. $2x^2 + 11x + 12 \geq 0$

23. $\dfrac{3}{x - 1} < 2$

24. $\dfrac{x}{(x - 1)(x + 2)} \geq 0$

25. $\dfrac{x - 2}{(x + 1)(x - 1)} \leq 0$

26. $\dfrac{1}{x} < 2$

27. $\dfrac{x}{2x - 1} \geq 1$

28. $\dfrac{x}{2x - 3} \leq 1$

29. $\dfrac{x}{2 - x} \leq -3$

30. $\dfrac{3}{x - 2} > \dfrac{2}{x + 2}$

31. $\dfrac{3}{x - 5} > \dfrac{1}{x + 1}$

APPLYING THE CONCEPTS

Graph the solution set.

32. $(x + 2)(x - 3)(x + 1)(x + 4) > 0$

33. $(x - 1)(x + 3)(x - 2)(x - 4) \geq 0$

34. $(x^2 + 2x - 8)(x^2 - 2x - 3) < 0$

35. $(x^2 + 2x - 3)(x^2 + 3x + 2) \geq 0$

36. $(x^2 + 1)(x^2 - 3x + 2) > 0$

37. $(x - 4)^2 > -2$

38. $x < x^2$

39. $x^3 > x$

Projects in Mathematics

Completing the Square

Essentially all of the investigations into mathematics before the Renaissance were geometric. The solutions of quadratic equations were calculated from a construction of a certain area. Proofs of theorems, even theorems about numbers, were based entirely on geometry. In this project, we will examine the geometric solution of a quadratic equation.

➡ Solve: $x^2 + 6x = 7$

Begin with a line of unknown length, x, and one of length 6, the coefficient of x. Using these lines, construct a rectangle as shown.

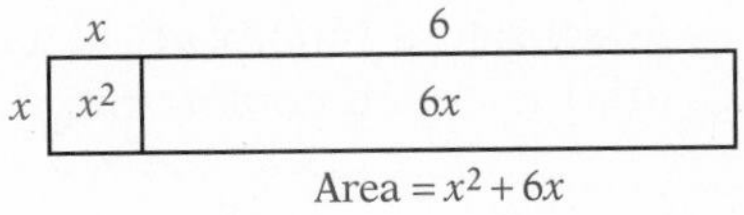

Fig. 1

Area $= x^2 + 6x$

Now draw another area that has exactly the same area as Figure 1 by cutting off one-half of the rectangle of area $6x$ and placing it on the bottom of the square labeled x^2. See Figure 2.

The unshaded area in Figure 2 has exactly the same area as Figure 1. However, when the shaded area is added to Figure 2 to make a square, the total area is 9 square units larger than that of Figure 1. In equation form,

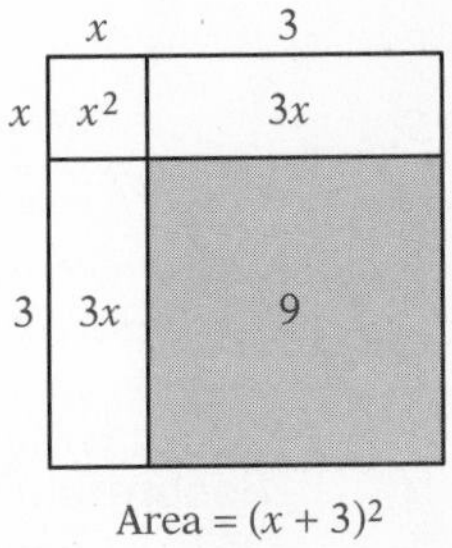

Area $= (x + 3)^2$

Fig. 2

(Area of Figure 1) + 9 = area of Figure 2

or

$$x^2 + 6x + 9 = (x + 3)^2$$

From the original equation, $x^2 + 6x = 7$. Thus,

$$x^2 + 6x + 9 = (x + 3)^2$$
$$7 + 9 = (x + 3)^2 \qquad \bullet \; x^2 + 6x = 7$$
$$16 = (x + 3)^2$$
$$4 = x + 3 \qquad \bullet \; \text{See note below.}$$
$$1 = x$$

Note: Although early mathematicians knew that a quadratic equation may have two solutions, both solutions were allowed only if they were positive. After all, a geometric construction could not have a negative length. Therefore, the solution of this equation was 1; the solution -7 would have been dismissed as *fictitious*, the actual word that was frequently used through the 15th century for negative-number solutions of an equation.

Try to solve the quadratic equation $x^2 + 4x = 12$ by geometrically completing the square.

Chapter Summary

Key Words   A *quadratic equation* is an equation of the form $ax^2 + bx + c = 0$, $a \neq 0$. A quadratic equation is also called a *second-degree equation*.

A quadratic equation is in *standard form* when the polynomial is in descending order and equal to zero.

When a quadratic equation has two solutions that are the same number, the solution is called a *double root* of the equation.

Adding to a binomial the constant term that makes it a perfect-square trinomial is called *completing the square*.

For an equation of the form $ax^2 + bx + c = 0$, the quantity $b^2 - 4ac$ is called the *discriminant*.

An equation is quadratic in form if it can be written as $au^2 + bu + c = 0$.

A *quadratic inequality* is one that can be written in the form $ax^2 + bx + c > 0$ or $ax^2 + bx + c < 0$, where $a \neq 0$. The symbols $\leq$ and $\geq$ can also be used.

Essential Rules   *The Principle of Zero Products*

If $ab = 0$, then $a = 0$ or $b = 0$.

Completing the Square

To complete the square on $x^2 + bx$,
add ($\frac{1}{2}$ the coefficient of x)^2 to the binomial.

The Quadratic Formula
$$x = \frac{-b \pm \sqrt{b^2 - 4ac}}{2a}$$

The Effect of the Discriminant on the Solutions of a Quadratic Equation

1. If $b^2 - 4ac = 0$, then the equation has one real number solution, a double root.
2. If $b^2 - 4ac > 0$, then the equation has two real number solutions that are not equal.
3. If $b^2 - 4ac < 0$, then the equation has two complex number solutions.

Chapter Review Exercises

. . . .

1. Solve by factoring: $3x^2 + 10x = 8$

2. Solve by factoring: $6x^2 - 5x - 6 = 0$

3. Write a quadratic equation that has integer coefficients and has solutions 3 and -3.

4. Write a quadratic equation that has integer coefficients and has solutions $\frac{1}{2}$ and -4.

5. Solve by taking square roots:
$3(x - 2)^2 - 24 = 0$

6. Solve by completing the square:
$x^2 - 6x - 2 = 0$

7. Solve by completing the square:
$3x^2 - 6x = 2$

8. Solve by using the quadratic formula:
$2x^2 - 2x = 1$

9. Solve by using the quadratic formula:
$x^2 + 4x + 12 = 0$

10. Use the discriminant to determine whether $3x^2 - 4x = 1$ has one real number solution, two real number solutions, or two complex number solutions.

11. Use the discriminant to determine whether $x^2 - 6x = -15$ has one real number solution, two real number solutions, or two complex number solutions.

12. Solve: $2x + 7x^{1/2} - 4 = 0$

13. Solve: $x^4 - 4x^2 + 3 = 0$

14. Solve: $\sqrt{2x + 1} + 5 = 2x$

15. Solve: $\sqrt{x - 2} = \sqrt{x} - 2$

16. Solve: $\dfrac{2x}{x - 3} + \dfrac{5}{x - 1} = 1$

17. Solve and graph the solution set of $(x - 2)(x + 4)(x - 4) < 0$.

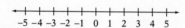

18. Solve and graph the solution set of $\dfrac{2x - 3}{x + 4} \le 0$.

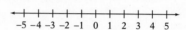

19. The base of a triangle is 3 ft more than three times the height. The area of the triangle is 30 ft². Find the base and height of the triangle.

20. The rate of a river's current is 2 mph. A canoe was paddled 6 mi down the river and back 6 mi in 4 h. Find the paddling rate in calm water.

Cumulative Review Exercises

1. Evaluate $2a^2 - b^2 \div c^2$ when $a = 3$, $b = -4$, and $c = -2$.

2. Solve: $|3x - 2| < 8$

3. Find the volume of a cylinder with a height of 6 m and a radius of 3 m. Give the exact measure.

4. Given $f(x) = \dfrac{2x - 3}{x^2 - 1}$, find $f(-2)$.

5. Find the slope of the line containing the points $(3, -4)$ and $(-1, 2)$.

6. Find the x- and y-intercepts of the graph of $6x - 5y = 15$.

7. Find the equation of the line that contains the point $(1, 2)$ and is parallel to the line $x - y = 1$.

8. Solve the system of equations.
$$x + y + z = 2$$
$$-x + 2y - 3z = -9$$
$$x - 2y - 2z = -1$$

9. Graph the solution set:
$$x + y \leq 3$$
$$2x - y < 4$$

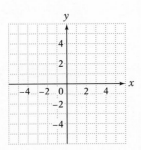

10. Triangles ABC and DEF are similar. Find the height of triangle DEF.

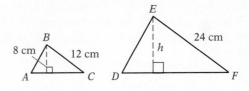

11. Divide: $(3x^3 - 13x^2 + 10) \div (3x - 4)$

12. Factor: $-3x^3y + 6x^2y^2 - 9xy^3$

13. Factor: $6x^2 - 7x - 20$

14. Multiply: $\dfrac{x^2 + 2x + 1}{8x^2 + 8x} \cdot \dfrac{4x^3 - 4x^2}{x^2 - 1}$

15. Solve: $\dfrac{x}{x + 2} - \dfrac{4x}{x + 3} = 1$

16. Solve $S = \dfrac{n}{2}(a + b)$ for b.

17. Multiply: $a^{-1/2}(a^{1/2} - a^{3/2})$

18. Multiply: $-2i(7 - 4i)$

19. Solve: $\sqrt{3x + 1} - 1 = x$

20. Solve: $x^4 - 6x^2 + 8 = 0$

21. A piston rod for an automobile is $9\dfrac{3}{8}$ in. with a tolerance of $\dfrac{1}{64}$ in. Find the lower and upper limits of the length of the piston rod.

22. The base of a triangle is $(x + 8)$ ft. The height is $(2x - 4)$ ft. Find the area of the triangle in terms of the variable x.

23. The graph shows the relationship between the value of a building and the time, in years, since depreciation began. Find the slope of the line between the two points shown on the graph. Write a sentence that states the meaning of the slope.

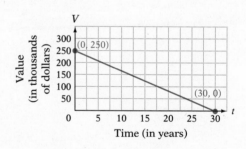

24. How high on a building will a 17-foot ladder reach when the bottom of the ladder is 8 ft from the building?

25. Use the discriminant to determine whether $2x^2 + 4x + 3 = 0$ has one real number solution, two real number solutions, or two complex number solutions.

Functions and Relations

Objectives

Conic Sections

The graphs of four curves—the circle, the ellipse, the parabola, and the hyperbola—are discussed in this chapter. These curves were studied by the Greeks and were known prior to 400 B.C. Their names were first used by Apollonius around 250 B.C. in *Conic Sections*, the most authoritative Greek discussion of these curves. Apollonius borrowed the names from a school founded by Pythagoras.

The diagram at the right shows the path of a planet around the sun. The curve traced out by the planet is an ellipse. The **aphelion** is the position of the planet when it is farthest from the sun. The **perihelion** is the planet's position when it is nearest to the sun.

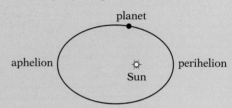

A telescope, like the one at the Palomar Observatory, has a cross section that is in the shape of a parabola. A parabolic mirror has the unusual property that all light rays parallel to the axis of symmetry that hit the mirror are reflected to the same point. This point is called the **focus of the parabola.**

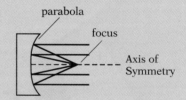

Some comets, unlike Halley's Comet, travel with such speed that they are not captured by the sun's gravitational field. The path of the comet as it comes around the sun is in the shape of a hyperbola.

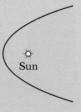

11.1 Linear Functions

Objective A   *To graph a linear function*

The graph of a function is the graph of the ordered pairs (x, y) that belong to the function. Because y and $f(x)$ are interchangeable, the ordered pairs of a function can be written as (x, y) or $(x, f(x))$. A function that can be written in the form $y = mx + b$ or $f(x) = mx + b$ is called a **linear function**. The graph of a linear function has certain characteristics. It is a straight line with slope m and y-intercept $(0, b)$.

➡ Graph: $f(x) = 2x + 1$.

This is a linear function. You can think of the function as the equation $y = 2x + 1$. The y-intercept is $(0, 1)$. The slope is 2.

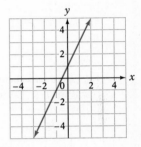

Beginning at the y-intercept, move right 1 and up 2. The point $(1, 3)$ is another point on the graph. Draw a straight line through the points $(0, 1)$ and $(1, 3)$.

When a function is given by an equation, the domain of the function is all real numbers for which the function evaluates to a real number. For instance,

• The domain of $f(x) = 2x + 1$ is all real numbers because the value of $2x + 1$ is a real number for any value of x.

• The domain of $g(x) = \dfrac{1}{x - 2}$ is all real numbers except 2; when $x = 2$, $g(2) = \dfrac{1}{2 - 2} = \dfrac{1}{0}$, which is not a real number.

Example 1   Graph: $f(x) = \dfrac{2}{3}x$

Solution

$b = 0$

y-intercept: $(0, 0)$

$m = \dfrac{2}{3}$

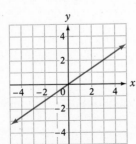

You Try It 1   Graph: $f(x) = \dfrac{3}{5}x - 4$

Your solution

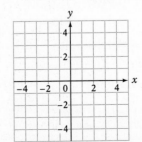

Solution on p. A51

Objective B   To solve application problems ....

Linear functions can be used to model a variety of applications in science and business. For each application, data are collected and the independent and dependent variables are selected. Then a linear function is determined that models the data.

Example 2

Suppose a manufacturer has determined that at a price of $115, consumers will purchase 1 million portable CD players and that at a price of $90, consumers will purchase 1.25 million portable CD players. Describe this situation with a linear function. Use this function to predict how many portable CD players consumers will purchase if the price is $80.

Strategy

• Select the independent and dependent variables. Because you are trying to determine the number of CD players, that quantity is the *dependent* variable, y. The price of CD players is the *independent* variable, x. From the given data, two ordered pairs are (115, 1) and (90, 1.25). (The ordinates are in millions of units.) Use these ordered pairs to determine the linear function.

• Evaluate the function for $x = 80$ to predict how many CD players consumers will purchase if the price is $80.

Solution

Let $(x_1, y_1) = (115, 1)$ and $(x_2, y_2) = (90, 1.25)$.

$$m = \frac{y_2 - y_1}{x_2 - x_1} = \frac{1.25 - 1}{90 - 115} = -\frac{0.25}{25} = -0.01$$

$$y - y_1 = m(x - x_1)$$
$$y - 1 = -0.01(x - 115)$$
$$y = -0.01x + 2.15$$

The linear function is
$f(x) = -0.01x + 2.15$.

$$f(80) = -0.01(80) + 2.15 = 1.35$$

Consumers will purchase 1.35 million CD players at a price of $80.

You Try It 2

Gabriel Daniel Fahrenheit invented the mercury thermometer in 1717. In terms of readings on this thermometer, water freezes at 32°F and boils at 212°F. In 1742 Anders Celsius invented the Celsius temperature scale. On this scale, water freezes at 0°C and boils at 100°C. Determine a linear function that can be used to predict the Celsius temperature when the Fahrenheit temperature is known.

Your strategy

Your solution

Solution on p. A51

Content and Format © 1996 HMCo.

11.1 Exercises

Objective A

Graph.

1. $f(x) = 3x - 4$

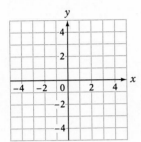

2. $f(x) = -2x + 3$

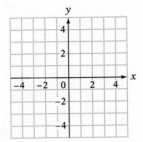

3. $f(x) = -\dfrac{2}{3}x$

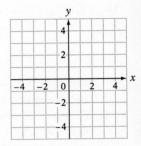

4. $f(x) = \dfrac{3}{2}x$

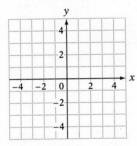

5. $f(x) = \dfrac{2}{3}x - 4$

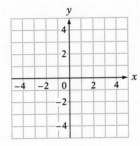

6. $f(x) = \dfrac{3}{4}x + 2$

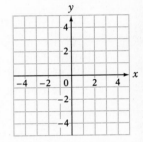

7. $f(x) = -\dfrac{1}{3}x + 2$

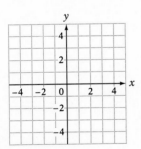

8. $f(x) = -\dfrac{3}{2}x - 3$

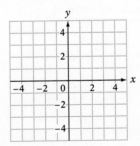

9. $f(x) = \dfrac{3}{5}x - 1$

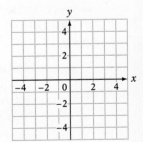

10. $f(x) = 2x - 3$

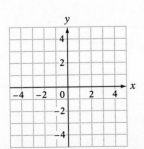

11. $f(x) = -2x - 3$

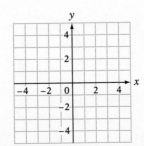

12. $f(x) = -\dfrac{2}{5}x + 2$

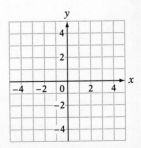

Objective B   Application Problems

Solve.

13. The sale price of an item is a function, s, of the original price, p, where $s(p) = 0.80p$. If an item's original price is $200, what is the sale price of the item?

14. The markup on an item is a function, m, of its cost, c, where $m(c) = 0.25c$. If the cost of an item is $150, what is the markup on the item?

15. A manufacturer of pickup trucks has determined that 50,000 trucks per month can be sold at a price of $9000. At a price of $8750, the number of trucks sold per month would increase to 55,000. Determine a linear function that will predict the number of trucks that would be sold at a given price. Use this model to predict the number of trucks that would be sold at a price of $8500.

16. A manufacturer of graphing calculators has determined that 10,000 calculators per week will be sold at a price of $95. At a price of $90, it is estimated that 12,000 calculators would be sold. Determine a linear function that will predict the number of calculators that would be sold at a given price. Use this model to predict the number of calculators per week that would be sold at a price of $75.

17. The operator of a hotel estimates that 500 rooms per night will be rented if the room rate per night is $75. For each $10 increase in the price of a room, 6 fewer rooms will be rented. Determine a linear function that will predict the number of rooms that will be rented for a given price per room. Use this model to predict the number of rooms that will be rented if the room rate is $100.

18. A general building contractor estimates that the cost to build a new home is $30,000 plus $85 for each square foot of floor space in the house. Determine a linear function that will give the cost of building a house that contains a given number of square feet. Use this model to determine the cost to build a house that contains 1800 ft^2.

APPLYING THE CONCEPTS

19. Let $f(x)$ be the digit in the xth decimal place of the repeating digit $0.\overline{387}$. For example, $f(3) = 7$ because 7 is the digit in the third decimal place. Find $f(14)$.

20. Let $f(x)$ be the digit in the xth decimal place of the repeating digit $0.\overline{018}$. For example, $f(1) = 0$ because 0 is the digit in the first decimal place. Find $f(21)$.

11.2 Quadratic Functions

Objective A *To graph a quadratic function*

Recall that a linear function is one that can be expressed by the equation $f(x) = mx + b$. The graph of a linear function has certain characteristics. It is a straight line with slope m and y-intercept $(0, b)$. A **quadratic function** is one that can be expressed by the equation $f(x) = ax^2 + bx + c$, $a \neq 0$. The graph of this function, called a **parabola**, also has certain characteristics. The graph of a quadratic function can be drawn by finding ordered pairs that belong to the function.

➡ Graph $f(x) = x^2 - 2x - 3$.

By evaluating the function for various values of x, find enough ordered pairs to determine the shape of the graph.

x	$f(x) = x^2 - 2x - 3$	$f(x)$	(x, y)
-2	$f(-2) = (-2)^2 - 2(-2) - 3$	5	$(-2, 5)$
-1	$f(-1) = (-1)^2 - 2(-1) - 3$	0	$(-1, 0)$
0	$f(0) = (0)^2 - 2(0) - 3$	-3	$(0, -3)$
1	$f(1) = (1)^2 - 2(1) - 3$	-4	$(1, -4)$
2	$f(2) = (2)^2 - 2(2) - 3$	-3	$(2, -3)$
3	$f(3) = (3)^2 - 2(3) - 3$	0	$(3, 0)$
4	$f(4) = (4)^2 - 2(4) - 3$	5	$(4, 5)$

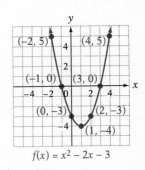

$f(x) = x^2 - 2x - 3$

Because the value of $f(x) = x^2 - 2x - 3$ is a real number for all values of x, the domain of f is all real numbers. From the graph, it appears that no value of y is less than -4. Thus the range is $\{y \mid y \geq -4\}$. The range can also be determined algebraically, as shown below, by completing the square.

CONSIDER THIS

In completing the square, 1 is both added and subtracted. Because $1 - 1 = 0$, the expression $x^2 - 2x - 3$ is not changed. Note that

$(x - 1)^2 - 4$
$= (x^2 - 2x + 1) - 4$
$= x^2 - 2x - 3$

which is the original expression.

$f(x) = x^2 - 2x - 3$
$\quad = (x^2 - 2x) - 3$ • Group the variable terms.
$\quad = (x^2 - 2x + 1) - 1 - 3$ • Complete the square of $x^2 - 2x$. Add and subtract $\left[\dfrac{1}{2}(-2)\right]^2 = 1$ to $x^2 - 2x$.
$\quad = (x - 1)^2 - 4$ • Factor and combine like terms.

Because the square of a positive number is always positive, we have

$(x - 1)^2 \geq 0$
$(x - 1)^2 - 4 \geq -4$ • Subtract 4 from each side of the inequality.
$f(x) \geq -4$ • $f(x) = x^2 - 2x - 3 = (x - 1)^2 - 4$
$y \geq -4$

From the last inequality, the range is $\{y \mid y \geq -4\}$.

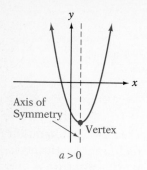

a > 0

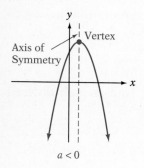

a < 0

In general, the graph of $f(x) = ax^2 + bx + c, a \neq 0$, resembles a "cup" shape as shown at the left. The parabola opens up when $a > 0$ (a is positive) and opens down when $a < 0$ (a is negative). When the parabola opens up, the **vertex** of the parabola is the point with the smallest y-coordinate. When the parabola opens down, the vertex is the point with the largest y-coordinate.

The **axis of symmetry** is a line that passes through the vertex of the parabola and is parallel to the y-axis. To understand the axis of symmetry, think of folding the graph along that line. The two portions of the graph will match up.

The following formulas can be used to find the axis of symmetry and the vertex of a parabola.

> **The Axis of Symmetry and Vertex of a Parabola**
>
> Let $f(x) = ax^2 + bx + c$ be the equation of a parabola.
> The equation of the axis of symmetry is $x = -\dfrac{b}{2a}$.
> The coordinates of the vertex are $\left(-\dfrac{b}{2a}, f\left(-\dfrac{b}{2a}\right)\right)$.

➡ Find the axis of symmetry and the vertex of the parabola whose equation is $g(x) = -2x^2 + 3x + 1$. Then graph the equation.

From the equation $g(x) = -2x^2 + 3x + 1$, $a = -2$, $b = 3$, and $c = 1$.

Axis of symmetry: $x = -\dfrac{b}{2a} = -\dfrac{3}{2(-2)} = \dfrac{3}{4}$

The axis of symmetry is a vertical line passing through the point $\left(\dfrac{3}{4}, 0\right)$.

The x-coordinate of the vertex is $\dfrac{3}{4}$.

Find the y-coordinate of the vertex by replacing x with $\dfrac{3}{4}$ and evaluating.

$$y = -2x^2 + 3x + 1$$
$$= -2\left(\frac{3}{4}\right)^2 + 3\left(\frac{3}{4}\right) + 1 = \frac{17}{8}$$

The vertex is $\left(\dfrac{3}{4}, \dfrac{17}{8}\right)$.

Because a is negative ($a = -2$), the graph opens down. Find a few ordered pairs that belong to the function, and then sketch the graph.

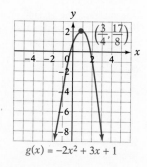

$g(x) = -2x^2 + 3x + 1$

CONSIDER THIS

Once the coordinates of the vertex are found, the range of a quadratic function can be determined.

Once the y-coordinate of the vertex is known, the range of the function can be determined. Here, the graph of g opens down, so the y-coordinate of the vertex is the largest value of y. Therefore, the range of g is $\left\{y \mid y \le \dfrac{17}{8}\right\}$. The value of $-2x^2 + 3x + 1$ is a real number for all values of x; the domain is all real numbers.

Example 1

Find the vertex and axis of symmetry of the parabola whose equation is $y = -x^2 + 4x + 1$. Then graph the equation.

Solution

x-coordinate of vertex:

$$-\frac{b}{2a} = -\frac{4}{2(-1)} = 2$$

y-coordinate of vertex:

$y = -x^2 + 4x + 1$
$\quad = -(2)^2 + 4(2) + 1$
$\quad = 5$

Vertex: $(2, 5)$
Axis of symmetry: $x = 2$

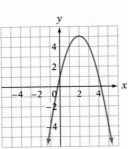

You Try It 1

Find the vertex and axis of symmetry of the parabola whose equation is $y = 4x^2 + 4x + 1$. Then graph the equation.

Your solution

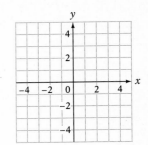

Solutions on p. A52

Objective B   *To find the x-intercepts of a parabola* ....

Recall that a point at which a graph crosses the x- or y-axis is called an *intercept* of the graph. The x-intercepts of the graph of an equation occur when $y = 0$; the y-intercepts occur when $x = 0$.

The graph of $y = x^2 + 3x - 4$ is shown at the right. The points whose coordinates are $(-4, 0)$ and $(1, 0)$ are x-intercepts of the graph.

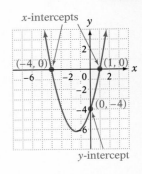

⟹ Find the x-intercepts for the parabola whose equation is $y = 4x^2 - 4x + 1$.

To find the x-intercepts, let $y = 0$ and then solve for x.

$y = 4x^2 - 4x + 1$
$0 = 4x^2 - 4x + 1$ • Let $y = 0$.

$0 = (2x - 1)(2x - 1)$ • Solve for x by factoring.

$\begin{array}{ll} 2x - 1 = 0 & 2x - 1 = 0 \\ \quad 2x = 1 & \quad 2x = 1 \\ \quad\; x = \dfrac{1}{2} & \quad\; x = \dfrac{1}{2} \end{array}$

The x-intercept is $\left(\dfrac{1}{2}, 0\right)$.

In this last example, the parabola has only one x-intercept. In this case, the parabola is said to be **tangent** to the x-axis at $x = \dfrac{1}{2}$.

➡️ Find the x-intercepts of $y = x^2 - 2x - 1$.

To find the x-intercepts, let $y = 0$ and solve for x.

$y = x^2 - 2x - 1$

$0 = x^2 - 2x - 1$

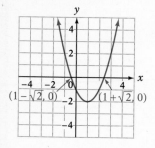

$$x = \frac{-b \pm \sqrt{b^2 - 4ac}}{2a}$$

$$= \frac{-(-2) \pm \sqrt{(-2)^2 - 4(1)(-1)}}{2(1)}$$

$$= \frac{2 \pm \sqrt{4 + 4}}{2} = \frac{2 \pm \sqrt{8}}{2}$$

$$= \frac{2 \pm 2\sqrt{2}}{2} = 1 \pm \sqrt{2}$$

- Because $x^2 - 2x - 1$ does not easily factor, use the quadratic formula to solve for x.

- $a = 1, b = -2, c = -1$

The x-intercepts are $(1 - \sqrt{2}, 0)$ and $(1 + \sqrt{2}, 0)$.

The graph of a parabola may not have x-intercepts. The graph of $y = -x^2 + 2x - 2$ is shown at the right. Note that the graph does not pass through the x-axis and thus there are no x-intercepts. This means there are no real number solutions of $-x^2 + 2x - 2 = 0$.

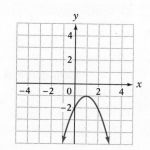

Using the quadratic formula, we find that the solutions of the equation $-x^2 + 2x - 2 = 0$ are the complex numbers $1 - i$ and $1 + i$.

Recall that the discriminant of the quadratic formula is the expression $b^2 - 4ac$ and that this expression can be used to determine whether $ax^2 + bx + c = 0$ has zero, one, or two real number solutions. Because there is a connection between the solutions of $ax^2 + bx + c = 0$ and the x-intercepts of the graph of $y = ax^2 + bx + c$, the discriminant can be used to determine the number of x-intercepts of a parabola.

The Effect of the Discriminant on the Number of x-Intercepts of a Parabola

1. If $b^2 - 4ac = 0$, the parabola has one x-intercept.
2. If $b^2 - 4ac > 0$, the parabola has two x-intercepts.
3. If $b^2 - 4ac < 0$, the parabola has no x-intercepts.

➡️ Use the discriminant to determine the number of x-intercepts of the parabola whose equation is $y = 2x^2 - x + 2$.

$b^2 - 4ac$

$(-1)^2 - 4(2)(2) = 1 - 16 = -15$

$-15 < 0$

- Evaluate the discriminant.
- $a = 2, b = -1, c = 2$

The discriminant is less than zero, so the parabola has no x-intercepts.

Example 2

Find the x-intercepts of
$y = 2x^2 - 5x + 2$.

Solution

$y = 2x^2 - 5x + 2$
$0 = 2x^2 - 5x + 2$
$0 = (2x - 1)(x - 2)$

$2x - 1 = 0 \qquad x - 2 = 0$
$\quad\ \ 2x = 1 \qquad\quad\ x = 2$
$\qquad\ x = \dfrac{1}{2}$

The x-intercepts are $\left(\dfrac{1}{2}, 0\right)$ and $(2, 0)$.

Example 3

Use the discriminant to determine the number of x-intercepts of
$y = x^2 - 6x + 9$.

Solution

$a = 1, b = -6, c = 9$
$b^2 - 4ac = (-6)^2 - 4(1)(9) = 36 - 36 = 0$

Because the discriminant is equal to zero, the parabola has one x-intercept.

You Try It 2

Find the x-intercepts of
$y = x^2 + 3x + 4$.

Your solution

You Try It 3

Use the discriminant to determine the number of x-intercepts of $y = x^2 - x - 6$.

Your solution

Solutions on p. A52

Objective C   *To find the minimum or maximum of a quadratic function* ················· 💻

POINT OF INTEREST

Calculus is a branch of mathematics that demonstrates, among other things, how to find the maximum or minimum of functions other than quadratic functions. These are very important problems in applied mathematics. For instance, an automotive engineer wants to design a car whose shape will *minimize* the effects of air flow. The same engineer tries to *maximize* the efficiency of a car's engine. Similarly, an economist may try to determine what business practices will *minimize* cost and *maximize* profit.

The graph of $f(x) = x^2 - 2x + 3$ is shown at the right. Because a is positive, the parabola opens up. The vertex of the parabola is the lowest point on the parabola. It is the point that has the minimum y-coordinate. Therefore, the value of the function at this point is a **minimum**.

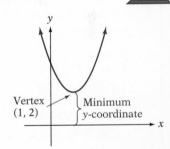

The graph of $f(x) = -x^2 + 2x + 1$ is shown at the right. Because a is negative, the parabola opens down. The vertex of the parabola is the highest point on the parabola. It is the point that has the maximum y-coordinate. Therefore, the value of the function at this point is a **maximum**.

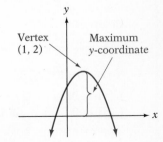

To find the minimum or maximum value of a quadratic function, first find the x-coordinate of the vertex. Then evaluate the function at that value.

Example 4

Find the minimum value of
$f(x) = 2x^2 - 3x + 1$.

Solution

$x = -\dfrac{b}{2a} = -\dfrac{-3}{2(2)} = \dfrac{3}{4}$

$f(x) = 2x^2 - 3x + 1$

$f\left(\dfrac{3}{4}\right) = 2\left(\dfrac{3}{4}\right)^2 - 3\left(\dfrac{3}{4}\right) + 1$

$\qquad = \dfrac{9}{8} - \dfrac{9}{4} + 1 = -\dfrac{1}{8}$

Because a is positive, the graph opens up.
The function has a minimum value.

The minimum value of the function is $-\dfrac{1}{8}$.

You Try It 4

Find the maximum value of
$f(x) = -3x^2 + 4x - 1$.

Your solution

Solution on p. A52

Objective D   *To solve application problems*

Example 5

A mining company has determined that
the cost (C) in dollars per ton of mining
a mineral is given by the equation
$C(x) = 0.2x^2 - 2x + 12$, where x is the
number of tons of the mineral that
are mined. Find the number of tons of
the mineral that should be mined to
minimize the cost. What is the
minimum cost?

Strategy

• To find the number of tons that will
  minimize the cost, find the
  x-coordinate of the vertex.
• To find the minimum cost, evaluate
  the function at the x-coordinate of the
  vertex.

Solution

$x = -\dfrac{b}{2a} = -\dfrac{-2}{2(0.2)} = 5$

To minimize cost, 5 tons should be mined.

$C(x) = 0.2x^2 - 2x + 12$

$C(5) = 0.2(5)^2 - 2(5) + 12 = 5 - 10 + 12 = 7$

The minimum cost per ton is \$7.

You Try It 5

The height (s) in feet of a ball thrown
straight up is given by the equation
$s(t) = -16t^2 + 64t$, where t is the time in
seconds. Find the time it takes the ball to
reach its maximum height. What is the
maximum height?

Your strategy

Your solution

Solution on p. A52

11.2 Exercises

Objective A

Find the vertex and axis of symmetry of the parabola. Then sketch its graph.

1. $y = x^2 - 2x - 4$

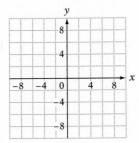

2. $y = x^2 + 4x - 4$

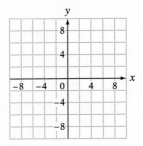

3. $y = -x^2 + 2x - 3$

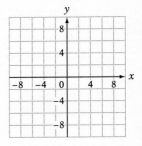

4. $y = -x^2 + 4x - 5$

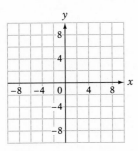

5. $f(x) = x^2 + 6x + 5$

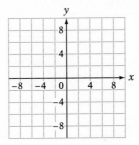

6. $f(x) = x^2 - x - 6$

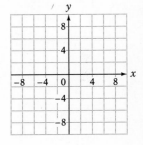

7. $G(x) = x^2 - x - 2$

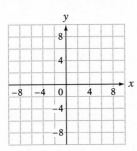

8. $F(x) = x^2 - 3x + 2$

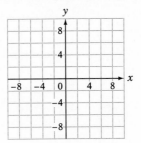

9. $y = -2x^2 + 6x$

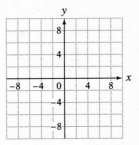

10. $y = -x^2 + 2x - 1$

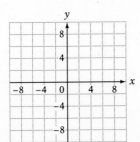

11. $h(x) = \frac{1}{2}x^2 - x + 1$

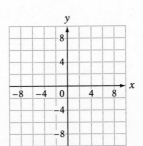

12. $P(x) = -\frac{1}{2}x^2 + 2x - 3$

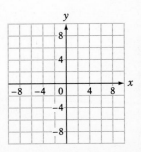

Objective B

Find the x-intercepts of the graph of the parabola.

13. $y = x^2 - 4$

14. $y = x^2 - 9$

15. $y = 2x^2 - 4x$

16. $y = 3x^2 + 6x$

17. $y = x^2 - x - 2$

18. $y = x^2 - 2x - 8$

19. $y = 2x^2 - x - 1$

20. $y = 2x^2 - 5x - 3$

21. $y = x^2 + 2x - 1$

22. $y = x^2 + 4x - 3$

23. $y = x^2 + 6x + 10$

24. $y = -x^2 - 4x - 5$

25. $y = x^2 - 2x - 2$

26. $y = -x^2 - 2x + 1$

27. $y = -x^2 + 4x + 1$

Use the discriminant to determine the number of x-intercepts of the graph.

28. $y = 2x^2 + x + 1$

29. $y = 2x^2 + 2x - 1$

30. $y = -x^2 - x + 3$

31. $y = -2x^2 + x + 1$

32. $y = x^2 - 8x + 16$

33. $y = x^2 - 10x + 25$

34. $y = -3x^2 - x - 2$

35. $y = -2x^2 + x - 1$

36. $y = 4x^2 - x - 2$

37. $y = 2x^2 + x + 4$

38. $y = -2x^2 - x - 5$

39. $y = -3x^2 + 4x - 5$

Objective C

Find the minimum or maximum value of the quadratic function.

40. $f(x) = x^2 - 2x + 3$

41. $f(x) = x^2 + 3x - 4$

42. $f(x) = -2x^2 + 4x - 3$

43. $f(x) = -2x^2 - 3x + 4$

44. $f(x) = 2x^2 + 4x$

45. $f(x) = -2x^2 - 3x$

46. $f(x) = -2x^2 + 4x - 5$

47. $f(x) = -3x^2 + x - 6$

48. $f(x) = 2x^2 + 3x - 8$

49. $f(x) = -x^2 - x + 2$

50. $f(x) = 3x^2 + 3x - 2$

51. $f(x) = x^2 - 5x + 3$

52. $f(x) = -3x^2 + 4x - 2$

53. $f(x) = -2x^2 - 5x + 1$

54. $f(x) = 3x^2 + 5x + 2$

Objective D   Application Problems

Solve.

55. The height in feet (s) of a rock thrown upward at an initial speed of 64 ft/s from a cliff 50 ft above an ocean beach is given by the function $s(t) = -16t^2 + 64t + 50$, where t is the time in seconds. Find the maximum height above the beach that the rock will attain.

56. The height in feet (s) of a ball thrown upward at an initial speed of 80 ft/s from a platform 50 ft high is given by the function $s(t) = -16t^2 + 80t + 50$, where t is the time in seconds. Find the maximum height above the ground that the ball will attain.

57. A pool is treated with a chemical to reduce the amount of algae. The amount of algae in the pool t days after the treatment can be approximated by the function $A(t) = 40t^2 - 400t + 1500$. How many days after treatment will the pool have the least amount of algae?

58. A manufacturer of microwave ovens believes that the revenue the company receives is related to the price (P) of an oven by the function $R(P) = 125P - \frac{1}{4}P^2$. What price will give the maximum revenue?

Solve.

59. The suspension cable that supports a footbridge hangs in the shape of a parabola. The height in feet of the cable above the bridge is given by the function $h(x) = 0.25x^2 - 0.8x + 25$, where x is the distance from one end of the bridge. What is the minimum height of the cable above the bridge?

60. A manufacturer of camera lenses estimated that the average monthly cost of production is given by $C(x) = 0.1x^2 - 20x + 2000$, where x is the number of lenses produced each month. Find the number of lenses the company should produce in order to minimize the average cost.

61. Find two numbers whose sum is 20 and whose product is a maximum.

62. Find two numbers whose sum is 50 and whose product is a maximum.

63. Find two numbers whose difference is 24 and whose product is a minimum.

64. Find two numbers whose difference is 14 and whose product is a minimum.

65. A rectangle has a perimeter of 60 ft. Find the dimensions of the rectangle that will yield the maximum area. What is the maximum area?

66. The perimeter of a rectangular window is 24 ft. Find the dimensions of the window that will enclose the largest area. What is the maximum area?

APPLYING THE CONCEPTS

67. Determine whether the statement is always true, sometimes true, or never true.
    a. An axis of symmetry of a parabola passes through the vertex.
    b. A parabola has two x-intercepts.
    c. A quadratic function has a minimum value.

68. An equation of the form $y = x^2 + bx + c$ can be written in the form $y = (x - h)^2 + k$, where (h, k) are the coordinates of the vertex of the parabola. Use the process of completing the square to rewrite the equation in the form $y = (x - h)^2 + k$. Find the vertex.
    a. $y = x^2 - 4x + 7$ **b.** $y = x^2 - 2x - 2$ **c.** $y = x^2 - 6x + 3$

69. A rancher has 200 ft of fencing to build a rectangular corral along the side of an existing fence. Determine the dimensions of the corral that will maximize the enclosed area.

11.3

Graphs of Functions

Objective A   *To graph functions*

The graphs of the polynomial functions $f(x) = mx + b$ (a straight line) and $f(x) = ax^2 + bx + c$, $a \neq 0$ (a parabola) have been discussed. The graphs of other functions can be drawn by finding ordered pairs that belong to the function, plotting the points that correspond to the ordered pairs, and then drawing a curve through the points.

CONSIDER THIS

The units along the x-axis are different from those along the y-axis. If the units along the x-axis were the same as those along the y-axis, the graph would appear narrower than the one shown.

➡ Graph: $F(x) = x^3$

Select several values of x and evaluate the function.

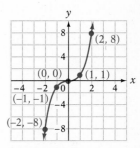

x	$F(x) = x^3$	$F(x)$	(x, y)
-2	$(-2)^3$	-8	$(-2, -8)$
-1	$(-1)^3$	-1	$(-1, -1)$
0	0^3	0	$(0, 0)$
1	1^3	1	$(1, 1)$
2	2^3	8	$(2, 8)$

Plot the ordered pairs and draw a graph through the points.

➡ Graph: $g(x) = x^3 - 4x + 5$

Select several values of x and evaluate the function.

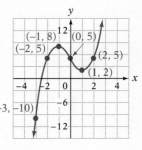

x	$g(x) = x^3 - 4x + 5$	$g(x)$	(x, y)
-3	$(-3)^3 - 4(-3) + 5$	-10	$(-3, -10)$
-2	$(-2)^3 - 4(-2) + 5$	5	$(-2, 5)$
-1	$(-1)^3 - 4(-1) + 5$	8	$(-1, 8)$
0	$(0)^3 - 4(0) + 5$	5	$(0, 5)$
1	$(1)^3 - 4(1) + 5$	2	$(1, 2)$
2	$(2)^3 - 4(2) + 5$	5	$(2, 5)$

Plot the ordered pairs and draw a graph through the points.

Note from the graphs of the two different cubic functions that the shapes of the graphs can be different. The following graphs of typical cubic polynomial functions show their general shapes.

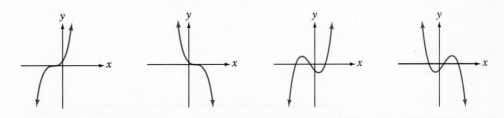

As the degree of a polynomial increases, the graph of the polynomial function can change significantly. In these cases, it may be necessary to plot many points before an accurate graph can be drawn. Only polynomials of degree 3 are considered here.

➡ Graph: $f(x) = |x + 2|$

This is an absolute value function.

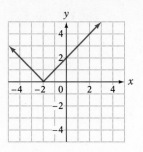

x	$f(x) = |x + 2|$	$f(x)$	(x, y)
-3	$|-3 + 2|$	1	$(-3, 1)$
-2	$|-2 + 2|$	0	$(-2, 0)$
-1	$|-1 + 2|$	1	$(-1, 1)$
0	$|0 + 2|$	2	$(0, 2)$
1	$|1 + 2|$	3	$(1, 3)$
2	$|2 + 2|$	4	$(2, 4)$

In general, the graph of the absolute value of a linear polynomial is V-shaped.

➡ Graph: $R(x) = \sqrt{2x - 4}$

This is a radical function. Because the square root of a negative number is not a real number, the domain of this function requires that $2x - 4 \geq 0$. Solve this inequality for x.

$$2x - 4 \geq 0$$
$$2x \geq 4$$
$$x \geq 2$$

The domain is $\{x \mid x \geq 2\}$. This means that only values of x that are greater than or equal to 2 can be chosen as values at which to evaluate the function. In this case, some of the y-coordinates must be approximated.

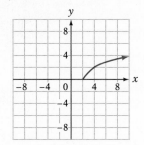

x	$R(x) = \sqrt{2x - 4}$	$R(x)$	(x, y)
2	$\sqrt{2(2) - 4}$	0	$(2, 0)$
3	$\sqrt{2(3) - 4}$	1.41	$(3, 1.41)$
4	$\sqrt{2(4) - 4}$	2	$(4, 2)$
5	$\sqrt{2(5) - 4}$	2.45	$(5, 2.45)$
6	$\sqrt{2(6) - 4}$	2.83	$(6, 2.83)$

Recall that a function is a special type of relation, one for which no two ordered pairs have the same first coordinate. Graphically, this means that the graph of a function cannot pass through two points that have the same x-coordinate and different y-coordinates. For instance, the graph at the right is not the graph of a function because there are ordered pairs with the same x-coordinate and different y-coordinates.

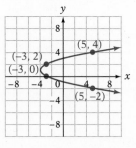

This last graph illustrates a general statement that can be made about whether a graph defines a function. It is called the *vertical-line test*.

Vertical-Line Test

A graph defines a function if any vertical line intersects the graph at no more than one point.

For example, the graph of a nonvertical straight line is the graph of a function. Any vertical line intersects the graph no more than once. The graph of a circle, however, is not the graph of a function. There are vertical lines that intersect the graph at more than one point.

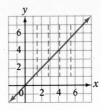
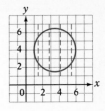

There may be practical situations in which a graph is not the graph of a function. The problem below is an example.

One of the causes of smog is an inversion layer where temperatures at higher altitudes are warmer than those at lower altitudes. The graph at the right shows the altitudes at which various temperatures were recorded. As shown by the dashed lines in the graph, there are two altitudes at which the temperature was 25°C. This means that there are two ordered pairs (shown in the graph) with the same first coordinate but different second coordinates. The graph does not define a function.

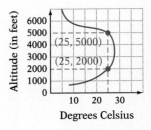

When a graph does define a function, the domain and range can be estimated from the graph.

➡ Determine the domain and range of the function given by the graph at the right.

The solid dots on the graph indicate its beginning and ending points.

The domain is the set of x-coordinates.
Domain: $\{x \mid 1 \leq x \leq 6\}$

The range is the set of y-coordinates.
Range: $\{y \mid 2 \leq y \leq 5\}$

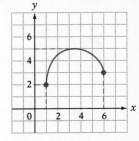

➡ Determine the domain and range of the function given by the graph at the right.

The arrows on the graph indicate that the graph continues in the same manner.

The domain is the set of x-coordinates.
Domain: $\{x \mid x$ is a real number$\}$

The range is the set of y-coordinates.
Range: $\{y \mid -4 \leq y \leq 4\}$

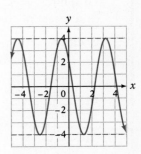

Example 1
Use the vertical-line test to determine whether the graph shown is the graph of a function.

Solution
A vertical line intersects the graph more than once. The graph is not the graph of a function.

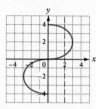

You Try It 1
Use the vertical-line test to determine whether the graph shown is the graph of a function.

Your solution

Example 2
Graph $f(x) = x^3 - 3x$. State the domain and range of the function.

Solution

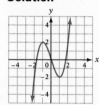

Domain:
$\{x \mid x \text{ is a real number}\}$
Range:
$\{y \mid y \text{ is a real number}\}$

You Try It 2
Graph $f(x) = -\frac{1}{2}x^3 + 2x$. State the domain and range of the function.

Your solution

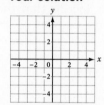

Example 3
Graph $f(x) = |x| + 2$. State the domain and range of the function.

Solution

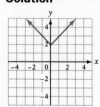

Domain:
$\{x \mid x \text{ is a real number}\}$
Range:
$\{y \mid y \geq 2\}$

You Try It 3
Graph $f(x) = |x - 1|$. State the domain and range of the function.

Your solution

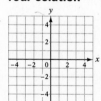

Example 4
Graph $f(x) = \sqrt{2 - x}$. State the domain and range of the function.

Solution

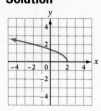

Domain: $\{x \mid x \leq 2\}$
Range: $\{y \mid y \geq 0\}$

You Try It 4
Graph $f(x) = -\sqrt{x - 1}$. State the domain and range of the function.

Your solution

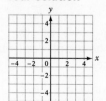

Solutions on p. A53

11.3 Exercises

. . . .

Objective A

Use the vertical-line test to determine whether the graph is the graph of a function.

1.

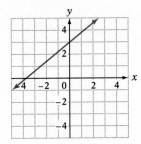

2.

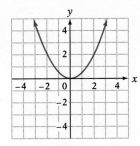

3.

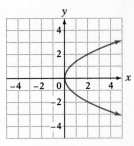

4.

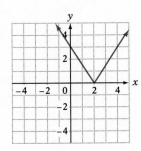

5.

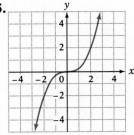

6.

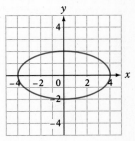

Graph the function and state its domain and range.

7. $f(x) = 3|2 - x|$

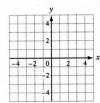

8. $f(x) = x^3 - 1$

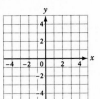

9. $f(x) = 1 - x^3$

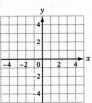

10. $f(x) = \sqrt{1 + x}$

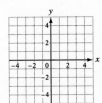

11. $f(x) = \sqrt{4 - x}$

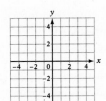

12. $f(x) = |2x - 1|$

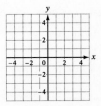

Graph the function and state its domain and range.

13. $f(x) = x^3 + 4x^2 + 4x$

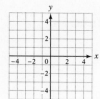

14. $f(x) = x^3 - x^2 - x + 1$

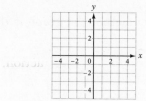

15. $f(x) = -\sqrt{x} + 2$

16. $f(x) = -\sqrt{x} - 3$

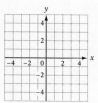

17. $f(x) = |2x + 2|$

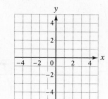

18. $f(x) = 2|x + 1|$

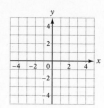

APPLYING THE CONCEPTS

19. If $f(x) = \sqrt{x} - 2$ and $f(a) = 4$, find a.

20. If $f(x) = \sqrt{x} + 5$ and $f(a) = 3$, find a.

21. $f(a, b) =$ the sum of a and b
$g(a, b) =$ the product of a and b
Find $f(2, 5) + g(2, 5)$.

22. $f(a, b) =$ the greatest common divisor of a and b
$g(a, b) =$ the least common multiple of a and b
Find $f(14, 35) + g(14, 35)$.

23. Given $f(x) = (x + 1)(x - 1)$, for what values of x is $f(x)$ negative?
Write your answer in set-builder notation.

24. Given $f(x) = (x + 2)(x - 2)$, for what values of x is $f(x)$ negative?
Write your answer in set-builder notation.

25. Given $f(x) = -|x + 3|$, for what value of x is $f(x)$ greatest?

26. Given $f(x) = |2x - 2|$, for what value of x is $f(x)$ smallest?

11.4

Algebra of Functions

Objective A   *To perform operations on functions* ....

The operations of addition, subtraction, multiplication, and division of functions are defined as follows.

Operations on Functions

If f and g are functions and x is an element of the domain of each function, then

$$(f + g)(x) = f(x) + g(x) \qquad (f \cdot g)(x) = f(x) \cdot g(x)$$

$$(f - g)(x) = f(x) - g(x) \qquad \left(\frac{f}{g}\right)(x) = \frac{f(x)}{g(x)}, \; g(x) \neq 0$$

➡ Given $f(x) = x^2 + 1$ and $g(x) = 3x - 2$, find $(f + g)(3)$ and $(f \cdot g)(-1)$.

$$\begin{aligned}
(f + g)(3) &= f(3) + g(3) \\
&= [(3)^2 + 1] + [3(3) - 2] \\
&= 10 + 7 = 17
\end{aligned}$$

$$\begin{aligned}
(f \cdot g)(-1) &= f(-1) \cdot g(-1) \\
&= [(-1)^2 + 1] \cdot [3(-1) - 2] \\
&= 2 \cdot (-5) = -10
\end{aligned}$$

Consider the functions f and g from the last example. Let $S(x)$ be the sum of the two functions. Then

$$\begin{aligned}
S(x) = (f + g)(x) &= f(x) + g(x) \\
&= (x^2 + 1) + (3x - 2)
\end{aligned}$$

$$S(x) = x^2 + 3x - 1$$

- The definition of addition of functions
- $f(x) = x^2 + 1$, $g(x) = 3x - 2$

Now evaluate $S(3)$.

$$\begin{aligned}
S(3) &= (3)^2 + 3(3) - 1 \\
&= 9 + 9 - 1 \\
&= 17 = (f + g)(3)
\end{aligned}$$

Note that $S(3) = 17$ and $(f + g)(3) = 17$. This shows that adding $f(x) + g(x)$ and then evaluating is the same as evaluating $f(x)$ and $g(x)$ and then adding. The same is true for the other operations on functions. For instance, let $P(x)$ be the product of the functions f and g. Then

$$\begin{aligned}
P(x) = (f \cdot g)(x) &= f(x) \cdot g(x) \\
&= (x^2 + 1)(3x - 2) \\
&= 3x^3 - 2x^2 + 3x - 2
\end{aligned}$$

$$\begin{aligned}
P(-1) &= 3(-1)^3 - 2(-1)^2 + 3(-1) - 2 \\
&= -3 - 2 - 3 - 2 \\
&= -10 = (f \cdot g)(-1)
\end{aligned}$$

➡️ Given $f(x) = 2x^2 - 5x + 3$ and $g(x) = x^2 - 1$, find $\left(\dfrac{f}{g}\right)(1)$.

$$\left(\frac{f}{g}\right)(1) = \frac{f(1)}{g(1)}$$

$$= \frac{2(1)^2 - 5(1) + 3}{(1)^2 - 1}$$

$$= \frac{0}{0} \qquad \bullet \text{ Not a real number}$$

Because $\dfrac{0}{0}$ is not defined, the expression $\left(\dfrac{f}{g}\right)(1)$ cannot be evaluated.

Example 1
Given $f(x) = x^2 - x + 1$ and
$g(x) = x^3 - 4$, find $(f - g)(3)$.

Solution
$$\begin{aligned}
(f - g)(3) &= f(3) - g(3) \\
&= (3^2 - 3 + 1) - (3^3 - 4) \\
&= 7 - 23 \\
&= -16
\end{aligned}$$

$(f - g)(3) = -16$

You Try It 1
Given $f(x) = x^2 + 2x$ and $g(x) = 5x - 2$,
find $(f + g)(-2)$.

Your solution

Example 2
Given $f(x) = x^2 + 2$ and $g(x) = 2x + 3$,
find $(f \cdot g)(-2)$.

Solution
$$\begin{aligned}
(f \cdot g)(-2) &= f(-2) \cdot g(-2) \\
&= [(-2)^2 + 2] \cdot [2(-2) + 3] \\
&= 6 \cdot (-1) \\
&= -6
\end{aligned}$$

$(f \cdot g)(-2) = -6$

You Try It 2
Given $f(x) = 4 - x^2$ and $g(x) = 3x - 4$,
find $(f \cdot g)(3)$.

Your solution

Example 3
Given $f(x) = x^2 + 4x + 4$ and
$g(x) = x^3 - 2$, find $\left(\dfrac{f}{g}\right)(3)$.

Solution
$$\left(\frac{f}{g}\right)(3) = \frac{f(3)}{g(3)}$$

$$= \frac{3^2 + 4(3) + 4}{3^3 - 2}$$

$$= \frac{25}{25}$$

$$= 1$$

$\left(\dfrac{f}{g}\right)(3) = 1$

You Try It 3
Given $f(x) = x^2 - 4$ and $g(x) = x^2 + 2x + 1$,
find $\left(\dfrac{f}{g}\right)(4)$.

Your solution

Solutions on p. A53

Objective B   *To find the composition of two functions* ....

A function can be evaluated at the value of another function. Consider

$$f(x) = 2x + 7 \quad \text{and} \quad g(x) = x^2 + 1$$

The expression $f[g(-2)]$ means to evaluate the function f at $g(-2)$.

$$g(-2) = (-2)^2 + 1 = 4 + 1 = 5$$
$$f[g(-2)] = f(5) = 2(5) + 7 = 10 + 7 = 17$$

Definition of the Composition of Two Functions

Let *f* and *g* be two functions such that *g*(*x*) is in the domain of *f* for all *x* in the domain of *g*. Then the **composition** of the two functions, denoted by *f* ∘ *g*, is the function whose value at *x* is given by $(f \circ g)(x) = f[g(x)]$.

The function defined by $f[g(x)]$ is called the **composite** of f and g.

The requirement in the definition of the composition of two functions that $g(x)$ be in the domain of f for all x in the domain of g is important. For instance, let

$$f(x) = \frac{1}{x - 1} \quad \text{and} \quad g(x) = 3x - 5$$

When $x = 2$,

$$g(2) = 3(2) - 5 = 1$$
$$f[g(2)] = f(1) = \frac{1}{1 - 1} = \frac{1}{0} \qquad \bullet \text{ This is not a real number.}$$

In this case, $g(2)$ is not in the domain of f. Thus the composition is not defined at 2.

➡ Given $f(x) = x^3 - x + 1$ and $g(x) = 2x^2 - 10$, evaluate $(g \circ f)(2)$.

$$f(2) = (2)^3 - (2) + 1 = 7$$
$$(g \circ f)(2) = g[f(2)]$$
$$= g(7)$$
$$= 2(7)^2 - 10 = 88$$
$$(g \circ f)(2) = 88$$

➡ Given $f(x) = 3x - 2$ and $g(x) = x^2 - 2x$, find $(f \circ g)(x)$.

$$(f \circ g)(x) = f[g(x)] = 3(x^2 - 2x) - 2$$
$$= 3x^2 - 6x - 2$$

In general, the composition of functions is not a commutative operation. That is, $(f \circ g)(x) \neq (g \circ f)(x)$. To show this, let $f(x) = x + 3$ and $g(x) = x^2 + 1$. Then

$$(f \circ g)(x) = f[g(x)] \qquad\qquad (g \circ f)(x) = g[f(x)]$$
$$= (x^2 + 1) + 3 \qquad\qquad = (x + 3)^2 + 1$$
$$= x^2 + 4 \qquad\qquad\qquad = x^2 + 6x + 10$$

Thus $(f \circ g)(x) \neq (g \circ f)(x)$.

Example 4

Given $f(x) = x^2 - x$ and $g(x) = 3x - 2$, find $f[g(3)]$.

Solution

$$g(x) = 3x - 2$$
$$g(3) = 3(3) - 2 = 9 - 2 = 7$$
$$f(x) = x^2 - x$$
$$f[g(3)] = f(7) = 7^2 - 7 = 42$$

You Try It 4

Given $f(x) = 1 - 2x$ and $g(x) = x^2$, find $f[g(-1)]$.

Your solution

Example 5

Given $f(x) = 2x^2 - x + 1$ and $g(x) = 1 - x^2$, find $(g \circ f)(-1)$.

Solution

$$f(x) = 2x^2 - x + 1$$
$$f(-1) = 2(-1)^2 - (-1) + 1 = 4$$
$$g(x) = 1 - x^2$$
$$(g \circ f)(-1) = g[f(-1)]$$
$$= g(4) = 1 - 4^2 = -15$$

You Try It 5

Given $f(x) = 3x - 1$ and $g(x) = \dfrac{x}{x^2 - 2}$, find $(g \circ f)(1)$.

Your solution

Example 6

Given $s(t) = t^2 + 3t - 1$ and $v(t) = 2t + 1$, determine $s[v(t)]$.

Solution

$$s(t) = t^2 + 3t - 1$$
$$s[v(t)] = (2t + 1)^2 + 3(2t + 1) - 1$$
$$= (4t^2 + 4t + 1) + 6t + 3 - 1$$
$$= 4t^2 + 10t + 3$$

You Try It 6

Given $L(s) = s + 1$ and $M(s) = s^3 + 1$, determine $M[L(s)]$.

Your solution

Example 7

Given $P(x) = x^2$ and $S(x) = \sqrt{x}$, determine $P[S(x)]$.

Solution

$$P(x) = x^2$$
$$P[S(x)] = P(\sqrt{x}) = (\sqrt{x})^2 = x$$

You Try It 7

Given $V(x) = x^2 + 1$ and $W(x) = -\sqrt{x - 1}$, find $V[W(x)]$.

Your solution

Solutions on p. A54

11.4 Exercises

* *

Objective A

For $f(x) = 2x^2 - 3$ and $g(x) = -2x + 4$, find:

1. $(f - g)(2)$

2. $(f - g)(3)$

3. $(f + g)(0)$

4. $(f + g)(1)$

5. $(f \cdot g)(2)$

6. $(f \cdot g)(-1)$

7. $\left(\dfrac{f}{g}\right)(4)$

8. $\left(\dfrac{f}{g}\right)(-1)$

9. $\left(\dfrac{g}{f}\right)(-3)$

For $f(x) = 2x^2 + 3x - 1$ and $g(x) = 2x - 4$, find:

10. $(f + g)(-3)$

11. $(f + g)(1)$

12. $(f - g)(-2)$

13. $(f - g)(4)$

14. $(f \cdot g)(-2)$

15. $(f \cdot g)(1)$

16. $\left(\dfrac{f}{g}\right)(2)$

17. $\left(\dfrac{f}{g}\right)(-3)$

18. $(f \cdot g)\left(\dfrac{1}{2}\right)$

For $f(x) = x^2 + 3x - 5$ and $g(x) = x^3 - 2x + 3$, find:

19. $(f - g)(2)$

20. $(f \cdot g)(-3)$

21. $\left(\dfrac{f}{g}\right)(-2)$

Objective B

Given $f(x) = 2x - 3$ and $g(x) = 4x - 1$, evaluate the composite function.

22. $f[g(0)]$

23. $g[f(0)]$

24. $f[g(2)]$

25. $g[f(-2)]$

26. $f[g(x)]$

27. $g[f(x)]$

Given $h(x) = 2x + 4$ and $f(x) = \frac{1}{2}x + 2$, evaluate the composite function.

28. $(h \circ f)(0)$ **29.** $(f \circ h)(0)$ **30.** $(h \circ f)(2)$

31. $(f \circ h)(-1)$ **32.** $(h \circ f)(x)$ **33.** $(f \circ h)(x)$

Given $f(x) = x^2 + x + 1$ and $h(x) = 3x + 2$, evaluate the composite function.

34. $(f \circ h)(0)$ **35.** $(h \circ f)(0)$ **36.** $(f \circ h)(-1)$

37. $(h \circ f)(-2)$ **38.** $(f \circ h)(x)$ **39.** $(h \circ f)(x)$

Given $f(x) = x - 2$ and $g(x) = x^3$, evaluate the composite function.

40. $(f \circ g)(2)$ **41.** $(f \circ g)(-1)$ **42.** $(g \circ f)(2)$

43. $(g \circ f)(-1)$ **44.** $(f \circ g)(x)$ **45.** $(g \circ f)(x)$

APPLYING THE CONCEPTS

For the function $g(x) = x^2 - 1$, find:

46. $g(2 + h)$ **47.** $g(3 + h) - g(3)$ **48.** $g(-1 + h) - g(-1)$

49. $\dfrac{g(1 + h) - g(1)}{h}$ **50.** $\dfrac{g(-2 + h) - g(-2)}{h}$ **51.** $\dfrac{g(a + h) - g(a)}{h}$

Given $f(x) = 2x$, $g(x) = 3x - 1$, and $h(x) = x - 2$, find:

52. $f(g[h(2)])$ **53.** $g(h[f(1)])$ **54.** $h(g[f(-1)])$

55. $f(h[g(0)])$ **56.** $f(g[h(x)])$ **57.** $g(f[h(x)])$

11.5

One-to-One and Inverse Functions

Objective A   *To determine whether a function is one-to-one*

Recall that a function is a set of ordered pairs in which no two ordered pairs that have the same first coordinate have different second coordinates. This means that given any x, there is only one y that can be paired with that x. A *one-to-one* function satisfies the additional condition that given any y, there is only one x that can be paired with the given y. One-to-one functions are commonly written as 1–1.

CONSIDER THIS

Recall that a function is a set of ordered pairs in which no two ordered pairs have the same first coordinate. A 1–1 function is a function in which no two ordered pairs have the same second coordinate.

> **One-to-One Function**
>
> A function f is a 1–1 function if, for any a and b in the domain of f, $f(a) = f(b)$ implies that $a = b$.

This definition states that if the y-coordinates of an ordered pair are equal, $f(a) = f(b)$, then the x-coordinates must be equal, $a = b$.

The function defined by $f(x) = 2x + 1$ is a 1–1 function. To show this, determine $f(a)$ and $f(b)$. Then form the equation $f(a) = f(b)$.

$$f(a) = 2a + 1 \qquad f(b) = 2b + 1$$

$$f(a) = f(b)$$
$$2a + 1 = 2b + 1$$
$$2a = 2b \qquad \text{• Subtract 1 from each side of the equation.}$$
$$a = b \qquad \text{• Divide each side of the equation by 2.}$$

Because $f(a) = f(b)$ implies that $a = b$, the function is a 1–1 function.

Consider the function defined by $g(x) = x^2 - x$. Evaluate the function at -2 and 3.

$$g(-2) = (-2)^2 - (-2) = 6 \qquad g(3) = 3^2 - 3 = 6$$

From this evaluation, $g(-2) = 6$ and $g(3) = 6$, but $-2 \neq 3$. Thus f is not a 1–1 function.

The graphs of $f(x) = 2x + 1$ and $g(x) = x^2 - x$ are shown below. Note that a horizontal line intersects the graph of f at no more than one point. However, a horizontal line intersects the graph of g at more than one point.

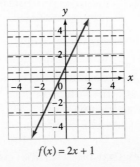

$f(x) = 2x + 1$

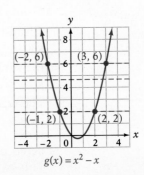

$g(x) = x^2 - x$

Looking at the graph of f on the previous page, note that for each y-coordinate there is only one x-coordinate. Thus f is a 1–1 function. On the graph of g, however, there are *two* x-coordinates for a given y-coordinate. For instance, $(-2, 6)$ and $(3, 6)$ are the coordinates of two points on the graph for which the y-coordinates are the same and the x-coordinates are different. Therefore, g is not a 1–1 function.

Horizontal-Line Test

The graph of a function represents the graph of a 1–1 function if any horizontal line intersects the graph at no more than one point.

Example 1

Determine whether the graph is the graph of a 1–1 function.

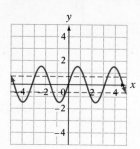

Solution

Because a horizontal line intersects the graph more than once, the graph is not the graph of a 1–1 function.

You Try It 1

Determine whether the graph is the graph of a 1–1 function.

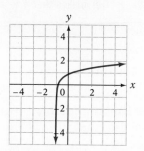

Your solution

Solution on p. A54

Objective B ***To find the inverse of a function***

The **inverse of a function** is the set of ordered pairs formed by reversing the coordinates of each ordered pair of the function.

For example, the set of ordered pairs of the function defined by $f(x) = 2x$ with domain $\{-2, -1, 0, 1, 2\}$ is $\{(-2, -4), (-1, -2), (0, 0), (1, 2), (2, 4)\}$. The set of ordered pairs of the inverse function is $\{(-4, -2), (-2, -1), (0, 0), (2, 1), (4, 2)\}$.

From the ordered pairs of f, we have

Domain = $\{-2, -1, 0, 1, 2\}$ and   Range = $\{-4, -2, 0, 2, 4\}$

From the ordered pairs of the inverse function, we have

Domain = $\{-4, -2, 0, 2, 4\}$ and   Range = $\{-2, -1, 0, 1, 2\}$

Note that the domain of the inverse function is the range of the function, and the range of the inverse function is the domain of the function.

Now consider the function defined by $g(x) = x^2$ with domain $\{-2, -1, 0, 1, 2\}$. The set of ordered pairs of this function is $\{(-2, 4), (-1, 1), (0, 0), (1, 1), (2, 4)\}$. Reversing the ordered pairs gives $\{(4, -2), (1, -1), (0, 0), (1, 1), (4, 2)\}$. These ordered pairs do not satisfy the condition of a function, because there are ordered pairs with the same first coordinate and different second coordinates. This example illustrates that not all functions have an inverse function.

The graphs of $f(x) = 2x$ and $g(x) = x^2$ with the set of real numbers as the domain are shown at the right.

By the horizontal-line test, f is a 1–1 function but g is not.

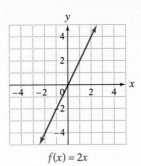

$f(x) = 2x$

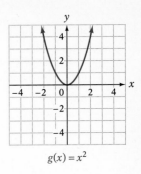

$g(x) = x^2$

Condition for an Inverse Function

A function f has an inverse function if and only if f is a 1–1 function.

CONSIDER THIS

It is important to note that f^{-1} is the symbol for the *inverse* function and does not mean reciprocal.

$$f^{-1}(x) \neq \frac{1}{f(x)}$$

The symbol f^{-1} is used to denote the inverse of the function f. The symbol $f^{-1}(x)$ is read "f inverse of x."

$f^{-1}(x)$ is *not* the reciprocal of $f(x)$ but is the notation for the inverse of a 1–1 function.

To find the inverse of a function, interchange x and y. Then solve for y.

➡ Find the inverse of the function defined by $f(x) = 3x + 6$.

CONSIDER THIS

The inverse of a function is the function with the coordinates of the ordered pairs reversed. This means interchanging the x and y.

$$f(x) = 3x + 6$$
$$y = 3x + 6 \qquad \bullet \text{ Replace } f(x) \text{ by } y.$$
$$x = 3y + 6 \qquad \bullet \text{ Interchange } x \text{ and } y.$$
$$x - 6 = 3y \qquad \bullet \text{ Solve for } y.$$
$$\frac{1}{3}x - 2 = y$$
$$f^{-1}(x) = \frac{1}{3}x - 2 \qquad \bullet \text{ Replace } y \text{ by } f^{-1}(x).$$

The inverse of the function is given by $f^{-1}(x) = \frac{1}{3}x - 2$.

The fact that the ordered pairs of the inverse of a function are the reverse of those of the function has a graphical interpretation. In the graph in the middle, the points with the coordinates reversed from the first graph are plotted. The inverse function is graphed by drawing a smooth curve through those points, as shown in the figure on the right below.

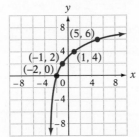

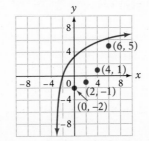

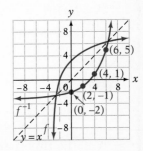

Note the dashed graph of $y = x$ that is shown in the figure on the right. If two functions are inverses of each other, their graphs are mirror images with respect to the graph of the line $y = x$.

The composition of a function and its inverse have a special property.

> **Property of the Composition of Inverse Functions**
>
> $$f^{-1}[f(x)] = x \qquad \text{and} \qquad f[f^{-1}(x)] = x$$

This property can be used to determine whether two functions are inverses of each other.

➡ Are $f(x) = 2x - 4$ and $g(x) = \frac{1}{2}x + 2$ inverses of each other?

To determine whether the functions are inverses, use the Property of the Composition of Inverse Functions.

$$f[g(x)] = 2\left(\frac{1}{2}x + 2\right) - 4 \qquad\qquad g[f(x)] = \frac{1}{2}(2x - 4) + 2$$
$$= x + 4 - 4 \qquad\qquad\qquad\qquad = x - 2 + 2$$
$$= x \qquad\qquad\qquad\qquad\qquad\quad = x$$

Because $f[g(x)] = x$ and $g[f(x)] = x$, the functions are inverses of each other.

Example 2

Find the inverse of the function defined by $f(x) = 2x - 3$.

Solution

$$f(x) = 2x - 3$$
$$y = 2x - 3 \qquad \bullet \text{ Replace } f(x) \text{ by } y.$$
$$x = 2y - 3 \qquad \bullet \text{ Interchange } x \text{ and } y.$$
$$x + 3 = 2y \qquad \bullet \text{ Solve for } y.$$
$$\frac{1}{2}x + \frac{3}{2} = y$$

$$f^{-1}(x) = \frac{1}{2}x + \frac{3}{2} \qquad \bullet \text{ Replace } y \text{ by } f^{-1}(x).$$

The inverse of the function is given by $f^{-1}(x) = \frac{1}{2}x + \frac{3}{2}$.

Example 3

Are $f(x) = 3x - 6$ and $g(x) = \frac{1}{3}x + 2$

inverses of each other?

Solution

$$f[g(x)] = 3\left(\frac{1}{3}x + 2\right) - 6 = x + 6 - 6 = x$$

$$g[f(x)] = \frac{1}{3}(3x - 6) + 2 = x - 2 + 2 = x$$

Yes, the functions are inverses of each other.

You Try It 2

Find the inverse of the function defined by
$$f(x) = \frac{1}{2}x + 4.$$

Your solution

You Try It 3

Are $f(x) = 2x - 6$ and $g(x) = \frac{1}{2}x - 3$

inverses of each other?

Your solution

Solutions on p. A54

11.5 Exercises

Objective A

Determine whether the graph represents the graph of a 1–1 function.

1.
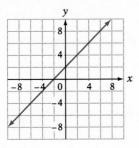

2.
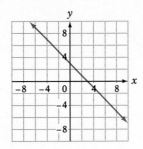

3.
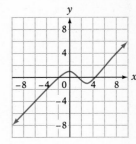

4.
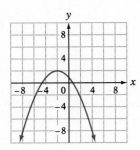

5.
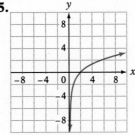

6.
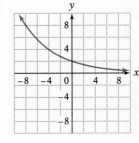

7.
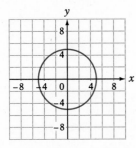

8.
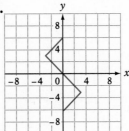

9.
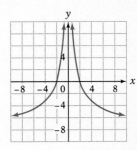

10.
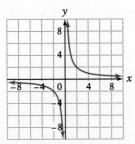

11.
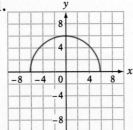

12.
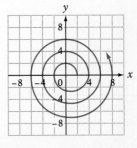

Objective B

Find the inverse of the function. If the function does not have an inverse function, write "no inverse."

13. $\{(1, 0), (2, 3), (3, 8), (4, 15)\}$

14. $\{(1, 0), (2, 1), (-1, 0), (-2, 0)\}$

15. $\{(3, 5), (-3, -5), (2, 5), (-2, -5)\}$

16. $\{(-5, -5), (-3, -1), (-1, 3), (1, 7)\}$

17. $\{(0, -2), (-1, 5), (3, 3), (-4, 6)\}$

18. $\{(-2, -2), (0, 0), (2, 2), (4, 4)\}$

19. $\{(-2, -3), (-1, 3), (0, 3), (1, 3)\}$

20. $\{(2, 0), (1, 0), (3, 0), (4, 0)\}$

Find $f^{-1}(x)$.

21. $f(x) = 4x - 8$

22. $f(x) = 3x + 6$

23. $f(x) = 2x + 4$

24. $f(x) = x - 5$

25. $f(x) = \dfrac{1}{2}x - 1$

26. $f(x) = \dfrac{1}{3}x + 2$

27. $f(x) = -2x + 2$

28. $f(x) = -3x - 9$

29. $f(x) = \dfrac{2}{3}x + 4$

30. $f(x) = \dfrac{3}{4}x - 4$

31. $f(x) = -\dfrac{1}{3}x + 1$

32. $f(x) = -\dfrac{1}{2}x + 2$

33. $f(x) = 2x - 5$

34. $f(x) = 3x + 4$

35. $f(x) = 5x - 2$

36. $f(x) = 4x - 2$

37. $f(x) = 6x - 3$

38. $f(x) = -8x + 4$

Find $f^{-1}(x)$.

39. $f(x) = -6x + 2$

40. $f(x) = 8x + 6$

41. $f(x) = 3x - 4$

42. $f(x) = \dfrac{5}{8}x + 5$

43. $f(x) = -\dfrac{2}{3}x + 4$

44. $f(x) = 5x + 5$

45. $f(x) = 3x - 2$

46. $f(x) = \dfrac{1}{4}x + 4$

47. $f(x) = -\dfrac{3}{4}x - 2$

Use the Property of the Composition of Inverse Functions to determine whether the functions are inverses of each other.

48. $f(x) = 4x; g(x) = \dfrac{x}{4}$

49. $g(x) = x + 5; h(x) = x - 5$

50. $f(x) = 3x; h(x) = \dfrac{1}{3x}$

51. $h(x) = x + 2; g(x) = 2 - x$

52. $g(x) = 3x + 2; f(x) = \dfrac{1}{3}x - \dfrac{2}{3}$

53. $h(x) = 4x - 1; f(x) = \dfrac{1}{4}x + \dfrac{1}{4}$

54. $f(x) = \dfrac{1}{2}x - \dfrac{3}{2}; g(x) = 2x + 3$

55. $g(x) = -\dfrac{1}{2}x - \dfrac{1}{2}; h(x) = -2x + 1$

56. $f(x) = \dfrac{3}{2}x + 2; g(x) = \dfrac{2}{3}x - \dfrac{4}{3}$

57. $f(x) = 3x - 2; g(x) = 2 - 3x$

58. $h(x) = \dfrac{3}{2}x + \dfrac{4}{3}; g(x) = \dfrac{2}{3}x + \dfrac{3}{4}$

59. $g(x) = -\dfrac{1}{2}x - 2; f(x) = -2x - 4$

60. $g(x) = -\dfrac{2}{3}x + 3; h(x) = \dfrac{3}{2}x - \dfrac{9}{2}$

61. $f(x) = -\dfrac{1}{2}x + \dfrac{3}{2}; h(x) = -2x + 3$

APPLYING THE CONCEPTS

Given the graph of the 1–1 function, draw the graph of the inverse of the function by using the technique shown on page 577.

62.
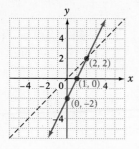

63.
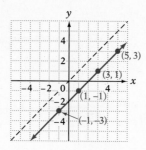

64.
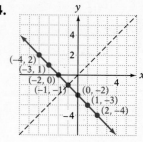

65.
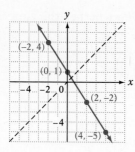

66.
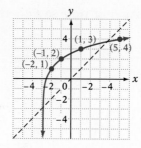

67.
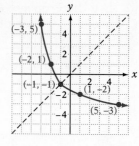

If f is a 1–1 function and $f(0) = 1$, $f(3) = -1$, and $f(5) = -3$, find:

68. $f^{-1}(-3)$

69. $f^{-1}(-1)$

70. $f^{-1}(1)$

If f is a 1–1 function and $f(-3) = 3$, $f(-4) = 7$, and $f(0) = 8$, find:

71. $f^{-1}(3)$

72. $f^{-1}(7)$

73. $f^{-1}(8)$

Each of the tables below defines a function. Is the inverse of the function a function? Explain your answer.

74. Grading Scale Table

[W]

Score	Grade
90–100	A
80–89	B
70–79	C
60–69	D
0–59	F

75. First-Class Postage

[W]

Weight	Cost
$0 < w \le 1$	$.32
$1 < w \le 2$	$.55
$2 < w \le 3$	$.78
$3 < w \le 4$	$1.01

76. Is the inverse of a constant function a function? Explain your answer.

[W]

77. The graphs of all functions given by $f(x) = mx + b$, $m \ne 0$, are straight lines. Are all of these functions 1–1 functions? If so, explain why. If not, give an example of a linear function that is not 1–1.

[W]

11.6 Conic Sections

Objective A   *To graph a parabola*

The **conic sections** are curves that can be constructed from the intersection of a plane and a right circular cone. The parabola, which was introduced earlier, is one of these curves. Here we will review some of that previous discussion and look at equations of parabolas that were not discussed before.

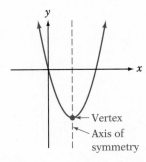

Every parabola has an axis of symmetry and a vertex that is on the axis of symmetry. To understand the axis of symmetry, think of folding the paper along that axis. The two halves of the curve will match up.

The graph of the equation $y = ax^2 + bx + c, a \neq 0$, is a parabola with the axis of symmetry parallel to the y-axis. The parabola opens up when $a > 0$ and opens down when $a < 0$. When the parabola opens up, the vertex is the lowest point on the parabola. When the parabola opens down, the vertex is the highest point on the parabola.

The coordinates of the vertex can be found by completing the square.

Find the vertex of the parabola whose equation is $y = x^2 - 4x + 5$.

$y = x^2 - 4x + 5$
$y = (x^2 - 4x) + 5$ • Group the terms involving x.
$y = (x^2 - 4x + 4) - 4 + 5$ • Complete the square of $x^2 - 4x$. Note that 4 is added and subtracted. Because $4 - 4 = 0$, the equation is not changed.
$y = (x - 2)^2 + 1$ • Factor the trinomial and combine like terms.

The coefficient of x^2 is positive, so the parabola opens up. The vertex is the lowest point on the parabola, or the point that has the least y-coordinate.

Because $(x - 2)^2 \geq 0$ for all x, the least y-coordinate occurs when $(x - 2)^2 = 0$, which occurs when $x = 2$. This means the x-coordinate of the vertex is 2.

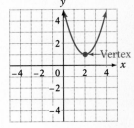

To find the y-coordinate of the vertex, replace x in $y = (x - 2)^2 + 1$ by 2 and solve for y.

$y = (x - 2)^2 + 1$
$\ = (2 - 2)^2 + 1 = 1$

The vertex is (2, 1).

POINT OF INTEREST

The suspension cables for some bridges, such as the Golden Gate bridge, hang in the shape of a parabola. Parabolic shapes are also used for mirrors in telescopes and in certain antenna designs.

By following the procedure of the last example and completing the square on the equation $y = ax^2 + bx + c$, we find that the **x-coordinate of the vertex** is $-\dfrac{b}{2a}$. The y-coordinate of the vertex can then be determined by substituting this value of x into $y = ax^2 + bx + c$ and solving for y.

Because the axis of symmetry is parallel to the y-axis and passes through the vertex, the equation of the **axis of symmetry** is $x = -\dfrac{b}{2a}$.

➡ Find the vertex and axis of symmetry of the parabola whose equation is $y = -3x^2 + 6x + 1$. Then sketch its graph.

Find the x-coordinate of the vertex and the axis of symmetry.

x-coordinate: $-\dfrac{b}{2a} = -\dfrac{6}{2(-3)} = 1$ • $a = -3, b = 6$

The x-coordinate of the vertex is 1.
The axis of symmetry is the line $x = 1$.

To find the y-coordinate of the vertex, replace x by 1 and solve for y.

$y = -3x^2 + 6x + 1$
$\quad = -3(1)^2 + 6(1) + 1 = 4$

The vertex is (1, 4).

Because a is negative, the parabola opens down.

Find a few ordered pairs and use symmetry to sketch the graph.

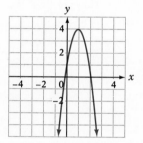

➡ Find the vertex and axis of symmetry of the parabola whose equation is $y = x^2 - 2$. Then sketch its graph.

Find the x-coordinate of the vertex and the axis of symmetry.

x-coordinate: $-\dfrac{b}{2a} = -\dfrac{0}{2(1)} = 0$ • $a = 1, b = 0$

The x-coordinate of the vertex is 0.
The axis of symmetry is the line $x = 0$.

To find the y-coordinate of the vertex, replace x by 0 and solve for y.

$y = x^2 - 2$
$\quad = 0^2 - 2 = -2$

The vertex is (0, −2).

Because a is positive, the parabola opens up.

Find a few ordered pairs and use symmetry to sketch the graph.

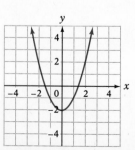

The graph of an equation of the form $x = ay^2 + by + c, a \neq 0$ is also a parabola. In this case, the parabola opens to the right when a is positive and opens to the left when a is negative.

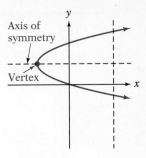

For a parabola of this form, the **y-coordinate of the vertex** is $-\dfrac{b}{2a}$. The **axis of symmetry** is the line $y = -\dfrac{b}{2a}$.

By the vertical-line test, the graph of a parabola of this form is not the graph of a function. The graph of $x = ay^2 + by + c$ is a relation.

⇒ Find the vertex and axis of symmetry of the parabola whose equation is $x = 2y^2 - 8y + 5$. Then sketch its graph.

Find the y-coordinate of the vertex and the axis of symmetry.

y-coordinate: $-\dfrac{b}{2a} = -\dfrac{-8}{2(2)} = 2$ • $a = 2, b = -8$

The y-coordinate of the vertex is 2.
The axis of symmetry is the line $y = 2$.

To find the x-coordinate of the vertex, replace y by 2 and solve for x.

$x = 2y^2 - 8y + 5$
$\quad = 2(2)^2 - 8(2) + 5 = -3$

The vertex is $(-3, 2)$.

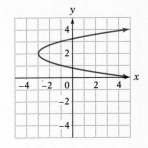

Because a is positive, the parabola opens to the right.

Find a few ordered pairs and use symmetry to sketch the graph.

⇒ Find the vertex and axis of symmetry of the parabola whose equation is $x = -2y^2 - 4y - 3$. Then sketch its graph.

Find the y-coordinate of the vertex and the axis of symmetry.

y-coordinate: $-\dfrac{b}{2a} = -\dfrac{-4}{2(-2)} = -1$ • $a = -2, b = -4$

The y-coordinate of the vertex is -1.
The axis of symmetry is the line $y = -1$.

To find the x-coordinate of the vertex, replace y by -1 and solve for x.

$x = -2y^2 - 4y - 3$
$\quad = -2(-1)^2 - 4(-1) - 3 = -1$

The vertex is $(-1, -1)$.

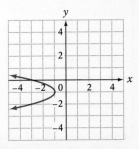

Because a is negative, the parabola opens to the left.

Find a few ordered pairs and use symmetry to sketch the graph.

Example 1

Find the vertex and axis of symmetry of the parabola whose equation is $y = x^2 - 4x + 3$. Then sketch its graph.

Solution

$-\dfrac{b}{2a} = -\dfrac{-4}{2(1)} = 2$

Axis of symmetry:
$\quad x = 2$

$y = 2^2 - 4(2) + 3$
$\quad = -1$

Vertex: $(2, -1)$

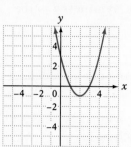

Example 2

Find the vertex and axis of symmetry of the parabola whose equation is $x = 2y^2 - 4y + 1$. Then sketch its graph.

Solution

$-\dfrac{b}{2a} = -\dfrac{-4}{2(2)} = 1$

Axis of symmetry:
$\quad y = 1$

$x = 2(1)^2 - 4(1) + 1$
$\quad = -1$

Vertex: $(-1, 1)$

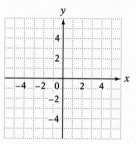

Example 3

Find the vertex and axis of symmetry of the parabola whose equation is $y = x^2 + 1$. Then sketch its graph.

Solution

$-\dfrac{b}{2a} = -\dfrac{0}{2(1)} = 0$

Axis of symmetry:
$\quad x = 0$

$y = 0^2 + 1 = 1$

Vertex: $(0, 1)$

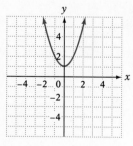

You Try It 1

Find the vertex and axis of symmetry of the parabola whose equation is $y = x^2 + 2x + 1$. Then sketch its graph.

Your solution

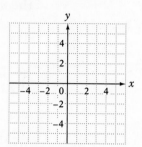

You Try It 2

Find the vertex and axis of symmetry of the parabola whose equation is $x = -y^2 - 2y + 2$. Then sketch its graph.

Your solution

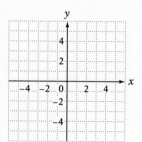

You Try It 3

Find the vertex and axis of symmetry of the parabola whose equation is $y = x^2 - 2x - 1$. Then sketch its graph.

Your solution

Solutions on pp. A55

Objective B   **To find the equation of a circle and to graph a circle** ....

A **circle** is a conic section formed by the intersection of a cone and a plane parallel to the base of the cone.

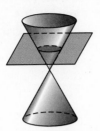

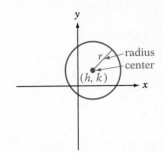

A **circle** can be defined as all the points (x, y) in the plane that are a fixed distance from a given point (h, k) called the **center.** The fixed distance is the **radius** of the circle.

The Standard Form of the Equation of a Circle

Let r be the radius of a circle and let (h, k) be the coordinates of the center of the circle. Then the equation of the circle is given by

$$(x - h)^2 + (y - k)^2 = r^2$$

➡ Sketch a graph of $(x - 1)^2 + (y + 2)^2 = 9$.

$(x - 1)^2 + [y - (-2)]^2 = 3^2$ • **Rewrite the equation in standard form.**

Center: $(1, -2)$ Radius: 3

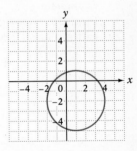

➡ Find the equation of the circle with radius 4 and center $(-1, 2)$. Then sketch its graph.

$(x - h)^2 + (y - k)^2 = r^2$ • **Use the standard form of the equation of a circle.**

$[x - (-1)]^2 + (y - 2)^2 = 4^2$ • **Replace r by 4, h by −1, and k by 2.**
$(x + 1)^2 + (y - 2)^2 = 16$

• **Sketch the graph by drawing a circle with center $(-1, 2)$ and radius 4.**

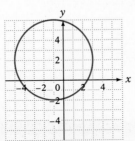

Applying the vertical-line test reveals that the graph of a circle is not the graph of a function. The graph of a circle is the graph of a relation.

Example 4

Sketch a graph of $(x + 2)^2 + (y - 1)^2 = 4$.

Solution

$$(x - h)^2 + (y - k)^2 = r^2$$
$$[x - (-2)]^2 + (y - 1)^2 = 2^2$$

Center: $(h, k) = (-2, 1)$

Radius: $r = 2$

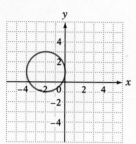

You Try It 4

Sketch a graph of $(x - 2)^2 + (y + 3)^2 = 9$.

Your solution

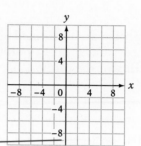

Example 5

Find the equation of the circle with radius 5 and center $(-1, 3)$. Then sketch its graph.

Solution

$$(x - h)^2 + (y - k)^2 = r^2$$
$$[x - (-1)]^2 + (y - 3)^2 = 5^2$$
$$(x + 1)^2 + (y - 3)^2 = 25$$

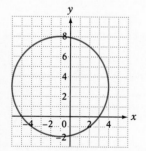

You Try It 5

Find the equation of the circle with radius 4 and center $(2, -3)$. Then sketch its graph.

Your solution

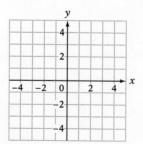

Solutions on p. A55

Objective C   **To graph an ellipse with center at the origin** ....

The orbits of the planets around the sun are "oval" shaped. This oval shape can be described as an **ellipse,** which is another of the conic sections.

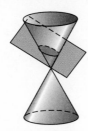

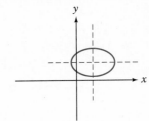

There are two **axes of symmetry** for an ellipse. The intersection of these two axes is the **center** of the ellipse.

An ellipse with center at the origin is shown at the right. Note that there are two x-intercepts and two y-intercepts.

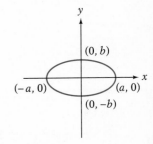

The Standard Form of the Equation of an Ellipse with Center at the Origin

The equation of an ellipse with center at the origin is $\dfrac{x^2}{a^2} + \dfrac{y^2}{b^2} = 1$.

The x-intercepts are $(a, 0)$ and $(-a, 0)$. The y-intercepts are $(0, b)$ and $(0, -b)$.

By finding the x- and y-intercepts for an ellipse and using the fact that the ellipse is "oval" shaped, we can sketch a graph of an ellipse.

➡ Sketch the graph of the ellipse whose equation is $\dfrac{x^2}{9} + \dfrac{y^2}{4} = 1$.

Comparing $\dfrac{x^2}{9} + \dfrac{y^2}{4} = 1$ with $\dfrac{x^2}{a^2} + \dfrac{y^2}{b^2} = 1$, we have $a^2 = 9$ and $b^2 = 4$.

Therefore, $a = 3$ and $b = 2$.

The x-intercepts are $(3, 0)$ and $(-3, 0)$.
The y-intercepts are $(0, 2)$ and $(0, -2)$.

Use the intercepts to sketch a graph of the ellipse.

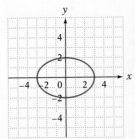

Using the vertical-line test, we find that the graph of an ellipse is not the graph of a function. The graph of an ellipse is the graph of a relation.

➡ Sketch a graph of the ellipse whose equation is $\dfrac{x^2}{16} + \dfrac{y^2}{16} = 1$.

The x-intercepts are $(4, 0)$ and $(-4, 0)$.

The y-intercepts are $(0, 4)$ and $(0, -4)$.

- $a^2 = 16$, $b^2 = 16$

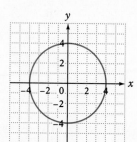

- Use the intercepts and symmetry to sketch the graph of the ellipse.

The graph in this example is the graph of a circle. A circle is a special case of an ellipse. It occurs when $a^2 = b^2$ in the equation $\dfrac{x^2}{a^2} + \dfrac{y^2}{b^2} = 1$.

Example 6

Sketch a graph of the ellipse whose equation is $\dfrac{x^2}{9} + \dfrac{y^2}{16} = 1$.

Solution

x-intercepts:
$(3, 0)$ and $(-3, 0)$

y-intercepts:
$(0, 4)$ and $(0, -4)$

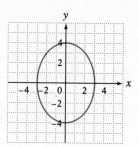

You Try It 6

Sketch a graph of the ellipse whose equation is $\dfrac{x^2}{4} + \dfrac{y^2}{25} = 1$.

Your solution

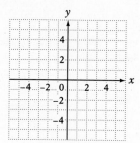

Example 7

Sketch a graph of the ellipse whose equation is $\dfrac{x^2}{16} + \dfrac{y^2}{12} = 1$.

Solution

x-intercepts:
$(4, 0)$ and $(-4, 0)$

y-intercepts:
$(0, 2\sqrt{3})$
and $(0, -2\sqrt{3})$
$(2\sqrt{3} \approx 3.5)$

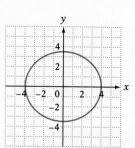

You Try It 7

Sketch a graph of the ellipse whose equation is $\dfrac{x^2}{18} + \dfrac{y^2}{9} = 1$.

Your solution

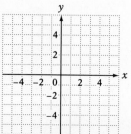

Solutions on p. A55

Objective D   *To graph a hyperbola with center at the origin* ....

A **hyperbola** is a conic section that is formed by the intersection of a cone and a plane perpendicular to the base of the cone.

 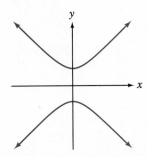

The hyperbola has two **vertices** and an **axis of symmetry** that passes through the vertices. The **center** of a hyperbola is the point halfway between the two vertices.

The graphs at the right show two possible graphs of a hyperbola with center at the origin.

In the first graph, an axis of symmetry is the *x*-axis and the vertices are *x*-intercepts.

In the second graph, an axis of symmetry is the *y*-axis and the vertices are *y*-intercepts.

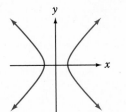 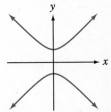

Note that in either case, the graph of a hyperbola is not the graph of a function. The graph of a hyperbola is the graph of a relation.

> **The Standard Form of the Equation of a Hyperbola with Center at the Origin**
>
> The equation of a hyperbola for which an axis of symmetry is the *x*-axis is $\dfrac{x^2}{a^2} - \dfrac{y^2}{b^2} = 1$. The vertices are $(a, 0)$ and $(-a, 0)$.
>
> The equation of a hyperbola for which an axis of symmetry is the *y*-axis is $\dfrac{y^2}{b^2} - \dfrac{x^2}{a^2} = 1$. The vertices are $(0, b)$ and $(0, -b)$.

To sketch a hyperbola, it is helpful to draw two lines that are "approached" by the hyperbola. These two lines are called **asymptotes**. As the hyperbola gets farther from the origin, the hyperbola "gets closer to" the asymptotes.

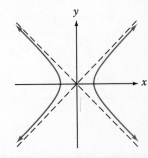

Because the asymptotes are straight lines, their equations are linear equations. The equations of the asymptotes for a hyperbola with center at the origin are $y = \dfrac{b}{a}x$ and $y = -\dfrac{b}{a}x$.

➡ Sketch a graph of the hyperbola whose equation is $\dfrac{y^2}{9} - \dfrac{x^2}{4} = 1$.

An axis of symmetry is the y-axis.

$b^2 = 9,\ a^2 = 4$

The vertices are $(0, 3)$ and $(0, -3)$.

The asymptotes are $y = \dfrac{3}{2}x$ and $y = -\dfrac{3}{2}x$.

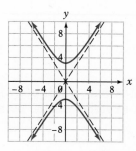

- **The vertices are (0, *b*) and (0, −*b*).**
- **The asymptotes are $y = \dfrac{b}{a}x$ and $y = -\dfrac{b}{a}x$.**
- **Sketch the asymptotes. Use symmetry and the fact that the hyperbola will approach the asymptotes to sketch its graph.**

Example 8
Sketch a graph of the hyperbola whose equation is $\dfrac{x^2}{16} - \dfrac{y^2}{4} = 1$.

Solution
Axis of symmetry:
x-axis

Vertices:
$(4, 0)$ and $(-4, 0)$

Asymptotes:
$y = \dfrac{1}{2}x$ and $y = -\dfrac{1}{2}x$

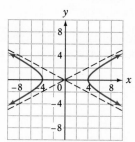

You Try It 8
Sketch a graph of the hyperbola whose equation is $\dfrac{x^2}{9} - \dfrac{y^2}{25} = 1$.

Your solution

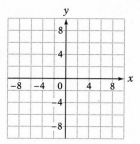

Example 9
Sketch a graph of the hyperbola whose equation is $\dfrac{y^2}{16} - \dfrac{x^2}{25} = 1$.

Solution
Axis of symmetry:
y-axis

Vertices:
$(0, 4)$ and $(0, -4)$

Asymptotes:
$y = \dfrac{4}{5}x$ and $y = -\dfrac{4}{5}x$

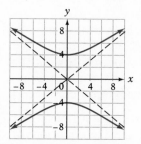

You Try It 9
Sketch a graph of the hyperbola whose equation is $\dfrac{y^2}{9} - \dfrac{x^2}{9} = 1$.

Your solution

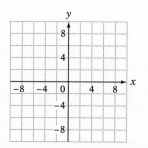

Solutions on p. A55

11.6 Exercises

Objective A

Find the vertex and axis of symmetry of the parabola given by the equation. Then sketch its graph.

1. $x = y^2 - 3y - 4$

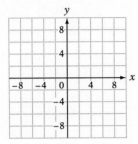

2. $y = x^2 - 2$

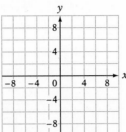

3. $y = x^2 + 2$

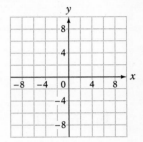

4. $x = -\frac{1}{2}y^2 + 4$

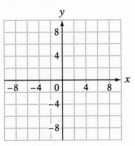

5. $x = -\frac{1}{4}y^2 - 1$

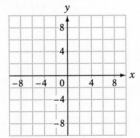

6. $x = \frac{1}{2}y^2 - y + 1$

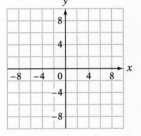

7. $x = -\frac{1}{2}y^2 + 2y - 3$

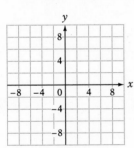

8. $y = -\frac{1}{2}x^2 + 2x + 6$

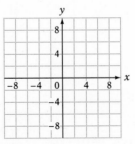

9. $y = \frac{1}{2}x^2 + x - 3$

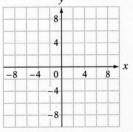

10. $x = y^2 + 6y + 5$

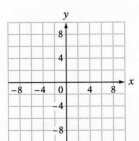

11. $x = y^2 - y - 6$

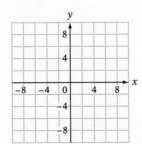

12. $x = -y^2 + 2y - 3$

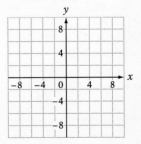

Objective B

Sketch a graph of the circle given by the equation.

13. $(x - 2)^2 + (y + 2)^2 = 9$ **14.** $(x + 2)^2 + (y - 3)^2 = 16$ **15.** $(x + 3)^2 + (y - 1)^2 = 25$

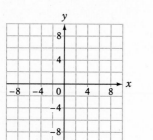 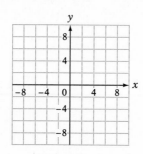 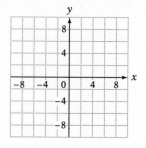

16. $(x - 2)^2 + (y + 3)^2 = 4$ **17.** $(x + 2)^2 + (y + 2)^2 = 4$ **18.** $(x - 1)^2 + (y - 2)^2 = 25$

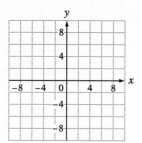 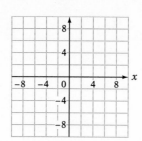 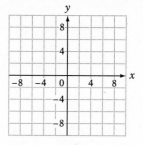

19. Find the equation of the circle with radius 2 and center $(2, -1)$. Then sketch its graph.

20. Find the equation of the circle with radius 3 and center $(-1, -2)$. Then sketch its graph.

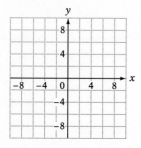 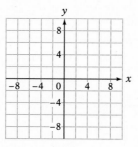

21. Find the equation of the circle with radius $\sqrt{5}$ and center $(-1, 1)$. Then sketch its graph.

22. Find the equation of the circle with radius $\sqrt{5}$ and center $(-2, 1)$. Then sketch its graph.

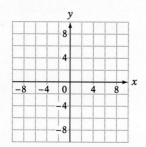 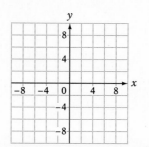

Objective C

Sketch a graph of the ellipse given by the equation.

23. $\dfrac{x^2}{4} + \dfrac{y^2}{9} = 1$

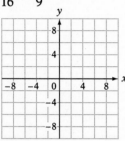

24. $\dfrac{x^2}{25} + \dfrac{y^2}{16} = 1$

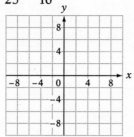

25. $\dfrac{x^2}{25} + \dfrac{y^2}{9} = 1$

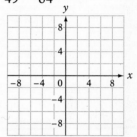

26. $\dfrac{x^2}{16} + \dfrac{y^2}{9} = 1$

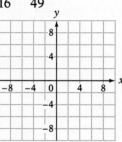

27. $\dfrac{x^2}{36} + \dfrac{y^2}{16} = 1$

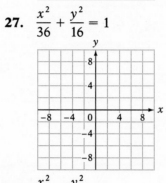

28. $\dfrac{x^2}{49} + \dfrac{y^2}{64} = 1$

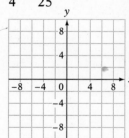

29. $\dfrac{x^2}{16} + \dfrac{y^2}{49} = 1$

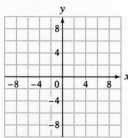

30. $\dfrac{x^2}{25} + \dfrac{y^2}{36} = 1$

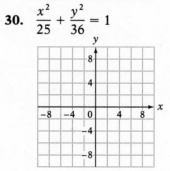

31. $\dfrac{x^2}{4} + \dfrac{y^2}{25} = 1$

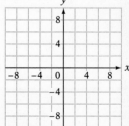

Objective D

Sketch a graph of the hyperbola given by the equation.

32. $\dfrac{x^2}{9} - \dfrac{y^2}{16} = 1$

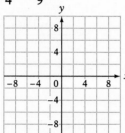

33. $\dfrac{x^2}{25} - \dfrac{y^2}{4} = 1$

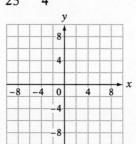

34. $\dfrac{y^2}{16} - \dfrac{x^2}{9} = 1$

Sketch a graph of the hyperbola given by the equation.

35. $\dfrac{y^2}{16} - \dfrac{x^2}{25} = 1$

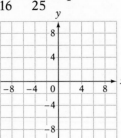

36. $\dfrac{x^2}{16} - \dfrac{y^2}{4} = 1$

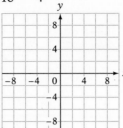

37. $\dfrac{x^2}{9} - \dfrac{y^2}{49} = 1$

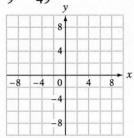

38. $\dfrac{y^2}{25} - \dfrac{x^2}{9} = 1$

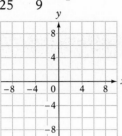

39. $\dfrac{y^2}{4} - \dfrac{x^2}{16} = 1$

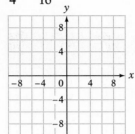

40. $\dfrac{x^2}{4} - \dfrac{y^2}{25} = 1$

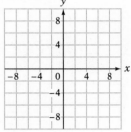

41. $\dfrac{x^2}{36} - \dfrac{y^2}{9} = 1$

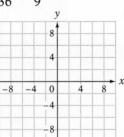

42. $\dfrac{y^2}{9} - \dfrac{x^2}{36} = 1$

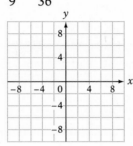

43. $\dfrac{y^2}{25} - \dfrac{x^2}{4} = 1$

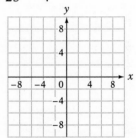

APPLYING THE CONCEPTS

44. Find the equation of a circle that has center (4, 0) and passes through the origin.

45. Find the equation of the circle with center at (3, 3) if the circle is tangent to the *x*-axis.

Write the equation in standard form. Identify the graph, and then graph the equation.

46. $4x^2 + 9y^2 = 36$

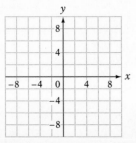

47. $16x^2 + 25y^2 = 400$

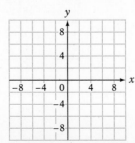

48. $9y^2 - 16x^2 = 144$

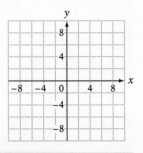

Projects in Mathematics

Exploring the Graphs of Fourth-Degree Polynomial Functions

A *turning point* in the graph of a function is a point where the graph changes direction. For instance, the graph of $f(x) = x^3 - 4x + 5$ shown at the right has two turning points. In general, the number of turning points that the graph of a polynomial function has depends on the degree of the polynomial. However, not all graphs of cubic functions have two turning points. For instance, the graph of $f(x) = x^3$ has no turning points.

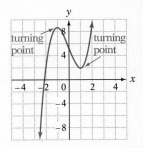

By using a graphing utility, graph the following and answer the questions. You may need to set Ymin and Ymax to fairly large numbers to accommodate these graphs.

1. Graph $f(x) = x^4 - 5$. How many turning points does the graph have?
2. Graph $f(x) = x^4 - 3x^3 - 7x^2 + x + 6$. How many turning points does the graph have?
3. Make up six more fourth-degree polynomial functions and graph each one. Record the number of turning points for each graph.
4. Make a conjecture about the number of turning points the graph of a fourth-degree polynomial will have. Graph a few more fourth-degree polynomials and determine whether your conjecture is correct for those graphs. If not, modify your conjecture and try a few more graphs.

Chapter Summary

Key Words

A function that can be written in the form $y = mx + b$ or $f(x) = mx + b$ is a *linear function*. The graph of a linear function is a straight line with slope m and y-intercept $(0, b)$.

A *quadratic function* is one that can be expressed by the equation $f(x) = ax^2 + bx + c, a \neq 0$. The graph of this function is a *parabola*.

The *inverse of a function* is the set of ordered pairs formed by reversing the coordinates of each ordered pair of the function.

The graph of a *conic section* can be represented by the intersection of a plane and a cone. The four conic sections are the *parabola, ellipse, hyperbola,* and *circle*.

A *circle* is the set of all points (x, y) in the plane that are a fixed distance from a given point (h, k) called the *center*. The fixed distance is the *radius* of the circle.

The *asymptotes* of a hyperbola are the two straight lines that are "approached" by the hyperbola. As the graph of the hyperbola gets farther from the origin, the hyperbola "gets closer to" the asymptotes.

Essential Rules *Vertical-Line Test* A graph defines a function if any vertical line intersects the graph at no more than one point.

Operations on Functions If f and g are functions and x is an element of the domain of each function, then

$$(f + g)(x) = f(x) + g(x) \qquad (f - g)(x) = f(x) - g(x)$$

$$(f \cdot g)(x) = f(x) \cdot g(x) \qquad \left(\frac{f}{g}\right)(x) = \frac{f(x)}{g(x)}, \; g(x) \neq 0$$

Composition of Two Functions The composition of two functions, $f \circ g$, is the function whose value at x is given by $(f \circ g)(x) = f[g(x)]$.

One-to-One Function A function f is a 1–1 function if, for any a and b in the domain of f, $f(a) = f(b)$ implies that $a = b$.

Horizontal-Line Test The graph of a function represents the graph of a 1–1 function if any horizontal line intersects the graph at no more than one point.

Condition for an Inverse Function A function f has an inverse function if and only if f is a 1–1 function.

Property of the Composition of Inverse Functions $f^{-1}[f(x)] = x$ and $f[f^{-1}(x)] = x$

Equation of a Parabola
$y = ax^2 + bx + c$

When $a > 0$, the parabola opens up.
When $a < 0$, the parabola opens down.

The x-coordinate of the vertex is $-\dfrac{b}{2a}$.

The axis of symmetry is the line $x = -\dfrac{b}{2a}$.

$x = ay^2 + by + c$

When $a > 0$, the parabola opens to the right.
When $a < 0$, the parabola opens to the left.

The y-coordinate of the vertex is $-\dfrac{b}{2a}$.

The axis of symmetry is the line $y = -\dfrac{b}{2a}$.

Equation of a Circle
$(x - h)^2 + (y - k)^2 = r^2$

The center is (h, k) and the radius is r.

Equation of an Ellipse
$\dfrac{x^2}{a^2} + \dfrac{y^2}{b^2} = 1$

The x-intercepts are $(a, 0)$ and $(-a, 0)$.
The y-intercepts are $(0, b)$ and $(0, -b)$.

Equation of a Hyperbola
$\dfrac{x^2}{a^2} - \dfrac{y^2}{b^2} = 1$

An axis of symmetry is the x-axis.
The vertices are $(a, 0)$ and $(-a, 0)$.

$\dfrac{y^2}{b^2} - \dfrac{x^2}{a^2} = 1$

An axis of symmetry is the y-axis.
The vertices are $(0, b)$ and $(0, -b)$.

The equations of the asymptotes are $y = \pm\dfrac{b}{a}x$.

Chapter Review Exercises

. . . .

1. Graph: $f(x) = \frac{1}{4}x - 2$

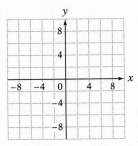

2. Sketch a graph of
$f(x) = \frac{1}{2}x^2 + x - 4.$

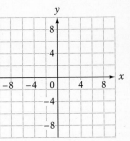

3. Find the x-intercepts of $y = 2x^2 - 3x + 4$.

4. Find the maximum value of the function $f(x) = -x^2 + 8x - 7$.

5. Graph $f(x) = -\sqrt{3 - x}$. State the domain and range.

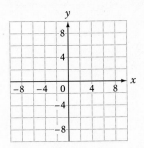

6. Graph $f(x) = \left|\frac{1}{2}x\right| - 2$. State the domain and range.

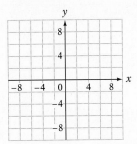

7. Graph $f(x) = x^3 - 3x + 2$. State the domain and range.

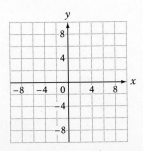

8. Determine whether the graph is the graph of a 1–1 function.

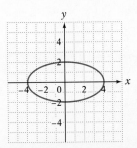

9. Given $f(x) = x^2 + 2x - 3$ and $g(x) = x^3 - 1$, find $(f - g)(2)$.

10. Given $f(x) = 4x - 5$ and $g(x) = x^2 + 3x + 4$, find $\left(\dfrac{f}{g}\right)(-2)$.

11. Given $f(x) = 4x + 2$ and $g(x) = \dfrac{x}{x + 1}$, find $f[g(3)]$.

12. Given $f(x) = 2x^2 - 7$ and $g(x) = x - 1$, find $f[g(x)]$.

13. Find the inverse of the function $\{(2, 6), (3, 5), (4, 4), (5, 3)\}$.

14. Find the inverse of the function $f(x) = \dfrac{1}{4}x - 4$.

15. Sketch a graph of $x = y^2 - y - 2$.

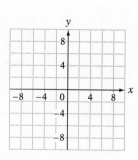

16. Sketch a graph of $(x - 2)^2 + (y + 1)^2 = 9$.

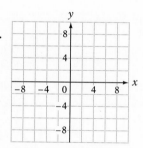

17. Sketch a graph of $\dfrac{x^2}{9} - \dfrac{y^2}{4} = 1$.

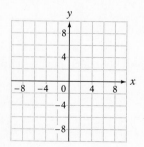

18. Sketch a graph of $\dfrac{x^2}{16} + \dfrac{y^2}{4} = 1$.

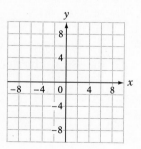

19. Find the equation of the circle with radius 4 and center $(-3, -3)$.

20. The perimeter of a rectangle is 200 cm. What dimensions would give the rectangle a maximum area? What is the maximum area?

Cumulative Review Exercises

. . . .

1. Graph the solution set of $\{x \,|\, x < 4\} \cap \{x \,|\, x > 2\}$.

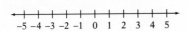
$$-5\;-4\;-3\;-2\;-1\;\;0\;\;1\;\;2\;\;3\;\;4\;\;5$$

2. Solve: $\dfrac{5x-2}{3} - \dfrac{1-x}{5} = \dfrac{x+4}{10}$

3. Evaluate the function $f(x) = -x^2 + 3x - 2$ at $x = -3$.

4. Find the equation of the line that contains the point $(2, -3)$ and has slope $-\dfrac{3}{2}$.

5. Find the equation of the line that contains the point $(4, -2)$ and is perpendicular to the line $y = -x + 5$.

6. Simplify: $\dfrac{ax - bx}{ax + ay - bx - by}$

7. Subtract: $\dfrac{x-4}{3x-2} - \dfrac{1+x}{3x^2 + x - 2}$

8. Solve: $\dfrac{6x}{2x-3} - \dfrac{1}{2x-3} = 7$

9. Simplify: $\left(\dfrac{12a^2b^2}{a^{-3}b^{-4}}\right)^{-1}\left(\dfrac{ab}{4^{-1}a^{-2}b^4}\right)^2$

10. Write $2\sqrt[4]{x^3}$ as an exponential expression.

11. Simplify: $\sqrt{18} - \sqrt{-25}$

12. Solve: $2x^2 + 2x - 3 = 0$

13. Solve: $x - \sqrt{2x-3} = 3$

14. Solve: $\dfrac{3x-2}{x+4} \le 1$

15. Find the maximum value of the function $f(x) = -2x^2 + 4x - 2$.

16. Find the inverse of the function $f(x) = 4x + 8$.

17. Graph the solution set of $5x + 2y \geq 10$.

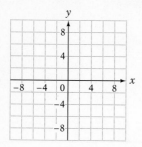

18. Graph: $x = y^2 - 2y + 3$

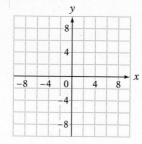

19. Graph: $\dfrac{x^2}{25} + \dfrac{y^2}{4} = 1$

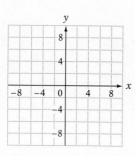

20. Graph: $\dfrac{y^2}{4} - \dfrac{x^2}{25} = 1$

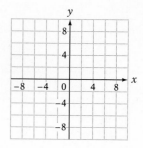

21. Tickets for a school play sold for $4.00 for each adult and $1.50 for each child. The total receipts for the 192 tickets sold were $493. Find the number of adult tickets sold.

22. A motorcycle travels 180 mi in the same amount of time that it takes a car to travel 144 mi. The rate of the motorcycle is 12 mph faster than the rate of the car. Find the rate of the motorcycle.

23. The rate of a river's current is 1.5 mph. A rowing crew can row 12 mi down this river and 12 mi back in 6 h. Find the rowing rate of the crew in calm water.

24. The speed (v) of a gear varies inversely as the number of teeth (t). If a gear that has 36 teeth makes 30 revolutions per minute, how many revolutions per minute will a gear that has 60 teeth make?

25. Find the maximum product of two numbers whose sum is 40.

Exponential and Logarithmic Functions

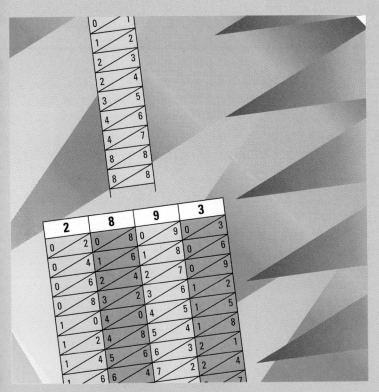

Napier Rods

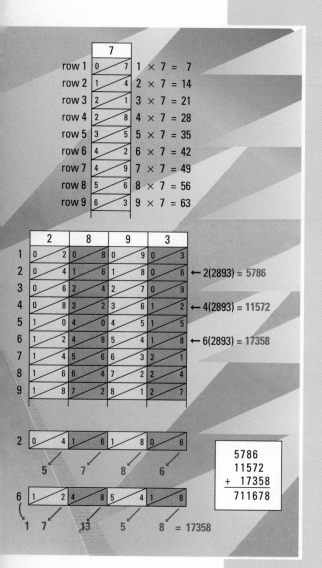

The labor involved in calculating the products of large numbers has led many people to devise ways to shorten the procedure. One such short-cut was first described in the early 1600s by John Napier and is based on Napier Rods.

Making a Napier Rod consists of placing a number and the first 9 multiples of that number on a rectangular piece of paper (Napier's Rod). This is done for the first 9 positive integers. It is necessary to have more than one rod for each number. The rod for 7 is shown at the left.

To illustrate how these rods were used to multiply, let's look at an example.

Multiply: $\begin{array}{r} 2893 \\ \times\ 246 \end{array}$

Place the rods for 2, 8, 9, and 3 next to one another. The products for 2, 4, and 6 are found by using the numbers along the 2nd, 4th, and 6th rows.

Each number is found by adding the digits diagonally downward on the diagonal and carrying to the next diagonal when necessary. The products for 2 and 6 are shown at the left.

Note that in the product for 6, the sum of the 3rd diagonal is 13, so carrying to the next diagonal is necessary.

The final product is then found by addition.

Before the invention of the electronic calculator, logarithms were used to ease the drudgery of lengthy calculations. John Napier is also credited with the invention of logarithms.

12.1

Exponential Functions

Objective A *To evaluate an exponential function* ....

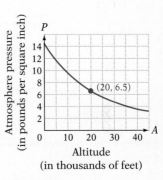

The growth of a $500 savings account that earns 5% annual interest compounded daily is shown at the right. In 14 years, the savings account contains approximately $1000, twice the initial amount. The growth of this savings account is an example of an exponential function.

V
Value of investment
2500
2000
1500 (14, 1000)
1000
500
0   5   15   25   t
Years

The pressure of the atmosphere at a certain height is shown in the graph at the right. This is another example of an exponential function. From the graph, we read that the air pressure is approximately 6.5 lb/in² at an altitude of 20,000 ft.

P
Atmosphere pressure (in pounds per square inch)
14
12
10
8   (20, 6.5)
6
4
2
0   10   20   30   40   A
Altitude (in thousands of feet)

CONSIDER THIS

It is important to distinguish between $F(x) = 2^x$ and $P(x) = x^2$. The first is an exponential function; the second is a polynomial function. Exponential functions are characterized by a constant base and a variable exponent. Polynomial functions have a variable base and a constant exponent.

> **Definition of an Exponential Function**
>
> The **exponential function** with base b is defined by
> $$f(x) = b^x$$
> where $b > 0$, $b \neq 1$, and x is any real number.

In the definition of an exponential function, b, the base, is required to be positive. If the base were a negative number, the value of the function would be a complex number for some values of x. For instance, the value of $f(x) = (-4)^x$ when $x = \frac{1}{2}$ is $f\left(\frac{1}{2}\right) = (-4)^{1/2} = \sqrt{-4} = 2i$. To avoid complex number values of a function, the base of the exponential function is a positive number.

➡ Evaluate $f(x) = 2^x$ at $x = 3$ and $x = -2$.

$f(3) = 2^3 = 8$ • Substitute 3 for x and simplify.

$f(-2) = 2^{-2} = \dfrac{1}{2^2} = \dfrac{1}{4}$ • Substitute −2 for x and simplify.

To evaluate an exponential expression for an irrational number such as $\sqrt{2}$, we obtain an approximation to the value of the function by approximating the irrational number. For instance, the value of $f(x) = 4^x$ when $x = \sqrt{2}$ can be approximated by using an approximation of $\sqrt{2}$.

$$f(\sqrt{2}) = 4^{\sqrt{2}} \approx 4^{1.4142} \approx 7.1029$$

Because $f(x) = b^x$ ($b > 0$, $b \neq 1$) can be evaluated at both rational and irrational numbers, the domain of f is all real numbers. And because $b^x > 0$ for all values of x, the range of f is the positive real numbers.

A frequently used base in applications of exponential functions is an irrational number designated by e. The number e is approximately 2.71828183. It is an irrational number, so it has a nonterminating, nonrepeating decimal representation.

POINT OF INTEREST

The natural exponential function is an extremely important function. It is used extensively in applied problems in virtually all disciplines from archeology to zoology. Leonard Euler (1707–1783) was the first to use the letter e as the base of the natural exponential function.

> **Natural Exponential Function**
>
> The function defined by $f(x) = e^x$ is called the **natural exponential function.**

The e^x key on a calculator can be used to evaluate the natural exponential function.

Example 1

Evaluate $f(x) = \left(\frac{1}{2}\right)^x$ at $x = 2$ and $x = -3$.

Solution

$$f(x) = \left(\frac{1}{2}\right)^x$$
$$f(2) = \left(\frac{1}{2}\right)^2 = \frac{1}{4}$$
$$f(-3) = \left(\frac{1}{2}\right)^{-3} = 2^3 = 8$$

You Try It 1

Evaluate $f(x) = \left(\frac{2}{3}\right)^x$ at $x = 3$ and $x = -2$.

Your solution

Example 2

Evaluate $f(x) = 2^{3x-1}$ at $x = 1$ and $x = -1$.

Solution

$$f(x) = 2^{3x-1}$$
$$f(1) = 2^{3(1)-1} = 2^2 = 4$$
$$f(-1) = 2^{3(-1)-1} = 2^{-4} = \frac{1}{2^4} = \frac{1}{16}$$

You Try It 2

Evaluate $f(x) = 2^{2x+1}$ at $x = 0$ and $x = -2$.

Your solution

Example 3

Evaluate $f(x) = e^{2x}$ at $x = 1$ and $x = -1$. Round to the nearest ten-thousandth.

Solution

$$f(x) = e^{2x}$$
$$f(1) = e^{2 \cdot 1} = e^2 \approx 7.3891$$
$$f(-1) = e^{2(-1)} = e^{-2} \approx 0.1353$$

You Try It 3

Evaluate $f(x) = e^{2x-1}$ at $x = 2$ and $x = -2$. Round to the nearest ten-thousandth.

Your solution

Solutions on p. A56

Objective B *To graph an exponential function* ....

Some properties of an exponential function can be seen in its graph.

➡ Graph $f(x) = 2^x$.

Think of this as the equation $y = 2^x$.

Choose values of x and find the corresponding values of y.

Graph the ordered pairs on a rectangular coordinate system.

Connect the points with a smooth curve.

x	$f(x) = y$
-2	$2^{-2} = \dfrac{1}{4}$
-1	$2^{-1} = \dfrac{1}{2}$
0	$2^0 = 1$
1	$2^1 = 2$
2	$2^2 = 4$
3	$2^3 = 8$

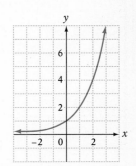

Note that a vertical line would intersect the graph at only one pont. Therefore, by the vertical-line test, the graph of $f(x) = 2^x$ is the graph of a function. Also note that a horizontal line would intersect the graph at only one point. Therefore, the graph of $f(x) = 2^x$ is the graph of a one-to-one function.

➡ Graph $f(x) = \left(\dfrac{1}{2}\right)^x$.

Think of this as the equation $y = \left(\dfrac{1}{2}\right)^x$.

Choose values of x and find the corresponding values of y.

Graph the ordered pairs on a rectangular coordinate system.

Connect the points with a smooth curve.

x	$f(x) = y$
-3	$\left(\dfrac{1}{2}\right)^{-3} = 8$
-2	$\left(\dfrac{1}{2}\right)^{-2} = 4$
-1	$\left(\dfrac{1}{2}\right)^{-1} = 2$
0	$\left(\dfrac{1}{2}\right)^{0} = 1$
1	$\left(\dfrac{1}{2}\right)^{1} = \dfrac{1}{2}$
2	$\left(\dfrac{1}{2}\right)^{2} = \dfrac{1}{4}$

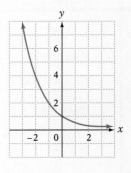

Applying the vertical-line and horizontal-line tests reveals that the graph of $f(x) = \left(\dfrac{1}{2}\right)^x$ is also the graph of a one-to-one function.

➡ Graph $f(x) = 2^{-x}$.

Think of this as the equation $y = 2^{-x}$.

Choose values of x and find the corresponding values of y.

Graph the ordered pairs on a rectangular coordinate system.

Connect the points with a smooth curve.

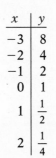

x	y
-3	8
-2	4
-1	2
0	1
1	$\dfrac{1}{2}$
2	$\dfrac{1}{4}$

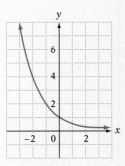

Note that because $2^{-x} = (2^{-1})^x = \left(\dfrac{1}{2}\right)^x$, the graphs of $f(x) = 2^{-x}$ and $f(x) = \left(\dfrac{1}{2}\right)^x$ are the same.

Example 4

Graph: $f(x) = 3^{\frac{1}{2}x - 1}$

Solution

x	y
-2	$\frac{1}{9}$
0	$\frac{1}{3}$
2	1
4	3

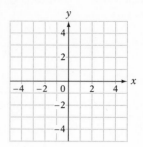

Example 5

Graph: $f(x) = 2^x - 1$

Solution

x	y
-2	$-\frac{3}{4}$
-1	$-\frac{1}{2}$
0	0
1	1
2	3
3	7

Example 6

Graph: $f(x) = \left(\frac{1}{3}\right)^x - 2$

Solution

x	y
-2	7
-1	1
0	-1
1	$-\frac{5}{3}$
2	$-\frac{17}{9}$

Example 7

Graph: $f(x) = 2^{-\frac{1}{2}x} - 1$

Solution

x	y
-6	7
-4	3
-2	1
0	0
2	$-\frac{1}{2}$
4	$-\frac{3}{4}$

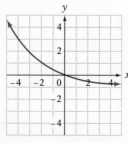

You Try It 4

Graph: $f(x) = 2^{-\frac{1}{2}x}$

Your solution

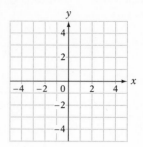

You Try It 5

Graph: $f(x) = 2^x + 1$

Your solution

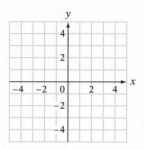

You Try It 6

Graph: $f(x) = 2^{-x} + 2$

Your solution

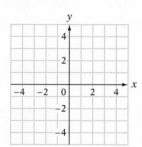

You Try It 7

Graph: $f(x) = \left(\frac{1}{2}\right)^{-\frac{1}{2}x} + 2$

Your solution

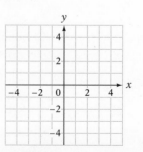

Solutions on p. A56

12.1 Exercises

· ·

Objective A

1. Given $f(x) = 3^x$, evaluate:
  a. $f(2)$　　**b.** $f(0)$　　**c.** $f(-2)$

2. Given $H(x) = 2^x$, evaluate:
  a. $H(-3)$　　**b.** $H(0)$　　**c.** $H(2)$

3. Given $g(x) = 2^{x+1}$, evaluate:
  a. $g(3)$　　**b.** $g(1)$　　**c.** $g(-3)$

4. Given $F(x) = 3^{x-2}$, evaluate:
  a. $F(-4)$　　**b.** $F(-1)$　　**c.** $F(0)$

5. Given $P(x) = \left(\dfrac{1}{2}\right)^{2x}$, evaluate:

  a. $P(0)$　　**b.** $P\left(\dfrac{3}{2}\right)$　　**c.** $P(-2)$

6. Given $R(t) = \left(\dfrac{1}{3}\right)^{3t}$, evaluate:

  a. $R\left(-\dfrac{1}{3}\right)$　　**b.** $R(1)$　　**c.** $R(-2)$

7. Given $G(x) = e^{\frac{x}{2}}$, evaluate the following. Round to the nearest ten-thousandth.
  a. $G(4)$　　**b.** $G(-2)$　　**c.** $G\left(\dfrac{1}{2}\right)$

8. Given $f(x) = e^{2x}$, evaluate the following. Round to the nearest ten-thousandth.
  a. $f(-2)$　　**b.** $f\left(-\dfrac{2}{3}\right)$　　**c.** $f(2)$

9. Given $H(r) = e^{-r+3}$, evaluate the following. Round to the nearest ten-thousandth.
  a. $H(-1)$　　**b.** $H(3)$　　**c.** $H(5)$

10. Given $P(t) = e^{-\frac{1}{2}t}$, evaluate the following. Round to the nearest ten-thousandth.
  a. $P(-3)$　　**b.** $P(4)$　　**c.** $P\left(\dfrac{1}{2}\right)$

11. Given $F(x) = 2^{x^2}$, evaluate the following. Round to the nearest ten-thousandth.
  a. $F(2)$　　**b.** $F(-2)$　　**c.** $F\left(\dfrac{3}{4}\right)$

12. Given $Q(x) = 2^{-x^2}$, evaluate:
  a. $Q(3)$　　**b.** $Q(-1)$　　**c.** $Q(-2)$

13. Given $f(x) = e^{-\frac{x^2}{2}}$, evaluate the following. Round to the nearest ten-thousandth.
  a. $f(-2)$　　**b.** $f(2)$　　**c.** $f(-3)$

14. Given $f(x) = e^{-2x} + 1$, evaluate the following. Round to the nearest ten-thousandth.
  a. $f(-1)$　　**b.** $f(3)$　　**c.** $f(-2)$

> *Objective B*

Graph.

15. $f(x) = 3^x$ **16.** $f(x) = 3^{-x}$ **17.** $f(x) = 2^{x+1}$ **18.** $f(x) = 2^{x-1}$

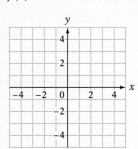 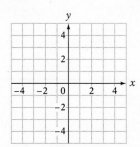 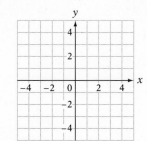 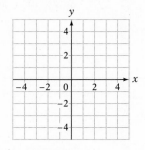

19. $f(x) = \left(\dfrac{1}{3}\right)^x$ **20.** $f(x) = \left(\dfrac{2}{3}\right)^x$ **21.** $f(x) = 2^{-x} + 1$ **22.** $f(x) = 2^x - 3$

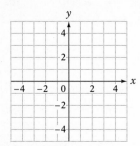 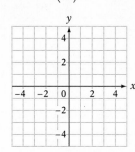 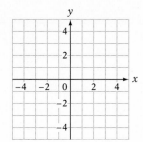 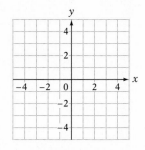

23. $f(x) = \left(\dfrac{1}{3}\right)^{-x}$ **24.** $f(x) = \left(\dfrac{3}{2}\right)^{-x}$ **25.** $f(x) = \left(\dfrac{1}{2}\right)^{-x} + 2$ **26.** $f(x) = \left(\dfrac{1}{2}\right)^x - 1$

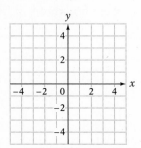 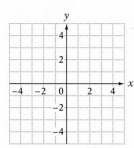 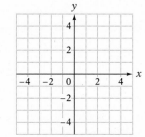 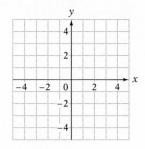

APPLYING THE CONCEPTS

27. Determine whether the following statements are always true, sometimes true, or never true for $f(x) = b^x$.
  a. If $b > 0$, $b \neq 1$, and $u < v$, then $f(u) < f(v)$.
  b. If $b > 0$ and $b \neq 1$, then $f(x) = b^x$ is a one-to-one function.
  c. If $b > 0$ and $b \neq 1$, then $f(x) = b^x$ is greater than zero.

28. Evaluate $\left(1 + \dfrac{1}{n}\right)^n$ for $n = 100$, 1000, 10,000, and 100,000 and compare the results with the value of e, the base of the natural exponential function. On the basis of your evaluation, complete the following sentence: As n increases, $\left(1 + \dfrac{1}{n}\right)^n$ becomes closer to _____ .

29. Evaluate $(1 + x)^{1/x}$ for $x = 0.1, 0.01, 0.001, 0.00001$, and 0.0000001 and compare the results with the value of e, the base of the natural exponential function. On the basis of your evaluation, complete the following sentence: As x gets closer to 0, $(1 + x)^{1/x}$ becomes closer to _____ .

Use a graphing calculator to graph the following.

30. $P(x) = (\sqrt{3})^x$

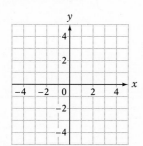

31. $F(x) = (\sqrt{5})^x$

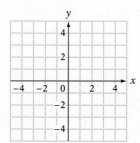

32. $Q(x) = (\sqrt{3})^{-x}$

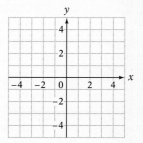

33. $A(x) = (\sqrt{2})^{-x}$

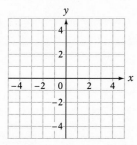

34. $\sinh(x) = \dfrac{e^x - e^{-x}}{2}$

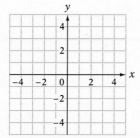

35. $\cosh(x) = \dfrac{e^x + e^{-x}}{2}$

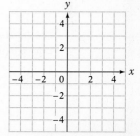

36. $N(x) = e^{-x^2/2}$

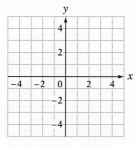

37. $M(x) = xe^{-x^2/2}$

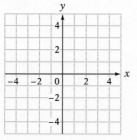

38. $f(x) = 2^x - x^2$

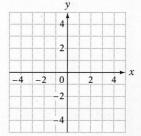

39. $h(x) = 3^x - x^3$

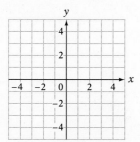

40. $f(x) = \pi^x$

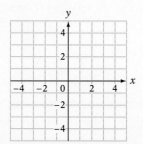

41. $g(x) = \pi^{-x}$

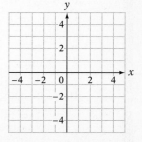

42. According to population studies, the population of China can be approximated by the equation $P(t) = 1.17(1.012)^t$, where $t = 0$ corresponds to 1993 and $P(t)$ is the population, in billions, of China in t years.

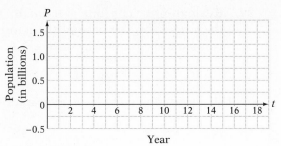

  a. Graph this equation. *Suggestion:* Use Xmin $= -0.5$, Xmax $= 18$, Xscl $= 1$, Ymin $= -0.5$, Ymax $= 1.5$, and Yscl $= 0.25$.

  b. The point whose approximate coordinates are (5, 1.242) is on the graph. Write a sentence that explains the meaning of these coordinates.

43. A model of the growth of the population of India can be approximated by the equation $P(t) = 0.883(1.017)^t$, where $t = 0$ corresponds to 1993 and $P(t)$ is the population, in billions, of India in t years.

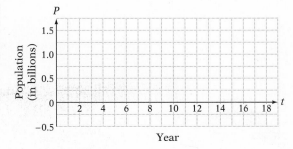

  a. Graph this equation. *Suggestion:* Use Xmin $= -0.5$, Xmax $= 18$, Xscl $= 1$, Ymin $= -0.5$, Ymax $= 1.5$, and Yscl $= 0.25$.

  b. The point whose approximate coordinates are (8, 1.01) is on the graph. Write a sentence that explains the meaning of these coordinates.

44. If air resistance is ignored, the speed v, in feet per second, of an object t seconds after it has been dropped is given by the equation $v = 32t$. This is true regardless of the mass of the object. However, if air resistance is considered, then the speed depends on the mass (and on other things). For a certain mass, the speed t seconds after it has been dropped is given by $v = 32(1 - e^{-t})$.

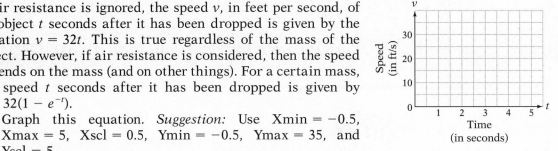

  a. Graph this equation. *Suggestion:* Use Xmin $= -0.5$, Xmax $= 5$, Xscl $= 0.5$, Ymin $= -0.5$, Ymax $= 35$, and Yscl $= 5$.

  b. The point whose approximate coordinates are (2, 27.7) is on the graph. Write a sentence that explains the meaning of these coordinates.

45. If air resistance is ignored, the speed v, in feet per second, of an object t seconds after it has been dropped is given by the equation $v = 32t$. This is true regardless of the mass of the object. However, if air resistance is considered, then the speed depends on the mass (and on other things). For a certain mass, the speed t seconds after it has been dropped is given by $v = 64(1 - e^{-t/2})$.

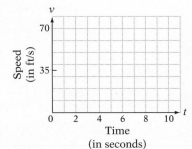

  a. Graph this equation. *Suggestion:* Use Xmin $= -0.5$, Xmax $= 10$, Xscl $= 1$, Ymin $= -0.5$, Ymax $= 70$, and Yscl $= 10$.

  b. The point whose approximate coordinates are (4, 55.3) is on the graph. Write a sentence that explains the meaning of these coordinates.

12.2 Introduction to Logarithms

Objective A *To write equivalent exponential and logarithmic equations*

CONSIDER THIS

Defining a logarithm as the inverse of the exponential function is similar to defining square root as the inverse of square.

"If $49 = x^2$, what is x?" The answer is the square root of 49.

Now consider "If $8 = 2^x$, what is x?" The answer is the logarithm, base 2, of 8.

Because the exponential function is a 1–1 function, it has an inverse function, which is called a *logarithm*. A logarithm is used to answer a question similar to the following one: "If $16 = 2^y$, what is the value of y?" Because $16 = 2^4$, the logarithm, base 2, of 16 is 4. This is written $\log_2 16 = 4$. Note that a logarithm is an exponent that solves a certain equation.

Definition of Logarithm

For $b > 0$, $b \neq 1$, $y = \log_b x$ is equivalent to $x = b^y$.

Read $\log_b x$ as "the logarithm of x, base b" or "log base b of x."

The table at the right shows equivalent statements written in both exponential and logarithmic form.

Exponential Form	Logarithmic Form
$2^4 = 16$	$\log_2 16 = 4$
$\left(\dfrac{2}{3}\right)^2 = \dfrac{4}{9}$	$\log_{2/3}\left(\dfrac{4}{9}\right) = 2$
$10^{-1} = 0.1$	$\log_{10}(0.1) = -1$

➡ Write $\log_3 81 = 4$ in exponential form.

$\log_3 81 = 4$ is equivalent to $3^4 = 81$.

➡ Write $10^{-2} = 0.01$ in logarithmic form.

$10^{-2} = 0.01$ is equivalent to $\log_{10}(0.01) = -2$.

The 1–1 property of exponential functions can be used to evaluate some logarithms.

1–1 Property of Exponential Functions

For $b > 0$, $b \neq 1$, if $b^u = b^v$, then $u = v$.

➡ Evaluate $\log_2 8$.

$\log_2 8 = x$	• Write an equation.
$8 = 2^x$	• Write the equation in its equivalent exponential form.
$2^3 = 2^x$	• Write 8 as 2^3.
$3 = x$	• Use the 1–1 Property of Exponential Functions.
$\log_2 8 = 3$	

⇒ Solve $\log_4 x = -2$ for x.

$$\log_4 x = -2$$

$\quad\quad 4^{-2} = x$ • Write the equation in its equivalent exponential form.

$\quad\quad \dfrac{1}{16} = x$ • Simplify.

The solution is $\dfrac{1}{16}$.

In this example, $\dfrac{1}{16}$ is called the *antilogarithm*, base 4, of -2.

Definition of Antilogarithm

If $\log_b M = N$, the **antilogarithm**, base b, of N is M. In exponential form, $M = b^N$.

The antilogarithm of a number can be determined by rewriting $\log_b M = N$ in exponential form, $b^N = M$. Thus M, the antilogarithm of N, is b^N. For instance, if $\log_5 x = 3$, then x (the antilogarithm, base 5, of 3) is $5^3 = x = 125$.

Logarithms, base 10, are called **common logarithms.** We usually omit the base, 10, when writing the common logarithm of a number. Therefore, $\log_{10} x$ is written $\log x$. To find the common logarithm of most numbers, a calculator is necessary. Because the logarithms of most numbers are irrational numbers, the value in the display of a calculator is an approximation of the number. Using a calculator reveals that

$$\log 384 \approx \overbrace{2.\underbrace{5843312}_{\text{Characteristic}}}^{\text{Mantissa}}$$

The decimal part of a *common logarithm* is called the **mantissa**; the integer part is called the **characteristic.**

When e (the base of the natural exponential function) is used as a base of a logarithm, the logarithm is referred to as the **natural logarithm** and is abbreviated $\ln x$. This is read "el en x." Using a calculator, we learn that

$$\ln 23 \approx 3.135494216$$

The integer and decimal parts of a natural logarithm do not have special names.

Example 1   Evaluate: $\log_3\left(\dfrac{1}{9}\right)$

Solution $\log_3\left(\dfrac{1}{9}\right) = x$

$$\dfrac{1}{9} = 3^x$$
$$3^{-2} = 3^x$$
$$-2 = x$$

$$\log_3\left(\dfrac{1}{9}\right) = -2$$

You Try It 1   Evaluate: $\log_4 64$

Your solution

Solution on p. A56

Example 2	Solve for x: $\log_5 x = 2$	You Try It 2	Solve for x: $\log_2 x = -4$
Solution	$\log_5 x = 2$ $5^2 = x$ $25 = x$ The solution is 25.	**Your solution**	
Example 3	Solve $\log x = -1.5$ for x. Round to the nearest ten-thousandth.	You Try It 3	Solve $\ln x = 3$ for x. Round to the nearest ten-thousandth.
Solution	$\log x = -1.5$ $10^{-1.5} = x$ $0.0316 \approx x$ • Use a calculator.	**Your solution**	

Solutions on pp. A56–A57

Objective B   *To use the Properties of Logarithms* ....

Because a logarithm is a special kind of exponent, the Properties of Logarithms are similar to the Properties of Exponents.

The property of logarithms that states that the logarithm of the product of two numbers equals the sum of the logarithms of the two numbers is similar to the property of exponents that states that to multiply two exponential expressions with the same base, we add the exponents.

CONSIDER THIS

Pay close attention to this theorem. Note, for instance, that this theorem states that

$\log_3(4 \cdot p) = \log_3 4 + \log_3 p$.

It also states that

$\log_5 9 + \log_5 z = \log_5(9z)$.

It does *not* state any relationship that involves $\log_b(x + y)$. **This expression cannot be simplified.**

> **The Logarithm Property of the Product of Two Numbers**
>
> For any positive real numbers x, y, and b, $b \neq 1$,
> $\log_b(xy) = \log_b x + \log_b y$.

A proof of this property can be found in the Appendix.

➡ Write $\log_b(6z)$ in expanded form.

$\log_b(6z) = \log_b 6 + \log_b z$ • Use the Logarithm Property of Products.

➡ Write $\log_b 12 + \log_b r$ as a single logarithm.

$\log_b 12 + \log_b r = \log_b(12r)$ • Use the Logarithm Property of Products.

The Logarithm Property of Products can be extended to include the logarithm of the product of more than two factors. For instance,

$$\log_b(xyz) = \log_b x + \log_b y + \log_b z$$
$$\log_b(7rt) = \log_b 7 + \log_b r + \log_b t$$

A second property of logarithms involves the logarithm of the quotient of two numbers. This property of logarithms is also based on the fact that a logarithm is an exponent and that to divide two exponential expressions with the same base, we subtract the exponents.

CONSIDER THIS

This theorem is used to rewrite expressions such as

$\log_5\left(\dfrac{m}{8}\right) = \log_5 m - \log_5 8$

It does *not* state any relationship that involves $\dfrac{\log_b x}{\log_b y}$. **This expression cannot be simplified.**

> **The Logarithm Property of the Quotient of Two Numbers**
>
> For any positive real numbers x, y, and b, $b \neq 1$,
>
> $\log_b \dfrac{x}{y} = \log_b x - \log_b y.$

A proof of this property can be found in the Appendix.

➡ Write $\log_b \dfrac{p}{8}$ in expanded form.

$\log_b \dfrac{p}{8} = \log_b p - \log_b 8$ ● **Use the Logarithm Property of Quotients.**

➡ Write $\log_b y - \log_b v$ as a single logarithm.

$\log_b y - \log_b v = \log_b \dfrac{y}{v}$ ● **Use the Logarithm Property of Quotients.**

A third property of logarithms is used to simplify powers of a number.

POINT OF INTEREST

Logarithms were developed independently by Jobst Burgi (1552–1632) and John Napier (1550–1617) as a means of simplifying the calculations of astronomers. The idea was to devise a method by which two numbers could be multiplied by performing additions. Napier is usually given credit for logarithms because he published his result first.

In Napier's original work, the logarithm of 10,000,000 was 0. After this work was published, Napier, in discussions with Henry Briggs (1561–1631), decided that tables of logarithms would be easier to use if the logarithm of 1 was 0. Napier died before new tables could be determined, and Briggs took on the task. His table consisted of logarithms accurate to 30 decimal places, all accomplished without a calculator!

The logarithms Briggs calculated are the common logarithms mentioned earlier.

> **The Logarithm Property of the Power of a Number**
>
> For any positive real numbers x and b, $b \neq 1$, and for any real number r, $\log_b x^r = r \log_b x.$

A proof of this property can be found in the Appendix.

➡ Rewrite $\log_b x^3$ in terms of $\log_b x$.

$\log_b x^3 = 3 \log_b x$ ● **Use the Logarithm Property of Powers.**

➡ Rewrite $\dfrac{2}{3} \log_b x$ with a coefficient of 1.

$\dfrac{2}{3} \log_b x = \log_b x^{2/3}$ ● **Use the Logarithm Property of Powers.**

The following table summarizes the properties of logarithms that we have discussed, along with two other properties.

> **Summary of the Properties of Logarithms**
>
> Let x, y, and b be positive real numbers with $b \neq 1$. Then
>
> | | |
> |---|---|
> | Product Property | $\log_b(x \cdot y) = \log_b x + \log_b y$ |
> | Quotient Property | $\log_b\left(\dfrac{x}{y}\right) = \log_b x - \log_b y$ |
> | Power Property | $\log_b x^r = r \log_b x$ |
> | Logarithm of One | $\log_b 1 = 0$ |
> | 1–1 Property | If $\log_b x = \log_b y$, then $x = y$. |

➡ Write $\log_b \dfrac{xy}{z}$ in expanded form.

$$\log_b \dfrac{xy}{z} = \log_b(xy) - \log_b z$$ • **Use the Logarithm Property of Quotients.**

$$= \log_b x + \log_b y - \log_b z$$ • **Use the Logarithm Property of Products.**

➡ Write $\log_b \dfrac{x^2}{y^3}$ in expanded form.

$$\log_b \dfrac{x^2}{y^3} = \log_b x^2 - \log_b y^3$$ • **Use the Logarithm Property of Quotients.**

$$= 2 \log_b x - 3 \log_b y$$ • **Use the Logarithm Property of Powers.**

➡ Write $2 \log_b x + 4 \log_b y$ as a single logarithm with a coefficient of 1.

$$2 \log_b x + 4 \log_b y = \log_b x^2 + \log_b y^4$$ • **Use the Logarithm Property of Powers.**

$$= \log_b x^2 y^4$$ • **Use the Logarithm Property of Products.**

Example 4
Write $\log \sqrt{x^3 y}$ in expanded form.

Solution

$$\log \sqrt{x^3 y} = \log(x^3 y)^{1/2} = \dfrac{1}{2} \log(x^3 y)$$

$$= \dfrac{1}{2}(\log x^3 + \log y)$$

$$= \dfrac{1}{2}(3 \log x + \log y)$$

$$= \dfrac{3}{2} \log x + \dfrac{1}{2} \log y$$

You Try It 4
Write $\log_8 \sqrt[3]{xy^2}$ in expanded form.

Your solution

Example 5
Write $\dfrac{1}{2}(\log_3 x - 3 \log_3 y + \log_3 z)$ as a single logarithm with a coefficient of 1.

Solution

$$\dfrac{1}{2}(\log_3 x - 3 \log_3 y + \log_3 z)$$

$$= \dfrac{1}{2}(\log_3 x - \log_3 y^3 + \log_3 z)$$

$$= \dfrac{1}{2}\left(\log_3 \dfrac{x}{y^3} + \log_3 z\right)$$

$$= \dfrac{1}{2}\left(\log_3 \dfrac{xz}{y^3}\right) = \log_3 \left(\dfrac{xz}{y^3}\right)^{1/2} = \log_3 \sqrt{\dfrac{xz}{y^3}}$$

You Try It 5
Write $\dfrac{1}{3}(\log_4 x - 2 \log_4 y + \log_4 z)$ as a single logarithm with a coefficient of 1.

Your solution

Example 6
Find $\log_4 1$.

Solution
$\log_4 1 = 0$ • **Logarithm of One**

You Try It 6
Find $\log_9 1$.

Your solution

Solutions on p. A57

Objective C *To use the change-of-base formula* ....

Although only common logarithms and natural logarithms are programmed into a calculator, the logarithms for other positive bases can be found.

➡ Evaluate $\log_5 22$.

$$\log_5 22 = x$$
$$5^x = 22$$
$$\log 5^x = \log 22$$
$$x \log 5 = \log 22$$
$$x = \frac{\log 22}{\log 5}$$
$$x \approx 1.9206$$
$$\log_5 22 \approx 1.9206$$

- Write an equation.
- Write the equation in its equivalent exponential form.
- Apply the common logarithm to each side of the equation.
- Use the Power Property of Logarithms.
- Divide each side by log 5. This is the exact answer.
- This is an approximate answer.

In the third step above, the natural logarithm, instead of the common logarithm, could have been applied to each side of the equation. The same result would have been obtained.

Using a procedure similar to the one for $\log_5 22$, a formula for changing bases can be derived.

Change-of-Base Formula

$$\log_a N = \frac{\log_b N}{\log_b a}$$

➡ Evaluate $\log_2 14$.

$$\log_2 14 = \frac{\log 14}{\log 2}$$
$$\approx 3.8074$$

- Use the Change-of-Base Formula with $N = 14$, $a = 2$, $b = 10$.

For the last example, common logarithms were used. Here is the same example using natural logarithms. Note that the answers are the same.

$$\log_2 14 = \frac{\ln 14}{\ln 2} \approx 3.8074$$

Example 7
Evaluate $\log_2 90.813$ by using common logarithms.

Solution

$$\log_2 90.813 = \frac{\log 90.813}{\log 2} \approx 6.5048$$

Example 8
Evaluate $\log_8 0.137$ by using natural logarithms.

Solution

$$\log_8 0.137 = \frac{\ln 0.137}{\ln 8} \approx -0.9559$$

You Try It 7
Evaluate $\log_7 6.45$ by using common logarithms.

Your solution

You Try It 8
Evaluate $\log_3 0.834$ by using natural logarithms.

Your solution

Solutions on p. A57

12.2 Exercises

Objective A

Write the exponential equation in logarithmic form.

1. $5^2 = 25$

2. $10^3 = 1000$

3. $4^{-2} = \dfrac{1}{16}$

4. $3^{-3} = \dfrac{1}{27}$

5. $10^y = x$

6. $e^y = x$

7. $a^x = w$

8. $b^y = c$

Write the logarithmic equation in exponential form.

9. $\log_3 9 = 2$

10. $\log_2 32 = 5$

11. $\log 0.01 = -2$

12. $\log_5 \dfrac{1}{5} = -1$

13. $\ln x = y$

14. $\log x = y$

15. $\log_b u = v$

16. $\log_c x = y$

Evaluate.

17. $\log_3 81$

18. $\log_7 49$

19. $\log_2 128$

20. $\log_5 125$

21. $\log 100$

22. $\log 0.001$

23. $\ln e^3$

24. $\ln e^2$

25. $\log_8 1$

26. $\log_3 243$

27. $\log_5 625$

28. $\log_2 64$

Solve for x.

29. $\log_3 x = 2$

30. $\log_5 x = 1$

31. $\log_4 x = 3$

32. $\log_2 x = 6$

33. $\log_7 x = -1$

34. $\log_8 x = -2$

35. $\log_6 x = 0$

36. $\log_4 x = 0$

Solve for x. Round to the nearest hundredth.

37. $\log x = 2.5$

38. $\log x = 3.2$

39. $\log x = -1.75$

40. $\log x = -2.1$

41. $\ln x = 2$

42. $\ln x = 1.4$

43. $\ln x = -\dfrac{1}{2}$

44. $\ln x = -1.7$

Objective B

Express as a single logarithm with a coefficient of 1.

45. $\log_3 x^3 + \log_3 y^2$ **46.** $\log_7 x + \log_7 z^2$ **47.** $\ln x^4 - \ln y^2$

48. $\ln x^2 - \ln y$ **49.** $3 \log_7 x$ **50.** $4 \log_8 y$

51. $3 \ln x + 4 \ln y$ **52.** $2 \ln x - 5 \ln y$ **53.** $2(\log_4 x + \log_4 y)$

54. $3(\log_5 r + \log_5 t)$ **55.** $\dfrac{1}{2}(\log_6 x - \log_6 y)$ **56.** $\dfrac{1}{3}(\log_8 x - \log_8 y)$

57. $2 \log_3 x - \log_3 y + 2 \log_3 z$ **58.** $4 \log_5 r - 3 \log_5 s + \log_5 t$

59. $\ln x - (2 \ln y + \ln z)$ **60.** $2 \log_b x - 3(\log_b y + \log_b z)$

61. $2(\log_4 s - 2 \log_4 t + \log_4 r)$ **62.** $3(\log_9 x + 2 \log_9 y - 2 \log_9 z)$

63. $\ln x - 2(\ln y + \ln z)$ **64.** $\ln t - 3(\ln u + \ln v)$

65. $3 \log_2 t - 2(\log_2 r - \log_2 v)$ **66.** $2 \log_{10} x - 3(\log_{10} y - \log_{10} z)$

67. $\dfrac{1}{2}(3 \log_4 x - 2 \log_4 y + \log_4 z)$ **68.** $\dfrac{1}{3}(4 \log_5 t - 5 \log_5 u - 7 \log_5 v)$

69. $\dfrac{1}{2}(\ln x - 3 \ln y)$ **70.** $\dfrac{1}{3} \ln a + \dfrac{2}{3} \ln b$

71. $\dfrac{1}{2} \log_2 x - \dfrac{2}{3} \log_2 y + \dfrac{1}{2} \log_2 z$ **72.** $\dfrac{2}{3} \log_3 x + \dfrac{1}{3} \log_3 y - \dfrac{1}{2} \log_3 z$

Write the logarithm in expanded form.

73. $\log_8(xz)$

74. $\log_7(rt)$

75. $\log_3 x^5$

76. $\log_2 y^7$

77. $\log_b\left(\dfrac{r}{s}\right)$

78. $\log_c\left(\dfrac{z}{4}\right)$

79. $\log_3(x^2 y^6)$

80. $\log_4(t^4 u^2)$

81. $\log_7\left(\dfrac{u^3}{v^4}\right)$

82. $\log\left(\dfrac{s^5}{t^2}\right)$

83. $\log_2(rs)^2$

84. $\log_3(x^2 y)^3$

85. $\ln(x^2 yz)$

86. $\ln(xy^2 z^3)$

87. $\log_5\left(\dfrac{xy^2}{z^4}\right)$

88. $\log_b\left(\dfrac{r^2 s}{t^3}\right)$

89. $\log_8\left(\dfrac{x^2}{yz^2}\right)$

90. $\log_9\left(\dfrac{x}{y^2 z^3}\right)$

91. $\log_4 \sqrt{x^3 y}$

92. $\log_3 \sqrt{x^5 y^3}$

93. $\log_7 \sqrt{\dfrac{x^3}{y}}$

94. $\log_b \sqrt[3]{\dfrac{r^2}{t}}$

95. $\ln\left(x \sqrt{\dfrac{y}{z}}\right)$

96. $\ln\left(y \sqrt[3]{\dfrac{r}{s}}\right)$

97. $\log_3 \dfrac{t}{\sqrt{x}}$

98. $\log_4 \dfrac{x}{\sqrt{y^2 z}}$

99. $\log_7\left(\dfrac{\sqrt{uv}}{x}\right)$

100. $\log_2\left(\dfrac{\sqrt{xy}}{z^2}\right)$

Objective C

Evaluate. Round to the nearest ten-thousandth.

101. $\log_{10} 7$

102. $\log_{10} 9$

103. $\log_{10}\left(\dfrac{3}{5}\right)$

104. $\log_{10}\left(\dfrac{13}{3}\right)$

105. $\ln 4$

106. $\ln 6$

107. $\ln\left(\dfrac{17}{6}\right)$

108. $\ln\left(\dfrac{13}{17}\right)$

109. $\log_8 6$

110. $\log_4 8$

111. $\log_5 30$

112. $\log_6 28$

113. $\log_3(0.5)$

114. $\log_5(0.6)$

115. $\log_7(1.7)$

116. $\log_6(3.2)$

117. $\log_5 15$

118. $\log_3 25$

119. $\log_{12} 120$

120. $\log_9 90$

121. $\log_4 2.55$

122. $\log_8 6.42$

123. $\log_5 67$

124. $\log_8 35$

APPLYING THE CONCEPTS

125. For each of the following, answer True or False. Assume all variables represent positive numbers.

a. $\log_3(-9) = -2$

b. $x^y = z$ and $\log_x z = y$ are equivalent equations.

c. $\log(x^{-1}) = \dfrac{1}{\log x}$

d. $\log(x + y) = \log x + \log y$

e. $\log(x \cdot y) = \log x \cdot \log y$

f. $\log\left(\dfrac{x}{y}\right) = \log x - \log y$

g. $\dfrac{\log x}{\log y} = \dfrac{x}{y}$

h. If $\log x = \log y$, then $x = y$.

126. Complete each statement using the equation $\log_a b = c$.

a. $a^c =$ _____

b. $\text{antilog}_a(\log_a b) =$ _____

12.3

Graphs of Logarithmic Functions

Objective A *To graph a logarithmic function*

The graph of a logarithmic function can be drawn by using the relationship between the exponential and logarithmic functions.

➡ Graph: $f(x) = \log_2 x$

POINT OF INTEREST
Although logarithms were originally developed to assist with computations, logarithmic functions have a much broader use today. These functions occur in geology, acoustics, chemistry, and economics, for example.

Think of this as the equation $y = \log_2 x$.

$$f(x) = \log_2 x$$
$$y = \log_2 x$$

Write the equivalent exponential equation.

$$x = 2^y$$

Because the equation is solved for x in terms of y, it is easier to choose values of y and find the corresponding values of x. The results can be recorded in a table.

Graph the ordered pairs on a rectangular coordinate system.

Connect the points with a smooth curve.

x	y
$\frac{1}{4}$	-2
$\frac{1}{2}$	-1
1	0
2	1
4	2

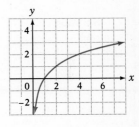

Applying the vertical-line and horizontal-line tests reveals that $f(x) = \log_2 x$ is a 1–1 function.

➡ Graph: $f(x) = \log_2(x) + 1$

Think of $f(x) = \log_2(x) + 1$ as the equation $y = \log_2(x) + 1$.

$$y = \log_2(x) + 1$$
$$y - 1 = \log_2(x)$$ • Solve for $\log_2(x)$.
$$2^{y-1} = x$$ • Write the equivalent exponential equation.

Choose values of y and find the corresponding values of x.

Graph the ordered pairs on a rectangular coordinate system.

Connect the points with a smooth curve.

x	y
$\frac{1}{4}$	-1
$\frac{1}{2}$	0
1	1
2	2
4	3

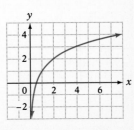

Example 1
Graph: $f(x) = \log_3 x$

Solution

$f(x) = \log_3 x$
$\quad y = \log_3 x$
$\quad 3^y = x$ • Write the equivalent exponential equation.

x	y
$\frac{1}{9}$	-2
$\frac{1}{3}$	-1
1	0
3	1

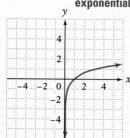

You Try It 1
Graph: $f(x) = \log_2(x - 1)$

Your solution

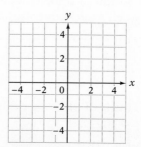

Example 2
Graph: $f(x) = 2 \log_3 x$

Solution

$f(x) = 2 \log_3 x$
$\quad y = 2 \log_3 x$
$\quad \dfrac{y}{2} = \log_3 x$
$\quad 3^{y/2} = x$ • Write the equivalent exponential equation.

x	y
$\frac{1}{9}$	-4
$\frac{1}{3}$	-2
1	0
3	2

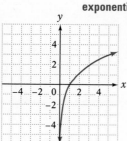

You Try It 2
Graph: $f(x) = \log_3(2x)$

Your solution

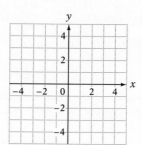

Example 3
Graph: $f(x) = -\log_2(x - 2)$

Solution

$\quad f(x) = -\log_2(x - 2)$
$\quad\quad y = -\log_2(x - 2)$
$\quad -y = \log_2(x - 2)$
$\quad 2^{-y} = x - 2$ • Write the equivalent
$2^{-y} + 2 = x$ exponential equation.

x	y
6	-2
4	-1
3	0
$\frac{5}{2}$	1
$\frac{9}{4}$	2
$\frac{17}{8}$	3

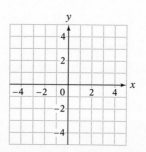

You Try It 3
Graph: $f(x) = -\log_3(x + 1)$

Your solution

Solutions on pp. A57–A58

12.3 Exercises

Objective A

Graph.

1. $f(x) = \log_4 x$

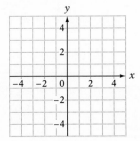

2. $f(x) = \log_2(x + 1)$

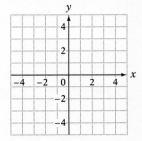

3. $f(x) = \log_3(2x - 1)$

4. $f(x) = \log_2\left(\dfrac{1}{2}x\right)$

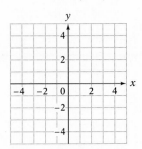

5. $f(x) = 3\log_2 x$

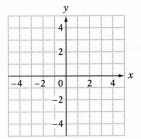

6. $f(x) = \dfrac{1}{2}\log_2 x$

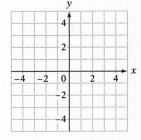

7. $f(x) = -\log_2 x$

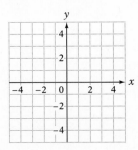

8. $f(x) = -\log_3 x$

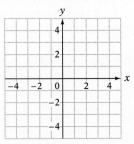

9. $f(x) = \log_2(x - 1)$

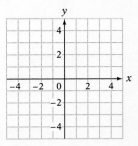

10. $f(x) = \log_3(2 - x)$

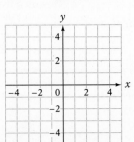

11. $f(x) = -\log_2(x - 1)$

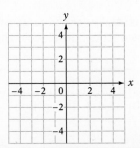

12. $f(x) = -\log_2(1 - x)$

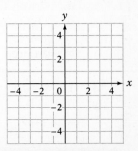

APPLYING THE CONCEPTS

 Use a graphing calculator to graph the following.

13. $f(x) = \log_2 x - 3$

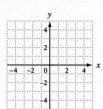

14. $f(x) = \log_3 x + 2$

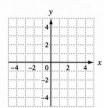

15. $f(x) = -\log_2 x + 2$

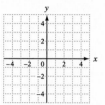

16. $f(x) = -\dfrac{1}{2} \log_2 x - 1$

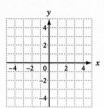

17. $f(x) = x - \log_2(1 - x)$

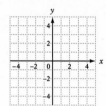

18. $f(x) = x + \log_3(2 - x)$

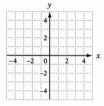

19. $f(x) = \dfrac{x}{2} - 2 \log_2(x + 1)$

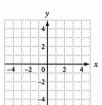

20. $f(x) = \dfrac{x}{3} - 3 \log_2(x + 3)$

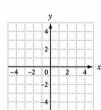

21. $f(x) = x^2 - 10 \ln(x - 1)$

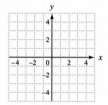

22. The proficiency of a typist decreases (without practice) over time. An equation that approximates this decrease is given by $S = 60 - 7 \ln(t + 1)$, where S is the typing speed in words per minute and t is the number of months without typing.
 a. Graph the equation.
 b. The point whose approximate coordinates are (4, 49) is on the graph. Write a sentence that describes the meaning of this ordered pair.

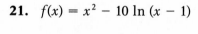

23. Astronomers use the *distance modulus* of a star as a method of determining the star's distance from the earth. The formula is $M = 5 \log s - 5$, where M is the distance modulus and s is the star's distance from the earth in parsecs. (One parsec $\approx 2.1 \times 10^{13}$ mi).
 a. Graph the equation.
 b. The point whose approximate coordinates are (25.1, 2) is on the graph. Write a sentence that describes the meaning of this ordered pair.

24. The Power Property of Logarithms tells us that $\ln x^2 = 2 \ln x$. Graph
[W] $f(x) = \ln x^2$ and $g(x) = 2 \ln x$. Are the graphs the same? Why or why not?

Solving Exponential and Logarithmic Equations

Objective A   *To solve an exponential equation* ....

An **exponential equation** is one in which the variable occurs in the exponent. The examples at the right are exponential equations.

$$6^{2x+1} = 6^{3x-2}$$
$$4^x = 3$$
$$2^{x+1} = 7$$

An exponential equation in which each side of the equation can be expressed in terms of the same base can be solved by using the 1–1 Property of Exponential Functions. Recall the 1–1 Property of Exponential Functions:

$$\text{If } b^x = b^y, \text{ then } x = y.$$

➡ Solve: $10^{3x+5} = 10^{x-3}$

$$10^{3x+5} = 10^{x-3}$$
$$3x + 5 = x - 3$$ • Use the 1–1 Property of Exponential Functions to equate the exponents.
$$2x + 5 = -3$$ • Solve the resulting equation.
$$2x = -8$$
$$x = -4$$

Check:
$$\frac{10^{3x+5} = 10^{x-3}}{\begin{array}{c|c} 10^{3(-4)+5} & 10^{-4-3} \\ 10^{-12+5} & 10^{-7} \\ 10^{-7} = & 10^{-7} \end{array}}$$

The solution is -4.

➡ Solve: $9^{x+1} = 27^{x-1}$

$$9^{x+1} = 27^{x-1}$$
$$(3^2)^{x+1} = (3^3)^{x-1}$$ • $3^2 = 9$; $3^3 = 27$
$$3^{2x+2} = 3^{3x-3}$$
$$2x + 2 = 3x - 3$$ • Use the 1–1 Property of Exponential Functions to equate the exponents.
$$2 = x - 3$$ • Solve for x.
$$5 = x$$

Check:
$$\frac{9^{x+1} = 27^{x-1}}{\begin{array}{c|c} 9^{5+1} & 27^{5-1} \\ 9^6 & 27^4 \\ 531{,}441 = & 531{,}441 \end{array}}$$

The solution is 5.

When each side of an exponential equation cannot easily be expressed in terms of the same base, logarithms are used to solve the exponential equation.

CONSIDER THIS

When evaluating $\dfrac{\log 7}{\log 4}$, first find the logarithm of each number.
Then divide the logarithms.
$\dfrac{\log 7}{\log 4} \approx \dfrac{0.845098}{0.602060} \approx 1.4037$

➡ Solve: $4^x = 7$

$$4^x = 7$$
$$\log 4^x = \log 7$$

$$x \log 4 = \log 7$$

$$x = \frac{\log 7}{\log 4} \approx 1.4037$$

The solution is 1.4037.

- Take the common logarithm of each side of the equation.
- Rewrite the equation using the Properties of Logarithms.

- Solve for x.

- Note that $\dfrac{\log 7}{\log 4} \neq \log 7 - \log 4$.

➡ Solve: $3^{x+1} = 5$

$$3^{x+1} = 5$$
$$\log 3^{x+1} = \log 5$$
$$(x + 1)\log 3 = \log 5$$

$$x + 1 = \frac{\log 5}{\log 3}$$
$$x = \frac{\log 5}{\log 3} - 1$$
$$x \approx 0.4650$$

The solution is 0.4650.

- Take the common logarithm of each side of the equation.
- Rewrite the equation using the Properties of Logarithms.

- Solve for x.

Example 1
Solve for x: $3^{2x} = 4$

Solution
$$3^{2x} = 4$$
$$\log 3^{2x} = \log 4$$
$$2x \log 3 = \log 4$$
$$2x = \frac{\log 4}{\log 3}$$
$$x = \frac{\log 4}{2 \log 3}$$
$$x \approx 0.6309$$

The solution is 0.6309.

You Try It 1
Solve for x: $4^{3x} = 25$

Your solution

Example 2
Solve for n: $(1.1)^n = 2$

Solution
$$(1.1)^n = 2$$
$$\log (1.1)^n = \log 2$$
$$n \log 1.1 = \log 2$$
$$n = \frac{\log 2}{\log 1.1}$$
$$n \approx 7.2725$$

The solution is 7.2725.

You Try It 2
Solve for n: $(1.06)^n = 1.5$

Your solution

Solutions on p. A58

Objective B   ***To solve a logarithmic equation*** ....

A logarithmic equation can be solved by using the Properties of Logarithms.

➡ Solve: $\log_9 x + \log_9(x - 8) = 1$

$\log_9 x + \log_9(x - 8) = 1$

$\log_9[x(x - 8)] = 1$ • Use the Logarithm Property of Products to rewrite the left side of the equation.

$9^1 = x(x - 8)$ • Write the equation in exponential form.

$9 = x^2 - 8x$ • Simplify.

$0 = x^2 - 8x - 9$ • Solve for x.

$0 = (x - 9)(x + 1)$ • Factor and use the Principle of Zero Products.

$x - 9 = 0 \qquad x + 1 = 0$

$x = 9 \qquad\quad x = -1$

Replacing x by 9 in the original equation reveals that 9 checks as a solution. Replacing x by -1 in the original equation results in the expression $\log_9(-1)$. Because the logarithm of a negative number is not a real number, -1 does not check as a solution.

The solution of the equation is 9.

➡ Solve: $\log_3 6 - \log_3(2x + 3) = \log_3(x + 1)$

$\log_3 6 - \log_3(2x + 3) = \log_3(x + 1)$

$\log_3 \dfrac{6}{2x + 3} = \log_3(x + 1)$ • Use the Quotient Property of Logarithms to rewrite the left side of the equation.

$\dfrac{6}{2x + 3} = x + 1$ • Use the 1–1 Property of Logarithms.

$6 = (2x + 3)(x + 1)$

$6 = 2x^2 + 5x + 3$ • Simplify.

$0 = 2x^2 + 5x - 3$

$0 = (2x - 1)(x + 3)$ • Solve for x.

$2x - 1 = 0 \qquad x + 3 = 0$

$x = \dfrac{1}{2} \qquad\quad x = -3$

Replacing x by $\dfrac{1}{2}$ in the original equation reveals that $\dfrac{1}{2}$ checks as a solution. Replacing x by -3 in the original equation results in the expression $\log_3(-2)$. Because the logarithm of a negative number is not a real number, -3 does not check as a solution.

The solution of the equation is $\dfrac{1}{2}$.

Example 3

Solve for x: $\log_3(2x - 1) = 2$

Solution

$\log_3(2x - 1) = 2$

$\quad\quad 3^2 = 2x - 1$ • Write in exponential form.
$\quad\quad\ 9 = 2x - 1$
$\quad\ 10 = 2x$
$\quad\quad\ 5 = x$

The solution is 5.

You Try It 3

Solve for x: $\log_4(x^2 - 3x) = 1$

Your solution

Example 4

Solve for x: $\log_2 x - \log_2(x - 1) = \log_2 2$

Solution

$\log_2 x - \log_2(x - 1) = \log_2 2$

$\log_2\left(\dfrac{x}{x - 1}\right) = \log_2 2$ • Use the Quotient Property of Logarithms.

$\quad\quad\quad \dfrac{x}{x - 1} = 2$ • Use the 1–1 Property of Logarithms.

$(x - 1)\left(\dfrac{x}{x - 1}\right) = (x - 1)2$

$\quad\quad\quad\quad\quad x = 2x - 2$
$\quad\quad\quad\quad -x = -2$
$\quad\quad\quad\quad\ \ x = 2$

The solution is 2.

You Try It 4

Solve for x: $\log_3 x + \log_3(x + 3) = \log_3 4$

Your solution

Example 5

Solve for x:
$\log_2(3x + 8) = \log_2(2x + 2) + \log_2(x - 2)$

Solution

$\log_2(3x + 8) = \log_2(2x + 2) + \log_2(x - 2)$
$\log_2(3x + 8) = \log_2[(2x + 2)(x - 2)]$
$\log_2(3x + 8) = \log_2(2x^2 - 2x - 4)$
$\quad\ 3x + 8 = 2x^2 - 2x - 4$ • Use the 1–1 Property of Logarithms.

$\quad\quad\quad 0 = 2x^2 - 5x - 12$
$\quad\quad\quad 0 = (2x + 3)(x - 4)$
$2x + 3 = 0 \quad\quad x - 4 = 0$
$\quad\quad x = -\dfrac{3}{2} \quad\quad\quad x = 4$

$-\dfrac{3}{2}$ does not check as a solution; 4

checks as a solution. The solution is 4.

You Try It 5

Solve for x: $\log_3 x + \log_3(x + 6) = 3$

Your solution

Solutions on p. A58

12.4 Exercises

· ·

Objective A

Solve for x. Round to the nearest ten-thousandth.

1. $5^{4x-1} = 5^{x-2}$

2. $7^{4x-3} = 7^{2x+1}$

3. $8^{x-4} = 8^{5x+8}$

4. $10^{4x-5} = 10^{x+4}$

5. $9^x = 3^{x+1}$

6. $2^{x-1} = 4^x$

7. $8^{x+2} = 16^x$

8. $9^{3x} = 81^{x-4}$

9. $16^{2-x} = 32^{2x}$

10. $27^{2x-3} = 81^{4-x}$

11. $25^{3-x} = 125^{2x-1}$

12. $8^{4x-7} = 64^{x-3}$

13. $5^x = 6$

14. $7^x = 10$

15. $e^x = 3$

16. $e^x = 2$

17. $10^x = 21$

18. $10^x = 37$

19. $2^{-x} = 7$

20. $3^{-x} = 14$

21. $2^{x-1} = 6$

22. $4^{x+1} = 9$

23. $3^{2x-1} = 4$

24. $4^{-x+2} = 12$

Objective B

Solve for x. Round to the nearest ten-thousandth.

25. $\log_2(2x - 3) = 3$

26. $\log_4(3x + 1) = 2$

27. $\log_2(x^2 + 2x) = 3$

28. $\log_3(x^2 + 6x) = 3$

29. $\log_5\left(\dfrac{2x}{x-1}\right) = 1$

30. $\log_6\left(\dfrac{3x}{x+1}\right) = 1$

31. $\log x = \log(1 - x)$

32. $\ln(3x - 2) = \ln(x + 1)$

33. $\ln 5 = \ln(4x - 13)$

34. $\log_3(x - 2) = \log_3(2x)$

35. $\ln(3x + 2) = 4$

36. $\ln(2x + 3) = -1$

Solve for x.

37. $\log_2(8x) - \log_2(x^2 - 1) = \log_2 3$

38. $\log_5(3x) - \log_5(x^2 - 1) = \log_5 2$

39. $\log_9 x + \log_9(2x - 3) = \log_9 2$

40. $\log_6 x + \log_6(3x - 5) = \log_6 2$

41. $\log_8(6x) = \log_8 2 + \log_8(x - 4)$

42. $\log_7(5x) = \log_7 3 + \log_7(2x + 1)$

APPLYING THE CONCEPTS

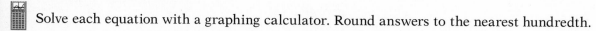 Solve each equation with a graphing calculator. Round answers to the nearest hundredth.

43. $3^x = -x$ **44.** $2^x = -x + 2$ **45.** $2^{-x} = x - 1$ **46.** $3^{-x} = 2x$

47. $\ln x = x^2$ **48.** $2\ln x = -x + 1$ **49.** $\log_3 x = -2x - 2$ **50.** $\log_5(2x) = -2x + 1$

51. A model for the distance s (in feet) an object that is experiencing air resistance will fall in t seconds is given by $s = 312.5 \ln\left(\dfrac{e^{0.32t} + e^{-0.32t}}{2}\right)$.
   a. Graph this equation. *Suggestion:* Use Xmin = −0.5, Xmax = 5, Xscl = 0.5, Ymin = −0.5, Ymax = 150, and Yscl = 20.
   b. Determine, to the nearest hundredth of a second, the time it takes the object to travel 100 ft.

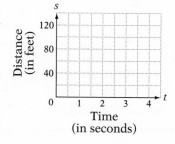

52. A model for the distance s (in feet) that an object that is experiencing air resistance will fall in t seconds is given by $s = 78 \ln\left(\dfrac{e^{0.8t} + e^{-0.8t}}{2}\right)$.
   a. Graph this equation. *Suggestion:* Use Xmin = −0.5, Xmax = 5, Xscl = 0.5, Ymin = −0.5, Ymax = 150, and Yscl = 20.
   b. Determine, to the nearest hundredth of a second, the time it takes the object to travel 125 ft.

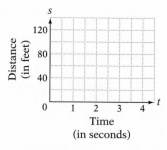

53. The following "proof" shows that $0.5 < 0.25$. Explain the error.
[W]

$$1 < 2$$
$$1 \cdot \log 0.5 < 2 \cdot \log 0.5$$
$$\log 0.5 < \log(0.5)^2$$
$$0.5 < (0.5)^2$$
$$0.5 < 0.25$$

12.5 Applications of Exponential and Logarithmic Functions

Objective A   *To solve application problems* ....

A biologist places one single-celled bacterium in a culture, and each hour that particular species of bacteria divides into two bacteria. After one hour there will be two bacteria. After two hours, each of the two bacteria will divide and there will be four bacteria. After three hours, each of the four bacteria will divide and there will be eight bacteria.

The table at the right shows the number of bacteria in the culture after various intervals of time, t, in hours. Values in this table could also be found by using the exponential equation $N = 2^t$.

Time, t	Number of Bacteria, N
0	1
1	2
2	4
3	8
4	16

The equation $N = 2^t$ is an example of an **exponential growth equation.** In general, any equation that can be written in the form $A = A_0 b^{kt}$, where A is the size at time t, A_0 is the initial size, $b > 1$, and k is a positive real number, is an exponential growth equation. These equations are important not only in population growth studies but also in physics, chemistry, psychology, and economics.

Recall that interest is the amount of money paid (or received) when borrowing (or investing) money. **Compound interest** is interest that is computed not only on the original principal, but also on the interest already earned. The compound interest formula is an exponential equation.

The **compound interest formula** is $P = A(1 + i)^n$, where A is the original value of an investment, i is the interest rate per compounding period, n is the total number of compounding periods, and P is the value of the investment after n periods.

POINT OF INTEREST

Another exponential function from finance is the one that enables us to calculate the amount of an amortized loan payment, such as a car loan.

$$P = B\left[\frac{i/12}{1 - [1 + (i/12)]^{-n}}\right]$$

B is the amount borrowed, i is the annual interest rate as a decimal, and n is the number of months to repay the loan.

➡ An investment broker deposits $1000 into an account that earns 12% annual interest compounded quarterly. What is the value of the investment after two years?

$i = \dfrac{12\%}{4} = \dfrac{0.12}{4} = 0.03$

• Find i, the interest rate per quarter. The quarterly rate is the annual rate divided by 4, the number of quarters in 1 year.

$n = 4 \cdot 2 = 8$

• Find n, the number of compounding periods. The investment is compounded quarterly, 4 times a year, for 2 years.

$P = A(1 + i)^n$

• Use the compound interest formula.

$P = 1000(1 + 0.03)^8$

• Replace A, i, and n by their values.

$P \approx 1267$

• Solve for P.

The value of the investment after two years is $1267.

Exponential decay offers another important example of an exponential equation. One of the most common illustrations is the decay of a radioactive substance.

A certain isotope of cobalt has a half-life of approximately 5 years. This means that one-half of any given amount of this cobalt isotope will disintegrate in 5 years.

The table at the right indicates the amount of the initial 10-milligram sample of a cobalt isotope that remains after various intervals of time, t, in years. Values in this table could also be found by using the exponential equation $A = 10\left(\dfrac{1}{2}\right)^{t/5}$.

Time, t	Amount, A
0	10
5	5
10	2.5
15	1.25
20	0.625

The equation $A = 10\left(\dfrac{1}{2}\right)^{t/5}$ is an example of an *exponential decay equation*. Comparing this equation to the exponential growth equation, note that for exponential growth, the base of the exponential equation is greater than 1, whereas for exponential decay, the base is between 0 and 1.

A method by which an archeologist can measure the age of a bone is called *carbon dating*. Carbon dating is based on a radioactive isotope of carbon called carbon-14, which has a half-life of approximately 5570 years. The exponential decay equation is given by $A = A_0\left(\dfrac{1}{2}\right)^{t/5570}$, where A_0 is the original amount of carbon-14 present in the bone, t is the age of the bone, and A is the amount present after t years.

➡ A bone that originally contained 100 mg of carbon-14 now has 70 mg of carbon-14. What is the approximate age of the bone? Round to the nearest year.

$$A = A_0\left(\dfrac{1}{2}\right)^{t/5570}$$ • Use the exponential decay equation.

$$70 = 100\left(\dfrac{1}{2}\right)^{t/5570}$$ • Replace A_0 and A by their given values and solve for t.

$$70 = 100(0.5)^{t/5570}$$

$$0.7 = (0.5)^{t/5570}$$ • Divide each side of the equation by 100.

$$\log 0.7 = \log (0.5)^{t/5570}$$ • Take the common logarithm of each side of the equation. Then simplify.

$$\log 0.7 = \dfrac{t}{5570} \log 0.5$$

$$\dfrac{5570 \log 0.7}{\log 0.5} = t$$

$$2866 \approx t$$

The age of the bone is approximately 2866 years.

Chemists measure the acidity or alkalinity of a solution in terms of the concentration of hydrogen ions, H^+, in the solution by the formula $pH = -\log(H^+)$. A neutral solution such as distilled water has a pH of 7, acids have a pH of less than 7, and alkaline (also called basic) solutions have a pH of greater than 7.

➡️ Find the pH of vinegar, for which $H^+ = 1.26 \times 10^{-3}$. Round to the nearest tenth.

$$\begin{aligned} pH &= -\log (H^+) \\ &= -\log (1.26 \times 10^{-3}) \\ &\approx 2.9 \end{aligned}$$

- Use the pH equation.
- $H^+ = 1.26 \times 10^{-3}$

The pH of vinegar is 2.9.

The *Richter scale* measures the magnitude, M, of an earthquake in terms of the intensity, I, of its shock waves. This can be expressed as the logarithmic equation $M = \log \dfrac{I}{I_0}$, where I_0 is a constant.

➡️ How many times stronger is an earthquake that has magnitude 4 on the Richter scale than one that has magnitude 2 on the scale?

Let I_1 represent the intensity of the earthquake that has magnitude 4, and let I_2 represent the intensity of the earthquake that has magnitude 2.

$$4 = \log \frac{I_1}{I_0}$$

$$2 = \log \frac{I_2}{I_0}$$

- Use the Richter equation to write a system of equations, one equation for magnitude 4 and one for magnitude 2. Then rewrite the system using the Properties of Logarithms.

$$4 = \log I_1 - \log I_0$$

$$2 = \log I_2 - \log I_0$$

$$2 = \log I_1 - \log I_2$$

- Use the addition method to eliminate $\log I_0$.

$$2 = \log \frac{I_1}{I_2}$$

- Rewrite the equation using the Properties of Logarithms.

$$\frac{I_1}{I_2} = 10^2 = 100$$

- Solve for the ratio using the relationship between logarithms and exponents.

An earthquake that has magnitude 4 on the Richter scale is 100 times stronger than an earthquake that has magnitude 2.

To calculate the percent of light that will pass through a substance, use the equation $\log P = -kd$, where P is the percent of light passing through the substance, k is a constant that depends on the substance, and d is the thickness of the substance.

➡️ Find the percent of light that will pass through glass for which $k = 0.4$ and d is 0.5 cm.

$$\begin{aligned} \log P &= -kd \\ \log P &= -(0.4)(0.5) \\ \log P &= -0.2 \end{aligned}$$

- Replace k and d in the equation by their given values and solve for P.

$$\begin{aligned} P &= 10^{-0.2} \\ P &\approx 0.631 \end{aligned}$$

- Use the relationship between the logarithmic and exponential functions.

Approximately 63.1% of the light will pass through the glass.

Example 1

An investment of $3000 is placed into an account that earns 12% annual interest compounded monthly. In approximately how many years will the investment be worth twice the original amount?

Strategy

To find the time, solve the compound interest formula for n. Use $P = 6000$, $A = 3000$, and $i = \frac{12\%}{12} = \frac{0.12}{12} = 0.01$.

Solution

$$P = A(1 + i)^n$$
$$6000 = 3000(1 + 0.01)^n$$
$$6000 = 3000(1.01)^n$$
$$2 = (1.01)^n$$
$$\log 2 = \log (1.01)^n$$
$$\log 2 = n \log 1.01$$
$$\frac{\log 2}{\log 1.01} = n$$
$$70 \approx n$$

70 months $\div$ 12 $\approx$ 5.8 years

In approximately 6 years, the investment will be worth $6000.

You Try It 1

Find the pH of sodium carbonate, for which $H^+ = 2.51 \times 10^{-12}$. Round to the nearest tenth.

Your strategy

Your solution

Example 2

The number of words per minute that a student can type will increase with practice and can be approximated by the equation $N = 100[1 - (0.9)^t]$, where N is the number of words typed per minute after t days of instruction. Find the number of words a student will type per minute after 8 days of instruction.

Strategy

To find the number of words per minute, replace t by its given value in the equation and solve for N.

Solution

$$N = 100[1 - (0.9)^t]$$
$$= 100[1 - (0.9)^8]$$
$$\approx 56.95$$

After 8 days of instruction, a student will type approximately 57 words per minute.

You Try It 2

An earthquake that measures 5 on the Richter scale can cause serious damage to buildings. The San Francisco earthquake of 1906 would have measured about 7.8 on this scale. How many times stronger was the San Francisco earthquake than the least forceful earthquake that can cause serious damage to buildings? Round to the nearest whole number.

Your strategy

Your solution

Solutions on p. A59

12.5 Exercises

· ·

Objective A *Application Problems*

Solve. Round to the nearest whole number.

For Exercises 1–6, use the compound interest formula $P = A(1 + i)^n$, where A is the original value of an investment, i is the interest rate per compounding period, n is the total number of compounding periods, and P is the value of the investment after n periods.

1. The Prime Investment Club deposits $5000 into an account that earns 9% annual interest compounded monthly. What is the value of the investment after 2 years?

2. What is the value of an $8000 investment after 2 years if it earns 8% annual interest compounded daily? Use a 365-day year.

3. An investment advisor deposits $12,000 into an account that earns 10% annual interest compounded semiannually. In approximately how many years will the investment double?

4. An insurance broker deposits $4000 into an account that earns 7% annual interest compounded monthly. In approximately how many years will the investment be worth $8000?

5. A manager estimates that in 4 years the company will need to purchase a new bottling machine at a cost of $25,000. How much money must be deposited in an account that earns 10% annual interest compounded monthly so that the value of the account in 4 years will be $25,000?

6. The comptroller of a company has determined that it will be necessary to purchase a new computer in 3 years. The estimated cost of the computer is $10,000. How much money must be deposited in an account that earns 9% annual interest compounded quarterly so that the value of the account in 3 years will be $10,000?

For Exercises 7–10, use the exponential decay equation $A = A_0 \left(\dfrac{1}{2}\right)^{t/k}$, where A is the amount of a radioactive material present after time t, k is the half-life, and A_0 is the original amount of radioactive substance. Round to the nearest tenth.

7. An isotope of technetium is used to prepare images of internal body organs. This isotope has a half-life of approximately 6 h. If a patient is injected with 30 mg of this isotope, how long (in hours) will it take for the technetium level to reach 20 mg?

8. Iodine-131 is an isotope of iodine that is used to study the functioning of the thyroid gland. This isotope has a half-life of approximately 8 days. If a patient is given an injection that contains 8 micrograms of iodine-131, how long (in days) will it take for the iodine level to reach 5 micrograms?

9. A sample of promethium-147 (used in some luminous paints) contains 25 mg. One year later, the sample contains 18.95 mg. What is the half-life of promethium-147, in years? Round to the nearest tenth.

10. Francium-223 is a very rare radioactive isotope discovered in 1939 by Marguerite Percy. In a laboratory, a 3-microgram sample of francium-223 will decay to 2.54 micrograms in 5 min. What is the half-life of francium-223, in minutes?

For Exercises 11 and 12, consider that the percent of correct welds a student can make will increase with practice and can be approximated by the equation $P = 1 - (0.75)^t$, where P is the percent of correct welds and t is the number of weeks of practice.

11. How many weeks of practice are necessary before a student will make 80% of the welds correctly?

12. Find the percent of correct welds a student will make after 4 weeks of practice.

For Exercises 13 and 14, use the equation $pH = -\log(H^+)$, where H^+ is the hydrogen ion concentration of a solution. Round to the nearest hundredth.

13. Find the pH of the digestive solution of the stomach, for which the hydrogen ion concentration is 0.045.

14. Find the pH of a morphine solution used to relieve pain, for which the hydrogen ion concentration is 3.2×10^{-10}.

The percent of light that will pass through a material is given by the equation $\log P = -kd$, where P is the percent of light passing through the material, k is a constant that depends on the material, and d is the thickness of the material in centimeters. Use this information in Exercises 15 and 16.

15. The constant k for a piece of opaque glass that is 0.5 cm thick is 0.2. Find the percent of light that will pass through the glass.

16. The constant k for a piece of tinted glass is 0.5. How thick is a piece of this glass that allows 60% of the light incident to the glass to pass through it?

For Exercises 17 and 18, use the Richter equation $M = \log \dfrac{I}{I_0}$, where M is the magnitude of an earthquake, I is the intensity of its shock waves, and I_0 is a constant. Round to the nearest tenth.

17. An earthquake that occurred in China in 1978 measured 8.2 on the Richter scale. In 1988, an earthquake in Armenia measured 6.9 on the Richter scale. How many times stronger was China's earthquake?

18. In October 1989, an earthquake of magnitude 7.1 on the Richter scale struck San Francisco, California. In January 1994, an earthquake measuring 6.6 on the Richter scale occurred in the San Fernando Valley in California. How many times stronger was the San Francisco earthquake than the San Fernando Valley earthquake?

Astronomers use the distance modulus formula $M = 5 \log r - 5$, where M is the distance modulus and r is the distance a star is from the earth in parsecs. Use this information in Exercises 19–22. Round to the nearest tenth.

19. The distance modulus of the star Betelgeuse is 5.89. How many parsecs from the earth is this star?

20. The distance modulus of Alpha Centauri is -1.11. How many parsecs from the earth is this star?

21. The distance modulus of the star Antares is 5.4, which is twice that of the star Pollux. Is Antares twice as far from the earth as Pollux? If not, how many times farther is Antares from the earth than Pollux is?

22. If one star has a distance modulus of 2 and a second star has a distance modulus of 6, is the second star three times as far from the earth as the first star?

The number of decibels, D, of a sound can be given by the equation $D = 10(\log I + 16)$, where I is the power of a sound measured in watts. Use this information in Exercises 23 and 24. Round to the nearest whole number.

23. Find the number of decibels of normal conversation. The power of the sound of normal conversation is approximately 3.2×10^{-10} watts.

24. The loudest sound made by any animal is made by the blue whale and can be heard over 500 mi away. The power of the sound made by the blue whale is 630 watts. Find the number of decibels of the sound emitted by the blue whale.

One model for the time it will take for the world's oil supply to be depleted is given by the equation $T = 14.29 \ln (0.00411r + 1)$, where r is the estimated world oil reserves in billions of barrels of oil and T is the time before that amount of oil is used up. Use this information in Exercises 25 and 26.

25. According to this model, how many barrels of oil are necessary to last 20 years?

26. According to this model, how many barrels of oil are necessary to last 50 years?

APPLYING THE CONCEPTS

27. Some banks now use *continuous compounding* of an amount invested. In this case, the equation that relates the value of an initial investment of A dollars in t years at an annual interest rate of $r\%$ is given by $P = Ae^{rt}$. Using this equation, find the value in 5 years of an investment of $2500 into an account that earns 5% annual interest.

28. Using the continuous compounding interest formula in Exercise 27, find the number of years it will take an investment to double if the annual interest rate is 6%. Does the time it takes to double in value depend on the amount of the initial investment?

29. Solve the equation $T = 14.29 \ln (0.00411r + 1)$ for r.

30. Solve the equation $M = 5 \log r - 5$ for r.

31. If $M = 5 \log r_1 - 5$ and $2M = 5 \log r_2 - 5$, express r_1 in terms of r_2.

Projects in Mathematics

. . . .

Fractals[1]

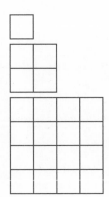

Fractals have a wide variety of applications. They have been used to create special effects for movies such as *Star Trek II: The Wrath of Kahn* and to explain the behavior of some biological and economic systems. One aspect of fractals that has fascinated mathematicians is that they apparently have fractional dimension.

To understand the idea of fractional dimension, one must first understand the terms "scale factor" and "size." Consider a unit square (a square of length one). By joining four of these squares, we can create another square, the length of which is two, and the size of which is four. (Here size = number of unit squares.) Four of these larger squares can in turn be put together to make a third square of length four and size sixteen. This process of grouping together four squares can in theory be done an infinite number of times; yet at each step, the following quantities will be the same:

$$\text{scale factor} = \frac{\text{new length}}{\text{old length}} \qquad \text{size ratio} = \frac{\text{new size}}{\text{old size}}$$

Consider the unit square as Step 1, the four unit squares as Step 2, etc.

1. Calculate the scale factor going (a) from Step 1 to Step 2, (b) from Step 2 to Step 3, and (c) from Step 3 to Step 4.

2. Calculate the size ratio going (a) from Step 1 to Step 2, (b) from Step 2 to Step 3, and (c) from Step 3 to Step 4.

3. What is (a) the scale factor and (b) the size ratio going from Step n to Step $n + 1$?

Mathematicians have defined dimension using the formula $d = \dfrac{\log(\text{size ratio})}{\log(\text{scale factor})}$.

For the squares discussed above, $d = \dfrac{\log(\text{size ratio})}{\log(\text{scale factor})} = \dfrac{\log 4}{\log 2} = 2$.

So by this definition of dimension, squares are two-dimensional figures.

Now consider a unit cube (Step 1). Group 8 unit cubes to form a cube that is 2 units on each side (Step 2). Group 8 of the cubes from Step 2 to form a cube that is 4 units on each side (Step 3).

4. Calculate (a) the scale factor and (b) the size ratio for this process.

5. Show that the cubes are three-dimensional figures.

In each of the above examples, if the process is continued indefinitely, we still have a square or a cube. Consider a process that is more difficult to envision. Let Step 1 be an equilateral triangle whose base has length one unit, and let Step 2 be a grouping of three of these equilateral triangles, such that the space between them is another equilateral triangle with a base of length one unit. Three shapes from Step 2 are arranged with an equilateral triangle in their center, and so on. It is hard to imagine the result if this is done an infinite number of times, but mathematicians have shown that the result is a single figure of fractional dimension. (Similar processes have been used to create fascinating artistic patterns and to explain scientific phenomenon.)

6. Show that for this process (a) the scale factor is 2 and (b) the size ratio is 3.

7. Calculate the dimension of the fractal. (Note that it is a *fractional* dimension!)

[1]This is adapted from Tami Martin, "Student Math Notes," *Mathematics Teacher* (Nov. 1991).

Chapter Summary

. . . .

Key Words   A function of the form $f(x) = b^x$ is an *exponential function*, where b is a positive real number not equal to one. The number b is the *base* of the exponential function.

For $b > 0$, $b \neq 1$, $y = \log_b x$ is equivalent to $x = b^y$.

$\log_b x$ is the logarithm of x to the base b.

Common logarithms are logarithms to the base 10. (We usually omit the base 10 when writing the common logarithm of a number.)

The *mantissa* is the decimal part of a common logarithm. The *characteristic* is the integer part of a common logarithm.

When e (the base of the natural exponential function) is used as a base of a logarithm, the logarithm is referred to as the natural logarithm and is abbreviated $\ln x$.

An *exponential equation* is an equation in which the variable occurs in the exponent.

Essential Rules   *The Logarithm Property of the Product of Two Numbers*

For any positive real numbers x, y, and b, $b \neq 1$,
$\log_b(xy) = \log_b x + \log_b y$.

The Logarithm Property of the Quotient of Two Numbers

For any positive real numbers x, y, and b, $b \neq 1$,
$\log_b \dfrac{x}{y} = \log_b x - \log_b y$.

The Logarithm Property of the Power of a Number

For any positive real numbers x and b, $b \neq 1$, and for any real number r, $\log_b x^r = r \log_b x$.

Additional Properties of Logarithms

Let x, y, and b be positive and real numbers with $b \neq 1$. Then
  $\log_b 1 = 0$
  If $\log_b x = \log_b y$, then $x = y$.

Change-of-Base Formula

$\log_a N = \dfrac{\log_b N}{\log_b a}$

1–1 Property of Exponential Functions

If $b^x = b^y$, then $x = y$.

Chapter Review Exercises

1. Evaluate $f(x) = \left(\dfrac{2}{3}\right)^x$ at $x = 0$.

2. Evaluate $f(x) = 3^{x+1}$ at $x = -2$.

3. Graph: $f(x) = 2^x - 3$

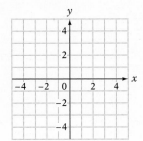

4. Graph: $f(x) = 2^x + 2$

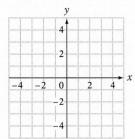

5. Evaluate: $\log_4 16$

6. Solve for x: $\log_3 x = -2$

7. Graph: $f(x) = \log_2(2x)$

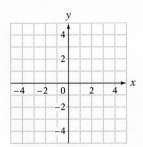

8. Graph: $f(x) = \log_3(x + 1)$

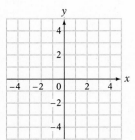

9. Write $\log_6 \sqrt{xy^3}$ in expanded form.

10. Write $\dfrac{1}{2}(\log_3 x - \log_3 y)$ as a single logarithm with a coefficient of 1.

11.   Write $\ln\left(\dfrac{x}{\sqrt{z}}\right)$ in expanded form.

12.   Write $3 \ln x - \ln y - \dfrac{1}{2} \ln z$ as a single logarithm with a coefficient of 1.

13.   Solve for x: $3^{7x+1} = 3^{4x-5}$

14.   Solve for x: $8^x = 2^{x-6}$

15.   Solve $3^x = 17$. Round to the nearest ten-thousandth.

16.   Solve for x: $\log x + \log(x - 4) = \log 12$

17.   Solve for x: $\log_6 x + \log_6(x - 1) = 1$

18.   Find $\log_5 9$. Round to the nearest ten-thousandth.

19.   Find $\log_3 19$. Round to the nearest ten-thousandth.

20.   Use the exponential decay equation $A = A_0\left(\dfrac{1}{2}\right)^{t/k}$, where A is the amount of a radioactive material present after time t, k is the half-life, and A_0 is the original amount of radioactive material, to find the half-life of a material that decays from 10 mg to 9 mg in 5 h. Round to the nearest whole number.

Cumulative Review Exercises

1. Solve: $4 - 2[x - 3(2 - 3x) - 4x] = 2x$

2. Find the equation of the line that contains the point $(2, -2)$ and is parallel to the line $2x - y = 5$.

3. Factor: $4x^{2n} + 7x^n + 3$

4. Simplify: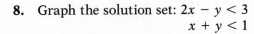

5. Simplify: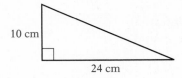

6. Solve by completing the square: $x^2 - 4x - 6 = 0$

7. Find the unknown side of the triangle in the figure below.

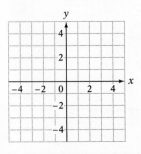

10 cm

24 cm

8. Graph the solution set: $2x - y < 3$
$\quad\quad\quad\quad\quad\quad\quad\quad\quad\quad x + y < 1$

9. Solve by the addition method.
$$3x - y + z = 3$$
$$x + y + 4z = 7$$
$$3x - 2y + 3z = 8$$

10. Subtract: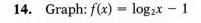

11. Solve: $x^2 + 4x - 5 \le 0$

12. Solve: $|2x - 5| \le 3$

13. Graph: $f(x) = \left(\dfrac{1}{2}\right)^x + 1$

14. Graph: $f(x) = \log_2 x - 1$

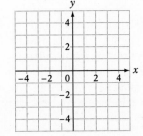

15. New carpet is installed in a room measuring 18 ft by 14 ft. Find the area of the room in square yards. ($9 \text{ ft}^2 = 1 \text{ yd}^2$)

16. Solve for x: $\log_5 x = 3$

17. Write $3 \log_b x - 5 \log_b y$ as a single logarithm with a coefficient of 1.

18. Find $\log_3 7$. Round to the nearest ten-thousandth.

19. Solve for x: $4^{5x-2} = 4^{3x+2}$

20. Solve for x: $\log x + \log(2x + 3) = \log 2$

21. A bank offers two types of checking accounts. One account has a charge of $5 per month plus 2 cents per check. The second account has a charge of $2 per month plus 8 cents per check. How many checks can a customer who has the second type of account write if it is to cost the customer less than the first type of checking account?

22. Find the cost per pound of a mixture made from 16 lb of chocolate that costs $4.00 per pound and 24 lb of chocolate that costs $2.50 per pound.

23. A plane can fly at a rate of 225 mph in calm air. Traveling with the wind, the plane flew 1000 mi in the same amount of time that it took to fly 800 mi against the wind. Find the rate of the wind.

24. The distance (d) that a spring stretches varies directly as the force (f) used to stretch the spring. If a force of 20 lb stretches a spring 6 in., how far will a force of 34 lb stretch the spring?

25. A carpenter purchased 80 ft of redwood and 140 ft of fir for a total cost of $67. A second purchase, at the same prices, included 140 ft of redwood and 100 ft of fir for a total cost of $81. Find the cost of redwood and of fir.

26. The compound interest formula is $P = A(1 + i)^n$, where A is the original value of an investment, i is the interest rate per compounding period, n is the total number of compounding periods, and P is the value of the investment after n periods. Use the compound interest formula to find the number of years in which an investment of $5000 will double in value. The investment earns 9% annual interest and is compounded semiannually. Round to the nearest whole number.

Final Exam

· ·

1. Simplify:
$12 - 8[3 - (-2)]^2 \div 5 - 3$

2. Evaluate $\dfrac{a^2 - b^2}{a - b}$ when $a = 3$ and $b = -4$.

3. Simplify: $5 - 2[3x - 7(2 - x) - 5x]$

4. Solve: $\dfrac{3}{4}x - 2 = 4$

5. Solve: $8 - |5 - 3x| = 1$

6. Find the volume of a sphere with a diameter of 8 ft. Round to the nearest tenth.

7. Graph $2x - 3y = 9$ using the x- and y-intercepts.

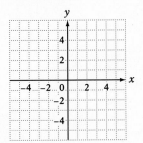

8. Find the equation of the line containing the points $(3, -2)$ and $(1, 4)$.

9. Find the equation of the line that contains the point $(-2, 1)$ and is perpendicular to the line $3x - 2y = 6$.

10. Simplify: $2a[5 - a(2 - 3a) - 2a] + 3a^2$

11. Factor: $8 - x^3y^3$

12. Factor: $x - y - x^3 + x^2y$

13. Divide: $(2x^3 - 7x^2 + 4) \div (2x - 3)$

14. Divide: $\dfrac{x^2 - 3x}{2x^2 - 3x - 5} \div \dfrac{4x - 12}{4x^2 - 4}$

15. Subtract: $\dfrac{x - 2}{x + 2} - \dfrac{x + 3}{x - 3}$

16. Simplify: $\dfrac{\dfrac{3}{x} + \dfrac{1}{x + 4}}{\dfrac{1}{x} + \dfrac{3}{x + 4}}$

17. Solve: $\dfrac{5}{x-2} - \dfrac{5}{x^2-4} = \dfrac{1}{x+2}$

18. Solve $a_n = a_1 + (n-1)d$ for d.

19. Simplify: $\left(\dfrac{4x^2y^{-1}}{3x^{-1}y}\right)^{-2}\left(\dfrac{2x^{-1}y^2}{9x^{-2}y^2}\right)^3$

20. Simplify: $\left(\dfrac{3x^{2/3}y^{1/2}}{6x^2y^{4/3}}\right)^6$

21. Subtract: $x\sqrt{18x^2y^3} - y\sqrt{50x^4y}$

22. Simplify: $\dfrac{\sqrt{16x^5y^4}}{\sqrt{32xy^7}}$

23. Simplify: $\dfrac{3}{2+i}$

24. Write a quadratic equation that has integer coefficients and has solutions $-\dfrac{1}{2}$ and 2.

25. Solve by using the quadratic formula: $2x^2 - 3x - 1 = 0$

26. Solve: $x^{2/3} - x^{1/3} - 6 = 0$

27. Graph: $f(x) = -x^2 + 4$

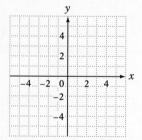

28. Graph: $\dfrac{x^2}{16} + \dfrac{y^2}{4} = 1$

29. Solve: $\dfrac{2}{x} - \dfrac{2}{2x+3} = 1$

30. Find the inverse of the function $f(x) = \dfrac{2}{3}x - 4$.

31. Solve by the addition method:
$3x - 2y = 1$
$5x - 3y = 3$

32. Evaluate the determinant:
$\begin{vmatrix} 3 & 4 \\ -1 & 2 \end{vmatrix}$

33. Solve:
$2 - 3x < 6$ and $2x + 1 > 4$

34. Solve: $|2x + 5| < 3$

35. Graph the solution set: $3x + 2y > 6$

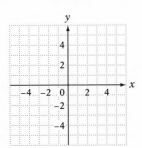

36. Graph: $f(x) = \left(\dfrac{2}{3}\right)^{x+1}$

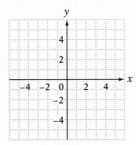

37. Graph: $f(x) = \log_2(x + 1)$

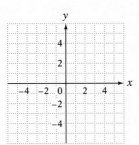

38. Write $2(\log_2 a - \log_2 b)$ as a single logarithm with a coefficient of 1.

39. Solve for x: $\log_3 x - \log_3(x - 3) = \log_3 2$

40. An average score of 70–79 in a history class receives a C grade. A student has grades of 64, 58, 82, and 77 on four history tests. Find the range of scores on the fifth test that will give the student a C grade for the course.

41. A jogger and a cyclist set out at 8 A.M. from the same point headed in the same direction. The average speed of the cyclist is two and a half times the average speed of the jogger. In 2 h, the cyclist is 24 mi ahead of the jogger. How far did the cyclist ride in that time?

42. You have a total of $12,000 invested in two simple interest accounts. On one account, a money market fund, the annual simple interest rate is 8.5%. On the other account, a tax free bond fund, the annual simple interest rate is 6.4%. The total annual interest earned by the two accounts is $936. How much do you have invested in each account?

43. The length of a rectangle is 1 ft less than three times the width. The area of the rectangle is 140 ft^2. Find the length and width of the rectangle.

44. Three hundred shares of a utility stock earn a yearly dividend of $486. How many additional shares of the utility stock would give a total dividend income of $810?

45. An account executive traveled 45 mi by car and then an additional 1050 mi by plane. The rate of the plane was seven times the rate of the car. The total time for the trip was $3\frac{1}{4}$ h. Find the rate of the plane.

46. An object is dropped from the top of a building. Find the distance the object has fallen when the speed reaches 75 ft/s. Use the equation $v = \sqrt{64d}$, where v is the speed of the object and d is the distance. Round to the nearest whole number.

47. A small plane made a trip of 660 mi in 5 h. The plane traveled the first 360 mi at a constant rate before increasing its speed by 30 mph. Then it traveled another 300 mi at the increased speed. Find the rate of the plane for the first 360 mi.

48. The intensity (L) of a light source is inversely proportional to the square of the distance (d) from the source. If the intensity is 8 lumens at a distance of 20 ft, what is the intensity when the distance is 4 ft?

49. A motorboat traveling with the current can go 30 mi in 2 h. Against the current, it takes 3 h to go the same distance. Find the rate of the motorboat in calm water and the rate of the current.

50. An investor deposits $4000 into an account that earns 9% annual interest compounded monthly. Use the compound interest formula $P = A(1 + i)^n$, where A is the original value of the investment, i is the interest rate per compounding period, n is the total number of compounding periods, and P is the value of the investment after n periods, to find the value of the investment after 2 years. Round to the nearest cent.

SOLUTIONS to Chapter 1 You Try It

SECTION 1.1 *pages 3–10*

You Try It 1 Replace y by each of the elements of the set and determine whether the inequality is true.

$$y > -1$$

$-5 > -1$ False

$-1 > -1$ False

$5 > -1$ True

The inequality is true for 5.

You Try It 2 Replace z by each element of the set and determine the value of the expression.

$-z$	$|z|$
$-(-11) = 11$	$|-11| = 11$
$-(0) = 0$	$|0| = 0$
$-(8) = -8$	$|8| = 8$

You Try It 3 $100 + (-43) = 57$

You Try It 4

$(-51) + 42 + 17 + (-102)$

$= -9 + 17 + (-102)$

$= 8 + (-102)$

$= -94$

You Try It 5 $19 - (-32) = 19 + 32$

$= 51$

You Try It 6

$-9 - (-12) - 17 - 4$

$= -9 + 12 + (-17) + (-4)$

$= 3 + (-17) + (-4)$

$= -14 + (-4)$

$= -18$

You Try It 7 $8(-9)10 = -72(10)$

$= -720$

You Try It 8 $(-2)3(-8)7 = -6(-8)7$

$= 48(7)$

$= 336$

You Try It 9 $(-135) \div (-9) = 15$

You Try It 10 $\dfrac{-72}{4} = -18$

You Try It 11 $-\dfrac{36}{-12} = -(-3)$

$= 3$

You Try It 12

Strategy

To find the average low temperature:
• Add the seven temperature readings.
• Divide the sum by 7.

Solution

$-6 + (-7) + 0 + (-5) + (-8) + (-1) + (-1) = -28$

$-28 \div 7 = -4$

The average daily low temperature was $-4°C$.

SECTION 1.2 *pages 17–26*

You Try It 1

$$9\overline{)4.000}$$ with quotient 0.444, $\dfrac{4}{9} = 0.\overline{4}$

$$\begin{array}{r} 0.444 \\ 9\overline{)4.000} \\ -3\,6 \\ \hline 40 \\ -36 \\ \hline 40 \\ -36 \\ \hline 4 \end{array}$$

You Try It 2

$$125\% = 125\left(\frac{1}{100}\right) = \frac{125}{100} = 1\frac{1}{4}$$

$$125\% = 125(0.01) = 1.25$$

You Try It 3

$$\frac{1}{3} = \frac{1}{3}(100\%)$$

$$= \frac{100}{3}\% = 33\frac{1}{3}\%$$

You Try It 4

$$0.043 = 0.043(100\%) = 4.3\%$$

You Try It 5 The LCM of 8, 6, and 4 is 24.

$$-\frac{7}{8} - \frac{5}{6} + \frac{3}{4} = -\frac{21}{24} - \frac{20}{24} + \frac{18}{24}$$

$$= \frac{-21}{24} + \frac{-20}{24} + \frac{18}{24}$$

$$= \frac{-21 - 20 + 18}{24}$$

$$= \frac{-23}{24} = -\frac{23}{24}$$

You Try It 6

$$16.127 - 67.91$$

$$= 16.127 + (-67.91)$$

$$= -51.783$$

You Try It 7 The quotient is positive.

$$-\frac{3}{8} \div \left(-\frac{5}{12}\right) = \frac{3}{8} \cdot \frac{12}{5}$$

$$= \frac{3 \cdot 12}{8 \cdot 5}$$

$$= \frac{3 \cdot \overset{1}{\cancel{2}} \cdot \overset{1}{\cancel{2}} \cdot 3}{\underset{1}{\cancel{2}} \cdot \underset{1}{\cancel{2}} \cdot 2 \cdot 5} = \frac{9}{10}$$

You Try It 8

$$\begin{array}{r} 5.44 \\ \times \ \ 3.8 \\ \hline 4352 \\ 1632 \ \ \\ \hline 20.672 \end{array}$$

$$-5.44(3.8) = -20.672$$

You Try It 9 $\quad -6^3 = -(6 \cdot 6 \cdot 6) = -216$

You Try It 10
$$(-3)^4 = (-3)(-3)(-3)(-3) = 81$$

You Try It 11
$$(3^3)(-2)^3 = (3)(3)(3) \cdot (-2)(-2)(-2)$$
$$= 27(-8) = -216$$

You Try It 12 $\quad \left(-\frac{2}{5}\right)^2 = \left(-\frac{2}{5}\right)\left(-\frac{2}{5}\right) = \frac{4}{25}$

You Try It 13 $\quad -3(0.3)^3 = -3(0.3)(0.3)(0.3)$
$$= -0.9(0.3)(0.3)$$
$$= -0.27(0.3) = -0.081$$

You Try It 14
$$-5\sqrt{32} = -5\sqrt{2^5} = -5\sqrt{2^4 \cdot 2} = -5\sqrt{2^4}\sqrt{2}$$
$$= -5 \cdot 2^2\sqrt{2} = -20\sqrt{2}$$

You Try It 15

$$\sqrt{216} = \sqrt{2^3 \cdot 3^3} = \sqrt{2^2 \cdot 3^2(2 \cdot 3)}$$
$$= \sqrt{2^2 \cdot 3^2}\sqrt{2 \cdot 3} = 2 \cdot 3\sqrt{2 \cdot 3} = 6\sqrt{6}$$

You Try It 16

Strategy To find the percent spent for Procardia:
- Add the 5 amounts shown in the circle graph to find the total spent.
- Divide the amount spent on Procardia (1.1 billion) by the total amount spent.
- Convert the decimal to a percent.

Solution $0.8 + 1.7 + 1.1 + 1.0 + 0.9 = 5.5$
The total amount spent was $5.5 billion.

$$\frac{1.1}{5.5} = 0.2$$

$$0.2 = 0.2(100\%) = 20\%$$

20% was spent for Procardia.

SECTION 1.3 *pages 33–34*

You Try It 1
$$18 - 5[8 - 2(2 - 5)] \div 10$$
$$= 18 - 5[8 - 2(-3)] \div 10$$
$$= 18 - 5[8 + 6] \div 10$$
$$= 18 - 5[14] \div 10$$
$$= 18 - 70 \div 10$$
$$= 18 - 7$$
$$= 11$$

You Try It 2
$$36 \div (8 - 5)^2 - (-3)^2 \cdot 2$$
$$= 36 \div (3)^2 - (-3)^2 \cdot 2$$
$$= 36 \div 9 - 9 \cdot 2$$
$$= 4 - 9 \cdot 2$$
$$= 4 - 18$$
$$= -14$$

You Try It 3
$$(6.97 - 4.72)^2 \cdot 4.5 \div 0.05$$
$$= (2.25)^2 \cdot 4.5 \div 0.05$$
$$= 5.0625 \cdot 4.5 \div 0.05$$
$$= 22.78125 \div 0.05$$
$$= 455.625$$

SECTION 1.4 *pages 37–46*

You Try It 1 $\dfrac{a^2 + b^2}{a + b}$

$$\dfrac{5^2 + (-3)^2}{5 + (-3)} = \dfrac{25 + 9}{5 + (-3)}$$
$$= \dfrac{34}{2}$$
$$= 17$$

You Try It 2 $x^3 - 2(x + y) + z^2$

$$(2)^3 - 2[2 + (-4)] + (-3)^2$$
$$= (2)^3 - 2(-2) + (-3)^2$$
$$= 8 - 2(-2) + 9$$
$$= 8 + 4 + 9$$
$$= 12 + 9$$
$$= 21$$

You Try It 3
$3a - 2b - 5a + 6b = -2a + 4b$

You Try It 4
$-3y^2 + 7 + 8y^2 - 14 = 5y^2 - 7$

You Try It 5 $-5(4y^2) = -20y^2$

You Try It 6 $-7(-2a) = 14a$

You Try It 7 $(-5x)(-2) = 10x$

You Try It 8 $-8(-2a + 7b) = 16a - 56b$

You Try It 9 $(3a - 1)5 = 15a - 5$

You Try It 10 $2(x^2 - x + 7) = 2x^2 - 2x + 14$

You Try It 11 $3y - 2(y - 7x) = 3y - 2y + 14x$
$$= 14x + y$$

You Try It 12
$-2(x - 2y) - (-x + 3y) = -2x + 4y + x - 3y$
$$= -x + y$$

You Try It 13
$3y - 2[x - 4(2 - 3y)] = 3y - 2[x - 8 + 12y]$
$$= 3y - 2x + 16 - 24y$$
$$= -2x - 21y + 16$$

You Try It 14 the unknown number: x
the difference between the number and sixty: $x - 60$

$5(x - 60);\ 5x - 300$

You Try It 15 the speed of the older model: s
the speed of the new model: $2s$

You Try It 16 the length of the longer piece: y
the length of the shorter piece: $6 - y$

SECTION 1.5 *pages 55–58*

You Try It 1 $A = \{-9, -7, -5, -3, -1\}$

You Try It 2 $A = \{1, 3, 5, \ldots\}$

You Try It 3 $A \cup B = \{-2, -1, 0, 1, 2, 3, 4\}$

You Try It 4 $C \cap D = \{10, 16\}$

You Try It 5 $A \cap B = \varnothing$

You Try It 6 $\{x \mid x < 59, x \in \text{positive even integers}\}$

You Try It 7 $\{x \mid x > -3, x \in \text{real numbers}\}$

You Try It 8 The solution set is the numbers greater than -2.

You Try It 9 The solution set is the numbers less than or equal to 4 and the numbers greater than or equal to -4.

You Try It 10 The solution set is the real numbers.

You Try It 11 The solution set is the numbers greater than -1 and the numbers less than -3.

SOLUTIONS to Chapter 2 You Try It

You Try It 1

$$5 - 4x = 8x + 2$$

$$5 - 4\left(\frac{1}{4}\right) \;\Big|\; 8\left(\frac{1}{4}\right) + 2$$

$$5 - 1 \;\Big|\; 2 + 2$$

$$4 = 4$$

Yes, $\frac{1}{4}$ is a solution.

You Try It 2

$$10x - x^2 = 3x - 10$$

$$10(5) - (5)^2 \;\Big|\; 3(5) - 10$$

$$50 - 25 \;\Big|\; 15 - 10$$

$$25 \neq 5$$

No, 5 is not a solution.

You Try It 3

$$\frac{5}{6} = y - \frac{3}{8}$$

$$\frac{5}{6} + \frac{3}{8} = y - \frac{3}{8} + \frac{3}{8}$$

$$\frac{29}{24} = y$$

The solution is $\frac{29}{24}$.

You Try It 4

$$-\frac{2x}{5} = 6$$

$$\left(-\frac{5}{2}\right)\left(-\frac{2}{5}x\right) = \left(-\frac{5}{2}\right)(6)$$

$$x = -15$$

The solution is -15.

You Try It 5

$$4x - 8x = 16$$

$$-4x = 16$$

$$\frac{-4x}{-4} = \frac{16}{-4}$$

$$x = -4$$

The solution is -4.

You Try It 1

Strategy
To find the amount, solve the basic percent equation.

Percent $= 33\frac{1}{3}\% = \frac{1}{3}$, base $= 45$, amount $= A$

Solution
Percent $\cdot$ base $=$ amount

$$\frac{1}{3}(45) = A$$

$$15 = A$$

15 is $33\frac{1}{3}\%$ of 45.

You Try It 2

Strategy
To find the percent, solve the basic percent equation. Percent $= P$, base $= 40$, amount $= 25$

Solution
Percent $\cdot$ base $=$ amount

$$P \cdot 40 = 25$$

$$P = \frac{25}{40} = 0.625$$

$$P = 62.5\%$$

25 is 62.5% of 40.

You Try It 3

Strategy

To find the base, solve the basic percent equation. Percent $= 16\frac{2}{3}\% = \frac{1}{6}$, base $= B$, amount $= 15$

Solution

Percent · base = amount

$$\frac{1}{6} \cdot B = 15$$

$$B = 15 \cdot 6$$

$$B = 90$$

$16\frac{2}{3}\%$ of 90 is 15.

You Try It 5

Strategy

To find the percent, use the basic percent equation. Percent $= P$, base $= 2165$, amount $= 324.75$

Solution

Percent · base = amount

$$P \cdot 2165 = 324.75$$

$$P = \frac{324.75}{2165}$$

$$P = 0.15 = 15\%$$

15% of the instructor's salary is deducted for income tax.

You Try It 7

Strategy

To find the increase in the hourly wage:

- Find last year's wage. Solve the basic percent equation. Percent $= 115\% = 1.15$, base $= B$, amount $= 20.01$
- Subtract last year's wage from this year's wage.

Solution

Percent · base = amount

$$1.15 \cdot B = 20.01$$

$$B = \frac{20.01}{1.15}$$

$$B = 17.40$$

$$20.01 - 17.40 = 2.61$$

The increase in the hourly wage was $2.61.

You Try It 4

Strategy

To find the amount budgeted for food, solve the basic percent equation for amount. Percent $= 8\% = 0.08$, base $= 26,000$, amount $= A$

Solution

Percent · base = amount

$$0.08 \cdot 26,000 = A$$

$$2080 = A$$

The amount budgeted for food is $2080.

You Try It 6

Strategy

To find the number polled, solve the base percent equation. Percent $= 70\% = 0.70$, base $= B$, amount $= 210$

Solution

Percent · base = amount

$$0.70 \cdot B = 210$$

$$B = \frac{210}{0.70}$$

$$B = 300$$

300 people were polled.

You Try It 1

$$5x + 7 = 10$$
$$5x + 7 - 7 = 10 - 7$$
$$5x = 3$$
$$\frac{5x}{5} = \frac{3}{5}$$
$$x = \frac{3}{5}$$

The solution is $\frac{3}{5}$.

You Try It 2

$$2 = 11 + 3x$$
$$2 - 11 = 11 - 11 + 3x$$
$$-9 = 3x$$
$$\frac{-9}{3} = \frac{3x}{3}$$
$$-3 = x$$

The solution is -3.

You Try It 3

$$x - 5 + 4x = 25$$
$$5x - 5 = 25$$
$$5x - 5 + 5 = 25 + 5$$
$$5x = 30$$
$$\frac{5x}{5} = \frac{30}{5}$$
$$x = 6$$

The solution is 6.

You Try It 4

$$5x + 4 = 6 + 10x$$
$$5x - 10x + 4 = 6 + 10x - 10x$$
$$-5x + 4 = 6$$
$$-5x + 4 - 4 = 6 - 4$$
$$-5x = 2$$
$$\frac{-5x}{-5} = \frac{2}{-5}$$
$$x = -\frac{2}{5}$$

The solution is $-\frac{2}{5}$.

You Try It 5

$$5x - 10 - 3x = 6 - 4x$$
$$2x - 10 = 6 - 4x$$
$$2x + 4x - 10 = 6 - 4x + 4x$$
$$6x - 10 = 6$$
$$6x - 10 + 10 = 6 + 10$$
$$6x = 16$$
$$\frac{6x}{6} = \frac{16}{6}$$
$$x = \frac{8}{3}$$

The solution is $\frac{8}{3}$.

You Try It 6

$$5x - 4(3 - 2x) = 2(3x - 2) + 6$$
$$5x - 12 + 8x = 6x - 4 + 6$$
$$13x - 12 = 6x + 2$$
$$13x - 6x - 12 = 6x - 6x + 2$$
$$7x - 12 = 2$$
$$7x - 12 + 12 = 2 + 12$$
$$7x = 14$$
$$\frac{7x}{7} = \frac{14}{7}$$
$$x = 2$$

The solution is 2.

You Try It 7

$$-2[3x - 5(2x - 3)] = 3x - 8$$
$$-2[3x - 10x + 15] = 3x - 8$$
$$-2[-7x + 15] = 3x - 8$$
$$14x - 30 = 3x - 8$$
$$14x - 3x - 30 = 3x - 3x - 8$$
$$11x - 30 = -8$$
$$11x - 30 + 30 = -8 + 30$$
$$11x = 22$$
$$\frac{11x}{11} = \frac{22}{11}$$
$$x = 2$$

The solution is 2.

You Try It 8

the unknown number: x

seven less than a number	is equal to	five more than three times the number

$$x - 7 = 3x + 5$$
$$x - 3x - 7 = 3x - 3x + 5$$
$$-2x - 7 = 5$$
$$-2x - 7 + 7 = 5 + 7$$
$$-2x = 12$$
$$\frac{-2x}{-2} = \frac{12}{-2}$$
$$x = -6$$

-6 checks as the solution.

The solution is -6.

You Try It 9

the smaller number: n
the larger number: $14 - n$

one more than three times the smaller number	equals	the sum of the larger number and three

$$3n + 1 = (14 - n) + 3$$
$$3n + 1 = 17 - n$$
$$3n + n + 1 = 17 - n + n$$
$$4n + 1 = 17$$
$$4n + 1 - 1 = 17 - 1$$
$$4n = 16$$
$$\frac{4n}{4} = \frac{16}{4}$$
$$n = 4$$

$14 - n = 14 - 4 = 10$

These numbers check as solutions.

The smaller number is 4.
The larger number is 10.

SECTION 2.4 *pages 99–106*

You Try It 1

Strategy
• Pounds of $.55 fertilizer: x

	Amount	Cost	Value
$.80 fertilizer	20	$.80	0.80(20)
$.55 fertilizer	x	$.55	0.55x
$.75 fertilizer	20 + x	$.75	0.75(20 + x)

• The sum of the values before mixing equals the value after mixing.

Solution
$$0.80(20) + 0.55x = 0.75(20 + x)$$
$$16 + 0.55x = 15 + 0.75x$$
$$16 - 0.20x = 15$$
$$-0.20x = -1$$
$$x = 5$$

5 lb of the $.55 fertilizer must be added.

You Try It 3

Strategy
• Additional amount: x

	Principal	Rate	Interest
8%	5000	0.08	0.08(5000)
11%	x	0.11	0.11x
9%	5000 + x	0.09	0.09(5000 + x)

• The sum of the interest earned by the two investments equals 9% of the total investment.

Solution
$$0.08(5000) + 0.11x = 0.09(5000 + x)$$
$$400 + 0.11x = 450 + 0.09x$$
$$400 + 0.02x = 450$$
$$0.02x = 50$$
$$x = 2500$$

$2500 more must be invested at 11%.

You Try It 2

Strategy
• Liters of the 6% solution: x

	Amount	Percent	Quantity
6%	x	0.06	0.06x
12%	5	0.12	5(0.12)
8%	x + 5	0.08	0.08(x + 5)

• The sum of the quantities before mixing equals the quantity after mixing.

Solution
$$0.06x + 5(0.12) = 0.08(x + 5)$$
$$0.06x + 0.60 = 0.08x + 0.40$$
$$0.06x + 0.20 = 0.08x$$
$$0.20 = 0.02x$$
$$10 = x$$

The pharmacist adds 10 L of the 6% solution to the 12% solution to get an 8% solution.

You Try It 4

Strategy

• Rate of the first train: r
  Rate of the second train: $2r$

	Rate	Time	Distance
1st train	r	3	$3r$
2nd train	$2r$	3	$3(2r)$

• The sum of the distances traveled by each train equals 288 mi.

Solution

$$3r + 3(2r) = 288$$
$$3r + 6r = 288$$
$$9r = 288$$
$$r = 32$$

$$2r = 2(32) = 64$$

The first train is traveling at 32 mph.
The second train is traveling at 64 mph.

You Try It 5

Strategy

• Time spent flying out: t
  Time spent flying back: $5 - t$

	Rate	Time	Distance
Out	150	t	$150t$
Back	100	$5 - t$	$100(5 - t)$

• The distance out equals the distance back.

Solution

$$150t = 100(5 - t)$$
$$150t = 500 - 100t$$
$$250t = 500$$
$$t = 2 \text{ (The time out was 2 h.)}$$

The distance $= 150t = 150(2) = 300$ mi.

The parcel of land was 300 mi away.

SECTION 2.5   *pages 115–120*

You Try It 1

$$2x - 1 < 6x + 7$$
$$-4x - 1 < 7$$
$$-4x < 8$$
$$\frac{-4x}{-4} > \frac{8}{-4}$$
$$x > -2$$
$$\{x \mid x > -2\}$$

You Try It 2

$$5x - 2 \le 4 - 3(x - 2)$$
$$5x - 2 \le 4 - 3x + 6$$
$$5x - 2 \le 10 - 3x$$
$$8x - 2 \le 10$$
$$8x \le 12$$
$$\frac{8x}{8} \le \frac{12}{8}$$
$$x \le \frac{3}{2}$$

$$\left\{ x \mid x \le \frac{3}{2} \right\}$$

You Try It 3

$$-2 \le 5x + 3 \le 13$$
$$-2 - 3 \le 5x + 3 - 3 \le 13 - 3$$
$$-5 \le 5x \le 10$$
$$\frac{-5}{5} \le \frac{5x}{5} \le \frac{10}{5}$$
$$-1 \le x \le 2$$
$$\{x \mid -1 \le x \le 2\}$$

You Try It 4

$$2 - 3x > 11 \quad \text{or} \quad 5 + 2x > 7$$
$$-3x > 9 \qquad\qquad 2x > 2$$
$$x < -3 \qquad\qquad x > 1$$
$$\{x \mid x < -3\} \qquad \{x \mid x > 1\}$$

$$\{x \mid x < -3\} \cup \{x \mid x > 1\}$$
$$= \{x \mid x < -3 \text{ or } x > 1\}$$

You Try It 5

Strategy   To find the maximum number of miles:
- Write an expression for the cost of each car, using x to represent the number of miles driven during the week.
- Write and solve an inequality.

Solution

Cost of a Company A car	is less than	Cost of a Company B car

$$8(7) + 0.10x < 10(7) + 0.08x$$
$$56 + 0.10x < 70 + 0.08x$$
$$56 + 0.10x - 0.08x < 70 + 0.08x - 0.08x$$
$$56 + 0.02x < 70$$
$$56 - 56 + 0.02x < 70 - 56$$
$$0.02x < 14$$
$$\frac{0.02x}{0.02} < \frac{14}{0.02}$$
$$x < 700$$

The maximum number of miles is 699.

SECTION 2.6   *pages 127–130*

You Try It 1

$|2x - 3| = 5$

$2x - 3 = 5$	$2x - 3 = -5$
$2x = 8$	$2x = -2$
$x = 4$	$x = -1$

The solutions are 4 and -1.

You Try It 2

$5 - |3x + 5| = 3$

$-|3x + 5| = -2$

$|3x + 5| = 2$

$3x + 5 = 2$	$3x + 5 = -2$
$3x = -3$	$3x = -7$
$x = -1$	$x = -\dfrac{7}{3}$

The solutions are -1 and $-\dfrac{7}{3}$.

You Try It 3

$|3x + 2| < 8$

$$-8 < 3x + 2 < 8$$
$$-8 - 2 < 3x + 2 - 2 < 8 - 2$$
$$-10 < 3x < 6$$
$$\frac{-10}{3} < \frac{3x}{3} < \frac{6}{3}$$
$$-\frac{10}{3} < x < 2$$

$$\left\{ x \,\middle|\, -\frac{10}{3} < x < 2 \right\}$$

You Try It 4

$|5x + 3| > 8$

$$5x + 3 < -8 \quad \text{or} \quad 5x + 3 > 8$$
$$5x < -11 \qquad\qquad 5x > 5$$
$$x < -\frac{11}{5} \qquad\qquad x > 1$$

$$\left\{ x \,\middle|\, x < -\frac{11}{5} \right\} \qquad \{x \,|\, x > 1\}$$

$$\left\{ x \,\middle|\, x < -\frac{11}{5} \right\} \cup \{x \,|\, x > 1\}$$

$$= \left\{ x \,\middle|\, x < -\frac{11}{5} \text{ or } x > 1 \right\}$$

You Try It 5

Strategy

Let b represent the diameter of the bushing, T the tolerance, and d the lower and upper limits of the diameter. Solve the absolute value inequality $|d - b| \le T$ for d.

Solution

$$|d - b| \le T$$
$$|d - 2.55| \le 0.003$$

$$-0.003 \le d - 2.55 \le 0.003$$
$$-0.003 + 2.55 \le d - 2.55 + 2.55 \le 0.003 + 2.55$$
$$2.547 \le d \le 2.553$$

The lower and upper limits of the diameter of the bushing are 2.547 in. and 2.553 in.

SOLUTIONS to Chapter 3 You Try It

SECTION 3.1   pages 143–152

You Try It 1 $QR + RS + ST = QT$
$24 + RS + 17 = 62$
$41 + RS = 62$
$RS = 21$

$RS = 21$ cm

You Try It 2 $AC = AB + BC$
$AC = \frac{1}{4}(BC) + BC$
$AC = \frac{1}{4}(16) + 16$
$AC = 4 + 16$
$AC = 20$

$AC = 20$ ft

You Try It 3

Strategy   Supplementary angles are two angles whose sum is 180°. To find the supplement, let x represent the supplement of a 129° angle. Write an equation and solve for x.

Solution $x + 129° = 180°$
$x = 51°$

The supplement of a 129° angle is a 51° angle.

You Try It 4

Strategy   To find the measure of $\angle a$, write an equation using the fact that the sum of the measure of $\angle a$ and 68° is 118°. Solve for $\angle a$.

Solution $\angle a + 68° = 118°$
$\angle a = 50°$

The measure of $\angle a$ is 50°.

You Try It 5

Strategy   The angles labeled are adjacent angles of intersecting lines and are, therefore, supplementary angles. To find x, write an equation and solve for x.

Solution $(x + 16°) + 3x = 180°$
$4x + 16° = 180°$
$4x = 164°$
$x = 41°$

You Try It 6

Strategy $3x = y$ because corresponding angles have the same measure. $y + (x + 40°) = 180°$ because adjacent angles of intersecting lines are supplementary angles. Substitute $3x$ for y and solve for x.

Solution $3x + (x + 40°) = 180°$
$4x + 40° = 180°$
$4x = 140°$
$x = 35°$

You Try It 7

Strategy   (1) To find the measure of angle b, use the fact that $\angle b$ and $\angle x$ are supplementary angles. (2) To find the measure of angle c, use the fact that the sum of the interior angles of a triangle is 180°. (3) To find the measure of angle y, use the fact that $\angle c$ and $\angle y$ are vertical angles.

Solution $\angle b + \angle x = 180°$
$\angle b + 100° = 180°$
$\angle b = 80°$

$\angle a + \angle b + \angle c = 180°$
$45° + 80° + \angle c = 180°$
$125° + \angle c = 180°$
$\angle c = 55°$
$\angle y = \angle c = 55°$

You Try It 8

Strategy   To find the measure of the third angle, use the facts that the measure of a right angle is 90° and the sum of the measures of the interior angles of a triangle is 180°. Write an equation using x to represent the measure of the third angle. Solve the equation for x.

Solution $x + 90° + 34° = 180°$
$x + 124° = 180°$
$x = 56°$

The measure of the third angle is 56°.

SECTION 3.2 *pages 159–168*

You Try It 1

Strategy
To find the perimeter, use the formula for the perimeter of a square. Substitute 60 for s and solve for P.

Solution
$P = 4s$
$P = 4(60)$
$P = 240$

The perimeter of the infield is 240 ft.

You Try It 2

Strategy
To find the perimeter, use the formula for the perimeter of a rectangle. Substitute 11 for L and $8\frac{1}{2}$ for W and solve for P.

Solution
$P = 2L + 2W$
$P = 2(11) + 2\left(8\frac{1}{2}\right)$
$P = 2(11) + 2\left(\frac{17}{2}\right)$
$P = 22 + 17$
$P = 39$

The perimeter of a standard piece of typing paper is 39 in.

You Try It 3

Strategy
To find the circumference, use the circumference formula that involves the diameter. Leave the answer in terms of π.

Solution
$C = \pi d$
$C = \pi(9)$
$C = 9\pi$

The circumference is 9π in.

You Try It 4

Strategy
To find the number of rolls of wallpaper to be purchased:
- Use the formula for the area of a rectangle to find the area of one wall.
- Multiply the area of one wall by the number of walls to be covered (2).
- Divide the area of wall to be covered by the area that one roll of wallpaper will cover (30).

Solution
$A = LW$
$A = 12 \cdot 8 = 96$ The area of one wall is 96 ft^2.

$2(96) = 192$ The area of the two two walls is 192 ft^2.

$192 \div 30 = 6.4$

Because a portion of a seventh roll is needed, 7 rolls of wallpaper should be purchased.

You Try It 5

Strategy
To find the area, use the formula for the area of a circle. An approximation is asked for; use the π key on a calculator.
$r = 11$.

Solution
$A = \pi r^2$
$A = \pi(11)^2$
$A = 121\pi$
$A \approx 380.13$

The area is approximately 380.13 cm^2.

You Try It 1

Strategy

To find the volume, use the formula for the volume of a cube. $s = 2.5$.

Solution

$V = s^3$

$V = (2.5)^3 = 15.625$

The volume of the cube is 15.625 m³.

You Try It 3

Strategy

To find the surface area of the cylinder:

- Find the radius of the base of the cylinder. $d = 6$.
- Use the formula for the surface area of a cylinder. An approximation is asked for; use the π key on a calculator.

Solution

$r = \dfrac{1}{2}d = \dfrac{1}{2}(6) = 3$

$SA = 2\pi r^2 + 2\pi rh$

$SA = 2\pi(3)^2 + 2\pi(3)(8)$

$\quad = 2\pi(9) + 2\pi(3)(8)$

$\quad = 18\pi + 48\pi$

$\quad = 66\pi$

$\quad \approx 207.35$

The surface area of the cylinder is approximately 207.35 ft².

You Try It 2

Strategy

To find the volume:

- Find the radius of the base of the cylinder. $d = 8$.
- Use the formula for the volume of a cylinder. Leave the answer in terms of π.

Solution

$r = \dfrac{1}{2}d = \dfrac{1}{2}(8) = 4$

$V = \pi r^2 h = \pi(4)^2(22) = \pi(16)(22) = 352\pi$

The volume of the cylinder is 352π ft³.

You Try It 4

Strategy

To find which solid has the larger surface area:

- Use the formula for the surface area of a cube to find the surface area of the cube. $s = 10$.
- Find the radius of the sphere. $d = 8$.
- Use the formula for the surface area of a sphere to find the surface area of the sphere. Because this number is to be compared to another number, use the π key on a calculator to approximate the surface area.
- Compare the two numbers.

Solution

$SA = 6s^2$

$SA = 6(10)^2 = 6(100) = 600$

The surface area of the cube is 600 cm².

$r = \dfrac{1}{2}d = \dfrac{1}{2}(8) = 4$

$SA = 4\pi r^2$

$SA = 4\pi(4)^2 = 4\pi(16) = 64\pi \approx 201.06$

The surface area of the sphere is approximately 201.06 cm².

$600 > 201.06$

The cube has a larger surface area than the sphere.

SOLUTIONS to Chapter 4 You Try It

SECTION 4.1   *pages 195–202*

You Try It 1

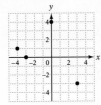

You Try It 2   The coordinates of A are $(4, -2)$.
The coordinates of B are $(-2, 4)$.
The coordinates of C are $(-2, 0)$.
The coordinates of D are $(0, -3)$.

You Try It 3

$$x - 3y = -14$$

$-2 - 3(4)$	-14
$-2 - 12$	-14
$-14 = -14$	

Yes, $(-2, 4)$ is a solution of $x - 3y = -14$.

You Try It 4   Replace x by -2 and solve for y.

$$y = \frac{3(-2)}{-2 + 1} = \frac{-6}{-1} = 6$$

The ordered-pair solution is $(-2, 6)$.

You Try It 5

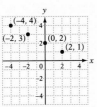

You Try It 6   {(82, 91), (78, 86), (81, 96), (87, 79), (81, 87)} The two ordered pairs (81, 96) and (81, 87) have the same first coordinate but different second coordinates.

No, the relation is not a function.

You Try It 7   Determine the ordered pairs defined by the equation. Replace x in $y = \frac{1}{2}x + 1$ by the given values and solve for y.
{$(-4, -1), (0, 1), (2, 2)$}

Yes, y is a function of x.

You Try It 8 $H(x) = \dfrac{x}{x - 4}$

$$H(8) = \frac{8}{8 - 4}$$

$$H(8) = \frac{8}{4} = 2$$

SECTION 4.2   *pages 207–214*

You Try It 1

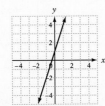

You Try It 2

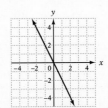

You Try It 3

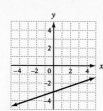

You Try It 4
$$5x - 2y = 10$$
$$-2y = -5x + 10$$
$$y = \frac{5}{2}x - 5$$

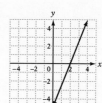

You Try It 5
x-intercept: $x - 4y = -4$
$$x - 4(0) = -4$$
$$x = -4$$
$(-4, 0)$
y-intercept: $x - 4y = -4$
$$0 - 4y = -4$$
$$-4y = -4$$
$$y = 1$$
$(0, 1)$

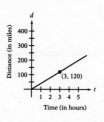

You Try It 6

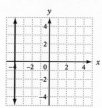

You Try It 7

The ordered pair (3, 120) means that in 3 h the car will travel 120 mi.

SECTION 4.3 *pages 219–224*

You Try It 1
Let $P_1 = (4, -3)$ and $P_2 = (2, 7)$.

$$m = \frac{y_2 - y_1}{x_2 - x_1} = \frac{7 - (-3)}{2 - 4} = \frac{10}{-2} = -5$$

The slope is -5.

You Try It 2
Let $P_1 = (6, -1)$ and $P_2 = (6, 7)$.

$$m = \frac{y_2 - y_1}{x_2 - x_1} = \frac{7 - (-1)}{6 - 6}$$

$$m = \frac{8}{0} \quad \text{Division by zero is not defined.}$$

The slope of the line is undefined.

You Try It 3

$$m = \frac{55{,}000 - 25{,}000}{2 - 5}$$

$$= \frac{30{,}000}{-3}$$

$$= -10{,}000$$

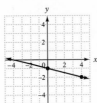

A slope of $-10{,}000$ means that the value of the printing press is decreasing by \$10,000 per year.

You Try It 4

y-intercept $= (0, b) = (0, -1)$

$$m = -\frac{1}{4}$$

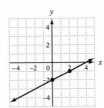

You Try It 5   Solve the equation for y.

$$x - 2y = 4$$
$$-2y = -x + 4$$
$$y = \frac{1}{2}x - 2$$

y-intercept $= (0, b) = (0, -2)$

$$m = \frac{1}{2}$$

SECTION 4.4   *pages 229–232*

You Try It 1

Because the slope and y-intercept are known, use the slope–intercept formula, $y = mx + b$.

$$y = mx + b$$
$$y = \frac{5}{3}x + 2$$

You Try It 2

$$m = -\frac{1}{3} \quad (x_1, y_1) = (-3, -2)$$

$$y - y_1 = m(x - x_1)$$
$$y - (-2) = -\frac{1}{3}[x - (-3)]$$
$$y + 2 = -\frac{1}{3}(x + 3)$$
$$y + 2 = -\frac{1}{3}x - 1$$
$$y = -\frac{1}{3}x - 1 - 2$$
$$y = -\frac{1}{3}x - 3$$

The equation of the line is $y = -\frac{1}{3}x - 3$.

You Try It 3 $m = -3 \quad (x_1, y_1) = (4, -3)$

$$y - y_1 = m(x - x_1)$$
$$y - (-3) = -3(x - 4)$$
$$y + 3 = -3x + 12$$
$$y = -3x + 12 - 3$$
$$y = -3x + 9$$

The equation of the line is $y = -3x + 9$.

You Try It 4 Find the slope of the line between the two points.

$$m = \frac{y_2 - y_1}{x_2 - x_1} = \frac{1 - (-1)}{3 - (-6)} = \frac{2}{9}$$

Use the point–slope formula.

$$y - y_1 = m(x - x_1)$$
$$y - (-1) = \frac{2}{9}[x - (-6)]$$
$$y + 1 = \frac{2}{9}x + \frac{4}{3}$$
$$y = \frac{2}{9}x + \frac{1}{3}$$

You Try It 5

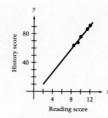

The slope of the line means that the grade on the history test increases 8.3 points for each 1-point increase in the grade on the reading test.

SECTION 4.5 *pages 237–240*

You Try It 1 $m_1 = \dfrac{1 - (-3)}{7 - (-2)} = \dfrac{4}{9}$

$$m_2 = \frac{-5 - 1}{6 - 4} = \frac{-6}{2} = -3$$

$$m_1 \cdot m_2 = \frac{4}{9} \cdot -3 = -\frac{4}{3}$$

No, the lines are not perpendicular.

You Try It 2 $5x + 2y = 2$
$$2y = -5x + 2$$
$$y = -\frac{5}{2}x + 1$$

$$m_1 = -\frac{5}{2}$$
$$5x + 2y = -6$$
$$2y = -5x - 6$$
$$y = -\frac{5}{2}x - 3$$

$$m_2 = -\frac{5}{2}$$

$$m_1 = m_2 = -\frac{5}{2}$$

Yes, the lines are parallel.

You Try It 3 $y = \dfrac{1}{4}x - 3$

$$m_1 = \frac{1}{4}$$

$$m_1 \cdot m_2 = -1$$
$$\frac{1}{4} \cdot m_2 = -1$$
$$m_2 = -4$$
$$y - y_1 = m(x - x_1)$$
$$y - 2 = -4[x - (-2)]$$
$$y - 2 = -4(x + 2)$$
$$y - 2 = -4x - 8$$
$$y = -4x - 6$$

The equation of the line is $y = -4x - 6$.

SECTION 4.6 *pages 243–244*

You Try It 1

$$x - 3y < 2$$
$$x - x - 3y < -x + 2$$
$$-3y < -x + 2$$
$$\frac{-3y}{-3} > \frac{-x + 2}{-3}$$
$$y > \frac{1}{3}x - \frac{2}{3}$$

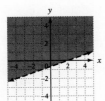

SOLUTIONS to Chapter 5 You Try It

You Try It 1

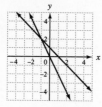

The solution is $(-1, 2)$.

You Try It 2

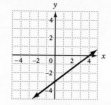

The two equations represent the same line. The system of equations is dependent. The solutions are the ordered pairs $\left(x, \frac{3}{4}x - 3\right)$.

You Try It 1

(1) $\quad 7x - y = 4$
(2) $\quad 3x + 2y = 9$

Solve Equation (1) for y.

$$7x - y = 4$$
$$-y = -7x + 4$$
$$y = 7x - 4$$

Substitute in Equation (2).

$$3x + 2y = 9$$
$$3x + 2(7x - 4) = 9$$
$$3x + 14x - 8 = 9$$
$$17x - 8 = 9$$
$$17x = 17$$
$$x = 1$$

Substitute in Equation (1).

$$7x - y = 4$$
$$7(1) - y = 4$$
$$7 - y = 4$$
$$-y = -3$$
$$y = 3$$

The solution is $(1, 3)$.

You Try It 2

(1) $\quad 3x - y = 4$
(2) $\quad\quad\quad y = 3x + 2$

$$3x - y = 4$$
$$3x - (3x + 2) = 4$$
$$3x - 3x - 2 = 4$$
$$-2 = 4$$

This is not a true equation. The system of equations is inconsistent and therefore does not have a solution.

You Try It 3

$$(1) \qquad y = -2x + 1$$
$$(2) \quad 6x + 3y = 3$$

$$6x + 3y = 3$$
$$6x + 3(-2x + 1) = 3$$
$$6x - 6x + 3 = 3$$
$$3 = 3$$

The system of equations is dependent. The solutions are the ordered pairs $(x, -2x + 1)$.

SECTION 5.3   *pages 267–274*

You Try It 1

$$(1) \quad 2x + 5y = 6$$
$$(2) \quad 3x - 2y = 6x + 2$$

Write Equation (2) in the form $Ax + By = C$.

$$3x - 2y = 6x + 2$$
$$-3x - 2y = 2$$

Solve the system:

$$2x + 5y = 6$$
$$-3x - 2y = 2$$

Eliminate y.

$$2(2x + 5y) = 2(6)$$
$$5(-3x - 2y) = 5(2)$$

$$4x + 10y = 12$$
$$-15x - 10y = 10$$

Add the equations.

$$-11x = 22$$
$$x = -2$$

Replace x in Equation (1).

$$2x + 5y = 6$$
$$2(-2) + 5y = 6$$
$$-4 + 5y = 6$$
$$5y = 10$$
$$y = 2$$

The solution is $(-2, 2)$.

You Try It 2

$$2x + y = 5$$
$$4x + 2y = 6$$

Eliminate y.

$$-2(2x + y) = -2(5)$$
$$4x + 2y = 6$$

$$-4x - 2y = -10$$
$$4x + 2y = 6$$

Add the equations.

$$0x + 0y = -4$$
$$0 = -4$$

This is not a true equation. The system is inconsistent and therefore has no solution.

You Try It 3

(1) $\quad x - y + z = 6$
(2) $2x + 3y - z = 1$
(3) $x + 2y + 2z = 5$

Eliminate z. Add Equations (1) and (2).

$$x - y + z = 6$$
$$2x + 3y - z = 1$$
(4) $\quad 3x + 2y = 7$

Multiply Equation (2) by 2 and add to Equation (3).

$$4x + 6y - 2z = 2$$
$$x + 2y + 2z = 5$$
(5) $\quad 5x + 8y = 7$

Solve the system of two equations.

(4) $3x + 2y = 7$
(5) $5x + 8y = 7$

Multiply Equation (4) by -4 and add to Equation (5).

$$-12x - 8y = -28$$
$$5x + 8y = 7$$
$$-7x = -21$$
$$x = 3$$

Replace x by 3 in Equation (4).

$$3x + 2y = 7$$
$$3(3) + 2y = 7$$
$$9 + 2y = 7$$
$$2y = -2$$
$$y = -1$$

Replace x by 3 and y by -1 in Equation (1).

$$x - y + z = 6$$
$$3 - (-1) + z = 6$$
$$4 + z = 6$$
$$z = 2$$

The solution is $(3, -1, 2)$.

SECTION 5.4   *pages 279–284*

You Try It 1

$$\begin{vmatrix} -1 & -4 \\ 3 & -5 \end{vmatrix} = -1(-5) - 3(-4) = 5 + 12 = 17$$

The value of the determinant is 17.

You Try It 2

Expand by cofactors of the first row.

$$\begin{vmatrix} 1 & 4 & -2 \\ 3 & 1 & 1 \\ 0 & -2 & 2 \end{vmatrix}$$

$$= 1 \begin{vmatrix} 1 & 1 \\ -2 & 2 \end{vmatrix} - 4 \begin{vmatrix} 3 & 1 \\ 0 & 2 \end{vmatrix} + (-2) \begin{vmatrix} 3 & 1 \\ 0 & -2 \end{vmatrix}$$

$$= 1(2 + 2) - 4(6 - 0) - 2(-6 - 0)$$
$$= 4 - 24 + 12$$
$$= -8$$

The value of the determinant is -8.

You Try It 3

$$\begin{vmatrix} 3 & -2 & 0 \\ 1 & 4 & 2 \\ -2 & 1 & 3 \end{vmatrix}$$

$$= 3 \begin{vmatrix} 4 & 2 \\ 1 & 3 \end{vmatrix} - (-2) \begin{vmatrix} 1 & 2 \\ -2 & 3 \end{vmatrix} + 0 \begin{vmatrix} 1 & 4 \\ -2 & 1 \end{vmatrix}$$

$$= 3(12 - 2) + 2(3 + 4) + 0$$
$$= 3(10) + 2(7)$$
$$= 30 + 14$$
$$= 44$$

The value of the determinant is 44.

You Try It 4

$$3x - y = 4$$
$$6x - 2y = 5$$

$$D = \begin{vmatrix} 3 & -1 \\ 6 & -2 \end{vmatrix} = -6 + 6 = 0$$

Because $D = 0$, $\dfrac{D_x}{D}$ is undefined. Therefore, the system of equations is dependent or inconsistent.

It is not possible to solve this system by Cramer's Rule.

You Try It 5

$$2x - y + z = -1$$
$$3x + 2y - z = 3$$
$$x + 3y + z = -2$$

$$D = \begin{vmatrix} 2 & -1 & 1 \\ 3 & 2 & -1 \\ 1 & 3 & 1 \end{vmatrix} = 21$$

$$D_x = \begin{vmatrix} -1 & -1 & 1 \\ 3 & 2 & -1 \\ -2 & 3 & 1 \end{vmatrix} = 9$$

$$D_y = \begin{vmatrix} 2 & -1 & 1 \\ 3 & 3 & -1 \\ 1 & -2 & 1 \end{vmatrix} = -3$$

$$D_z = \begin{vmatrix} 2 & -1 & -1 \\ 3 & 2 & 3 \\ 1 & 3 & -2 \end{vmatrix} = -42$$

$$x = \frac{D_x}{D} = \frac{9}{21} = \frac{3}{7}$$

$$y = \frac{D_y}{D} = \frac{-3}{21} = -\frac{1}{7}$$

$$z = \frac{D_z}{D} = \frac{-42}{21} = -2$$

The solution is $\left(\dfrac{3}{7}, -\dfrac{1}{7}, -2 \right)$.

SECTION 5.5 *pages 287–290*

You Try It 1

Strategy

- Rate of the rowing team in calm water: t
  Rate of the current: c

	Rate	Time	Distance
With current	$t + c$	2	$2(t + c)$
Against current	$t - c$	2	$2(t - c)$

- The distance traveled with the current is 18 mi.
  The distance traveled against the current is 10 mi.

Solution

$2(t + c) = 18$ $\qquad \dfrac{1}{2} \cdot 2(t + c) = \dfrac{1}{2} \cdot 18$

$2(t - c) = 10$ $\qquad \dfrac{1}{2} \cdot 2(t - c) = \dfrac{1}{2} \cdot 10$

$$t + c = 9$$
$$t - c = 5$$

$$2t = 14$$
$$t = 7$$

$t + c = 9$
$7 + c = 9$
$\quad c = 2$

The rate of the rowing team in calm water is 7 mph.

The rate of the current is 2 mph.

You Try It 2

Strategy

- Amount invested at 8%: x
  Amounted invested at 7%: y

Investment	Amount	Interest Rate	Interest Earned
8%	x	0.08	$0.08x$
7%	y	0.07	$0.07y$

- The total amount invested was \$15,000.
  The total amount of interest earned was \$1170.

Solution

$$x + y = 15{,}000$$
$$0.08x + 0.07y = 1170$$

$$-8x - 8y = -120{,}000$$
$$8x + 7y = 117{,}000$$
$$-y = -3000$$
$$y = 3000$$

$$x + y = 15{,}000$$
$$x + 3000 = 15{,}000$$
$$x = 12{,}000$$

\$12,000 was invested in the 8% account.

\$3000 was invested in the 7% account.

You Try It 1

Shade above the solid line $y = 2x - 3$.
Shade above the dashed line $y = -3x$.

The solution set of the system is the intersection of the solution sets of the individual inequalities.

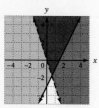

You Try It 2

$$3x + 4y > 12$$
$$4y > -3x + 12$$
$$y > -\frac{3}{4}x + 3$$

Shade above the dashed line $y = -\frac{3}{4}x + 3$.

Shade below the dashed line $y = \frac{3}{4}x - 1$.

The solution set of the system is the intersection of the solution sets of the individual inequalities.

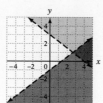

SOLUTIONS to Chapter 6 You Try It

SECTION 6.1   *pages 307–314*

You Try It 1
$$(-3a^2b^4)(-2ab^3)^4 = (-3a^2b^4)[(-2)^4a^4b^{12}]$$
$$= (-3a^2b^4)(16a^4b^{12})$$
$$= -48a^6b^{16}$$

You Try It 2
$$(y^{n-3})^2 = y^{(n-3)2} = y^{2n-6}$$

You Try It 3
$$[(ab^3)^3]^4 = [a^3b^9]^4 = a^{12}b^{36}$$

You Try It 4
$$\frac{20r^{-2}t^{-5}}{-16r^{-3}s^{-2}} = -\frac{4 \cdot 5r^{-2-(-3)}s^2t^{-5}}{4 \cdot 4}$$
$$= -\frac{5rs^2}{4t^5}$$

You Try It 5
$$\frac{(9u^{-6}v^4)^{-1}}{(6u^{-3}v^{-2})^{-2}} = \frac{9^{-1}u^6v^{-4}}{6^{-2}u^6v^4}$$
$$= 9^{-1} \cdot 6^2u^0v^{-8}$$
$$= \frac{36}{9v^8}$$
$$= \frac{4}{v^8}$$

You Try It 6
$$\frac{a^{2n+1}}{a^{n+3}} = a^{2n+1-(n+3)} = a^{2n+1-n-3} = a^{n-2}$$

You Try It 7
$$942{,}000{,}000 = 9.42 \times 10^8$$

You Try It 8
$$2.7 \times 10^{-5} = 0.000027$$

You Try It 9
$$\frac{5{,}600{,}000 \times 0.000000081}{900 \times 0.000000028}$$
$$= \frac{5.6 \times 10^6 \times 8.1 \times 10^{-8}}{9 \times 10^2 \times 2.8 \times 10^{-8}}$$
$$= \frac{(5.6)(8.1) \times 10^{6+(-8)-2-(-8)}}{(9)(2.8)}$$
$$= 1.8 \times 10^4 = 18{,}000$$

You Try It 10

Strategy
To find the number of arithmetic operations:
• Find the reciprocal of 1×10^{-7}, which is the number of operations performed in one second.
• Write the number of seconds in one minute (60) in scientific notation.
• Multiply the number of arithmetic operations per second by the number of seconds in one minute.

Solution
$$\frac{1}{1 \times 10^{-7}} = 10^7$$

$$60 = 6 \times 10$$

$$6 \times 10 \times 10^7 = 6 \times 10^8$$

The computer can perform 6×10^8 operations in one minute.

SECTION 6.2   *pages 319–324*

You Try It 1

$R(x) = -2x^4 - 5x^3 + 2x - 8$

$R(2) = -2(2^4) - 5(2^3) + 2(2) - 8$

$ = -2(16) - 5(8) + 2(2) - 8$

$ = -32 - 40 + 4 - 8$

$ = -76$

You Try It 2

The leading coefficient is -3, the constant term is -12, and the degree is 4.

You Try It 3

a. This is a polynomial function.

b. This is not a polynomial function. A polynomial function does not have a variable expression raised to a negative power.

c. This is not a polynomial function. A polynomial function does not have a variable expression within a radical.

You Try It 4

x	y
-2	9
-1	3
0	1
1	3
2	9

You Try It 5

x	y
-3	28
-2	9
-1	2
0	1
1	0
2	-7
3	-26

You Try It 6

$$\begin{array}{r} -3x^2 - 4x + 9 \\ -5x^2 - 7x + 1 \\ \hline -8x^2 - 11x + 10 \end{array}$$

You Try It 7

Add the additive inverse of $6x^2 + 3x - 7$ to $-5x^2 + 2x - 3$.

$$\begin{array}{r} -5x^2 + 2x - 3 \\ -6x^2 - 3x + 7 \\ \hline -11x^2 - x + 4 \end{array}$$

You Try It 8

$S(x) = P(x) + R(x)$

$ = (4x^3 - 3x^2 + 2) + (-2x^2 + 2x - 3)$

$ = 4x^3 - 5x^2 + 2x - 1$

$S(-1) = 4(-1)^3 - 5(-1)^2 + 2(-1) - 1$

$ = 4(-1) - 5(1) + 2(-1) - 1$

$ = -4 - 5 - 2 - 1$

$ = -12$

You Try It 9

$D(x) = P(x) - R(x)$

$D(x) = (5x^{2n} - 3x^n - 7) - (-2x^{2n} - 5x^n + 8)$

$ = (5x^{2n} - 3x^n - 7) + (2x^{2n} + 5x^n - 8)$

$ = 7x^{2n} + 2x^n - 15$

SECTION 6.3   *pages 327–330*

You Try It 1
$(-2y + 3)(-4y) = 8y^2 - 12y$

You Try It 2
$-a^2(3a^2 + 2a - 7) = -3a^4 - 2a^3 + 7a^2$

You Try It 3

$$
\begin{array}{r}
2y^3 + 2y^2 \qquad\quad - 3 \\
3y - 1 \\
\hline
-2y^3 - 2y^2 \qquad\;\; + 3 \\
6y^4 + 6y^3 \qquad\quad - 9y \\
\hline
6y^4 + 4y^3 - 2y^2 - 9y + 3
\end{array}
$$

You Try It 4
$$
\begin{aligned}
(4y - 5)(2y - 3) &= 8y^2 - 12y - 10y + 15 \\
&= 8y^2 - 22y + 15
\end{aligned}
$$

You Try It 5
$$
\begin{aligned}
(3b + 2)(3b - 5) &= 9b^2 - 15b + 6b - 10 \\
&= 9b^2 - 9b - 10
\end{aligned}
$$

You Try It 6
$(2a + 5c)(2a - 5c) = 4a^2 - 25c^2$

You Try It 7 $(3x + 2y)^2 = 9x^2 + 12xy + 4y^2$

You Try It 8

Strategy

To find the area, replace the variables b and h in the equation $A = \frac{1}{2}bh$ by the given values and solve for A.

Solution

$A = \frac{1}{2}bh$

$A = \frac{1}{2}(2x + 6)(x - 4)$

$A = (x + 3)(x - 4)$

$A = x^2 - 4x + 3x - 12$

$A = x^2 - x - 12$

The area is $(x^2 - x - 12)$ ft^2.

You Try It 1

$$
\begin{array}{r}
5x - 1 \\
3x + 4\overline{)15x^2 + 17x - 20} \\
\underline{15x^2 + 20x} \\
-3x - 20 \\
\underline{-3x - 4} \\
-16
\end{array}
$$

$$\frac{15x^2 + 17x - 20}{3x + 4} = 5x - 1 - \frac{16}{3x + 4}$$

You Try It 2

$$
\begin{array}{r}
x^2 + 3x - 1 \\
3x - 1\overline{)3x^3 + 8x^2 - 6x + 2} \\
\underline{3x^3 - x^2} \\
9x^2 - 6x \\
\underline{9x^2 - 3x} \\
-3x + 2 \\
\underline{-3x + 1} \\
1
\end{array}
$$

$$\frac{3x^3 + 8x^2 - 6x + 2}{3x - 1} = x^2 + 3x - 1 + \frac{1}{3x - 1}$$

You Try It 3

$$
\begin{array}{r}
3x^2 - 2x + 4 \\
x^2 - 3x + 2\overline{)3x^4 - 11x^3 + 16x^2 - 16x + 8} \\
\underline{3x^4 - 9x^3 + 6x^2} \\
-2x^3 + 10x^2 - 16x \\
\underline{-2x^3 + 6x^2 - 4x} \\
4x^2 - 12x + 8 \\
\underline{4x^2 - 12x + 8} \\
0
\end{array}
$$

$$\frac{3x^4 - 11x^3 + 16x^2 - 16x + 8}{x^2 - 3x + 2} = 3x^2 - 2x + 4$$

You Try It 4

$$
\begin{array}{r|rrr}
-2 & 6 & 8 & -5 \\
& & -12 & 8 \\
\hline
& 6 & -4 & 3
\end{array}
$$

$$(6x^2 + 8x - 5) \div (x + 2) = 6x - 4 + \frac{3}{x + 2}$$

You Try It 5

$$
\begin{array}{r|rrrr}
2 & 5 & -12 & -8 & 16 \\
& & 10 & -4 & -24 \\
\hline
& 5 & -2 & -12 & -8
\end{array}
$$

$$(5x^3 - 12x^2 - 8x + 16) \div (x - 2)$$
$$= 5x^2 - 2x - 12 - \frac{8}{x - 2}$$

You Try It 6

$$
\begin{array}{r|rrrrr}
3 & 2 & -3 & -8 & 0 & -2 \\
& & 6 & 9 & 3 & 9 \\
\hline
& 2 & 3 & 1 & 3 & 7
\end{array}
$$

$$(2x^4 - 3x^3 - 8x^2 - 2) \div (x - 3)$$
$$= 2x^3 + 3x^2 + x + 3 + \frac{7}{x - 3}$$

You Try It 7

$$
\begin{array}{r|rrrr}
-3 & 2 & -5 & 0 & 7 \\
& & -6 & 33 & -99 \\
\hline
& 2 & -11 & 33 & -92
\end{array}
$$

$$P(-3) = -92$$

SOLUTIONS to Chapter 7 You Try It

SECTION 7.1   *pages 353–356*

You Try It 1
The GCF is $7a^2$.

$$14a^2 - 21a^4b = 7a^2(2) + 7a^2(-3a^2b)$$
$$= 7a^2(2 - 3a^2b)$$

You Try It 2
The GCF is 9.

$$27b^2 + 18b + 9 = 9(3b^2) + 9(2b) + 9(1)$$
$$= 9(3b^2 + 2b + 1)$$

You Try It 3
The GCF is $3x^2y^2$.

$$6x^4y^2 - 9x^3y^2 + 12x^2y^4$$
$$= 3x^2y^2(2x^2) + 3x^2y^2(-3x) + 3x^2y^2(4y^2)$$
$$= 3x^2y^2(2x^2 - 3x + 4y^2)$$

You Try It 4
The GCF is a^2.

$$a^{n+4} + a^2 = a^2(a^{n+2}) + a^2(1)$$
$$= a^2(a^{n+2} + 1)$$

You Try It 5
$$2y(5x - 2) - 3(2 - 5x)$$
$$= 2y(5x - 2) + 3(5x - 2)$$
$$= (5x - 2)(2y + 3)$$

You Try It 6
$$a^2 - 3a + 2ab - 6b$$
$$= (a^2 - 3a) + (2ab - 6b)$$
$$= a(a - 3) + 2b(a - 3)$$
$$= (a - 3)(a + 2b)$$

You Try It 7
$$2mn^2 - n + 8mn - 4 = (2mn^2 - n) + (8mn - 4)$$
$$= n(2mn - 1) + 4(2mn - 1)$$
$$= (2mn - 1)(n + 4)$$

SECTION 7.2   *pages 359–362*

You Try It 1

Factors	Sum
1, 20	21
2, 10	12
4, 5	9

● Find the positive factors of 20 whose sum is 9.

$$x^2 + 9x + 20 = (x + 4)(x + 5)$$

You Try It 2

Factors	Sum
+1, −18	−17
−1, +18	17
+2, −9	−7
−2, +9	7
+3, −6	−3
−3, +6	+3

● Find the factors of −18 whose sum is 7.

$$x^2 + 7x - 18 = (x + 9)(x - 2)$$

You Try It 3
The GCF is $-2x$.

$$-2x^3 + 14x^2 - 12x = -2x(x^2 - 7x + 6)$$

Factor the trinomial $x^2 - 7x + 6$. Find two negative factors of 6 whose sum is -7.

Factors	Sum
−1, −6	−7
−2, −3	−5

$$-2x^3 + 14x^2 - 12x = -2x(x - 6)(x - 1)$$

You Try It 4
The GCF is 3.

$$3x^2 - 9xy - 12y^2 = 3(x^2 - 3xy - 4y^2)$$

Factor the trinomial. Find the factors of −4 whose sum is -3.

Factors	Sum
+1, −4	−3
−1, +4	3
+2, −2	0

$$3x^2 - 9xy - 12y^2 = 3(x + y)(x - 4y)$$

You Try It 1

Factor the trinomial $2x^2 - x - 3$.

Positive factors of 2: 1, 2	Factors of -3: 1, -3
	-1, 3

Trial Factors	Middle Term
$(1x + 1)(2x - 3)$	$-3x + 2x = -x$
$(1x - 3)(2x + 1)$	$x - 6x = -5x$
$(1x - 1)(2x + 3)$	$3x - 2x = x$
$(1x + 3)(2x - 1)$	$-x + 6x = 5x$

$2x^2 - x - 3 = (x + 1)(2x - 3)$

You Try It 2

The GCF is $-3y$.

$-45y^3 + 12y^2 + 12y = -3y(15y^2 - 4y - 4)$

Factor the trinomial $15y^2 - 4y - 4$.

Positive factors of 15: 1, 15	Factors of -4: -1, 4
3, 5	1, -4
	-2, 2

Trial Factors	Middle Term
$(1y - 1)(15y + 4)$	$4y - 15y = -11y$
$(1y + 4)(15y - 1)$	$-y + 60y = 59y$
$(1y + 1)(15y - 4)$	$-4y + 15y = 11y$
$(1y - 4)(15y + 1)$	$y - 60y = -59y$
$(1y - 2)(15y + 2)$	$2y - 30y = -28y$
$(1y + 2)(15y - 2)$	$-2y + 30y = 28y$
$(3y - 1)(5y + 4)$	$12y - 5y = 7y$
$(3y + 4)(5y - 1)$	$-3y + 20y = 17y$
$(3y + 1)(5y - 4)$	$-12y + 5y = -7y$
$(3y - 4)(5y + 1)$	$3y - 20y = -17y$
$(3y - 2)(5y + 2)$	$6y - 10y = -4y$
$(3y + 2)(5y - 2)$	$-6y + 10y = 4y$

$-45y^3 + 12y^2 + 12y = -3y(3y - 2)(5y + 2)$

You Try It 3

Factors of -14 [2(-7)]	Sum
-1, 14	13
1, -14	-13
2, -7	-5
-2, 7	5

$$2a^2 + 13a - 7 = 2a^2 - a + 14a - 7$$
$$= (2a^2 - a) + (14a - 7)$$
$$= a(2a - 1) + 7(2a - 1)$$
$$= (2a - 1)(a + 7)$$

$2a^2 + 13a - 7 = (2a - 1)(a + 7)$

You Try It 4

The GCF is $5x$.

$15x^3 + 40x^2 - 80x = 5x(3x^2 + 8x - 16)$

Factors of -48 [3(-16)]	Sum
-1, 48	47
1, -48	-47
-2, 24	22
2, -24	-22
-3, 16	13
3, -16	-13
-4, 12	8

$$3x^2 + 8x - 16 = 3x^2 - 4x + 12x - 16$$
$$= (3x^2 - 4x) + (12x - 16)$$
$$= x(3x - 4) + 4(3x - 4)$$
$$= (3x - 4)(x + 4)$$

$$15x^3 + 40x^2 - 80x = 5x(3x^2 + 8x - 16)$$
$$= 5x(3x - 4)(x + 4)$$

SECTION 7.4   *pages 375–380*

You Try It 1

$x^2 - 36y^4 = x^2 - (6y^2)^2$

$\qquad\quad = (x + 6y^2)(x - 6y^2)$

You Try It 2

$9x^2 + 12x + 4 = (3x + 2)^2$

You Try It 3

$(a + b)^2 - (a - b)^2$

$\quad = [(a + b) + (a - b)][(a + b) - (a - b)]$

$\quad = (a + b + a - b)(a + b - a + b)$

$\quad = (2a)(2b) = 4ab$

You Try It 4

$8x^3 + y^3z^3 = (2x)^3 + (yz)^3$

$\qquad\quad = (2x + yz)(4x^2 - 2xyz + y^2z^2)$

You Try It 5

$(x - y)^3 + (x + y)^3$

$\quad = [(x - y) + (x + y)][(x - y)^2 - (x - y)(x + y) + (x + y)^2]$

$\quad = 2x[x^2 - 2xy + y^2 - (x^2 - y^2) + x^2 + 2xy + y^2]$

$\quad = 2x(x^2 - 2xy + y^2 - x^2 + y^2 + x^2 + 2xy + y^2)$

$\quad = 2x(x^2 + 3y^2)$

You Try It 6

Let $u = xy$.

$6x^2y^2 - 19xy + 10 = 6u^2 - 19u + 10$

$\qquad\qquad\qquad\quad\; = (2u - 5)(3u - 2)$

$\qquad\qquad\qquad\quad\; = (2xy - 5)(3xy - 2)$

You Try It 7

Let $u = x^2$.

$3x^4 + 4x^2 - 4 = 3u^2 + 4u - 4$

$\qquad\qquad\quad\; = (u + 2)(3u - 2)$

$\qquad\qquad\quad\; = (x^2 + 2)(3x^2 - 2)$

You Try It 8

Let $u = a^2b^2$.

$a^4b^4 + 6a^2b^2 - 7 = u^2 + 6u - 7$

$\qquad\qquad\qquad\;\; = (u + 7)(u - 1)$

$\qquad\qquad\qquad\;\; = (a^2b^2 + 7)(a^2b^2 - 1)$

You Try It 9

$18x^3 - 6x^2 - 60x = 6x(3x^2 - x - 10)$

$\qquad\qquad\qquad\quad\; = 6x(3x + 5)(x - 2)$

You Try It 10

$4x - 4y - x^3 + x^2y$

$\quad = (4x - 4y) - (x^3 - x^2y)$

$\quad = 4(x - y) - x^2(x - y)$

$\quad = (x - y)(4 - x^2)$

$\quad = (x - y)(2 + x)(2 - x)$

You Try It 11

$x^{4n} - x^{2n}y^{2n} = x^{2n+2n} - x^{2n}y^{2n}$

$\qquad\qquad\quad\; = x^{2n}(x^{2n} - y^{2n})$

$\qquad\qquad\quad\; = x^{2n}[(x^n)^2 - (y^n)^2]$

$\qquad\qquad\quad\; = x^{2n}(x^n + y^n)(x^n - y^n)$

You Try It 12

$ax^5 - ax^2y^6 = ax^2(x^3 - y^6)$

$\qquad\qquad\quad\; = ax^2(x - y^2)(x^2 + xy^2 + y^4)$

You Try It 1

$2x(x + 7) = 0$

$2x = 0 \qquad x + 7 = 0$
$\;x = 0 \qquad\qquad x = -7$

The solutions are 0 and −7.

You Try It 2

$\qquad\quad 4x^2 - 9 = 0$
$(2x - 3)(2x + 3) = 0$

$2x - 3 = 0 \qquad 2x + 3 = 0$
$\quad 2x = 3 \qquad\qquad 2x = -3$
$\qquad x = \dfrac{3}{2} \qquad\qquad x = -\dfrac{3}{2}$

The solutions are $\dfrac{3}{2}$ and $-\dfrac{3}{2}$.

You Try It 3

$(x + 2)(x - 7) = 52$
$\quad x^2 - 5x - 14 = 52$
$\quad x^2 - 5x - 66 = 0$
$(x + 6)(x - 11) = 0$

$x + 6 = 0 \qquad x - 11 = 0$
$\quad x = -6 \qquad\qquad x = 11$

The solutions are −6 and 11.

You Try It 4

Strategy
First positive consecutive integer: n
Second positive consecutive integer: $n + 1$

The sum of the squares of two positive consecutive integers is 61.

Solution
$\qquad\quad n^2 + (n + 1)^2 = 61$
$\quad n^2 + n^2 + 2n + 1 = 61$
$\qquad\quad 2n^2 + 2n + 1 = 61$
$\qquad\quad 2n^2 + 2n - 60 = 0$
$\qquad\quad 2(n^2 + n - 30) = 0$
$\qquad\qquad n^2 + n - 30 = 0$
$\qquad (n - 5)(n + 6) = 0$

$n - 5 = 0 \qquad n + 6 = 0$
$\quad n = 5 \qquad\qquad n = -6$

Since −6 is not a positive integer, it is not a solution.

$n = 5$
$n + 1 = 5 + 1 = 6$

The two integers are 5 and 6.

You Try It 5

Strategy
Width = x
Length = $2x + 4$

The area of the rectangle is 96 in². Use the equation $A = L \cdot W$.

Solution
$A = L \cdot W$
$96 = (2x + 4)x$
$96 = 2x^2 + 4x$
$\;0 = 2x^2 + 4x - 96$
$\;0 = 2(x^2 + 2x - 48)$
$\;0 = x^2 + 2x - 48$
$\;0 = (x + 8)(x - 6)$

$x + 8 = 0 \qquad x - 6 = 0$
$\quad x = -8 \qquad\qquad x = 6$

Since the width cannot be a negative number, −8 is not a solution.

$x = 6$
$2x + 4 = 2(6) + 4 = 12 + 4 = 16$

The width is 6 in.
The length is 16 in.

SOLUTIONS to Chapter 8 You Try It

SECTION 8.1 *pages 401–404*

You Try It 1

$$\frac{6x^5y}{12x^2y^3} = \frac{\overset{1}{\cancel{2}} \cdot \overset{1}{\cancel{3}} \cdot x^5y}{\underset{1}{\cancel{2}} \cdot 2 \cdot \underset{1}{\cancel{3}} \cdot x^2y^3} = \frac{x^3}{2y^2}$$

You Try It 2

$$\frac{x^2 + 2x - 24}{16 - x^2} = \frac{\overset{-1}{\cancel{(x - 4)}}(x + 6)}{\underset{1}{\cancel{(4 - x)}}(4 + x)} = -\frac{x + 6}{x + 4}$$

You Try It 3

$$\frac{x^2 + 4x - 12}{x^2 - 3x + 2} = \frac{\overset{1}{\cancel{(x - 2)}}(x + 6)}{(x - 1)\underset{1}{\cancel{(x - 2)}}} = \frac{x + 6}{x - 1}$$

You Try It 4

$$\frac{12x^2 + 3x}{10x - 15} \cdot \frac{8x - 12}{9x + 18} = \frac{3x(4x + 1)}{5(2x - 3)} \cdot \frac{4(2x - 3)}{9(x + 2)}$$

$$= \frac{\overset{1}{\cancel{3}}x(4x + 1) \cdot 2 \cdot 2\overset{1}{\cancel{(2x - 3)}}}{5\underset{1}{\cancel{(2x - 3)}} \cdot \underset{1}{\cancel{3}} \cdot 3(x + 2)}$$

$$= \frac{4x(4x + 1)}{15(x + 2)}$$

You Try It 5

$$\frac{x^2 + 2x - 15}{9 - x^2} \cdot \frac{x^2 - 3x - 18}{x^2 - 7x + 6}$$

$$= \frac{(x - 3)(x + 5)}{(3 - x)(3 + x)} \cdot \frac{(x + 3)(x - 6)}{(x - 1)(x - 6)}$$

$$= \frac{\overset{-1}{\cancel{(x - 3)}}(x + 5) \cdot \overset{1}{\cancel{(x + 3)}}\overset{1}{\cancel{(x - 6)}}}{\underset{1}{\cancel{(3 - x)}}\underset{1}{\cancel{(3 + x)}} \cdot (x - 1)\underset{1}{\cancel{(x - 6)}}} = -\frac{x + 5}{x - 1}$$

You Try It 6

$$\frac{a^2}{4bc^2 - 2b^2c} \div \frac{a}{6bc - 3b^2} = \frac{a^2}{4bc^2 - 2b^2c} \cdot \frac{6bc - 3b^2}{a}$$

$$= \frac{a^2 \cdot 3b\overset{1}{\cancel{(2c - b)}}}{2bc\underset{1}{\cancel{(2c - b)}} \cdot a} = \frac{3a}{2c}$$

You Try It 7

$$\frac{3x^2 + 26x + 16}{3x^2 - 7x - 6} \div \frac{2x^2 + 9x - 5}{x^2 + 2x - 15}$$

$$= \frac{3x^2 + 26x + 16}{3x^2 - 7x - 6} \cdot \frac{x^2 + 2x - 15}{2x^2 + 9x - 5}$$

$$= \frac{\overset{1}{\cancel{(3x + 2)}}(x + 8) \cdot \overset{1}{\cancel{(x + 5)}}\overset{1}{\cancel{(x - 3)}}}{\underset{1}{\cancel{(3x + 2)}}\underset{1}{\cancel{(x - 3)}} \cdot (2x - 1)\underset{1}{\cancel{(x + 5)}}} = \frac{x + 8}{2x - 1}$$

SECTION 8.2 *pages 409–414*

You Try It 1
$8uv^2 = 2 \cdot 2 \cdot 2 \cdot u \cdot v \cdot v$
$12uw = 2 \cdot 2 \cdot 3 \cdot u \cdot w$
$\text{LCM} = 2 \cdot 2 \cdot 2 \cdot 3 \cdot u \cdot v \cdot v \cdot w = 24uv^2w$

You Try It 2
$m^2 - 6m + 9 = (m - 3)(m - 3)$
$m^2 - 2m - 3 = (m + 1)(m - 3)$
$\text{LCM} = (m - 3)(m - 3)(m + 1)$

You Try It 3
The LCM is $36xy^2z$.

$$\frac{x - 3}{4xy^2} = \frac{x - 3}{4xy^2} \cdot \frac{9z}{9z} = \frac{9xz - 27z}{36xy^2z}$$

$$\frac{2x + 1}{9y^2z} = \frac{2x + 1}{9y^2z} \cdot \frac{4x}{4x} = \frac{8x^2 + 4x}{36xy^2z}$$

You Try It 4
The LCM is $(x + 2)(x - 5)(x + 5)$.

$$\frac{x + 4}{x^2 - 3x - 10} = \frac{x + 4}{(x + 2)(x - 5)} \cdot \frac{x + 5}{x + 5}$$

$$= \frac{x^2 + 9x + 20}{(x + 2)(x - 5)(x + 5)}$$

$$\frac{2x}{25 - x^2} = \frac{2x}{-(x^2 - 25)} = -\frac{2x}{(x - 5)(x + 5)} \cdot \frac{x + 2}{x + 2}$$

$$= -\frac{2x^2 + 4x}{(x + 2)(x - 5)(x + 5)}$$

You Try It 5

$$\frac{3}{xy} + \frac{12}{xy} = \frac{3 + 12}{xy} = \frac{15}{xy}$$

You Try It 6

$$\frac{2x^2}{x^2 - x - 12} - \frac{7x + 4}{x^2 - x - 12}$$

$$= \frac{2x^2 - (7x + 4)}{x^2 - x - 12} = \frac{2x^2 - 7x - 4}{x^2 - x - 12}$$

$$= \frac{(2x + 1)\overset{1}{\cancel{(x - 4)}}}{(x + 3)\underset{1}{\cancel{(x - 4)}}} = \frac{2x + 1}{x + 3}$$

You Try It 7

$$\frac{x^2 - 1}{x^2 - 8x + 12} - \frac{2x + 1}{x^2 - 8x + 12} + \frac{x}{x^2 - 8x + 12}$$

$$= \frac{(x^2 - 1) - (2x + 1) + x}{x^2 - 8x + 12} = \frac{x^2 - 1 - 2x - 1 + x}{x^2 - 8x + 12}$$

$$= \frac{x^2 - x - 2}{x^2 - 8x + 12} = \frac{(x + 1)\overset{1}{\cancel{(x - 2)}}}{\underset{1}{\cancel{(x - 2)}}(x - 6)} = \frac{x + 1}{x - 6}$$

You Try It 8

The LCM of the denominators is $24y$.

$$\frac{z}{8y} - \frac{4z}{3y} + \frac{5z}{4y} = \frac{z}{8y} \cdot \frac{3}{3} - \frac{4z}{3y} \cdot \frac{8}{8} + \frac{5z}{4y} \cdot \frac{6}{6}$$

$$= \frac{3z}{24y} - \frac{32z}{24y} + \frac{30z}{24y}$$

$$= \frac{3z - 32z + 30z}{24y} = \frac{z}{24y}$$

You Try It 9

$$2 - x = -(x - 2)$$

Therefore, $\dfrac{3}{2 - x} = \dfrac{-3}{x - 2}$.

The LCM is $x - 2$.

$$\frac{5x}{x - 2} - \frac{3}{2 - x} = \frac{5x}{x - 2} - \frac{-3}{x - 2}$$

$$= \frac{5x - (-3)}{x - 2} = \frac{5x + 3}{x - 2}$$

You Try It 10

The LCM is $(3x - 1)(x + 4)$.

$$\frac{4x}{3x - 1} - \frac{9}{x + 4} = \frac{4x}{3x - 1} \cdot \frac{x + 4}{x + 4} - \frac{9}{x + 4} \cdot \frac{3x - 1}{3x - 1}$$

$$= \frac{4x^2 + 16x}{(3x - 1)(x + 4)} - \frac{27x - 9}{(3x - 1)(x + 4)}$$

$$= \frac{(4x^2 + 16x) - (27x - 9)}{(3x - 1)(x + 4)}$$

$$= \frac{4x^2 + 16x - 27x + 9}{(3x - 1)(x + 4)}$$

$$= \frac{4x^2 - 11x + 9}{(3x - 1)(x + 4)}$$

You Try It 11

The LCM is $x - 3$.

$$2 - \frac{1}{x - 3} = 2 \cdot \frac{x - 3}{x - 3} - \frac{1}{x - 3}$$

$$= \frac{2x - 6}{x - 3} - \frac{1}{x - 3}$$

$$= \frac{2x - 6 - 1}{x - 3}$$

$$= \frac{2x - 7}{x - 3}$$

You Try It 12

The LCM is $a(a - 5)(a + 5)$.

$$\frac{a - 3}{a^2 - 5a} + \frac{a - 9}{a^2 - 25}$$

$$= \frac{a - 3}{a(a - 5)} \cdot \frac{a + 5}{a + 5} + \frac{a - 9}{(a - 5)(a + 5)} \cdot \frac{a}{a}$$

$$= \frac{(a - 3)(a + 5) + a(a - 9)}{a(a - 5)(a + 5)}$$

$$= \frac{(a^2 + 2a - 15) + (a^2 - 9a)}{a(a - 5)(a + 5)}$$

$$= \frac{a^2 + 2a - 15 + a^2 - 9a}{a(a - 5)(a + 5)}$$

$$= \frac{2a^2 - 7a - 15}{a(a - 5)(a + 5)} = \frac{(2a + 3)(a - 5)}{a(a - 5)(a + 5)}$$

$$= \frac{(2a + 3)\overset{1}{\cancel{(a - 5)}}}{a\underset{1}{\cancel{(a - 5)}}(a + 5)} = \frac{2a + 3}{a(a + 5)}$$

You Try It 13
The LCM is $(3x + 2)(x - 1)$.

$$\frac{2x - 3}{3x^2 - x - 2} + \frac{5}{3x + 2} - \frac{1}{x - 1}$$

$$= \frac{2x - 3}{(3x + 2)(x - 1)} + \frac{5}{3x + 2} \cdot \frac{x - 1}{x - 1} - \frac{1}{x - 1} \cdot \frac{3x + 2}{3x + 2}$$

$$= \frac{2x - 3}{(3x + 2)(x - 1)} + \frac{5x - 5}{(3x + 2)(x - 1)} - \frac{3x + 2}{(3x + 2)(x - 1)}$$

$$= \frac{(2x - 3) + (5x - 5) - (3x + 2)}{(3x + 2)(x - 1)}$$

$$= \frac{2x - 3 + 5x - 5 - 3x - 2}{(3x + 2)(x - 1)}$$

$$= \frac{4x - 10}{(3x + 2)(x - 1)} = \frac{2(2x - 5)}{(3x + 2)(x - 1)}$$

SECTION 8.3   *pages 421–422*

You Try It 1
The LCM of 3, x, 9, and x^2 is $9x^2$.

$$\frac{\dfrac{1}{3} - \dfrac{1}{x}}{\dfrac{1}{9} - \dfrac{1}{x^2}} = \frac{\dfrac{1}{3} - \dfrac{1}{x}}{\dfrac{1}{9} - \dfrac{1}{x^2}} \cdot \frac{9x^2}{9x^2} = \frac{\dfrac{1}{3} \cdot 9x^2 - \dfrac{1}{x} \cdot 9x^2}{\dfrac{1}{9} \cdot 9x^2 - \dfrac{1}{x^2} \cdot 9x^2}$$

$$= \frac{3x^2 - 9x}{x^2 - 9} = \frac{3x(x - 3)}{(x - 3)(x + 3)} = \frac{3x}{x + 3}$$

You Try It 2
The LCM of x and x^2 is x^2.

$$\frac{1 + \dfrac{4}{x} + \dfrac{3}{x^2}}{1 + \dfrac{10}{x} + \dfrac{21}{x^2}} = \frac{1 + \dfrac{4}{x} + \dfrac{3}{x^2}}{1 + \dfrac{10}{x} + \dfrac{21}{x^2}} \cdot \frac{x^2}{x^2}$$

$$= \frac{1 \cdot x^2 + \dfrac{4}{x} \cdot x^2 + \dfrac{3}{x^2} \cdot x^2}{1 \cdot x^2 + \dfrac{10}{x} \cdot x^2 + \dfrac{21}{x^2} \cdot x^2}$$

$$= \frac{x^2 + 4x + 3}{x^2 + 10x + 21} = \frac{(x + 1)(x + 3)}{(x + 3)(x + 7)} = \frac{x + 1}{x + 7}$$

You Try It 3
The LCM is $x - 5$.

$$\frac{x + 3 - \dfrac{20}{x - 5}}{x + 8 + \dfrac{30}{x - 5}} = \frac{x + 3 - \dfrac{20}{x - 5}}{x + 8 + \dfrac{30}{x - 5}} \cdot \frac{x - 5}{x - 5}$$

$$= \frac{x(x - 5) + 3(x - 5) - \dfrac{20}{x - 5} \cdot (x - 5)}{x(x - 5) + 8(x - 5) + \dfrac{30}{x - 5} \cdot (x - 5)}$$

$$= \frac{x^2 - 5x + 3x - 15 - 20}{x^2 - 5x + 8x - 40 + 30}$$

$$= \frac{x^2 - 2x - 35}{x^2 + 3x - 10}$$

$$= \frac{(x + 5)(x - 7)}{(x - 2)(x + 5)} = \frac{x - 7}{x - 2}$$

You Try It 1

$$\frac{x}{x+6} = \frac{3}{x} \quad \text{The LCM is } x(x+6).$$

$$\frac{x(x+6)}{1} \cdot \frac{x}{x+6} = \frac{x(x+6)}{1} \cdot \frac{3}{x}$$

$$x^2 = (x+6)3$$
$$x^2 = 3x + 18$$
$$x^2 - 3x - 18 = 0$$
$$(x+3)(x-6) = 0$$

$$x + 3 = 0 \qquad x - 6 = 0$$
$$x = -3 \qquad x = 6$$

Both −3 and 6 check as solutions.
The solutions are −3 and 6.

You Try It 2

$$\frac{5x}{x+2} = 3 - \frac{10}{x+2} \quad \text{The LCM is } x + 2.$$

$$\frac{(x+2)}{1} \cdot \frac{5x}{x+2} = \frac{(x+2)}{1}\left(3 - \frac{10}{x+2}\right)$$

$$\frac{x+2}{1} \cdot \frac{5x}{x+2} = (x+2)3 - \frac{x+2}{1} \cdot \frac{10}{x+2}$$

$$5x = (x+2)3 - 10$$
$$5x = 3x + 6 - 10$$
$$5x = 3x - 4$$
$$2x = -4$$
$$x = -2$$

−2 does not check as a solution.
The equation has no solution.

You Try It 3

$$\frac{2}{x+3} = \frac{6}{5x+5}$$

$$\frac{(x+3)(5x+5)}{1} \cdot \frac{2}{x+3} = \frac{(x+3)(5x+5)}{1} \cdot \frac{6}{5x+5}$$

$$\frac{(x+3)(5x+5)}{1} \cdot \frac{2}{x+3} = \frac{(x+3)(5x+5)}{1} \cdot \frac{6}{5x+5}$$

$$(5x+5)2 = (x+3)6$$
$$10x + 10 = 6x + 18$$
$$4x + 10 = 18$$
$$4x = 8$$
$$x = 2$$

The solution is 2.

You Try It 4

Strategy

To find *FG*, write a proportion using the fact that in similar triangles, the ratio of corresponding sides equals the ratio of corresponding heights. Solve the proportion for *FG*.

Solution

$$\frac{AC}{DF} = \frac{CH}{FG}$$

$$\frac{10}{15} = \frac{7}{FG}$$

$$15(FG)\left(\frac{10}{15}\right) = 15(FG)\left(\frac{7}{FG}\right)$$

$$10(FG) = 15(7)$$
$$10(FG) = 105$$
$$FG = 10.5$$

The height *FG* of triangle *DEF* is 10.5 m.

You Try It 5

Strategy
To find the total area that 256 ceramic tiles will cover, write and solve a proportion using x to represent the number of square feet that 256 tiles will cover.

Solution
$$\frac{9}{16} = \frac{x}{256}$$
$$256\left(\frac{9}{16}\right) = 256\left(\frac{x}{256}\right)$$
$$144 = x$$

A 144-square-foot area can be tiled using 256 ceramic tiles.

You Try It 6

Strategy
To find the additional amount of medication required for a 200-pound adult, write and solve a proportion using x to represent the additional medication. Then $3 + x$ is the total amount required for a 200-pound adult.

Solution
$$\frac{150}{3} = \frac{200}{3 + x}$$
$$\frac{50}{1} = \frac{200}{3 + x}$$
$$(3 + x) \cdot 50 = (3 + x) \cdot \frac{200}{3 + x}$$
$$(3 + x) \cdot 50 = 200$$
$$150 + 50x = 200$$
$$50x = 50$$
$$x = 1$$

One additional ounce is required for a 200-pound adult.

SECTION 8.5   *pages 437–438*

You Try It 1
$$A = P + Prt$$
$$A - P = P - P + Prt$$
$$A - P = Prt$$
$$\frac{A - P}{Pt} = \frac{Prt}{Pt}$$
$$\frac{A - P}{Pt} = r$$

You Try It 2
$$s = \frac{A + L}{2}$$
$$2 \cdot s = 2\left(\frac{A + L}{2}\right)$$
$$2s = A + L$$
$$2s - A = A - A + L$$
$$2s - A = L$$

You Try It 3

$$S = a + (n - 1)d$$
$$S = a + nd - d$$
$$S - a = a - a + nd - d$$
$$S - a = nd - d$$
$$S - a + d = nd - d + d$$
$$S - a + d = nd$$
$$\frac{S - a + d}{d} = \frac{nd}{d}$$
$$\frac{S - a + d}{d} = n$$

You Try It 4

$$S = C + rC$$
$$S = (1 + r)C$$
$$\frac{S}{1 + r} = \frac{(1 + r)C}{1 + r}$$
$$\frac{S}{1 + r} = C$$

SECTION 8.6   *pages 441–444*

You Try It 1

Strategy
• Time for one printer to complete the job: t

	Rate	*Time*	*Part*
1st printer	$\frac{1}{t}$	2	$\frac{2}{t}$
2nd printer	$\frac{1}{t}$	5	$\frac{5}{t}$

• The sum of the parts of the task completed must equal 1.

Solution

$$\frac{2}{t} + \frac{5}{t} = 1$$
$$t\left(\frac{2}{t} + \frac{5}{t}\right) = t \cdot 1$$
$$2 + 5 = t$$
$$7 = t$$

Working alone, one printer takes 7 h to print the payroll.

You Try It 2

Strategy
• Rate sailing across the lake: r
  Rate sailing back: $3r$

	Distance	*Rate*	*Time*
Across	6	r	$\frac{6}{r}$
Back	6	$3r$	$\frac{6}{3r}$

• The total time for the trip was 2 h.

Solution

$$\frac{6}{r} + \frac{6}{3r} = 2$$
$$3r\left(\frac{6}{r} + \frac{6}{3r}\right) = 3r(2)$$
$$3r \cdot \frac{6}{r} + 3r \cdot \frac{6}{3r} = 6r$$
$$18 + 6 = 6r$$
$$24 = 6r$$
$$4 = r$$

The rate across the lake was 4 km/h.

SECTION 8.7 *pages 449–452*

You Try It 1

Strategy
To find the distance:

- Write the basic direct variation equation, replace the variables by the given values, and solve for k.
- Write the direct variation equation, replacing k by its value. Substitute 5 for t and solve for s.

Solution

$$s = kt^2$$
$$64 = k(2)^2$$
$$64 = k \cdot 4$$
$$16 = k$$

$$s = 16t^2 = 16(5)^2 = 400$$

The object will fall 400 ft in 5 s.

You Try It 2

Strategy
To find the resistance:

- Write the basic inverse variation equation, replace the variables by the given values, and solve for k.
- Write the inverse variation equation, replacing k by its value. Substitute 0.02 for d and solve for R.

Solution

$$R = \frac{k}{d^2}$$
$$0.5 = \frac{k}{(0.01)^2}$$
$$0.5 = \frac{k}{0.0001}$$
$$0.00005 = k$$

$$R = \frac{0.00005}{d^2} = \frac{0.00005}{(0.02)^2} = 0.125$$

The resistance is 0.125 ohm.

You Try It 3

Strategy
To find the strength:

- Write the basic combined variation equation, replace the variables by the given values, and solve for k.
- Write the combined variation equation, replacing k by its value. Substitute 4 for w, 8 for d, and 16 for L, and solve for s.

Solution

$$s = \frac{kwd^2}{L}$$
$$1200 = \frac{k \cdot 2 \cdot (12)^2}{12}$$
$$1200 = \frac{k \cdot 2 \cdot 144}{12}$$
$$1200 = 24k$$
$$50 = k$$

$$s = \frac{50wd^2}{L} = \frac{50 \cdot 4 \cdot (8)^2}{16} = 800$$

The strength is 800 lb.

SOLUTIONS to Chapter 9 You Try It

SECTION 9.1 *pages 463–468*

You Try It 1
$$16^{-3/4} = (2^4)^{-3/4}$$
$$= 2^{-3}$$
$$= \frac{1}{2^3} = \frac{1}{8}$$

You Try It 2 $(-81)^{3/4}$
The base of the exponential expression is negative, and the denominator of the exponent is a positive even number.

Therefore, $(-81)^{3/4}$ is not a real number.

You Try It 3
$$(x^{3/4}y^{1/2}z^{-2/3})^{-4/3} = x^{-1}y^{-2/3}z^{8/9}$$
$$= \frac{z^{8/9}}{xy^{2/3}}$$

You Try It 4
$$\left(\frac{16a^{-2}b^{4/3}}{9a^4b^{-2/3}}\right)^{-1/2} = \left(\frac{2^4a^{-6}b^2}{3^2}\right)^{-1/2}$$
$$= \frac{2^{-2}a^3b^{-1}}{3^{-1}}$$
$$= \frac{3a^3}{2^2b} = \frac{3a^3}{4b}$$

You Try It 5
$$(2x^3)^{3/4} = \sqrt[4]{(2x^3)^3}$$
$$= \sqrt[4]{8x^9}$$

You Try It 6
$$-5a^{5/6} = -5(a^5)^{1/6}$$
$$= -5\sqrt[6]{a^5}$$

You Try It 7 $\sqrt[3]{3ab} = (3ab)^{1/3}$

You Try It 8 $\sqrt[4]{x^4 + y^4} = (x^4 + y^4)^{1/4}$

You Try It 9
$$\sqrt{121x^{10}y^4} = \sqrt{11^2x^{10}y^4}$$
$$= (11^2x^{10}y^4)^{1/2}$$
$$= 11x^5y^2$$

You Try It 10
$$\sqrt[3]{-8x^{12}y^3} = \sqrt[3]{(-2)^3x^{12}y^3}$$
$$= [(-2)^3x^{12}y^3]^{1/3}$$
$$= -2x^4y$$

You Try It 11
$$-\sqrt[4]{81x^{12}y^8} = -\sqrt[4]{3^4x^{12}y^8}$$
$$= -(3^4x^{12}y^8)^{1/4}$$
$$= -3x^3y^2$$

SECTION 9.2 *pages 473–478*

You Try It 1
$$\sqrt[5]{x^7} = \sqrt[5]{x^5 \cdot x^2}$$
$$= \sqrt[5]{x^5}\sqrt[5]{x^2}$$
$$= x\sqrt[5]{x^2}$$

You Try It 2
$$\sqrt[3]{-64x^8y^{18}} = \sqrt[3]{(-4)^3x^8y^{18}}$$
$$= \sqrt[3]{(-4)^3x^6y^{18}(x^2)}$$
$$= \sqrt[3]{(-4)^3x^6y^{18}}\sqrt[3]{x^2}$$
$$= -4x^2y^6\sqrt[3]{x^2}$$

You Try It 3
$$3xy\sqrt[3]{81x^5y} - \sqrt[3]{192x^8y^4}$$
$$= 3xy\sqrt[3]{3^4x^5y} - \sqrt[3]{2^6 \cdot 3x^8y^4}$$
$$= 3xy\sqrt[3]{3^3x^3}\sqrt[3]{3x^2y} - \sqrt[3]{2^6x^6y^3}\sqrt[3]{3x^2y}$$
$$= 3xy \cdot 3x\sqrt[3]{3x^2y} - 2^2x^2y\sqrt[3]{3x^2y}$$
$$= 9x^2y\sqrt[3]{3x^2y} - 4x^2y\sqrt[3]{3x^2y} = 5x^2y\sqrt[3]{3x^2y}$$

You Try It 4
$$\sqrt{5b}(\sqrt{3b} - \sqrt{10})$$
$$= \sqrt{15b^2} - \sqrt{50b}$$
$$= \sqrt{3 \cdot 5b^2} - \sqrt{2 \cdot 5^2b}$$
$$= \sqrt{b^2}\sqrt{3 \cdot 5} - \sqrt{5^2}\sqrt{2b}$$
$$= b\sqrt{15} - 5\sqrt{2b}$$

You Try It 5

$(2\sqrt[3]{2x} - 3)(\sqrt[3]{2x} - 5)$

$\quad = 2\sqrt[3]{4x^2} - 10\sqrt[3]{2x} - 3\sqrt[3]{2x} + 15$

$\quad = 2\sqrt[3]{4x^2} - 13\sqrt[3]{2x} + 15$

You Try It 6

$(\sqrt{a} - 3\sqrt{y})(\sqrt{a} + 3\sqrt{y})$

$\quad = (\sqrt{a})^2 - (3\sqrt{y})^2$

$\quad = a - 9y$

You Try It 7

$\dfrac{y}{\sqrt{3y}} = \dfrac{y}{\sqrt{3y}} \cdot \dfrac{\sqrt{3y}}{\sqrt{3y}} = \dfrac{y\sqrt{3y}}{\sqrt{3^2y^2}} = \dfrac{y\sqrt{3y}}{3y} = \dfrac{\sqrt{3y}}{3}$

You Try It 8

$\dfrac{3}{\sqrt[3]{3x^2}} = \dfrac{3}{\sqrt[3]{3x^2}} \cdot \dfrac{\sqrt[3]{3^2x}}{\sqrt[3]{3^2x}} = \dfrac{3\sqrt[3]{9x}}{\sqrt[3]{3^3x^3}}$

$\quad = \dfrac{3\sqrt[3]{9x}}{3x} = \dfrac{\sqrt[3]{9x}}{x}$

You Try It 9

$\dfrac{3 + \sqrt{6}}{2 - \sqrt{6}} = \dfrac{3 + \sqrt{6}}{2 - \sqrt{6}} \cdot \dfrac{2 + \sqrt{6}}{2 + \sqrt{6}} = \dfrac{6 + 3\sqrt{6} + 2\sqrt{6} + (\sqrt{6})^2}{2^2 - (\sqrt{6})^2}$

$\quad = \dfrac{6 + 5\sqrt{6} + 6}{4 - 6} = \dfrac{12 + 5\sqrt{6}}{-2} = -\dfrac{12 + 5\sqrt{6}}{2}$

SECTION 9.3   *pages 483–488*

You Try It 1

$\sqrt{-45} = i\sqrt{45} = i\sqrt{3^2 \cdot 5} = 3i\sqrt{5}$

You Try It 2

$\sqrt{98} - \sqrt{-60} = \sqrt{98} - i\sqrt{60}$

$\quad = \sqrt{2 \cdot 7^2} - i\sqrt{2^2 \cdot 3 \cdot 5}$

$\quad = 7\sqrt{2} - 2i\sqrt{15}$

You Try It 3

$(-4 + 2i) - (6 - 8i) = -10 + 10i$

You Try It 4

$(16 - \sqrt{-45}) - (3 + \sqrt{-20})$

$\quad = (16 - i\sqrt{45}) - (3 + i\sqrt{20})$

$\quad = (16 - i\sqrt{3^2 \cdot 5}) - (3 + i\sqrt{2^2 \cdot 5})$

$\quad = (16 - 3i\sqrt{5}) - (3 + 2i\sqrt{5})$

$\quad = 13 - 5i\sqrt{5}$

You Try It 5

$(3 - 2i) + (-3 + 2i) = 0 + 0i = 0$

You Try It 6

$(-3i)(-10i) = 30i^2 = 30(-1) = -30$

You Try It 7

$-\sqrt{-8} \cdot \sqrt{-5} = -i\sqrt{8} \cdot i\sqrt{5} = -i^2\sqrt{40}$

$\quad = -(-1)\sqrt{40} = \sqrt{2^3 \cdot 5} = 2\sqrt{10}$

You Try It 8

$-6i(3 + 4i) = -18i - 24i^2$

$\quad = -18i - 24(-1) = 24 - 18i$

You Try It 9

$\sqrt{-3}(\sqrt{27} - \sqrt{-6}) = i\sqrt{3}(\sqrt{27} - i\sqrt{6})$

$\quad = i\sqrt{81} - i^2\sqrt{18}$

$\quad = i\sqrt{3^4} - (-1)\sqrt{2 \cdot 3^2}$

$\quad = 9i + 3\sqrt{2}$

$\quad = 3\sqrt{2} + 9i$

You Try It 10

$(4 - 3i)(2 - i) = 8 - 4i - 6i + 3i^2$

$\quad = 8 - 10i + 3i^2$

$\quad = 8 - 10i + 3(-1)$

$\quad = 5 - 10i$

You Try It 11

$$(3 + 6i)(3 - 6i) = 3^2 + 6^2$$
$$= 9 + 36$$
$$= 45$$

You Try It 12

$$(3 - i)\left(\frac{3}{10} + \frac{1}{10}i\right) = \frac{9}{10} + \frac{3}{10}i - \frac{3}{10}i - \frac{1}{10}i^2$$
$$= \frac{9}{10} - \frac{1}{10}i^2 = \frac{9}{10} - \frac{1}{10}(-1)$$
$$= \frac{9}{10} + \frac{1}{10} = 1$$

You Try It 13

$$\frac{2 - 3i}{4i} = \frac{2 - 3i}{4i} \cdot \frac{i}{i}$$
$$= \frac{2i - 3i^2}{4i^2}$$
$$= \frac{2i - 3(-1)}{4(-1)}$$
$$= \frac{3 + 2i}{-4} = -\frac{3}{4} - \frac{1}{2}i$$

You Try It 14

$$\frac{2 + 5i}{3 - 2i} = \frac{2 + 5i}{3 - 2i} \cdot \frac{3 + 2i}{3 + 2i} = \frac{6 + 4i + 15i + 10i^2}{3^2 + 2^2}$$
$$= \frac{6 + 19i + 10(-1)}{13} = \frac{-4 + 19i}{13}$$
$$= -\frac{4}{13} + \frac{19}{13}i$$

SECTION 9.4   *pages 491–494*

You Try It 1

$$\sqrt[4]{x - 8} = 3$$
$$(\sqrt[4]{x - 8})^4 = 3^4$$
$$x - 8 = 81$$
$$x = 89$$

Check:

$$\sqrt[4]{x - 8} = 3$$

$\sqrt[4]{89 - 8}$	3
$\sqrt[4]{81}$	3
$3 = 3$	

The solution is 89.

You Try It 2

$$\sqrt{x} - \sqrt{x + 5} = 1$$
$$\sqrt{x} = 1 + \sqrt{x + 5}$$
$$(\sqrt{x})^2 = (1 + \sqrt{x + 5})^2$$
$$x = 1 + 2\sqrt{x + 5} + x + 5$$
$$0 = 6 + 2\sqrt{x + 5}$$
$$-6 = 2\sqrt{x + 5}$$
$$-3 = \sqrt{x + 5}$$
$$(-3)^2 = (\sqrt{x + 5})^2$$
$$9 = x + 5$$
$$4 = x$$

4 does not check as a solution. The equation has no solution.

You Try It 3

Strategy
To find the diagonal, use the Pythagorean Theorem. One leg is the length of the rectangle. The second leg is the width of the rectangle. The hypotenuse is the diagonal of the rectangle.

Solution
$$c^2 = a^2 + b^2$$
$$c^2 = (6)^2 + (3)^2$$
$$c^2 = 36 + 9$$
$$c^2 = 45$$
$$(c^2)^{1/2} = (45)^{1/2}$$
$$c = \sqrt{45}$$
$$c \approx 6.7$$

The diagonal is 6.7 cm.

You Try It 5

Strategy
To find the distance, replace the variables v and a in the equation by their given values and solve for s.

Solution
$$v = \sqrt{2as}$$
$$88 = \sqrt{2 \cdot 22s}$$
$$88 = \sqrt{44s}$$
$$(88)^2 = (\sqrt{44s})^2$$
$$7744 = 44s$$
$$176 = s$$

The distance required is 176 ft.

You Try It 4

Strategy
To find the height, replace d in the equation with the given value and solve for h.

Solution
$$d = \sqrt{1.5h}$$
$$5.5 = \sqrt{1.5h}$$
$$(5.5)^2 = (\sqrt{1.5h})^2$$
$$30.25 = 1.5h$$
$$20.17 \approx h$$

The periscope must be approximately 20.17 ft above the water.

SOLUTIONS to Chapter 10 You Try It

You Try It 1
$$2x^2 = 7x - 3$$
$$2x^2 - 7x + 3 = 0$$
$$(2x - 1)(x - 3) = 0$$

$$2x - 1 = 0 \qquad x - 3 = 0$$
$$2x = 1 \qquad\qquad x = 3$$
$$x = \frac{1}{2}$$

The solutions are $\frac{1}{2}$ and 3.

You Try It 2 $x^2 - 3ax - 4a^2 = 0$
$$(x + a)(x - 4a) = 0$$

$$x + a = 0 \qquad x - 4a = 0$$
$$x = -a \qquad\qquad x = 4a$$

The solutions are $-a$ and $4a$.

You Try It 3
$$(x - r_1)(x - r_2) = 0$$
$$(x - 3)\left[x - \left(-\frac{1}{2}\right)\right] = 0$$
$$(x - 3)\left(x + \frac{1}{2}\right) = 0$$
$$x^2 - \frac{5}{2}x - \frac{3}{2} = 0$$
$$2\left(x^2 - \frac{5}{2}x - \frac{3}{2}\right) = 2 \cdot 0$$
$$2x^2 - 5x - 3 = 0$$

You Try It 4
$$2(x + 1)^2 - 24 = 0$$
$$2(x + 1)^2 = 24$$
$$(x + 1)^2 = 12$$
$$\sqrt{(x + 1)^2} = \pm\sqrt{12}$$
$$x + 1 = \pm 2\sqrt{3}$$

$$x + 1 = 2\sqrt{3} \qquad x + 1 = -2\sqrt{3}$$
$$x = -1 + 2\sqrt{3} \qquad x = -1 - 2\sqrt{3}$$

The solutions are
$-1 + 2\sqrt{3}$ and $-1 - 2\sqrt{3}$.

You Try It 1
$$4x^2 - 4x - 1 = 0$$
$$4x^2 - 4x = 1$$
$$\frac{1}{4}(4x^2 - 4x) = \frac{1}{4} \cdot 1$$
$$x^2 - x = \frac{1}{4}$$

Complete the square.

$$x^2 - x + \frac{1}{4} = \frac{1}{4} + \frac{1}{4}$$
$$\left(x - \frac{1}{2}\right)^2 = \frac{2}{4}$$
$$\sqrt{\left(x - \frac{1}{2}\right)^2} = \pm\sqrt{\frac{2}{4}}$$
$$x - \frac{1}{2} = \pm\frac{\sqrt{2}}{2}$$

$$x - \frac{1}{2} = \frac{\sqrt{2}}{2} \qquad x - \frac{1}{2} = -\frac{\sqrt{2}}{2}$$
$$x = \frac{1}{2} + \frac{\sqrt{2}}{2} \qquad x = \frac{1}{2} - \frac{\sqrt{2}}{2}$$

The solutions are $\dfrac{1 + \sqrt{2}}{2}$ and $\dfrac{1 - \sqrt{2}}{2}$.

You Try It 2
$$2x^2 + x - 5 = 0$$
$$2x^2 + x = 5$$
$$\frac{1}{2}(2x^2 + x) = \frac{1}{2} \cdot 5$$
$$x^2 + \frac{1}{2}x = \frac{5}{2}$$

Complete the square.

$$x^2 + \frac{1}{2}x + \frac{1}{16} = \frac{5}{2} + \frac{1}{16}$$
$$\left(x + \frac{1}{4}\right)^2 = \frac{41}{16}$$
$$\sqrt{\left(x + \frac{1}{4}\right)^2} = \pm\sqrt{\frac{41}{16}}$$
$$x + \frac{1}{4} = \pm\frac{\sqrt{41}}{4}$$

$$x + \frac{1}{4} = \frac{\sqrt{41}}{4} \qquad x + \frac{1}{4} = -\frac{\sqrt{41}}{4}$$
$$x = -\frac{1}{4} + \frac{\sqrt{41}}{4} \qquad x = -\frac{1}{4} - \frac{\sqrt{41}}{4}$$

The solutions are $\dfrac{-1 + \sqrt{41}}{4}$ and $\dfrac{-1 - \sqrt{41}}{4}$.

SECTION 10.3   *pages 521–524*

You Try It 1 $x^2 - 2x + 10 = 0$
$a = 1, b = -2, c = 10$
$x = \dfrac{-b \pm \sqrt{b^2 - 4ac}}{2a}$

$= \dfrac{-(-2) \pm \sqrt{(-2)^2 - 4(1)(10)}}{2 \cdot 1}$

$= \dfrac{2 \pm \sqrt{4 - 40}}{2} = \dfrac{2 \pm \sqrt{-36}}{2}$

$= \dfrac{2 \pm 6i}{2} = 1 \pm 3i$

The solutions are $1 + 3i$
and $1 - 3i$.

You Try It 2 $4x^2 = 4x - 1$
$4x^2 - 4x + 1 = 0$
$a = 4, b = -4, c = 1$
$x = \dfrac{-b \pm \sqrt{b^2 - 4ac}}{2a}$

$= \dfrac{-(-4) \pm \sqrt{(-4)^2 - 4(4)(1)}}{2 \cdot 4}$

$= \dfrac{4 \pm \sqrt{16 - 16}}{8} = \dfrac{4 \pm \sqrt{0}}{8}$

$= \dfrac{4}{8} = \dfrac{1}{2}$

The solution is $\dfrac{1}{2}$.

You Try It 3 $3x^2 - x - 1 = 0$
$a = 3, b = -1, c = -1$
$b^2 - 4ac = (-1)^2 - 4(3)(-1)$
$\qquad\qquad = 1 + 12 = 13$

$13 > 0$

Because the discriminant is
greater than zero, the equation
has two real number solutions.

SECTION 10.4   *pages 527–530*

You Try It 1 $x - 5x^{1/2} + 6 = 0$
$(x^{1/2})^2 - 5(x^{1/2}) + 6 = 0$
$u^2 - 5u + 6 = 0$
$(u - 2)(u - 3) = 0$

$u - 2 = 0 \qquad u - 3 = 0$
$u = 2 \qquad\quad u = 3$

Replace u by $x^{1/2}$.

$x^{1/2} = 2 \qquad\quad x^{1/2} = 3$
$\sqrt{x} = 2 \qquad\quad \sqrt{x} = 3$
$(\sqrt{x})^2 = 2^2 \qquad (\sqrt{x})^2 = 3^2$
$x = 4 \qquad\qquad x = 9$

The solutions are 4 and 9.

You Try It 2 $\sqrt{2x + 1} + x = 7$
$\sqrt{2x + 1} = 7 - x$
$(\sqrt{2x + 1})^2 = (7 - x)^2$
$2x + 1 = 49 - 14x + x^2$
$0 = x^2 - 16x + 48$
$0 = (x - 4)(x - 12)$

$x - 4 = 0 \qquad x - 12 = 0$
$x = 4 \qquad\qquad x = 12$

4 checks as a solution.
12 does not check as a solution.

The solution is 4.

You Try It 3

$\sqrt{2x - 1} + \sqrt{x} = 2$

Solve for one of the radical expressions.

$\sqrt{2x - 1} = 2 - \sqrt{x}$

$(\sqrt{2x - 1})^2 = (2 - \sqrt{x})^2$

$\quad 2x - 1 = 4 - 4\sqrt{x} + x$

$\quad\quad x - 5 = -4\sqrt{x}$

Square each side of the equation.

$\quad (x - 5)^2 = (-4\sqrt{x})^2$

$x^2 - 10x + 25 = 16x$

$x^2 - 26x + 25 = 0$

$(x - 1)(x - 25) = 0$

$x - 1 = 0 \qquad x - 25 = 0$

$\quad x = 1 \qquad\quad\; x = 25$

1 checks as a solution.
25 does not check as a solution.

The solution is 1.

You Try It 4

$$3y + \frac{25}{3y - 2} = -8$$

$$(3y - 2)\left(3y + \frac{25}{3y - 2}\right) = (3y - 2)(-8)$$

$$(3y - 2)(3y) + (3y - 2)\left(\frac{25}{3y - 2}\right) = (3y - 2)(-8)$$

$$9y^2 - 6y + 25 = -24y + 16$$

$$9y^2 + 18y + 9 = 0$$

$$9(y^2 + 2y + 1) = 0$$

$$y^2 + 2y + 1 = 0$$

$$(y + 1)(y + 1) = 0$$

$$y + 1 = 0 \qquad y + 1 = 0$$

$$y = -1 \qquad\quad y = -1$$

The solution is -1.

SECTION 10.5   *pages 533–534*

You Try It 1

Strategy
- This is a geometry problem.
- Width of the rectangle: W
  Length of the rectangle: $W + 3$
- Use the equation $A = L \cdot W$.

Solution

$A = L \cdot W$

$54 = (W + 3)(W)$

$54 = W^2 + 3W$

$0 = W^2 + 3W - 54$

$0 = (W + 9)(W - 6)$

$W + 9 = 0 \qquad W - 6 = 0$

$\quad W = -9 \qquad\quad W = 6$

The solution -9 is not possible.

$W + 3 = 6 + 3 = 9$

The length is 9 m.

You Try It 1

$$2x^2 - x - 10 \le 0$$
$$(2x - 5)(x + 2) \le 0$$

$$\left\{ x \mid -2 \le x \le \frac{5}{2} \right\}$$

SOLUTIONS to Chapter 11 You Try It

SECTION 11.1   *pages 549–550*

You Try It 1

$b = -4$

y-intercept: $(0, -4)$

$m = \dfrac{3}{5}$

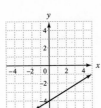

You Try It 2

Strategy

- Select the independent and dependent variables. The function is to predict the Celsius temperature, so that quantity is the dependent variable, y. The Fahrenheit variable is the independent variable, x.
- From the given data, two ordered pairs are $(212, 100)$ and $(32, 0)$. Use these ordered pairs to determine the linear function.

Solution

Let $(x_1, y_1) = (32, 0)$ and $(x_2, y_2) = (212, 100)$.

$$m = \frac{y_2 - y_1}{x_2 - x_1} = \frac{100 - 0}{212 - 32} = \frac{100}{180} = \frac{5}{9}$$

$$y - y_1 = m(x - x_1)$$

$$y - 0 = \frac{5}{9}(x - 32)$$

$$y = \frac{5}{9}(x - 32), \text{ or } C = \frac{5}{9}(F - 32).$$

The linear function is $f(F) = \dfrac{5}{9}(F - 32)$.

SECTION 11.2   *pages 553–558*

You Try It 1

x-coordinate of vertex:

$$-\frac{b}{2a} = -\frac{4}{2(4)} = -\frac{1}{2}$$

y-coordinate of vertex:

$$y = 4x^2 + 4x + 1$$
$$= 4\left(-\frac{1}{2}\right)^2 + 4\left(-\frac{1}{2}\right) + 1$$
$$= 1 - 2 + 1$$
$$= 0$$

Vertex: $\left(-\frac{1}{2}, 0\right)$

Axis of symmetry: $x = -\frac{1}{2}$.

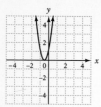

You Try It 3

$$y = x^2 - x - 6$$
$$a = 1, b = -1, c = -6$$

$$b^2 - 4ac$$
$$(-1)^2 - 4(1)(-6) = 1 + 24 = 25$$

Because the discriminant is greater than zero, the parabola has two x-intercepts.

You Try It 5

Strategy

- To find the time it takes the ball to reach its maximum height, find the t-coordinate of the vertex.
- To find the maximum height, evaluate the function at the t-coordinate of the vertex.

You Try It 2

$$y = x^2 + 3x + 4$$
$$0 = x^2 + 3x + 4$$

$$x = \frac{-b \pm \sqrt{b^2 - 4ac}}{2a} \qquad \bullet \ a = 1, b = 3, c = 4$$

$$= \frac{-3 \pm \sqrt{3^2 - 4(1)(4)}}{2 \cdot 1}$$

$$= \frac{-3 \pm \sqrt{-7}}{2}$$

$$= \frac{-3 \pm i\sqrt{7}}{2} = -\frac{3}{2} \pm \frac{\sqrt{7}}{2}i$$

The equation has no real number solutions. There are no x-intercepts.

You Try It 4

$$f(x) = -3x^2 + 4x - 1$$
$$x = -\frac{b}{2a} = -\frac{4}{2(-3)} = \frac{2}{3}$$
$$f(x) = -3x^2 + 4x - 1$$
$$f\left(\frac{2}{3}\right) = -3\left(\frac{2}{3}\right)^2 + 4\left(\frac{2}{3}\right) - 1$$
$$= -\frac{4}{3} + \frac{8}{3} - 1 = \frac{1}{3}$$

Because a is negative, the function has a maximum value.

The maximum value of the function is $\frac{1}{3}$.

Solution

$$t = -\frac{b}{2a} = -\frac{64}{2(-16)} = 2$$

The ball reaches its maximum height in 2 s.

$$s(t) = -16t^2 + 64t$$
$$s(2) = -16(2)^2 + 64(2) = -64 + 128 = 64$$

The maximum height is 64 ft.

SECTION 11.3 *pages 563–566*

You Try It 1
Any vertical line intersects the graph only once. The graph is the graph of a function.

You Try It 2
Domain: $\{x \mid x \text{ is a real number}\}$
Range: $\{y \mid y \text{ is a real number}\}$

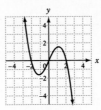

You Try It 3
Domain: $\{x \mid x \text{ is a real number}\}$
Range: $\{y \mid y \geq 0\}$

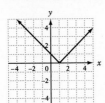

You Try It 4
Domain: $\{x \mid x \geq 1\}$
Range: $\{y \mid y \leq 0\}$

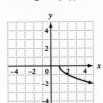

SECTION 11.4 *pages 569–572*

You Try It 1
$$(f + g)(-2) = f(-2) + g(-2)$$
$$= [(-2)^2 + 2(-2)] + [5(-2) - 2]$$
$$= (4 - 4) + (-10 - 2)$$
$$= -12$$

$$(f + g)(-2) = -12$$

You Try It 2
$$(f \cdot g)(3) = f(3) \cdot g(3)$$
$$= (4 - 3^2) \cdot [3(3) - 4]$$
$$= (4 - 9) \cdot (9 - 4)$$
$$= (-5)(5)$$
$$= -25$$

$$(f \cdot g)(3) = -25$$

You Try It 3
$$\left(\frac{f}{g}\right)(4) = \frac{f(4)}{g(4)}$$
$$= \frac{4^2 - 4}{4^2 + 2 \cdot 4 + 1}$$
$$= \frac{16 - 4}{16 + 8 + 1}$$
$$= \frac{12}{25}$$

$$\left(\frac{f}{g}\right)(4) = \frac{12}{25}$$

You Try It 4
$$g(x) = x^2$$
$$g(-1) = (-1)^2 = 1$$

$$f(x) = 1 - 2x$$
$$f[g(-1)] = f(1) = 1 - 2(1) = -1$$

You Try It 5

$f(x) = 3x - 1$

$f(1) = 3(1) - 1 = 2$

$g(x) = \dfrac{x}{x^2 - 2}$

$(g \circ f)(1) = g[f(1)] = g(2) = \dfrac{2}{2^2 - 2} = \dfrac{2}{2} = 1$

You Try It 7

$V(x) = x^2 + 1$

$$V[W(x)] = V(-\sqrt{x + 1}) = (-\sqrt{x - 1})^2 + 1$$
$$= x - 1 + 1$$
$$= x$$

You Try It 6

$M(s) = s^3 + 1$

$$M[L(s)] = M(s + 1) = (s + 1)^3 + 1$$
$$= s^3 + 3s^2 + 3s + 1 + 1$$
$$= s^3 + 3s^2 + 3s + 2$$

SECTION 11.5 *pages 575–578*

You Try It 1

Because any horizontal line intersects the graph at most once, the graph is the graph of a 1–1 function.

You Try It 3

$$f[g(x)] = 2\left(\frac{1}{2}x - 3\right) - 6$$
$$= x - 6 - 6 = x - 12$$

No, $g(x)$ is not the inverse of $f(x)$.

You Try It 2

$$f(x) = \frac{1}{2}x + 4$$

$$y = \frac{1}{2}x + 4$$

$$x = \frac{1}{2}y + 4$$

$$x - 4 = \frac{1}{2}y$$

$$2x - 8 = y$$

$$f^{-1}(x) = 2x - 8$$

The inverse of the function is given by $f^{-1}(x) = 2x - 8$.

SECTION 11.6   *pages 583–592*

You Try It 1

$y = x^2 + 2x + 1$

$-\dfrac{b}{2a} = -\dfrac{2}{2(1)} = -1$

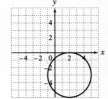

Axis of symmetry:
$$x = -1$$
$$y = (-1)^2 + 2(-1) + 1$$
$$= 0$$

Vertex: $(-1, 0)$

You Try It 2

$x = -y^2 - 2y + 2$

$-\dfrac{b}{2a} = -\dfrac{-2}{2(-1)} = -1$

Axis of symmetry:
$$y = -1$$
$$x = -(-1)^2 - 2(-1) + 2$$
$$= 3$$

Vertex: $(3, -1)$

You Try It 3

$y = x^2 - 2x - 1$

$-\dfrac{b}{2a} = -\dfrac{-2}{2(1)} = 1$

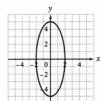

Axis of symmetry:
$$x = 1$$
$$y = 1^2 - 2(1) - 1$$
$$= -2$$

Vertex: $(1, -2)$

You Try It 4

$$(x - h)^2 + (y - k)^2 = r^2$$
$$(x - 2)^2 + [y - (-3)]^2 = 3^2$$
Center: $(h, k) = (2, -3)$
Radius: $r = 3$

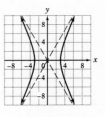

You Try It 5

$$(x - h)^2 + (y - k)^2 = r^2$$
$$(x - 2)^2 + [y - (-3)]^2 = 4^2$$
$$(x - 2)^2 + (y + 3)^2 = 16$$

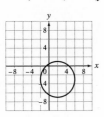

You Try It 6

x-intercepts:
$(2, 0)$ and $(-2, 0)$

y-intercepts:
$(0, 5)$ and $(0, -5)$

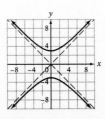

You Try It 7

x-intercepts:
$(3\sqrt{2}, 0)$ and $(-3\sqrt{2}, 0)$

y-intercepts:
$(0, 3)$ and $(0, -3)$

$\left(3\sqrt{2} \approx 4\dfrac{1}{4}\right)$

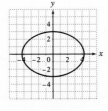

You Try It 8

Axis of symmetry:
x-axis

Vertices:
$(3, 0)$ and $(-3, 0)$

Asymptotes:
$y = \dfrac{5}{3}x$ and $y = -\dfrac{5}{3}x$

You Try It 9

Axis of symmetry:
y-axis

Vertices:
$(0, 3)$ and $(0, -3)$

Asymptotes:
$y = x$ and $y = -x$

SOLUTIONS to Chapter 12 You Try It

You Try It 1

$$f(x) = \left(\frac{2}{3}\right)^x$$

$$f(3) = \left(\frac{2}{3}\right)^3 = \frac{8}{27}$$

$$f(-2) = \left(\frac{2}{3}\right)^{-2} = \left(\frac{3}{2}\right)^2 = \frac{9}{4}$$

You Try It 2

$$f(x) = 2^{2x+1}$$

$$f(0) = 2^{2(0)+1} = 2^1 = 2$$

$$f(-2) = 2^{2(-2)+1} = 2^{-3} = \frac{1}{2^3} = \frac{1}{8}$$

You Try It 3

$$f(x) = e^{2x-1}$$

$$f(2) = e^{2 \cdot 2 - 1} = e^3 \approx 20.0855$$

$$f(-2) = e^{2(-2)-1} = e^{-5} \approx 0.0067$$

You Try It 4

x	y
-4	4
-2	2
0	1
2	$\frac{1}{2}$
4	$\frac{1}{4}$

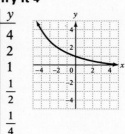

You Try It 5

x	y
-2	$\frac{5}{4}$
-1	$\frac{3}{2}$
0	2
1	3
2	5

You Try It 6

x	y
-2	6
-1	4
0	3
1	$\frac{5}{2}$
2	$\frac{9}{4}$

You Try It 7

x	y
-4	$\frac{9}{4}$
-2	$\frac{5}{2}$
0	3
2	4
4	6

You Try It 1

$$\log_4 64 = x$$
$$64 = 4^x$$
$$4^3 = 4^x$$
$$3 = x$$

$$\log_4 64 = 3$$

You Try It 2

$$\log_2 x = -4$$
$$2^{-4} = x$$

$$\frac{1}{2^4} = x$$

$$\frac{1}{16} = x$$

The solution is $\frac{1}{16}$.

You Try It 3

$$\ln x = 3$$
$$e^3 = x$$
$$20.0855 \approx x$$

You Try It 4

$$\log_8 \sqrt[3]{xy^2} = \log_8 (xy^2)^{1/3} = \frac{1}{3} \log_8 (xy^2)$$

$$= \frac{1}{3} (\log_8 x + \log_8 y^2)$$

$$= \frac{1}{3} (\log_8 x + 2 \log_8 y)$$

$$= \frac{1}{3} \log_8 x + \frac{2}{3} \log_8 y$$

You Try It 5

$$\frac{1}{3} (\log_4 x - 2 \log_4 y + \log_4 z)$$

$$= \frac{1}{3} (\log_4 x - \log_4 y^2 + \log_4 z)$$

$$= \frac{1}{3} \left(\log_4 \frac{x}{y^2} + \log_4 z \right) = \frac{1}{3} \left(\log_4 \frac{xz}{y^2} \right)$$

$$= \log_4 \left(\frac{xz}{y^2} \right)^{1/3} = \log_4 \sqrt[3]{\frac{xz}{y^2}}$$

You Try It 6

Because $\log_b 1 = 0$, $\log_9 1 = 0$.

You Try It 7

$$\log_7 6.45 = \frac{\log 6.45}{\log 7} \approx 0.9579$$

You Try It 8

$$\log_3 0.834 = \frac{\ln 0.834}{\ln 3} \approx -0.1652$$

SECTION 12.3   *pages 623–624*

You Try It 1

$$f(x) = \log_2 (x - 1)$$
$$y = \log_2 (x - 1)$$
$$2^y = x - 1$$
$$2^y + 1 = x$$

x	y
$\frac{5}{4}$	-2
$\frac{3}{2}$	-1
2	0
3	1
5	2

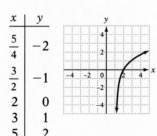

You Try It 2

$$f(x) = \log_3 (2x)$$
$$y = \log_3 (2x)$$
$$3^y = 2x$$
$$\frac{3^y}{2} = x$$

x	y
$\frac{1}{18}$	-2
$\frac{1}{6}$	-1
$\frac{1}{2}$	0
$\frac{3}{2}$	1
$\frac{9}{2}$	2

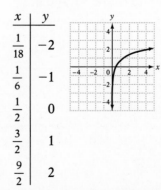

You Try It 3

$$f(x) = -\log_3(x + 1)$$
$$y = -\log_3(x + 1)$$
$$-y = \log_3(x + 1)$$
$$3^{-y} = x + 1$$
$$3^{-y} - 1 = x$$

x	y
8	−2
2	−1
0	0
$-\dfrac{2}{3}$	1
$-\dfrac{8}{9}$	2

SECTION 12.4 *pages 627–630*

You Try It 1

$$4^{3x} = 25$$
$$\log 4^{3x} = \log 25$$
$$3x \log 4 = \log 25$$
$$3x = \frac{\log 25}{\log 4}$$
$$x = \frac{\log 25}{3 \log 4}$$
$$x \approx 0.7740$$

The solution is 0.7740.

You Try It 2

$$(1.06)^n = 1.5$$
$$\log(1.06)^n = \log 1.5$$
$$n \log 1.06 = \log 1.5$$
$$n = \frac{\log 1.5}{\log 1.06}$$
$$n \approx 6.9585$$

The solution is 6.9585.

You Try It 3 $\log_4(x^2 - 3x) = 1$

Rewrite in exponential form.

$$4^1 = x^2 - 3x$$
$$4 = x^2 - 3x$$
$$0 = x^2 - 3x - 4$$
$$0 = (x + 1)(x - 4)$$

$$x + 1 = 0 \qquad x - 4 = 0$$
$$x = -1 \qquad x = 4$$

The solutions are −1 and 4.

You Try It 4 $\log_3 x + \log_3(x + 3) = \log_3 4$
$$\log_3[x(x + 3)] = \log_3 4$$

Use the fact that if
$\log_b u = \log_b v$, then $u = v$.

$$x(x + 3) = 4$$
$$x^2 + 3x = 4$$
$$x^2 + 3x - 4 = 0$$
$$(x + 4)(x - 1) = 0$$

$$x + 4 = 0 \qquad x - 1 = 0$$
$$x = -4 \qquad x = 1$$

−4 does not check as a
solution. The solution is 1.

You Try It 5

$$\log_3 x + \log_3(x + 6) = 3$$
$$\log_3[x(x + 6)] = 3$$
$$x(x + 6) = 3^3$$
$$x^2 + 6x = 27$$
$$(x + 9)(x - 3) = 0$$

$$x + 9 = 0 \qquad x - 3 = 0$$
$$x = -9 \qquad x = 3$$

−9 does not check as a solution. The
solution is 3.

SECTION 12.5   *pages 633–636*

You Try It 1

Strategy
To find the pH, replace H^+ by 2.51×10^{-12} in the equation $pH = -\log(H^+)$ and solve for pH.

Solution
$$pH = -\log(H^+)$$
$$= -\log(2.51 \times 10^{-12})$$
$$\approx 11.6$$

The pH of sodium carbonate is 11.6.

You Try It 2

Strategy
To find how many times stronger the San Francisco earthquake was, use the Richter equation to write a system of equations. Solve the system of equations for the ratio $\dfrac{I_1}{I_2}$.

Solution

$$7.8 = \log \frac{I_1}{I_0}$$

$$5 = \log \frac{I_2}{I_0}$$

$$7.8 = \log I_1 - \log I_0$$
$$5 = \log I_2 - \log I_0$$
$$2.8 = \log I_1 - \log I_2$$

$$2.8 = \log \frac{I_1}{I_2}$$

$$\frac{I_1}{I_2} = 10^{2.8} \approx 631$$

The San Francisco earthquake was 631 times stronger than the least forceful earthquake that can cause serious damage.

ANSWERS to Chapter 1 Selected Exercises

SECTION 1.1 *pages 11–16*

1. $8 > -6$ **3.** $-12 < 1$ **5.** $42 > 19$ **7.** $0 > -31$ **9.** $53 > -46$ **11.** False
13. True **15.** False **17.** True **19.** False **21.** -23 and -18 **23.** 21 and 37
25. -23 **27.** -4 **29.** 9 **31.** 28 **33.** 14 **35.** -77 **37.** 0 **39.** 74 **41.** -82
43. -81 **45.** $|-83| > |58|$ **47.** $|43| < |-52|$ **49.** $|-68| > |-42|$ **51.** $|-45| < |-61|$
53. $19, 0, -28$ **55.** $-45, 0, -17$ **57.** -11 **59.** -5 **61.** -83 **63.** -46 **65.** 0
67. -5 **69.** 9 **71.** 1 **73.** 8 **75.** -7 **77.** -9 **79.** 9 **81.** -3 **83.** 18
85. -10 **87.** -18 **89.** -41 **91.** -12 **93.** 0 **95.** -34 **97.** 0 **99.** -9
101. 11 **103.** -18 **105.** 0 **107.** 2 **109.** -138 **111.** -8 **113.** -12 **115.** -20
117. 42 **119.** -28 **121.** 60 **123.** -253 **125.** -238 **127.** -114 **129.** -2
131. 8 **133.** -7 **135.** -12 **137.** -6 **139.** -7 **141.** 11 **143.** -14 **145.** 15
147. -16 **149.** 0 **151.** -29 **153.** undefined **155.** -11 **157.** undefined
159. -105 **161.** 252 **163.** -240 **165.** 96 **167.** -216 **169.** -315 **171.** 420
173. 2880 **175.** -2772 **177.** 0 **179.** The difference is 20,602 ft.
181. The difference is 30,314 ft. **183.** The difference is 35°F.
185. No. The largest decrease is 7°F. **187.** The temperature dropped 100°F.
189. The average daily low temperature was -2°F. **191.** The student's score is 93. **193.** 6
195. positive

SECTION 1.2 *pages 27–32*

1. 0.125 **3.** $0.\overline{2}$ **5.** $0.1\overline{6}$ **7.** 0.5625 **9.** $0.58\overline{3}$ **11.** 0.24 **13.** 0.225 **15.** $0.\overline{45}$

17. $\frac{2}{5}, 0.4$ **19.** $\frac{22}{25}, 0.88$ **21.** $\frac{8}{5}, 1.6$ **23.** $\frac{87}{100}, 0.87$ **25.** $\frac{9}{2}, 4.5$ **27.** $\frac{3}{70}$ **29.** $\frac{3}{8}$

31. $\frac{1}{400}$ **33.** $\frac{1}{16}$ **35.** $\frac{23}{400}$ **37.** 0.091 **39.** 0.167 **41.** 0.009 **43.** 0.0915

45. 0.1823 **47.** 37% **49.** 2% **51.** 12.5% **53.** 136% **55.** 0.4% **57.** 83%

59. 37.5% **61.** $44\frac{4}{9}\%$ **63.** 45% **65.** 250% **67.** $-\frac{5}{24}$ **69.** $-\frac{19}{24}$ **71.** $\frac{5}{26}$ **73.** $\frac{11}{8}$

75. $\frac{1}{12}$ **77.** $\frac{7}{24}$ **79.** 0 **81.** $\frac{3}{8}$ **83.** $-\frac{7}{60}$ **85.** $-\frac{1}{16}$ **87.** 7.29 **89.** -3.049

91. -1.06 **93.** -23.845 **95.** -10.7893 **97.** -37.19 **99.** -17.5 **101.** 19.61

103. $-\frac{3}{8}$ **105.** $\frac{1}{10}$ **107.** $-\frac{4}{9}$ **109.** $-\frac{7}{30}$ **111.** $\frac{15}{64}$ **113.** $-\frac{10}{9}$ **115.** $-\frac{147}{32}$

117. $\frac{25}{8}$ **119.** $\frac{2}{3}$ **121.** 4.164 **123.** 4.347 **125.** -4.028 **127.** -2.22 **129.** -1.104

131. 0.506 **133.** -0.2376 **135.** $-274.\overline{4}$ **137.** -2.59 **139.** -5.11 **141.** -2060.55

143. 2401 **145.** -64 **147.** -8 **149.** -125 **151.** $-\frac{27}{64}$ **153.** 3.375 **155.** -1

157. 6.75 **159.** -405 **161.** -8 **163.** -6750 **165.** -144 **167.** -18 **169.** $\frac{4}{}$
171. 7 **173.** $4\sqrt{2}$ **175.** $2\sqrt{2}$ **177.** $18\sqrt{2}$ **179.** $10\sqrt{10}$ **181.** $\sqrt{15}$ **183.** $\sqrt{29}$
185. $-54\sqrt{2}$ **187.** $3\sqrt{5}$ **189.** 0 **191.** $48\sqrt{2}$ **193.** 15.492 **195.** 16.971 **197.** 16
199. 16.583 **201.** 15.652 **203.** 18.762

205. **a.** San Antonio came closest to winning $\frac{2}{3}$ of its games.

 b. Minnesota and Dallas lost more than $\frac{3}{5}$ of their games.

 c. Dallas lost 84.1% of its games.
207. **a.** The difference is $31.0 million. **b.** The difference is $30.8 million.
209. The total tax paid is $335,506,050.00. **211.** **a.** 36 **b.** 4 and 14

SECTION 1.3 *pages 35–36*

1. 0 **3.** -11 **5.** 20 **7.** -10 **9.** 20 **11.** 29 **13.** 11 **15.** 7 **17.** -11

19. 6 **21.** 15 **23.** 4 **25.** 5 **27.** -1 **29.** 4 **31.** 0.51 **33.** 1.7 **35.** $-\frac{1}{16}$

37. Answers will vary, for example, $\frac{17}{24}, \frac{33}{48}$

39. Answers will vary. For example, $a.\ r = \frac{1}{2}, b.\ r = 1, c.\ r = 2.$ **41.** $a = 2, b = 3, c = 6$

SECTION 1.4 *pages 47–54*

1. -9 **3.** 41 **5.** -7 **7.** 13 **9.** -15 **11.** 41 **13.** 1 **15.** 5 **17.** 1 **19.** 57
21. 5 **23.** 8 **25.** -3 **27.** -2 **29.** -4 **31.** 10 **33.** -25 **35.** $25x$ **37.** $9a$

39. $2y$ **41.** $-12y - 3$ **43.** $9a$ **45.** $6ab$ **47.** $-12xy$ **49.** 0 **51.** $-\frac{1}{10}y$ **53.** $\frac{2}{9}y^2$

55. $20x$ **57.** $-4a$ **59.** $-2y^2$ **61.** $-2x + 8y$ **63.** $8x$ **65.** $19a - 12b$
67. $-12x - 2y$ **69.** $-7x^2 - 5x$ **71.** $60x$ **73.** $-10a$ **75.** $30y$ **77.** $72x$ **79.** $-28a$
81. $108b$ **83.** $-56x^2$ **85.** x^2 **87.** x **89.** a **91.** b **93.** x **95.** n **97.** $2x$
99. $-2x$ **101.** $-15a^2$ **103.** $6y$ **105.** $3y$ **107.** $-2x$ **109.** $-9y$ **111.** $-x - 7$
113. $10x - 35$ **115.** $-5a - 80$ **117.** $-15y + 35$ **119.** $20 - 14b$ **121.** $-4x + 2y$
123. $18x^2 + 12x$ **125.** $10x - 35$ **127.** $-14x + 49$ **129.** $-30x^2 - 15$ **131.** $-24y^2 + 96$

133. $5x^2 + 5y^2$ **135.** $-\frac{1}{2}x + 2y$ **137.** $3x^2 + 6x - 18$ **139.** $-2y^2 + 4y - 8$

141. $-2x + 3y - \frac{1}{3}$ **143.** $10x^2 + 15x - 35$ **145.** $6x^2 + 3xy - 9y^2$ **147.** $-3a^2 - 5a + 4$

149. $-2x - 16$ **151.** $-12y - 9$ **153.** $7n - 7$ **155.** $-2x + 41$ **157.** $3y - 3$
159. $2a - 4b$ **161.** $-4x + 24$ **163.** $-2x - 16$ **165.** $-3x + 21$ **167.** $-4x + 12$

169. $-x + 50$ **171.** $\frac{x}{18}$ **173.** $x + 20$ **175.** $10(x - 50);\ 10x - 500$ **177.** $\frac{5}{8}x + 6$

179. $x - (x + 3);\ -3$ **181.** $4(x + 19);\ 4x + 76$ **183.** $\frac{15}{x + 12}$ **185.** $\frac{2}{3}(x + 7);\ \frac{2}{3}x + \frac{14}{3}$

187. $40 - \frac{x}{20}$ **189.** $x^2 + 2x$ **191.** $(x + 8) + \frac{1}{3}x;\ \frac{4}{3}x + 8$ **193.** $x + (x + 9);\ 2x + 9$

195. $x - (8 - x);\ 2x - 8$ **197.** $\frac{1}{3}x - \frac{5}{8}x;\ -\frac{7}{24}x$ **199.** $(x + 5) + 2;\ x + 7$

201. $2(6x + 7);\ 12x + 14$ **203.** width of the football field: w; length of the football field: $w + 30$
205. amount of gold in 12-carat gold chain: g; amount of gold in 18-carat gold chain: $2g - 6$

207. rate of the ball: r; rate of the bat: $\frac{2}{3}r$

209. length of the shorter piece: s; length of the longer piece: $3s - 2$
211. distance traveled by the faster car: d; distance traveled by the slower car: $200 - d$
213. No; 0.
215. **a.** False. $8 \div 4 \neq 4 \div 8$ **b.** False. $(8 \div 4) \div 2 \neq 8 \div (4 \div 2)$
    c. False. $(7 - 5) - 1 \neq 7 - (5 - 1)$ **d.** False. $6 - 3 \neq 3 - 6$ **e.** True
217. **a.** 4 **b.** 5 **c.** 6 **d.** $7; n^x > x^n$ if $x \geq n + 1$.

SECTION 1.5 *pages 59–60*

1. $\{16, 17, 18, 19, 20, 21\}$ **3.** $\{9, 11, 13, 15, 17\}$ **5.** $\{b, c\}$ **7.** $A \cup B = \{3, 4, 5, 6\}$
9. $A \cup B = \{-10, -9, -8, 8, 9, 10\}$ **11.** $A \cup B = \{a, b, c, d, e, f\}$ **13.** $A \cup B = \{1, 3, 7, 9, 11, 13\}$
15. $A \cap B = \{4, 5\}$ **17.** $A \cap B = \varnothing$ **19.** $A \cap B = \{c, d, e\}$
21. $\{x \mid x > -5, x \in \text{negative integers}\}$ **23.** $\{x \mid x > 30, x \in \text{integers}\}$
25. $\{x \mid x > 5, x \in \text{even integers}\}$ **27.** $\{x \mid x > 8, x \in \text{real numbers}\}$
29. **31.** **33.** **35.**

37. **39.** **a.** never true **b.** always true **c.** always true

41. **a.** yes **b.** yes

CHAPTER REVIEW EXERCISES   *pages 65–66*

1. $-4, 0$ [1.1A]   **2.** 4 [1.1B]   **3.** -5 [1.1B]   **4.** -13 [1.1C]   **5.** 1 [1.1C]
6. -42 [1.1D]   **7.** -20 [1.1D]   **8.** 0.28 [1.2A]   **9.** 0.062 [1.2B]   **10.** 62.5% [1.2B]
11. $\frac{7}{12}$ [1.2C]   **12.** -1.068 [1.2C]   **13.** $-\frac{72}{85}$ [1.2D]   **14.** -4.6224 [1.2D]   **15.** $\frac{16}{81}$ [1.2E]
16. 12 [1.2F]   **17.** $-6\sqrt{30}$ [1.2F]   **18.** 31 [1.3A]   **19.** 29 [1.4A]   **20.** $8a - 4b$ [1.4B]
21. $36y$ [1.4C]   **22.** $10x - 35$ [1.4D]   **23.** $7x + 46$ [1.4D]   **24.** $-90x + 25$ [1.4D]
25. $\{1, 3, 5, 7\}$ [1.5A]   **26.** $A \cap B = \{1, 5, 9\}$ [1.5A]
27. [1.5C]   **28.** [1.5C]

29. The student's score was 98.   [1.1E]   **30.** The value of the purchase is $7958.81.   [1.2G]
31. $2x - \frac{1}{2}x; \frac{3}{2}x$ [1.4E]
32. number of American League players' cards: n; number of National League players' cards: $5n$ [1.4E]
33. number of ten-dollar bills: n; number of five-dollar bills: $35 - n$ [1.4E]

ANSWERS to Chapter 2 Selected Exercises

SECTION 2.1   *pages 73–76*

1. yes   **3.** no   **5.** no   **7.** yes   **9.** yes   **11.** yes   **13.** no   **15.** yes   **17.** yes
19. yes   **21.** no   **23.** $y = 6$ **25.** $z = 16$ **27.** $x = 7$ **29.** $t = -2$ **31.** $x = 1$
33. $y = 0$ **35.** $x = 3$ **37.** $n = -10$ **39.** $x = -3$ **41.** $y = -14$ **43.** $t = 2$
45. $a = 11$ **47.** $n = -9$ **49.** $y = -1$ **51.** $x = -14$ **53.** $x = -5$ **55.** $m = -1$
57. $x = 1$ **59.** $y = -\frac{1}{2}$ **61.** $m = -\frac{3}{4}$ **63.** $x = \frac{1}{12}$ **65.** $x = -\frac{7}{12}$ **67.** $d = 0.6529$
69. $x = -0.283$ **71.** $z = 9.257$ **73.** $x = -3$ **75.** $b = 0$ **77.** $x = -2$ **79.** $x = 9$
81. $c = 80$ **83.** $w = -4$ **85.** $x = 0$ **87.** $y = 8$ **89.** $t = -7$ **91.** $x = 12$
93. $b = -18$ **95.** $x = 15$ **97.** $m = -20$ **99.** $x = 0$ **101.** $x = 15$ **103.** $x = 75$
105. $x = \frac{8}{3}$ **107.** $y = \frac{1}{3}$ **109.** $x = -\frac{1}{2}$ **111.** $n = -\frac{3}{2}$ **113.** $m = \frac{15}{7}$ **115.** $n = 4$
117. $y = 3$ **119.** $x = 4.745$ **121.** $a = 2.06$ **123.** $x = -2.13$ **125.** $x = \frac{b}{a}, a \neq 0$
127. For example, $3x = -3$.   **129. a.** $x = \frac{6}{5}$ **b.** $a = 2$

SECTION 2.2   *pages 83–86*

1. 8 **3.** 0.075 **5.** $16\frac{2}{3}\%$ **7.** 37.5% **9.** 100 **11.** 1200 **13.** 51.895 **15.** 13
17. 2.7% **19.** 400% **21.** 7.5 **23.** 200 **25.** 2.5% **27.** 37.5% **29.** 80 **31.** 9
33. The programmer's salary increased by 8%.   **35.** The estimated life of the brakes is 50,000 mi.
37. The amount deducted for income tax is $403.20.
39. 32% of last month's income was spent for rent.
41. 20.9% of the total injuries happen on slides.   **43.** The previous year's snowfall was 165 in.
45. The price at the competitor's store was $49.82.
47. The city's population 5 years ago was 56,000.
49. The farmer would receive a tax credit of $12,750.
51. 9.375% of the units are taken in mathematics.
53. 7944 of the computer boards tested were not defective.
55. The amount of money offered is $1,518,000.
57. A student must answer correctly $83\frac{1}{3}\%$ of the questions.   **59.** yes
61. The new value is two times the original value.

SECTION 2.3 *pages 93–98*

1. $x = 3$ **3.** $a = 6$ **5.** $x = -1$ **7.** $x = -3$ **9.** $w = 2$ **11.** $t = 2$ **13.** $a = 5$
15. $b = -3$ **17.** $x = 6$ **19.** $x = 3$ **21.** $a = 1$ **23.** $b = 6$ **25.** $m = -7$ **27.** $y = 0$
29. $x = \dfrac{3}{4}$ **31.** $x = \dfrac{4}{9}$ **33.** $x = \dfrac{1}{3}$ **35.** $w = -\dfrac{1}{2}$ **37.** $b = -\dfrac{3}{4}$ **39.** $a = \dfrac{1}{3}$
41. $x = -\dfrac{1}{6}$ **43.** $a = 1$ **45.** $x = 1$ **47.** $x = 0$ **49.** $x = \dfrac{13}{10}$ **51.** $a = \dfrac{2}{5}$
53. $x = -\dfrac{4}{3}$ **55.** $x = -\dfrac{3}{2}$ **57.** $m = 18$ **59.** $n = 8$ **61.** $b = -16$ **63.** $y = 25$
65. $c = 21$ **67.** $w = 15$ **69.** $x = -16$ **71.** $x = -21$ **73.** $x = \dfrac{15}{2}$ **75.** $y = -\dfrac{18}{5}$
77. $y = 2$ **79.** $z = 3$ **81.** $b = 1$ **83.** $m = -2$ **85.** $x = 2$ **87.** $x = 3$ **89.** $x = -1$
91. $x = 2$ **93.** $x = -2$ **95.** $b = -3$ **97.** $y = 0$ **99.** $x = -1$ **101.** $x = -3$
103. $x = -1$ **105.** $m = 4$ **107.** $b = \dfrac{2}{3}$ **109.** $x = \dfrac{5}{6}$ **111.** $m = \dfrac{3}{4}$ **113.** $y = 1$
115. $x = 4$ **117.** $m = -1$ **119.** $x = -1$ **121.** $b = -\dfrac{2}{3}$ **123.** $a = \dfrac{4}{3}$ **125.** $x = \dfrac{1}{2}$
127. $y = -\dfrac{1}{3}$ **129.** $a = \dfrac{10}{3}$ **131.** $b = -\dfrac{1}{4}$ **133.** $x + 12 = 20;\ 8$ **135.** $\dfrac{3}{5}x = -30;\ -50$
137. $3x + 4 = 13;\ 3$ **139.** $9x - 6 = 12;\ 2$ **141.** $x + 2x = 9;\ 3$ **143.** $5x - 17 = 2;\ \dfrac{19}{5}$
145. $6x + 7 = 3x - 8;\ -5$ **147.** $30 = 7x - 9;\ \dfrac{39}{7}$ **149.** $2x = (21 - x) + 3;\ 8,\ 13$
151. $23 - x = 2x + 5;\ 6,\ 17$ **153.** 19 **155.** -14 **157.** 41,493
159. It is not an equation and cannot be solved. **161.** No solution **163.** $m = -21$

SECTION 2.4 *pages 107–114*

1. The mixture contains 2 lb of diet supplement and 3 lb of vitamin supplement.
3. The cost is $6.98 per pound. **5.** The combination contained 56 oz at $4.30 and 144 oz at $1.80.
7. The cost is $2.90 per lb. **9.** There must be 10 kg of hard candy.
11. The mixture contains 30 lb at $2.20 and 20 lb at $4.20.
13. The caterer should use 25 gal of ice cream and 75 gal of fruit juice.
15. The cost is $2.75 per gallon.
17. The solution must contain 20 ml of 13% acid and 30 ml of 18% acid.
19. There is a 50% concentration of silver. **21.** There were 30 lb of 60% mixture used.
23. There is 0.74% of hydrocortisone. **25.** The hair dye contains 100 ml of 7% and 200 ml of 4%.
27. There must be 25 oz of pure water added. **29.** The percent concentration is 27%.
31. There must be 10 oz of pure bran flakes added.
33. There must be an additional $5000 added.
35. There was $9000 invested at 7% and $6000 at 6.5%.
37. There was $2500 deposited in the mutual fund.
39. The university deposited $200,000 at 10% and $100,000 at 8.5%.
41. The mechanic invests $3000 in additional bonds.
43. $40,500 was invested at 8%. $13,500 was invested at 12%.
45. The total amount invested was $650,000. **47.** The total amount invested was $500,000.
49. The plane flew 2 h at 105 mph and 3 h at 115 mph. **51.** The sailboat traveled 36 mi.
53. The rate of the passenger train is 50 mph, and the rate of the freight train is 30 mph.
55. The rate of the cyclist is 16 mph.
57. The rate of the first plane is 95 mph, and the rate of the second plane is 120 mph.
59. They will meet after 1 h. **61.** The second runner will overtake the first runner after 3 h.
63. It took 20 min downstream. **65.** The cost is $3.65 per ounce.
67. 10 oz of water must be evaporated to produce a 15% salt solution.
69. 3.75 gal must be drained and replaced by pure antifreeze.
71. $12,000 was invested in bonds, $21,000 was invested in a simple interest account, and $27,000 was invested in corporate bonds.
73. The campers turned around at 10:15 A.M. **75.** The cyclist's average speed is $13\dfrac{1}{3}$ mph.

Content and Format © 1996 HMCo.

SECTION 2.5 *pages 121–126*

1. $\{x \mid x < 5\}$ **3.** $\{x \mid x \le 2\}$ **5.** $\{x \mid x < -4\}$ **7.** $\{x \mid x > 3\}$ **9.** $\{x \mid x > 4\}$
11. $\{x \mid x \le 2\}$ **13.** $\{x \mid x > -2\}$ **15.** $\{x \mid x \ge 2\}$ **17.** $\{x \mid x > -2\}$ **19.** $\{x \mid x \le 3\}$

21. $\{x \mid x < 2\}$ **23.** $\{x \mid x < -3\}$ **25.** $\{x \mid x \le 5\}$ **27.** $\left\{x \mid x \ge -\dfrac{1}{2}\right\}$ **29.** $\left\{x \mid x < \dfrac{23}{16}\right\}$

31. $\left\{x \mid x < \dfrac{8}{3}\right\}$ **33.** $\{x \mid x > 1\}$ **35.** $\left\{x \mid x > \dfrac{14}{11}\right\}$ **37.** $\{x \mid x \le 1\}$ **39.** $\left\{x \mid x \le \dfrac{3}{8}\right\}$

41. $\left\{x \mid x \le \dfrac{7}{4}\right\}$ **43.** $\left\{x \mid x \ge -\dfrac{5}{4}\right\}$ **45.** $\{x \mid x \le 2\}$ **47.** $\{x \mid -2 \le x \le 4\}$

49. $\{x \mid x < 3 \text{ or } x > 5\}$ **51.** $\{x \mid -4 < x < 2\}$ **53.** $\{x \mid x > 6 \text{ or } x < -4\}$
55. $\{x \mid x < -3\}$ **57.** $\{x \mid -3 < x < 2\}$ **59.** $\{x \mid -2 < x < 1\}$ **61.** $\{x \mid x < -2 \text{ or } x > 2\}$

63. $\{x \mid 2 < x < 6\}$ **65.** $\{x \mid -3 < x < -2\}$ **67.** $\left\{x \mid x > 5 \text{ or } x < -\dfrac{5}{3}\right\}$ **69.** $\{x \mid x < 3\}$

71. The solution set is the empty set. **73.** The solution set is the set of real numbers.
75. $\left\{x \mid \dfrac{17}{7} \le x \le \dfrac{45}{7}\right\}$ **77.** $\left\{x \mid -5 < x < \dfrac{17}{3}\right\}$ **79.** The solution set is the set of real numbers.
81. The solution set is the empty set. **83.** The smallest number is -12.
85. The maximum width as an integer is 11 cm.
87. You can use the cellular phone for no more than 45 min. **89.** You can drive less than 350 mi.
91. The temperature range is $32° < F < 86°$. **93.** The amount of sales is $44,000 or more.
95. The customer can write fewer than 50 checks. **97.** The range of scores is $58 \le x \le 100$.
99. The ski area is more than 38 mi away. **101.** $\{1, 2\}$ **103.** $\varnothing$

SECTION 2.6 *pages 131–134*

1. 7 and -7 **3.** 4 and -4 **5.** 6 and -6 **7.** 7 and -7 **9.** There is no solution.
11. There is no solution. **13.** -5 and 1 **15.** 2 and 8 **17.** 2 **19.** There is no solution.

21. 3 and $-\dfrac{3}{2}$ **23.** $\dfrac{3}{2}$ **25.** There is no solution. **27.** -3 and 7 **29.** $-\dfrac{10}{3}$ and 2

31. 3 and 1 **33.** $\dfrac{3}{2}$ **35.** There is no solution. **37.** $-\dfrac{1}{6}$ and $\dfrac{11}{6}$ **39.** -1 and $-\dfrac{1}{3}$

41. There is no solution. **43.** 0 and 3 **45.** $-\dfrac{7}{3}$ and $\dfrac{11}{3}$ **47.** $\dfrac{13}{3}$ and 1

49. There is no solution. **51.** $\dfrac{1}{3}$ and $\dfrac{7}{3}$ **53.** $-\dfrac{1}{2}$ **55.** $-\dfrac{7}{2}$ and $-\dfrac{1}{2}$ **57.** $\dfrac{10}{3}$ and $-\dfrac{8}{3}$

59. There is no solution. **61.** $\{x \mid x > 3 \text{ or } x < -3\}$ **63.** $\{x \mid x > 1 \text{ or } x < -3\}$

65. $\{x \mid 4 \le x \le 6\}$ **67.** $\{x \mid x \ge 5 \text{ or } x \le -1\}$ **69.** $\{x \mid -3 < x < 2\}$ **71.** $\left\{x \mid x > 2 \text{ or } x < -\dfrac{14}{5}\right\}$

73. $\left\{x \mid \dfrac{1}{4} \le x \le \dfrac{5}{4}\right\}$ **75.** $\{x \mid x < -6 \text{ or } x > -1\}$ **77.** $\left\{x \mid x \le -\dfrac{1}{3} \text{ or } x \ge 3\right\}$

79. $\left\{x \mid -2 \le x \le \dfrac{9}{2}\right\}$ **81.** $\{x \mid x = 2\}$ **83.** $\left\{x \mid x < -2 \text{ or } x > \dfrac{22}{9}\right\}$ **85.** $\left\{x \mid -\dfrac{3}{2} < x < \dfrac{9}{2}\right\}$

87. $\left\{x \mid x < 0 \text{ or } x > \dfrac{4}{5}\right\}$ **89.** $\{x \mid x > 5 \text{ or } x < 0\}$

91. The lower limit is 3.95 cc. The upper limit is 4.05 cc.
93. The lower limit is 2.648 in. The upper limit is 2.652 in.
95. The lower limit is $9\dfrac{19}{32}$ in. The upper limit is $9\dfrac{21}{32}$ in.
97. The lower limit is 28,420 ohms. The upper limit is 29,580 ohms.
99. The lower limit is 23,750 ohms. The upper limit is 26,250 ohms.
101. **a.** Sometimes **b.** Sometimes **103.** **a.** $\le$ **b.** $\ge$ **c.** $\ge$ **d.** $=$

CHAPTER REVIEW EXERCISES *pages 137–138*

1. no [2.1A] **2.** $x = 21$ [2.1B] **3.** $a = 20$ [2.1C] **4.** $x = -3$ [2.3A]
5. $x = -\dfrac{6}{7}$ [2.3A] **6.** $y = \dfrac{1}{3}$ [2.3B] **7.** $x = 4$ [2.3B] **8.** $x = -2$ [2.3C]

9. $x = 10$ [2.3C]   **10.** $\left\{x \mid x > \frac{5}{3}\right\}$ [2.5A]   **11.** $\{x \mid x > -1\}$ [2.5A]

12. $\left\{x \mid -3 < x < \frac{4}{3}\right\}$ [2.5B]   **13.**   The solution set is the real numbers.   [2.5B]

14. $-\frac{9}{5}$ and 3   [2.6A]   **15.** -1 and 9   [2.6A]   **16.** $\{x \mid 1 < x < 4\}$ [2.6B]

17. $\left\{x \mid x \le \frac{1}{2} \text{ or } x \ge 2\right\}$ [2.6B]   **18.**   30 is 250% of 12.   [2.2A]   **19.** $\frac{1}{2}\%$ of 1600 is 8.   [2.2A]

20. $5x - 4 = 16; x = 4$ [2.3D]   **21.** $3x = 2(21 - x) - 2$; 8 and 13   [2.3D]

22.   The airline would sell 177 tickets.   [2.2B]   **23.**   The rebate is 6.1% of the cost.   [2.2B]

24.   7 qt of cranberry juice and 3 qt of apple juice were used.   [2.4A]

25.   The cost is $1.84 per pound.   [2.4A]

26.   The percent concentration of butterfat is 14%.   [2.4B]

27.   375 lb of the alloy containing 30% tin and 125 lb of the alloy containing 70% tin were used.   [2.4B]

28.   The total amount invested was $450,000.   [2.4C]

29. $1400 was deposited at 6.75%. $1000 was deposited at 9.45%.   [2.4C]

30.   The jet overtakes the propeller-driven plane 600 mi from the starting point.   [2.4D]

31.   The average speed on the winding road was 32 mph.   [2.4D]

32.   The range of scores is $82 \le x \le 100$.   [2.5C]

33.   The lower limit is 1.75 cc. The upper limit is 2.25 cc.   [2.6C]

CUMULATIVE REVIEW EXERCISES   *pages 139–140*

1.   6   [1.1C]   **2.** -48 [1.1D]   **3.** $-\frac{19}{48}$ [1.2C]   **4.**   54   [1.2E]   **5.** $\frac{49}{40}$ [1.3A]

6.   6   [1.4A]   **7.** $-17x$ [1.4B]   **8.** $-5a - 4b$ [1.4B]   **9.** $2x$ [1.4C]   **10.** $36y$ [1.4C]

11. $2x^2 + 6x - 4$ [1.4D]   **12.** $-4x + 14$ [1.4D]   **13.** $6x - 34$ [1.4D]

14. $A \cap B = \{-4, 0\}$ [1.5B]   **15.**   [1.5C]   **16.**   yes   [2.1A]

17. $x = -25$ [2.1C]   **18.** $x = -3$ [2.3A]   **19.** $x = 3$ [2.3A]   **20.** $x = -3$ [2.3B]

21. $x = \frac{1}{2}$ [2.3B]   **22.** $x = 13$ [2.3C]   **23.** $\{x \mid x \le 1\}$ [2.5A]   **24.** $\{x \mid -4 \le x \le 1\}$ [2.5B]

25. -1 and 4   [2.6A]   **26.** $\left\{x \mid x < -\frac{4}{3} \text{ or } x > 2\right\}$ [2.6B]   **27.** $\frac{11}{20}$ [1.2B]   **28.**   103%   [1.2B]

29.   25% of 120 is 30.   [2.2A]   **30.** $6x + 13 = 3x - 5; x = -6$ [2.3D]

31.   20 lb of oat flour must be used.   [2.4A]   **32.**   25 g of pure gold must be added.   [2.4B]

33.   The length of the track is 120 m.   [2.4D]

ANSWERS to Chapter 3 Selected Exercises

SECTION 3.1   *pages 153–158*

1.   40°; acute   **3.**   115°; obtuse   **5.**   90°; right   **7.**   The complement is 28°.

9.   The supplement is 18°.   **11.**   The length of BC is 14 cm.   **13.**   The length of QS is 28 ft.

15.   The length of EG is 30 m.   **17.**   The measure of $\angle MON$ is 86°.   **19.**   71°   **21.**   30°

23.   36°   **25.**   127°   **27.**   116°   **29.**   20°   **31.**   20°   **33.**   20°   **35.**   141°   **37.**   106°

39.   11°   **41.** $\angle a = 38°, \angle b = 142°$ **43.** $\angle a = 47°, \angle b = 133°$ **45.**   20°   **47.**   47°

49. $\angle x = 155°, \angle y = 70°$ **51.** $\angle a = 45°, \angle b = 135°$ **53.** $90° - x$

55.   The measure of the third angle is 60°.   **57.**   The measure of the third angle is 35°.

59.   The measure of the third angle is 102°.   **61. a.**   1°   **b.**   179°   **65.**   360°

SECTION 3.2   *pages 169–176*

1.   hexagon   **3.**   pentagon   **5.**   scalene   **7.**   equilateral   **9.**   obtuse   **11.**   acute

13.   56 in.   **15.**   14 ft   **17.**   47 mi   **19.** 8π cm or approximately 25.13 cm

21. 11π mi or approximately 34.56 mi   **23.** 17π ft or approximately 53.41 ft
25. The perimeter is 17.4 cm.   **27.** The perimeter is 8 cm.   **29.** The perimeter is 24 m.
31. The perimeter is 48.8 cm.   **33.** The perimeter is 17.5 in.
35. The length of a diameter is 8.4 cm.   **37.** The circumference is 1.5π in.
39. The circumference is 226.19 cm.   **41.** 60 ft of fencing should be purchased.
43. The carpet must be nailed down along 44 ft.   **45.** The length is 120 ft.
47. The length of the third side is 10 in.   **49.** The length of each side is 12 in.
51. The length of a diameter is 2.55 cm.   **53.** The length is 13.19 ft.
55. The bicycle travels 50.27 ft.   **57.** The circumference is 39,935.93 km.   **59.** 60 ft^2
61. 20.25 in^2 **63.** 546 ft^2 **65.** 16π cm^2 or approximately 50.27 cm^2
67. 30.25π mi^2 or approximately 95.03 mi^2 **69.** 72.25π ft^2 or approximately 226.98 ft^2
71. The area is 156.25 cm^2.   **73.** The area is 570 in^2.   **75.** The area is 192 in^2.
77. The area is 13.5 ft^2.   **79.** The area is 330 cm^2.   **81.** The area is 25π in^2.
83. The area is 9.08 ft^2.   **85.** The area is $10{,}000\pi$ in^2.   **87.** The area is 126 ft^2.
89. 7500 yd^2 must be purchased.   **91.** The width is 10 in.   **93.** The length of the base is 20 m.
95. You should buy 2 qt.   **97.** It will cost \$74.   **99.** The increase in area is 113.10 in^2.
101. The cost will be \$638.   **103.** The area is 216 m^2.   **105.** 4:9   **109.** $A = \dfrac{1}{4}\pi d^2$

SECTION 3.3   *pages 183–186*

1. 840 in^3 **3.** 15 ft^3 **5.** 4.5π cm^3 or approximately 14.14 cm^3 **7.** The volume is 34 m^3.
9. The volume is 15.625 in^3.   **11.** The volume is 36π ft^3.   **13.** The volume is 8143.01 cm^3.
15. The volume is 75π in^3.   **17.** The volume is 120 in^3.   **19.** The width is 2.5 ft.
21. The radius of the base is 4.00 in.   **23.** The length is 5 in. The width is 5 in.
25. There are 75.40 m^3 in the tank.   **27.** 94 m^2 **29.** 56 m^2
31. 96π in^2 or approximately 301.59 in^2 **33.** The surface area is 184 ft^2.
35. The surface area is 69.36 m^2.   **37.** The surface area is 225π cm^2.
39. The surface area is 402.12 in^2.   **41.** The surface area is 6π ft^2.
43. The surface area is 297 in^2.   **45.** The width is 3 cm.
47. 11 cans of paint should be purchased.   **49.** 456 in^2 of glass are needed.
51. The surface area of the pyramid is 22.53 cm^2 larger.
53. **a.** always true   **b.** never true   **c.** sometimes true

CHAPTER REVIEW EXERCISES   *pages 189–190*

1. $\angle x = 22°$, $\angle y = 158°$ [3.1C]   **2.** $\angle x = 68°$ [3.1B]   **3.** The length of AC is 44 cm. [3.1A]
4. $x = 19°$ [3.1A]   **5.** The volume is 96 cm^3. [3.3A]   **6.** $\angle a = 138°$, $\angle b = 42°$ [3.1B]
7. The surface area is 220 ft^2. [3.3B]   **8.** The supplement is 148°. [3.1A]
9. The area is 78 cm^2. [3.2B]   **10.** The area is 63 m^2. [3.2B]
11. The volume is 39 ft^3. [3.3A]   **12.** The measure of the third angle is 95°. [3.1C]
13. The length of the base is 8 cm. [3.2B]   **14.** The volume is 288π mm^3. [3.3A]
15. The volume is $\dfrac{784\pi}{3}$ cm^3. [3.3A]   **16.** Each side measures 21.5 cm. [3.2A]
17. 4 cans of paint should be purchased. [3.3B]   **18.** 208 yd of fencing are needed. [3.2A]
19. The area is 90.25 m^2. [3.2B]   **20.** The area is 276 m^2. [3.2B]

CUMULATIVE REVIEW EXERCISES   *pages 191–192*

1. -3, 0, and 1 [1.1A]   **2.** 0.089 [1.2B]   **3.** 35% [1.2B]   **4.** $-\dfrac{2}{3}$ [1.2D]
5. -24.51 [1.2D]   **6.** $-5\sqrt{5}$ [1.2F]   **7.** -28 [1.3A]   **8.** -8 [1.4A]
9. $-3m + 3n$ [1.4B]   **10.** $21y$ [1.4C]   **11.** $7x + 9$ [1.4D]   **12.** $\{-2, -1\}$ [1.5A]
13. $\{-10, 0, 10, 20, 30\}$ [1.5A]   **14.** $\begin{array}{c} \text{(number line)} \\ \underset{-5\,-4\,-3\,-2\,-1\ 0\ 1\ 2\ 3\ 4\ 5}{} \end{array}$ [1.5C]   **15.** $x = 5$ [2.3B]
16. $x = \dfrac{1}{2}$ [2.3C]   **17.** $\{y \mid y \le -4\}$ [2.5A]   **18.** $\{x \mid x \ge 2\}$ [2.5A]
19. $\{x \mid x < -3 \text{ or } x > 4\}$ [2.5B]   **20.** $\{x \mid 2 \le x \le 6\}$ [2.5B]   **21.** $-\dfrac{1}{3}$ and 1 [2.6A]

22. $\{x \mid 6 \leq x \leq 10\}$ [2.6B] **23.** $\angle x = 131°$ [3.1B] **24.** $4x - 10 = 2; x = 3$ [2.3D]
25. The third angle measures 122°. [3.1C] **26.** \$5000 is deposited in the 9.5% account. [2.4C]
27. The third side measures 4.5 m. [3.2A] **28.** 80% of the students went on to college. [2.2B]
29. The area is 20.25π cm². [3.2B] **30.** The height of the box is 3 ft. [3.3A]

ANSWERS to Chapter 4 Selected Exercises

SECTION 4.1 *pages 203–206*

1.

3.

5.

7. The coordinates of A are $(2, 3)$. The coordinates of B are $(4, 0)$.
The coordinates of C are $(-4, 1)$. The coordinates of D are $(-2, -2)$.
9. The coordinates of A are $(-2, 5)$. The coordinates of B are $(3, 4)$.
The coordinates of C are $(0, 0)$. The coordinates of D are $(-3, -2)$.
11. The coordinates of A are $(2, 4)$. The coordinates of B are $(0, 1)$.
The coordinates of C are $(-4, 0)$. The coordinates of D are $(3, -3)$.
13. Yes **15.** No **17.** No **19.** No **21.** $(3, 7)$ **23.** $(6, 3)$ **25.** $(0, 1)$ **27.** $(-5, 0)$
29.

31.

33.

35. $\{(4, L), (6, W), (4, W), (2, W), (1, L), (6, W)\}$ No, the relation is not a function.
37. $\{(11, 200), (9, 150), (7, 200), (12, 175), (12, 250)\}$ No, the relation is not a function. **39.** Yes
41. No **43.** Yes **45.** 8 **47.** 9 **49.** 2 **51.** -1 **53.** 22 **55.** $-\dfrac{3}{2}$ **57.** -7

SECTION 4.2 *pages 215–218*

1.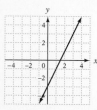

3.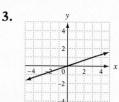

5.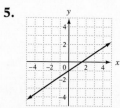

7.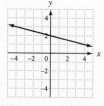

9.

11.

13.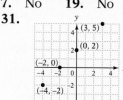

15.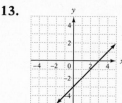

17.

19. $(3, 0)$ and $(0, -3)$ **21.** $(10, 0)$ and $(0, -2)$ **23.** $(0, 0)$ and $(0, 0)$

25. **27.** 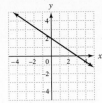 **29.** 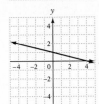 **31.**

33. 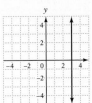 **35.** 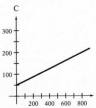 **37.** 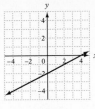 **39.**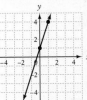

41. 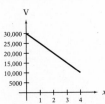 **43.**

The ordered pair (1, 25,000) means that in 1 year, the value of the computer will be $25,000.

The ordered pair (500, 140) means that if the car is driven 500 mi, the total cost to rent the car will be $140.

45. The value of y decreases by 2. The change in y is -2.

SECTION 4.3 *pages 225–228*

1. -1 **3.** $\frac{1}{3}$ **5.** $-\frac{2}{3}$ **7.** $-\frac{3}{4}$ **9.** The slope is undefined. **11.** $\frac{7}{5}$

13. The line has zero slope. **15.** $-\frac{1}{2}$ **17.** The slope is undefined.

19. The slope is 40. The slope is the average speed of the motorist in miles per hour.
21. The slope is 0.28. The slope is the tax rate.
23. The slope is 385.5. The slope is the average speed of the runner in meters per minute.

25. **27.** **29.** **31.**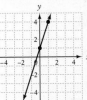

33. 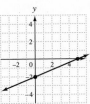 **35.** 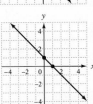 **37.** 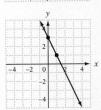 **39.**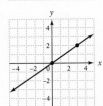

41. 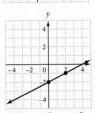 **43.** No. For instance, the graph of $x = 3$ does not have a y-intercept.

45. The value of y increases by 2 as the value of x increases by 1.
47. The value of y increases by 2 as the value of x decreases by 1.
49. The value of y increases by $\frac{1}{2}$ as the value of x increases by 1. **51.** $k = 10$ **53.** $k = -4$

SECTION 4.4 *pages 233–236*

1. $y = 2x + 5$ **3.** $y = \frac{1}{2}x + 2$ **5.** $y = \frac{5}{4}x + \frac{21}{4}$ **7.** $y = -\frac{5}{3}x + 5$ **9.** $y = -3x + 9$

11. $y = -3x + 4$ **13.** $y = \frac{2}{3}x - \frac{7}{3}$ **15.** $y = \frac{1}{2}x$ **17.** $y = 3x - 9$ **19.** $y = -\frac{2}{3}x + 7$

21. $y = -x - 3$ **23.** $y = \frac{7}{5}x - \frac{27}{5}$ **25.** $y = -\frac{2}{5}x + \frac{3}{5}$ **27.** $x = 3$ **29.** $y = -\frac{5}{4}x - \frac{15}{2}$

31. $y = -3$ **33.** $y = -2x + 3$ **35.** $x = -5$ **37.** $y = x + 2$ **39.** $y = -2x - 3$

41. $y = \frac{2}{3}x + \frac{5}{3}$ **43.** $y = \frac{1}{3}x + \frac{10}{3}$ **45.** $y = \frac{3}{2}x - \frac{1}{2}$ **47.** $y = -\frac{3}{2}x + 3$ **49.** $y = -1$

51. $y = x - 1$ **53.** $y = -x + 1$ **55.** $y = \frac{2}{3}x + \frac{5}{3}$ **57.** $y = \frac{1}{2}x - 1$ **59.** $y = -4$

61. $y = \frac{3}{4}x$ **63.** $y = -\frac{4}{3}x + \frac{5}{3}$ **65.** $x = -2$ **67.** $y = x - 1$ **69.** $y = \frac{4}{3}x + \frac{7}{3}$

71. $y = -x + 3$ **73.**

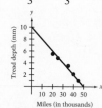

The slope represents the loss of tread depth per thousand miles.

75.

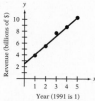

The slope represents the increase in sales per year.

77. $-\frac{3}{2}$ **79.** $y = -\frac{2}{3}x + \frac{5}{3}$

SECTION 4.5 *pages 241–242*

1. The lines are perpendicular.   **3.** No, the lines are not parallel.
5. No, the lines are not parallel.   **7.** The lines are perpendicular.   **9.** The lines are parallel.
11. The lines are perpendicular.   **13.** No, the lines are not parallel.

15. The lines are perpendicular.   **17.** The lines are parallel.   **19.** $y = -\frac{3}{4}x$

21. $y = \frac{2}{3}x - \frac{8}{3}$ **23.** $y = \frac{1}{3}x - \frac{1}{3}$ **25.** $y = -\frac{1}{4}x - 2$ **27.** $y = -\frac{5}{3}x - \frac{14}{3}$ **29.** $k = -3$

31. $k = -3$ **33.** Any equation of the form $y = 2x + b$, $b \neq -13$ or $y = -\frac{3}{2}x + c$, $c \neq 8$.

SECTION 4.6 *pages 245–246*

1. 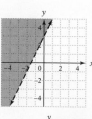   **3.** 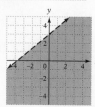   **5.** 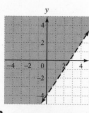   **7.**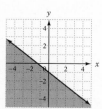

9. 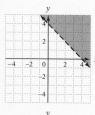   **11.**   **13.**   **15.**

17. 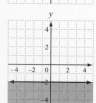   **19.** 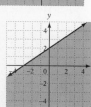   **21.** The inequality is $y \geq 2x + 2$.

CHAPTER REVIEW EXERCISES   *pages 249–250*

1. [4.1A]   **2.** 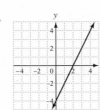 [4.1B]   **3.** $(-3, 0)$ [4.1B]   **4.** Yes [4.1C]

5. $f(-1) = -1$ [4.1D]   **6.** $(8, 0)$ and $(0, -12)$ [4.2B]   **7.** 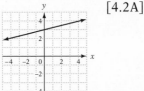 [4.2A]

8. 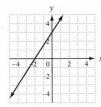 [4.2B]   **9.** [4.3B]   **10.** 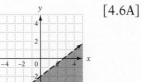 [4.6A]

11. $\dfrac{7}{11}$ [4.3A]   **12.** $y = -3x + 5$ [4.4A]   **13.** $y = \dfrac{2}{3}x + 6$ [4.4A]

14. $y = -\dfrac{2}{3}x + \dfrac{11}{3}$ [4.4B]   **15.** $y = -\dfrac{3}{2}x + \dfrac{7}{2}$ [4.5A]   **16.** $y = 2x + 1$ [4.5A]

17. 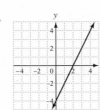   The cost of 100 min of access time during one month is \$30.   [4.2C]

18. $\{(55, 95), (57, 101), (53, 94), (57, 98), (60, 100), (61, 105), (58, 97), (54, 95)\}$ No, the relation is not a function.   [4.1C]

19. The slope is 1.5. The tuition is increasing \$1500 per year.   [4.3A]

20. 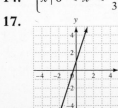   As the age of the machine increases by one year, the number of hours of down time per month increases by 10.4 h.   [4.4C]

CUMULATIVE REVIEW EXERCISES   *pages 251–252*

1. $-5, -3$ [1.1A]   **2.** 0.85 [1.2B]   **3.** $9\sqrt{5}$ [1.2F]   **4.** -12 [1.3A]   **5.** $-\dfrac{5}{8}$ [1.4A]

6. $-4d - 9$ [1.4B]   **7.** $-32z$ [1.4C]   **8.** $-13x + 7y$ [1.4D]   **9.** [1.5C]

10. $\dfrac{3}{2}$ [2.3A]   **11.** 1 [2.3C]   **12.** $\{x \mid x > -1\}$ [2.5A]   **13.** $\varnothing$ [2.5B]

14. $\left\{x \mid 0 < x < \dfrac{10}{3}\right\}$ [2.6B]   **15.** $f(2) = 6$ [4.1D]   **16.** The slope is 2.   [4.3A]

17. [4.2A]   **18.** [4.2B]   **19.** [4.3B]

20. 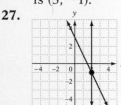   [4.6A]   **21.** $y = -\frac{1}{3}x - 2$ [4.4B]   **22.** $y = -\frac{3}{2}x + 7$ [4.5A]

23. The first plane is traveling at 400 mph. The second plane is traveling at 200 mph.   [2.4D]

24. 32 lb of \$8 coffee and 48 lb of \$3 coffee should be used.   [2.4A]

25. The slope is $-\frac{10,000}{3}$. The value of the house decreases by \$3333.33 each year.   [4.3A]

ANSWERS to Chapter 5 Selected Exercises

SECTION 5.1   *pages 257–260*

1. Yes, (2, 3) is a solution of the system of equations.

3. Yes, (1, −2) is a solution of the system of equations.

5. No, (4, 3) is not a solution of the system of equations.

7. No, (−1, 3) is not a solution of the system of equations.

9. No, (0, 0) is not a solution of the system of equations.

11. Yes, (2, −3) is a solution of the system of equations.

13. Yes, (5, 2) is a solution of the system of equations.

15. Yes, (−2, −3) is a solution of the system of equations.

17. No, (0, −3) is not a solution of the system of equations.

19.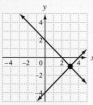
The solution is (3, −1).

21.
The solution is (2, 4).

23.
The solution is (4, 3).

25.
The solution is (4, −1).

27.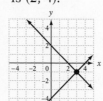
The solution is (2, −1).

29.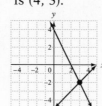
The solution is (3, −1).

31.
The solution is (3, −2).

33.
The lines are parallel and therefore do not intersect. The system of equations has no solution.

35.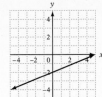

The equations are dependent. The solutions are the ordered pairs $\left(x, \frac{2}{5}x - 2\right)$.

37.

The solution is $(0, -3)$.

39.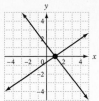

The solution is $(1, 0)$.

41. The solution is $(1.20, 1.40)$. **43.** The solution is $(0.64, -0.10)$.
45. For example: **a.** $2x + y = -1$ **b.** $3x + 2y = 7$ **c.** $x - 3y = 4$
$x + 2y = 7$ $\quad y = -\frac{3}{2}x + \frac{7}{3}$ $\quad y = \frac{1}{3}x - \frac{4}{3}$

SECTION 5.2 *pages 265–266*

1. The solution is $(2, 1)$. **3.** The solution is $(4, 1)$. **5.** The solution is $(-1, 1)$.
7. The system of equations is inconsistent and has no solution.
9. The system of equations is inconsistent and has no solution. **11.** The solution is $\left(-\frac{3}{4}, -\frac{3}{4}\right)$.
13. The solution is $(1, 1)$. **15.** The solution is $(2, 0)$. **17.** The solution is $(1, -2)$.
19. The solution is $(0, 0)$.
21. The system of equations is dependent. The solutions are the ordered pairs $(x, 2x - 2)$.
23. The solution is $(-4, -2)$. **25.** The solution is $(10, 31)$. **27.** The solution is $(3, -10)$.
29. The solution is $(-22, -5)$. **31.** $k = 2$ **33.** $k = 2$

SECTION 5.3 *pages 275–278*

1. The solution is $(6, 1)$. **3.** The solution is $(1, 1)$. **5.** The solution is $(2, 1)$.
7. The solution is $(-2, 1)$.
9. The equations are dependent. The solutions are the ordered pairs $(x, 3x - 4)$.
11. The solution is $\left(-\frac{1}{2}, 2\right)$. **13.** The system is inconsistent and therefore has no solution.
15. The solution is $(-1, -2)$. **17.** The solution is $(-5, 4)$. **19.** The solution is $(2, 5)$.
21. The solution is $\left(\frac{1}{2}, \frac{3}{4}\right)$. **23.** The solution is $(0, 0)$. **25.** The solution is $(-1, 3)$.
27. The solution is $\left(\frac{2}{3}, -\frac{2}{3}\right)$. **29.** The solution is $(-2, 3)$. **31.** The solution is $(2, -1)$.
33. The solution is $(10, -5)$. **35.** The solution is $\left(-\frac{1}{2}, \frac{2}{3}\right)$. **37.** The solution is $\left(\frac{5}{3}, \frac{1}{3}\right)$.
39. The system is inconsistent and therefore has no solution. **41.** The solution is $(1, -1)$.
43. The solution is $(2, 1, 3)$. **45.** The solution is $(1, -1, 2)$. **47.** The solution is $(1, 2, 4)$.
49. The solution is $(2, -1, -2)$. **51.** The solution is $(-2, -1, 3)$.
53. The system is inconsistent and therefore has no solution. **55.** The solution is $(1, 4, 1)$.
57. The solution is $(1, 3, 2)$. **59.** The solution is $(1, -1, 3)$. **61.** The solution is $(0, 2, 0)$.
63. The solution is $(1, 5, 2)$. **65.** The solution is $(-2, 1, 1)$. **67.** $A = 3, B = -3$
69. **a.** 14 **b.** $\frac{2}{3}$ **c.** $\frac{1}{2}$ **71.** $A = 1$

SECTION 5.4 *pages 285–286*

1. The value of the determinant is 11. **3.** The value of the determinant is 18.
5. The value of the determinant is 0. **7.** The value of the determinant is 15.
9. The value of the determinant is -30. **11.** The value of the determinant is 0.

13. The solution is $(3, -4)$.   **15.** The solution is $(4, -1)$.   **17.** The solution is $\left(\frac{11}{14}, \frac{17}{21}\right)$.

19. The solution is $\left(\frac{1}{2}, 1\right)$.   **21.** The system of equations is not independent.

23. The solution is $(-1, 0)$.   **25.** The solution is $(1, -1, 2)$.   **27.** The solution is $(2, -2, 3)$.

29. The system of equations is not independent.   **31.** The solution is $\left(\frac{68}{25}, \frac{56}{25}, -\frac{8}{25}\right)$.

33. The solution is $\left(\frac{2}{3}, -1, \frac{1}{2}\right)$.   **35.** 0   **37. a.** 0   **b.** 0

SECTION 5.5   *pages 291–294*

1. The rate of the boat is 15 mph. The rate of the current is 3 mph.
3. The rate of the plane is 502.5 mph. The rate of the wind is 47.5 mph.
5. The rate of the team is 8 km/h. The rate of the current is 2 km/h.
7. The rate of the plane is 180 mph. The rate of the wind is 20 mph.
9. The rate of the plane is 110 mph. The rate of the wind is 10 mph.
11. The rate of the motorboat is 13 mph. The rate of the current is 3 mph.
13. The rate of the cabin cruiser is 15 mph. The rate of the current is 3 mph.
15. The cost of the cinnamon tea is $2.50/lb. The cost of the spice tea is $3/lb.
17. The cost of the wool carpet is $26/yd.
19. The company plans to manufacture 25 ten-speed bicycles.
21. There are 480 g of the first alloy and 40 g of the second alloy.
23. The measure of the smaller angle is 9°. The measure of the larger angle is 81°.
25. There were 155 adult tickets and 90 student tickets sold.
27. You have $1500 in the 6% account and $3500 in the 8% account.
29. The original number is 84.   **31.** There are 30 nickels in the bank.

SECTION 5.6   *pages 297–298*

1. 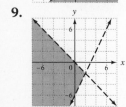   **3.** 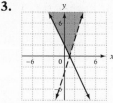   **5.** 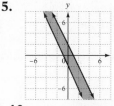   **7.**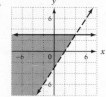

9. 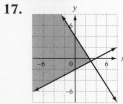   **11.** 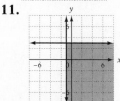   **13.** 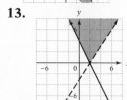   **15.**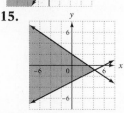

17. 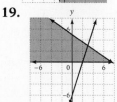   **19.** 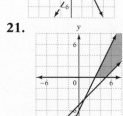   **21.**   **23.**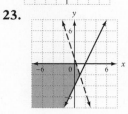

CHAPTER REVIEW EXERCISES   *pages 301–302*

1. The solution is $(3, 4)$.   [5.1A]

2. 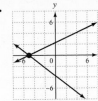 The solution is $(-5, 0)$. [5.1A]   **3.** 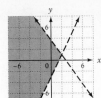 [5.6A]

4. 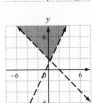 [5.6A]   **5.** The solution is $\left(\dfrac{3}{4}, \dfrac{7}{8}\right)$. [5.2A]

6. The solution is $(-3, -4)$. [5.2A]   **7.** The solution is $(2, -1)$. [5.2A]
8. The solution is $(-2, 1)$. [5.3A]   **9.** The system is inconsistent and therefore has no solution. [5.3A]   **10.** The solution is $(1, 1)$. [5.3A]
11. The system is inconsistent and therefore has no solution. [5.3B]
12. The solution is $(2, -1, -2)$. [5.3B]   **13.** The value of the determinant is 10. [5.4A]
14. The value of the determinant is -32. [5.4A]   **15.** The solution is $\left(-\dfrac{1}{3}, -\dfrac{10}{3}\right)$. [5.4B]
16. The solution is $\left(\dfrac{59}{19}, -\dfrac{62}{19}\right)$. [5.4B]   **17.** The solution is $\left(\dfrac{1}{5}, -\dfrac{6}{5}, \dfrac{3}{5}\right)$. [5.4B]
18. The solution is $(0, -2, 3)$. [5.4B]
19. The rate of the plane is 150 mph. The rate of the wind is 25 mph. [5.5A]
20. The cost of the cotton is \$4.50/yd. The cost of the wool is \$7/yd. [5.5B]

CUMULATIVE REVIEW EXERCISES   *pages 303–304*

1. $-6\sqrt{10}$ [1.2F]   **2.** 22 [2.3C]   **3.** $3x - 24$ [1.4D]   **4.** -4 [1.4A]
5. $\{x \mid x < 6\}$ [2.5B]   **6.** $\{x \mid -4 < x < 8\}$ [2.6B]   **7.** $\{x \mid x > 4 \text{ or } x < -1\}$ [2.6B]
8. $F(2) = 1$ [4.1D]   **9.** 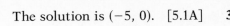 [1.5C]   **10.** $y = -\dfrac{2}{3}x + \dfrac{5}{3}$ [4.4A]

11. $y = 5x - 11$ [4.4B]   **12.** $y = -\dfrac{3}{2}x + \dfrac{1}{2}$ [4.5A]   **13.** 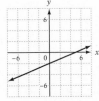 [4.3B]

14. 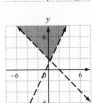 [4.6A]   **15.** $(2, 0)$ [5.1A]   **16.** $(-5, -11)$ [5.2A]   **17.** $(1, 0, -1)$ [5.3B]

18. 3 [5.4A]   **19.** $(2, -3)$ [5.4B]   **20.** 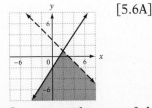 [5.6A]

21. 60 ml of pure water must be used. [2.4B]   **22.** The rate of the wind is 12.5 mph. [5.5A]
23. One pound of steak costs \$3. [5.5B]
24. The lower limit is 10,200 ohms. The upper limit is 13,800 ohms. [2.6C]
25. The slope is 40. The slope is the commission rate of the executive divided by 100. [4.3A]

ANSWERS to Chapter 6 Selected Exercises

SECTION 6.1   *pages 315–318*

1. a^4b^4 **3.** $-18x^3y^4$ **5.** x^8y^{16} **7.** $81x^8y^{12}$ **9.** $729a^{10}b^6$ **11.** x^5y^{11} **13.** $729x^6$
15. $a^{18}b^{18}$ **17.** $4096x^{12}y^{12}$ **19.** $64a^{24}b^{18}$ **21.** x^{2n+1} **23.** y^{6n-2} **25.** a^{2n^2-6n} **27.** x^{15n+10}
29. $-6x^5y^5z^4$ **31.** $-12a^2b^9c^2$ **33.** $-6x^4y^4z^5$ **35.** $-432a^7b^{11}$ **37.** $54a^{13}b^{17}$ **39.** 243

41. y^3 **43.** $\dfrac{a^3b^2}{4}$ **45.** $\dfrac{x}{y^4}$ **47.** $\dfrac{1}{2}$ **49.** $-\dfrac{1}{9}$ **51.** $\dfrac{1}{y^8}$ **53.** $\dfrac{1}{x^6y^{10}}$ **55.** x^5y^5

57. $\dfrac{a^8}{b^9}$ **59.** $\dfrac{1}{2187a}$ **61.** $\dfrac{y^6}{x^3}$ **63.** $\dfrac{y^4}{x^3}$ **65.** $-\dfrac{1}{2x^2}$ **67.** $-\dfrac{1}{243a^5b^{10}}$ **69.** $\dfrac{16x^8}{81y^4z^{12}}$

71. $\dfrac{3a^4}{4b^3}$ **73.** $-\dfrac{9a}{8b^6}$ **75.** $\dfrac{16x^2}{y^6}$ **77.** $\dfrac{1}{b^{4n}}$ **79.** $-\dfrac{1}{y^{6n}}$ **81.** y^{n-2} **83.** $\dfrac{1}{y^{2n}}$ **85.** $\dfrac{x^{n-5}}{y^6}$

87. $\dfrac{8b^{15}}{3a^{18}}$ **89.** $\dfrac{8}{5}$ **91.** 4.67×10^{-6} **93.** 1.7×10^{-10} **95.** 2×10^{11} **97.** 0.000000123

99. 8,200,000,000,000,000   **101.** 0.039   **103.** 150,000   **105.** 20,800,000
107. 0.000000015   **109.** 0.0000000000178   **111.** 140,000,000   **113.** 11,000,000
115. 0.000008   **117.** Light travels 2.592×10^{10} km in one day.
119. The computer can perform 4×10^9 operations in one hour.
121. The centrifuge makes one revolution in $6.\overline{6} \times 10^{-7}$ s.
123. The sun is 3.38983×10^5 times heavier.
125. The weight of one orchid seed is approximately 3.225806×10^{-8} oz.
127. Alpha Centauri is approximately 2.522249×10^{13} mi from the earth.

SECTION 6.2   *pages 325–326*

1. $P(3) = 13$ **3.** $R(2) = 10$ **5.** $f(-1) = -11$
7. a. The leading coefficient is -1.   **b.** The constant term is 8.   **c.** The degree is 2.
9. The expression is not a polynomial function.   **11.** The expression is not a polynomial function.
13. a. The leading coefficient is 3.   **b.** The constant term is π.   **c.** The degree is 5.
15. a. The leading coefficient is -5.   **b.** The constant term is 2.   **c.** The degree is 3.
17. a. The leading coefficient is 14.   **b.** The constant term is 14.   **c.** The degree is 0.
19.   **21.**   **23.**   **25.** $6x^2 - 6x + 5$

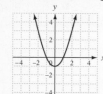
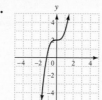

27. $-x^2 + 1$ **29.** $5y^2 - 15y + 2$ **31.** $7a^2 - a + 2$ **33.** $P(x) + R(x) = 3x^2 - 3xy - 2y^2$
35. $P(x) - R(x) = 8x^2 - 2xy + 5y^2$ **37.** $S(x) = 3x^4 - 8x^2 + 2x; \;\; S(2) = 20$
39. a. $k = 8$ **b.** $k = -4$

SECTION 6.3   *pages 331–334*

1. $x^2 - 2x$ **3.** $-x^2 - 7x$ **5.** $3a^3 - 6a^2$ **7.** $-5x^4 + 5x^3$ **9.** $-3x^5 + 7x^3$
11. $12x^3 - 6x^2$ **13.** $6x^2 - 12x$ **15.** $3x^2 + 4x$ **17.** $-x^3y + xy^3$ **19.** $2x^4 - 3x^2 + 2x$
21. $2a^3 + 3a^2 + 2a$ **23.** $3x^6 - 3x^4 - 2x^2$ **25.** $-6y^4 - 12y^3 + 14y^2$ **27.** $-2a^3 - 6a^2 + 8a$
29. $6y^4 - 3y^3 + 6y^2$ **31.** $x^3y - 3x^2y^2 + xy^3$ **33.** $x^3 + 4x^2 + 5x + 2$
35. $a^3 - 6a^2 + 13a - 12$ **37.** $-2b^3 + 7b^2 + 19b - 20$ **39.** $-6x^3 + 31x^2 - 41x + 10$
41. $x^3 - 3x^2 + 5x - 15$ **43.** $x^4 - 4x^3 - 3x^2 + 14x - 8$ **45.** $15y^3 - 16y^2 - 70y + 16$
47. $5a^4 - 20a^3 - 5a^2 + 22a - 8$ **49.** $y^4 + 4y^3 + y^2 - 5y + 2$ **51.** $x^2 + 4x + 3$
53. $a^2 + a - 12$ **55.** $y^2 - 5y - 24$ **57.** $y^2 - 10y + 21$ **59.** $2x^2 + 15x + 7$
61. $3x^2 + 11x - 4$ **63.** $4x^2 - 31x + 21$ **65.** $3y^2 - 2y - 16$ **67.** $9x^2 + 54x + 77$
69. $21a^2 - 83a + 80$ **71.** $6a^2 - 25ab + 14b^2$ **73.** $2a^2 - 11ab - 63b^2$
75. $100a^2 - 100ab + 21b^2$ **77.** $15x^2 + 56xy + 48y^2$ **79.** $14x^2 - 97xy - 60y^2$
81. $56x^2 - 61xy + 15y^2$ **83.** $y^2 - 25$ **85.** $4x^2 - 9$ **87.** $x^2 + 2x + 1$ **89.** $9a^2 - 30a + 25$

91. $9x^2 - 49$ **93.** $4a^2 + 4ab + b^2$ **95.** $x^2 - 4xy + 4y^2$ **97.** $16 - 9y^2$
99. $25x^2 + 20xy + 4y^2$ **101.** The area is $(10x^2 - 35x)$ ft^2.
103. The area is $(4x^2 + 4x + 1)$ km^2.   **105.** The area is $(4x^2 + 10x)$ m^2.
107. The area is $(18x^2 + 12x + 2)$ in^2.   **109.** The total area is $(90x + 2025)$ ft^2.
111. $4ab$ **113.** $a^3 + 9a^2 + 27a + 27$ **115.** **a.** $7x^2 - 11x - 8$ **b.** $x^3 - 7x^2 - 7$

SECTION 6.4   *pages 341–344*

1. $x + 8$ **3.** $x^2 + \dfrac{2}{x - 3}$ **5.** $3x + 5 + \dfrac{3}{2x + 1}$ **7.** $5x + 7 + \dfrac{2}{2x - 1}$

9. $4x^2 + 6x + 9 + \dfrac{18}{2x - 3}$ **11.** $3x^2 + 1 + \dfrac{1}{2x^2 - 5}$ **13.** $x^2 - 3x - 10$

15. $x^2 - 2x + 1 - \dfrac{1}{x - 3}$ **17.** $2x^3 - 3x^2 + x - 4$ **19.** $2x + \dfrac{x + 2}{x^2 + 2x - 1}$

21. $x^2 + 4x + 6 + \dfrac{10x + 8}{x^2 - 2x - 1}$ **23.** $x^2 - 2x + 4 - \dfrac{6x + 7}{x^2 + 2x + 3}$ **25.** $2x - 8$ **27.** $3x - 8$

29. $3x + 3 - \dfrac{1}{x - 1}$ **31.** $2x^2 - 3x + 9$ **33.** $4x^2 + 8x + 15 + \dfrac{12}{x - 2}$ **35.** $2x^2 - 3x + 7 - \dfrac{8}{x + 4}$

37. $3x^3 + 2x^2 + 12x + 19 + \dfrac{33}{x - 2}$ **39.** $3x^3 - x + 4 - \dfrac{2}{x + 1}$ **41.** $2x^3 + 6x^2 + 17x + 51 + \dfrac{155}{x - 3}$

43. $P(3) = 8$ **45.** $R(4) = 43$ **47.** $P(-2) = -39$ **49.** $Z(-3) = -60$ **51.** $Q(2) = 31$
53. $F(-3) = 178$ **55.** $P(5) = 122$ **57.** $R(-3) = 302$ **59.** $Q(2) = 0$ **61.** $R(-2) = -65$
63. **a.** $a^2 - ab + b^2$ **b.** $x^4 - x^3y + x^2y^2 - xy^3 + y^4$ **c.** $x^5 - x^4y + x^3y^2 - x^2y^3 + xy^4 - y^5$
65. **a.** $\dfrac{1}{2}x^2 + \dfrac{3}{4}x - 1$ **b.** $\dfrac{2}{3}x^2 + \dfrac{7}{2}x + \dfrac{10}{3} + \dfrac{64}{6x - 12}$ **c.** $2x^2 - x + 1 + \dfrac{3x - 11}{x^2 - x + 1}$

d. $x^2 + 6x + 17 + \dfrac{51x + 56}{x^2 - 2x - 3}$

CHAPTER REVIEW EXERCISES   *pages 347–348*

1. $-6a^3b^6$ [6.1A]   **2.** $9x^4y^6$ [6.1A]   **3.** $\dfrac{c^{10}}{2b^{17}}$ [6.1B]   **4.** 0.00254 [6.1C]   **5.** -7 [6.2A]

6. $4x^2 - 8xy + 5y^2$ [6.2B]   **7.** $\dfrac{18x^5y^4}{z}$ [6.1B]   **8.** $-54a^{13}b^5c^7$ [6.1A]

9.   [6.2A]   **10.** **a.** 3   **b.** 8   **c.** 5 [6.2A]   **11.** 1 [6.4C]

12. $-8x^3 - 14x^2 + 18x$ [6.3A]   **13.** $8a^3b^3 - 4a^2b^4 + 6ab^5$ [6.3A]
14. $6y^3 + 17y^2 - 2y - 21$ [6.3B]   **15.** $10a^2 + 31a - 63$ [6.3C]   **16.** 1.27×10^{-7} [6.1C]
17. $25y^2 - 70y + 49$ [6.3D]   **18.** $5x + 4 + \dfrac{6}{3x - 2}$ [6.4A]   **19.** $4x^2 + 3x - 8 + \dfrac{50}{x + 6}$ [6.4B]
20. $x^3 + 4x^2 + 16x + 64 + \dfrac{252}{x - 4}$ [6.4B]   **21.** $8{,}100{,}000{,}000$ [6.1C]   **22.** $-\dfrac{2a^3}{3b^3}$ [6.1B]
23. $25a^2 - 4b^2$ [6.3D]   **24.** 68 [6.4C]   **25.** $21y^2 + 4y - 1$ [6.2B]
26. $12b^5 - 4b^4 - 6b^3 - 8b^2 + 5$ [6.3B]   **27.** $a^2 - 49$ [6.3D]   **28.** 7.65×10^{11} [6.1C]
29. The mass of the moon is 8.103×10^{19} tons. [6.1D]
30. The Great Galaxy of Andromeda is 1.291224×10^{19} mi from the earth. [6.1D]
31. There are 6.048×10^5 s in one week. [6.1D]   **32.** The area is $(9x^2 - 12x + 4)$ in^2. [6.3E]
33. The area is $(10x^2 - 29x - 21)$ cm^2. [6.3E]

CUMULATIVE REVIEW EXERCISES   *pages 349–350*

1. -3 and 3 [1.1A]   **2.** -83 [1.1B]   **3.** 6 [1.3A]   **4.** $-\dfrac{5}{4}$ [1.4A]   **5.** $-50\sqrt{3}$ [1.2F]

6. The Inverse Property of Addition   [1.4B]   **7.** $-186x + 8$ [1.4D]   **8.** $-\frac{1}{6}$ [2.3A]

9. $-\frac{11}{4}$ [2.3B]   **10.** -1 and $\frac{7}{3}$ [2.6A]   **11.** 18 [4.1D]   **12.** Yes [4.1C]   **13.** $-\frac{1}{6}$ [4.3A]

14. $y = -\frac{3}{2}x + \frac{1}{2}$ [4.4A]   **15.** $y = \frac{2}{3}x + \frac{16}{3}$ [4.5A]   **16.** $\left(-\frac{7}{5}, -\frac{8}{5}\right)$ [5.4B]

17. $\left(-\frac{9}{7}, \frac{2}{7}, \frac{11}{7}\right)$ [5.3B]   **18.** $5x - 3xy$ [1.4B]   **19.** $4x^3 - 7x + 3$ [6.3B]

20. 5.01×10^{-6} [6.1C]   **21.** 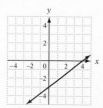 [4.2B]   **22.** 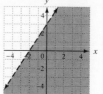 [4.6A]

23. 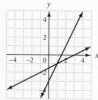   **24.** 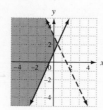 [5.6A]   **25.** $\frac{b^5}{a^8}$ [6.1B]

The solution is $(1, -1)$. [5.1A]

26. $\frac{y^2}{25x^6}$ [6.1B]   **27.** The two integers are 9 and 15. [2.3D]

28. 40 oz of pure gold must be used. [2.4A]

29. The cyclists are traveling at 5 mph and 7.5 mph. [2.4D]

30. \$4500 must be invested at an annual simple interest rate of 10%. [2.4C]

31. The slope is 50. The slope represents the average speed in miles per hour. [4.3B]

32. The length is 15 m. The width is 6 m. [3.2A]   **33.** The area is $(4x^2 + 12x + 9)$ m^2. [6.3E]

ANSWERS to Chapter 7 Selected Exercises

SECTION 7.1   *pages 357–358*

1. $5(a + 1)$ **3.** $8(2 - a^2)$ **5.** $4(2x + 3)$ **7.** $6(5a - 1)$ **9.** $x(7x - 3)$ **11.** $a^2(3 + 5a^3)$
13. $y(14y + 11)$ **15.** $2x(x^3 - 2)$ **17.** $2x^2(5x^2 - 6)$ **19.** $4a^5(2a^3 - 1)$ **21.** $xy(xy - 1)$
23. $3xy(xy^3 - 2)$ **25.** $xy(x - y^2)$ **27.** $5y(y^2 - 4y + 2)$ **29.** $3y^2(y^2 - 3y - 2)$
31. $3y(y^2 - 3y + 8)$ **33.** $a^2(6a^3 - 3a - 2)$ **35.** $ab(2a - 5ab + 7b)$ **37.** $2b(2b^4 + 3b^2 - 6)$
39. $x^2(8y^2 - 4y + 1)$ **41.** $a^{2n}(2a^{3n} + 1)$ **43.** $y^{2n}(y^{2n} + 1)$ **45.** $b^5(b^n - 1)$
47. $(y + 7)(a + z)$ **49.** $(3r + s)(a - b)$ **51.** $(t - 7)(m - 7)$ **53.** $(2y + 1)(4a - b)$
55. $(x + 2)(x + 2y)$ **57.** $(p - 2)(p - 3r)$ **59.** $(b - 4)(a + 6)$ **61.** $(z + y)(2z - 1)$
63. $(4v + 7)(2v - 3y)$ **65.** $(2x - 5)(x - 3y)$ **67.** $(3y - a)(y - 2)$ **69.** $(y + 1)(3x - y)$
71. $(t - 2)(t + 3s)$ **73.** 28   **75.** P doubles when $L + W$ doubles.

SECTION 7.2   *pages 363–366*

1. $(x + 1)(x + 2)$ **3.** $(x - 2)(x + 1)$ **5.** $(a + 4)(a - 3)$ **7.** $(a - 1)(a - 2)$
9. $(a + 2)(a - 1)$ **11.** $(b - 3)^2$ **13.** $(b + 8)(b - 1)$ **15.** $(y + 11)(y - 5)$
17. $(y - 3)(y - 2)$ **19.** $(z - 5)(z - 9)$ **21.** $(z - 20)(z + 8)$ **23.** $(p + 3)(p + 9)$
25. $(x + 10)^2$ **27.** $(b + 4)(b + 5)$ **29.** $(x + 3)(x - 14)$ **31.** $(b - 5)(b + 4)$
33. $(y - 17)(y + 3)$ **35.** $(p - 7)(p + 3)$ **37.** Nonfactorable over the integers
39. $(x - 5)(x - 15)$ **41.** $(x - 7)(x - 8)$ **43.** $(x + 8)(x - 7)$ **45.** $(a - 24)(a + 3)$
47. $(a - 3)(a - 12)$ **49.** $(z - 17)(z + 8)$ **51.** $(c - 10)(c + 9)$ **53.** $(z + 11)(z + 4)$
55. $(c + 2)(c + 17)$ **57.** $(x - 12)(x + 8)$ **59.** $(x - 8)(x - 14)$ **61.** $(b + 15)(b - 7)$
63. $(a - 12)(a + 3)$ **65.** $(b - 6)(b - 17)$ **67.** $(a + 3)(a + 24)$ **69.** $(x + 12)(x + 13)$
71. $(x - 16)(x + 6)$ **73.** $2(x + 1)(x + 2)$ **75.** $-(x + 2)(x - 9)$ **77.** $a(b + 5)(b - 3)$
79. $x(y - 2)(y - 3)$ **81.** $z(z - 4)(z - 3)$ **83.** $-3y(y - 2)(y - 3)$ **85.** $3(x + 4)(x - 3)$

87. $5(z - 7)(z + 4)$ **89.** $2a(a + 8)(a - 4)$ **91.** $(x - 3y)(x - 2y)$ **93.** $(a - 4b)(a - 5b)$
95. $(x - 7y)(x + 4y)$ **97.** Nonfactorable over the integers **99.** $z^2(z - 5)(z - 7)$
101. $b^2(b - 10)(b - 12)$ **103.** $2y^2(y - 16)(y + 3)$ **105.** $-x^2(x + 8)(x - 1)$
107. $4y(x + 7)(x - 2)$ **109.** $c(c + 20)(c - 2)$ **111.** $-4x(x + 3)(x - 2)$
113. $(y - 8x)(y + x)$ **115.** $(y + 7z)(y - 3z)$ **117.** $(y - z)(y - 15z)$ **119.** $4y(x - 18)(x + 1)$
121. $4x(x + 8)(x - 5)$ **123.** $5z(z - 12)(z + 2)$ **125.** $5x(x + 2)(x + 4)$
127. $4(p - 15)(p + 8)$ **129.** $p^2(p + 8)(p - 7)$ **131.** $(a - 5b)^2$ **133.** $(x + 10y)(x - 6y)$
135. $6x(x - 5)(x + 4)$ **137.** $y(x - 9)(x + 6)$ **139.** $k = -36, -12, 12, 36$
141. $-22, -10, 10, 22$ **143.** $k = 6, 10, 12$ **145.** $k = 6, 10, 12$ **147.** $k = 4, 6$

SECTION 7.3 *pages 371–374*

1. $(2x + 1)(x + 1)$ **3.** $(2y + 1)(y + 3)$ **5.** $(2a - 1)(a - 1)$ **7.** $(2b - 1)(b - 5)$
9. $(2x - 1)(x + 1)$ **11.** $(2x + 1)(x - 3)$ **13.** $(2t - 5)(t + 2)$ **15.** $(3p - 1)(p - 5)$
17. $(4y - 1)(3y - 1)$ **19.** Nonfactorable over the integers **21.** $(2t - 1)(3t - 4)$
23. $(8x + 1)(x + 4)$ **25.** Nonfactorable over the integers **27.** $(4y + 5)(3y + 1)$
29. $(7a - 2)(a + 7)$ **31.** $(3b - 4)(b - 4)$ **33.** $(2z + 1)(z - 14)$ **35.** $(3p - 2)(p + 8)$
37. $2(2x + 1)(x + 1)$ **39.** $5(3y - 7)(y - 1)$ **41.** $x(2x - 1)(x - 5)$ **43.** $b(3a - 4)(a - 4)$
45. Nonfactorable over the integers **47.** $-3x(x + 4)(x - 3)$ **49.** $4(4y - 1)(5y - 1)$
51. $z(4z + 1)(2z + 3)$ **53.** $y(3x + 2)(2x - 5)$ **55.** $5(2t - 5)(t + 2)$ **57.** $p(3p - 1)(p - 5)$
59. $2(13z - 3)(z + 4)$ **61.** $2y(5y - 2)(y - 4)$ **63.** $yz(4z - 3)(z + 2)$
65. $3a(2a + 3)(7a - 3)$ **67.** $y(3x - 5y)^2$ **69.** $xy(3x - 4y)^2$ **71.** $(3x - 4)(2x - 3)$
73. $(5b - 2)(b + 7)$ **75.** $(3a + 8)(2a - 3)$ **77.** $(4z + 3)(z + 2)$ **79.** $(2p + 5)(11p - 2)$
81. $(8y + 9)(y + 1)$ **83.** $(3t + 1)(6t - 5)$ **85.** $(6b - 1)(b + 12)$ **87.** $(3x + 2)^2$
89. $(3b - 2)(2b - 3)$ **91.** $(11b - 7)(3b + 5)$ **93.** $(3y - 4)(6y - 5)$ **95.** $(3a + 7)(5a - 3)$
97. $(2y - 5)(4y - 3)$ **99.** $(4z - 5)(2z + 3)$ **101.** Nonfactorable over the integers
103. $(2z - 5)(5z - 2)$ **105.** $(6z + 7)(6z + 5)$ **107.** $(3x - 2y)(x + y)$ **109.** $(3a - b)(a + 2b)$
111. $(4y - 3z)(y - 2z)$ **113.** $-(z - 7)(z + 4)$ **115.** $-(x - 1)(x + 8)$ **117.** $3(3x - 4)(x + 5)$
119. $4(2x - 3)(3x - 2)$ **121.** $a^2(7a - 1)(5a + 2)$ **123.** $5(3b - 2)(b - 7)$
125. $(3x - 5y)(x - 7y)$ **127.** $3(8y - 1)(9y + 1)$ **129.** $-(x - 1)(x + 21)$
131. $(3a - 2b)(5a + 7b)$ **133.** $-z(z + 11)(z - 3)$ **135.** $(3a + 2)(a + 3)$
137. $k = -7, -5, 5, 7$ **139.** $k = -7, -5, 5, 7$

SECTION 7.4 *pages 381–384*

1. $(x + 4)(x - 4)$ **3.** $(2x + 1)(2x - 1)$ **5.** $(4x + 11)(4x - 11)$ **7.** $(1 + 3a)(1 - 3a)$
9. $(xy + 10)(xy - 10)$ **11.** Nonfactorable over the integers **13.** $(5 + ab)(5 - ab)$
15. $(a^n + 1)(a^n - 1)$ **17.** $(x - 6)^2$ **19.** $(b - 1)^2$ **21.** $(4x - 5)^2$
23. Nonfactorable over the integers **25.** Nonfactorable over the integers **27.** $(x + 3y)^2$
29. $(5a - 4b)^2$ **31.** $(x^n + 3)^2$ **33.** $(x - 7)(x - 1)$ **35.** $(x - y + a + b)(x - y - a - b)$
37. $(x - 3)(x^2 + 3x + 9)$ **39.** $(2x - 1)(4x^2 + 2x + 1)$ **41.** $(x - y)(x^2 + xy + y^2)$
43. $(m + n)(m^2 - mn + n^2)$ **45.** $(4x + 1)(16x^2 - 4x + 1)$ **47.** $(3x - 2y)(9x^2 + 6xy + 4y^2)$
49. $(xy + 4)(x^2y^2 - 4xy + 16)$ **51.** Nonfactorable over the integers
53. Nonfactorable over the integers **55.** $(a - 2b)(a^2 - ab + b^2)$
57. $(x^{2n} + y^n)(x^{4n} - x^{2n}y^n + y^{2n})$ **59.** $(x^n + 2)(x^{2n} - 2x^n + 4)$ **61.** $(xy - 3)(xy - 5)$
63. $(xy - 5)(xy - 12)$ **65.** $(x^2 - 3)(x^2 - 6)$ **67.** $(b^2 + 5)(b^2 - 18)$ **69.** $(x^2y^2 - 2)(x^2y^2 - 6)$
71. $(x^n + 1)(x^n + 2)$ **73.** $(3xy - 5)(xy - 3)$ **75.** $(2ab - 3)(3ab - 7)$ **77.** $(2x^2 - 15)(x^2 + 1)$
79. $(2x^n - 1)(x^n - 3)$ **81.** $(2a^n + 5)(3a^n + 2)$ **83.** $3(2x - 3)^2$ **85.** $a(3a - 1)(9a^2 + 3a + 1)$
87. $5(2x + 1)(2x - 1)$ **89.** $y^3(y + 11)(y - 5)$ **91.** $(4x^2 + 9)(2x + 3)(2x - 3)$
93. $2a(2 - a)(4 + 2a + a^2)$ **95.** $b^3(ab - 1)(a^2b^2 + ab + 1)$ **97.** $2x^2(2x - 5)^2$
99. $(x^2 + y^2)(x + y)(x - y)$ **101.** $(x^2 + y^2)(x^4 - x^2y^2 + y^4)$
103. Nonfactorable over the integers **105.** $2a(2a - 1)(4a^2 + 2a + 1)$
107. $a^2b^2(a + 4b)(a - 12b)$ **109.** $2b^2(3a + 5b)(4a - 9b)$ **111.** $(x - 2)^2(x + 2)$
113. $(x + y)(x - y)(2x + 1)(2x - 1)$ **115.** $(x - 1)(x^2 + x + 1)(xy + 1)(x^2y^2 - xy + 1)$
117. $x(x^n + 1)^2$ **119.** $b^n(3b - 2)(b + 2)$ **121.** $(x - 3)(x - 2)(x - 1)$
123. The dimensions are $(4x + 3)$ m by $(4x + 3)$ m. Yes, x can equal 0. The possible values of x are
$x > -\dfrac{3}{4}$.

SECTION 7.5 *pages 389–392*

1. −3 and −2 **3.** 7 and 3 **5.** 0 and 5 **7.** 0 and 9 **9.** 0 and $-\frac{3}{2}$ **11.** 0 and $\frac{2}{3}$

13. −2 and 5 **15.** 9 and −9 **17.** $\frac{7}{2}$ and $-\frac{7}{2}$ **19.** $\frac{1}{3}$ and $-\frac{1}{3}$ **21.** −2 and −4

23. −7 and 2 **25.** 2 and 3 **27.** −7 and 3 **29.** $-\frac{1}{2}$ and 5 **31.** $-\frac{1}{3}$ and $-\frac{1}{2}$

33. 0 and 3 **35.** 0 and 7 **37.** −1 and −4 **39.** 2 and 3 **41.** $\frac{1}{2}$ and −4 **43.** $\frac{1}{3}$ and 4

45. 3 and 9 **47.** 9 and −2 **49.** −1 and −2 **51.** −9 and 5 **53.** −7 and 4
55. −2 and −3 **57.** −8 and 9 **59.** 1 and 4 **61.** −5 and 2 **63.** 3 and 4
65. −4 and $-\frac{1}{3}$ **67.** The number is −2. **69.** There is no solution.
71. The length is 16 ft. The width is 11 ft.
73. The length of a side of the original square is 3 in. **75.** The width of the border is 2 ft.
77. The increase in area is 138.23 in^2. **79.** There will be 15 consecutive numbers.
81. There are 10 teams. **83.** The object will hit 4 s later. **85.** It will be 6 s later.
87. The length is 20 in. The width is 10 in. **89.** 1 and 18

CHAPTER REVIEW EXERCISES *pages 395–396*

1. $5x(x^2 + 2x + 7)$ [7.1A] **2.** $3ab(4a + b)$ [7.1A] **3.** $7y^3(2y^6 − 7y^3 + 1)$ [7.1A]
4. $(x − 3)(4x + 5)$ [7.1B] **5.** $(2x + 5)(5x + 2y)$ [7.1B] **6.** $(3a − 5b)(7x + 2y)$ [7.1B]
7. $(b − 3)(b − 10)$ [7.2A] **8.** $(c + 6)(c + 2)$ [7.2A] **9.** $(y − 4)(y + 9)$ [7.2A]
10. $3(a + 2)(a − 7)$ [7.2B] **11.** $4x(x − 6)(x + 1)$ [7.2B] **12.** $n^2(n + 1)(n − 3)$ [7.2B]
13. $(2x − 7)(3x − 4)$ [7.3A] **14.** $(6y − 1)(2y + 3)$ [7.3A]
15. Nonfactorable over the integers [7.3A] **16.** $(3x − 2)(x − 5)$ [7.3B]
17. $(2a + 5)(a − 12)$ [7.3B] **18.** $(6a − 5)(3a + 2)$ [7.3B] **19.** $(xy + 3)(xy − 3)$ [7.4A]
20. $(2x + 3y)^2$ [7.4A] **21.** $(x^n − 6)^2$ [7.4A] **22.** $(4a − 3b)(16a^2 + 12ab + 9b^2)$ [7.4B]
23. $(3x^2 + 2)(5x^2 − 3)$ [7.4C] **24.** $(7x^2y^2 + 3)(3x^2y^2 + 2)$ [7.4C]

25. $3a^2(a^2 + 1)(a^2 − 6)$ [7.4D] **26.** $\frac{1}{4}$ and −7 [7.5A] **27.** −3 and 7 [7.5A]

28. The length is 100 yd. The width is 60 yd. [7.5B]

29. The width of the frame is $\frac{3}{2}$ in. or 1.5 in. or $1\frac{1}{2}$ in. [7.5B]

30. The length of a side is 20 ft. [7.5B]

CUMULATIVE REVIEW EXERCISES *pages 397–398*

1. 7 [1.1C] **2.** 4 [1.3A] **3.** −7 [1.4A] **4.** $15x^2$ [1.4C] **5.** 12 [1.4D] **6.** $\frac{2}{3}$ [2.1C]

7. $\frac{7}{4}$ [2.3B] **8.** 3 [2.3C] **9.** 45 [2.2A] **10.** 1 [4.1D] **11.** 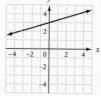 [4.2A]

12. 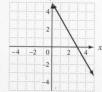 [4.2B] **13.** $y = \frac{2}{3}x + 6$ [4.4A] **14.** (1, 6) [5.2A]

15. (−1, −2) [5.3A] **16.** $9a^6b^4$ [6.1A] **17.** $x^3 − 3x^2 − 6x + 8$ [6.3B]

18. $4x + 8 + \dfrac{21}{2x - 3}$ [6.4A]   **19.** $\dfrac{y^6}{x^8}$ [6.1B]   **20.** $(a - b)(3 - x)$ [7.1B]

21. $5xy^2(3 - 4y^2)$ [7.1A]   **22.** $(x - 7y)(x + 2y)$ [7.2A]   **23.** $\dfrac{2}{3}$ and -7 [7.5A]

24. $\dfrac{5}{2}$ and 4 [7.5A]   **25.** The third angle measures $59°$. [3.1C]   **26.** The width is 15 ft. [3.2A]

27. The pieces are 4 ft long and 6 ft long. [2.3D]   **28.** $6500 must be invested at 11%. [2.4C]

29. The distance to the resort is 168 mi. [2.4D]

30. The length of the base of the triangle is 12 in. [7.5B]

ANSWERS to Chapter 8 Selected Exercises

SECTION 8.1   *pages 405–408*

1. $\dfrac{3}{4x}$ **3.** $\dfrac{1}{x + 3}$ **5.** -1 **7.** $\dfrac{2}{3y}$ **9.** $-\dfrac{3}{4x}$ **11.** $\dfrac{a}{b}$ **13.** $-\dfrac{2}{x}$ **15.** $\dfrac{y - 2}{y - 3}$ **17.** $\dfrac{x + 5}{x + 4}$

19. $\dfrac{x + 4}{x - 3}$ **21.** $-\dfrac{x + 2}{x + 5}$ **23.** $\dfrac{2(x + 2)}{x + 3}$ **25.** $\dfrac{2x - 1}{2x + 3}$ **27.** $-\dfrac{x + 7}{x + 6}$ **29.** $\dfrac{35ab^2}{24x^2y}$ **31.** $\dfrac{4x^3y^3}{3a^2}$

33. $\dfrac{3}{4}$ **35.** ab^2 **37.** $\dfrac{x^2(x - 1)}{y(x + 3)}$ **39.** $\dfrac{y(x - 1)}{x^2(x + 10)}$ **41.** $-ab^2$ **43.** $\dfrac{x + 5}{x + 4}$ **45.** 1

47. $-\dfrac{n - 10}{n - 7}$ **49.** $\dfrac{x(x + 2)}{2(x - 1)}$ **51.** $-\dfrac{x + 2}{x - 6}$ **53.** $\dfrac{x + 5}{x - 12}$ **55.** $\dfrac{3y + 2}{3y + 1}$ **57.** $\dfrac{7a^3y^2}{40bx}$ **59.** $\dfrac{4}{3}$

61. $\dfrac{3a}{2}$ **63.** $\dfrac{x^2(x + 4)}{y^2(x + 2)}$ **65.** $\dfrac{x(x - 2)}{y(x - 6)}$ **67.** $-\dfrac{3by}{ax}$ **69.** $\dfrac{(x + 6)(x - 3)}{(x + 7)(x - 6)}$ **71.** 1 **73.** $-\dfrac{x + 8}{x - 4}$

75. $\dfrac{2n + 1}{2n - 3}$ **77.** $-\dfrac{3x + 1}{2x - 3}$ **79.** yes; $3 + 10^{-8}$ **81.** $5, -5$ **83.** $\dfrac{4x}{3y}$ **85.** $-\dfrac{1}{y^3}$

SECTION 8.2   *pages 415–420*

1. $24x^3y^2$ **3.** $30x^4y^2$ **5.** $8x^2(x + 2)$ **7.** $6x^2y(x + 4)$ **9.** $36x(x + 2)^2$ **11.** $6(x + 1)^2$

13. $(x - 1)(x + 2)(x + 3)$ **15.** $(2x + 3)^2(x - 5)$ **17.** $(x - 1)(x - 2)$

19. $(x - 3)(x + 2)(x + 4)$ **21.** $(x + 4)(x + 1)(x - 7)$ **23.** $(x - 6)(x + 6)(x + 4)$

25. $(x - 10)(x - 8)(x + 3)$ **27.** $(3x - 2)(x - 3)(x + 2)$ **29.** $(x + 2)(x - 3)$

31. $(x - 5)(x + 1)$ **33.** $(x - 3)(x - 2)(x - 1)(x - 6)$ **35.** $\dfrac{5}{ab^2}$; $\dfrac{6b}{ab^2}$ **37.** $\dfrac{15y^2}{18x^2y}$; $\dfrac{14x}{18x^2y}$

39. $\dfrac{ay + 5a}{y^2(y + 5)}$; $\dfrac{6y}{y^2(y + 5)}$ **41.** $\dfrac{a^2y + 7a^2}{y(y + 7)^2}$; $\dfrac{ay}{y(y + 7)^2}$ **43.** $\dfrac{b}{y(y - 4)}$; $\dfrac{-b^2y}{y(y - 4)}$ **45.** $-\dfrac{3y - 21}{(y - 7)^2}$; $\dfrac{2}{(y - 7)^2}$

47. $\dfrac{2y^2}{y^2(y - 3)}$; $\dfrac{3}{y^2(y - 3)}$ **49.** $\dfrac{x^3 + 4x^2}{(2x - 1)(x + 4)}$; $\dfrac{2x^2 + x - 1}{(2x - 1)(x + 4)}$ **51.** $\dfrac{3x^2 + 15x}{(x - 5)(x + 5)}$; $\dfrac{4}{(x - 5)(x + 5)}$

53. $\dfrac{x - 3}{(3x - 2)(x + 2)}$; $\dfrac{6x - 4}{(3x - 2)(x + 2)}$ **55.** $\dfrac{x^2 - 1}{(x + 5)(x - 3)(x + 1)}$; $\dfrac{x^2 - 3x}{(x + 5)(x - 3)(x + 1)}$ **57.** $\dfrac{4}{ab}$ **59.** $\dfrac{x - 2}{x + 6}$

61. $-\dfrac{5y}{4y + 1}$ **63.** $\dfrac{7x - 5}{2x - 7}$ **65.** $\dfrac{3x - 1}{x - 10}$ **67.** $\dfrac{-3n + 3}{3n + 4}$ **69.** $\dfrac{3}{x + 5}$ **71.** $\dfrac{1}{x - 1}$ **73.** $\dfrac{1}{x - 4}$

75. $\dfrac{x - 2}{x - 9}$ **77.** $\dfrac{7b + 5a}{ab}$ **79.** $\dfrac{11}{12a}$ **81.** $\dfrac{11}{12y}$ **83.** $\dfrac{120 + 7y}{20y^2}$ **85.** $\dfrac{21b + 4a}{12ab}$ **87.** $\dfrac{14x + 3}{12x}$

89. $\dfrac{8x - 3}{6x}$ **91.** $\dfrac{7}{36}$ **93.** $\dfrac{-3x^2 - 11x - 2}{3x^2}$ **95.** $\dfrac{6x^2 + 5x + 10}{6x^2}$ **97.** $\dfrac{2x^2 + 1}{x}$ **99.** $\dfrac{4x + 2}{x + 1}$

101. $\dfrac{x^2 - x + 2}{x^2y}$ **103.** $\dfrac{16xy - 12y + 6x^2 + 3x}{12x^2y^2}$ **105.** $\dfrac{3xy - 6y - 2x^2 - 14x}{24x^2y}$ **107.** $\dfrac{9x + 2}{(x - 2)(x + 3)}$

109. $\dfrac{2(x + 23)}{(x - 7)(x + 3)}$ **111.** $\dfrac{2x^2 - 5x + 1}{(x + 1)(x - 3)}$ **113.** $\dfrac{4x^2 - 34x + 5}{(2x - 1)(x - 6)}$ **115.** $\dfrac{2a - 5}{a - 7}$ **117.** $\dfrac{4x + 9}{(x + 3)(x - 3)}$

119. $\dfrac{-x + 9}{(x + 2)(x - 3)}$ **121.** $\dfrac{8y}{(y + 2)^2}$ **123.** $\dfrac{3x - y}{x - y}$ **125.** $\dfrac{11}{(a - 3)(a - 4)}$ **127.** $\dfrac{-x - 1}{(x + 2)(x - 5)}$

129. $\dfrac{x + 3}{x + 5}$ **131.** $\dfrac{x - 2}{x + 4}$ **133.** $\dfrac{4(2x - 1)}{(x + 5)(x - 2)}$ **135. a.** $\dfrac{3}{y} + \dfrac{6}{x}$ **b.** $\dfrac{4}{b^2} + \dfrac{3}{ab}$ **c.** $\dfrac{1}{4mn} + \dfrac{1}{6m^2}$

137. The sum of 50 terms is $\frac{50}{51}$. The sum of 100 terms is $\frac{100}{101}$. The sum of 1000 terms is $\frac{1000}{1001}$.

SECTION 8.3   *pages 423–424*

1. $\frac{x}{x-3}$ **3.** $\frac{2}{3}$ **5.** $\frac{y+3}{y-4}$ **7.** $\frac{4x+26}{5x+36}$ **9.** $\frac{x+2}{x+3}$ **11.** $\frac{x-6}{x+5}$ **13.** $-\frac{x-2}{x+1}$ **15.** $x-1$

17. $\frac{1}{2x-1}$ **19.** $\frac{x-3}{x+5}$ **21.** $\frac{x-7}{x-8}$ **23.** $\frac{2y-1}{2y+1}$ **25.** $\frac{x-2}{2x-5}$ **27.** $-\frac{x+1}{4x-3}$ **29.** $\frac{x+1}{2(5x-2)}$

31. $\frac{5}{3}$ **33.** $-\frac{1}{x-1}$ **35.** $\frac{y+4}{2(y-2)}$

SECTION 8.4   *pages 431–436*

1. 3 **3.** 1 **5.** 9 **7.** 1 **9.** $\frac{1}{4}$ **11.** 1 **13.** −3 **15.** $\frac{1}{2}$ **17.** 8 **19.** 5

21. −1 **23.** 5 **25.** The equation has no solution. **27.** 2 and 4 **29.** 4 and $-\frac{3}{2}$ **31.** 3

33. 4 **35.** 9 **37.** 36 **39.** 10 **41.** 113 **43.** −2 **45.** 15 **47.** 4 **49.** $\frac{1}{2}$

51. $\frac{3}{4}$ **53.** The length of *DE* is 7.2 cm. **55.** The height is 3.3 m.

57. The perimeter is 12 m. **59.** The perimeter is 12 in. **61.** The area is 56.3 cm².
63. The height is 18 ft. **65.** The height is 16 m. **67.** 6 c of sugar are required.
69. 20,000 voters voted in favor of the amendment. **71.** Yes, the shipment will be accepted.
73. 180 panels are necessary. **75.** The cities are 750 mi apart.
77. There are approximately 800 fish in the lake. **79.** Two additional gallons will be necessary.

81. $-\frac{2}{5}$ **83.** The first person's share is $1.25 million.

85. The basketball player made 210 shots.

SECTION 8.5   *pages 439–440*

1. $h = \frac{2A}{b}$ **3.** $t = \frac{d}{r}$ **5.** $T = \frac{PV}{nR}$ **7.** $l = \frac{P-2w}{2}$ **9.** $b_1 = \frac{2A - hb_2}{h}$ **11.** $h = \frac{3V}{A}$

13. $S = C - Rt$ **15.** $P = \frac{A}{1+rt}$ **17.** $w = \frac{A}{S+1}$

19. a. $h = \frac{S - 2\pi r^2}{2\pi r}$ **b.** The height is 5 in. **c.** The height is 4 in.

21. a. $S = \frac{F + BV}{B}$ **b.** The required selling price is $180. **c.** The required selling price is $75.

SECTION 8.6   *pages 445–448*

1. It would take the experienced painter 6 h.
3. It would take both skiploaders 3 h working together.
5. It would take 30 h with both computers working.
7. It would take 6 min with both air conditioners working.
9. It would take the second welder 15 h. **11.** It would take the second harvester 3 h.
13. It would take the second mason 6 h. **15.** It would take the second technician 3 h.

17. It would take the small unit $14\frac{2}{3}$ h. **19.** The camper hiked at a rate of 5 mph.

21. The rate of the jogger is 8 mph. The rate of the cyclist is 20 mph.
23. The rate of the jet is 360 mph. **25.** The rate of the second plane is 150 mph.
27. The rate of the car is 48 mph.

29. The rate at which the hiker walked the first 9 mi was 3 mph.
31. The rate of the current is 2 mph.   **33.** The rate of the gulf current is 6 mph.
35. The rate of the trucker for the first 330 mi was 55 mph.

37. It will take $1\frac{1}{19}$ h to fill the tank.   **39.** The bus usually travels 60 mph.

SECTION 8.7   *pages 453–454*

1. The profit is \$80,000.   **3.** The pressure is 6.75 lb/in^2.
5. The distance to the horizon is 24.03 mi.   **7.** The ball will roll 54 ft in 3 s.
9. The length is 10 ft when the width is 4 ft.   **11.** The pressure will be 75 lb/in^2.
13. The pressure will be 112.5 lb/in^2.   **15.** The repulsive force will be 80 lb.
17. The resistance will be 56.25 ohms.   **19.** y is doubled.   **21.** inversely   **23.** inversely

CHAPTER REVIEW EXERCISES   *pages 457–458*

1. $\frac{2x^3}{3y^3}$ [8.1A]   **2.** $-\frac{x+5}{x+1}$ [8.1A]   **3.** $\frac{x+1}{x^3(x-2)}$ [8.1B]   **4.** $\frac{(x-5)(2x-1)}{(x+3)(2x+5)}$ [8.1B]

5. $\frac{x+5}{x+4}$ [8.1C]   **6.** $3(2x-1)(x+1)$ [8.2A]   **7.** $\frac{3(x+2)}{x(x+2)(x-2)};\ \frac{x^2}{x(x+2)(x-2)}$ [8.2B]

8. $\frac{2}{x+5}$ [8.2C]   **9.** $\frac{5}{(2x-1)(3x+1)}$ [8.2D]   **10.** $\frac{x^2-4x+5}{(x-2)(x+3)}$ [8.2D]   **11.** $\frac{x-3}{x-2}$ [8.3A]

12. 2 [8.4A]   **13.** The equation has no solution. [8.4A]   **14.** -1 [8.4B]

15. The area is 64.8 m^2. [8.4C]   **16.** $t = \frac{d-s}{r}$ [8.5A]

17. 14 rolls of wallpaper are needed. [8.4D]
18. It would take 10 min with both landscapers working. [8.6A]
19. The rate of the cyclist is 10 mph. [8.6B]   **20.** The resistance is 0.4 ohm. [8.7A]

CUMULATIVE REVIEW EXERCISES   *pages 459–460*

1. $\frac{31}{30}$ [1.3B]   **2.** 21 [1.4A]   **3.** $5x - 2y$ [1.4B]   **4.** $-8x + 26$ [1.4D]   **5.** $-\frac{9}{2}$ [2.3A]

6. -12 [2.3C]   **7.** 10 [2.2A]   **8.** $\{x\,|\,x \le 8\}$ [2.5A]   **9.** The volume is 200 ft^3. [3.3A]

10.

[4.2B]   **11.** $\frac{3}{7}$ [4.1D]   **12.** $y = \frac{3}{2}x + 2$ [4.5A]   **13.** -28 [5.4A]

14. a^3b^7 [6.1A]   **15.** $\frac{a^5}{b^6}$ [6.1B]   **16.** 3.5×10^{-8} [6.1C]   **17.** $-4a^4 + 6a^3 - 2a^2$ [6.3A]
18. $a^2 + ab - 12b^2$ [6.3C]   **19.** $x^2 + 2x + 4$ [6.4B]   **20.** $(y-6)(y-1)$ [7.2A]
21. $(4x+1)(3x-1)$ [7.3A/7.3B]   **22.** $a(2a-3)(a+5)$ [7.3A/7.3B]
23. $4(b+5)(b-5)$ [7.4D]   **24.** -3 and $\frac{5}{2}$ [7.5A]   **25.** $\frac{2x^3}{3y^5}$ [8.1A]   **26.** $-\frac{x-2}{x+5}$ [8.1A]
27. 1 [8.1C]   **28.** $\frac{3}{(2x-1)(x+1)}$ [8.2D]   **29.** $\frac{x+3}{x+5}$ [8.3A]   **30.** 4 [8.4A]
31. The alloy contains 70% silver. [2.4B]   **32.** It would cost \$80. [8.4D]
33. It would take 6 min to fill the tank. [8.6A]

ANSWERS to Chapter 9 Selected Exercises

SECTION 9.1 *pages 469–472*

1. 2   **3.** 27   **5.** $\dfrac{1}{9}$ **7.** 4   **9.** $(-25)^{5/2}$ is not a real number.   **11.** $\dfrac{343}{125}$ **13.** x

15. $y^{1/2}$ **17.** $x^{1/12}$ **19.** $a^{7/12}$ **21.** $\dfrac{1}{a}$ **23.** $\dfrac{1}{y}$ **25.** $y^{3/2}$ **27.** $\dfrac{1}{x}$ **29.** $\dfrac{1}{x^4}$ **31.** a

33. $x^{3/10}$ **35.** a^3 **37.** $\dfrac{1}{x^{1/2}}$ **39.** $y^{1/9}$ **41.** x^4y **43.** $x^6y^3z^9$ **45.** $\dfrac{x}{y^2}$ **47.** $\dfrac{x^{3/2}}{y^{1/4}}$

49. $\dfrac{x^2}{y^8}$ **51.** $\dfrac{1}{x^{11/12}}$ **53.** $\dfrac{1}{y^{5/2}}$ **55.** $\dfrac{1}{b^{7/8}}$ **57.** a^5b^{13} **59.** $\dfrac{m^2}{4n^{3/2}}$ **61.** $\dfrac{y^{17/2}}{x^3}$ **63.** $\dfrac{16b^2}{a^{1/3}}$

65. $y^2 - y$ **67.** $a - a^2$ **69.** x^{4n} **71.** $x^{3n/2}$ **73.** $y^{3n/2}$ **75.** x^{2n^2} **77.** $x^{2n}y^n$

79. $x^{4n}y^{2n}$ **81.** $\sqrt[4]{3}$ **83.** $\sqrt{a^3}$ **85.** $\sqrt{32t^5}$ **87.** $-2\sqrt[3]{x^2}$ **89.** $\sqrt[3]{a^4b^2}$ **91.** $\sqrt[5]{a^6b^{12}}$

93. $\sqrt[4]{(4x-3)^3}$ **95.** $\dfrac{1}{\sqrt[3]{x^2}}$ **97.** $14^{1/2}$ **99.** $x^{1/3}$ **101.** $x^{4/3}$ **103.** $b^{3/5}$ **105.** $(2x^2)^{1/3}$

107. $-(3x^5)^{1/2}$ **109.** $3xy^{2/3}$ **111.** $(a^2-2)^{1/2}$ **113.** x^8 **115.** $-x^4$ **117.** xy^3

119. $-x^5y$ **121.** $4a^2b^6$ **123.** $\sqrt{-16x^4y^2}$ is not a real number.   **125.** $3x^3$ **127.** $-4x^3y^4$

129. $-x^2y^3$ **131.** x^4y^2 **133.** $3xy^5$ **135.** $2ab^2$

137. **a.** False; 2   **b.** True   **c.** True   **d.** False; $(a^n + n^n)^{1/n}$ **e.** False; $a + 2(ab)^{1/2} + b$
f. False; $a^{n/m}$

SECTION 9.2 *pages 479–482*

1. $x^2yz^2\sqrt{yz}$ **3.** $2ab^4\sqrt{2a}$ **5.** $3xyz^2\sqrt{5yz}$ **7.** $\sqrt{-9x^3}$ is not a real number.   **9.** $a^5b^2\sqrt[3]{ab^2}$

11. $-5y\sqrt[3]{x^2y}$ **13.** $abc^2\sqrt[4]{ab^2}$ **15.** $2x^2y\sqrt[4]{xy}$ **17.** $-6\sqrt{x}$ **19.** $-2\sqrt{2}$

21. $3\sqrt{2b} + 5\sqrt{3b}$ **23.** $-2xy\sqrt{2y}$ **25.** $6ab^2\sqrt{3ab} + 3ab\sqrt{3ab}$ **27.** $-\sqrt[3]{2}$ **29.** $8b\sqrt[3]{2b^2}$

31. $3a\sqrt[4]{2a}$ **33.** $17\sqrt{2} - 15\sqrt{5}$ **35.** $5b\sqrt{b}$ **37.** $-8xy\sqrt{2x} + 2xy\sqrt{xy}$ **39.** $2y\sqrt[3]{2x}$

41. $-4ab\sqrt[4]{2b}$ **43.** 16   **45.** $2\sqrt[3]{4}$ **47.** $xy^3\sqrt{x}$ **49.** $8xy\sqrt{x}$ **51.** $2x^2y\sqrt[3]{2}$

53. $2ab\sqrt[4]{3a^2b}$ **55.** 6   **57.** $x - \sqrt{2x}$ **59.** $4x - 8\sqrt{x}$ **61.** $x - 6\sqrt{x} + 9$

63. $84 + 16\sqrt{5}$ **65.** $672x^2y^2$ **67.** $4a^3b^3\sqrt[3]{a}$ **69.** $-8\sqrt{5}$ **71.** $x - y^2$ **73.** $12x - y$

75. $y\sqrt{5y}$ **77.** $b\sqrt{13b}$ **79.** $\dfrac{\sqrt{2}}{2}$ **81.** $\dfrac{2\sqrt{3y}}{3y}$ **83.** $\dfrac{3\sqrt{3a}}{a}$ **85.** $\dfrac{\sqrt{2y}}{2}$ **87.** $\dfrac{5\sqrt[3]{3}}{3}$

89. $\dfrac{5\sqrt[3]{9y^2}}{3y}$ **91.** $\dfrac{b\sqrt{2a}}{2a^2}$ **93.** $\dfrac{\sqrt{15x}}{5x}$ **95.** $2 + 2\sqrt{2}$ **97.** $\dfrac{-12 - 4\sqrt{2}}{7}$ **99.** $-\dfrac{10 + 5\sqrt{7}}{3}$

101. $-\dfrac{7\sqrt{x} + 21}{x - 9}$ **103.** $-\sqrt{6} + 3 - 2\sqrt{2} + 2\sqrt{3}$ **105.** $-\dfrac{17 + 5\sqrt{5}}{4}$ **107.** $\dfrac{8a - 10\sqrt{ab} + 3b}{16a - 9b}$

109. $\dfrac{3 - 7\sqrt{y} + 2y}{1 - 4y}$

111. **a.** False; $\sqrt[6]{432}$ **b.** True   **c.** False; $\sqrt[3]{x} \cdot \sqrt[3]{x} = x^{2/3}$ **d.** False; $\sqrt{x} + \sqrt{y}$
e. False; $\sqrt[2]{2} + \sqrt[3]{3}$ **f.** True

113. $a + b$

SECTION 9.3 *pages 489–490*

1. $2i$ **3.** $7i\sqrt{2}$ **5.** $3i\sqrt{3}$ **7.** $4 + 2i$ **9.** $2\sqrt{3} - 3i\sqrt{2}$ **11.** $4\sqrt{10} - 7i\sqrt{3}$

13. $8 - i$ **15.** $-8 + 4i$ **17.** $6 - 6i$ **19.** $19 - 7i\sqrt{2}$ **21.** $6\sqrt{2} - 3i\sqrt{2}$ **23.** 63

25. -4 **27.** $-3\sqrt{2}$ **29.** $-4 + 12i$ **31.** $-2 + 4i$ **33.** $17 - i$ **35.** $8 + 27i$ **37.** 1

39. 1   **41.** $-3i$ **43.** $\dfrac{3}{4} + \dfrac{1}{2}i$ **45.** $\dfrac{10}{13} - \dfrac{2}{13}i$ **47.** $\dfrac{4}{5} + \dfrac{2}{5}i$ **49.** $-i$ **51.** $-\dfrac{\sqrt{5}}{5} + \dfrac{2\sqrt{5}}{5}i$

53. $\dfrac{3}{10} - \left(\dfrac{11}{10}\right)i$ **55.** $\dfrac{6}{5} + \dfrac{7}{5}i$

57. **a.** $(y + i)(y - i)$ **b.** $(7x + 4i)(7x - 4i)$ **c.** $(3a + 8i)(3a - 8i)$ **59.** $i^{76} = 1$

SECTION 9.4 *pages 495–498*

1. 25 **3.** 27 **5.** 48 **7.** −2 **9.** The equation has no solution. **11.** 9 **13.** −23

15. −16 **17.** $\dfrac{9}{4}$ **19.** 14 **21.** 7 **23.** 7 **25.** −122 **27.** 35 **29.** −5 and −2

31. 45 **33.** 23 **35.** −4 **37.** 10 **39.** 1 **41.** 3 **43.** 21 **45.** $\dfrac{25}{72}$ **47.** 5

49. −2 and 4 **51.** 6 **53.** 0 and 1 **55.** −2 and 2 **57.** −2 and 1 **59.** 2 and 11
61. The width is 6 ft. **63.** The bottom of the ladder is 10 ft from the wall.
65. The object has fallen 2500 ft. **67.** The periscope must be 8.64 ft above the water.
69. The length of the pendulum is 4.67 ft. **71.** The distance is 180 m. **73.** 2
75. $a = \sqrt{c^2 - b^2}$ **77.** The length is $\sqrt{6}$.
79. The first cyclist had traveled 40 mi. The second cyclist had traveled 30 mi.

CHAPTER REVIEW EXERCISES *pages 501–502*

1. $r^{1/6}$ [9.1A] **2.** $\dfrac{64x^3}{y^6}$ [9.1A] **3.** $\dfrac{b^3}{8a^6}$ [9.1A] **4.** $3\sqrt[5]{y^2}$ [9.1B] **5.** $\dfrac{1}{2}x^{3/4}$ [9.1B]

6. $2xy^2$ [9.1C] **7.** $4x^2y^3\sqrt{2y}$ [9.2A] **8.** $3abc^2\sqrt[3]{ac}$ [9.2A] **9.** $8a\sqrt{2a}$ [9.2B]
10. $-2x^2y\sqrt[3]{2x}$ [9.2B] **11.** $-4x\sqrt{3}$ [9.2C] **12.** $14 + 10\sqrt{3}$ [9.2C]
13. $2a - \sqrt{ab} - 15b$ [9.2C] **14.** $4x + 4\sqrt{xy} + y$ [9.2C] **15.** $\dfrac{4x^2}{y}$ [9.2D] **16.** 2 [9.2D]

17. $\dfrac{x + \sqrt{xy}}{x - y}$ [9.2D] **18.** $5i\sqrt{2}$ [9.3A] **19.** $-3 + 2i$ [9.3B] **20.** $18 + 16i$ [9.3C]

21. $-\dfrac{4}{5} + \dfrac{7}{5}i$ [9.3D] **22.** −4 [9.3C] **23.** 4 [9.4A] **24.** −3 [9.4A]
25. The object has fallen 576 ft. [9.4B]

CUMULATIVE REVIEW EXERCISES *pages 503–504*

1. 92 [1.3B] **2.** 56 [1.4A] **3.** $-10x + 1$ [1.4D] **4.** $\dfrac{3}{2}$ [2.3A] **5.** $\dfrac{2}{3}$ [2.3C]

6. $\{x \mid x > 1\}$ [2.5A] **7.** $\dfrac{1}{3}$ and $\dfrac{7}{3}$ [2.6A] **8.** $\{x \mid -6 \le x \le 3\}$ [2.6B]

9. The area is 187.5 cm². [3.2B] **10.** 3 [5.4A] **11.** $m = \dfrac{3}{2}$, $b = 3$ [4.2B]

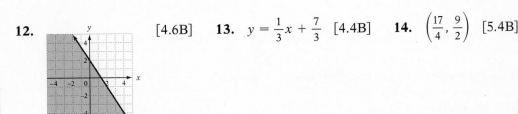

12. [4.6B] **13.** $y = \dfrac{1}{3}x + \dfrac{7}{3}$ [4.4B] **14.** $\left(\dfrac{17}{4}, \dfrac{9}{2}\right)$ [5.4B]

15. $2x^2y^2$ [6.1B] **16.** $(9x + y)(9x - y)$ [7.4A] **17.** $x(x^2 + 3)(x + 1)(x - 1)$ [7.4D]
18. $C = R - nP$ [8.5A] **19.** $\dfrac{y^5}{x^4}$ [9.1A] **20.** $-x\sqrt{10x}$ [9.2B] **21.** $13 - 7\sqrt{3}$ [9.2C]

22. $\sqrt{6} + \sqrt{2}$ [9.2D] **23.** $-\dfrac{1}{5} + \dfrac{3}{5}i$ [9.3D] **24.** −20 [9.4A]

25. The length of side *DE* is 27 m. [8.4C] **26.** $4000 must be invested at 8.4%. [2.4C]

27. The rate of the plane is 250 mph.   [8.6B]
28. It takes 1.25 s for light to travel to the earth from the moon.   [6.1D]
29. The slope is 0.08. The slope indicates that the annual income is 8% of the investment.   [4.3B]
30. The periscope must be 32.7 ft above the water.   [9.4B]

ANSWERS to Chapter 10 Selected Exercises

SECTION 10.1   *pages 511–514*

1. 0 and 4   3. 5 and −5   5. 3 and −2   7. 3   9. 0 and 2   11. 5 and −2

13. 2 and 5   15. 6 and $-\frac{3}{2}$ 17. 2 and $\frac{1}{4}$ 19. $\frac{1}{3}$ and −4   21. $\frac{9}{2}$ and $-\frac{2}{3}$

23. $\frac{1}{4}$ and −4   25. 9 and −2   27. −2 and $-\frac{3}{4}$ 29. 2 and −5   31. −4 and $-\frac{3}{2}$

33. $2b$ and $7b$ 35. $7c$ and $-c$ 37. $-\frac{b}{2}$ and $-b$ 39. $4a$ and $\frac{2a}{3}$ 41. $-\frac{a}{3}$ and $3a$

43. $-\frac{3y}{2}$ and $-\frac{y}{2}$ 45. $-\frac{a}{2}$ and $-\frac{4a}{3}$ 47. $x^2 - 7x + 10 = 0$ 49. $x^2 + 6x + 8 = 0$

51. $x^2 - 5x - 6 = 0$ 53. $x^2 - 9 = 0$ 55. $x^2 - 8x + 16 = 0$ 57. $x^2 - 5x = 0$

59. $x^2 - 3x = 0$ 61. $2x^2 - 7x + 3 = 0$ 63. $4x^2 - 5x - 6 = 0$ 65. $3x^2 + 11x + 10 = 0$

67. $9x^2 - 4 = 0$ 69. $6x^2 - 5x + 1 = 0$ 71. $10x^2 - 7x - 6 = 0$ 73. $8x^2 + 6x + 1 = 0$

75. $50x^2 - 25x - 3 = 0$ 77. 7 and −7   79. $2i$ and $-2i$ 81. 2 and −2   83. $\frac{9}{2}$ and $-\frac{9}{2}$

85. $7i$ and $-7i$ 87. $4\sqrt{3}$ and $-4\sqrt{3}$ 89. $5\sqrt{3}$ and $-5\sqrt{3}$ 91. $3i\sqrt{2}$ and $-3i\sqrt{2}$

93. 7 and −5   95. 0 and −6   97. −7 and 3   99. 1 and 0   101. $-5 + \sqrt{6}$ and $-5 - \sqrt{6}$

103. $3 + 3i\sqrt{5}$ and $3 - 3i\sqrt{5}$ 105. $-\frac{2 - 9\sqrt{2}}{3}$ and $-\frac{2 + 9\sqrt{2}}{3}$ 107. $x^2 - 2 = 0$

109. $x^2 - 18 = 0$ 111. $\frac{3b}{a}$ and $-\frac{3b}{a}$ 113. $-\frac{1}{2}$

SECTION 10.2   *pages 519–520*

1. 5 and −1   3. −9 and 1   5. 3   7. $-2 + \sqrt{11}$ and $-2 - \sqrt{11}$ 9. $3 + \sqrt{2}$ and $3 - \sqrt{2}$

11. $1 + i$ and $1 - i$ 13. 8 and −3   15. 4 and −9   17. $\frac{3 + \sqrt{5}}{2}$ and $\frac{3 - \sqrt{5}}{2}$

19. $\frac{1 + \sqrt{5}}{2}$ and $\frac{1 - \sqrt{5}}{2}$ 21. $3 + \sqrt{13}$ and $3 - \sqrt{13}$ 23. 5 and 3   25. $2 + 3i$ and $2 - 3i$

27. $-3 + 2i$ and $-3 - 2i$ 29. $1 + 3\sqrt{2}$ and $1 - 3\sqrt{2}$ 31. $\frac{1 + \sqrt{17}}{2}$ and $\frac{1 - \sqrt{17}}{2}$

33. $1 + 2i\sqrt{3}$ and $1 - 2i\sqrt{3}$ 35. $\frac{1}{2} + i$ and $\frac{1}{2} - i$ 37. $\frac{1}{3} + \frac{1}{3}i$ and $\frac{1}{3} - \frac{1}{3}i$

39. $\frac{2 + \sqrt{14}}{2}$ and $\frac{2 - \sqrt{14}}{2}$ 41. 1 and $-\frac{3}{2}$ 43. $1 + \sqrt{5}$ and $1 - \sqrt{5}$ 45. $\frac{1}{2}$ and 5

47. $2 + \sqrt{5}$ and $2 - \sqrt{5}$ 49. $-a$ and $2a$ 51. $-5a$ and $2a$

53. No. The ball will have gone only 197.2 ft when it hits the ground.

SECTION 10.3   *pages 525–526*

1. 5 and −2   3. 4 and −9   5. $4 + 2\sqrt{22}$ and $4 - 2\sqrt{22}$ 7. 3 and −8   9. −3 and $\frac{1}{2}$

11. $\frac{3}{2}$ and $-\frac{1}{4}$ 13. 2 and 12   15. $\frac{1 + \sqrt{3}}{2}$ and $\frac{1 - \sqrt{3}}{2}$ 17. $-1 + i$ and $-1 - i$

19. $1 + 2i$ and $1 - 2i$ 21. $2 + 3i$ and $2 - 3i$ 23. $\frac{1}{2} + \frac{3}{2}i$ and $\frac{1}{2} - \frac{3}{2}i$

25. $-\frac{3}{2} + \frac{1}{2}i$ and $-\frac{3}{2} - \frac{1}{2}i$ **27.** $\frac{3}{4} + \frac{3\sqrt{3}}{4}i$ and $\frac{3}{4} - \frac{3\sqrt{3}}{4}i$

29. The quadratic equation has two complex number solutions.
31. The quadratic equation has one real number solution.
33. The quadratic equation has two real number solutions.   **35.** $\{p \mid p < 25\}$ **37.** $\{p \mid p > 4\}$
39. a. $x = 1.05$; will increase   **b.** $x = 0.95$; will decrease

SECTION 10.4   *pages 531–532*

1. 3, −3, 2, and −2   **3.** $\sqrt{2}$, −$\sqrt{2}$, 2, and −2   **5.** 1 and 4   **7.** 16   **9.** $2i$, $-2i$, 1, and −1

11. $4i$, $-4i$, 2, and −2   **13.** 16   **15.** 1 and 512   **17.** $\frac{2}{3}$, $-\frac{2}{3}$, 1, and −1   **19.** 3   **21.** 9

23. 2 and −1   **25.** 0 and 2   **27.** 2 and $-\frac{1}{2}$ **29.** −2   **31.** 1   **33.** 1   **35.** −3

37. 10 and −1   **39.** $-\frac{1}{2} + \frac{\sqrt{7}}{2}i$ and $-\frac{1}{2} - \frac{\sqrt{7}}{2}i$ **41.** 1 and −3   **43.** 0 and −1

45. $\frac{1}{2}$ and $-\frac{1}{3}$ **47.** $-\frac{2}{3}$ and 6   **49.** $\frac{4}{3}$ and 3   **51.** $-\frac{1}{4}$ and 3   **53.** 4

SECTION 10.5   *pages 535–536*

1. The height is 3 cm. The base is 14 cm.   **3.** The length is 13 ft. The width is 5 ft.
5. The integers are 3 and 5 or −5 and −3.   **7.** The integer is 3 or −8.
9. The time for a projectile to return is 12.5 s.   **11.** The distance is 2.3 in.
13. The rate of the cruise ship is 10 mph.
15. The speed of the first car is 30 mph. The speed of the second car is 40 mph.
17. a. The maximum height of the arch is 27 ft.   **b.** The height is 24 ft.
   c. The distance is 20.13 ft.

SECTION 10.6   *pages 539–540*

1. $\{x \mid x < -2 \text{ or } x > 4\}$ **3.** $\{x \mid x \leq 1 \text{ or } x \geq 2\}$

5. $\{x \mid -3 < x < 4\}$ **7.** $\{x \mid x < -2 \text{ or } 1 < x < 3\}$

9. $\{x \mid -4 \leq x \leq 1 \text{ or } x \geq 2\}$ **11.** $\{x \mid x < -2 \text{ or } x > 4\}$

13. $\{x \mid -1 < x \leq 3\}$ **15.** $\{x \mid x \leq -2 \text{ or } 1 \leq x < 3\}$

17. $\{x \mid x > 4 \text{ or } x < -4\}$ **19.** $\{x \mid -3 \leq x \leq 12\}$ **21.** $\left\{x \mid \frac{1}{2} < x < \frac{3}{2}\right\}$

23. $\left\{x \mid x < 1 \text{ or } x > \frac{5}{2}\right\}$ **25.** $\{x \mid x < -1 \text{ or } 1 < x \leq 2\}$ **27.** $\left\{x \mid \frac{1}{2} < x \leq 1\right\}$

29. $\{x \mid 2 < x \leq 3\}$ **31.** $\{x \mid x > 5 \text{ or } -4 < x < -1\}$ **33.**

35.   **37.**   **39.**

CHAPTER REVIEW EXERCISES   *pages 543–544*

1. $\frac{2}{3}$ and −4 [10.1A]   **2.** $\frac{3}{2}$ and $-\frac{2}{3}$ [10.1A]   **3.** $x^2 - 9 = 0$ [10.1B]

4. $2x^2 + 7x - 4 = 0$ [10.1B]   **5.** $2 + 2\sqrt{2}$ and $2 - 2\sqrt{2}$ [10.1C]

6. $3 + \sqrt{11}$ and $3 - \sqrt{11}$ [10.2A]   **7.** $\frac{3 + \sqrt{15}}{3}$ and $\frac{3 - \sqrt{15}}{3}$ [10.2A]

8. $\frac{1 + \sqrt{3}}{2}$ and $\frac{1 - \sqrt{3}}{2}$ [10.3A]   **9.** $-2 + 2i\sqrt{2}$ and $-2 - 2i\sqrt{2}$ [10.3A]

10. The quadratic equation has two real number solutions. [10.3A]

11. The quadratic equation has two complex number solutions. [10.3A]   **12.** $\frac{1}{4}$ [10.4A]

13. $1, -1, \sqrt{3}$, and $-\sqrt{3}$ [10.4A]   **14.** 4 [10.4B]   **15.** The equation has no solution. [10.4B]
16. 2 and -9 [10.4C]   **17.** $\{x \mid x < -4 \text{ or } 2 < x < 4\}$ [10.6A]
18. $\left\{ x \mid -4 < x \le \dfrac{3}{2} \right\}$ [10.6A]   **19.** The base is 15 ft. The height is 4 ft. [10.5A]
20. The paddling rate in calm water is 4 mph. [10.5A]

CUMULATIVE REVIEW EXERCISES   *pages 545–546*

1. 14 [1.4A]   **2.** $\left\{ x \mid -2 < x < \dfrac{10}{3} \right\}$ [2.6B]   **3.** The volume is 54π m³. [3.3A]

4. $-\dfrac{7}{3}$ [4.1D]   **5.** $-\dfrac{3}{2}$ [4.3B]   **6.** $\left(\dfrac{5}{2}, 0 \right)$ and $(0, -3)$ [4.2B]   **7.** $y = x + 1$ [4.5A]

8. $(1, -1, 2)$ [5.3B]   **9.**   [5.6A]

10. The height of triangle *DEF* is 16 cm. [8.4C]   **11.** $x^2 - 3x - 4 - \dfrac{6}{3x - 4}$ [6.4A]

12. $-3xy(x^2 - 2xy + 3y^2)$ [7.1A]   **13.** $(2x - 5)(3x + 4)$ [7.3A/7.3B]   **14.** $\dfrac{x}{2}$ [8.1B]

15. $-\dfrac{3}{2}$ and -1 [8.4A]   **16.** $b = \dfrac{2S - an}{n}$ [8.5A]   **17.** $1 - a$ [9.1A]   **18.** $-8 - 14i$ [9.3C]

19. 0 and 1 [9.4A]   **20.** $2, -2, \sqrt{2}$, and $-\sqrt{2}$ [10.4A]

21. The lower limit is $9\dfrac{23}{64}$ in. The upper limit is $9\dfrac{25}{64}$ in. [2.6C]

22. The area is $(x^2 + 6x - 16)$ ft². [6.3E]

23. The slope is $-\dfrac{25{,}000}{3}$. The building decreases $\$\dfrac{25{,}000}{3}$ in value each year. [4.3A]

24. The ladder will reach 15 ft up on the building. [9.4B]
25. There are two complex number solutions. [10.3A]

ANSWERS to Chapter 11 Selected Exercises

SECTION 11.1   *pages 551–552*

1. 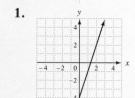   **3.** 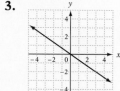   **5.** 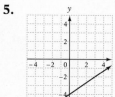   **7.**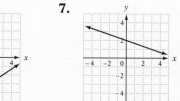

9. 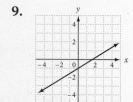   **11.** 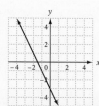   **13.** The sale price is $160.

15. $f(x) = -20x + 230{,}000$. At a price of $8500, 60,000 trucks would be sold.
17. $f(x) = -\dfrac{3}{5}x + 545$. If the room rate is $100, 485 rooms will be rented.   **19.** 8

SECTION 11.2   *pages 559–562*

1.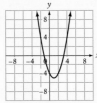
Vertex: $(1, -5)$
Axis of symmetry: $x = 1$

3.
Vertex: $(1, -2)$
Axis of symmetry: $x = 1$

5.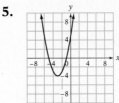
Vertex: $(-3, -4)$
Axis of symmetry: $x = -3$

7.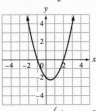
Vertex: $\left(\frac{1}{2}, -\frac{9}{4}\right)$
Axis of symmetry: $x = \frac{1}{2}$

9.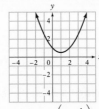
Vertex: $\left(\frac{3}{2}, \frac{9}{2}\right)$
Axis of symmetry: $x = \frac{3}{2}$

11.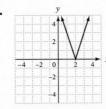
Vertex: $\left(1, \frac{1}{2}\right)$
Axis of symmetry: $x = 1$

13. $(2, 0)$ and $(-2, 0)$ **15.** $(0, 0)$ and $(2, 0)$ **17.** $(2, 0)$ and $(-1, 0)$

19. $\left(-\frac{1}{2}, 0\right)$ and $(1, 0)$ **21.** $(-1 + \sqrt{2}, 0)$ and $(-1 - \sqrt{2}, 0)$

23. The parabola has no x-intercepts.   **25.** $(1 + \sqrt{3}, 0)$ and $(1 - \sqrt{3}, 0)$.
27. $(2 + \sqrt{5}, 0)$ and $(2 - \sqrt{5}, 0)$ **29.** The parabola has two x-intercepts.
31. The parabola has two x-intercepts.   **33.** The parabola has one x-intercept.
35. The parabola has no x-intercepts.   **37.** The parabola has no x-intercepts.
39. The parabola has no x-intercepts.   **41.** The minimum value of the function is $-\frac{25}{4}$.

43. The maximum value of the function is $\frac{41}{8}$.   **45.** The maximum value of the function is $\frac{9}{8}$.

47. The maximum value of the function is $-\frac{71}{12}$.   **49.** The maximum value of the function is $\frac{9}{4}$.

51. The minimum value of the function is $-\frac{13}{4}$.   **53.** The maximum value of the function is $\frac{33}{8}$.

55. The maximum height is 114 ft.   **57.** The pool will have the least amount of algae after 5 days.
59. The minimum height of the cable is 24.36 ft.   **61.** The two numbers are 10 and 10.
63. The two numbers are 12 and −12.   **65.** The dimensions are 15 ft by 15 ft. The area is 225 ft².
67. a. Always true   **b.** Sometimes true   **c.** Sometimes true
69. The dimensions are 100 ft by 50 ft.

SECTION 11.3   *pages 567–568*

1. Yes   **3.** No   **5.** Yes   **7.**
Domain: $\{x \mid x \text{ is a real number}\}$;
Range: $\{y \mid y \geq 0\}$

9.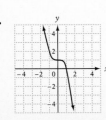
Domain: $\{x \mid x \text{ is a real number}\}$;
Range: $\{y \mid y \text{ is a real number}\}$

11.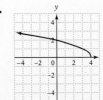
Domain: $\{x \mid x \leq 4\}$;
Range: $\{y \mid y \geq 0\}$

13. 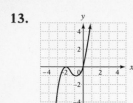 Domain: $\{x \mid x \text{ is a real number}\}$;
Range: $\{y \mid y \text{ is a real number}\}$

15. 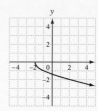 Domain: $\{x \mid x \geq -2\}$;
Range: $\{y \mid y \leq 0\}$

17. 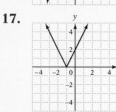 Domain: $\{x \mid x \text{ is a real number}\}$;
Range: $\{y \mid y \geq 0\}$

19. $a = 18$ **21.** $f(2, 5) + g(2, 5) = 17$

23. $\{x \mid -1 < x < 1\}$ **25.** $f(x)$ is greatest when $x = -3$.

SECTION 11.4 *pages 573–574*

1. 5 **3.** 1 **5.** 0 **7.** $-\dfrac{29}{4}$ **9.** $\dfrac{2}{3}$ **11.** 2 **13.** 39 **15.** -8 **17.** $-\dfrac{4}{5}$ **19.** -2
21. 7 **23.** -13 **25.** -29 **27.** $8x - 13$ **29.** 4 **31.** 3 **33.** $x + 4$ **35.** 5
37. 11 **39.** $3x^2 + 3x + 5$ **41.** -3 **43.** -27 **45.** $x^3 - 6x^2 + 12x - 8$ **47.** $6h + h^2$
49. $2 + h$ **51.** $h + 2a$ **53.** -1 **55.** -6 **57.** $6x - 13$

SECTION 11.5 *pages 579–582*

1. Yes **3.** No **5.** Yes **7.** No **9.** No **11.** No **13.** $\{(0, 1), (3, 2), (8, 3), (15, 4)\}$

15. No inverse **17.** $\{(-2, 0), (5, -1), (3, 3), (6, -4)\}$ **19.** No inverse **21.** $f^{-1}(x) = \dfrac{1}{4}x + 2$

23. $f^{-1}(x) = \dfrac{1}{2}x - 2$ **25.** $f^{-1}(x) = 2x + 2$ **27.** $f^{-1}(x) = -\dfrac{1}{2}x + 1$ **29.** $f^{-1}(x) = \dfrac{3}{2}x - 6$

31. $f^{-1}(x) = -3x + 3$ **33.** $f^{-1}(x) = \dfrac{1}{2}x + \dfrac{5}{2}$ **35.** $f^{-1}(x) = \dfrac{1}{5}x + \dfrac{2}{5}$ **37.** $f^{-1}(x) = \dfrac{1}{6}x + \dfrac{1}{2}$

39. $f^{-1}(x) = -\dfrac{1}{6}x + \dfrac{1}{3}$ **41.** $f^{-1}(x) = \dfrac{1}{3}x + \dfrac{4}{3}$ **43.** $f^{-1}(x) = -\dfrac{3}{2}x + 6$ **45.** $f^{-1}(x) = \dfrac{1}{3}x + \dfrac{2}{3}$

47. $f^{-1}(x) = -\dfrac{4}{3}x - \dfrac{8}{3}$ **49.** Yes **51.** No **53.** Yes **55.** No **57.** No **59.** Yes

61. Yes **63.** 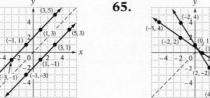 **65.** 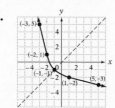 **67.**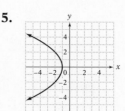

69. $f^{-1}(-1) = 3$ **71.** $f^{-1}(3) = -3$ **73.** $f^{-1}(8) = 0$

SECTION 11.6 *pages 593–596*

1.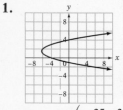

Vertex: $\left(-\dfrac{25}{4}, \dfrac{3}{2}\right)$

Axis of symmetry: $y = \dfrac{3}{2}$

3.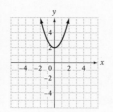

Vertex: $(0, 2)$
Axis of symmetry: $x = 0$

5.

Vertex: $(-1, 0)$
Axis of symmetry: $y = 0$

7.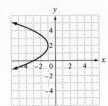

Vertex: $(-1, 2)$
Axis of symmetry: $y = 2$

9.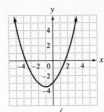

Vertex: $\left(-1, -\dfrac{7}{2}\right)$
Axis of symmetry: $x = -1$

11.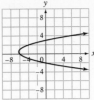

Vertex: $\left(-\dfrac{25}{4}, \dfrac{1}{2}\right)$
Axis of symmetry: $y = \dfrac{1}{2}$

13.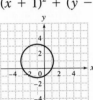

15.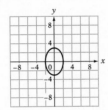

17.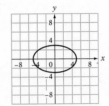

19. $(x - 2)^2 + (y + 1)^2 = 4$

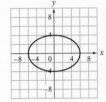

21. $(x + 1)^2 + (y - 1)^2 = 5$

23.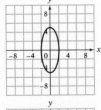

25.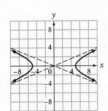

27.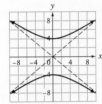

29.

31.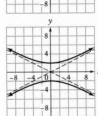

33.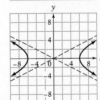

35.

37.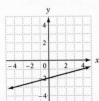

39.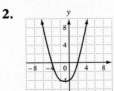

41.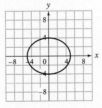

43.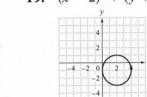

45. $(x - 3)^2 + (y - 3)^2 = 9$

47. $\dfrac{x^2}{25} + \dfrac{y^2}{16} = 1$, ellipse

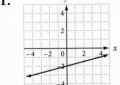

CHAPTER REVIEW EXERCISES   *pages 599–600*

1.   [11.1A]

2.   [11.2A]

3. There are no x-intercepts.   [11.2B]
4. The maximum value of the function is 9.   [11.2C]

5. 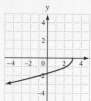   Domain: $\{x \mid x \le 3\}$; Range: $\{y \mid y \le 0\}$ [11.3A]

6. 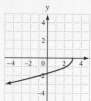   Domain: $\{x \mid x \text{ is a real number}\}$; Range: $\{y \mid y \ge -2\}$ [11.3A]

7. 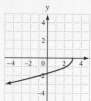   Domain: $\{x \mid x \text{ is a real number}\}$; Range: $\{y \mid y \text{ is a real number}\}$ [11.3A]

8. No [11.5A]   **9.** -2 [11.4A]   **10.** $-\dfrac{13}{2}$ [11.4A]   **11.** 5 [11.4B]

12. $2x^2 - 4x - 5$ [11.4B]   **13.** $\{(6, 2), (5, 3), (4, 4), (3, 5)\}$ [11.5B]   **14.** $f^{-1}(x) = 4x + 16$ [11.5B]

15.   **16.** 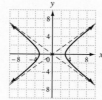   **17.** 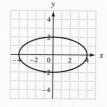   **18.**

[11.6A]   [11.6B]   [11.6D]   [11.6C]

19. $(x + 3)^2 + (y + 3)^2 = 16$ [11.6B]

20. The dimensions are 50 cm by 50 cm. The area is 2500 cm². [11.2D]

CUMULATIVE REVIEW EXERCISES   *pages 601–602*

1. [1.5C]   **2.** $\dfrac{38}{53}$ [8.4A]   **3.** -20 [4.1D]   **4.** $y = -\dfrac{3}{2}x$ [4.4A]

5. $y = x - 6$ [4.5A]   **6.** $\dfrac{x}{x + y}$ [8.1A]   **7.** $\dfrac{x - 5}{3x - 2}$ [8.2D]   **8.** $\dfrac{5}{2}$ [8.4A]   **9.** $\dfrac{4a}{3b^{12}}$ [6.1B]

10. $2x^{3/4}$ [9.1B]   **11.** $3\sqrt{2} - 5i$ [9.3A]   **12.** $\dfrac{-1 + \sqrt{7}}{2}$ and $\dfrac{-1 - \sqrt{7}}{2}$ [10.2A/10.3A]

13. 6 [10.4B]   **14.** $\{x \mid -4 < x \le 3\}$ [10.6A]   **15.** 0 [11.2C]   **16.** $f^{-1}(x) = \dfrac{1}{4}x - 2$ [11.5B]

17. 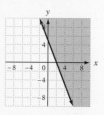   **18.**   **19.**   **20.**

[4.6A]   [11.6A]   [11.6C]   [11.6D]

21. There were 82 adult tickets sold. [5.5B]   **22.** The rate of the motorcycle is 60 mph. [8.6B]

23. The rate of the crew in calm water is 4.5 mph. [5.5A]

24. A gear with 60 teeth will make 18 revolutions/min. [8.7A]

25. The maximum product is 400. [11.2D]

ANSWERS to Chapter 12 Selected Exercises

SECTION 12.1   *pages 609–612*

1. a. $f(2) = 9$ **b.** $f(0) = 1$ **c.** $f(-2) = \dfrac{1}{9}$

3. a. $g(3) = 16$ **b.** $g(1) = 4$ **c.** $g(-3) = \dfrac{1}{4}$

5. a. $P(0) = 1$ **b.** $P\left(\dfrac{3}{2}\right) = \dfrac{1}{8}$ **c.** $P(-2) = 16$

7. a. $G(4) \approx 7.3891$ **b.** $G(-2) \approx 0.3679$ **c.** $G\left(\dfrac{1}{2}\right) \approx 1.2840$

9. a. $H(-1) \approx 54.5982$ **b.** $H(3) = 1$ **c.** $H(5) \approx 0.1353$

11. a. $F(2) = 16$ **b.** $F(-2) = 16$ **c.** $F\left(\dfrac{3}{4}\right) \approx 1.4768$

13. a. $f(-2) \approx 0.1353$ **b.** $f(2) \approx 0.1353$ **c.** $f(-3) \approx 0.0111$ **15.**

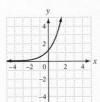

17. 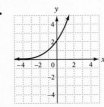   **19.** 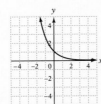   **21.** 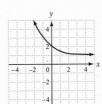   **23.**

25.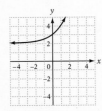

27. a. Sometimes true   **b.** Always true   **c.** Always true   **29.** e

31. 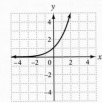   **33.** 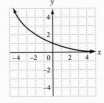   **35.**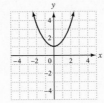

37. 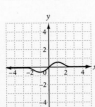   **39.** 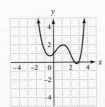   **41.**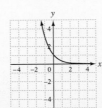

43.   **a.**

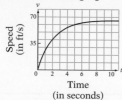

Year

   b.   In 2001 the population of India will be 1.01 billion people.

45.   **a.**

   b.   4 s after the object is dropped, it will be falling at 55.3 ft/s.

SECTION 12.2   *pages 619–622*

1. $\log_5 25 = 2$ **3.** $\log_4 \frac{1}{16} = -2$ **5.** $\log_{10} x = y$ **7.** $\log_a w = x$ **9.** $3^2 = 9$

11. $10^{-2} = 0.01$ **13.** $e^y = x$ **15.** $b^v = u$ **17.** $\log_3 81 = 4$ **19.** $\log_2 128 = 7$

21. $\log 100 = 2$ **23.** $\ln e^3 = 3$ **25.** $\log_8 1 = 0$ **27.** $\log_5 625 = 4$ **29.** 9

31. 64   **33.** $\frac{1}{7}$ **35.** 1   **37.** 316.23   **39.** 0.02   **41.** 7.39   **43.** 0.61   **45.** $\log_3 x^3 y^2$

47. $\ln\left(\frac{x^4}{y^2}\right)$ **49.** $\log_7 x^3$ **51.** $\ln x^3 y^4$ **53.** $\log_4 x^2 y^2$ **55.** $\log_6 \sqrt{\frac{x}{y}}$ **57.** $\log_3\left(\frac{x^2 z^2}{y}\right)$

59. $\ln\left(\frac{x}{y^2 z}\right)$ **61.** $\log_4 \frac{s^2 r^2}{t^4}$ **63.** $\ln\left(\frac{x}{y^2 z^2}\right)$ **65.** $\log_2\left(\frac{t^3 v^2}{r^2}\right)$ **67.** $\log_4 \sqrt{\frac{x^3 z}{y^2}}$ **69.** $\ln \sqrt{\frac{x}{y^3}}$

71. $\log_2 \frac{\sqrt{xy}}{\sqrt[3]{y^2}}$ **73.** $\log_8 x + \log_8 z$ **75.** $5 \log_3 x$ **77.** $\log_b r - \log_b s$ **79.** $2 \log_3 x + 6 \log_3 y$

81. $3 \log_7 u - 4 \log_7 v$ **83.** $2 \log_2 r + 2 \log_2 s$ **85.** $2 \ln x + \ln y + \ln z$

87. $\log_5 x + 2 \log_5 y - 4 \log_5 z$ **89.** $2 \log_8 x - \log_8 y - 2 \log_8 z$ **91.** $\frac{3}{2} \log_4 x + \frac{1}{2} \log_4 y$

93. $\frac{3}{2} \log_7 x - \frac{1}{2} \log_7 y$ **95.** $\ln x + \frac{1}{2} \ln y - \frac{1}{2} \ln z$ **97.** $\log_3 t - \frac{1}{2} \log_3 x$

99. $\frac{1}{2} \log_7 u + \frac{1}{2} \log_7 v - \log_7 x$ **101.** 0.8451   **103.** -0.2218 **105.** 1.3863   **107.** 1.0415

109. 0.8617   **111.** 2.1133   **113.** -0.6309 **115.** 0.2727   **117.** 1.6826   **119.** 1.9266

121. 0.6752   **123.** 2.6125

125.   **a.** False   **b.** True   **c.** False   **d.** False   **e.** False   **f.** True   **g.** False

     h. True

SECTION 12.3   *pages 625–626*

1. 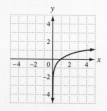   **3.** 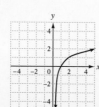   **5.** 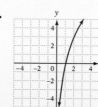   **7.**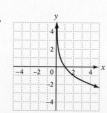

9. 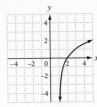   **11.** 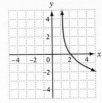   **13.** 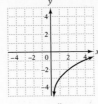   **15.**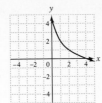

17. 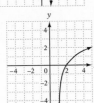   **19.** 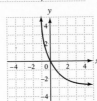   **21.**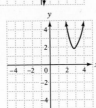

23. **a.**

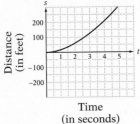

Distance from Earth
(in parsecs)

b. The point (25.1, 2) means that a star that is 25.1 parsecs from the earth has a distance modulus of 2.

SECTION 12.4   *pages 631–632*

1. $-\dfrac{1}{3}$ **3.** -3 **5.** 1 **7.** 6 **9.** $\dfrac{4}{7}$ **11.** $\dfrac{9}{8}$ **13.** 1.1133 **15.** 1.0986 **17.** 1.3222

19. -2.8074 **21.** 3.5850 **23.** 1.1309 **25.** $\dfrac{11}{2}$ **27.** -4 and 2 **29.** $\dfrac{5}{3}$ **31.** $\dfrac{1}{2}$

33. $\dfrac{9}{2}$ **35.** 17.5327 **37.** 3 **39.** 2 **41.** The equation has no solution.   **43.** -0.55

45. 1.38 **47.** The equation has no solution.   **49.** 0.09

51. **a.**

Distance
(in feet)

Time
(in seconds)

b. It will take the object approximately 2.64 s to fall 100 ft.

SECTION 12.5   *pages 637–640*

1. The value of the investment is $5982 after 2 years.
3. The investment should double in 7 years.
5. The company must deposit $16,786 in the account.
7. It will take 3.5 h to decay to 20 mg.   **9.** The half-life of the material is 2.5 years.
11. It will take 6 weeks of practice to make 80% of the welds correctly.
13. The pH of the solution is 1.35.   **15.** 79% of the light will pass through the glass.
17. The earthquake in China was 20 times stronger.
19. The star is 150.7 parsecs from the earth.
21. No, Antares is 3.5 times farther.   **23.** There are 65 decibels in normal conversation.
25. 742.9 billion barrels will last 20 years.

27. The value of the investment will be $3210.06 in 5 years.   **29.** $\dfrac{e^{T/14.29} - 1}{0.00411}$

31. $r_1 = \sqrt{10 r_2}$

CHAPTER REVIEW EXERCISES   *pages 643–644*

1. 1 [12.1A]   **2.** $\frac{1}{3}$ [12.1A]   **3.** [12.1B]

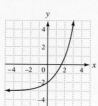

4. [12.1B]   **5.** 2 [12.2A]   **6.** $\frac{1}{9}$ [12.2A]

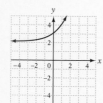

7. [12.3A]   **8.** [12.3A]   **9.** $-\frac{1}{2}\log_6 x + \frac{3}{2}\log_6 y$ [12.2B]

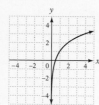
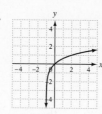

10. $\log_3 \sqrt{\dfrac{x}{y}}$ [12.2B]   **11.** $\ln x - \frac{1}{2}\ln z$ [12.2B]   **12.** $\ln \dfrac{x^3}{y\sqrt{z}}$ [12.2B]   **13.** -2 [12.4A]

14. -3 [12.4A]   **15.** 2.5789 [12.4A]   **16.** 6 [12.4B]   **17.** 3 [12.4B]

18. 1.3652 [12.2C]   **19.** 2.6801 [12.2C]   **20.** The half-life is 33 h. [12.5A]

CUMULATIVE REVIEW EXERCISES   *pages 645–646*

1. $\frac{8}{7}$ [2.3C]   **2.** $y = 2x - 6$ [4.5A]   **3.** $(4x^n + 3)(x^n + 1)$ [7.3A/7.3B]   **4.** $\dfrac{x-3}{x+3}$ [8.3A]

5. $\dfrac{x\sqrt{y} + y\sqrt{x}}{x - y}$ [9.2D]   **6.** $2 + \sqrt{10}$ and $2 - \sqrt{10}$ [10.2A]

7. The length of the side is 26 cm. [9.4B]   **8.** [5.6A]   **9.** $(0, -1, 2)$ [5.3B]

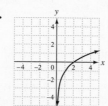

10. $-\dfrac{2x^2 - 17x + 13}{(x-2)(2x-3)}$ [8.2D]   **11.** $\{x \mid -5 \le x \le 1\}$ [10.6A]   **12.** $\{x \mid 1 \le x \le 4\}$ [2.6B]

13. [12.1B]   **14.** [12.3A]   **15.** The area is 28 yd². [3.2B]

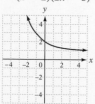

16. 125 [12.2A]   **17.** $\log_b \dfrac{x^3}{y^5}$ [12.2B]   **18.** 1.7712 [12.2C]   **19.** 2 [12.4A]

20. $\frac{1}{2}$ [12.4B]   **21.** The customer can write fewer than 50 checks. [2.5C]

22. The mixture costs \$3.10 per pound. [2.4A]   **23.** The rate of the wind is 25 mph. [5.5A]

24. The force will stretch the spring 10.2 in. [8.7A]

25. The redwood costs $.40 per foot. The fir costs $.25 per foot.   [5.5B]
26. The investment should double in 8 years.   [12.5A]

FINAL EXAM   *pages 647–650*

1. -31 [1.3A]   **2.** -1 [1.4A]   **3.** $-10x + 33$ [1.4D]   **4.** 8 [2.3A]

5. 4 and $-\dfrac{2}{3}$ [2.6A]   **6.** The volume is 268.1 ft^3.   [3.3A]

7. 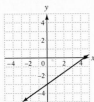   [4.2B]   **8.** $y = -3x + 7$ [4.4B]   **9.** $y = -\dfrac{2}{3}x - \dfrac{1}{3}$ [4.5A]

10. $6a^3 - 5a^2 + 10a$ [6.3A]   **11.** $(2 - xy)(4 + 2xy + x^2y^2)$ [7.4B]

12. $(x - y)(1 + x)(1 - x)$ [7.4D]   **13.** $x^2 - 2x - 3 - \dfrac{5}{2x - 3}$ [6.4A]   **14.** $\dfrac{x(x - 1)}{2x - 5}$ [8.1C]

15. $\dfrac{-10x}{(x + 2)(x - 3)}$ [8.2D]   **16.** $\dfrac{x + 3}{x + 1}$ [8.3A]   **17.** $-\dfrac{7}{4}$ [8.4A]   **18.** $d = \dfrac{a_n - a_1}{n - 1}$ [8.5A]

19. $\dfrac{y^4}{162x^3}$ [6.1B]   **20.** $\dfrac{1}{64x^8y^5}$ [9.1A]   **21.** $-2x^2y\sqrt{2y}$ [9.2B]   **22.** $\dfrac{x^2\sqrt{2y}}{2y^2}$ [9.2D]

23. $\dfrac{6}{5} - \dfrac{3}{5}i$ [9.3D]   **24.** $2x^2 - 3x - 2 = 0$ [10.1B]   **25.** $\dfrac{3 + \sqrt{17}}{4}$ and $\dfrac{3 - \sqrt{17}}{4}$ [10.3A]

26. -8 and 27 [10.4A]   **27.** 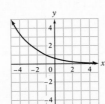   [11.2A]   **28.** 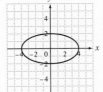   [11.6C]

29. -2 and $\dfrac{3}{2}$ [10.4C]   **30.** $f^{-1}(x) = \dfrac{3}{2}x + 6$ [11.5B]   **31.** $(3, 4)$ [5.3A]   **32.** 10 [5.4A]

33. $\left\{x \mid x > \dfrac{3}{2}\right\}$ [2.5B]   **34.** $\{x \mid -4 < x < -1\}$ [2.6B]

35. 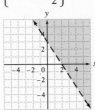   [4.6A]   **36.** 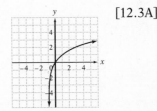   [12.1B]   **37.** 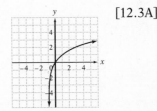   [12.3A]

38. $\log_2 \dfrac{a^2}{b^2}$ [12.2B]   **39.** 6 [12.4B]   **40.** The range of scores is $69 \le x \le 100$.   [2.5C]

41. The cyclist rode 40 mi.   [2.4D]
42. There is $8000 invested at 8.5% and $4000 invested at 6.4%.   [2.4C]
43. The length is 20 ft. The width is 7 ft.   [10.5A]
44. An additional 200 shares are needed.   [8.4B]
45. The rate of the plane was 420 mph.   [8.6B]
46. The object has fallen 88 ft when the speed reaches 75 ft/s.   [9.4B]
47. The rate of the plane for the first 360 mi was 120 mph.   [8.6B]
48. The intensity is 200 lumens.   [8.7A]
49. The rate of the boat is 12.5 mph. The rate of the current is 2.5 mph.   [5.5A]
50. The value of the investment after 2 years will be $4785.65.   [12.5A]

APPENDIX: Guidelines for Using Graphing Calculators

Texas Instruments *TI-82*

To evaluate an expression

a. Press the $\boxed{\text{Y=}}$ key. A menu showing y_1 through y_8 will be displayed vertically with the cursor on y_1. Press $\boxed{\text{CLEAR}}$, if necessary, to delete an unwanted expression.

b. Input the expression to be evaluated. For example, to input the expression $-3a^2b - 4c$, use the following keystrokes:

$\boxed{\text{Y=}}$ $\boxed{\text{CLEAR}}$ $\boxed{(-)}$ 3 $\boxed{\text{ALPHA}}$ A $\boxed{\wedge}$ 2 $\boxed{\text{ALPHA}}$ B $\boxed{-}$ 4 $\boxed{\text{ALPHA}}$ C $\boxed{\text{2nd}}$ QUIT

Note the difference between the keys for a *negative* sign $\boxed{(-)}$ and a *minus* sign $\boxed{-}$.

c. Store the value of each variable that will be used in the expression. For example, to evaluate the expression above when $a = 3$, $b = -2$, and $c = -4$, use the following keystrokes:

3 $\boxed{\text{STO}\triangleright}$ $\boxed{\text{ALPHA}}$ A $\boxed{\text{ENTER}}$ $\boxed{(-)}$ 2 $\boxed{\text{STO}\triangleright}$ $\boxed{\text{ALPHA}}$ B $\boxed{\text{ENTER}}$ $\boxed{(-)}$ 4 $\boxed{\text{STO}\triangleright}$ $\boxed{\text{ALPHA}}$ C $\boxed{\text{ENTER}}$

These steps store the value of each variable.

d. Press $\boxed{\text{2nd}}$ Y-VARS $\boxed{1}$ $\boxed{1}$ $\boxed{\text{ENTER}}$. The value of the expression, y_1, for the given values is displayed, in this case, $y_1 = 70$.

To graph a function

a. Press the $\boxed{\text{Y=}}$ key. A menu showing y_1 through y_8 will be displayed vertically with the cursor on y_1. Press $\boxed{\text{CLEAR}}$, if necessary, to delete an unwanted expression.

b. Input the expression for each function that is to be graphed. Press $\boxed{\text{X,T,}\theta}$ to input x. For example, to input $f(x) = x^3 + 2x^2 - 5x - 6$, use the following keystrokes:

$\boxed{\text{Y=}}$ $\boxed{\text{X,T,}\theta}$ $\boxed{\wedge}$ 3 $\boxed{+}$ 2 $\boxed{\text{X,T,}\theta}$ $\boxed{\wedge}$ 2 $\boxed{-}$ 5 $\boxed{\text{X,T,}\theta}$ $\boxed{-}$ 6

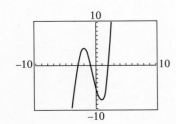

c. Set the domain and range by pressing $\boxed{\text{WINDOW}}$. Enter the values for the minimum x-value (Xmin), the maximum x-value (Xmax), the distance between tick marks on the x-axis (Xscl), the minimum y-value (Ymin), the maximum y-value (Ymax), and the distance between tick marks on the y-axis (Yscl). Now press $\boxed{\text{GRAPH}}$. For the graph shown at the left, Xmin $= -10$, Xmax $= 10$, Xscl $= 1$, Ymin $= -10$, Ymax $= 10$, and Yscl $= 1$. This is called the standard viewing rectangle. Pressing $\boxed{\text{ZOOM}}$ $\boxed{6}$ is a quick way to set the calculator to the standard viewing rectangle. *Note:* This will also immediately graph the function in that window.

d. Press the $\boxed{\text{Y=}}$ key. The equal sign has a black rectangle around it. This indicates that the function is *active* and will be graphed when the $\boxed{\text{GRAPH}}$ key is pressed. A function is deactivated by using the arrow keys. Move the cursor over the equal sign and press $\boxed{\text{ENTER}}$. When the cursor is moved to the right, the black rectangle will not be present and that equation will not be active.

e. Graphing some radical equations requires special care. To graph the function $f(x) = x^{2/3} - 1$, enter the following keystrokes:

$\boxed{\text{Y=}}$ $\boxed{(}$ $\boxed{\text{X,T,}\theta}$ $\boxed{\wedge}$ $\boxed{(}$ 1 $\boxed{\div}$ 3 $\boxed{)}$ $\boxed{)}$ $\boxed{\wedge}$ 2 $\boxed{-}$ 1

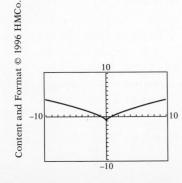

You are entering $x^{2/3}$ as $(x^{1/3})^2$. The graph is shown at the left.

To display the *x*-coordinates of rectangular coordinates as integers

a. Set the viewing window as follows: Xmin $= -47$, Xmax $= 47$, Xscl $= 10$, Ymin $= -32$, Ymax $= 32$, Yscl $= 10$.

b. Graph the function and use the TRACE feature. Press ⌈TRACE⌉ and then move the cursor with the ⌈◁⌉ and ⌈▷⌉ keys. The values of *x* and $y = f(x)$ displayed on the bottom of the screen are the coordinates of a point on the graph.

To display the *x*-coordinates of rectangular coordinates in tenths

a. Set the viewing window as follows: ⌈ZOOM⌉ ⌈4⌉

b. Graph the function and use the TRACE feature. Press ⌈TRACE⌉ and then move the cursor with the ⌈◁⌉ and ⌈▷⌉ keys. The values of *x* and $y = f(x)$ displayed on the bottom of the screen are the coordinates of a point on the graph.

To evaluate a function for a given value of *x*, or to produce a pair of rectangular coordinates

a. Input the equation; for example, input $y_1 = 2x^3 - 3x + 2$.

b. Press ⌈2nd⌉QUIT.

c. Input a value for *x*; for example, to input 3 press 3 ⌈STO▷⌉ ⌈X,T,θ⌉ ⌈ENTER⌉.

d. Press ⌈2nd⌉Y-VARS⌈1⌉⌈1⌉ ⌈ENTER⌉. The value of the function, y_1, for the given *x*-value is shown; in this case, $y_1 = 47$. An ordered pair of the function is (3, 47).

e. Repeat steps (c)–(d) to produce as many pairs as desired. The TABLE feature of the *TI-82* can also be used to determine ordered pairs.

To graph the sum (or difference) of two functions

a. Input the first function as y_1 and the second function as y_2.

b. After the second function is entered, press ⌈ENTER⌉ to move the cursor to y_3.

c. Press ⌈2nd⌉Y-VARS⌈1⌉⌈1⌉⌈+⌉⌈2ND⌉Y-VARS⌈1⌉⌈2⌉. (Enter ⌈−⌉ instead of ⌈+⌉ to graph $y_1 - y_2$.)

d. Graph y_3. If you want to show only the graph of the sum (or difference), be sure that the black rectangle around the equal sign has been removed from y_1 and y_2.

Zoom Features of the *TI-82*

To zoom in or out on a graph

a. Here are two methods of using ZOOM. The first method uses the built-in features of the calculator. Move the cursor to a point on the graph that is of interest. Press ⌈ZOOM⌉. The ZOOM menu will appear. Press ⌈2⌉ ⌈ENTER⌉ to zoom in on the graph by the amount shown under the SET FACTORS menu. The center of the new graph is the location at which you placed the cursor. Press ⌈3⌉ ⌈ENTER⌉ to zoom out on the graph by the amount under the SET FACTORS menu. (The SET FACTORS menu is accessed by pressing ⌈ZOOM⌉ ⌈▷⌉ ⌈4⌉.)

b. The second method uses the ZBOX option under the ZOOM menu. To use this method, press ⌈ZOOM⌉ ⌈1⌉. A cursor will appear on the graph. Use the arrow keys to move the cursor to a portion of the graph that is of interest. Press ⌈ENTER⌉. Now use the arrow keys to draw a box around the portion of the graph you wish to see. Press ⌈ENTER⌉. The portion of the graph defined by the box will be drawn.

c. Pressing ⌈ZOOM⌉ ⌈6⌉ resets the window to the standard 10 × 10 viewing window.

Solving Equations with the *TI-82*

This discussion is based on the fact that the solution of an equation can be related to the x-intercepts of a graph. For instance, the solutions of the equation $x^3 = x + 1$ are the x-intercepts of the graph of $f(x) = x^3 - x - 1$, which are the zeros of f.

To solve $x^3 = x + 1$, rewrite the equation with all terms on one side. The equation is now $x^3 - x - 1 = 0$. Think of this equation as $y_1 = x^3 - x - 1$. The x-intercepts of the graph of y_1 are the solutions of the equation $x^3 = x + 1$.

a. Enter the function into y_1.

b. Graph the equation. You may need to adjust the viewing window so that the x-intercept is visible.

c. Press $\boxed{\text{2ND}}$ $\boxed{\text{CALC}}$ $\boxed{2}$.

d. Move the cursor to a point on the curve that is to the left of the x-intercept. Press $\boxed{\text{ENTER}}$.

e. Move the cursor to a point on the curve that is to the right of the x-intercept. Press $\boxed{\text{ENTER}}$.

f. Press $\boxed{\text{ENTER}}$.

g. The root is shown as the x-coordinate on the bottom of the screen; in this case, the root is approximately 1.324718. The SOLVE feature under the MATH menu can also be used to find solutions of equations.

Solving Systems of Equations in Two Variables with the *TI-82*

To solve a system of equations

a. Solve each equation for y.

b. Enter the first equation as y_1. For instance, let $y_1 = x^2 - 1$.

c. Enter the second equation as y_2. For instance, let $y_2 = 1 - x$.

d. Graph both equations. (*Note:* The points of intersection must appear on the screen. It may be necessary to adjust the viewing window so that the point(s) of intersection are displayed.)

e. Press $\boxed{\text{2nd}}$ $\boxed{\text{CALC}}$ $\boxed{5}$.

f. Move the cursor to the left of the first point of intersection. Press $\boxed{\text{ENTER}}$.

g. Move the cursor to the right of the first point of intersection. Press $\boxed{\text{ENTER}}$.

h. Press $\boxed{\text{ENTER}}$.

i. The first point of intersection is $(-2, 3)$.

j. Repeat this procedure for each point of intersection.

Finding Minimum or Maximum Values of a Function with the *TI-82*

a. Enter the function into y_1. The equation $y_1 = x^3 - x - 1$ is used here.

b. Graph the equation. You may need to adjust the viewing window so that the maximum or minimum points are visible.

c. Press $\boxed{\text{2nd}}$ $\boxed{\text{CALC}}$ $\boxed{3}$ to determine a minimum value or press $\boxed{\text{2nd}}$ $\boxed{\text{CALC}}$ $\boxed{4}$ to determine a maximum value.

d. Move the cursor to a point on the curve that is to the left of the minimum (maximum). Press $\boxed{\text{ENTER}}$.

e. Move the cursor to a point on the curve that is to the right of the minimum (maximum). Press ENTER.

f. Press ENTER.

g. The minimum (maximum) is shown as the y-coordinate on the bottom of the screen; in this case the minimum value is -1.3849 and the maximum value is -0.6150998.

CASIO *fx-7700GB*

To evaluate an expression

a. For example, to input the expression $-3a^2b - 4c$, use the keystrokes shown below. Note the difference between the keys for a *negative* sign $(-)$ and a *minus* sign $\boxed{-}$. To enter $(-)$, press $\boxed{\text{SHIFT}}$ $(-)$.

$\boxed{\text{SHIFT}}$ $(-)$ 3 $\boxed{\text{ALPHA}}$ A $\boxed{x^y}$ 2 $\boxed{\text{ALPHA}}$ B $\boxed{-}$ 4 $\boxed{\text{ALPHA}}$ C $\boxed{\text{SHIFT}}$ $\boxed{\text{F}}$MEM $\boxed{\text{F1}}$ $\boxed{1}$ $\boxed{\text{EXE}}$

The number 1 entered here can be any number from 1 to 6.

b. Store the value of each variable that will be used in the expression. For example, to evaluate the expression above when $a = 3$, $b = -2$, and $c = -4$, use the following keystrokes:

3 $\boxed{\rightarrow}$ $\boxed{\text{ALPHA}}$ A $\boxed{\text{EXE}}$ $\boxed{\text{SHIFT}}$ $(-)$ 2 $\boxed{\rightarrow}$ $\boxed{\text{ALPHA}}$ B $\boxed{\text{EXE}}$ $\boxed{\text{SHIFT}}$ $(-)$ 4 $\boxed{\rightarrow}$ $\boxed{\text{ALPHA}}$ C $\boxed{\text{EXE}}$

These steps store the values of each variable.

c. To evaluate the expression, recall the expression from the function menu. Press $\boxed{\text{SHIFT}}$ $\boxed{\text{F}}$MEM $\boxed{\text{F2}}$ $\boxed{1}$ $\boxed{\text{EXE}}$. The value of the expression is displayed as 70.

To graph a function

a. Ensure the calculator is in graphics mode. Press $\boxed{\text{MODE}}$ $\boxed{1}$ $\boxed{\text{MODE}}$ $\boxed{+}$ $\boxed{\text{MODE}}$ $\boxed{\text{SHIFT}}$ $\boxed{+}$.

b. To graph $f(x) = x^3 + 2x^2 - 5x - 6$, use the following keystrokes:

$\boxed{\text{Graph}}$ $\boxed{\text{X,}\theta\text{,T}}$ $\boxed{x^y}$ 3 $\boxed{+}$ 2 $\boxed{\text{X,}\theta\text{,T}}$ $\boxed{\text{SHIFT}}$ $\boxed{x^2}$ $\boxed{-}$ 5 $\boxed{\text{X,}\theta\text{,T}}$ $\boxed{-}$ $\boxed{6}$ $\boxed{\text{EXE}}$

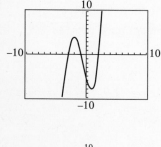

c. Set the domain and range by pressing $\boxed{\text{Range}}$. Enter the values for the minimum x-value (Xmin), the maximum x-value (Xmax), the distance between tick marks on the x-axis (Xscl), the minimum y-value (Ymin), the maximum y-value (Ymax), and the distance between tick marks on the y-axis (Yscl). Press $\boxed{\text{EXE}}$ after each entry. For the graph shown at the left, Xmin $= -10$, Xmax $= 10$, Xscl $= 1$, Ymin $= -10$, Ymax $= 10$, and Yscl $= 1$. Press the $\boxed{\text{Range}}$ key until you return to the display of the graph. Press $\boxed{\text{EXE}}$.

d. Graphing some radical equations requires special care. To graph the function $f(x) = x^{2/3} - 1$, enter the following keystrokes:

$\boxed{\text{Graph}}$ $\boxed{(}$ $\boxed{\text{X,}\theta\text{,T}}$ $\boxed{x^y}$ $\boxed{(}$ 1 $\boxed{\div}$ 3 $\boxed{)}$ $\boxed{)}$ $\boxed{x^y}$ 2 $\boxed{-}$ 1 $\boxed{\text{EXE}}$

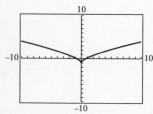

You are entering $x^{2/3}$ as $(x^{1/3})^2$. The graph is shown at the left.

e. If you graph a function and then return to the regular screen (for example, by pressing $\boxed{\text{AC}}$), you can view the graph again by pressing $\boxed{\text{G}\leftrightarrow\text{T}}$. To clear a graph, press $\boxed{\text{SHIFT}}$ $\boxed{\text{F5}}$ $\boxed{\text{EXE}}$.

To display the x-coordinates of rectangular coordinates as integers

a. Set the viewing window as follows: Xmin $= -47$, Xmax $= 47$, Xscl $= 10$, Ymin $= -31$, Ymax $= 31$, Yscl $= 10$.

b. Graph the function and use the TRACE feature. Press $\boxed{\text{F1}}$ and then use the cursor keys to move along the graph. The values of x and $y = f(x)$ displayed on the bottom of the screen are the coordinates of a point on the graph.

To display the *x*-coordinates of rectangular coordinates in tenths

a. Set the viewing window as follows: $\boxed{\text{Range}}$ $\boxed{\text{F1}}$ $\boxed{\text{Range}}$ $\boxed{\text{Range}}$

b. Graph the function and use the TRACE feature. Press $\boxed{\text{F1}}$ and then use the cursor keys to move along the graph.

To produce a pair of rectangular coordinates

a. Input the function; for example, input $3x - 4$.

b. Press $\boxed{\text{SHIFT}}$ $\boxed{\text{F}\,\text{MEM}}$ $\boxed{\text{F1}}$ $\boxed{1}$ $\boxed{\text{AC}}$ to store the function in f_1.

c. Input any value for x; for example, input 3 using the keystrokes 3 $\boxed{\rightarrow}$ $\boxed{\text{X},\theta,\text{T}}$ $\boxed{\text{EXE}}$ to store the value in x.

d. Press $\boxed{\text{SHIFT}}$ $\boxed{\text{F}\,\text{MEM}}$ $\boxed{\text{F2}}$ $\boxed{1}$ $\boxed{\text{EXE}}$ to find the corresponding function value for the stored x-value; in this example you should get 5. The point is (3, 5).

e. Repeat steps (c)–(d) to produce as many pairs as desired.

To graph the sum (or difference) of two functions

a. Input the first function.

b. Press $\boxed{\text{SHIFT}}$ $\boxed{\text{F}\,\text{MEM}}$ $\boxed{\text{F1}}$ $\boxed{1}$ $\boxed{\text{AC}}$.

c. Input the second function.

d. Press $\boxed{\text{SHIFT}}$ $\boxed{\text{F}\,\text{MEM}}$ $\boxed{\text{F1}}$ $\boxed{2}$ $\boxed{\text{AC}}$.

e. Press $\boxed{\text{Graph}}$ $\boxed{\text{SHIFT}}$ $\boxed{\text{F}\,\text{MEM}}$ $\boxed{\text{F3}}$ $\boxed{1}$ $\boxed{+}$ $\boxed{\text{F3}}$ $\boxed{2}$ $\boxed{\text{EXE}}$. (Enter $\boxed{-}$ instead of $\boxed{+}$ to graph the difference of the two functions.)

Zoom Features of the CASIO *fx-7700GB*

To zoom in or out on a graph

a. Here are two methods of using ZOOM. The first method uses the built-in features of the calculator. Press $\boxed{\text{F1}}$. This activates the cursor. Use the arrow keys to move the cursor to a portion of the graph that is of interest. Press $\boxed{\text{F2}}$. The ZOOM menu will appear. Press $\boxed{\text{F3}}$ to zoom in on the graph by the amount shown in FCT (factor). The center of the new graph is the location at which you placed the cursor. Press $\boxed{\text{F2}}$ followed by $\boxed{\text{F4}}$ to zoom out on the graph by the amount shown in FCT.

b. The second method uses the BOX option under the ZOOM menu. If necessary, press $\boxed{\text{G}\leftrightarrow\text{T}}$ to view the graph. Press $\boxed{\text{F2}}$. Press $\boxed{\text{F1}}$ to select BOX. Use the arrow keys to move the cursor to a portion of the graph that is of interest. Press $\boxed{\text{EXE}}$. Now use the arrow keys to draw a box around the portion of the graph you wish to see. Press $\boxed{\text{EXE}}$. The portion of the graph defined by the box will be drawn.

Solving Equations with the CASIO *fx-7700GB*

This discussion is based on the fact that the solution of an equation can be related to the x-intercepts of a graph. For instance, the solutions of the equation $x^2 - x = 1$ are the x-intercepts of the graph of $f(x) = x^2 - x - 1$, which are the zeros of f.

To solve $x^3 = x + 1$, rewrite the equation with all terms on one side. The equation is now $x^3 - x - 1 = 0$. Think of this equation as $y_1 = x^3 - x - 1$. The x-intercepts of the graph of y_1 are the solutions of the equation $x^3 = x + 1$.

a. Press $\boxed{\texttt{Graph}}$, enter the function, and press $\boxed{\texttt{EXE}}$. You may need to adjust the viewing window so that the x-intercept is visible. Note that if you adjust the viewing window after graphing the function, the equation must be reentered.

b. Press $\boxed{\texttt{F1}}$ and move the cursor to the approximate x-intercept.

c. Press $\boxed{\texttt{F2}}$ $\boxed{\texttt{F3}}$.

d. Repeat steps (c) and (d) until you can approximate the x-intercept to the desired degree of accuracy. The root is about 1.324718.

Solving Systems of Equations in Two Variables with the CASIO *fx-7700GB*

To solve a system of equations

a. Solve each equation for y.

b. Enter the first equation as f_1. For instance, let $y = x^2 - 1$.

c. Enter the second equation as f_2. For instance, let $y = 1 - x$.

d. Graph both equations. (*Note:* The points of intersection must appear on the screen. It may be necessary to adjust the viewing window so that the point(s) of intersection are displayed.)

e. Press $\boxed{\texttt{F1}}$ and then move the cursor to an approximate intersection point.

f. Press $\boxed{\texttt{F2}}$ $\boxed{\texttt{F3}}$.

g. Repeat steps (e) and (f) until you can approximate the point of intersection to the desired degree of accuracy. The first point of intersection is $(-2, 3)$.

Finding Minimum or Maximum Values of a Function with the CASIO *fx-7700GB*

To find the minimum (maximum) value of a function

a. Graph the function as the equation $y = x^3 - x - 1$.

b. Press $\boxed{\texttt{F1}}$ and then move the cursor to the approximate minimum (maximum).

c. Press $\boxed{\texttt{F2}}$ $\boxed{\texttt{F3}}$.

d. Repeat steps (b) and (c) until you can approximate the minimum (maximum) value to the desired degree of accuracy. The minimum (maximum) value is shown as the y-coordinate on the bottom of the screen; in this case the minimum value is -1.3849 and the maximum value is -0.6150998.

SHARP EL-9300

To evaluate an expression

a. The SOLVER mode of the calculator is used to evaluate expressions. To enter SOLVER mode, press $\boxed{\texttt{2ndF}}$ $\texttt{SOLVER}$ $\boxed{\texttt{CL}}$. The expression $-3a^2b - 4c$ must be entered as the equation $-3a^2b - 4c = t$. The letter t can be any letter other than one used in the expression. When entering an expression in SOLVER mode, the variables appear on the screen in lower case. Use the following keystrokes to input $-3a^2b - 4c = t$:

$\boxed{\texttt{(−)}}$ 3 $\boxed{\texttt{ALPHA}}$ $\boxed{\texttt{A}}$ $\boxed{a^b}$ 2 $\boxed{\triangleright}$ $\boxed{\texttt{ALPHA}}$ $\boxed{\texttt{B}}$ $\boxed{-}$ 4 $\boxed{\texttt{ALPHA}}$ $\texttt{C}$ $\boxed{\texttt{ALPHA}}$ $\boxed{=}$ $\boxed{\texttt{ALPHA}}$ $\boxed{\texttt{T}}$ $\boxed{\texttt{ENTER}}$

Note the difference between the keys for a *negative* sign $\boxed{\texttt{(−)}}$ and a *minus* sign $\boxed{-}$.

b. After you press ENTER, the variables used in the equation will be displayed on the screen. To evaluate the expression for $a = 3$, $b = -2$, and $c = -4$, input each value, pressing ENTER after each number. When the cursor moves to t, press ENTER. A small window will appear. Press ENTER. In this case you will see "$t = 70$" on the screen.

c. Pressing ENTER again will allow you to evaluate the expression for new values of a, b, and c. Press 🔳 to return to normal operation.

To graph a function

a. Press the ▱ key. The screen will show Y1= .

b. Input the expression for a function that is to be graphed. Press X/θ/T to enter x. For example, to input $f(x) = x^3 + 2x^2 - 5x - 6$, use the following keystrokes:

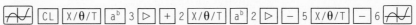

c. Set the viewing window by pressing RANGE. Enter the values for the minimum x-value (Xmin), the maximum x-value (Xmax), the distance between tick marks on the x-axis (Xscl), minimum y-value (Ymin), the maximum y-value (Ymax), and the distance between tick marks on the y-axis (Yscl). Press ENTER after each entry. Press ▱. For the graph shown at the left, enter Xmin = −10, Xmax = 10, Ymin = −10, and Ymax = 10. Press ▱.

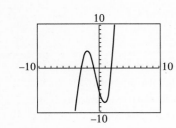

d. Press EQTN to return to the equation. The equal sign has a black rectangle around it. This indicates that the function is *active* and will be graphed when the ▱ key is pressed. A function is deactivated by using the arrow keys. Move the cursor over the equal sign and press ENTER. When the cursor is moved to the right, the black rectangle will not be present and that equation will not be active.

e. Graphing some radical equations requires special care. To graph the function $f(x) = x^{2/3} - 1$, enter the following keystrokes:

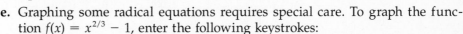

You are entering the exponent in the form $(x^{1/3})^2$. The graph is shown at the left.

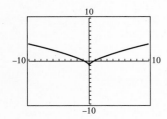

To display the x-coordinates of rectangular coordinates as integers

a. Set the viewing window as follows: Xmin = −47, Xmax = 47, Xscl = 10, Ymin = −31, Ymax = 31, Yscl = 10.

b. Graph the function and use the right and left arrow keys to trace the function. The values of x and $y = f(x)$ displayed on the bottom of the screen are the coordinates of a point on the graph.

To display the x-coordinates of rectangular coordinates in tenths

a. Set the viewing window as follows: Xmin = −4.7, Xmax = 4.7, Xscl = 1, Ymin = −3.1, Ymax = 3.1, Yscl = 1. This is accomplished by pressing RANGE MENU A ENTER RANGE .

b. Graph the function and use the arrow keys to move along the graph of the function. The coordinates are displayed at the bottom of the screen.

To evaluate a function for a given value of x, or to produce a pair of rectangular coordinates

a. Enter SOLVER mode. Press 2ndF SOLVER CL .

b. Input the expression; for instance, input $x^3 - 4x^2 + 1 = y$. Press ENTER .

c. Move the cursor to the x variable (if it is not already there). Input any value for x; for example, input 3 and then press ENTER . The cursor will now be over y. Press ENTER twice to evaluate the function. In this case, the value is -8. An ordered pair of the function is $(3, -8)$.

d. Repeat step (c) to produce as many pairs as desired.

To graph the sum (or difference) of two functions

a. Press $\boxed{\sim\!\!\!\!\vee\!\!\!\!}$. Input the first function as y_1, press ENTER, and input the second function as y_2. Press ENTER to move to y_3. ·

b. Press MATH E 1 + MATH E 2 $\boxed{\sim\!\!\!\!\vee\!\!\!\!}$. (To graph the difference $y_1 - y_2$, replace + with −.)

c. If you want to show only the graph of the sum (or difference), be sure that the black rectangle around the equal sign has been removed from y_1 and y_2. To return to the first equation, press MENU A 1. To return to the second equation, press MENU A 2.

Zoom Features of the SHARP EL-9300

To zoom in or out on a graph

a. Here are two methods of using ZOOM. The first method uses the built-in features of the calculator. Move the cursor to a point on the graph that is of interest. Press ZOOM. The ZOOM menu will appear. Press 2 to zoom in on the graph by the amount shown by FACTOR. The center of the new graph is the location at which you placed the cursor. Press ZOOM 3 to zoom out on the graph by the amount shown in FACTOR.

b. The second method uses the BOX option under the ZOOM menu. To use this method, press ZOOM 1. A cursor will appear on the graph. Use the arrow keys to move the cursor to a portion of the graph that is of interest. Press ENTER. Use the arrow keys to draw a box around the portion of the graph you wish to see. Press ENTER.

Solving Equations or Systems of Equations in Two Variables with the SHARP EL-9300

a. The x-intercept, y-intercept, and the point of intersection of the two graphs can be determined by using the JUMP command. Graph the functions of interest. Using the arrow keys, place the cursor on the graph of one of the functions. Press 2ndF JUMP; the JUMP menu will appear. Press 1 to jump to the intersection of two graphs, press 4 to jump to the x-intercept, or press 5 to jump to the y-intercept. If there is more than one intercept or intersection, pressing 2ndF JUMP again will allow you to find the remaining points. *Important:* The intersection must be a point in the viewing window.

Finding Maximum and Minimum Values of a Function with the SHARP EL-9300

a. The maximum and minimum values of a function can be determined by using the JUMP command. Graph the function of interest. Press 2ndF JUMP; the JUMP menu will appear. Press 2 to jump to the minimum value or press 3 to jump to the maximum value of the function. If there is more than one minimum or maximum, pressing 2ndF JUMP again will allow you to find the remaining points. *Important:* The minimum or maximum must be a point in the viewing window.

b. The minimum (maximum) value is shown as the y-coordinate on the bottom of the screen. For the equation $y_1 = x^3 - x - 1$, the minimum value is -1.3849 and the maximum value is -0.615099.

Table of Symbols

+	add
−	subtract
·, ×, $(a)(b)$	multiply
$\dfrac{a}{b}$, ÷	divide
( )	parentheses, a grouping symbol
[ ]	brackets, a grouping symbol
π	pi, a number approximately equal to $\dfrac{22}{7}$ or 3.14
$-a$	the opposite, or additive inverse, of a
$\dfrac{1}{a}$	the reciprocal, or multiplicative inverse, of a
=	is equal to
≈	is approximately equal to
≠	is not equal to

<	is less than
≤	is less than or equal to
>	is greater than
≥	is greater than or equal to
(a, b)	an ordered pair whose first component is a and whose second component is b
°	degree (for angles)
$\sqrt{a}$	the principal square root of a
∅, { }	the empty set
$|a|$	the absolute value of a
∪	union of two sets
∩	intersection of two sets
∈	is an element of (for sets)
∉	is not an element of (for sets)

Table of Measurement Abbreviations

U.S. Customary System

Length		Capacity		Weight		Area	
in.	inches	oz	ounces	oz	ounces	in^2	square inches
ft	feet	c	cups	lb	pounds	ft^2	square feet
yd	yards	qt	quarts				
mi	miles	gal	gallons				

Metric System

Length		Capacity		Weight/Mass		Area	
mm	millimeter (0.001 m)	ml	milliliter (0.001 L)	mg	milligram (0.001 g)	cm^2	square centimeters
cm	centimeter (0.01 m)	cl	centiliter (0.01 L)	cg	centigram (0.01 g)	m^2	square meters
dm	decimeter (0.1 m)	dl	deciliter (0.1 L)	dg	decigram (0.1 g)		
m	meter	L	liter	g	gram		
dam	decameter (10 m)	dal	decaliter (10 L)	dag	decagram (10 g)		
hm	hectometer (100 m)	hl	hectoliter (100 L)	hg	hectogram (100 g)		
km	kilometer (1000 m)	kl	kiloliter (1000 L)	kg	kilogram (1000 g)		

Time

h	hours	min	minutes	s	seconds

Proofs of Logarithmic Properties

In each of the following proofs of logarithmic properties, it is assumed that the Properties of Exponents are true for all real number exponents.

The Logarithm Property of the Product of Two Numbers

For any positive real numbers x, y, and b, $b \neq 1$, $\log_b xy = \log_b x + \log_b y$.

Proof: Let $\log_b x = m$ and $\log_b y = n$.

Write each equation in its equivalent exponential form. $x = b^m \qquad y = b^n$

Use substitution and the Properties of Exponents. $xy = b^m b^n$
$xy = b^{m+n}$

Write the equation in its equivalent logarithmic form. $\log_b xy = m + n$

Substitute $\log_b x$ for m and $\log_b y$ for n. $\log_b xy = \log_b x + \log_b y$

The Logarithm Property of the Quotient of Two Numbers

For any positive real numbers x, y, and b, $b \neq 1$, $\log_b \dfrac{x}{y} = \log_b x - \log_b y$.

Proof: Let $\log_b x = m$ and $\log_b y = n$.

Write each equation in its equivalent exponential form. $x = b^m \qquad y = b^n$

Use substitution and the Properties of Exponents. $\dfrac{x}{y} = \dfrac{b^m}{b^n}$
$\dfrac{x}{y} = b^{m-n}$

Write the equation in its equivalent logarithmic form. $\log_b \dfrac{x}{y} = m - n$

Substitute $\log_b x$ for m and $\log_b y$ for n. $\log_b \dfrac{x}{y} = \log_b x - \log_b y$

The Logarithm Property of the Power of a Number

For any real numbers x and b, $b \neq 1$, and for any real number r, $\log_b x^r = r \log_b x$.

Proof: Let $\log_b x = m$.

Write the equation in its equivalent exponential form. $x = b^m$

Raise both sides to the r power. $x^r = (b^m)^r$
$x^r = b^{mr}$

Write the equation in its equivalent logarithmic form. $\log_b x^r = mr$

Substitute $\log_b x$ for m. $\log_b x^r = r \log_b x$

Glossary

· ·

abscissa   The first number of an ordered pair; it measures a horizontal distance and is also called the first coordinate of an ordered pair. (Sec. 4.1)

absolute value of a number   The distance of the number from zero on the number line. (Sec. 1.1)

absolute-value equation   An equation containing an absolute-value symbol. (Sec. 2.6)

acute angle   An angle whose measure is between 0° and 90°. (Sec. 3.1)

acute triangle   A triangle that has three acute angles. (Sec. 3.1)

addition method   An algebraic method of finding an exact solution of a system of linear equations wherein we use the Addition Property of Equations. (Sec. 5.3)

additive inverses   Numbers that are the same distance from zero on the number line but on different sides of zero; also called opposites. (Sec. 1.1)

adjacent angles   Two angles that share a common side. (Sec. 3.1)

alternate exterior angles   Two angles that are on opposite sides of the transversal and outside the parallel lines. (Sec. 3.1)

alternate interior angles   Two angles that are on opposite sides of the transversal and between the parallel lines. (Sec. 3.1)

analytic geometry   Geometry in which a coordinate system is used to study relationships between variables. (Sec. 4.1)

angle   An angle is formed when two rays start at the same point; it is measured in degrees. (Sec. 3.1)

antilogarithm   If $\log_b M = N$, then the antilogarithm, base b, of N is M. (Sec. 12.2)

area   A measure of the amount of surface in a region. (Sec. 3.2)

asymptotes   The two straight lines that a hyperbola "approaches." (Sec. 11.6)

axes   The two number lines that form a rectangular coordinate system; also called coordinate axes. (Sec. 4.1)

axis of symmetry of a parabola   A line of symmetry that passes through the vertex of the parabola. (Sec. 11.2)

base   In an exponential expression, the number that is taken as a factor as many times as indicated by the exponent. (Sec. 1.2)

binomial   A polynomial of two terms. (Sec. 6.2)

characteristic   The integer part of a common logarithm. (Sec. 12.2)

circle   A plane figure in which all points are the same distance from point O, which is called the center of the circle. (Sec. 3.2)

circumference   The distance around a circle. (Sec. 3.2)

clearing denominators   Multiplying each side of an equation by the least common multiple of the denominators. (Sec. 8.4)

cofactor of an element of a matrix $(-1)^{i+j}$ times the minor of that element, where i is the row number of the element and j is its column number. (Sec. 5.4)

combined variation   A variation in which two or more types of variation occur at the same time. (Sec. 8.7)

combining like terms   Using the Distributive Property to add the coefficients of like variable terms. (Sec. 1.4)

common logarithms   Logarithms to the base 10. (Sec. 12.2)

complementary angles   Two angles whose measures have the sum 90°. (Sec. 3.1)

completing the square   Adding to a binomial the constant term that makes it a perfect-square trinomial. (Sec. 10.2)

complex fraction   A fraction whose numerator or denominator contains one or more fractions. (Sec. 8.3)

complex number   A number of the form $a + bi$, where a and b are real numbers and $i = \sqrt{-1}$. (Sec. 9.3)

composition of functions   The operation on two functions f and g denoted by $f \circ g$. The value of the composition of f and g is given by $(f \circ g)(x) = f[g(x)]$. (Sec. 11.4)

compound inequality Two inequalities joined with a connective word such as "and" or "or." (Sec. 2.5)

compound interest Interest that is computed not only on the original principal but also on the interest already earned. (Sec. 12.5)

conic section A curve that can be constructed from the intersection of a plane and a right circular cone. The four conic sections are the parabola, hyperbola, ellipse, and circle. (Sec. 11.6)

conjugates Binomial expressions that differ only in the sign of a term. The expressions $a + b$ and $a - b$ are conjugates. (Sec. 9.2)

consecutive even integers Even integers that follow one another in order. (Sec. 7.5)

consecutive integers Integers that follow one another in order. (Sec. 7.5)

consecutive odd integers Odd integers that follow one another in order. (Sec. 7.5)

constant of proportionality k in a variation equation; also called the constant of variation. (Sec. 10.3)

constant term A term that includes no variable part; also called a constant. (Sec. 1.4)

constant of variation k in a variation equation; also called the constant of proportionality. (Sec. 10.3)

coordinate axes The two number lines that form a rectangular coordinate system; also called simply axes. (Sec. 4.1)

coordinates of a point The numbers in the ordered pair that is associated with the point. (Sec. 4.1)

corresponding angles Two angles that are on the same side of the transversal and are both acute angles or are both obtuse angles. (Sec. 3.1)

cube A rectangular solid in which all six faces are square. (Sec. 3.3)

cube root of a perfect cube One of the three equal factors of the perfect cube. (Sec. 7.4)

cubic function A third-degree polynomial function. (Sec. 6.2)

degree The unit used to measure angles; one complete revolution is 360°. (Sec. 3.1)

degree of a monomial The sum of the exponents of the variables. (Sec. 6.1)

degree of a polynomial The greatest of the degrees of any of its terms. (Sec. 6.2)

dependent system of equations A system of equations that has an infinite number of solutions. (Sec. 5.1)

dependent variable In a function, the variable whose value depends on the value of another variable known as the independent variable. (Sec. 4.1)

descending order The terms of a polynomial in one variable are arranged in descending order when the exponents of the variable decrease from left to right. (Sec. 6.2)

determinant A number associated with a square matrix. (Sec. 5.4)

diameter of a circle A line segment with endpoints on the circle and going through the center. (Sec. 3.2)

diameter of a sphere A line segment with endpoints on the sphere and going through the center. (Sec. 3.3)

direct variation A function that can be expressed as the equation $y = kx$, where k is a constant value called the constant of variation or the constant of proportionality. (Sec. 10.3)

discriminant For an equation of the form $ax^2 + bx + c = 0$, the quantity $b^2 - 4ac$ is called the discriminant. (Sec. 10.3)

domain The set of the first coordinates of the ordered pairs in a relation. (Sec. 4.1)

double root When a quadratic equation has two solutions that are the same number, the solution is called a double root of the equation. (Sec. 10.1)

element of a matrix A number in a matrix. (Sec. 5.4)

element of a set One of the objects in a set. (Sec. 1.1)

ellipse An oval shape that is one of the conic sections. (Sec. 11.6)

empty set The set that contains no elements; also called the null set. (Sec. 1.5)

equation A statement of the equality of two mathematical expressions. (Sec. 2.1)

equilateral triangle A triangle that has three sides of equal length; the three angles are of equal measure. (Sec. 3.2)

equivalent equations Equations that have the same solution. (Sec. 2.1)

evaluating a function Determining $f(x)$ for a given value of x. (Sec. 4.1)

evaluating a variable expression Replacing each variable by its value and then simplifying the resulting numerical expression. (Sec. 1.4)

even integer An integer that is divisible by 2. (Sec. 7.5)

expanding by cofactors A technique for finding the value of a 3×3 or larger determinant. (Sec. 5.4)

exponent In an exponential expression, the raised number that indicates how many times the factor, or base, occurs in the multiplication. (Sec. 1.2)

exponential equation An equation in which the variable occurs in the exponent. (Sec. 12.4)

exponential form The expression 2^6 is in exponential form. Compare *factored form*. (Sec. 1.2)

exponential function The exponential function with base b is defined by $f(x) = b^x$, where b is a positive real number not equal to one. (Sec. 12.1)

exterior angle of a triangle An angle adjacent to an interior angle of the triangle. (Sec. 3.1)

extraneous solution When each side of an equation is raised to an even power, the resulting equation may have a solution that is not a solution of the original equation. Such a solution is called an extraneous solution. (Sec. 9.4)

factor a polynomial To write the polynomial as a product of other polynomials. (Sec. 7.1)

factor a trinomial of the form $ax^2 + bx + c$ To express the trinomial as the product of two binomials. (Sec. 7.2)

factored form The multiplication $2 \cdot 2 \cdot 2 \cdot 2 \cdot 2 \cdot 2$ is in factored form. Compare *exponential form*. (Sec. 1.2)

factors In multiplication, the numbers that are multiplied. (Sec. 1.1)

FOIL A method of finding the product of two binomials. The letters stand for First, Outer, Inner, and Last. (Sec. 6.3)

formula A literal equation that states rules about measurement. (Sec. 8.5)

function A relation in which no two ordered pairs that have the same first coordinate have different second coordinates. (Sec. 4.1)

functional notation Notation used for those equations that define functions. The letter f is commonly used to name a function. (Sec. 4.1)

geometric solid A figure in space. (Sec. 3.3)

graph of a function A graph of the ordered pairs that belong to the function. (Sec. 11.1)

graph of an equation in two variables A graph of the ordered-pair solutions of an equation. (Sec. 4.2)

graph of an integer A heavy dot directly above that number on the number line. (Sec. 1.1)

graph of an ordered pair The dot drawn at the coordinates of the point in the plane. (Sec. 4.1)

graphing a point in the plane Placing a dot at the location given by the ordered pair; also called plotting a point in the plane. (Sec. 4.1)

graph of a relation The graph of the ordered pairs that belong to the relation. (Sec. 4.1)

greater than A number a is greater than another number b, written $a > b$, if a is to the right of b on the number line. (Sec. 1.1)

greater than or equal to The symbol $\geq$ means "is greater than or equal to." (Sec. 1.1)

greatest common factor The greatest common factor (GCF) of two or more integers is the greatest integer that is a factor of all the integers. The greatest common factor of two or more monomials is the product of the GCF of the coefficients and the common variable factors. (Sec. 7.1)

half-plane The solution set of a linear inequality in two variables. (Sec. 4.6)

hyperbola A conic section formed by the intersection of a cone and a plane perpendicular to the base of the cone. (Sec. 11.6)

hypotenuse In a right triangle, the side opposite the 90° angle. (Sec. 9.4)

imaginary number A number of the form ai, where a is a real number and $i = \sqrt{-1}$. (Sec. 9.3)

imaginary part of a complex number For the complex number $a + bi$, b is the imaginary part. (Sec. 9.3)

inconsistent system of equations A system of equations that has no solution. (Sec. 5.1)

independent system of equations A system of equations that has one solution. (Sec. 5.1)

independent variable In a function, the variable that varies independently and whose value determines the value of the dependent variable. (Sec. 4.1)

index In the expression $\sqrt[n]{a}$, n is the index of the radical. (Sec. 9.1)

inequality An expression that contains the symbol $>$, $<$, $\geq$ (is greater than or equal to), or $\leq$ (is less than or equal to). (Sec. 1.5)

integers The numbers $\ldots, -3, -2, -1, 0, 1, 2, 3, \ldots$. (Sec. 1.1)

interior angle of a triangle One of the angles within the region enclosed by the triangle. (Sec. 3.1)

intersecting lines Lines that cross at a point in the plane. (Sec. 3.1)

intersection of sets A and B The set that contains the elements that are common to both A and B. (Sec. 1.5)

inverse of a function The set of ordered pairs formed by reversing the coordinates of each ordered pair of the function. (Sec. 11.5)

inverse variation A function that can be expressed as the equation $y = k/x$, where k is a constant value. (Sec. 8.7)

irrational number The decimal representation of an irrational number never terminates or repeats and can only be approximated. (Sec. 1.2)

isosceles triangle A triangle that has two sides of equal length; the angles opposite the equal sides are of equal measure. (Sec. 3.2)

joint variation A variation in which a variable varies directly as the product of two or more variables. A joint variation can be expressed as the equation $z = kxy$, where k is a constant value. (Sec. 8.7)

leading coefficient In a polynomial function, the coefficient of the variable with the largest exponent. (Sec. 6.2)

least common denominator The smallest number that is a multiple of each denominator in question. (Sec. 1.2)

least common multiple (LCM) The LCM of two or more numbers is the smallest number that is a multiple of each of those numbers. (Sec. 1.2)

least common multiple of two polynomials The simplest polynomial of least degree that contains the factors of each polynomial. (Sec. 8.2)

legs In a right triangle, the sides opposite the acute angles. (Sec. 11.3)

less than A number a is less than another number b, written $a < b$, if a is to the left of b on the number line. (Sec. 1.1)

less than or equal to The symbol $\leq$ means "is less than or equal to." (Sec. 1.1)

like terms Terms of a variable expression that have the same variable part. Having no variable part, constant terms are like terms. (Sec. 1.4)

line A line extends indefinitely in two directions in a plane; it has no width. (Sec. 3.1)

line of best fit A line drawn to approximate data that are graphed as points in a coordinate system. (Sec. 4.4)

linear equation in three variables An equation of the form $Ax + By + Cz = D$, where A, B, C, and D are constants. (Sec. 5.3)

linear equation in two variables An equation of the form $y = mx + b$, where m is the coefficient of x and b is a constant; also called a linear function. (Sec. 4.6)

linear function A function that can be expressed in the form $f(x) = mx + b$. Its graph is a straight line. (Sec. 4.2)

linear inequality in two variables An inequality of the form $y > mx + b$ or $Ax + By > C$. (The symbol $>$ could be replaced by $\geq$, $<$, or $\leq$.) (Sec. 4.6)

linear model A first-degree equation that is used to describe a relationship between quantities. (Sec. 4.4)

line segment Part of a line; it has two endpoints. (Sec. 3.1)

literal equation An equation that contains more than one variable. (Sec. 8.5)

logarithm For b greater than zero and not equal to one, the statement $y = \log_b x$ (the logarithm of x, base b) is equivalent to $x = b^y$. (Sec. 12.2)

mantissa The decimal part of a common logarithm. (Sec. 12.2)

matrix A rectangular array of numbers. (Sec. 5.4)

minor of an element The minor of an element in a 3×3 determinant is the 2×2 determinant obtained by eliminating the row and column that contain that element. (Sec. 5.4)

monomial A number, a variable, or a product of a number and variables; a polynomial of one term. (Sec. 6.1)

multiplicative inverse The multiplicative inverse of a nonzero real number a is $1/a$; also called the reciprocal. (Sec. 1.4)

natural exponential function The function defined by $f(x) = e^x$, where $e \approx 2.71828$. (Sec. 12.1)

natural logarithm When e (the base of the natural exponential function) is used as the base of a logarithm, the logarithm is referred to as the natural logarithm and is abbreviated $\ln x$. (Sec. 12.2)

natural numbers The numbers $1, 2, 3, \ldots$; also called the positive integers. (Sec. 1.1)

negative integers The numbers $\ldots, -3, -2, -1$. (Sec. 1.1)

negative slope The slope of a line that slants downward to the right. (Sec. 4.3)

nonfactorable over the integers A polynomial is nonfactorable over the integers if it does not factor using only integers. (Sec. 7.2)

null set The set that contains no elements; also called the empty set. (Sec. 1.5)

numerical coefficient The number part of a variable term. When the numerical coefficient is 1 or -1, the 1 is usually not written. (Sec. 1.4)

obtuse angle An angle whose measure is between $90°$ and $180°$. (Sec. 3.1)

obtuse triangle A triangle that has one obtuse angle. (Sec. 3.2)

odd integer An integer that is not divisible by 2. (Sec. 7.5)

one-to-one function In a one-to-one function, given any y, there is only one x that can be paired with the given y. (Sec. 11.5)

opposites Numbers that are the same distance from zero on the number line but on different sides of zero; also called additive inverses. (Sec. 1.1)

order $m \times n$ A matrix of m rows and n columns is of order $m \times n$. (Sec. 5.4)

Order of Operations Agreement A set of rules that specifies the order in which we perform operations in simplifying numerical expressions. (Sec. 1.3)

ordered pair A pair of numbers expressed in the form (a, b) and used to locate a point in the plane determined by a rectangular coordinate system. (Sec. 4.1)

ordinate The second number of an ordered pair; it measures a vertical distance and is also called the second coordinate of an ordered pair. (Sec. 4.1)

origin The point of intersection of the two number lines that form a rectangular coordinate system. (Sec. 4.1)

parabola The graph of a quadratic equation in two variables. (Sec. 11.2)

parallel lines Lines that never meet; the distance between them is always the same. (Sec. 3.1)

parallelogram A quadrilateral that has opposite sides equal and parallel. (Sec. 3.2)

percent Parts of 100. (Sec. 1.2)

perfect cube The product of the same three factors. (Sec. 7.4)

perfect square The square of an integer or a variable. (Sec. 1.2)

perfect-square trinomial A trinomial that is a product of a binomial and itself. (Sec. 7.4)

perimeter The distance around a plane figure. (Sec. 3.2)

perpendicular lines Intersecting lines that form right angles. (Sec. 3.1)

plane A flat surface that extends in all directions. (Sec. 3.1)

plane figure A figure that lies totally in a plane. (Sec. 3.1)

plotting a point in the plane Placing a dot at the location given by the ordered pair; also called graphing a point in the plane. (Sec. 4.1)

point–slope formula The equation $y - y_1 = m(x - x_1)$, where m is the slope of a line and (x_1, y_1) is a point on the line. (Sec. 4.4)

polygon A closed figure determined by three or more line segments that lie in a plane. (Sec. 3.2)

polynomial A variable expression in which the terms are monomials. (Sec. 6.2)

positive integers The numbers $1, 2, 3, \ldots$; also called the natural numbers. (Sec. 1.1)

positive slope The slope of a line that slants upward to the right. (Sec. 4.3)

prime polynomial A polynomial that is nonfactorable over the integers. (Sec. 7.2)

principal square root The positive square root of a number. (Sec. 1.2)

product In multiplication, the result of multiplying two numbers. (Sec. 1.1)

proportion An equation that states the equality of two ratios or rates. (Sec. 8.4)

Pythagorean Theorem The square of the hypotenuse of a right triangle is equal to the sum of the squares of the two legs. (Sec. 9.4)

quadrant One of the four regions into which the two axes of a rectangular coordinate system divide the plane. (Sec. 4.1)

quadratic equation An equation of the form $ax^2 + bx + c = 0$, where a is not equal to zero; also called a second-degree equation. (Sec. 7.5)

quadratic formula A general formula for solving quadratic equations derived by applying the method of completing the square to the standard form of a quadratic equation. (Sec. 10.3)

quadratic function A function that can be expressed by the equation $f(x) = ax^2 + bx + c$, where a is not equal to zero. (Sec. 6.2)

quadratic inequality An inequality that can be written in the form $ax^2 + bx + c < 0$ or $ax^2 + bx + c > 0$, where a is not equal to zero. The symbols $\leq$ and $\geq$ can also be used. (Sec. 10.6)

quadrilateral A four-sided polygon. (Sec. 3.2)

radical sign The symbol $\sqrt{}$, which is used to indicate the positive, or principal, square root of a number. (Sec. 1.2)

radical equation An equation that contains a variable expression in a radicand. (Sec. 9.4)

radicand In a radical expression, the expression under the radical sign. (Sec. 1.2)

radius of a circle A line segment going from the center to a point on the circle. (Sec. 3.2)

radius of a sphere A line segment going from the center to a point on the sphere. (Sec. 3.3)

range The set of second coordinates of the ordered pairs in a relation. (Sec. 4.1)

rate The quotient of two quantities that have different units. (Sec. 8.4)

rate of work That part of a task which is completed in one unit of time. (Sec. 8.6)

ratio The quotient of two quantities that have the same unit. (Sec. 8.4)

rational expression A fraction in which the numerator or denominator is a polynomial. (Sec. 8.1)

rational number A number that can be written in the form a/b, where a and b are integers and b is not equal to zero. (Sec. 1.2)

rationalizing the denominator The procedure used to remove a radical from the denominator of a fraction. (Sec. 9.2)

ray A ray starts at a point and extends indefinitely in one direction. (Sec. 3.1)

real numbers The rational numbers and the irrational numbers taken together. (Sec. 1.2)

real part of a complex number For the complex number $a + bi$, a is the real part. (Sec. 9.3)

reciprocal The reciprocal of a nonzero real number a is $1/a$; also called the multiplicative inverse. (Sec. 1.4)

reciprocal of a rational expression The rational expression with the numerator and denominator interchanged. (Sec. 8.1)

rectangle A parallelogram that has four right angles. (Sec. 3.2)

rectangular coordinate system A coordinate system formed by two number lines, one horizontal and one vertical, that intersect at the zero point of each line. (Sec. 4.1)

rectangular solid A solid in which all six faces are rectangles. (Sec. 3.3)

regular polygon A polygon in which each side has the same length and each angle has the same measure. (Sec. 3.2)

relation A set of ordered pairs. (Sec. 4.2)

repeating decimal A decimal formed when dividing the numerator of a fraction by its denominator, in which one or more of the digits in the decimal repeat infinitely. (Sec. 1.1)

right angle An angle whose measure is 90 degrees. (Sec. 3.1)

right triangle A triangle that contains one right angle. (Sec. 3.2)

roster method A method of designating a set by enclosing a list of its elements in braces. (Sec. 1.5)

scalene triangle A triangle that has no sides of equal length; no two of its angles are of equal measure. (Sec. 3.2)

scatter diagram A graph of collected data as points in a coordinate system. (Sec. 4.3)

scientific notation Notation in which a number is expressed as the product of a number between 1 and 10 and a power of 10. (Sec. 6.1)

second-degree equation An equation of the form $ax^2 + bx + c = 0$, where a is not equal to zero; also called a quadratic equation. (Sec. 10.1)

set A collection of objects. (Sec. 1.1)

set-builder notation A method of designating a set that makes use of a variable and a certain property that only elements of that set possess. (Sec. 1.5)

similar objects Objects that have the same shape but not necessarily the same size. (Sec. 8.4)

slope–intercept form of a straight line The equation $y = mx + b$, where m is the slope of the line and $(0, b)$ is the y-intercept. (Sec. 4.3)

slope of a line A measure of the slant of a line. The symbol for slope is m. The formula for the slope is

$$m = \frac{y_2 - y_1}{x_2 - x_1}$$

where (x_1, y_1) and (x_2, y_2) are the coordinates of two points on the line and $x_1 \neq x_2$. (Sec. 4.3)

solid An object in space. (Sec. 3.3)

solution of a system of equations in three variables An ordered triple that is a solution of each equation of the system. (Sec. 5.3)

solution of a system of equations in two variables An ordered pair that is a solution of each equation of the system. (Sec. 5.1)

solution of an equation in two variables An ordered pair whose coordinates make the equation a true statement. (Sec. 4.1)

solution set of a system of inequalities The intersection of the solution sets of the individual inequalities. (Sec. 5.6)

solution set of an inequality A set of numbers, each element of which, when substituted for the variable, results in a true inequality. (Sec. 2.5)

solution(s) of an equation The replacement value(s) of the variable that will make the equation true; also called the root(s) of the equation. (Sec. 2.1)

solving an equation Finding a root, or solution, of the equation. (Sec. 2.1)

sphere A solid in which all points are the same distance from point O, which is called the center of the sphere. (Sec. 3.3)

square A rectangle that has four equal sides. (Sec. 3.2)

square matrix A matrix that has the same number of rows as columns. (Sec. 5.4)

square root A square root of a positive number x is a number a for which $a^2 = x$. (Sec. 1.2)

straight angle A 180° angle. (Sec. 3.1)

standard form A quadratic equation is in standard form when the polynomial is in descending order and equal to zero. $ax^2 + bx + c = 0$ is in standard form. (Sec. 7.5)

substitution method An algebraic method of finding an exact solution of a system of linear equations wherein we use the Substitution Property of Equality. (Sec. 5.2)

supplementary angles Two angles whose measures have the sum 180°. (Sec. 3.1)

surface area The total area on the surface of a solid. (Sec. 3.3)

synthetic division A shorter method of dividing a polynomial by a binomial of the form $x - a$. This method uses only the coefficients of the variable terms. (Sec. 6.4)

system of equations Two or more equations considered together. (Sec. 5.1)

system of inequalities Two or more inequalities considered together. (Sec. 5.6)

terminating decimal A decimal formed when dividing the numerator of a fraction by its denominator and the remainder is zero. (Sec. 1.1)

terms of a variable expression The addends of the expression. (Sec. 1.4)

tolerance of a component The acceptable amount by which the component may vary from a given measurement. (Sec. 2.6)

transversal A line intersecting two other lines at two different points. (Sec. 3.1)

triangle A three-sided polygon. (Sec. 3.2)

trinomial A polynomial of three terms. (Sec. 6.2)

undefined slope The slope of a vertical line is undefined. (Sec. 4.3)

uniform motion The motion of a moving object whose speed and direction do not change. (Sec. 2.4)

union of sets A and B The set that contains all elements of A and all elements of B. (Sec. 1.5)

value of a function The value of the dependent variable for a given value of the independent variable. (Sec. 4.1)

variable A letter of the alphabet used to stand for a quantity that is unknown or can change. (Sec. 1.1)

variable expression An expression that contains one or more variables. (Sec. 1.4)

variable term A term composed of a numerical coefficient and a variable part. When the numerical coefficient is 1 or -1, the 1 is not usually written. (Sec. 1.4)

vertex of an angle The common endpoint of the two rays that form the angle. (Sec. 3.1)

vertex of a quadratic function The lowest point on a parabola that opens up; the highest point on a parabola that opens down. (Sec. 11.2)

vertical angles Two angles that are on opposite sides of the intersection of two lines. (Sec. 3.1)

volume A measure of the amount of space inside a closed surface. (Sec. 3.3)

x-coordinate The abscissa in an ordered pair. (Sec. 4.1)

x-intercept The point at which a graph crosses the x-axis. (Sec. 4.2)

xy-coordinate system A rectangular coordinate system in which the horizontal axis is labeled x and the vertical axis is labeled y. (Sec. 4.1)

y-coordinate The ordinate in an ordered pair. (Sec. 4.1)

y-intercept The point at which a graph crosses the y-axis. (Sec. 4.2)

zero slope The slope of a horizontal line. (Sec. 4.3)

Index

Content and Format © 1996 HMCo.

Content and Format © 1996 HMCo.